PLEASE STAMP DATE DUE, BOTH BELOW AND ON CARD

| DATE DUE | DATE DUE | DATE DUE |

Geology
QE517.5 .E94 1999
Exhumation processes : normal
faulting, ductile flow, and
erosion

# Exhumation Processes: Normal Faulting, Ductile Flow and Erosion

Geological Society Special Publications
*Series Editors*
A. J. FLEET
R. E. HOLDSWORTH
A. C. MORTON
M. S. STOKER

It is recommended that reference to all or part of this book should be made in one of the following ways.

RING, U., BRANDON, M. T., LISTER, G. S. & WILLETT, S. D. (eds) 1999. *Exhumation Processes: Normal Faulting, Ductile Flow and Erosion*. Geological Society, London, Special Publications, **154**.

THOMSON, S. N., STÖCKHERT, B. & BRIX, M. R. 1999. Miocene high-pressure metamorphic rocks of Crete, Greece: rapid exhumation by buoyant escape. *In*: RING, U., BRANDON, M. T., LISTER, G. S. & WILLETT, S. D. (eds) *Exhumation Processes: Normal Faulting, Ductile Flow and Erosion*. Geological Society, London, Special Publications, **154**.

GEOLOGICAL SOCIETY SPECIAL PUBLICATION NO. 154

# Exhumation Processes: Normal Faulting, Ductile Flow and Erosion

EDITED BY

UWE RING
(Johannes Gutenberg-Universität, Mainz, Germany)

MARK T. BRANDON
(Yale University, New Haven, USA)

GORDON S. LISTER
(Monash University, Clayton, Australia)

and

SEAN D. WILLETT
(Pennsylvania State University, USA)

1999
Published by
The Geological Society
London

# THE GEOLOGICAL SOCIETY

The Society was founded in 1807 as The Geological Society of London and is the oldest geological society in the world. It received its Royal Charter in 1825 for the purpose of 'investigating the mineral structure of the Earth'. The Society is Britain's national society for geology with a membership of around 8500. It has country-wide coverage and approximately 1500 members reside overseas. The Society is responsible for all aspects of the geological sciences including professional matters. The Society has its own publishing house, which produces the Society's international journals, books and maps, and which acts as the European distributor for publications of the American Association of Petroleum Geologists, SEPM and the Geological Society of America.

Fellowship is open to those holding a recognized honours degree in geology or cognate subject and who have at least two years' relevant postgraduate experience, or who have not less than six years' relevant experience in geology or a cognate subject. A Fellow who has not less than five years' relevant postgraduate experience in the practice of geology may apply for validation and, subject to approval, may be able to use the designatory letters C Geol (Chartered Geologist).

Further information about the Society is available from the Membership Manager, The Geological Society, Burlington House, Piccadilly, London W1V 0JU, UK. The Society is a Registered Charity, No. 210161.

Published by The Geological Society from:
The Geological Society Publishing House
Unit 7, Brassmill Enterprise Centre
Brassmill Lane
Bath BA1 3JN
UK
(*Orders*: Tel. 01225 445046
Fax 01225 442836)

First published 1999

The publishers make no representation, express or implied, with regard to the accuracy of the information contained in this book and cannot accept any legal responsibility for any errors or omissions that may be made.

© The Geological Society 1998. All rights reserved. No reproduction, copy or transmission of this publication may be made without written permission. No paragraph of this publication may be reproduced, copied or transmitted save with the provisions of the Copyright Licensing Agency, 90 Tottenham Court Road, London W1P 9HE. Users registered with the Copyright Clearance Center, 27 Congress Street, Salem, MA 01970, USA: the item-fee code for this publication is 0305–8719/98/$10.00.

**British Library Cataloguing in Publication Data**
A catalogue record for this book is available from the British Library.

ISBN 1–86239–025–8

Typeset by Type Study, Scarborough, UK

Printed in Great Britain by
Cambridge University Press, Cambridge, UK.

**Distributors**
*USA*
AAPG Bookstore
PO Box 979
Tulsa
OK 74101–0979
USA
(*Orders*: Tel. (918) 584–2555
Fax (918) 560–2652)

*Australia*
Australian Mineral Foundation
63 Conyngham Street
Glenside
South Australia 5065
Australia
(*Orders*: Tel. (08) 379–0444
Fax (08) 379–4634)

*India*
Affiliated East–West Press PVT Ltd
G–1/16 Ansari Road
New Delhi 110 002
India
(*Orders*: Tel. (11) 327–9113
Fax (11) 326–0538)

*Japan*
Kanda Book Trading Co.
Cityhouse Tama 204
Tsurumaki 1-3-10
Tama-Shi
Tokyo 206–0034
Japan
(*Orders*: Tel. (0423) 57–7650
Fax (0423) 57–7651)

# Contents

| | |
|---|---|
| Preface | vii |
| Acknowledgements | viii |
| RING, U., BRANDON, M. T., WILLETT, S. D. & LISTER, G. S. Exhumation Processes | 1 |

**Subduction-related accretionary wedges (B-type subduction)**

| | |
|---|---|
| SEDLOCK, R. L. Evaluation of exhumation mechanisms for coherent blueschists in western Baja California, Mexico | 29 |
| RING, U. & BRANDON, M. T. Ductile deformation and mass loss in the Franciscan Subduction Complex: implications for exhumation processes in accretionary wedges | 55 |
| THOMSON, S. N., STÖCKHERT, B. & BRIX, M. R. Miocene high-pressure metamorphic rocks of Crete, Greece: rapid exhumation by buoyant escape | 87 |
| RAWLING, T. J. & LISTER, G. S. Oscillating modes of orogeny in the Southwest Pacific and the tectonic evolution of New Caledonia | 109 |
| WINTSCH, R. P., BYRNE, T. & TORIUMI, M. Exhumation of the Sanbagawa blueschist belt, SW Japan, by lateral flow and extrusion: evidence from structural kinematics and retrograde $P$–$T$–$t$ paths | 129 |

**Collisional belts and intra-continental convergence (A-type subduction)**

| | |
|---|---|
| SCHLUNEGGER, F. & WILLETT, S. Spatial and temporal variations in exhumation of the central Swiss Alps and implications for exhumation mechanisms | 157 |
| VANDERHAEGHE, O., BURG, J.-P. & TEYSSIER, C. Exhumation of migmatites in two collapsed orogens: Canadian Cordillera and French Variscides | 181 |
| CALVERT, A. T., GANS, P. B. & AMATO, J. M. Diapiric ascent and cooling of a sillimanite gneiss dome revealed by $^{40}Ar/^{39}Ar$ thermochronology: the Kigluaik Mountains, Seward Peninsula, Alaska | 205 |
| GLAZNER, A. F. Exposure of deep, dense rocks: interplay between erosion and sinking | 233 |
| MILLER, J. McL., GREGORY, R. T., GRAY, D. R. & FOSTER, D. A. Geological and geochronological constraints on the exhumation of a high-pressure metamorphic terrane, Oman | 241 |
| BATT, G. E., KOHN, B. P., BRAUN, J., MCDOUGALL, I. & IRELAND, T. R. New insight into the dynamic development of the Southern Alps, New Zealand, from detailed thermochronological investigation of the Mataketake Range pegmatites | 261 |
| GARVER, J. I., BRANDON, M. T., RODEN-TICE, M. & KAMP, P. J. J. Exhumation history of orogenic highlands determined by detrital fission-track thermochronology | 283 |

**Lithospheric extension: divergent plate motions (rifting)**

| | |
|---|---|
| FORSTER, M. A. & LISTER, G. S. Detachment faults in the Aegean core complex of Ios, Cyclades, Greece | 305 |

GOODWIN, L. B. Controls on pseudotachylyte formation during tectonic exhumation in the South Mountains metamorphic core complex, Arizona 325

FOSTER, D. A. & JOHN, B. E. Quantifying tectonic exhumation in an extensional orogen with thermochronology: examples from the southern Basin and Range Province 343

# Preface

The general idea of this book is to provide an overview of exhumation processes. The idea was conceived during a conference on the same topic. The papers were organized to provide a broad sampling of frontier research on all processes that contribute to exhumation of metamorphic rocks in ancient and modern orogens. Examples from a variety of tectonic settings, including continental rifts, oceanic subduction zones and continental collision zones are given. The papers provide innovative applications of structural geology, metamorphic petrology and thermochronology to resolve the rates and geometry of normal faulting, and the interaction between erosion and tectonics. One of the weaknesses of the volume is that the role that surface processes play in the exhumation of deeply seated rocks is not well presented.

The volume contains 16 papers. The opening contribution by **Ring** *et al.* is a review aimed at providing an umbrella for the more focused papers that follow. The remaining papers are arranged into three sections, divided according to tectonic setting. The first section, *Subduction-related accretionary wedges (B-type subduction)*, starts with studies of the Franciscan subduction complex by **Sedlock** and **Ring & Brandon**. **Sedlock** documents normal faulting as the primary exhumation mechanism for blueschists in Baja California. **Ring & Brandon** examine the role of deformation, primarily ductile flow in tectonically exhuming the high-grade Eastern Franciscan Belt, and show that these processes played only a minor role. The papers of **Thomson** *et al.* and **Rawling & Lister** are examples of dominantly normal-faulting-controlled exhumation at retreating subduction zones. The contribution of **Wintsch** *et al.* calls for margin-parallel shear as the principal exhumation process for Cretaceous high-pressure metamorphic rocks at the Japan convergent margin.

The second section is *Collisional belts and intra-continental convergence (A-type subduction)*. The first three papers address the exhumation of high-temperature rocks in the internides of collisional orogens. The paper of **Schlunegger & Willett** examines how exhumation influenced the postcollisional growth and destruction of the Alps. The paper illustrates how difficult it is to distinguish between tectonic exhumation and erosion. **Vanderhaeghe** *et al.* also highlight the interplay between various exhumation mechanisms during the destruction of the migmatitic core in two collisional orogens. **Calvert** *et al.* report on an interesting example of exhumation by diapirism. **Glazner** reviews how magmatism might influence extensional deformation. The paper focuses on buoyancy of magmas and their level of emplacement in the crust.

**Miller** *et al.* report thermochronological aspects of exhumation of eclogite beneath the Samail ophiolite in Oman. The time-temperature history of exhumation is also addressed in **Batt** *et al.* and **Garver** *et al*. **Batt** *et al.* concentrate on constraining cooling and exhumation rates and their temporal variations, whereas **Garver** *et al.* use fission-track ages preserved in detrital sediments to resolve the progressive evolution of exhumation in several orogens.

The last section is called *Lithospheric extension and divergent plate motions (rifting)*. **Forster & Lister** describe multiple generations of low-angle normal faults in the back-arc region of the active Hellenic subduction zone in the eastern Aegean. They also examine differences between extensional faulting in the Aegean and the Basin-and-Range province. The remaining two papers by **Goodwin** and **Foster & John** deal with the highly attenuated crustal sections in the Basin-and-Range province. **Goodwin** concentrates on a microstructural and microchemical study of pseudotachylyte formation associated with the later stages of crustal extension in the South Mountains metamorphic core complex. **Foster & John** use thermochronological techniques to place constraints on the rate of extensional faulting, the change in thermal gradients during detachment faulting, and the original dip of the detachments surfaces in the Colorado River extensional corridor.

The editors are grateful to the Geological Society of London for making the publication of this book possible and to Angharad Hills, Staff Editor of the Geological Society Publishing House, and Alan Roberts as the corresponding editor for their continued support and patience.

Uwe Ring
Mark T. Brandon
Gordon S. Lister
Sean D. Willett

# Acknowledgements

The following colleagues and two anonymous referees kindly reviewed the articles submitted to this book:

| | | |
|---|---|---|
| John Aitchinson | Ron Harris | Ray Price |
| Gary Axen | Rod Holcomb | Steve Ralser |
| Sue Baldwin | Simon Inger | Steve Reddy |
| John Bartley | Rebecca Jamieson | Mike Sandiford |
| Geoff Batt | Christopher Johnson | Elizabeth Schermer |
| Andy Bobyarchick | Laurent Jolivet | Stefan Schmid |
| Mark Brandon | Peter Koons | Jane Selverstone |
| Martin Burkhart | Paul Layer | Jinny Sisson |
| Darrel Cowan | Gordon Lister | John Spray |
| Allen Dennis | Neil Mancktelow | Bernhard Stöckhert |
| David Dinter | Dieter Mertz | Stuart Thomson |
| Mihai Ducea | Nick Mortimer | Janos Urai |
| Michel Faure | Roland Oberhänsli | John Wakabayashi |
| David Foster | Onno Oncken | Simon Wallis |
| Kevin Furlong | Cees Passchier | Dave Waters |
| Arthur Goldstein | Terry Pavlis | Sean Willett |

# Exhumation processes

UWE RING[1], MARK T. BRANDON[2], SEAN D. WILLETT[3] & GORDON S. LISTER[4]

[1] *Institut für Geowissenschaften, Johannes Gutenberg-Universität, 55099 Mainz, Germany*
[2] *Department of Geology and Geophysics, Yale University, New Haven, CT 06520, USA*
[3] *Department of Geosciences, Pennsylvania State University, University Park, PA 16802, USA*
*Present address: Department of Geological Sciences, University of Washington, Seattle, WA 98125, USA*
[4] *Department of Earth Sciences, Monash University, Clayton, Victoria VIC 3168, Australia*

**Abstract:** Deep-seated metamorphic rocks are commonly found in the interior of many divergent and convergent orogens. Plate tectonics can account for high-pressure metamorphism by subduction and crustal thickening, but the return of these metamorphosed crustal rocks back to the surface is a more complicated problem. In particular, we seek to know how various processes, such as normal faulting, ductile thinning, and erosion, contribute to the exhumation of metamorphic rocks, and what evidence can be used to distinguish between these different exhumation processes.

In this paper, we provide a selective overview of the issues associated with the exhumation problem. We start with a discussion of the terms *exhumation*, *denudation* and *erosion*, and follow with a summary of relevant tectonic parameters. Then, we review the characteristics of exhumation in different tectonic settings. For instance, continental rifts, such as the severely extended Basin-and-Range province, appear to exhume only middle and upper crustal rocks, whereas continental collision zones expose rocks from 125 km and greater. Mantle rocks are locally exhumed in oceanic rifts and transform zones, probably due to the relatively thin crust associated with oceanic lithosphere.

Another topic is the use of $P–T–t$ data to distinguish between different exhumation processes. We conclude that this approach is generally not very diagnostic since erosion and normal faulting show the same range of exhumation rates, reaching maximum rates of $>5–10$ km Ma$^{-1}$ for both processes. In contrast, ductile thinning appears to operate at significantly slower rates. The pattern of cooling ages can be used to distinguish between different exhumation processes. Normal faulting generally shows an asymmetric distribution of cooling ages, with an abrupt discontinuity at the causative fault, whereas erosional exhumation is typically characterized by a smoothly varying cooling-age pattern with few to no structural breaks.

Last, we consider the challenging problem of ultrahigh-pressure crustal rocks, which indicate metamorphism at depths greater than 100–125 km. Understanding the exhumation of these rocks requires that we first know where and how they were formed. One explanation is that metamorphism occurred within a thickened crustal root, but it does seem unlikely that the crust, including an eclogitized mafic lower crust, could get much thicker than *c*. 110 km while maintaining a reasonable Moho depth (<70 km, assuming that the seismically defined Moho would be observed to lie above the eclogitized lower crust). Diamond-bearing crustal rocks cannot be explained by this scheme. The alternative is to accrete the upper 10–40 km of lithospheric mantle into the orogenic root. This scenario will provide sufficient pressures for both coesite- and diamond-bearing eclogite-facies metamorphism, while maintaining a reasonable Moho depth (<70 km) and reasonable mean topography (≤3 km). We speculate that the detachment and foundering of the mantle root may contribute to the exhumation of any crustal rocks contained within the mantle root.

Exhumation is a generic term describing the return of once deep-seated metamorphic rocks to the Earth's surface. Field geology is, by definition, the geology of exhumed rocks. In fact, most of our understanding of crustal deformation and metamorphism is based on studies of exhumed rocks. Exhumation occurs by three processes: normal faulting, ductile thinning, and erosion. These processes are important, not only for the exhumation that they cause, but also for their influence on the formation of orogenic topography and the contribution to production of synorogenic sediments.

Exhumation can occur in virtually any geological setting, regardless of age or tectonic

RING, U., BRANDON, M. T., WILLETT, S. D. & LISTER, G. S. 1999. Exhumation processes. *In:* RING, U., BRANDON, M. T., LISTER, G. S. & WILLETT, S. D. (eds) *Exhumation Processes: Normal Faulting, Ductile Flow and Erosion.* Geological Society, London, Special Publications, **154**, 1–27.

regime. Even in the earliest studies of alpine tectonics, erosion was recognized as an important process for unroofing the internal metamorphic zones of convergent mountain belts. Early geologists observed that mountainous regions eroded faster than adjacent lowlands, and that ancient mountain belts were commonly flanked by thick synorogenic deposits that could be traced by provenance to erosional sources within the orogen.

The term 'tectonic denudation' (Moores *et al.* 1968; Armstrong 1972) made its way into the literature with the discovery of metamorphic core complexes in the Basin-and-Range province of western United States. Early workers recognised that normal faulting (Fig. 1) was capable of unroofing mid-crustal rocks, and that the hallmark of this type of exhumation was the 'resetting' of footwall rocks to a common isotopic age. In fact, we now understand that the common isotopic age is caused by rapid cooling as the hanging wall is stripped away. This observation has lead to the widely held view that rapid cooling is a diagnostic feature of tectonic exhumation. Work over the last ten years has demonstrated that exhumation by normal faulting often occurs in convergent as well as divergent orogens.

A third exhumation process is ductile thinning (Fig. 1), which can contribute to unroofing of metamorphic rocks. This idea was at the centre of the debate about diapiric emplacement of migmatites and gneiss domes (Ramberg 1967, 1972, 1980, 1981). In this sense, diapiric emplacement of a pluton can also be viewed as a type of exhumation, given that the pluton is 'exhumed' by thinning of its cover. The role of ductile thinning has received less attention than other exhumation mechanisms, but it appears to be important in some cases. For example, there has been much debate recently about the possibility of buoyant rise of high-pressure and ultra-high-pressure quartzofeldspathic rocks, an idea that has close similarity to the diapiric model for gneiss dome emplacement (Calvert *et al.* this volume).

Our objective is to provide a selective review of the exhumation problem. We focus on five topics: (1) a review of the terminology used to discuss exhumation and its relationship to orogenesis, (2) identification of tectonic parameters relevant to the exhumation processes, (3) a summary of how exhumation varies as a function of tectonic setting, (4) the critical review of evidence that might be diagnostic of specific exhumation processes, and (5) a discussion of the origin and exhumation of ultra-high-pressure metamorphic rocks, which represent a particularly challenging example of deep exhumation.

## Terminology

The exhumation problem is surrounded by a confusing and inconsistent terminology, which can leave even simple concepts, such as uplift (England & Molnar 1990) and extension (Wheeler & Butler 1994; Butler & Freeman 1996), difficult to follow. In this section, we examine the terminology and provide some simple definitions and suggestions for consistent usage.

The term *orogen* has broadened over the years, and now is commonly used to refer to any mountainous topography at the Earth's surface resulting from localized deformation. This usage includes convergent orogens like the European Alps and the Cascadia accretionary wedge of northwestern North America, and divergent orogens like the Basin-and-Range province and

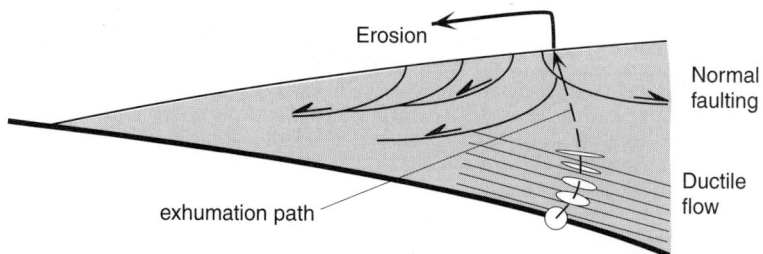

**Fig. 1.** Schematic illustration of the three exhumation processes: normal faulting, ductile flow, and erosion. Normal faulting covers both brittle normal faulting in the upper crust and normal-sense ductile shear zones in the deeper crust. Ductile thinning refers to wholesale vertical shortening in the orogenic wedge. The circle shows an undeformed particle accreted at the base of the wedge, which becomes deformed (indicated by ellipses) along the exhumation path.

the East African rift. Orogens are commonly described as *compressional* and *extensional*, which emphasises the horizontal stress regime within the orogen. Unfortunately, these terms are not complementary, given that compression refers to stress and extension to strain. More importantly, the simple reference to a dominant horizontal strain ignores the fact that some orogens are characterized by coeval horizontal extension and horizontal contraction of different levels within the same orogen (mixed-mode flow field in Fig. 2). Such mixed-mode deformation fields have been recognized in a number of convergent orogens, such as the Apennines of Italy (Royden 1993), the Betic Cordillera of Spain (Platt & Vissers 1989, Vissers *et al.* 1995), and the Himalayas (Burchfiel & Royden 1985).

Plate boundary terminology provides a better classification of an orogen. The terms *convergent*, *divergent* and *translational* refer to the kinematic relationship of the plates that bound the orogen, and carry no implications about the dominant state of stress or strain within the orogen. For example, one can describe the Apennines as a convergent orogen, even though the deformation field is mixed with some parts of

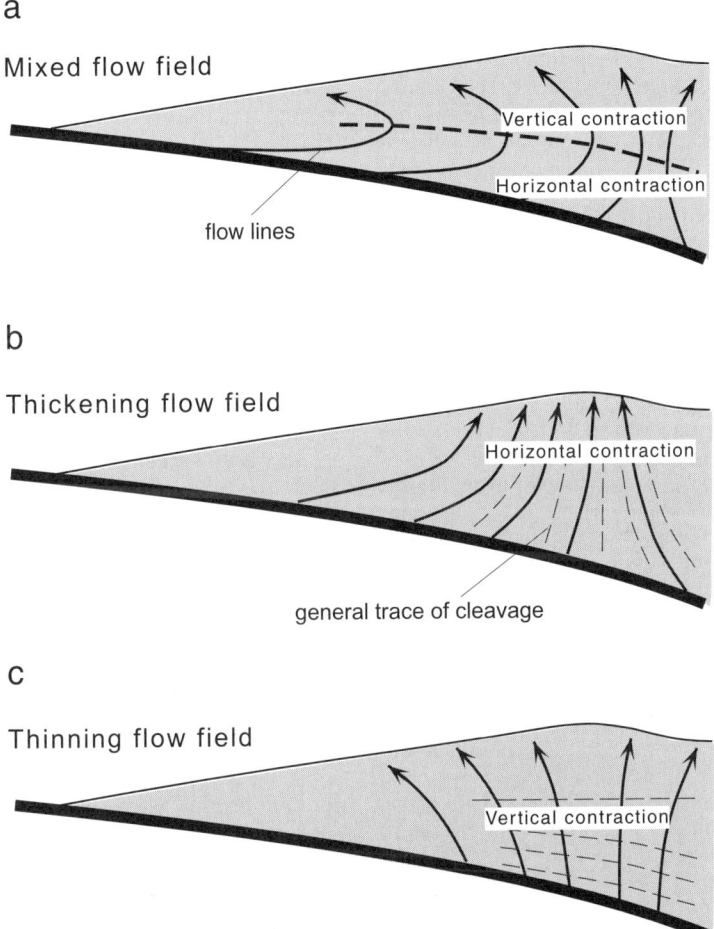

**Fig. 2.** Schematic illustrations of flow fields for three end-member convergent wedges (modified from Feehan & Brandon 1999). (**a**) A mixed-mode flow field characterized by horizontal contraction at the base of the wedge and vertical contraction near the top. This situation was proposed by Platt (1987). The dashed line marks the level where the principle strain-rate directions reverse. (**b**) A thickening flow field is characterized by converging flow lines and widespread horizontal contraction. (**c**) A thinning flow field is characterized by diverging flow lines, which indicates widespread vertical contraction and horizontal extension. Note that horizontal extension can be achieved by either ductile flow or normal faulting.

the orogen showing horizontal contraction and others, horizontal extension.

Horizontal extension is commonly taken as evidence of tectonic exhumation. The implicit assumption is that horizontal extension indicates vertical shortening, which in turn indicates tectonic thinning of the overburden. There are two problems with this linkage of horizontal extension to tectonic exhumation. The first is that the strain geometry is not always as assumed. For instance, broad shear zones at translational plate boundaries are characterized by plane strain with the maximum extension and maximum shortening directions lying in the horizontal, and little to no strain in the vertical. The result is little to no vertical thinning or thickening of the overburden. In another example, Ring & Brandon (this volume) show that the Franciscan subduction complex was shortened by about 30% in the vertical, but that shortening was balanced by a nearly equal volume strain. The result was that the other principal strains were nearly zero. This example shows that vertical thinning can occur without any appreciable horizontal extension.

This ambiguity can be avoided by focusing on the vertical strain – either brittle or ductile – when assessing the role of tectonic exhumation. In this regard, the terms *thickening* and *thinning* are unambiguous descriptions of the change in overburden thickness caused by vertical strain. Vertical thickening causes tectonic burial and vertical thinning, tectonic exhumation.

There has been some discussion about the suitability of *exhumation* as a term for unroofing of rocks. In a strict sense, *exhumation* means 'to dig up or disinter' (Summerfield & Brown 1998). As such, it refers to the motion between a rock relative to the Earth's surface (England & Molnar 1990). An alternative term is *denudation*, which has a general meaning 'to make bare' and a geological meaning 'to expose rock strata by erosion'. Denudation and erosion are, in fact, widely used synonyms, although there is a tendency to restrict denudation to mean erosion at the regional scale (Summerfield & Brown 1998). The use of denudation in a more generic sense has clear precedent given the term *tectonic denudation,* which means to unroof by normal faulting (Moores *et al.* 1968; Armstrong 1972).

In our review of the literature, *exhumation* and *denudation* are used almost interchangeably, although a subtle distinction is sometimes made about the frame of reference. Exhumation is used to refer to the unroofing history of a rock, defined by the vertical distance transversed by the rock relative to the Earth's surface. On the other hand, denudation, and erosion as well, are frequently used to refer to the removal of material at a particular point on the Earth's surface. This distinction is not universally accepted, but we think it is useful and encourage its adoption. Walcott (1998) emphasized this usage in his synthesis of the Southern Alps orogen of New Zealand. The Southern Alps are characterized by a large gradient in surface erosion rates across the orogen, from $c.$ 1 km $Ma^{-1}$ on the eastern dry side of the range to $c.$ 10 km $Ma^{-1}$ on the western rainy side of the range. Most particle paths in the orogen have a significant westward component of motion, causing them to move horizontally through the orogen from the slowly eroding east side to the rapidly eroding west side. As stated by Walcott (1998) (our italics): 'In two dimensions the amount of rock removed by erosion at a *spatial* point on the Earth's surface may be vastly greater than the exhumation *experienced by the rock* and is better referred to as denudation *or erosion*'.

England & Molnar (1990) made the useful distinction between *surface uplift*, meaning the vertical motion of the Earth's surface relative to sea level, and *rock uplift*, meaning the vertical motion of rock relative to sea level. Erosion and denudation are then defined as the difference between rock uplift and surface uplift at a single spatial point. Since erosion is measured in the vertical, it is probably best viewed as a flux (i.e., velocity normal to the approximately horizontal surface of the Earth) rather than a true velocity. This usage is consistent with estimates of drainage-scale erosion rates, which are calculated by dividing sediment yield by drainage area.

As used here, exhumation is measured by integrating the difference between the rock velocity and surface velocity while following the rock along its material path (cf. England & Molnar 1990). Barometers and thermochronometers provide information about exhumation since they tell us about the unroofing history specific to a rock sample. For the simple case of a vertical particle path, exhumation and erosion are equivalent, but this situation cannot be expected in general. When exhumation is divided by time, the resulting exhumation rate can be viewed as a spatially and temporally averaged erosion rate, but this equivalence should be viewed with caution. The approximation becomes increasingly flawed as the closure depth for the relevant barometric and thermochronometric data increases. For instance, cosmogenic isotopes provide estimates of exhumation rates within meters of the surface (Cerling & Craig 1994), which means that the measured rates are essentially equal to erosion rates. Low-temperature thermochronometers have deeper closure depths, and thus provide a more averaged estimate of the erosion rate. For instance, the apatite (U–Th)/He thermochronometer (Lippolt et al. 1994; Wolff et al. 1997) has a very low closure temperature (c. 75°C), which implies closure depths of 2–3 km, assuming slow erosion rates. This depth will be even shallower in areas of fast erosion because of the upward advection of heat caused by erosion. Apatite fission-track ages have a higher closure temperature (c. 110°C) and deeper closure depths (<4–5 km), so exhumation rates determined from those data will have a more approximate relationship to surface erosion rates. Zircon fission-track ages (c. 240°C) and $^{40}$Ar–$^{39}$Ar ages (>300°C) have greater closure temperatures, and an even more distant relationship to surface erosion rates.

To summarize, we recommend the following definitions.

*Exhumation*: the unroofing history of a rock, as caused by tectonic and/or surficial processes.

*Erosion*: the surficial removal of mass at a spatial point in the landscape by both mechanical and chemical processes.

*Denudation*: the removal of rock by tectonic and/or surficial processes at a specified point at or under the Earth's surface.

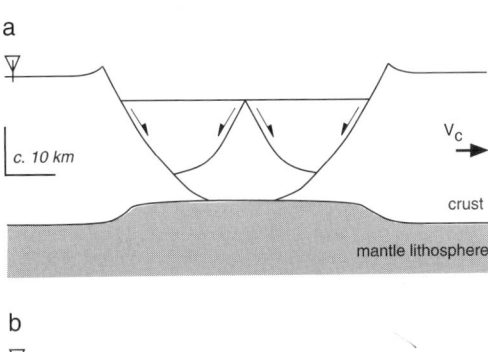

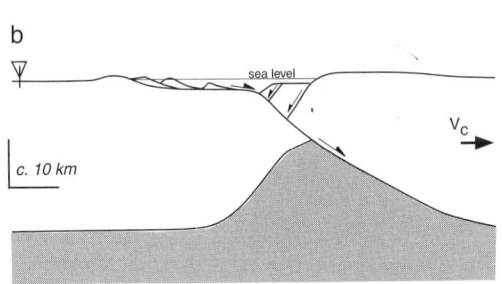

**Fig. 3.** Schematic illustration of divergent settings. (**a**) Symmetric rifting, resulting from coaxial extension of the lithosphere. (**b**) Asymmetric rifting, due to non-coaxial deformation of the lithosphere.

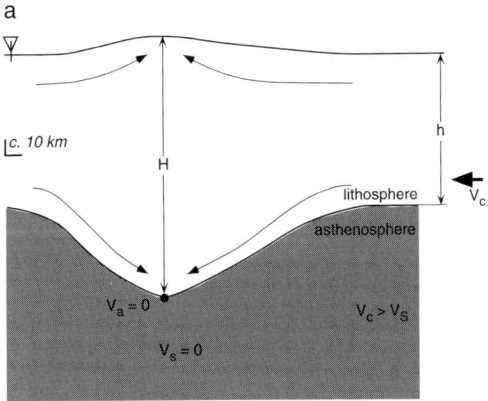

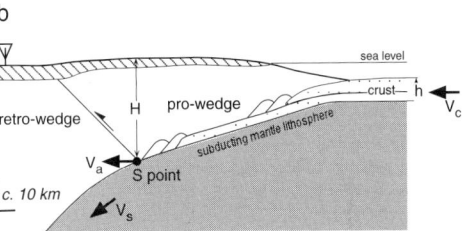

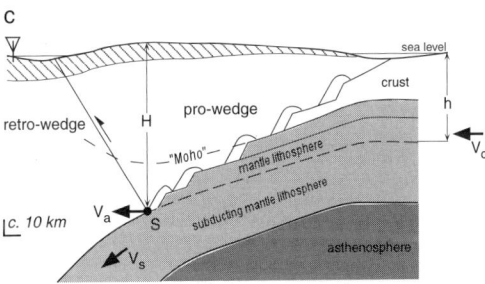

**Fig. 4.** Schematic illustration of convergent wedges showing the relationship between $V_c$, $V_a$, $V_S$ and the S point. The vector for $V_a$ shows the case for subduction-zone advance where $V_a > 0$. (**a**) Extreme case where both lithospheric plates are completely accreted into the orogen. $V_S$ and the S point are undefined in this case. (**b**) Subduction-related accretionary wedge. The S point is at a depth of about 30 km. Incoming crustal section is about 6 km thick. (**c**) Convergent continental wedge. The S point is at a depth of about 120 km, which means that imbrication and accretion include both crust and lithospheric mantle from the underthrust plate.

## Orogenic deformation and tectonic models

Plate-tectonic theory dictates that the velocity field at the Earth's surface is dominated by the motion of rigid plates. Deformation occurs mainly at the velocity gradients within the zones that define the plate boundaries. To a first order, orogens are either divergent (Fig. 3) or convergent (Fig. 4) depending on the sign of the relative velocity, $V_c$, defined here to be positive for convergence.

Divergent orogens display considerable complexity depending on the symmetry of internal structure (Fig. 3). In continental settings, divergence results in symmetric or asymmetric stretching and rifting. Symmetric stretching can result in significant amounts of ductile thinning. A more asymmetric style of stretching is influenced by the development of strongly localized normal faults, which impart a sense of vergence to the deformation. In both cases, the deformation, whether brittle or ductile, involves a strong component of horizontal extension and tectonic denudation of the footwall beneath major normal faults.

Similarly, convergent orogens show a broad diversity depending on both the boundary conditions and the internal response. The general expectation is that convergence causes pervasive horizontal contraction throughout the orogen. However, horizontal extension has been recognized as an important feature in a number of convergent orogens. For instance, in the special case of a retreating plate boundary (Royden 1993; Waschbusch & Beaumont 1996), oceanic subduction is commonly characterized by horizontal contraction in the accretionary wedge and simultaneous horizontal extension within the arc and back arc.

The complexity of deformation within convergent orogens, as well as the diversity of tectonic style and exhumation processes, has led to considerable effort in describing and modelling these orogenic systems. Much of the deformational response to convergence is dictated by how much of the lithosphere is involved in the deformation. At one limit, the entire lithosphere of one or both bounding plates is forced to contract and thicken (Fig. 4a). Whole-lithosphere thickening has been proposed as an important feature of large-scale continental collisions, such as the India–Asia collision (England & McKenzie 1982; England & Houseman 1986). England & Houseman (1986) have concluded that tectonic exhumation will occur when the gravitationally unstable mantle lithospheric root detaches, but they argue that the orogen otherwise shows no strong tendency for vertical thinning during convergence.

The other end member for convergent orogens is oceanic subduction zones, which are characterized by accretion of a thin layer, often limited to only the sedimentary cover of the downing plate (Fig. 4b). In this case, the orogen is the accretionary wedge, which grows slowly by accretion of sediments. Intermediate to these end-members are relatively small convergent orogens that form by accretion and contraction of sediment, crustal basement rocks, and sometimes lithospheric mantle as well (Fig. 4c). Well-studied examples include the European Alps, the Pyrenees, the Apennines, and the arc–continent collision of Taiwan.

For a subduction/accretion model, there is always a point or, more precisely, a locus of points at the base of the orogen, that mark the lower limit of mass transfer (i.e. accretion) across the plate boundary (Willett et al. 1993; Beaumont et al. 1994; Waschbusch & Beaumont 1996). Below this point, which we denote as S, subduction of the downgoing plate occurs beneath a discrete shear zone. The velocity of S, designated as $V_a$, describes retreat or advance of the subduction zone. As shown in Fig. 4, our convention is to define $V_c$ and $V_a$ relative to a fixed overriding plate, with positive towards the overriding plate. A positive $V_a$ implies advance of the subduction zone towards the overriding plate; a negative $V_a$ represents slab rollback and motion of the subduction boundary away from the overriding plate. Waschbusch & Beaumont (1996) used a slightly different convention. In their notation, $V_c = V_P - V_R$ and $V_a = V_S - V_R$, where $V_P$ and $V_R$ are the absolute velocities of the subducting (pro-) and overriding (retro-) plates, and $V_S$ is the horizontal migration velocity of the subducted slab relative to the deep mantle.

Subduction-zone advance ($V_a > 0$) describes the motion of the S point towards the overriding plate. Advance must be accompanied by contraction of the overriding plate. A modern example is western South America where crustal thickening and growth of the Andean orogenic belt has occurred without any significant accretion from the subducting Nazca plate. Without accretion, shortening of the South American plate must be accompanied by eastward motion of the Nazca slab (Pardo-Casas & Molnar 1987; Isacks 1988; Pope & Willett 1998). Note that outward growth of an orogen is a natural consequence of the addition of mass by accretion and will occur independent of the velocity of S. As a result, the advance of the retro-side deformation front does not require the advance of the underlying subduction zone.

When $V_a < 0$, the S point migrates away from the overriding plate, defining the case of subduction retreat. Slab retreat or rollback has long been recognized as an important factor influencing the style of deformation in convergent orogens (Uyeda & Kanamori 1979; Dewey 1980; Jarrard 1986; Royden & Burchfiel 1989; Waschbusch & Beaumont 1996). Examples of retreating convergent orogens include the Mariana margin in the western Pacific and the Hellenic margin of Greece. In an oceanic setting, slab retreat leads to extension of the upper plate and formation of a back-arc basin. In a continental setting, the consequences are less clear, but as the slab retreats away from the overriding plate, the position of active contraction migrates away from the upper plate, creating additional space to accommodate accreted material. If retreat creates space faster than accretion can fill it, then the upper plate might respond by horizontal extension.

The deformation within a convergent orogen is also controlled by the distribution of fluxes around the orogen. Accretion and surface erosion control the fluxes into and out of the orogen. Accretionary fluxes generally enter into the orogen on its pro-side, either by frontal accretion (offscraping) or basal accretion (underplating). Accretion can also occur on the retro-side of the wedge (Willett et al. 1993), but the fluxes are usually much smaller than those on the pro-side, except for the case of subduction-zone advance. A significant erosional flux only occurs when there is extensive subaerial topography, and when that topography has sufficient runoff to develop an integrated river system to carry out eroded material.

As a general rule, frontal accretion and erosion both tend to promote horizontal contraction across an orogen. A fast rate of frontal accretion will tend to cause pervasive contraction throughout the orogen. This case is nicely illustrated by the strain calculations in Dahlen & Suppe (1988). Waschbusch & Beaumont (1996) show that at retreating subduction zones, frontal accretion will also result in horizontal contraction in the accreted material. They also show that when rates of accretion are low or nil, subduction-zone retreat can cause mixed contraction and extension in pre-existing rocks that border the subduction zone. A fast rate of erosion in the Olympic Mountains of the Cascadia fore-arc high has caused pronounced vertical extension in the most deeply exhumed part of the accretionary wedge (Brandon et al. 1998; Brandon & Fletcher 1998). In contrast, a fast rate of underplating combined with a slow rate of erosion should cause horizontal extension in the upper rear part of the orogen (Platt 1986, 1993; Brandon & Fletcher 1998). A possible example might be the Apennine thrust belt, which shows active extension in the internal part of the orogen (Elter et al. 1975; Patacca et al. 1993). Fast underplating might be expected given that the Apennines saw a transition over the last 5 Ma from oceanic subduction to subduction of the passive margin of the Adriatic continental block (Dewey et al. 1989). An alternative explanation (Royden 1993) is that extension is caused by subduction-zone retreat.

## What gets exhumed?

Exhumation occurs at a variety of tectonic settings (Fig. 5), but mainly at oceanic rifts and transform faults, continental rift zones, subduction zones, and at continent–continent collision zones. Here we summarize the types of rocks that are exhumed in these settings. In general, we find that each tectonic setting has a maximum exhumation depth, with oceanic rifts and transforms showing the shallowest exhumation depths (c. 10 km) and continental collision zones, the deepest (>125 km).

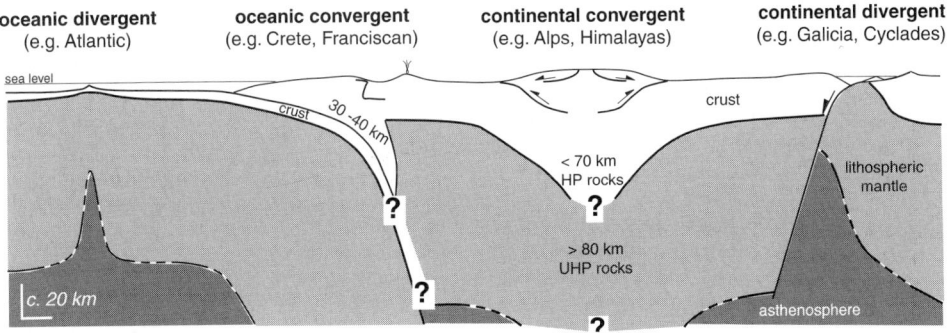

**Fig. 5.** Idealized cross section showing depth range of metamorphism in ocean-convergent, continental-convergent, and continental-rift and oceanic-rift settings.

## Oceanic rifts and transforms

This setting is charaterized by shallow exhumation from depths of *c.* 10 km. Given the thin crust for oceanic lithosphere, this depth is sufficient to expose the mantle, as evidenced by spectacular local exposure on the seafloor of the serpentinized peridotites. In the modern oceans, exhumed mantle is most commonly found along long-offset transform faults and 'under-fed' ridge crests. The exposed peridotites are plagioclase- and spinel-bearing, indicating shallow mantle rocks.

Exhumed mantle has also been observed in the deep ocean in association with extreme rifting. Off the Spanish/Portuguese coast between the supposedly oceanic crust (Grau *et al.* 1973) of the Iberian abyssal plain and the Galicia margin, serpentinized peridotite derived from the upper mantle is largely buried beneath sediments but also crops out locally (Boillot *et al.* 1980). The Galicia margin is made up of a number of tilted blocks formed during continental rifting. According to Boillot *et al.* (1980), the serpentinized peridotite is thought to be the result of serpentinite diapirism and tectonic unroofing of mantle rocks along the rift axis of the margin just before sea-floor spreading started between Galicia and Newfoundland. The setting would thus be transitional from continental to oceanic rifting.

Ophicalcites record ancient examples of exhumed oceanic lithosphere. An ophicalcite is a sedimentary breccia made up of mafic and ultramafic clasts set in a pelagic carbonate matrix (e.g. Lemoine 1980; Bernoulli & Weissert 1985). They are thought to form in intra-oceanic settings. The clasts provide a clear record that all levels of the underlying oceanic lithosphere were exposed at the seafloor.

## Continental rifts

The rift shoulders of the East African rift expose mainly middle to lower crustal Precambrian gneisses, but those rocks are known to have been mostly exhumed prior to the onset of Cenozoic rifting. The rift process itself appears to have caused only minor exhumation. Continental rifting in the Basin-and-Range province of western North America locally resulted in exhumation of the upper and middle crust (e.g. Applegate & Hodges 1995). However, despite large-magnitude extension (about 100%, Wernicke *et al.* 1988), no high-pressure metamorphic rocks (i.e. >10 kbar) are found at the surface.

Deeper exhumation in continental rift zones has occurred in the Cyclades of Greece, where blueschist and eclogite from *c.* 50 km depth are locally exposed (Schliestedt *et al.* 1987; Okrusch & Bröcker 1990). Divergence started there sometime after the middle Oligocene (Raouzaios *et al.* 1996; Thomson *et al.* 1998; Ring *et al.* 1999), but yet we know that much of the exhumation of the Cycladic high-pressure rocks occurred earlier, during the Eocene and Early Oligocene, shortly after the rocks were subducted and accreted (Avigad *et al.* 1997; Ring *et al.* 1999). Forster & Lister (this volume) argue that deep exhumation in the Cyclades is a result of multiple episodes of normal faulting, with any single event involving only a modest amount of exhumation.

Collectively, these examples suggest that individual episodes of continental rifting will exhume rocks from depths no greater than *c.* 25 km. Nevertheless, it may be difficult to tell whether or not tectonic exhumation was caused by divergent plate motions (rifting) or by synconvergent extension, as highlighted by the discussion between Andersen (1993) and Fossen (1993) concerning exhumation in the Norwegian Caledonides.

## Subduction zones

Subduction zones expose a wider variety of deeply exhumed rocks. In a number of cases, the metamorphic grade is no greater than blueschist facies, indicating exhumation from depths of about 30–40 km, which is the maximum thickness of modern subduction-related accretionary wedges.

The Mariana convergent margin (Fryer 1996) displays an intriguing example of deep exhumation. Cold intrusions of serpentinite are found as conical volcano-like features littering the seafloor of the Mariana forearc. Rare blueschist minerals are locally found in association with these serpentinite diapirs. For this case, exhumation of mantle and high-pressure rocks is thought to be driven by the buoyancy caused by serpentinization of mantle peridotites. Serpentinization can cause a decrease in density from 3300 kg m$^{-3}$ to as low as 2600 kg m$^{-3}$ (−22%; Christensen & Mooney 1995).

Eclogite-facies mafic rocks (i.e. eclogites) have been recognized at many ancient subduction zones. These rocks may occur as isolated blocks, in association with lenses and blocks of serpentinized peridotite. The evidence usually indicates that the eclogites were severely disrupted and dismembered after metamorphism. Pressure estimates indicate exhumation from 50 km and deeper. The Central Belt of the Franciscan subduction complex of coastal California serves as a classic example of disrupted eclogites in a subduction zone setting. The 'knockers' are made up of typical oceanic basalts. Some are surrounded by an actinolitic rind, indicating that they resided for some time in a serpentinite matrix (Coleman & Lanphere 1971; Moore 1984).

The hanging wall of the subduction zone is, at least in some cases, invoked as a source for these deep rocks (e.g. Coleman & Lanphere 1971; Platt 1975; Moore 1984 for the Franciscan subduction complex). The reason is that modern subduction zones have well-defined Benioff zones indicating that the lower-plate mantle is subducted, not accreted. This observation suggests that eclogite blocks were formed by accretion of mafic crust from the subducting slab into the mantle of the overriding plate. The exhumation of these rocks remains poorly understood, although the observation of exhumation of high-pressure metamorphic minerals in the Mariana forearc suggests that serpentinite diapirs might be involved in exhumation of eclogite blocks as well.

## Continental collision zones

The internal zones of collisional belts expose the widest variety of exhumed rocks. Schist and gneiss from the upper and middle crust are commonly exhumed in this setting. Some convergent orogens, such as the Delamerian orogen of south Australia, the Mount Isa orogen of northeast Australia, and the Rocky Mountains of the western US, are almost entirely made up of upper and middle crustal rocks and generally lack exposed high-pressure granulite, blueschist, or eclogite (i.e. rocks characterized by metamorphic pressures <10 kbar). We suspect that high-pressure metamorphic rocks are formed in these settings but the exhumation processes operating there were somehow unable to bring those rocks to the surface.

Other collisional belts show clear evidence of crustal thickening and deep exhumation during orogenesis, as indicated by the general occurrence of metamorphic rocks from > 40 km depth. A classic example is the Sesia zone in the Italian Alps (Compagnoni & Maffeo 1973), which contains high-pressure continental crustal rocks that were metamorphosed at the base of a thickened crustal root, and then subsequently exhumed. Pressure estimates indicate metamorphism at c. 70 km, which is consistent with the maximum Moho depths in modern orogens (e.g. Meissner 1986; Christensen & Mooney 1995).

Less common within collisional orogens are deeply exhumed oceanic assemblages. An interesting example is the Zermatt-Saas zone of the Swiss-Italian Alps (e.g. Bearth 1956, 1967, 1976), which consists of metabasalts, metacherts, and ultramafite. Metamorphic assemblages indicate both high pressure and ultrahigh-pressure conditions (locally >100 km depth). The lithological association suggests a subduction-zone environment, with initial accretion of crust and mantle from an oceanic lower plate, followed by low-temperature/high-pressure metamorphism. However, age constraints (Reddy *et al.* 1998) suggest that exhumation occurred during continental collision, which distinguishes these rocks from subduction-zone eclogites, like those of the Franciscan complex.

The most challenging examples of deeply exhumed rocks are ultrahigh-pressure (UHP) metamorphic rocks, which are usually found exhumed in continental collisional zones. These rocks were first discovered in the Cima-Lunga nappe (Ernst 1977; Evans & Trommsdorff 1978), the Dora-Maira massif (Chopin 1984), and the Zermatt-Saas zone (Reinecke 1991) of the Alps, and the Western Gneiss region of Norway (Griffin 1987; Smith & Lappin 1989). They are now known from a number of collisional orogens (Coleman & Wang 1995). Ultrahigh-pressure rocks are continental or oceanic crustal rocks that were metamorphosed within the stability field of coesite or diamond (Schreyer 1995; Coleman & Wang 1995). The exposure of these rocks at the Earth's surface demonstrates that continental and oceanic crust can be subducted to depths >100 km and then returned to the surface. A common debate is whether or not UHP rocks formed within highly overthickened crust or within the mantle (see section on UHP rocks below).

Garnet peridotites are also exposed in several collisional orogens such as the Alps (Alpe Arami, Ernst 1977; Evans & Trommsdorff 1978) and the Betic-Rif orogen (e.g. Ronda and Beni Bousera peridotites, Loomis 1975; Pearson *et al.* 1989). For the Ronda and Beni Bousera peridotites, graphite pseudomorphs after diamond indicate depths >125 km (Pearson *et al.* 1989; Tabit *et al.* 1990). The initial stages of exhumation are attributed to Mesozoic rifting (Vissers *et al.* 1995). However, the Alpe Arami peridotite was subjected to metamorphism at depths of >70–100 km during the Alpine orogeny (e.g. Ernst 1977; Evans & Trommsdorff 1978; Becker 1993), which would have postdated Mesozoic rifting. The presence of deep-seated ultramafic rocks at the Earth's surface indicates that orogenic wedges must have sufficient upward flow to overcome the negative buoyancy of these rocks.

## General remarks

Our summary suggests a counter-intuitive result: continental rift zones seem to have only modest potential for deep exhumation, whereas continental collision zones seem to have the greatest potential. We see no simple explanation for this result.

Another highlight is the evidence from the Mariana subduction zone indicating that some high-pressure metamorphic rocks might be brought back to the surface by buoyancy-driven exhumation, triggered by sepentinization of upper-plate mantle. If such serpentinization does occur beneath the forearc region, then we will probably need to reconsider what the seismic Moho means in those settings. Serpentinization can cause a decrease in compressional wave velocities from about 7.7–8.2 km s$^{-1}$ to 5.3–5.5 km s$^{-1}$ (Christensen & Mooney 1995).

## Diagnostic features of different exhumation processes

A difficult question in most orogenic belts, especially the older ones, is the relative contributions of different exhumation processes. Here, we assess features that might be diagnostic of erosion, normal faulting, and ductile thinning. We also evaluate some problems with quantifying the rates of these exhumation processes.

### Erosion

The large volumes of detrital deposits found adjacent to almost all convergent continental orogens provides ample evidence that erosion is a significant exhumation process. The Himalayan foreland and the offshore Bengal and Indus fans preserve an important record of erosional exhumation of the Himalayas (Cerveny et al. 1988; Copeland & Harrison 1990; Burbank et al. 1993). Nie et al. (1994) proposed that the voluminous Songpan–Ganzi flysch was produced during exhumation of the Dabie Shan ultrahigh-pressure rocks. None of this evidence requires that erosion is the sole exhumation process or even the dominant process, but clearly erosion cannot be ignored as a contributing factor.

Erosion is often assumed to be a fairly slow exhumation process, but there is little support for this view. Figure 6 summarizes Milliman & Syvitski's (1992) compilation of modern sediment yield from a global distribution of 280 river drainages. The conclusion is that the average rate of mechanical erosion of the drained part of the continents is $c.$ 0.052 km Ma$^{-1}$. However, more important is the fact that recorded erosion rates reach local values of 5–13 km Ma$^{-1}$ (e.g. Southern Alps of New Zealand and the Taiwan Alps). In fact, 2% of the drained area of the continents has erosion rates > 0.5 km Ma$^{-1}$. It is important to stress that these rates represent drainage-scale averages. Thus, local rates within a drainage could be much greater. Erosion rates associated with warm-based alpine glaciers, such as those of southern Alaska, have rates of 1–100 km Ma$^{-1}$ (Hallet et al. 1996). It is interesting to speculate that an increase in Alpine glaciation, caused by global cooling, increased precipitation, and/or growth of mountainous topography, might play a major role in exhuming metamorphic rocks and in limiting the maximum height of mountains. The important conclusion is that relative to tectonic exhumation, which may be no greater than $c.$ 5–10 km Ma$^{-1}$, surficial erosion can locally be a very fast process.

Fast eroding regions tend to be mountainous, tectonically active, and wet. Conversely, arid climates tend to have slow erosion rates regardless of the amount of topography. At present, arid landscapes cover about one-third of the continents (p. 278 in Bloom 1998). Arid high plateaus, such as Tibet and the Altiplano in the

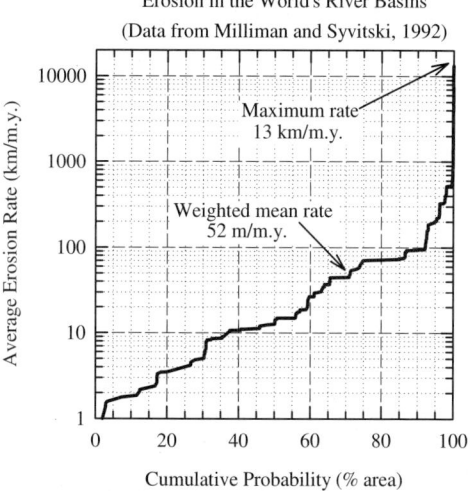

**Fig. 6.** Average erosion rate from the world's river basins plotted against cumulative probability (data compiled from Milliman & Syvitski 1992). The diagram shows that mechanical erosion is typically slow (weighted mean rate $c.$ 0.05 km Ma$^{-1}$), but in some orogens, mechanical erosion can be extremely fast with rates locally exceeding 10 km Ma$^{-1}$.

Andes, serve as examples of mountainous landscapes with little to no erosion. The transition from an arid to a humid climate appears to be marked by a dramatic increase in erosion rates, primarily due to the enhanced transport capacity that comes with a well-drained landscape. Drainage-scale erosion rates are commonly assumed to scale with precipitation or runoff, but regional compilations do not provide clear support for this inference (e.g. Pinet & Souriau 1988; p. 134 in Allen 1997). In other words, erosion rates do not appear to be strongly influenced by climate as long as the climate is not arid. The rate of tectonically driven uplift appears to be a much more important factor in controlling local erosion rates. Recent work by Burbank et al. (1996) and Hovius et al. (1997) have called attention to the importance of bedrock landslides in producing transportable material in well-drained tectonically active landscapes. In this type of setting, the rate of erosion is probably directly driven by the rate of the tectonically-driven uplift, given the usual conclusion that in well-drained humid landscapes, river incision keeps pace with tectonic uplift.

An important question is whether modern erosion rates are representative of long-term erosion rates. We ignore human effects here; most compilations of modern erosion rates have included corrections to remove this factor. Perhaps the most important factor influencing erosion rates on the short-term is climate-induced changes in vegetation cover, which controls the amount of regolith and soil that can be stored on the hill slopes (Bull 1991). A loss of vegetation cover can result in very high short-term erosion rates, but those rates can only be sustained for a short period of time until the stored regolith is striped away. The major climate cycles have periods ranging from 21 to 100 ka, with the 100 ka glacial cycle being especially pronounced during the Quaternary. In this context, long-term erosion rates can be usefully defined as the average rate of erosion for a period of time >100 ka, to ensure that 'short-term' climate variations are averaged out.

Long-term erosion rates can be estimated by reconstructing eroded geological features, by isopach analysis of synorogenic sediments, and by thermochronologic dating of exhumed bedrock or detrital grains in synorogenic sediments. A number of studies have documented long-term syn-orogenic erosion rates ranging from 1 to 15 km Ma$^{-1}$ (e.g. Kamp et al. 1989 for the Southern Alps of New Zealand; Copeland & Harrison 1990 for the Himalayas; Burbank & Beck 1991 for the Salt Range of Pakistan; Fitzgerald et al. 1995 for the central Alaska range; Johnson 1997 for the Betic Cordillera of southern Spain; Brandon et al. 1998 for the Olympic Mountains).

Our conclusion is that erosion cannot be excluded as an exhumation process on the basis of rate alone. For modern orogens, lack of subaerial relief (e.g. submarine accretionary wedge) or an arid climate or pronounced rain shadow (e.g. Tibet) would qualify as credible arguments for generally slow erosion rates, but these arguments are fairly difficult to apply to ancient settings. Isopach analysis provides a direct measure of erosional exhumation, although it is commonly found that synorogenic sediments end up being dispersed to great distances from the orogen. In the Himalayan system, the dispersal is over distances of thousands of kilometres, reaching distant ocean basins and subduction zones. The reason for the wide dispersal is that a collisional foreland basin can never hold more than about half of the sediment produced by erosional lowering of the orogen (England 1981). Wide dispersal makes it difficult, even in the best-preserved orogens, to make a useful comparison between the volume of eroded sediment and the depth of exhumation.

Another factor that has to be taken into account in subduction zone settings is the recycling of eroded material back into the subduction zone. In a similar manner, sedimentary basins adjacent to collision orogens might be overridden and hidden underneath large-slip faults.

Another method that has much promise for distinguishing between normal faulting and erosional exhumation is the use of isotopic cooling ages in synorogenic sediments (e.g. Garver et al. this volume). Detrital ages provide information about the lag-time distribution of the sediment, where lag time is defined as the difference between the cooling age of a detrital grain and its depositional age. For instance, the growth and decay of an erosion-dominated orogen should show a parallel decrease and increase in lag time for the sediment shed from the orogen. The changes observed in the lag-time distribution will be delayed relative to the actual growth and decay of the topography because some time is needed to erode away rocks that have pre-orogenic cooling ages after initial orogenic uplift, and to erode away rocks that have syn-orogenic cooling ages during the post-orogenic decay of the topography. In contrast, after orogenic uplift has stopped normal faulting should be marked by the rapid appearance of sediment with short lag time, and then long periods of time where the lag time increases at the same rate as stratigraphic age. These two features of the lag-time evolution mark the fact that normal faulting brings hot rocks to the surface without any need for erosion. Furthermore, if normal faulting is rapid, much of the upper crust will be reset to a common isotopic cooling age. Brandon & Vance (1992) reported an example like this for sediments shed from the Eocene metamorphic core complexes in the Cordilleran thrust belt of eastern Washington State (USA) and southeast British Columbia (Canada). Sandstone samples spanning from the Eocene through the Early Miocene show peaks in the fission-track grain-age distribution for detrital zircons that remain fixed at c. 43 and c. 57 Ma. These peaks were called static peaks because they did not move with depositional age. They can be related to the slow erosion of the Eocene core complexes after they formed.

## Normal faulting

There is abundant evidence that normal faulting aids the exhumation of metamorphic rocks. The Basin-and-Range province (Armstrong 1972; Crittenden et al. 1980; Davis 1988; Foster & John this volume), the Aegean (Lister et al. 1984; Thomson et al. this volume; Forster & Lister this volume), the Betic Cordillera (Platt & Vissers 1989; Vissers et al. 1995) and the Alps (Mancktelow 1985; Selverstone 1985; Behrmann 1988; Ratschbacher et al. 1991; Ring & Merle 1992; Reddy et al. 1999) are well-documented settings where normal faulting contributed to exhumation.

The most commonly cited evidence for normal faulting is the presence of 'younger over older' or 'low-grade on high-grade' tectonic contacts, where large faults, with low to moderate dips, have placed younger above older rocks or low-grade rocks on high-grade rocks, and in the process have cut out a significant thickness of stratigraphic or metamorphic section. A more diagnostic feature for large-slip normal faults is the juxtaposition of a ductilely deformed footwall with a brittlely deformed hanging wall (e.g. Lister & Davis 1989; Foster & John this volume). The evolution of superimposed sedimentary basins can be used to determine the relative displacement of footwall and hanging wall. These 'piggy-back' style basins also provide a record palaeohorizontal, which can be used to restore underlying faults back to their original orientation. This information, together with the shear sense of a fault, is essential for resolving the nature (reverse or normal in relative sense) of a fault.

The pattern of cooling ages in the field can also provide clues about the exhumation process. Exhumation due to normal faulting commonly results in abrupt breaks in the cooling-age pattern with younger ages in the footwall of the normal fault (Fig. 7a) (Wheeler & Butler 1994). Johnson (1997), Foster & John (this volume) and Thomson et al. (this volume) provide examples from the Betic Cordillera, the Basin-and-Range province and Crete, respectively. In contrast, erosional exhumation should be characterized by a relatively smooth variation in cooling ages across the eroded region (Fig. 7b). Brandon et al. (1998) report a useful example for the erosional exhumed fore-arc high above the Cascadia subduction zone.

Attenuation of stratigraphic or metamorphic units, by itself, is not diagnostic of tectonic thinning by normal faulting (Wheeler & Butler 1994; Ring & Brandon 1994; Ring 1995; Butler & Freeman 1996) (Fig. 8). Thrust faults can thin a stratigraphic or metamorphic section when that section is tilted rearward prior to thrusting. In this case, the thrusts can cut down-section in the direction of transport, while cutting up towards the Earth's surface as thrust faults should.

In some cases thrusts may be rotated by

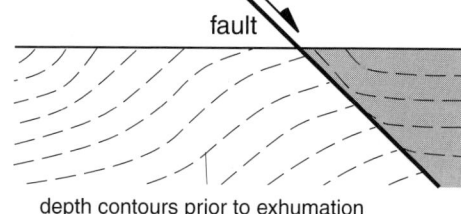

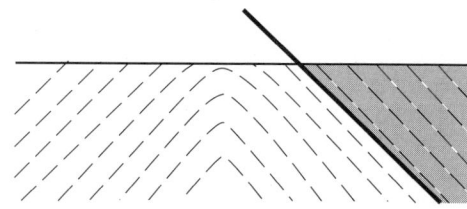

**Fig. 7.** Exhumation pattern as a function of process. Dashed lines show depth contours prior to exhumation. (**a**) Normal faulting will create an asymmetric exhumation pattern. Cooling ages will get younger as one approaches the normal fault (see Foster & John and Thomson et al. this volume). (**b**) Exhumation by erosion alone will commonly result in a broadly domed pattern, which should vary smoothly, independent of faults and local structures.

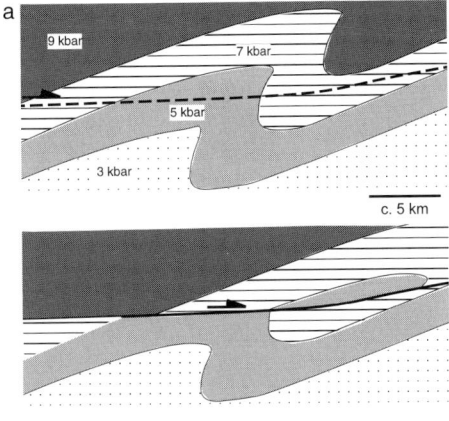

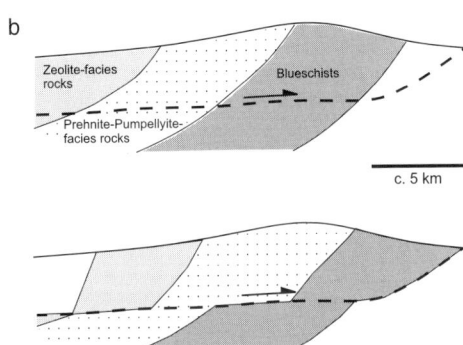

**Fig. 8.** Examples of attenuation of metamorphic section by contractional faults. (**a**) Normal nappe sequence characterized by the tectonic juxtaposition of higher-grade rocks over lower-grade rocks. Nappe sequence is further affected by post-nappe folding. Re-imbrication of the thrust and fold sequence; the re-imbricating contractional fault climbs upwards towards the foreland, but cuts downward through the tilted section resulting in faulted contacts where lower-grade rocks are thrust over higher-grade rocks. Such a multiphase contractional history is commonly found in the internal zones of orogens. (**b**) Contractional fault cutting down through an uplifted metamorphic section, while cutting upward relative to the foreland in the direction of tectonic transport, and thereby attenuating metamorphic or structural section.

folding, block tilting or differential exhumation until their geometry classifies them as normal faults at the front of a thrust culmination (e.g. Price 1981). Conversely, normal faults are sometimes domed by isostatic rebound due to unloading of their footwalls until the fault locally has a thrust sense of offset (Buck 1988; Lister & Davis 1989). Therefore, the kinematic development of a fault has to be related to the $P$–$T$–$t$ evolution in its foot- and hanging wall at the time of tectonic transport, or displacement along a fault relative to the palaeosurface at the time of fault activity has to be demonstrated in order to distinguish between normal faults and thrusts (cf. Wheeler & Butler 1994).

A problem for quantifying the relative contribution of normal faulting to exhumation of deep-seated rocks is that the total offset and original dip of most crustal-scale normal faults, especially those involving deep crustal rocks, are typically poorly resolved. The offset is commonly approximated with barometric estimates from above and below the fault. In this case, it has to be demonstrated that faulting is actually contemporaneous with the pressure break across the fault. Reddy *et al.* (1999) report an example from the Zermatt–Saas zone where the break in metamorphic pressure across a shear zone allowed them to constrain a total of 30 km of exhumation by normal faulting within 9 Ma, indicating an average exhumation rate of 3.5 km $Ma^{-1}$.

Only a few studies have attempted to measure slip rates for normal faults. The greatest reported slip rates come from some extensional detachments in the Basin-and-Range province and are of the order of 7–9 km $Ma^{-1}$ (Davis & Lister 1988; Spencer & Reynolds 1991; Foster *et al.* 1993; Scott *et al.* 1999; Foster & John this volume). There is growing evidence that many of these 'high-speed' extensional detachments had an initially gentle dip (<30°) (Scott & Lister 1992; Foster & John this volume). Other estimates for exhumation by extensional detachments in the Basin and Range indicate lower rates of about 1–2 km $Ma^{-1}$ (see review in Foster & John this volume). Average rates from other orogens appear to be generally slow: (1) <2 km $Ma^{-1}$ for a 10 Ma interval for the Tauern window in the Eastern Alps (Frisch *et al.* 1998) and, (2) 0.6–0.9 km $Ma^{-1}$ and <0.6–0.7 km $Ma^{-1}$, respectively, for a 20 Ma interval for the Menderes Massif in western Turkey (Hetzel *et al.* unpublished data; Ring unpublished data).

Normal faults in long-lived continental rifts also appear to have relatively low slip rates and slip generally occurred along steeply dipping (*c.* 60°) normal faults. In the Miocene to Recent East African rift system, apatite fission-track ages are, in general, not reset by the young rifting events (Foster & Gleadow 1996), which indicates a slip rate for the normal faults of <0.3 km $Ma^{-1}$. At the Livingstone escarpment of the northern Malawi sector of the East African rift system, the Miocene to Recent offset at the escarpment yields similar low slip rates (Ring unpublished data).

The exhumation rate during normal faulting is a function of both slip rate and fault dip. A normal fault with a 30° dip that slipped at a rate of 7–9 km Ma$^{-1}$ would exhume rocks at a rate of 3.5–4.5 km Ma$^{-1}$. If this detachment had an original dip as gentle as 20°, it would exhume rocks at a rate of 2.4–3.1 km Ma$^{-1}$. It is important to note that most slip rates are averaged over relatively long periods of time (see cited time intervals above for the Tauern window and the central Menderes Massif). It is conceivable that normal faulting occurred in pulses at faster rates. Nonetheless, the long-term average appears to have been relatively slow.

## Ductile thinning

Penetrative deformation fabrics present in most exhumed mountain belts indicate that ductile flow is an important process. This process can either aid or hinder exhumation, depending upon whether ductile flow causes vertical thinning as associated with the formation of a subhorizontal foliation, or vertical thickening as associated with the formation of a subvertical foliation (Fig. 2b and c). The presence of a subhorizontal foliation is generally diagnostic of ductile thinning. The general observation of subhorizontal foliations in the internal zones of many orogens shows that ductile thinning commonly aids exhumation (Selverstone 1985; Wallis 1992, 1995; Wallis *et al.* 1993; Platt 1993; Mortimer 1993; Ring 1995; Krabbendam & Dewey 1998; Ring *et al.* 1999). Ductile thinning by itself cannot fully exhume rocks and an additional exhumation process is required to bring rocks back to the Earth's surface (Platt *et al.* 1998; Feehan & Brandon 1999).

In the simplest case, where exhumation is entirely controlled by ductile thinning, exhumation by ductile thinning is given by the average stretch in the vertical because this tells how much the vertical has changed in thickness. However, if exhumation occurs by additional processes as well, it is more difficult to quantify the contribution of ductile thinning to exhumation. In this case, the vertical rate at which a rock moved through its overburden and the rate of thinning of the remaining overburden at each step along the exhumation path have to be considered (Feehan & Brandon 1999). The general conclusion is that the contribution of ductile thinning to exhumation will always be less than that estimated from the vertical stretch only. Simple one-dimensional calculations show that the contribution of ductile thinning in convergent wedges from western North America is less than half that indicated by the estimated finite vertical shortening in the rock (Feehan & Brandon 1999; Ring & Brandon this volume). It follows that huge vertical strains are needed for ductile thinning to make a significant contribution to exhumation. Vertical contraction on the order of 70% has indeed been documented, for example, by Norris & Bishop (1990) and Maxelon *et al.* (1998) from the interior of the Otago accretionary wedge, exposed on the South Island of New Zealand. Dewey *et al.* (1993) also reported vertical contraction on the order of 70–80% from the Western Gneiss region. The studies by Feehan & Brandon (1999) and Ring & Brandon (this volume) show that ductile thinning operated at low rates of <0.3 km Ma$^{-1}$. Recent modelling by Platt *et al.* (1998) in the Betic Cordillera argues for vertical shortening of 75% and ductile thinning operating at a rate of 4.5 km Ma$^{-1}$.

Platt (1993) pointed out that thrusting alone can not tectonically exhume rocks. This is certainly true if thrusting is confined to a thin zone at the base of a sequence of nappes, which is commonly the case in the upper crust. However, in the internal zones of convergent orogens, nappes usually show a pervasive degree of internal deformation and generally flat-lying foliations. The foliations show that nappe stacking was associated with pronounced vertical thinning, which would ultimately contribute to exhumation of the nappes. In this regard, it is interesting to note that high-pressure belts, as a rule, occur above lower grade units indicating that the overburden of the high-pressure nappes must have been reduced prior to their emplacement above the lower pressure units. Pervasive ductile flow in the hanging wall associated with thrusting of the high-pressure nappe might have aided the exhumation of the latter. Note that vertical ductile thinning might be coupled with erosion and/or normal faulting in upper parts of the nappe pile.

## *P–T–t* data and exhumation processes

### *P–T* paths

The shape of *P–T* paths is sometimes used to distinguish exhumation processes. The shape of a *P–T* path is typically difficult to resolve, especially the youngest part of the path, which is probably the most diagnostic of process (J. Selverstone, pers. comm. 1996). Nonetheless, fast initial exhumation rates of deeply buried rocks should generally result in near-isothermal decompression. Such *P–T* paths are known, for instance, from the garnet-oligoclase-facies rocks of the Southern Alps of New Zealand (Holm et al. 1989; Grapes 1995) and the Mulhacen nappe of the Betic Cordillera (e.g. Gomez-Pugnaire & Fernandez-Soler 1987) (paths 1 and 2 in Fig. 9). The Southern Alps were dominantly exhumed by erosion and the Mulhacen nappe is thought to be exhumed primarily by normal faulting. Nonetheless, the shapes of the P–T paths from both units are virtually indistinguishable (Fig. 9).

T–t *paths*

A number of isotopic studies have used evidence of rapid cooling to argue for rapid exhumation (e.g. Copeland et al. 1987; Zeck et al. 1992). Rapid exhumation of deep-seated rocks causes upward advection of heat and thus retards cooling and promotes isothermal decompression. The result is a relatively high thermal gradient and compressed isotherms in the upper crust (Fig. 10). In this situation, rocks can quickly move through the closely spaced isotherms without being much exhumed. The closure-temperature concept (Dodson 1973) indicates that a radiogenic isotopic system will record the time when a dated mineral cools below its effective closer temperature. However, there is no 'closure pressure'. Thus, for cases of near-isothermal decompression typical for fast exhumation (path 3 in Fig. 9), it is difficult to estimate the depth where the isotopic system closed. As a result, exhumation rates will be poorly resolved. Only hairpin P–T

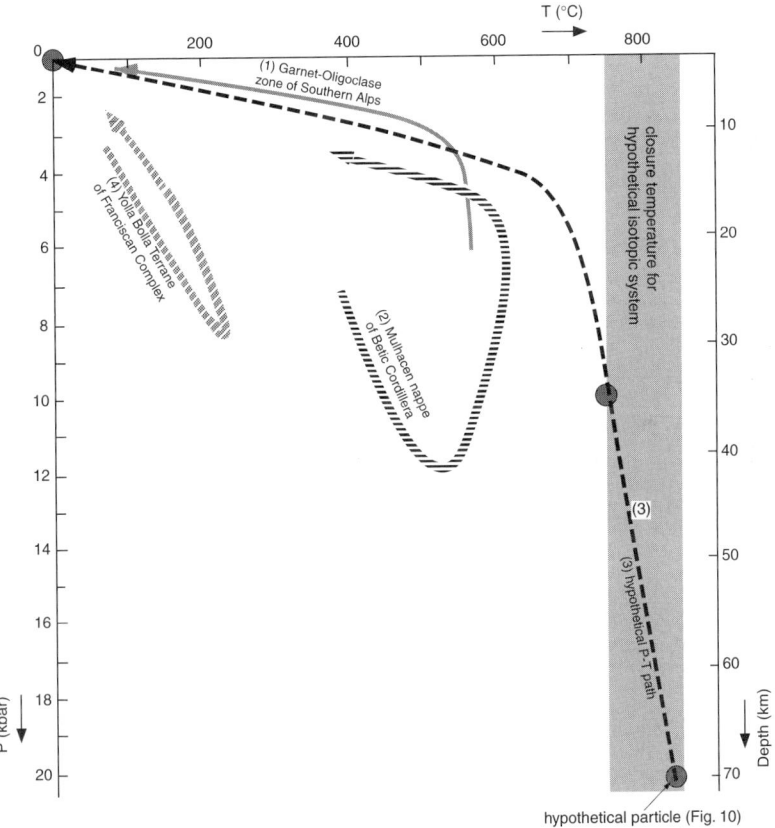

**Fig. 9.** Selected *P–T* paths from exhumation settings that are controlled (1) mainly by erosion, e.g. Southern Alps of New Zealand (Holm et al. 1989; Grapes & Wantanabe 1995), and (2) mainly by crustal extension, e.g. Mulhacen nappe of Betic Cordillera in southern Spain (Gomez-Pugnaire & Fernandez-Soler 1987). Path (3) shows a hypothetical isothermal decompression curve, which illustrates the problem of applying the 'closure-temperature concept' for estimating exhumation rates. Dark grey dot shows *P–T* evolution of the hypothetical particle shown in Fig. 10. The hairpin shape of path (4) from the Eastern Belt of the Franciscan subduction complex of California (e.g. Ernst 1993) is probably caused by slow exhumation in a subduction-zone setting.

paths like that for curve 4 in Fig. 9 from the Franciscan subduction complex have cooling rates that can be simply related to exhumation rates. This situation seems to hold only when exchumation rates are <c. 1 km Ma$^{-1}$.

## P–t data

P–t information provides probably the most powerful tool for resolving exhumation rates. This is because metamorphic pressure can be directly related to depth by assuming an average density for the crustal column. P–t data are relatively easy to obtain in granitoids (e.g. magmatic pressure using the Al-in-hornblende barometer and magmatic age from U-Pb dating of zircon). In metamorphic rocks, the relation between pressure and time can only be determined by correlating barometric data to the thermal history data, as obtained from petrology and isotopic dating. Alternatively, dating of the crystallization age of minerals such as phengite, from which minimum metamorphic pressure can be obtained (Massone 1995), may provide a means for constraining a P–t history.

Reddy et al. (1999) showed that minimum pressure estimates for phengite crystallization show no correlation with age within a shear zone in the hanging wall of the Zermatt–Saas eclogites. The ages for phengite crystallization range from 36 to 45 Ma (Reddy et al. 1999). They concluded that the hanging wall of the shear zone was not exhumed while the shear zone was active (duration c. 9 Ma). These data, in conjunction with the break in pressure across the shear zone, allowed them to constrain a rate of 3.5 km Ma$^{-1}$ for exhumation of the Zermatt–Sass eclogites by normal faulting.

## Ultrahigh-pressure rocks

Surface exposures of UHP metamorphic rocks are a showcase for exhumation in the extreme. The mode of exhumation remains uncertain, but there seems to be a growing consensus that several processes, operating at different levels in the lithosphere, are involved. The evidence suggests that UHP metamorphism mainly occurs within collisional orogens, but where within the orogen? More specifically, does

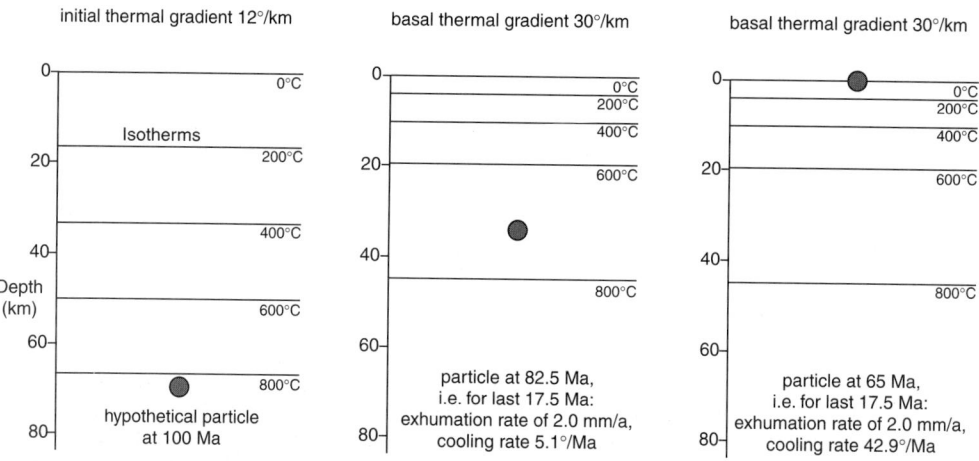

**Fig. 10.** Illustration of possible effects of rapid exhumation. A P–T path for the hypothetical particle considered in this example is shown in Fig. 9. (**a**) Depth arrangement of isotherms during rapid underthrusting of a particle to a depth of 70 km, assuming an initial thermal gradient of 12°C km$^{-1}$. (**b**) Accretion is followed by exhumation at a constant rate of 2 km Ma$^{-1}$, which initiates a new, relatively steep thermal gradient (i.e. rapid exhumation results in a very modest cooling rate of 5°C Ma$^{-1}$ averaged over the first 17.5 Ma of exhumation). The development of the isotherms in this example has been modelled using a one-dimensional steady-state model where accretion is balanced by erosion. The model assumes that temperatures at the surface and at the base of a layer remain fixed at their set values of 0 °C and 840 °C, respectively. These set values can be viewed as describing a basal geotherm of 30°C km$^{-1}$. The thermal diffusivity ($\kappa$) in our example is 32 km$^2$ Ma$^{-1}$, which is typical for collisional belts such as the Himalayas (Zeitler 1985). For model details, see appendix of Brandon et al. (1998). (**c**) Final exhumation of particle at the same rate as in (b), but average cooling rate is now more than eight times higher as in (b) (43°C Ma$^{-1}$, averaged over 17.5 Ma).

metamorphism occur within the crust or the mantle?

Ernst *et al.* (1997) argues that ultrahigh-pressure metamorphism is caused by subduction of continental crust along an oceanic subduction zone. The continental crust is embedded in the oceanic lithosphere, so that slab pull is able to drag the crust down to depths of >100–125 km. The crust somehow detaches and returns to the surface, driven mainly by its strong positive buoyancy relative to the surrounding mantle.

This interpretation is certainly plausible, but it relies heavily on the early presence of an oceanic subduction zone to account for deep burial and metamophism. Early subduction of oceanic lithosphere is supported by the occurrences of ophiolitic rocks in some ultrahigh pressure metamorphic terrains (i.e. Zermatt–Sass zone described above). However, ultrahigh-pressure metamorphic rocks are generally found in Mediterranean-style continental collision zones with little to no evidence of a pre-collisional magmatic arc (Ernst *et al.* 1997). Thus, one is left to question if there was a well-organized pre-collisional oceanic subduction zone during UHP metamorphism. This problem has caused us to consider if it might be possible to form ultrahigh-pressure metamorphic rocks within the collisional orogen itself, and without appealing to an early oceanic subduction zone.

## Orogenic root models

The pressures associated with UHP metamorphism seem incompatible with metamorphism in thickened continental crust of an average crustal density (2830 kg m$^{-3}$, Christensen & Mooney 1995). Assuming a reasonable metamorphic temperature of 800°C, this crust would have to be 100 km thick to produce the 29 kbar conditions needed to convert quartz into coesite and a thickness of >120 km, corresponding to pressures of >35 kbar, to stabilize diamond. At 100 km crustal thickness, the resulting orogenic topography would have a mean elevation of >10 km, which is twice the height of the highest orogenic topography on present Earth. It seems unlikely that such high topography could be supported for any geologically reasonable length of time (e.g. Bird 1991). Some other explanation is needed. We explore two ideas (Figs 11 and 12), both of which invoke a thicker and denser orogenic root, either due to the inclusion of lithospheric mantle (e.g. Molnar *et al.* 1993) or eclogitized lower crust (Dewey *et al.* 1993).

For the first option, the orogenic root is made up of crust and a fraction δ of the incoming lithospheric mantle (Fig. 11a). Deeper lithospheric mantle does not thicken and is thus passively depressed or subducted into the asthenosphere. UHP metamorphism is inferred to take place beneath the Moho, within the mantle part of the orogenic root. Some type of fault imbrication is required to interleave continental and oceanic crust rocks within this mantle root. This seems plausible because the lithospheric mantle would be expected to behave in a brittle fashion given the relatively low temperatures of typical UHP metamorphism (<800 to 900°C).

To explore the implications of the model, we assume the following simplified scenario. An incoming continental platform with an initial crustal thickness $h_c = 40$ km and a mean elevation $z = 0$ km (sea level), and a thickened crust $H_c = 70$ km within the orogen. The 70 km thickness would be representative of the Moho in a collisional orogen (Christensen & Mooney 1995). The initial thickness of the lithospheric mantle is set at $h_m = 100$ km (Molnar *et al.* 1993). The crust and lithospheric mantle are constrained to thicken by the same amount,

$$S_v = H_c/h_c = \delta H_m/\delta h_m = 1.75$$

assuming the representative values above for $H_c$ and $h_c$. Assuming isostatic equilibrium, we can predict the maximum thickness of the orogenic root,

$$(H_c + \delta H_m) = S_v (h_c + \delta h_m),$$

the mean elevation of the orogen above sea level,

$$z = [((\rho_a - \rho_c)/\rho_a) h_c + ((\rho_a - \rho_m)/\rho_a) \delta h_m] (S_v - 1),$$

and the pressure at the base of the orogen,

$$P_{base} = (\rho_c h_c + \rho_m \delta h_m) S_v g,$$

where g is the acceleration of gravity.

These results are illustrated in Fig. 11b–d, using $S_v = 1.75$. About 10–40% of the incoming lithospheric mantle would have to be accreted to the orogenic root to get pressures sufficient for UHP metamorphism. The resulting mean elevation would be c. 3 km. For this model, the metamorphic pressure needed for UHP metamorphism is provided by the negative buoyancy of the mantle portion of the root. Note, however, that the amount of interleaved crustal rocks in the mantle part of the root must remain small to retain sufficient negative buoyancy within the root.

If this explanation is correct, we are still left to explain how the crustal rocks separate from the mantle part of the root and return to the surface. Houseman *et al.* (1981) and Molnar *et al.* (1993)

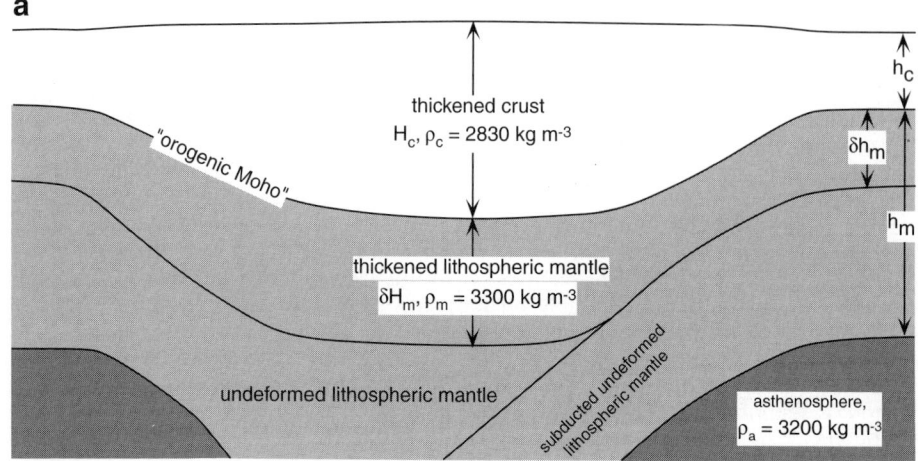

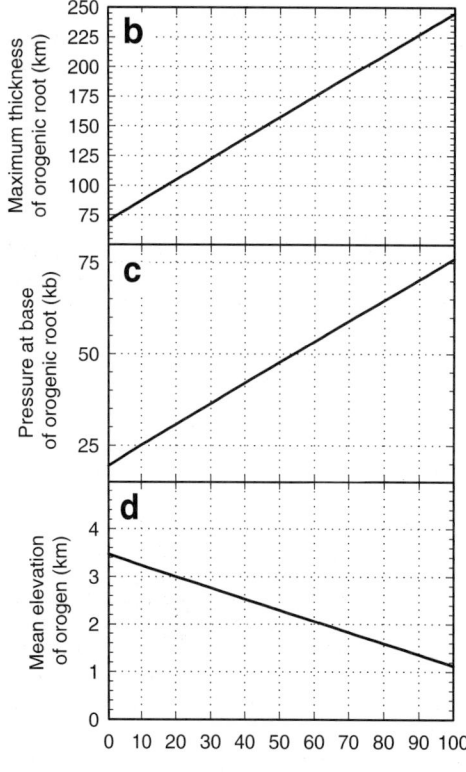

**Fig. 11.** A thickened lithospheric mantle model for the origin of UHP metamorphic rocks. (**a**) The orogenic root is formed by accretion of both crust and a specified fraction of lithospheric mantle. Average densities are from Christensen & Mooney (1995), Molnar *et al.* (1993) and Houseman & Molnar (1997). (**b–d**) The results of isostatic calculations for the maximum thickness, basal pressure and elevation of the orogen. This calculation assumes uniform thickening of the crust and accreted lithospheric mantle by a factor $S_V = 1.75$, which gives a final crustal thickness of 70 km, comparable to maximum Moho depths in modern collisional orogens. Note that mean elevation is greatest when no lithospheric mantle is accreted because of the greater density of the lithospheric mantle relative to asthenospheric mantle.

have argued that the gravitationally unstable lithospheric mantle in the root will ultimately detach and sink into the asthenosphere. The rise time for this detachment process depends on time constants for thermal relaxation of the root and for viscous flow associated with the Raleigh–Taylor instability (see Molnar *et al.* 1993 for a recent analysis). A similar rise time is probably associated with separation of the more buoyant UHP crustal rocks from the mantle lithosphere (Wallis *et al.* 1998). Buoyant rise or diapirism may account for how UHP rocks reach

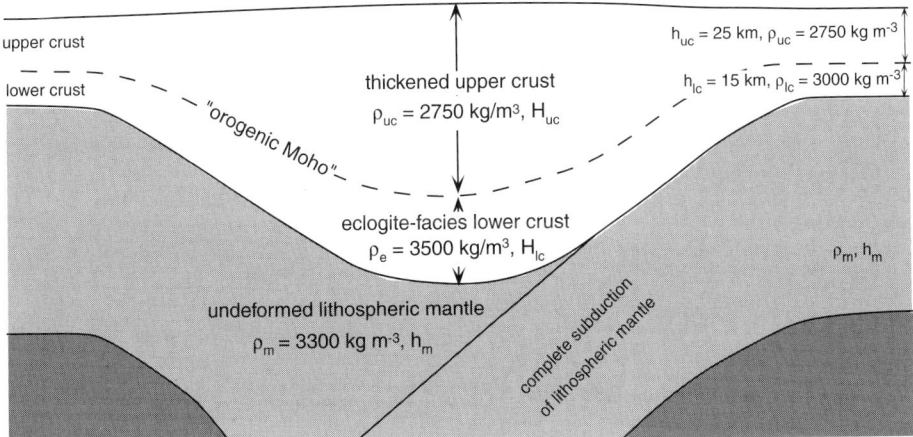

**Fig. 12.** An eclogitized lower crust model for the origin of UHP metamorphic rocks. See text for details.

the base of the crust, perhaps at the same time as the mantle root descends into the asthenosphere. This interpretation remains speculative but provides a stimulating view of how UHP crustal rocks might start to make their way back to the surface.

We consider a second option, that eclogization of the mafic lower crust provides the thickness and pressure needed for UHP metamorphism (Fig. 12). According to Christensen & Mooney (1995), the lower crust of the continents, from 25 to 40 km on average, is mainly mafic in composition with a density of 2900–3100 kg m$^{-3}$. At depths $>c.$ 50 km, corresponding to pressures $>c.$ 14 kbar, these rocks would be converted to eclogite with an average density ($\rho_e$) of $c.$ 3500 kg m$^{-3}$ and P-wave velocities of $c.$ 8 km s$^{-1}$ (see Christensen & Mooney 1995, Table 4, p. 9775). In this case, the Moho would no longer mark the top of the mantle, but rather the transition into eclogite-facies mafic crustal rocks (Dewey et al. 1993). The high density cited above only applies to eclogite, and not to the more silicic coesite- and diamond-bearing rocks that have come to define the UHP metamorphic problem. The UHP silicic rocks are not eclogites. None the less, they belong to the higher pressure part of the eclogite facies, and can be referred to as the coesite or diamond subfacies. As in the previous interpretation, we must argue that the eclogite-rich crustal root would contain a small fraction of structurally interleaved silicic UHP metamorphic rocks.

To explore the implications of the eclogite model, we again start with a 40 km thick crust with a mean elevation at sea level, an upper crustal thickness $h_{uc} = 25$ km, and a lower mafic crustal thickness $h_{lc} = 15$ km. We examine the extreme case where the crust above the Moho is formed from the upper crust by itself. $H_{uc}$ is set to 70 km, which we use again as our representative depth for Moho in a collisional orogen. The upper and lower crust are assumed to thicken by the same amount, which means that

$$S_v = H_{uc}/h_{uc} = H_{lc}/h_{lc} = 2.8$$

assuming the representative values for $H_{uc}$ and $h_{uc}$. Isostatic balance gives the following equations for the maximum thickness of the orogenic root:

$$(H_{uc} + H_{lc}) = S_v (h_{uc} + h_{lc}),$$

mean elevation:

$$z = [((\rho_a - \rho_{uc}) / \rho_a) h_{uc} + ((\rho_a - \rho_{lc}) / \rho_a) h_{lc}] \\ (S_v - 1) - ((\rho_e - \rho_{lc}) / \rho_a) S_v h_{lc}$$

and pressure at the base of the orogenic root:

$$P_{base} = [(h_{uc} \rho_{uc}) + (h_{lc} \rho_e)] S_v g.$$

For this model, our equations predict an orogenic root that is 112 km thick with an average elevation of only 1.5 km and a maximum pressure at the base of the root of 33 kbar. The pressure is sufficient for coesite stability but diamond would only be stable if temperatures were well below 700°C. A thicker orogen would be needed to account for the full range of pressures observed for UHP metamorphic suites but we cannot see how this model could be used to generate a thicker orogen without making the Moho deeper than 70 km.

For this model, the orogenic root is composed entirely of crust, but the ecologitized lower crust is gravitationally unstable and thus prone to detachment as described above for the thickened lithospheric mantle model. A critical problem is that the crustal root described here may be so weak that it could not persist for any significant time before detaching and sinking into the asthenosphere. Lithospheric mantle is stronger and should be able to persist in a thickened form for a longer period of time. Given this factor, plus the greater range of possible metamorphic pressures, the thickened lithospheric mantle model seems to provide a better explanation for the origin of UHP metamorphism. Note however that neither model is exclusive and that a mafic lower crust would become eclogitized if it was thickened to depths $>c.$ 50 km.

## Other unresolved issues

There are many additional unresolved issues regarding ultrahigh-pressure rocks, some of which are listed here.

(1) *What is the thickness of UHP units?* It has been proposed that ultrahigh-pressure rocks are typically found as relatively large (i.e. up to hundreds of $km^2$) internally coherent nappes, which are only a few kilometres thick (Ernst *et al.* 1997). Such a statement is certainly true for the well-mapped ultrahigh-pressure Brossasco-Isasca unit of the Dora Maira Massif which is sandwiched between lower pressure nappes (Chopin 1984; Chopin *et al.* 1991). However, in other UHP nappes the regional extent and the thickness of the nappes are poorly constrained. For example, diamond-bearing eclogites in the Erzgebirge of eastern Germany apparently represent very small lenses (< hundreds of metres in diameter) within high-pressure gneiss (Massone 1999). The average thickness of UHP nappes is crucial for understanding their thermal history and also their rheologic behaviour during exhumation.

(2) *What is the relationship of ultrahigh-pressure metamorphism to magmatism?* The high radiogenic heat production typical of continental rocks should lead to thermally-induced melting since ultrahigh-pressure metamorphism is generally at conditions above the 'wet' solidus for granitic melts (Huang & Wyllie 1975). The local presence of melt has been discussed by Schreyer *et al.* (1987), Schreyer (1995), Phillipot (1993) and Sharp *et al.* (1993) but clear evidence for melting remains scarce. Most people argue that the exhumation path of ultrahigh-pressure rocks is characterized by cooling during decompression (Roberto Compagnoni, pers. comm. 1996, 1997; Ernst *et al.* 1997). The exhumation-related cooling suggest that the ultrahigh-pressure rocks were thrust onto fairly cold foreland units, which were capable of cooling down hundreds of square kilometre-sized nappes quickly. Rapid exhumation and quick cooling might have prevented melting. However, the initial size of an UHP nappe is critical in this regard.

(3) *Is the preservation of ultrahigh-pressure assemblages in continental basement consistent with dry metamorphic conditions during exhumation?* A relatively dry granitic basement rock (e.g. Dora Maira Massif, Western Gneiss region) typically contains about 2 wt% structurally bound water. To transform dry mineral assemblages to high- and ultrahigh-pressure assemblages, one has to hydrate the rocks to stabilize minerals like talc and phengite in such rocks. The presence of a fluid phase during ultrahigh-pressure metamorphism has also been inferred from mass-transfer processes and fluid-inclusion evidence (Schreyer 1995; Harley & Carswell 1995; Phillipot *et al.* 1995). Oxygen-isotope studies in the ultrahigh-pressure rocks of Dabie Shan, China, show that the granitoids where heavily hydrothermally altered by near-surface waters before metamorphism (Rumble *et al.* 1998). This summary suggests that the continental basement is in general not dry before being converted to an ultrahigh-pressure rock. The preservation of ultrahigh-pressure assemblages during exhumation would then suggest that fluid was channelized in shear zones. Localized deformation and fluid flow may have allowed the preservation of ultrahigh-pressure assemblages in shielded blocks.

(4) *How much of the exhumation history of an ultrahigh-pressure terrain is preserved in its present structural and stratigraphic setting?* Standard structural field studies, in conjunction with $P-T$ work, are capable of constraining aspects of the last 20–30 km of the exhumation path. Van der Klauw *et al.* (1997) and Stöckhert & Renner (1998) demonstrated, for instance, that quartz microfabrics in UHP rocks record greenschist-facies deformation. Structures that formed at deeper crustal levels are commonly thought to have been highly or completely obliterated. The preserved high- or even ultrahigh-pressure deformation relics are hard to relate to mapable large-scale nappe contacts or accretion-related structural discontinuities.

A remarkable feature of at least some UHP nappes is the lack of pronounced deformation. The virtually undeformed Variscan Brossasco granite of the Dora Maira Massif preserves its original igneous texture and its intrusive relationship with a surrounding metasedimentary unit (Biino & Compagnoni 1992). In Dabie Shan, some of the rocks preserve a pre-metamorphic hydrothermal alteration by near-surface meteoric waters (Rumble et al. 1998). In both cases, ultrahigh pressure metamorphism caused virtually no change to rock texture or isotopic composition. Stöckhert & Renner (1998) show that the undeformed Brossasco granite indicates that differential stress was too low to cause plastic flow during burial, accretion and exhumation. The only evidence for significant deformation by dislocation creep under ultrahigh-pressure conditions comes from ultrahigh-pressure eclogite from the Zermatt–Saas zone (omphacite microstructures reported by van der Klauw et al. 1997).

(5) *Are there transient accelerations in the rate of the processes involved in exhumation (e.g. Hill et al. 1995)?* Many of the arguments that concern the sustainability of topography and the maintenance of steady-state geotherms, critically depend on the time constants of the processes involved. For example, it is relatively easy to produce a depressed geotherm in the overriding plate, above a subduction zone, as the result of large-scale overthrusting of a back-arc basin, providing an alternative explanation for the synchroneity of ophiolite emplacement and the formation of eclogites and blueschists (see Rawling & Lister this volume). Since relatively thick crust might exist for less than 1 Ma, rapid oscillations in tectonic mode may provide an explanation for many of the paradoxes outlined in the discussions above.

## Concluding remarks – outstanding problems

In most active mountain belts, erosion and tectonism are dynamically coupled to the point where it may be difficult to separate cause and effect. This makes it difficult to distinguish between different exhumation processes. Nonetheless, exhumation typically occurs by multiple processes and there is a need to quantify the relative contributions of the different exhumation processes, using information from metamorphic petrology, isotope thermochronology, structural and kinematic analysis, synorogenic stratigraphy, geomorphology, and palaeo-elevation analysis.

To highlight some of our conclusion, we ask the following four questions.

(1) *How diagnostic are exhumation rates for distinguishing exhumation processes?* We believe that exhumation rates alone cannot distinguish between exhumation processes. Likewise, net exhumation rates do not supply much information on the rates of specific processes. The following example may help to illustrates this point: an erosion rate of 1 km Ma$^{-1}$, a fast slip rate of 3 km Ma$^{-1}$ for a 20°-dipping normal fault (which equals an exhumation rate of 1 km Ma$^{-1}$), and a rate of 1 km Ma$^{-1}$ for ductile thinning would combine to give a total exhumation rate of 3 km Ma$^{-1}$. The fast net rate is not useful in distinguishing between different exhumation processes.

Fast exhumation rates inhibits fast cooling. However, fast cooling may commonly follow rapid exhumation because of the upward advection of heat. There appears to be a serious need to constrain well-defined $T$–$t$ and especially $P$–$t$ paths for exhumed rocks.

(2) *How important is tectonic exhumation?* Tectonic exhumation, especially in the form of normal faulting has been recognized as a common factor in continental orogenesis. A widely held view is that early crustal thickening in convergent continental orogens will generally lead to normal faulting and crustal thinning. The first salient problem is to diagnose unequivocally horizontal crustal extension in orogens. Other problems associated with normal faulting appear to be the depth range of normal faults and the maximum throw on normal faults.

The few data on ductile thinning suggest that it is a slow exhumation process. More quantitative work on ductile thinning, especially in continental collision zones, is needed to demonstrate that this process contributes in a significant way to the exhumation of metamorphic rocks.

(3) *What is the role of erosion?* Erosion appears to be able to operate at very fast rates, perhaps as high as 15 km Ma$^{-1}$, given sufficient precipitation, steep terrain, and comparable uplift rates. There is no reason to indicate that these rates could not be sustained for long periods of time, as long as uplift rates continued to match erosion rates and climate conditions remained favourable for fast erosion. Alpine glaciation appears to be the most aggressive agent of erosion and one that is particularly sensitive to global climate. In this regard, the relatively high sediment production rate of the Quaternary may be the result of a cooler climate and more extensive alpine glaciation. An outstanding problem is that much of our current understanding of erosion rates is based on relatively short records. There is a serious need for better long-term estimates using sediment inventories or thermochronometry.

(4) *Where do UHP rocks form and how are they exhumed?* Ultrahigh-pressure rocks occur only in collisional belts, but they otherwise appear to form below the seismically determined Moho. Coesite-bearing ultrahigh-pressure rocks can form in the root of a highly overthickened

crust when large parts of the more mafic lower crust have been eclogitized. We have shown an example where eclogitized lower crust would be placed beneath the Moho in which case the crust might be about 110 km thick (with the Moho at a depth of c. 70 km). The predicted mean elevation would only be about 1.5 km. Ultrahigh-pressure rocks may also form within a mantle-rich orogenic root. The involvement of lithospheric mantle limits the mean elevation of the orogen to about 3 km. The question that remains unanswered is how do these rocks make their way back to the surface?

This introduction benefited from comments by Christopher Beaumont, Sören Dürr, Gary Ernst, Thomas Flöttmann, Jessica Graybill, Kit Johnson, Bernhard Stöckhert and Simon Turner and a formal review by Alan Roberts.

## References:

ALLEN, P. 1997. *Earth Surface Processes*. Blackwell Science, Oxford.

ANDERSEN, T. B. 1993. The role of extensional tectonics in the Caledonides of south Norway: Discussion. *Journal of Structural Geology*, **15**, 1379–1380.

APPELGATE, J. D. R. & HODGES, K. V. 1995. Mesozoic and Cenozoic extension recorded by metamorphic rocks in the Funeral Mountains, California. *Geological Society of America Bulletin*, **107**, 1063–1076.

ARMSTRONG, R. L. 1972. Low-Angle (Denudation) Faults, Hinterland of the Sevier Orogenic Belt, Eastern Nevada and Western Utah. *Geological Society of America Bulletin*, **83**, 1729–1754.

AVIGAD, D., GARFUNKEL, Z., JOLIVET, L. & AZANON, J. M. 1997. Back arc extension and denudation of Mediterranean eclogites. *Tectonics*, **16**, 924–941.

BECKER, H. 1993. Garnet peridotite and eclogite Sm-Nd mineral ages from the Lepontine dome (Swiss Alps). New evidence for Eocene high pressure metamorphism in the Central Alps. *Geology*, **21**, 599–602.

BEARTH, P. 1956. Geologische Beobachtungen im Grenzgebiet der lepontinischen und penninischen Alpen. *Eclogae Geologicae Helvetiae*, **49**, 279–290.

—— 1967. Die Ophiolithe der Zone von Zermatt-Saas Fee. *Beiträge zur Geologischen Karte der Schweiz*, **130**, 1–132.

—— 1976. Zur Gliederung der Bündnerschiefer in der Region von Zermatt. *Eclogae Geologicae Helvetiae*, **69**, 149–161.

BEAUMONT, C., FULLSACK, P. & HAMILTON, J. 1994. Styles of crustal deformation in compressional orogens caused by subduction of the underlying lithosphere. *Tectonophysics*, **232**, 119–132.

BEHRMANN, J. H. 1988. Crustal-scale extension in a convergent orogen: The Sterzing-Steinach mylonite zone in the Eastern Alps. *Geodinamica Acta*, **2**, 63–73.

BERNOULLI, D. & WEISSERT, H. 1985. Sedimentary fabrics in Alpine ophicalcites, South Pennine Arosa Zone, Switzerland. *Geology*, **13**, 755–758.

BIINO, G. & COMPAGNONI, R. 1992. Very-high pressure metamorphism of the Brossasco coronite metagranite, southern Dora Maira Massif, Western Alps. *Schweizerische Mineralogische und Petrographische Mitteilungen*, **72**, 347–363.

BIRD, P. 1991. Lateral extrusion of lower crust from under high topography in the isostatic limit: *Journal of Geophysical Research*, **96**, 10275–10286.

BOILLOT, G., GRIMAUD, S., MAUFFRET, A., MOUGENOT, D., MERGOIL-DANIEL, J., KORNPROBST, J. & TORRENT, G. 1980. Ocean-continent boundary off the Iberian margin: a serpentinite diapir west of the Galicia Bank. *Earth and Planetary Science Letters*, **48**, 23–34.

BLOOM, A. 1998. *Geomorphology: A systematic analysis of Late Cretaceous landforms*, 3rd edition. Prentice-Hall, New Jersey.

BRANDON, M. T. & FLETCHER, R. C. 1998. Accretion and exhumation at a steady-state wedge; a new analytical model with comparisons to geologic examples. *Geological Society of America, Abstracts with Programs*, **29**(6), 120.

—— & VANCE, J. A. 1992. Tectonic evolution of the Cenozoic Olympic subduction complex, Washington State, as deduced from fission track ages for detrital zircons. *American Journal of Science*, **292**, 565–636.

——, RODEN-TICE, M. K. & GARVER, J. I. 1998. Late Cenozoic exhumation of the Cascadia accretionary wedge in the Olympic Mountains, NW Washington State. *Geological Society of America Bulletin*, **110**, 985–1009.

BUCK, R. 1988. Flexural rotation of normal faults. *Tectonics*, **7**, 959–973.

BULL, W. 1991. *Geomorphic responses to climatic change*. Oxford Press, New York.

BURBANK, D. & BECK, R. 1991. Rapid, long-term rates of denudation. *Geology*, **19**, 1169–1172.

——, DERRY, L. A. & FRANCE-LANORD, C. 1993. Reduced Himalayan sediment production 8 Myr ago despite an intensified monsoon. *Nature*, **364**, 48–50.

——, LELAND, J., FIELDING, E., ANDERSON, R., BROZOVIC, N., REID, M. & DUNCAN, C. 1996. Bedrock incision, rock uplift and threshold hillslopes in the northwestern Himalayas. *Nature*, **379**, 505–510.

BURCHFIEL, B. C. & ROYDEN, L. H. 1985. North-south extension within the convergent Himalayan region. *Geology*, **13**, 679–682.

BUTLER, R. W. H. & FREEMAN, S. 1996. Can crustal extension be distinguished from thrusting in the internal parts of mountain belts? A case histoty o the Entrelor shear zone, Western Alps. *Journal of Structural Geology*, **18**, 909–923.

CALVERT, A. T., GANS, P. B. & AMATO, J. M. 1999. Diapiric ascent and cooling of a sillimanite gneiss dome revealed by $^{40}Ar/^{39}Ar$ thermochronology: the Kigluaik Mountains, Seward Peninsula, Alaska. *This volume*.

CERLING, T. E. & CRAIG, H. 1994. Geomorphology and

in-situ cosmogenic isotopes. *Annual Review of Earth and Planetary Sciences*, **22**, 273–317.

CERVENY, P.F., NAESER, N. D., ZEITLER, P. K., NAESER, C. W. & JOHNSON, N. M. 1988. History of uplift and relief of the Himalaya during the past 18 million years: Evidence from fission-track ages of detrital zircons from sandstones of the Siwalik group. *In*: KLEINSPEHN, K. L. & PAOLA, C. (eds) *New Perspectives in Basin Analysis*. Springer-Verlag, Berlin, 43–61.

CHOPIN, C. 1984. Coesite and pure pyrope in high-grade blueschists of the Western Alps: a first record and some consequences. *Contributions to Mineralogy and Petrology*, **86**, 107–118.

——, HENRY, C. & MICHARD, A. 1991. Geology and petrology o the coesite-bearing terrain, Dora Maira massif, Western Alps. *European Journal of Mineralogy*, **3**, 263–291.

CHRISTENSEN, N. I. & MOONEY, W. D. 1995. Seismic velocity structure and composition of the continental crust: A global view. *Journal of Geophysical Research*, **100**, 9761–9788.

COLEMAN, R. G. & LANPHERE, M. A. 1971. Distribution and age of high-grade blueschists, associated eclogites, and amphibolites from Oregon and California. *Geological Society of America Bulletin*, **82**, 2397–2412.

—— & WANG, X. 1995. *Ultrahigh-pressure metamorphism*. Cambridge University Press.

COMPAGNONGI, R. & MAFFEO, B. 1973. Jadeite-bearing metagranite s.l. and related rock in the Monte Mucrone area (Sesia-Lanzo Zone, Western Italian Alps). *Schweizerische Mineralogische und Petrographische Mitteilungen*, **53**, 355–378.

COPELAND, P. & HARRISON, M.T. 1990. Episodic rapid uplift in the Himalayas revealed by $^{40}Ar/^{39}Ar$ analysis of detrital K-feldspar and muscovite, Bengal fan. *Geology*, **18**, 354–357.

——, HARRISON, M. T., KIDD, W. S. F., RONGHUA, X. & YUQUAN, Z. 1987. Rapid early Miocene acceleration of uplift in the Gandese belt, Xizang (southern Tibet), and ist bearing on accommodation mechanisms o the India-Asia collision. *Earth and Planetary Science Letters*, **86**, 240–252.

CRITTENDEN, M. D., CONEY, P. J. & DAVIS, G. H. 1980. Cordilleran metamorphic core complexes. *Geological Society of America Memoir*, **153**, 490p.

DAHLEN, F. A. & SUPPE, J. 1988. Mechanics, growth, and erosion of mountain belts. *In*: CLARK, S. P., BURCHFIEL, B. C. & SUPPE, J. (eds) *Processes in continental lithospheric deformation*. Geological Society of America, Special Paper **218**, 161–208.

DAVIS, G. A. 1988. Rapid upward transport of mid-crustal mylonitic gneisses in the footwall of a Miocene detchment fault, Whipple Mountains, southeastern California. *Geologische Rundschau*, **77**, 191–209.

—— & LISTER, G. S. 1988. Detachment faulting on continental extension: perspectives from the southwestern U.S. Cordillera. *In*: CLARK, S. P., BURCHFIEL, B. C. & SUPPE, J. (eds) *Processes in continental lithospheric deformation*. Geological Society of America, Special Papers, **218**, 133–159.

DEWEY, J. F. 1980. Episodicity, sequence, and style at convergent plate boundaries. *In*: STRANGWAY, D. W. (ed.) *The continental crust and its mineral deposits*. Special Papers of the Geological Association of Canada, **20**, 553–573.

——, HELMAN, M. L., TURCO, E., HUTTON, D. H. W. & KNOTT, S. D. 1989. Kinematics of the western Mediterranean. *In*: COWARD, M. P., DIETRICH, D. & PARK, R. G. (eds) *Alpine Tectonics*. Geological Society, London, Special Publications, **45**, 265–283.

——, RYAN, P. D. & ANDERSEN, T. B. 1993. Orogenic uplift and collapse, crustal thickness, fabrics and metamorphic phase changes: the role of eclogites. *In*: ALABASTER, H. M., HARRIS, N. B. W. & NEARY, C. R. (eds) *Magmatic Processes and Plate Tectonics*. Geological Society, London, Special Publications, **76**, 325–343.

DODSON, M. H. 1973. Closure temperature in cooling geochronological and petrological systems. *Contributions to Mineralogy and Petrology*, **40**, 259–274.

ELTER, P., GIGLIA, G., TONGIORGI, M. & TREVISAN, L. 1975. Tensional and contractional areas in the recent (Tortonian to present) evolution of the Northern Apennines. *Bolletini Geofisica Teorica ed Applicata*, **17**, 3–18.

ENGLAND, P. 1981. Metamorphic pressure estimates and sediment volumes for the Alpine orogeny: An independent control on geobarometers? *Earth and Planetary Science Letters*, **56**, 387–397.

—— & McKENZIE, D. P. 1982. A thin viscous sheet model for continental deformation. *Geophysical Journal of the Royal Astronomical Society*, **70**, 295–321.

—— & HOUSEMAN, G. 1986. Finite strain calculations of continental deformation, 2, Comparison with the India-Asia collision. *Journal of Geophysical Research*, **91**, 3664–3676.

—— & MOLNAR, P. 1990. Surface uplift, uplift of rocks, and exhumation of rocks. *Geology*, **18**, 1173–1177.

ERNST, W. G. 1977. Mineralogic study of eclogitic rocks from Alpe Arami, Lepontine Alps, Southern Switzerland. *Journal of Petrology*, **18**, 317–398.

—— 1993. Metamorphism of Franciscan tectonostratigraphic assemblage, Pacheco Pass area, east-central Diablo Range, California Coast Ranges. *Geological Society of America Bulletin*, **105**, 618–636.

——, MARUYAMA, S. & WALLIS, S. R. 1997. Buoyancy-driven, rapid exhumation of ultrahigh-pressure metamorphosed continental crust. *Proceedings of the National Academy of Science*, **94**, 9532–9537.

EVANS, B. W. & TROMMSDORFF, V. 1978. Petrogenesis of garnet peridotite, Cima di Gagnone, Lepontine Alps: *Earth and Planetary Science Letters*, **40**, 333–348.

FEEHAN, J. G. & BRANDON, M. T. 1999. Contribution of ductile flow to exhumation of low T– high P metamorphic rocks: San Juan – Cascade Nappes, NW Washington State. *Journal of Geophysical Research* (in press).

FITZGERALD, P. G., SORKHABI, R. B., REFIELD, T. F. & STUMP, E. 1995. Uplift and denudation of the central Alaska Range: A case study in the use of apatite fission track thermochronology to determine absolute uplift parameters. *Journal of Geophysical Research*, **100**, 20175–20191.

FORSTER, M. & LISTER, G. S. 1999. Detachment faults in the Aegean core complex of Ios, Cyclades, Greece. *This volume*.

FOSSEN, H. 1993. The role of extensional tectonics in the Caledonides of south Norway: Reply. *Journal of Structural Geology*, **15**, 1381–1383.

FOSTER, D. A. & GLEADOW, A. J. W. 1996. Structural framework and denudation history of the flanks of the Kenya and Anza Rifts, East Africa. *Tectonics*, **15**, 258–271.

—— & JOHN, B. E. 1999. Quantifying tectonic exhumation in an extensional orogen with thermochronology: examples from the southern Basin and Range province. *This volume*.

——, GLEADOW, A. J. W., REYNOLDS, S. J. & FITZGERALD, P. G. 1993. The denudation of metamorphic core complexes and the reconstruction of the Transition Zone, west-central Arizona: constraints from apatite fission-track thermochronology. *Journal of Geophysical Research*, **98**, 2167–2185.

FRISCH, W., DUNKL, I. & KUHLEMANN, J. 1998. Large-scale extension in the Alps: Tectonic versus erosional denudation. *Terra Nostra*, **98**, 9–10.

FRYER, P. 1996. Evolution of the Mariana convergent plate margin system. *Reviews of Geophysics*, **34**, 89–125.

GARVER, J. I., BRANDON, M. T., RODEN-TICE, M. & KAMP, P. J. J. 1999. Exhumation history of orogenic highlands determined by detrital fission-track thermochronology. *This volume*.

GOMEZ-PUGNAIRE, M. T. & FERNANDEZ-SOLER, J. 1987. High-pressure metamorphism in metabasites from the Betic Cordilleras (S.E. Spain) and ist evolution during the Alpine orogeny. *Contributions to Mineralogy and Petrology*, **95**, 231–244.

GRAPES, R. & WANTANABE, T. 1995. Metamorphism and uplift of Alpine schist in the Franz Josef-Fox Glacier area of the Southern Alp, New Zealand. *Journal of Metamorphic Geology*, **10**, 171–180.

GRAU, G., MONTADERT, L., DELTEIL, R. & WINNOCK, E. 1973. Structure of the European continental margin between Portugal and Ireland from seismic data. *Tectonophysics*, **20**, 319–339.

GRIFFIN, W. L. 1987. On the eclogites of Norway, 65 years later. *Mineralogical Magazine*, **51**, 333–343.

HALLET, B., HUNTER, L. & BOGEN, J. 1996. Rates of erosion and sediment evacuation by glaciers; a review of field data and their implications. *Global and Planetary Change*, **12**, 213–235.

HARLEY, S. L. & CARSWELL, D. A. 1995. Ultradeep crustal metamorphism: A prospective view. *Journal of Geophysical Reserach*, **100**, 8367–8380.

HILL, E., BALDWIN, S. & LISTER, G. 1995. Magmatism as an essential driving force for formation of active metamorphic core complexes in eastern Papua New Guinea. *Journal of Geophysical Research*, **100**, 10441–10451.

HOLM, D. K., NORRIS, R. J. & CRAW, D. 1989. Brittle/ductile deformation in a zone of rapid uplift: Central Southern Alps, New Zealand. *Tectonics*, **8**, 153–168.

HOUSEMAN, G. A. & MOLNAR, P. 1997. Gravitational (Rayleigh-Taylor) instability of a layer with non-linear viscosity and convective thinning of continental lithosphere. *Geophysical Journal International*, **128**, 125–150.

——, MCKENZIE, D. P. & MOLNAR, P. 1981. Convective instability of a thickened boundary layer and its relevance for the thermal evolution of continental convergent belts. *Journal of Geophysical Research*, **86**, 6115–6132.

HOVIUS, N., STARK, C. & ALLEN, P. 1997. Sediment flux from a mountain belt derived by landslide mapping. *Geology*, **25**, 231–234.

HUANG, W.-L. & WYLLIE, P. J. 1975. Melting and subsolidus phase relations for $CaSiO_3$ to 35 kilobars pressure. *American Mineralogist*, **60**, 213–217.

ISACKS, B. L. 1988. Uplift of the central Andean plateau and bending of the Bolivian Orocline. *Journal of Geophysical Research*, **93**, 3211–3231.

JARRAD, M. 1986. Relations among subduction parameters. *Reviews of Geophysics*, **24**, 217–284.

JOHNSON, C. 1997. Resolving dendudational histories in orogenic belts with apatite fission-track thermochronology and structural data: An example from southern Spain. *Geology*, **25**, 623–626.

KAMP, P. J. J., GREEN, P. F. & WHITE, S. H. 1989. Fission track analysis reveals character of collisional tectonics in New Zealand. *Tectonics*, **8**, 169–195.

KRABBENDAM, M. & DEWEY, J. 1998. Exhumation of UHP rocks by transtension in the Western Gneiss Region, Scandinavian Caledonides. *In*: HOLDSWORTH, R. E., STRACHAN, R. A. & DEWEY, J. F. (eds) *Continental Transpressional and Transtensional Tectonics*. Geological Society of London, Special Publications, **135**, 159–181.

LEMOINE, M. 1980. Serpentinites, gabbros and ophicalcites in the Piemont-Ligurian domain of the Western Alps: possible indicators of oceanic fracture zones and associated serpentinite protrusions in the Jurassic-Cretaceous Tethys. *In*: BERTRAND, J. & DEFERNE, J. (eds) *Proceedings of the International Symposium on tectonic inclusions and associated rocks in serpentinites*. Archives des Sciences, Geneva, **33**, 105–115.

LIPPOLT, H. J., LEITZ, M., WERNICKE, R. S. & HAGEDORN, B. 1994. (U-Th)/He dating of apatite: Experience with samples from different geochemical environments. *Chemical Geology*, **112**, 179–191.

LISTER, G. S., BANGA, G. & FEENSTRA, A. 1984. Metamorphic core complexes of Cordilleran type in the Cyclades, Aegean Sea, Greece. *Geology*, **12**, 221–225.

—— & DAVIS, G. A. 1989. The origin of metamorphic core complexes and detachment faults formed during continental extension in the northern Colorado River region, U.S.A. *Journal of Structural Geology*, **11**, 65–94.

LOOMIS, T. P. 1975. Tertiary mantle diapirism, orogeny, and plate tectonics east of the Strait of Gibraltar. *American Journal of Science*, **275**, 1–30.

MANCKTELOW, N. 1985. The Simplon Line, a major

displacement zone in the western Lepontine Alps. *Eclogae Geologicae Helvetiae*, **78**, 73–96.

MASSONNE, H.-J. 1995. Experimental and petrogenetic study of UHPM. *In:* COLEMAN R. G. & WANG, X. (eds) *Ultrahigh Pressure Metamorphism*. Cambridge University Press, 33–95.

—— 1999. The gneiss–eclogite unit of the central Erzgebirge as a natural laboratory for understanding processes at orogenic roots. *Terra Nostra*, **99**, 143–144.

MAXELON, M., WOHLERS, A., HALAMA, R., RING, U., MORTIMER, N. & BRANDON, M. T. 1998. Ductile strain in the Torlesse wedge, South Island, New Zealand. *EOS*, **79**(45), 889.

MEISSNER, R. 1986. *The continental crust, a geophysical appraoch*. Academic Press, London.

MILLIMAN, J. D. & SYVITSKI, J. P. M. 1992. Geomorphic/tectonic control of sediment discharge to the ocean: The importance of small mountainous rivers. *Journal of Geology*, **100**, 525–544.

MOLNAR, P., ENGLAND, R. & MARTINOD, J. 1993. Mantle dynamics, uplift of the Tibetan plateau, and the Indian monsoon. *Reviews of Geophysics*, **31**, 357–396.

MOORE, D. E. 1984. Metamorphic history of a high-grade blueschist exotic block from the Franciscan Complex, California. *Journal of Petrology*, **25**, 126–150.

MOORES, E. M., SCOTT, R. B. & LUMSDEN, W. W. 1968. Tertiary tectonics of the White Pine-Grant Range region, east-central Nevada, and some regional implications. *Geological Society of America Bulletin*, **79**, 1703–1726.

MORTIMER, N. 1993. Geology of the Otago schist and adjacent rocks, 1: 500 000; Map 7. Institute of Geological and Nuclear Sciences, Lower Hutt, New Zealand.

NIE, S., YIN, A., ROWLEY, D. B. & JIN, Y. 1994. Exhumation of the Dabie Shan ultra-high-pressure rocks and accumulation of the Songpan-Ganzi flysch sequence, central China. *Geology*, **22**, 999–1002.

NORRIS, R. J. & BISHOP, D. G. 1990. Deformed conglomerates and textural zones in the Otago Schists, South Island, New Zealand. *Tectonophysics*, **174**, 331–349.

OKRUSCH, M. & BRÖCKER, M. 1990. Eclogites associated with high-grade blueschists in the Cyclades archipelago, Greece: A review. *Eurupean Journal of Mineralogy*, **2**, 451–478.

PARDO-CASAS, F. & MOLNAR, P. 1987. Relative motion of the Nazca (Farallon) and South American plates since late Cretaceous time. *Tectonics*, **6**, 233–248.

PATACCA, E., SARTORI, R. & SCANDONE, P. 1993. Tyrrhenian Sea basin and Apenninic arcs: kinematic relations since Late Tortonian times. *Memorie della Società Geologica Italiana*, **45**, 425–451.

PEARSON, D. G., DAVIES, G. R., NIXON, P. H. & MILLEDGE, H. J. 1989. Graphitized diamonds from a peridotite massif in Morocco and implications for anomalous diamond occurrences. *Nature*, **338**, 60–62.

PHILLIPOT, P. 1993. Fluid-melt-rock interaction in mafic eclogites and coesite-bearing metasediment. Contributions on volatile recycling during subduction. *Chemical Geology*, **108**, 93–112.

——, CHEVALLIER, P., CHOPIN, C. & DUBESSY, J. 1995. Fluid composition and evolution in coesite-bearing rocks (Dora-Maira Massif, Western Alps): Implications for element recycling during subduction. *Contributions to Mineralogy and Petrology*, **121**, 29–44.

PINET, P. & SOURIAU, M. 1988. Continental erosion and large-scale relief. *Tectonics*, **7**, 563–582.

PLATT, J. P. 1975. Metamorphic and deformational processes in the Franciscan Complex, California: Some insights from the Catalina schist terrane. *Geological Society of America Bulletin*, **86**, 1337–1347.

—— 1986. Dynamics of orogenic wedges and the uplift of high-pressure metamorphic rocks. *Geological Society of America Bulletin*, **97**, 1037–1053.

—— 1987. The uplift of high-pressure–low-temperature metamorphic rocks. *Philosophical Transactions of the Royal Society London*, **A321**, 87–103.

—— 1993. Exhumation of high-pressure rocks: a review of concepts and processes. *Terra Nova*, **5**, 119–133.

——, SOTO J. I., WHITEHOUSE, M. J., HURFORD, A. J. & KELLEY, S. P. 1998. Thermal evolution, rate of exhumation, and tectonic significance of metamorphic rocks from the floor of the Alboran extensional basin, western Mediterranean. *Tectonics*, **17**, 671–689.

—— & VISSERS 1989. Extensional collapse of thickened continental lithospere: A working hypothesis for the Alboran Sea and Gibralter Arc. *Geology*, **17**, 540–545.

POPE, D. & WILLETT, S. 1998. A thermo-mechanical model for crustal thickening in the Central Andes driven by ablative subduction. *Geology*, **26**, 511–514.

PRICE, R. A. 1981. The Cordilleran thrust and fold belt in the southern Canadian Rocky Mountains. *In*: MCCLAY K. R. & PRICE N. J. (eds) *Thrust and nappe tectonics*. Geological Society, London, Special Publications, **9**, 427–448.

RAMBERG, H. 1967. *Gravity, deformation and the Earth's crust*. Academic Press.

—— 1972. Theoretical models of density stratification and diapirism in the Earth. *Journal of Geophysical Research*, **77**, 877–889.

—— 1980. Diapirism and gravity collapse in the Scandinavian Caledonides. *Journal of the Geological Society London*, **137**, 261–270.

—— 1981. *Gravity, deformation and the Earth's crust: Theory, experiments, and geological application*. Academic Press, London.

RATSCHBACHER, L., FRISCH, W., LINZER, H.-G. & MERLE, O. 1991. Lateral extrusion in the Eastern Alps. Part 2: Structural analysis. *Tectonics*, **10**, 257–271.

RAOUZAIOS, A., LISTER, G. S. & FOSTER, D. A. 1996. Oligocene exhumation and metamorphism of eclogite-blueschists from the island Sifnos, Cyclades, Greece. *Geological Society of Australia abstracts*, **41**, 358.

RAWLING, T. J. & LISTER, G. S. 1999. Oscillating modes of orogeny in the Southwest Pacific and the tectonic evolution of New Caledonia. *This volume*.

REDDY, S. M., WHEELER, J. & CLIFF, R. A. 1999. The geometry and timing of orogenic extension: An example from the western Italian Alps. *Journal of Metamorphic Geology*, in press.

REINECKE, T. 1991. Very-high pressure metamorphism and uplift of coesite-bearing metasediments from the Zermatt-Saas zone, Western Alps: *European Journal of Mineralogy*, **3**, 7–17.

RING, U. 1995. Horizontal contraction or horizontal extension: Heterogeneous Late Eocene and Early Oligocene general shearing during blueschist- and greenschist-facies metamorphism at the Pennine-Austroalpine boundary zone in the Western Alps. *Geologische Rundschau*, **84**, 843–859.

—— & BRANDON, M. T. 1994. Kinematic data for the Coast Range fault zone zone and implications for the exhumation of the Franciscan Subduction Complex. *Geology*, **22**, 735–738.

—— & —— 1999. Ductile deformation and mass loss in the Franciscan subduction complex: implications for exhumation processes in accretionary wedges. *This volume*.

——, LAWS, S. & BERNET, M. 1999. Structural analysis of a complex nappe sequence and late-orogenic basins from the Aegean Island of Samos, Greece. *Journal of Structural Geology*, in press.

ROYDEN, L.H. 1993. The tectonic expression of slab pull at continental convergent boundaries. *Tectonics*, **12**, 303–325.

—— & BURCHFIEL, B. C. 1989. Are systematic variations in thrust belt style related to plate boundary processes? The Western Alps versus the Carpathians. *Tectonics*, **8**, 51–61.

RUMBLE, D., XU, H. & YUI, T.-F. 1998. Subduction, UHP metamorphism, and exhumation of an intact meteoric water-hydrothermal system. *EOS*, **79** (45), 982.

SCHLIESTEDT, M., ALTHERR, R. & MATTHEWS, M. 1987. Evolution of the Cycladic crystalline complex: petrology, isotope geochemistry and geochronology. *In:* HELGESON, H. C. (ed.) *Chemical transport in metasomatic processes*. NATO ASI Series, 389–428.

SCHREYER, W. 1995. Ultradeep metamorphic rocks: The retrospective view. *Journal of Geophysical Research*, **100**, 8353–8366.

——, MASSONE, H.-J. & CHOPIN, C. 1987. Continental crust subducted to mantle depth near 100 km: Implications for magma and fluid genesis in collision zones. *In:* MYSEN, B. O. (ed.) *Magmatic Processes: Physiochemical principles*. Special Publications of the Gechemical Society, **1**, 155–163.

SCOTT, R. S. & LISTER, G. S. 1992. Detachment faults: evidence for a low-angle origin. *Geology*, **20**, 833–836.

——, FOSTER, D. A. & LISTER, G. S. 1999. Rapid cooling of denuded lower plate rocks from the Buksin-Rawhide metamorphic core complex, west-central Arizona. *Geological Society of America Bulletin*, in press.

SELVERSTONE, J. 1985. Petrologic constraints on imbrication, metamorphism, and uplift in the SW Tauern window, Eastern Alps. *Tectonics*, **4**, 687–704.

SHARP, Z. D., ESENE, E. J. & HUNZIKER, J. C. 1993. Stable isotope geochemistry and phase equilibria of coesite-bearing whiteschists, Dora Maira Massif, Western Alps. *Contributions to Mineralogy and Petrology*, **114**, 1–12.

SMITH, D. C. & LAPPIN, M. A. 1989. Coesite in the Straumen kyanite-eclogite pod, Norway. *Terra Nova*, **1**, 47–56.

SPENCER, J. E. & REYNOLDS, S. J. 1991. Tectonics of mid-Tertiary extension along a transect through west-central Arizona. *Tectonics*, **10**, 1204–1221.

STÖCKHERT, B. & RENNER, J. 1998. Rheology of crustal rocks at ultrahigh pressure. *In:* HACKER, B. & LIOU, G. (eds) *When continents collide: Geodynamics and geochemistry of ultrahigh-pressure rocks*. Kluwer, Dordrecht, 57–95.

SUMMERFIELD, M. A. & BROWN, R. W. 1998. Geomorphic factors in the interpretation of fission-track data. *In:* VAN DEN HAUTE, P. & DE CORTE, F. (eds) *Advances in fission-track geochronology*. Kluwer, Dordrecht, 19–32.

TABIT, A., KORNPROBST, J., LI, J. P. & WOODLAND, A. B. 1990. Origin and evolution of the massif at Beni Bousera, Morocco: petrological evidence. *In. International workshop on orogenic lherzolites and mantle processes*. Blackwell Science, Oxford.

THOMSON, S. N., STÖCKHERT, B. & BRIX, M. A. 1998. Thermochronology of the high-pressure metamorphic rocks of Crete, Greece: Implications for the speed of tectonic processes. *Geology*, **26**, 259–262.

——, —— & —— 1999. Miocene high-pressure metamorphic rocks of Crete, Greece: rapid exhumation by buoyant escape. *This volume*.

UYEDA, S. & KANAMORI, H. 1979. Back-arc opening and the mode of subduction. *Journal of Geophysical Research*, **84**, 1049–1061.

VAN DER KLAUW, S. N., REINECKE, T. & STÖCKHERT, B. 1997. Exhumation of ultrahigh-pressure metamorphic oceanic crust from Lago di Cignana, Piemontese zone, Western Alps: the structural record in metabasites. *Lithos*, **41**, 79–102.

VISSERS, R. L. M., PLATT, J. P. & VAN DER WAL, D. 1995. Late orogenic extension of the Betic Cordillera and the Alboran domain. *Tectonics*, **14**, 786–803.

WALCOTT, R. I. 1998. Modes of oblique compression: Late Cenozoic tectonics of the South Island of New Zealand. *Reviews in Geophysics*, **36**, 1–26.

WALLIS, S. R. 1992. Vorticity analysis in a metachert from the Sanbagawa Belt, SW Japan. *Journal of Structural Geology*, **14**, 271–280.

—— 1995. Vorticity analysis and recognition of ductile extension in the Sanbagawa belt, SW Japan. *Journal of Structural Geology*, **17**, 1077–1093.

——, NAKAMURA, D. & TAKAO, H. 1998. Rapid exhumation of the Su Lu UHP terrain. *EOS*, **79**(45), 983.

——, PLATT, J. P. & KNOTT, S. D. 1993. Recognition of syn-convergence extension in accretionary wedges with examples from the Calabrian arc and the Eastern Alps. *American Journal of Science*, **293**, 463–495.

WASCHBUSCH, P. & BEAUMONT, C. 1996. Effect of a retreating subduction zone on deformation in simple regions of plate convergence. *Journal of Geophysical Research*, **101**, 28133–28148.

WERNICKE, B., AXEN, G. & SNOW, J., 1988, Basin and Range extensional tectonics at the latitude of Las Vegas, Nevada. *Geological Society of America Bulletin*, **100**, 1738–1757.

WHEELER, J. & BUTLER, R. W. H. 1994. Criteria for identifying structures related to true crustal extension in orogens. *Journal of Structural Geology*, **16**, 1023–1027.

WILLETT, S. D., BEAUMONT, C. & FULLSACK, P. 1993. Mechanical model for the tectonics of doubly vergent compressional orogens. *Geology*, **21**, 371–374.

WOLFF, R. A., FARLEY, K. A. & SILVER, L. T. 1997. Assessment of (U-Th)/He thermochronometry: The low-temperature history of the San Jacinto mountains, California. *Geology*, **25**, 65–68.

ZECK, H. P., MONIE, P., VILLA, I. M. & HANSEN, B. T. 1992. Very high rates of cooling and uplift in the Alpine belt of the Betic Cordillera, southern Spain. *Geology*, **20**, 79–82.

ZEITLER, P. K. 1985. Cooling history of the NW Himalaya, Pakistan. *Tectonics*, **4**, 127–151.

# Evaluation of exhumation mechanisms for coherent blueschists in western Baja California, Mexico

RICHARD L. SEDLOCK

*Department of Geology, San José State University, San José, CA 95192–0102, USA*
*(e-mail: sedlock@geosun1.sjsu.edu)*

**Abstract:** Structural data and field observations from Franciscan-type coherent blueschists in western Baja California, Mexico, are used to test proposed mechanisms for exhumation of these high-pressure, low-temperature (high $P/T$) metamorphic rocks. Analysis of specific observations that are predicted by, consistent with, or inconsistent with specific mechanisms supports extension and normal faulting as the dominant mechanism in Baja. Erosion and strike-slip faulting may have been active, but their contributions cannot be documented and are inferred to be minor. Buoyancy forces and ductile flow probably played a negligible role. Extension and exhumation probably occurred in Late Cretaceous and Paleogene time, while subduction continued at the convergent margin west of North America. Blueschists probably were exhumed to shallow crustal levels by 30 Ma, and possibly before a decrease in convergence rate at about 40 Ma. Coherent blueschists lie in the footwalls of major normal fault zones that contain serpentinite-matrix mélange with exotic blocks of greenschist, amphibolite, eclogite, and blueschist. Mélange probably formed before normal slip on the fault zones. Most mélange components are consistent with tectonic thinning of the mantle wedge and adjacent parts of the accretionary prism that once lay structurally higher than the lower-plate blueschists. Many of these findings may be applicable to similar rocks in California, including Franciscan blueschists.

This paper addresses the exhumation of high-pressure, low-temperature (high $P/T$) metamorphic rocks in western Baja California, Mexico. Protoliths, metamorphic and structural history, and tectonic relations indicate that these blueschists and rare eclogites formed in a Cretaceous subduction zone on the western margin of North America (Sedlock 1988b; Sedlock & Isozaki 1990). High $P/T$ rocks in Baja lack a higher-temperature overprint, indicating one or both of the following: (1) exhumation was so rapid that, though blueschists may have experienced conditions of higher temperature or lower $P/T$, the kinetics of re-equilibration reactions were too slow to affect the original assemblages; (2) continual depression of isotherms during exhumation enabled blueschists to avoid higher temperatures or lower $P/T$ conditions. Similar conditions have been deduced in a few other blueschist belts (e.g. Hokkaido, Sakakibara & Kanisawa 1994) that are widely referred to as 'Franciscan-type' after the Franciscan Complex of California, USA (Ernst 1988). Such belts are distinct from higher-pressure, higher-temperature belts that form in areas of continental collision.

Surface exposures of blueschist-facies metamorphic rocks are widely inferred to be exhumed parts of ancient accretionary prisms because their high $P/T$ metamorphic assemblages require conditions found in subduction zones (pressures of 0.4–1.2 GPa, corresponding to depths of about 12–40 km, and temperatures of about 170–350°C). The presence of these rocks at the surface poses several difficult questions. What forces drove exhumation? What mechanisms operated during exhumation? What happened to the overburden?

Exhumation mechanisms for tectonic blocks of high $P/T$ rocks, typically metres or tens of metres across and immersed in a matrix of mud or serpentinite, include diapiric rise of low-density matrix (Lockwood 1972; Moore 1984), corner or channel flow of matrix (Cloos & Shreve 1988), and geometric irregularities in strike-slip fault systems (Karig 1980). Less clearly understood are the exhumation mechanisms for regionally metamorphosed high $P/T$ rocks in coherent tracts that range from kilometres to hundreds of kilometres long. Proposed mechanisms include erosion, ductile flow, normal faulting, strike-slip faulting, and buoyancy (see Platt (1993) for summary).

Tests of proposed exhumation mechanisms of coherent Franciscan-type (e.g. eastern belt) blueschists are likely to be more straightforward in western Baja than in California because the quality of exposure is far superior, and because western Baja has undergone comparatively minor late Cenozoic tectonism. This paper examines proposed exhumation mechanisms and specific observations predicted by each;

SEDLOCK, R. L. 1999. Evaluation of exhumation mechanisms for coherent blueschists in western Baja California, Mexico. *In*: RING, U., BRANDON, M. T., LISTER, G. S. & WILLETT, S. D. (eds) *Exhumation Processes: Normal Faulting, Ductile Flow and Erosion.* Geological Society, London, Special Publications, **154**, 29–54.

describes relevant geological data from western Baja, including new structural and field observations; and evaluates blueschist exhumation in western Baja.

## Proposed exhumation mechanisms

Proposed exhumation mechanisms of high $P/T$ rocks from depths of $\geq 20$ km include the following: (A) erosion, (B) ductile flow, (C) normal faulting, (D) strike-slip faulting, and (E) buoyancy. Two or more of these processes may have acted in sequence or together, so evidence for a particular process does not necessarily disprove the others.

Each proposed mechanism predicts specific observations that can be used to test for applicability of the mechanism. However, observations differ in their effectiveness in validating and eliminating possible mechanisms. In this section, I briefly summarize the proposed mechanisms and evaluate the observations predicted by each. In Table 1, each predicted observation is assigned a letter-number code; the letter refers to mechanisms A–E, and the number describes the following roles of the observation in the evaluation process, with lower numbers serving as more effective tools: (1) predicted by the mechanism; inconsistent with other mechanisms (though other mechanisms may be indicated by other observations); (2) rules out the mechanism; (3) predicted by the mechanism, but absence does not rule out the mechanism; (4) predicted by the mechanism; absence indicates it was not the sole mechanism, but may have contributed; (5) typically associated with the mechanism, but consistent with other mechanism(s); (6) not characteristic of any specific mechanism. A given exhumation mechanism is specifically validated only by observations rated 1 or 3; it is specifically eliminated only by observations rated 2. Thus, efforts to assess the exhumation of high $P/T$ rocks should focus on these observations, rather than on observations that are consistent with more than one mechanism.

*(A) Erosion.* One obvious way to exhume high $P/T$ rocks is by erosion of the overburden as the deeper-level rocks rise through the crust. Surface uplift and erosion rate may be controlled by orogen-thickening mechanisms such as underplating, thrusting, and contractional pure shear (Platt 1993; Thompson *et al.* 1997).

Erosion acting as the sole exhumation mechanism should expose high $P/T$ rocks as part of a semi-complete crustal section (Table 1, prediction 1). If this is not the case, at least one other exhumation mechanism must have been active, and erosion may or may not have been active.

Erosion also will produce a large volume of eroded sediment, which will equal the volume of exhumed rock if erosion was the sole exhumation mechanism (Table 1, prediction 2). For instance, the volume of Triassic flysch in the Songpan–Ganzi basin of central China is roughly similar to the calculated overburden that must have been removed to exhume ultra-high-pressure metamorphic rocks in the Dabie Shan (Nie *et al.* 1994). Unfortunately, the preservation potential of such sediment will be low, particularly if delivered to trench or trench-slope basins, so estimates of sediment volume based on existing outcrops may be much lower than the actual volume that was eroded (Hsu 1991). A shortfall in sediment volume, and inappropriate composition of preserved sediment, may not necessarily indicate that other mechanisms were operative (Table 1, predictions 3 and 4).

*(B) Ductile flow.* High $P/T$ rocks and their overburden may be internally thinned or thickened by ductile flow at depths below the brittle–plastic transition. Platt (1993) suggested that extension in an accretionary prism will be plastic, rather than brittle, below about 10 km. However, geothermal gradients of 10–15°C km$^{-1}$ in long-lived Franciscan-type accretionary prisms (Ernst 1988) imply that brittle behaviour in quartz-rich rocks, which typically change to plastic behaviour at about 300°C (Sibson 1982), should persist to 20–25 km or even 30 km.

Ductile flow may have contributed to exhumation if a penetrative fabric is present (Table 1, predictions 5 and 6). Ductile thinning may account for up to 25% of the 18 km of unroofing required to exhume the San Juan–Cascades high $P/T$ rocks in Washington state (Feehan & Brandon 1993); similar studies of the eastern belt of the Franciscan Complex recognized <1% exhumation because of ductile flow (Ring & Brandon 1994; Bohlar *et al.* 1997). Vertical flattening and subhorizontal stretching may characterize the deep levels of orogenic belts that undergo gravitational collapse (Dewey 1988). Such fabrics have been used to attribute exhumation of high $P/T$ rocks to ductile shortening in some continental collision zones (Jun *et al.* 1995). However, thickening by ductile flow would require that more exhumation be accommodated by other mechanisms.

*(C) Normal faulting.* High $P/T$ rocks may be exhumed in the footwalls of faults that have a large-magnitude component of normal

**Table 1.** *Evaluation of observations predicted by proposed exhumation mechanisms*

| Predicted observation | Code* | Comments |
|---|---|---|
| 1, high $P/T$ rocks form part of continuous crustal section | A-4 | if not the case, other mechanism(s) must have been active, with or without erosion |
| 2, sediment volume = exhumed volume | A-1 | difficult to pinpoint source of sediment, thus difficult to prove that its volume equals exhumed volume |
| 3, sediment volume < exhumed volume | A-5 | poor preservation potential of sediment, thus difficult to prove that its volume is less than exhumed volume |
| 4, sediment contains detritus typical of the exhumed volume | A-3 | poor preservation potential of sediment; appropriate strata may not be exposed |
| 5, absence of a ductile fabric | B-2 | no ductile flow if no ductile fabric |
| 6, presence of a ductile fabric | B-1 | fabric must be penetrative, rather than limited to margins of normal (C) or strike-slip (D) faults |
| 7, exhumed rocks in footwalls of normal faults | C-1 | sense of slip must be proven to be normal |
| 8, metamorphic break and tectonic thinning at fault | C-3 | sense of slip must be proven to be normal; beware thermal re-equilibration that may obscure metamorphic differences |
| 9, high $P/T$ rocks bounded by strike-slip faults | D-1 | indicates strike-slip was final mechanism; other mechanisms may have been active earlier |
| 10, high $P/T$ rocks near strike-slip faults | D-5 | spatial proximity could be coincidental |
| 11, exhumed rocks within higher-density rocks | E-5 | consistent with all mechanisms |
| 12, exhumed rocks within lower-density rocks | E-2 | indicates buoyancy did not play a role |
| 13, contact of high $P/T$ unit displays melange with exotic blocks | C-5 D-5 E-5 | may have been emplaced as tectonic breccia along fault (C,D) or buoyant diapir (E); hard to explain by erosion (A) |
| 14, fast or slow exhumation rate | all-6 | similar rates possible by different mechanisms |
| 15, shape of $P$–$T$–$t$ path | all-6 | similar paths have been attributed to different mechanisms |
| 16, thrusting or other contractional deformation | all-6 | contributes to crustal thickening, not thinning and exhumation |

*A, erosion; B, ductile flow; C, normal faulting; D, strike-slip faulting; E, buoyancy; 1–6: see text.

displacement (Table 1, prediction 7). A widely perceived driving force for such normal faulting is underplating of the forearc, possibly by a subducted bathymetric high, which leads to overthickening, gravitational collapse, and extension of a tapered accretionary prism (Lister *et al.* 1984; Platt 1986). Normal faulting may also be driven by oblique convergence (Avé Lallemant & Guth 1990). Many studies of active convergent margins have documented underplating (e.g. Westbrook & Smith 1983; Clowes *et al.* 1987; Fuis & Plafker 1991) and normal faulting (e.g. Crespi *et al.* 1996; McNeill *et al.* 1997).

Normal faulting has been suggested as the exhumation mechanism of high $P/T$ rocks from modern and ancient convergent margins around the world, including the Franciscan Complex of California (Jayko *et al.* 1987; Harms *et al.* 1992), the Dora Maira ultra-high-pressure metamorphic rocks of the western Alps during the later (crustal) stage of exhumation (Avigad 1992;

Michard et al. 1993), the Tauern Window of the eastern Alps (Selverstone 1988), the Betic Cordillera of southern Spain (Platt & Vissers 1989; Johnson et al. 1997; Martínez-Martínez & Azañón 1997), high $P/T$ rocks in Crete (Jolivet et al. 1996) and throughout the Cyclades (Avigad & Garfunkel 1991), blueschists in eastern Australia (Little et al. 1992), and the Kamuikotan unit on Hokkaido, Japan (Kimura 1994). Caution must be taken in inferring normal slip on faults and shear zones in such orogens; for instance, shallower-level rocks can be emplaced atop deeper-level rocks by thrusting of previously deformed rocks (Table 1, prediction 8).

*(D) Strike-slip faults.* Strike-slip faults at obliquely convergent margins may be partly responsible for the distribution of high $P/T$ rocks. Modern strike-slip fault systems show surface uplift at restraining bends or steps and where they are oblique to relative plate motion. In some orogens, such areas expose high $P/T$ rocks, implying a causal effect between strike slip and exhumation (Karig 1980; Mann & Gordon 1996) (Table 1, predictions 9 and 10). However, two factors limit the applicability of strike-slip faulting as an exhumation mechanism: (1) effects are likely to be localized near the fault, and thus cannot be responsible for exhuming large, coherent tracts; (2) vertical separation along such faults is unlikely to exceed several kilometres unless obliquity is high and persists for many millions of years. Nevertheless, strike-slip faulting may assist in exhuming deep-level rocks, especially those already brought to near-surface levels by other mechanisms.

*(E) Buoyancy.* If high $P/T$ rocks are sufficiently less dense than surrounding rock, they may return to the surface along the subduction zone or through the accretionary prism as low-density diapirs or masses (Ernst 1971) (Table 1, prediction 11). Buoyancy may account for the exhumation of small blocks of high $P/T$ rocks within a lower-density matrix material (England & Holland 1979), but cannot be applied to large tracts such as the Franciscan Complex, whose average density of 2.8–3.0 g cm$^{-3}$ is similar to or greater than that of typical continental crust (Table 1, prediction 12). However, buoyancy may provide the chief means of exhuming ultra-high-pressure rocks from the mantle to the base of the crust (Michard et al. 1993; Platt 1993).

Several observations are consistent with several or all of the proposed mechanisms. Serpentinite-matrix mélange (Table 1, prediction 13) may have been emplaced during normal faulting or strike-slip faulting, or may have been buoyantly emplaced by diapiric forces. A particular rate of exhumation (prediction 14) or shape of $P–T–t$ path (prediction 15) probably can result from different mechanisms (e.g. Ruppel et al. 1988), so these are not likely to be diagnostic of a particular mechanism. Thrusting or other contractional deformation (prediction 16) may thicken the crust and increase the surface slope, thereby preparing the crust for various exhumation mechanisms, but by itself cannot exhume rock (Platt 1993).

## Regional geology and tectonics of western Baja California

Baja California (Fig. 1) can be divided into rocks of chiefly continental affinity that make up most of the peninsula (the Cortes and Santa Ana terranes of Coney & Campa-Uranga (1987) or the Serí and Yuma terranes of Sedlock et al. (1993)), and oceanic rocks that underlie westernmost Baja and offshore islands (the Vizcaíno terrane of Coney & Campa-Uranga (1987) or the Cochimí composite terrane of Sedlock et al. (1993)). The boundary between the two assemblages is covered by younger sedimentary rocks, but differences in lithology and metamorphic and structural history imply that it is a major fault zone. Its position in the subsurface is known within a few tens of kilometres from the distribution of outcrops and subcrops and from distinctive positive gravity anomalies (Couch et al. 1991; Sedlock et al. 1993).

Continental rocks of 'mainland Baja' include Palaeozoic and Mesozoic metasedimentary rocks deposited atop or west of North America, the Early Cretaceous Alisitos–Santiago Peak volcanic arc, and the Cretaceous (130–90 Ma) Peninsular Ranges batholith (Gastil et al. 1975; Silver & Chappell 1988). The latitude of formation of mainland Baja is controversial: paleomagnetic and geobarometric studies infer a position 10–15° farther south in Late Cretaceous time (Beck 1991; Lund & Bottjer 1991; Ague & Brandon 1992), but geological correlations between Baja and Sonora are difficult to reconcile with long-distance transport, and a fault or faults that accommodated displacement have not been identified (Gastil 1991). The western flank of the continental assemblage is onlapped by Upper Cretaceous to Paleogene siliciclastic marine strata, chiefly turbidites, that probably were deposited in a forearc basin (Gastil et al. 1975; Bottjer & Link 1984).

Oceanic rocks in western Baja consist of Mesozoic blueschist, ophiolite, and island-arc terranes overlain by Cretaceous forearc basin strata. Oceanic rocks crop out on Cedros Island,

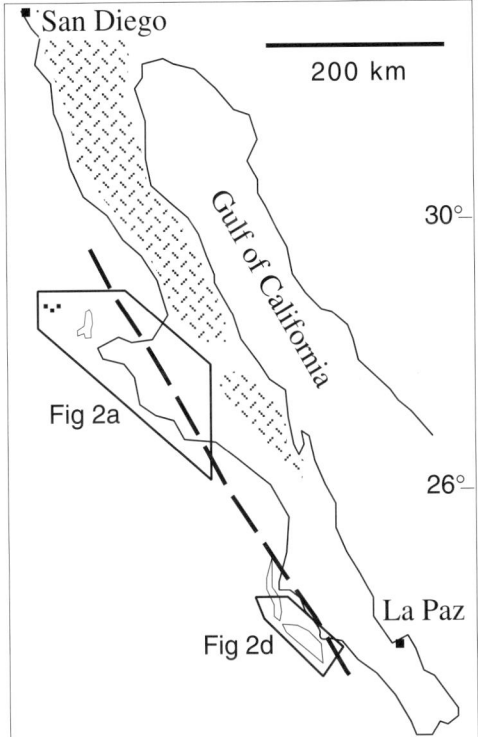

**Fig. 1.** Sketch map of Baja California, Mexico, showing locations of Fig. 2a and d. Dashed thick line is boundary between oceanic terranes to west and crust underlain by Peninsular Ranges batholith (stipple pattern) to east.

the San Benito Islands, and the Vizcaíno Peninsula near 28°N, and on Magdalena and Santa Margarita Islands near 24°30′N (Figs 1 and 2a–d). Similar rocks probably underlie most of the continental shelf between 30°N and 23°N, judging from gravity and magnetic anomalies (Couch *et al.* 1991). These rocks are described in detail in the next section.

Baja lay entirely within the North American forearc from about 80 Ma to 25 Ma (see references in Sedlock *et al.* (1993)). An abrupt westward jump of arc magmatism led to the eruption of thick arc volcanic rocks in eastern Baja from about 25 Ma to 12 Ma (Lonsdale 1991). Diachronous north-to-south change from a convergent to transform plate boundary occurred from about 20 Ma to 5 Ma, with development of dextral strike-slip faults in parts of the borderland. Since about 5 Ma, Baja has been part of the Pacific plate, and tectonism has been limited to widely spaced, steep normal faults that contribute to the topographic expression of the offshore islands (Yeats & Haq 1981; Sedlock *et al.* 1993).

## Mesozoic oceanic rocks in western Baja

Mesozoic oceanic rocks that crop out in western Baja (Fig. 2a–d) are divided into three structural components: a blueschist-facies subduction complex, serpentinite-matrix mélange, and magmatic and sedimentary rocks referred to as the 'upper plate'. Many aspects of these rocks have been described elsewhere (Sedlock 1988*a*, *b*, 1993, 1996; Baldwin & Harrison 1989, 1992; Sedlock & Isozaki 1990); relevant details are included here with new structural and field data.

### Blueschist-facies subduction complex

The structurally lowest unit in western Baja consists of large tracts of regionally metamorphosed blueschists (Western Baja terrane of Sedlock (1988*b*)). The rocks are assigned to four subterranes that differ in protolith proportions, metamorphic assemblages, and structural style. All subterranes (hereafter st1, st2, st3, st4) consist of fault-bounded domains that contain part of an originally continuous stratigraphic sequence. Stratal disruption locally produces broken formation, but the term mélange is inappropriate because block-in-matrix fabric and exotic lithotypes are absent.

Blueschist protoliths include mafic volcanic rocks, and pelagic and siliciclastic sedimentary rocks (Sedlock & Isozaki 1990). Mafic pillow lavas, porphyritic effusive rocks, volcaniclastic breccias, and thin-bedded tuff or fine-grained effusive rocks have whole-rock and relict clinopyroxene compositions that imply an ocean-floor or possibly intraplate origin (R. Sedlock, unpublished data). Chert forms Triassic to Cretaceous radiolarian-rich sequences up to 40 m thick atop mafic volcanic rocks, and thin (<1–20 cm) beds interbedded with tuff and siliceous argillite. Limestone is present as interpillow deposits and in beds up to 2 m thick interbedded with argillite and ferruginous chert or ironstone. Siliciclastic turbidites with sedimentary structures and facies characteristic of submarine fan deposition constitute most of st2, depositionally overlie radiolarian chert in st1 and st3, and are absent from st4. These protoliths are interpreted as Triassic oceanic crust, Triassic to Cretaceous pelagic sedimentary rocks, and Cretaceous terrigenous turbidites that were deposited as the oceanic rocks neared North America (Sedlock & Isozaki 1990).

All subterranes were subducted and exhumed under high-pressure, low-temperature conditions

(Fig. 3). The extent of recrystallization during subduction-related peak metamorphism (M1) is greatest in st1 (100%) and least in st3 and st4 (<5%). Index phases that developed during M1 include jadeitic–acmitic sodic clinopyroxene, sodic amphibole, lawsonite, quartz, and aragonite. Peak metamorphic conditions were determined from index mineral occurrence, sodic clinopyroxene composition, vitrinite reflectance data, and the absence of higher-temperature overprinting:

st1: $T = 230–300°C$, $P > 0.8$ GPa (8 kbar);
st2: $T = 170–220°C$, $P = 0.7–0.8$ GPa;
st3: $T = 170–200°C$, $P = 0.6–0.7$ GPa;
st4: $T = 170–200°C$, $P = 0.7–0.8$ GPa.

Estimates for st1, st2, and st3 are from Sedlock (1988a), except st3 pressure, which was revised in light of the aragonite–calcite equilibrium determined by Redfern *et al.* (1989). Estimates for st4 are estimated from mineral assemblages and the absence of higher-temperature overprinting.

Metamorphic conditions during exhumation are deduced chiefly from veins, which truncate structures produced during M1 in all subterranes (see below). Vein phases (M2) include aegirine–jadeite or aegirine, albite, aragonite, lawsonite, sodic amphibole, quartz, and pumpellyite. Pressures lower than those of M1 are indicated by (1) the low Al content of vein sodic clinopyroxene in rocks that grew higher-pressure jadeite during M1, and (2) albite veins in rocks that grew jadeite + quartz from albite during M1. Sedlock (1988a), following Carlson & Rosenfeld (1981), argued that aragonite–bearing rocks in st2 and st3 were exhumed across the aragonite–calcite stability curve at 125–175°C. However, the work of Liu & Yund (1993) raises that temperature limit as high as 235°C. Nevertheless, the absence of greenschist-facies minerals such as actinolite indicates constant or decreasing temperature during decompression (Fig. 3).

The deformation history of each subterrane includes early shortening, D1, and later extension, D2. Petrographic relations indicate that D1 foliation, folds, and thrust faults were coeval with peak metamorphism and thus probably formed during duplexing and underplating at the base of the accretionary prism. Structural style indicates that D1 occurred while the subterranes were at depths that straddle the brittle–plastic transition.

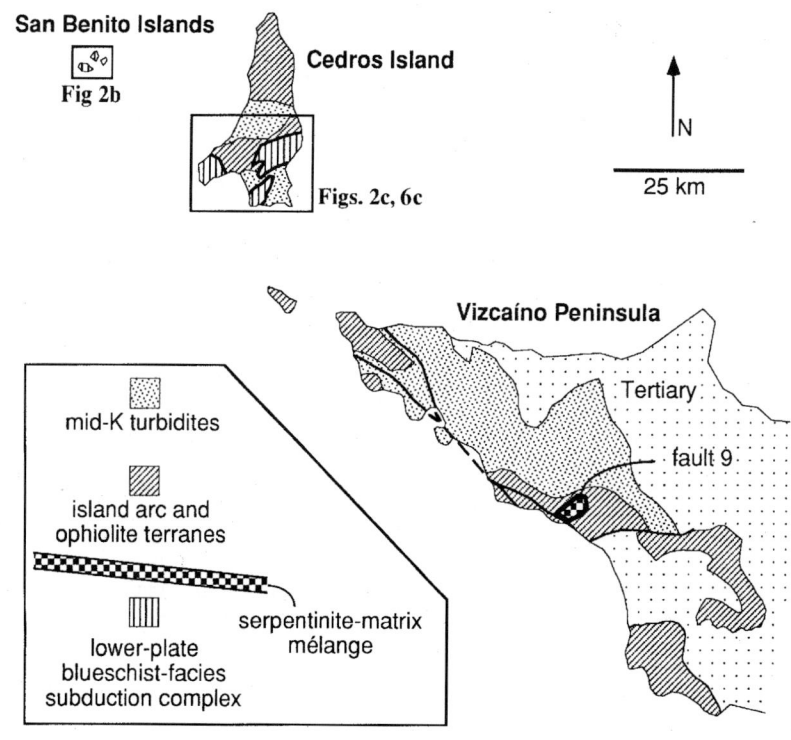

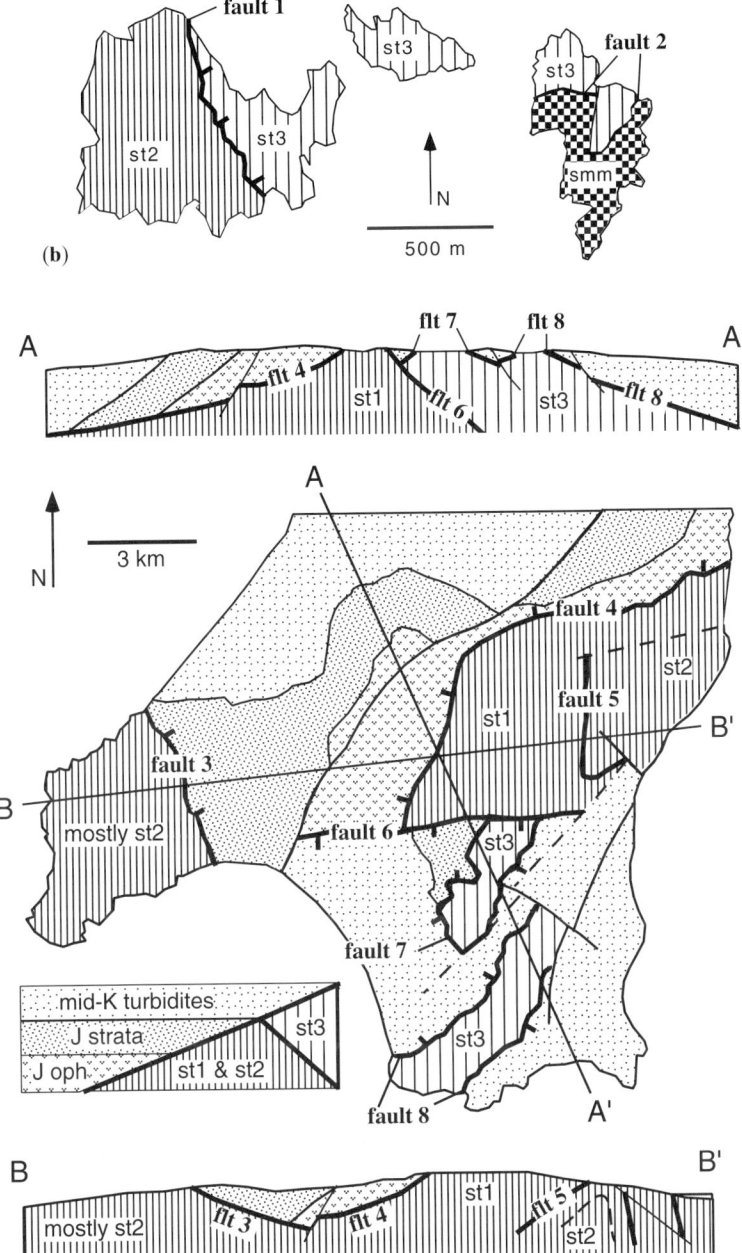

**Fig. 2.** (a) Generalized geological map of Mesozoic rocks in west–central Baja California. Upper-plate rocks include ophiolite and island-arc terranes on Cedros Island and Vizcaíno Peninsula and unconformably overlying middle Cretaceous turbidites. Lower-plate blueschist-facies subduction complex crops out on Cedros and San Benito Islands. Serpentinite-matrix mélange crops out along all faults between upper and lower plates on Cedros and within fault 9 on Vizcaíno. Boxes show locations of (b), (c) and Fig. 6c. (b) Geological map of San Benito Islands (Sedlock, unpublished mapping). Patterns as in (a) and (c); smm, serpentinite-matrix mélange. (c) Geological map and structure sections of Mesozoic rocks on southern Cedros Island; parts of map modified from Kilmer (1979) and Smith & Busby (1993). Heavy lines: faults; lighter lines, depositional contacts. Box shows structural disposition of major units. st1, st2 and st3 are subterranes of lower-plate blueschists.

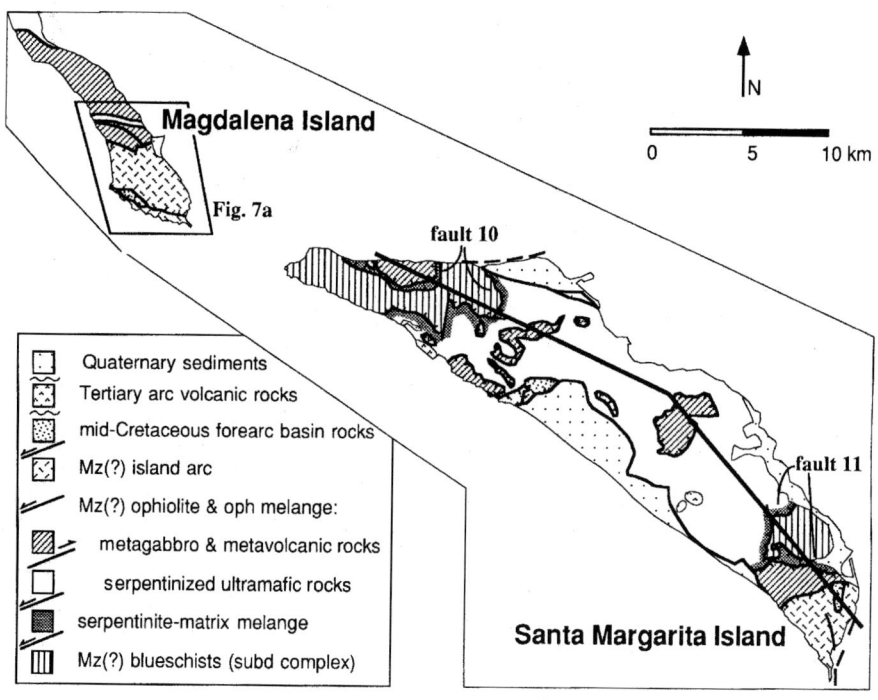

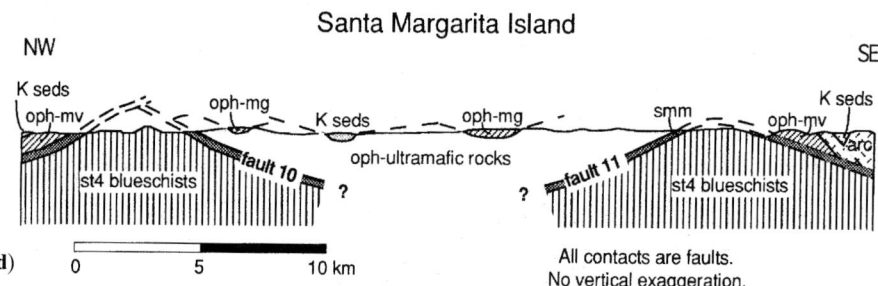

**Fig. 2.** (**d**) Top: geological map of Isla Santa Margarita and southern Isla Magdalena; geology of northern Magdalena Island from Blake *et al.* (1984). Bottom: structure section through Isla Santa Margarita. oph, ophiolite; mv, metavolcanic rocks; mg, metagabbro; smm, serpentinite-matrix mélange; K seds, Cretaceous forearc basin sedimentary rocks. Faults 10 and 11 may comprise a single unbroken surface or several discrete faults. Box shows location of Fig. 6a.

Layer-parallel foliation, tight to isoclinal folding, and boudinage accompanied complete metamorphic reconstitution of st1 rocks at depths greater than 25 km ($P > 0.8$ GPa), but brittle features such as discrete faults, fault zones, and spaced foliation predominated during very weak recrystallization of st3 rocks at about 20 km depth ($P = 0.6$–$0.7$ GPa). In st2 and st4 ($0.7$–$0.8$ GPa), variably developed discontinuous foliation, local pinch-and-swell structure and boudinage, open to tight folds, and abundant faults suggest that D1 was transitional from brittle to plastic behaviour at depths of about 22–25 km. Because quartz-rich rocks generally begin to behave plastically at about 300°C (e.g. Sibson 1982), these observations imply a geothermal gradient of 12–14°C km$^{-1}$ for subduction-related M1 and D1.

D2 is represented by normal fault and vein systems that truncate D1 structures (Fig. 4). Normal faults are uncommon and few fault surfaces crop out, so D2 kinematics are

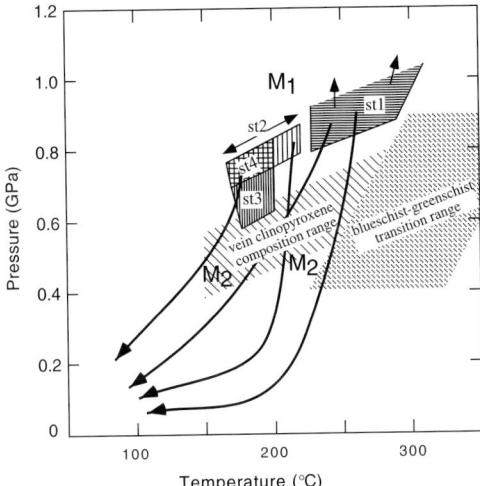

**Fig. 3.** Estimated conditions of peak metamorphism (M1 labelled st1, st2, st3, st4), and constraints on metamorphism during exhumation (M2) in western Baja. Upper pressure limit of st1 is indeterminate. Patterned fields indicate range of aegirine–jadeite composition in late veins (part of M2), and estimates by other workers of P–T conditions of transition from blueschist facies to greenschist facies (Sedlock 1988a). Filled arrows schematically show range of possible trajectories during exhumation.

unknown. Normal separations range from a few centimetres to at least 30 m; larger separations are likely but cannot be determined because of a dearth of marker beds. Most veins cut earlier foliation and contain M2 assemblages, and thus are assigned to this younger event; however, some veins in st1 and st2 may have formed as a result of the 10–15% volume loss that accompanied the breakdown of albite to jadeite + quartz during M1.

$^{40}Ar/^{39}Ar$ studies of the metamorphic minerals in the subterranes are hindered by fine grain size and low potassium concentrations. D1 and M1 in st2 and st3 are probably of late Early Cretaceous age, based on 112–100 Ma $^{40}Ar/^{39}Ar$ ages from metamorphic white mica in metasandstones (Baldwin & Harrison 1989; Baldwin 1996). The ages of D1 and M1 in st1 and st4 are unknown. The ages of D2 and M2, interpreted to represent exhumation, are problematic. Plagioclase in an st2 conglomerate clast yielded a saddle-shaped age spectrum with a minimum age of 20 Ma (at 145°C) (Baldwin & Harrison 1989). The significance of this date is discussed in a later section.

Blueschists were exposed at the surface by about 4 Ma in the Cedros–Vizcaíno region, where their detritus is present in Pliocene shallow-marine rocks, and probably by about 6 Ma in the Santa Margarita region, where they crop out within 1 km of undisturbed subaerial volcanic rocks.

## Serpentinite-matrix mélange

Serpentinite-matrix mélange occupies fault zones up to 500 m thick at all contacts between the blueschist-facies subduction complex and structurally higher rocks. Similar mélange also crops out along a few faults within the footwalls and hanging walls of these thicker fault zones. The matrix of the mélange consists chiefly of strongly foliated antigorite, chrysotile, and lizardite, with subordinate magnetite, chlorite, magnesite, and anthophyllite.

Most blocks in the mélange are competent, rounded to nodular bodies of serpentinized ultramafic rocks; gradational textures and petrographic similarities suggest that the mélange matrix formed by shearing of similar rocks. Protoliths are difficult to determine due to complete serpentinization, but probably include dunite, wehrlite, olivine clinopyroxenite, clinopyroxenite, and very rare chromitite.

About half of the non-ultramafic blocks are fragments of hanging-wall or footwall units, up to 250 m across, that usually crop out within a few kilometres of their source. The remaining blocks comprise diverse metamorphic rocks that were derived from cryptic source terranes not exposed in western Baja; these are hereafter referred to as exotic blocks. Many exotic blocks are mantled by rinds up to 1 m thick of actinolite schist that probably formed by chemical interaction with the enclosing serpentinite host (Moore 1984). Actinolite–tremolite is the most abundant phase, with randomly oriented crystals up to 15 cm. Other phases include chlorite, magnetite, muscovite, talc, and rare fuchsite.

Exotic blocks include greenschist, eclogite, two types of blueschist, and amphibolite. Greenschists derived from a variety of sedimentary and volcanic protoliths are sparsely but widely distributed in fault zones on southern Cedros Island, San Benito East Island, the Vizcaíno Peninsula, and Santa Margarita Island. Metapelitic schists on Santa Margarita yielded $^{40}Ar/^{39}Ar$ ages of about 120 Ma (Bonini 1994; Bonini & Baldwin 1994).

Eclogite crops out in mélange on Vizcaíno Peninsula (Moore 1986) and along the margins of st1 on Cedros Island. Peak metamorphic conditions were calculated at 1.2–1.5 GPa, 560–640°C (Sedlock, unpublished data). One Cedros eclogite was overprinted by a blueschist-facies assemblage at 115–105 Ma (Baldwin & Harrison 1989; Baldwin & Harrison 1992).

**Fig. 4.** Photographs of D2 normal faults cutting D1 fabrics in blueschist-facies subduction complex. (**a**) Normal fault cutting radiolarian chert (light bed above hammer), argillite, and tuff, Santa Margarita Island. (**b**) Normal faults cutting thin-bedded turbidites, Cedros Island; lens cap in lower centre of photo.

Type 1 blueschist blocks are foliated, very coarse-grained rocks that crop out in most mélange zones on Cedros. Type 1 blueschists contain sodic amphibole and white mica to 0.8 cm, almandine garnet to 2.5 cm, and poikiloblastic lawsonite to 1.5 cm, and lack epidote, actinolite, and any other indication of a higher-temperature overprint. Several blocks yielded $^{40}Ar/^{39}Ar$ isochron ages of 103–94 Ma on sodic amphibole and age gradients of 115–95 Ma on white mica (Baldwin & Harrison 1992).

Type 2 blueschist blocks contain peak assemblages that crystallized at higher temperatures than coherent footwall blueschists and type 1 blueschist blocks, as indicated by the presence of epidote rather than lawsonite and local actinolite and pumpellyite (Evans 1990). Type 2 blueschists crop out only on the Vizcaíno Peninsula and San

Benito East Island. A few blocks with younger greenschist-facies assemblages are the only Baja blueschists with a lower $P/T$ overprint.

Amphibolite blocks of mafic composition are widely distributed in all areas. Blocks contain amphibolite facies and epidote–amphibolite facies assemblages, which in some cases probably replaced eclogite assemblages. Garnet is locally abundant. Calcic amphibole and white mica from amphibolite blocks on Cedros yielded $^{40}$Ar/$^{39}$Ar isochron and plateau ages of 170–163 Ma; some blocks were overprinted by mid-Cretaceous blueschist-facies metamorphism (Baldwin & Harrison 1992). Hornblende from amphibolite blocks on the Vizcaíno Peninsula and central Magdalena Island (Cabo San Lázaro) yielded 138 Ma $^{40}$Ar/$^{39}$Ar plateau and isochron ages, respectively (Bonini 1994; S. Baldwin, unpublished data). Hornblende from amphibolite blocks on Santa Margarita yielded a K–Ar age of 138 Ma (Forman *et al.* 1971) and an $^{40}$Ar/$^{39}$Ar plateau age of 126 Ma (Bonini 1994).

*Upper plate*

The structurally highest rocks in western Baja, informally referred to as the upper plate, include Mesozoic island arcs, ophiolites, and volcanogenic to terrigenous cover strata that crop out in all areas except the San Benito Islands (Boles & Landis 1984; Kimbrough 1984; Moore 1985; Busby-Spera 1988; Sedlock 1993). Island-arc and ophiolite terranes evolved in the Pacific oceanic realm but were accreted to the western edge of North America in Late Jurassic or earliest Cretaceous time. After attachment, magmatism shifted eastward to the Cretaceous arc in mainland Baja, and the accreted terranes formed part of the forearc above an east-dipping subduction zone that fed this arc. The accreted terranes were overlapped by forearc basin turbidites that range from lower Aptian to uppermost Cretaceous age (Smith & Busby 1993; D. Smith, unpublished data; R. Sedlock, unpublished data).

Most upper-plate rocks have been confined to shallow crustal levels throughout their history, and all were at depths of <10 km before exhumation of footwall blueschists. Metamorphism typically ranges from absent to very low-grade, and primary igneous and sedimentary structures are well preserved. Combined with regional stratigraphic relations, these observations indicate maximum structural depths of <4 km for volcanogenic and terrigenous cover strata, and <10 km for island-arc terranes and the Cedros–Vizcaíno ophiolites. Ophiolites on Magdalena and Santa Margarita Islands may have been more deeply buried during (locally mylonitic) shortening and amphibolite-facies metamorphism of probable Late Jurassic age (Bonini 1994), but later were raised to a position above or adjacent to the island arcs (i.e. <10 km) when intruded by dykes and sills of these arcs.

All upper-plate units are cut by brittle extensional structures. These extensional features cut early foliation, folds, and thrusts in the Magdalena and Santa Margarita ophiolite, and represent the only strain event recognized in all other upper-plate units. The next section summarizes the style, geometry, kinematics, and timing of this extension.

## Brittle extension in western Baja

Brittle extension is the youngest or only deformation in every Mesozoic unit in western Baja. Similar style and geometry of brittle extension is shared by faults within all units and by faults between upper-plate units. Kinematic studies focused on the upper-plate units that lack older shortening strain, specifically the island-arc units and Jurassic and Cretaceous sedimentary rocks.

*Style*

All Mesozoic units are cut by sharp fault surfaces that are planar to subplanar at outcrop scale. Faults with separations of ≤10 m typically are knife-sharp and lack mesoscopically visible cataclasite. Fault gouge, scaly mudstone, and fault breccia are up to 20 cm thick at a few individual faults, and up to 3 m thick in fault zones that comprise several subparallel strands. Over 95% of measured faults show normal separation that ranges from a few millimetres to at least 2 km, and individual beds are extended up to 65% along sets of subparallel normal faults. Fault dips are typically 25° to 80°. Normal drag affects hanging-wall strata at some faults.

Widespread carbonate–quartz vein systems are up to 3 m thick and record up to 15% bulk extension. Many normal faults and veins terminate downward within serpentinite-matrix mélange at the base of the upper plate.

Faults between upper-plate units always place younger or shallower-level units atop older or deeper-level ones. Most depositional contacts have been cut out along such faults. Crustal thinning across such faults ranges from several hundred metres to at least 5 km (Fig. 5). Faults on Santa Margarita Island and the Vizcaíno Peninsula juxtapose Cretaceous turbidites with ultramafic units of subjacent ophiolites, indicating nearly complete distension of the upper plate.

Many fault zones contain horses that are remnants of tectonically excised units. For example, a major fault zone on the Vizcaíno Peninsula that places Cretaceous turbidites atop ophiolitic serpentinite contains slivers of ophiolitic gabbro. A nearby fault that places the Cretaceous turbidites atop ophiolitic gabbro contains horses of ophiolitic volcanic rocks (Fig. 5c). A major fault zone on Santa Margarita that places Cretaceous turbidites on ophiolitic ultramafic rocks contains dispersed fragments of island-arc dyke complex and ophiolitic metagabbro.

## Kinematic analysis

Palaeostresses that produced the extension can be estimated by kinematic analysis of the orientations of fault surfaces and striations. Data from the following areas indicate that palaeostress orientations are consistent over areas of several to tens of square kilometres, but are variable at a regional scale.

On southern Magdalena Island, data were collected from Cretaceous turbidites, island-arc rocks, and the fault boundary between them. Kinematic analysis suggests a single phase of

**Fig. 5.** Photographs of faults between rock units within upper plate. (**a**) Shallowly dipping normal fault on Cedros Island between Cretaceous turbidites in the hanging wall (bedding dips to left) and subhorizontal Jurassic shale in footwall; hanging wall is cut by synthetic normal faults (planar, dip to right). (**b**) Subhorizontal, undulatory normal fault on Vizcaíno Peninsula between Cretaceous turbidites in hanging wall and serpentinized ophiolitic harzburgite in footwall; cliff in foreground is about 25 m high. (**c**) Tectonic horse of brecciated ophiolitic metavolcanic rocks within a fault zone that separates hanging-wall Cretaceous turbidites and footwall ophiolitic gabbro, Vizcaíno Peninsula. Students at lower left for scale.

subhorizontal northeast–southwest tension (Fig. 6a).

On the northwestern Vizcaíno Peninsula, data were collected from Cretaceous turbidites, Jurassic strata, and Triassic ophiolite (Fig. 6b). Near the serpentinite-matrix melange of Puerto Nuevo ('fault 9' in Fig. 6b), subhorizontal east–west to northwest–southeast tension is inferred for faults within Cretaceous sedimentary rocks and for large-magnitude faults that place the Cretaceous strata atop ophiolitic rocks. Subhorizontal north–south tension is inferred for faults in Cretaceous and Jurassic sedimentary rocks about 10 km north of Puerto Nuevo and about 40 km to the northwest.

On Cedros, data were collected from Cretaceous turbidites in the southern and east–central parts of the island. Kinematic analysis indicates subhorizontal, roughly north–south tension in all areas (Fig. 6c). Subhorizontal east–west tension of late Cenozoic age affects Miocene–Pliocene rocks (Fig. 6c, 'faulted Tertiary rocks') and probably overprints older structures in Cretaceous turbidites on the east coast of Cedros (Fig. 6c, 'younger faults').

The uniformity of palaeostress orientations over large areas supports the field-based interpretation that most Mesozoic rocks underwent a single, though possibly protracted, phase of extension. Variations in palaeostress orientation between these areas indicate one or more of the following: (1) faults comprise superimposed systems that developed under different stress regimes; (2) faults were rotated or tilted after the main phase of extension; (3) stress or strain fields were heterogeneous during the main phase of extension. Possibility 1 is unlikely given the paucity of field and kinematic evidence. Possibility 2 is limited to rotations about a subvertical axis to account for the consistency of subvertical $\sigma 1$ in all areas. Possibility 3 seems the most likely, but is the hardest to test directly. Whatever the case, the current orientations make it impossible to judge whether tension was oblique or parallel to the plate boundary.

## Age of extension

Geological and geochronological data indicate that normal faulting in western Baja was active in Late Cretaceous and Paleogene time. Syndepositional normal faulting of forearc basin turbidites began by early Cenomanian time (97 Ma), as indicated by asymmetric basin

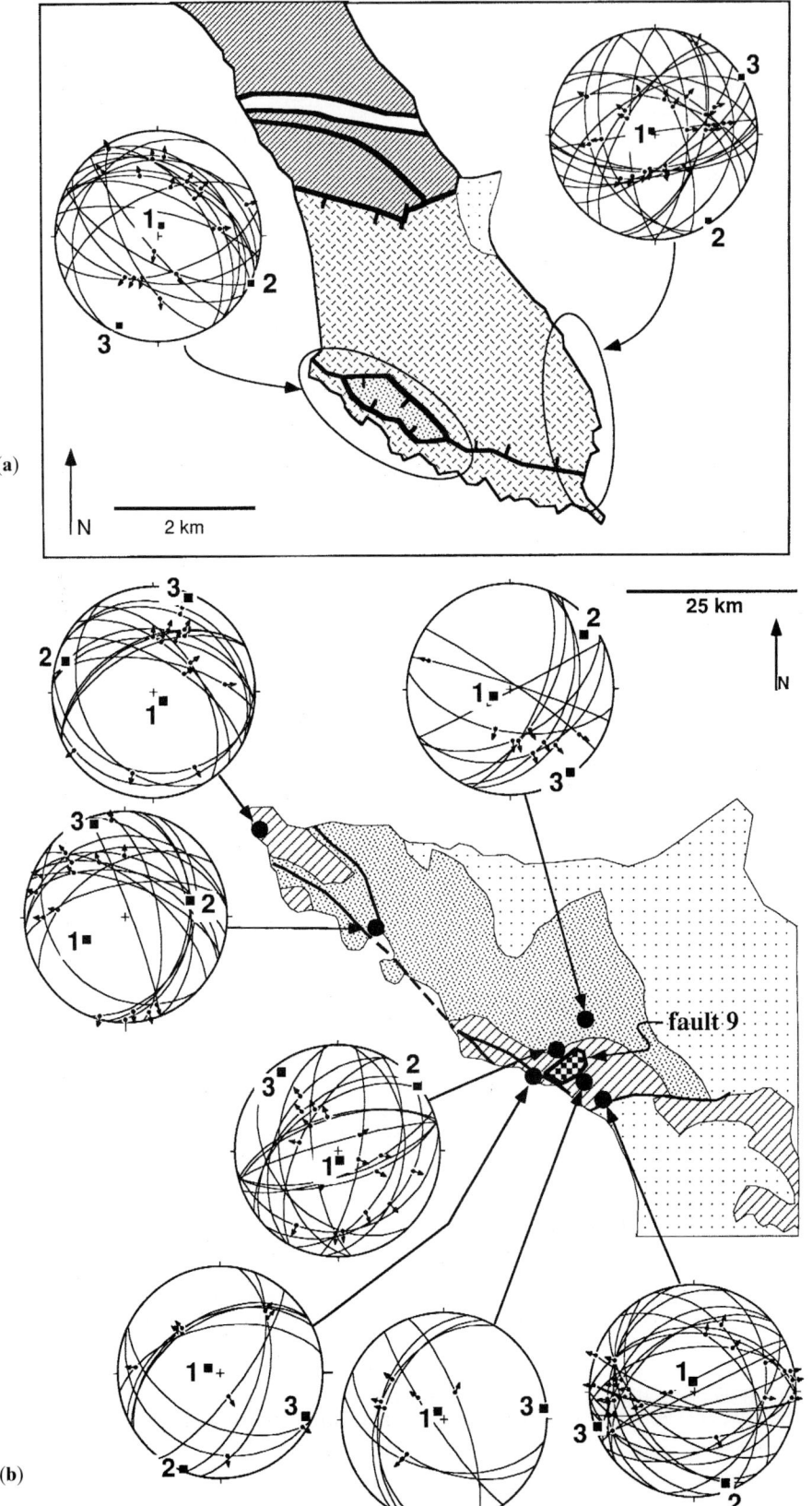

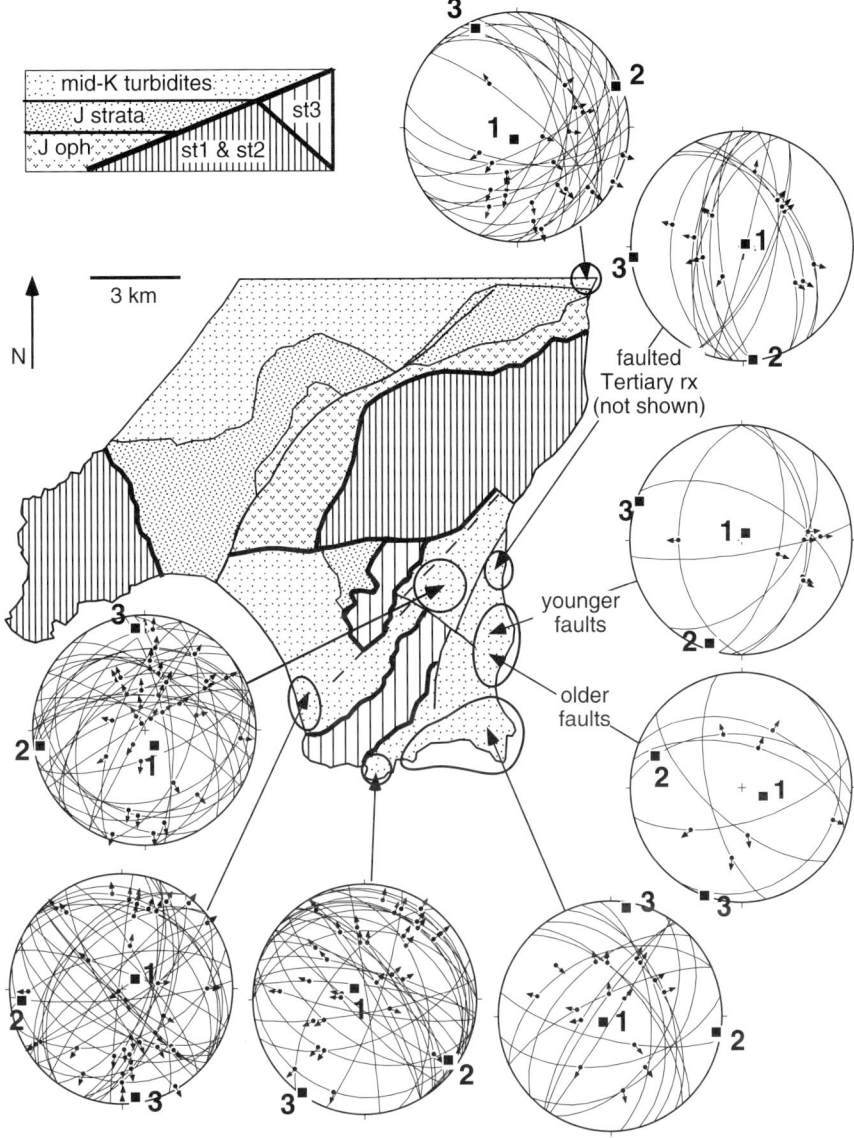

**Fig. 6.** Kinematic results of fault-striation data from upper-plate units in western Baja. Lower-hemisphere, equal-area projections; 1, 2, 3 indicate calculated palaeostress axes. Geological units keyed to Fig. 2. (**a**) southern Magdalena Island; (**b**) Vizcaíno Peninsula; (**c**) southern Cedros Island.

subsidence and olistostromes that formed at the base of normal fault scarps (Smith & Busby 1993). Extension must have persisted into early Cenozoic time because forearc basin strata as young as Maastrichtian date are cut by normal faults. Lamprophyre dykes on Santa Margarita Island that have yielded K–Ar and $^{40}Ar/^{39}Ar$ ages of 30 Ma (Forman et al. 1971; Bonini 1994) intrude and are not cut by upper-plate normal faults, indicating that this phase of extension ceased on Santa Margarita before 30 Ma. I infer that extension also ceased by this time elsewhere in western Baja, on the basis of region-wide strong similarities in the structural history of hanging-wall rocks, the structural and metamorphic history of footwall blueschists, and the composition, thickness, and mode of occurrence of serpentinite-matrix mélange.

High-angle normal faults with unknown magnitudes of net slip currently are active on or offshore Cedros, Magdalena, and Santa Margarita Islands (e.g. Yeats & Haq 1981). These faults are easily distinguished from Cretaceous–Paleogene normal faults by their planarity and lack of mélange, and because they truncate the older faults.

## Major faults that bound coherent blueschists

Subterranes of coherent blueschist-facies subduction complex rocks in western Baja are bounded by major faults that juxtapose them with upper-plate rocks or with other blueschist-facies subterranes. Most boundaries are fault zones that contain serpentinite-matrix mélange of varying thickness. These fault zones are fundamental to understanding the extensional component of blueschist exhumation, and possibly to understanding analogous, but poorly exposed, faults in other Franciscan-type orogens. Faults are numbered sequentially from northwest to southeast across the area (Fig. 2) and described individually below.

Nearly all deformation during faulting was accommodated by the serpentinite matrix of mélange within the fault zones. Foliation in serpentinite generally parallels the margins of enclosed blocks and, to a lesser degree, the margins of the fault zone. Striations on foliation surfaces typically rake 45–90°, suggesting that the predominant displacement on these faults was dip-slip or oblique-slip, but their geometric complexity prevents systematic kinematic analysis. For instance, striations on parallel foliation planes a few centimetres apart in the same fault zone rake 70–90° and 0–10°, respectively. The variability of striation orientations may reflect non-laminar motion of serpentinite during emplacement.

At most faults, strain gradients in the footwall and hanging wall are negligible, i.e. strain does not perceptibly increase toward the fault contact. However, upper-plate rocks within about 100 m of a few faults display broken formation produced by brittle disruption on faults that merge downward into mélange of the major fault. Bedding orientation in the broken formation is typically homoclinal, indicating disruption without folding, tilting, or rotation.

Emplacement of serpentinite caused varied chemical changes in rocks along the fault margins. At many faults, the hanging wall and, less commonly, the footwall exhibit a dark brown 'blackwall' effect ranging in thickness from <25 cm to 100 m that probably resulted from magnesium metasomatism. At most faults, upper-plate rocks and (to a lesser extent) footwall blueschists within 2–10 m of the serpentinite-matrix mélange are cut by veins of magnesite ± serpentinite that emanate from the mélange unit. In a few fault zones, the upper 10 cm of serpentinite is a very resistant crust formed by serpentine fibres growing at high angles to the fault surface.

### San Benito Islands

Fault 1 (Fig. 2a) is a nearly planar, 45°-west-dipping fault that juxtaposes hanging-wall st3 blueschists (depth at peak metamorphism 19–22 km) with footwall st2 blueschists (22–25 km) on San Benito West Island. The fault is marked by a discontinuous zone of serpentinite-matrix mélange, up to 4 m thick, with rare exotic blocks of blueschist. Fault 1 coincides with the western limit of the San Benito shear zone of Cohen et al. (1963).

Fault 2 is a >100-m-thick zone of serpentinite-matrix mélange that crops out over much of San Benito East Island. St3 blueschists occupy the footwall, but no upper-plate rocks are preserved in the hanging wall here or elsewhere on the San Benito Islands. The contact between the mélange and footwall blueschists dips about 50° south. Blocks within mélange in the fault zone include greenschists, type 2 blueschists, and amphibolites.

### Cedros Island

Fault 3 (Fig. 2b) juxtaposes hanging-wall Jurassic volcanogenic strata (3–4 km depth) with footwall st2 and st3 blueschists (19–25 km). The fault is marked by a 5–25-m-thick belt of serpentinite-matrix mélange that dips 5–30° east and truncates at a small angle the subhorizontal bedding in hanging-wall strata. Sparse exotic blocks include type 1 blueschists and amphibolite. These observations contradict the assignment by Kilmer (1979) of the footwall blueschists to the >5-km-wide San Augustin shear zone, which he speculated might be a continuation of fault 1 on San Benito West Island. Fault 3 may connect eastward in the subsurface with fault 4 (Fig. 2b).

Fault 4, the San Carlos fault of Kilmer (1979), juxtaposes hanging-wall ophiolite (<10 km depth) with footwall st1 (>25 km). The fault zone consists of strongly foliated serpentinite-matrix mélange up to 30 m thick with exotic blocks of type 1 blueschist and rare eclogite. Kimbrough (1982) interpreted the serpentinite as an integral part of the Cedros ophiolite, and

Kilmer (1979) inferred steep (to 80°) dips of the fault zone, but detailed mapping clearly shows that the serpentinite-matrix mélange is a structurally discrete, fairly tabular unit that dips 10–45° north to west off the north and west flanks of Monte Cedros. Fault 4 may connect westward in the subsurface with fault 3 (Fig. 2b).

Fault 5 juxtaposes hanging-wall st1 blueschists (>25 km depth) with footwall st2 blueschists (22–25 km). This poorly exposed fault probably dips moderately west except at its southern limit, where it and underlying st2 rocks are folded into a south-plunging anticline. To the north, fault 5 is truncated by an inferred high-angle fault. Very sparse outcrops of serpentinite and resistant exotic blocks of type 1 blueschist suggest that fault 5 is marked by serpentinite-matrix mélange.

Fault 6 juxtaposes hanging-wall st3 blueschists (19–22 km depth) and Jurassic and Cretaceous strata (1–4 km) with footwall st1 blueschists (>25 km). Fault 6 strikes roughly east–west and dips 50–60° south. Serpentinite-matrix mélange, which lacks exotic blocks here, is typically 3–5 m thick but ranges from 0 m (absent) to 25 m thick. Fault 6 truncates faults 4 and 7. The fault is covered by extensive landslide deposits west of fault 4, and is probably truncated by a high-angle late Cenozoic fault east of fault 7.

Faults 7 and 8 (Fig. 2a) juxtapose hanging-wall Cretaceous and Jurassic strata (1–4 km depth) with ridge-forming footwall st3 blueschists (19–22 km). Both faults are marked by serpentinite-matrix mélange with exotic blocks of greenschist, type 1 blueschist, and amphibolite. At the outcrop scale, mélange typically forms roughly tabular, 5–15-m-thick belts. However, the fault zones clearly are not planar surfaces. Mélange generally dips east on the east sides of the ridges, dips west on the west sides, and forms klippe atop the ridges, indicating that faults 7 and 8 are roughly antiformal. Faults 7 and 8 may be part of a single undulating fault surface; where the two ridges are en echelon, the faults dip < 20° towards each other and probably intersect at a depth of less than 1 km. Faults 7 and 8 both are cut out by younger, higher-angle normal faults of late Cenozoic age on the southeast sides of the ridges.

Faults 7 and 8 display a number of geometric irregularities that may be due to the mobility of serpentinite-matrix mélange, including a kilometre-long, 100-m-thick tabular splay into the footwall of fault 7; bulbous diapirs, 100–500 m across, into the footwall of fault 8; undulations that produce hanging-wall klippe and footwall windows; and locally overturned contacts of the mélange with hanging-wall or footwall rocks.

Exotic blocks are abundant, particularly at the south tip of the island, where slabs of amphibolite at least 500 m long occupy a zone at least 200 m thick. Serpentinite-matrix mélange in the southern half of fault 8 consists of a structurally lower member rich in exotic blocks and a structurally higher member poor in exotic blocks. Foliation in the serpentinite matrix sometimes parallels fault boundaries, but in many areas it strikes at moderate to high angles to the fault zone.

## Vizcaíno Peninsula

Fault 9 (Fig. 2a) is a complex fault zone at the base of upper-plate rocks in the Puerto Nuevo region of the Vizcaíno Peninsula. Serpentinite-matrix mélange is at least 250 m thick and contains greenschist, eclogite, blueschist, and amphibolite blocks (Moore 1985, 1986; San José State University Geological Survey, unpublished data). Coherent high $P/T$ rocks are not exposed on the Vizcaíno Peninsula.

## Santa Margarita Island

Fault 10 (Fig. 2d) juxtaposes hanging-wall ophiolitic rocks (pre-exhumation depth <10 km) with footwall st4 blueschists (22–25 km) on northern Santa Margarita Island. Fault 10 dips 35–50° east and is marked by serpentinite-matrix mélange, typically 50–100 m thick, that contains exotic blocks of striped gneiss, amphibolite, and greenschist, and abundant slivers and slabs of adjacent wall rocks (st4, ophiolite). Mélange also occupies splays of fault 10 that cut into the footwall and hanging wall. Fault 10 may connect in the subsurface with fault 11 to the southeast (Fig. 2d).

Fault 11 juxtaposes hanging-wall ophiolitic rocks (pre-exhumation depth <10 km) with footwall st4 blueschists (22–25 km) on southern Santa Margarita Island. The fault is marked by a belt of serpentinite-matrix mélange about 100 m thick that dips 10–30° to the north, west, and south off Monte Santa Margarita. Thus, fault 11 either has a domal geometry or is a composite of variably oriented faults. Fault 11 may connect in the subsurface with fault 10 to the northwest (Fig. 2d).

## Sense and amount of slip

Earlier workers interpreted the fault zones between the upper-plate rocks and the blueschist-facies subduction complex as thrusts (Rangin 1978) or possible strike-slip faults (Kilmer 1979), but I have reinterpreted all but fault 5 as crustal-scale normal faults (Sedlock 1988b, 1993, 1996). Faults dip < 60° and generally < 40°, and most striations rake 45–90°, indicating dip-slip or oblique-slip. Brittle extension is the only or youngest strain event in every unit within the hanging wall of the major fault zones, and many hanging-wall normal faults and vein systems terminate downward within serpentinite-matrix mélange, suggesting a contemporaneous origin.

Wheeler & Butler (1994) discussed ways to identify true normal slip on major faults and shear zones in orogenic belts. Metamorphic and thermobarometric criteria have limited, if any, utility in western Baja, where hanging-wall rocks of most major fault zones are weakly metamorphosed to unmetamorphosed. However, two structural criteria can be tested in western Baja: (1) the surface of a true normal fault will intersect the Earth's surface when traced in a direction opposite that of hanging-wall transport, even after subsequent erosion; (2) tectonic thinning across true normal faults can be demonstrated only if that layering was subhorizontal when slip on the surface began. Application of both criteria in western Baja confirms that the major fault zones are normal. With respect to the first criterion, all faults project into space above the Earth's surface. With respect to the second, brittle extension was the sole deformation to affect most upper-plate units, including all island-arc rocks and all Jurassic–Cretaceous volcanogenic and terrigenous sedimentary rocks. Thus, at the onset of exhumation (c. 100 Ma), bedding in these units must have been horizontal or nearly horizontal, eliminating concern that normal separation might be caused by thrusting of strata of appropriate dip (Wheeler & Butler 1994). The interpretation of pressure differences at the major faults (Wheeler & Butler 1994) is straightforward because constant refrigeration of the long-lived Mesozoic subduction zone west of North America probably prohibited thermal re-equilibration (e.g. Ward 1991).

Geobarometric and stratigraphic data require tectonic thinning of 12–25 km or more across most of the major faults. Net normal slip was calculated using the amount of crustal thinning and the maximum and average surface dips of the fault zones, and projecting surface dip downward (Table 2). These slip estimates are minimum values of the total slip because strike-slip components of displacement cannot be determined, owing to a lack of piercing points. Normal sense and slip cannot be determined for faults 2 and 9 because they lack hanging-wall and footwall rocks, respectively. However, the overall style and structural setting of these faults suggest strong similarities with the tabulated faults. Fault 5, which places slightly deeper-level blueschists atop other blueschists, is interpreted as a syn-subduction thrust that was folded during duplexing and underplating at the base of the accretionary prism.

## Age of exhumation

Stratigraphic ages of hanging-wall rocks (e.g. Smith & Busby 1993) and peak metamorphic ages of footwall blueschists (Baldwin & Harrison 1989) indicate that exhumation must have started after c. 100 Ma and must have been active during Late Cretaceous and Paleogene time (Fig. 7). Lamprophyre dykes that intrude normal faults on Santa Margarita show that extension ceased by 30 Ma there and, based on striking region-wide similarities discussed earlier, probably elsewhere in western Baja.

A mid-Tertiary age for the cessation of major exhumation is also supported by plate tectonic considerations (Ward 1991). The preservation of blueschists in western Baja California probably reflects protracted exhumation in the continuously refrigerated forearc of a steady-state subduction zone. Geophysical and geological

**Table 2.** *Tectonic thinning and net normal slip on major faults*

| Fault | Tectonic thinning (km) | Fault dip (degrees) | | Net normal slip (km) | |
|---|---|---|---|---|---|
| | | Maximum | Average | Minimum | Probable |
| 1 | 0–6 | 45 | 45 | 0–8 | 0–8 |
| 3 | 15–22 | 30 | 15 | 30–44 | 58–85 |
| 4 | >15 | 45 | 25 | >21 | >35 |
| 6 | >21 | 60 | 55 | >24 | >26 |
| 7 | 15–21 | ≈35 | 25 | 26–37 | 36–50 |
| 8 | 15–21 | ≈60 | 35 | 17–24 | 26–37 |
| 10 | >12–15 | 50 | 40 | >16–20 | >19–23 |
| 11 | >12–15 | 30 | 20 | >24–30 | >35–44 |

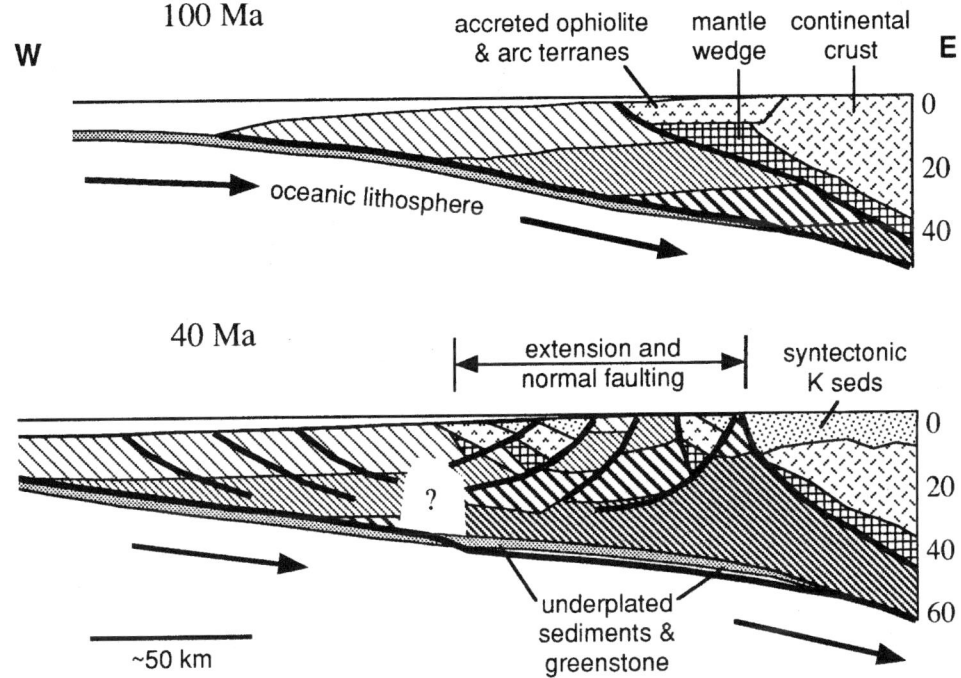

**Fig. 7.** Generalized, unbalanced cross-sections of western margin of Baja at 100 Ma (top) and about 40 Ma (bottom). Accretionary prism consists of four diagonally lined units divided at intervals of 0.4 GPa, 0.8 GPa, and 1.2 GPa. Underplating led to extension and normal faulting during Late Cretaceous and Early Tertiary time (see text). Queried blank area is transition between extensional and contractional regimes.

evidence shows that oceanic lithosphere (Farallon, Kula, or both) was subducted beneath North America from at least 100 Ma until c. 30 Ma. Continuous subduction of cold lithosphere would have maintained low geothermal gradients and prevented blueschists from warming during exhumation. However, by c. 40 Ma, young Farallon lithosphere entering the subduction zone probably slowed the convergence rate (Ward 1991) and led to heating of the forearc and overprinting of blueschists not already at near-surface depths. Thus, extension and exhumation may have been completed by c. 40 Ma.

The age of exhumation is not effectively constrained by geochronological data. Baldwin & Harrison (1989) interpreted $^{40}Ar/^{39}Ar$ data from an st2 blueschist to indicate cooling below 145°C at c. 20 Ma. Assuming exhumation of these aragonite-bearing rocks into the calcite stability field at a temperature of 125–175°C (Carlson & Rosenfeld 1981), Baldwin (1996) inferred average exhumation rates of 0.1 mm $a^{-1}$ from about 100 Ma to 20 Ma and >0.8 mm $a^{-1}$ from 20 Ma to 5 Ma. The rocks thus would have been exhumed from depths of 12–15 km during Neogene time. However, the same geochronological data, and revised aragonite–calcite stability relations (Liu & Yund 1993), are consistent with exhumation at about 0.3 mm $a^{-1}$ from 100 Ma to 30 Ma and at <0.1 mm $a^{-1}$ since 30 Ma. High $P/T$ rocks would have been within a few kilometres of the surface by 30 Ma, allowing little subsequent exhumation. This second alternative resembles the exhumation history inferred for the high $P/T$ Catalina Schist in southern California (Grove & Bebout 1995), and is consistent with protracted plate subduction in late Cretaceous and early Tertiary time followed by a decrease or stalling of convergence in Oligocene time. In any case, the exhumation history cannot be uniquely determined from geochronological data.

## Evaluation of exhumation mechanisms in western Baja

What role did different mechanisms play in the exhumation of blueschists in western Baja? In this section, I consider each of the five mechanisms with emphasis on observations that validate or eliminate particular mechanisms.

*Erosion.* Erosion probably contributes to the exhumation of high *P/T* rocks in all orogens that raise rocks above sea level. However, western Baja high *P/T* rocks and their overburden may have been submerged throughout the exhumation phase. Exhumation began after peak blueschist metamorphism at 115–100 Ma, so erosion must have been of Cenomanian age or younger. Most of western Baja, including the subsurface, contains shallow to deep marine terrigenous clastic rocks that range from Aptian to early Eocene age (López-Ramos 1985; D. Smith, unpublished data). This suggests, but does not prove, that the forearc above the blueschist-facies subduction complex was submarine and thus not eroded.

The composition and volume of the Cenomanian–Eocene marine strata do not support erosion as an exhumation mechanism (Table 1, predictions 3 and 4). Provenance studies concur that the strata were derived from the arc or continent to the east; the absence of blueschist detritus indicates that no high *P/T* rocks had been exhumed to the Earth's surface as late as early Eocene time. Volume estimates can be indirectly estimated from the two-dimensional case of a margin-normal transect, where the cumulative thickness of Cenomanian–Eocene marine strata in Baja is about 5 km (López-Ramos 1985; D. Smith, unpublished data), or less than one-third of the mean thickness of the exhumed overburden. A similar volumetric difference applies if the widths of the exhumed forearc and the depositional basins were roughly similar. Another observation that is inconsistent with erosion as the sole or chief exhumation mechanism is that the high *P/T* rocks do not form part of a continuous crustal section (Table 1, prediction 1).

Nevertheless, erosion cannot be completely invalidated as a contributing mechanism for exhumation. Detritus of the subduction complex may not be preserved or exposed in current outcrop areas, or they could have been transported away from western Baja by margin-parallel faults. Perhaps Cenomanian–Eocene marine strata were partly derived from forearc rocks that are petrographically indistinguishable from arc or terrigenous sources.

*Ductile flow.* Ductile processes such as diffusional mass transfer probably produced early shortening strain features in st1, st2, and st4 blueschists. Thus, ductile flow may have operated during synmetamorphic shortening, probably shortly before or during underplating of the blueschists. However, deformation that accompanied exhumation, although imperfectly understood, appears to consist solely of brittle (frictional) extensional structures such as normal faults and vein systems. Ductile deformation is absent from most upper-plate units, and where it is present, as in the Magdalena and Santa Margarita ophiolites, it is probably syn-accretion and pre-exhumation. Ductile flow probably played a negligible role in the exhumation of the blueschists in western Baja.

*Normal faulting.* As discussed in preceding sections of the paper, boundaries between upper-plate rocks and the coherent blueschists are probably major normal fault zones. The simple tectonic history of upper-plate rocks in the hanging walls of these faults indicates that slip was normal and that metamorphic breaks across the fault zones have not been modified (Table 1, predictions 7 and 8). Tectonic thinning of 12–25 km or more and net normal slip of 20–45 km or more (Table 2) are comparable with values in many metamorphic core complexes (e.g. Spencer & Reynolds 1991). Thus, this paper affirms the interpretation that normal faulting was the principal mechanism for exhuming Baja blueschists (Sedlock 1988*b*, 1993, 1996). The timing of extension and exhumation, discussed in the previous section, indicates that extension occurred while subduction continued (Fig. 7).

*Strike-slip faulting.* Plate tectonic history suggests that western Baja has accommodated significant dextral slip since about 100 Ma. The margin-parallel component of right-oblique convergence in Late Cretaceous time and, to a lesser extent, in Paleogene time was probably accommodated by dextral slip faults in the forearc. Such faults may have hosted northward translation of the subduction complex and possibly upper-plate rocks in western Baja (Hagstrum & Sedlock 1990; Lund & Bottjer 1991). Modern subduction margins with obliquity as low as 30° reveal complex shortening, extension, and translation (Ryan & Scholl 1989).

Western Baja also was cut by dextral slip faults in late Tertiary time, when the transform Pacific–North American plate boundary lay offshore between about 18 Ma and 5 Ma (e.g. Lonsdale 1991). For example, Miocene slip is believed to have occurred in the borderland on the submarine Tosco–Abreojos fault zone, which lies 50 km southwest of Santa Margarita Island and dies out northward near the Vizcaíno Peninsula. Mann & Gordon (1996) suggested that such Neogene strike-slip faults may have played a major role in exhuming the high *P/T* rocks in western Baja (Table 1, predictions 9 and 10). To do so, such fault systems must involve significant shortening and surface uplift at a restraining bend or stepover. However, the exhumed Baja blueschists are not within nor bounded by a strike-slip fault zone with such a geometry, and offshore candidates are 20–50 km distant (Normark et al. 1987). The compilation maps of Fenby & Gastil (1991) show through-going strike-slip faults near southwestern Cedros Island and the San Benito Islands, but field relations documented in this paper show that the supposed onland continuations of such faults (faults 1 and 3) are not strike-slip faults at all, but rather shallowly to moderately dipping normal faults. Furthermore, detailed seismic reflection studies in this part of the borderland, as well as near Magdalena and Santa Margarita Islands, recognize late Cenozoic and active normal faults but no throughgoing strike-slip faults (Normark et al. 1987). Thus, I conclude that late Cenozoic strike-slip faulting probably played a minor role in exhuming the Baja blueschists.

*Buoyancy.* Buoyancy may have contributed to exhumation of blueschists if they are less dense than the crustal rocks through which they passed. Density measurements are not available for Mesozoic rocks in western Baja, but useful proxies are provided by similar rocks elsewhere. Subduction-complex blueschists probably have Franciscan-like densities of 2.8–3.0 g cm$^{-3}$, and upper-plate terrigenous sedimentary rocks, island arcs, and ophiolites probably have average densities of 2.7–3.0 g cm$^{-3}$, so buoyancy probably contributed little to exhumation.

## Discussion

### Significance of mélange

What can be learned of blueschist exhumation from the serpentinite-matrix mélange within the major fault zones between upper-plate rocks and footwall blueschists? Was the mélange present before faulting, perhaps as a remnant of the mantle wedge that was tectonized during exhumation-related fault slip? Or was the mélange emplaced later along these existing weak surfaces after slip had ceased?

Several observations support the interpretation that mélange had already formed by the time slip began on the major normal fault zones. (1) Most of the major normal faults have hosted 20–35 km of slip or more (Table 2), yet wall rocks lack strain gradients that are typical of faults with such slip magnitudes. In comparison, footwall and hanging-wall rocks in most metamorphic core complexes are affected by strong ductile and especially brittle strain hundreds or thousands of metres from the detachment fault (e.g. Coney 1980). In western Baja, the strongly sheared matrix of the mélange appears to have accommodated the bulk of slip-related strain. (2) Most mélange-occupied fault zones are tabular; even where curved, fault zones generally have parallel upper and lower boundaries. This geometry suggests that the shape of the mélange bodies was controlled by fault slip, and thus that mélange emplacement is not post-kinematic. Local irregularities in some fault zone margins probably reflect minor post-slip mélange remobilization. (3) Serpentinite-matrix mélange also crops out along folded faults, probably thrusts, within the coherent blueschists (fault 5) and within upper-plate ophiolites on Santa Margarita Island. This mélange is similar in appearance, composition, block type, and structural style to mélange in the major normal faults between the upper plate and the blueschists. Mélange was probably emplaced before folding, because post-folding emplacement would require intricate navigation of 1-km-amplitude tight to open folds with almost no intrusion of wall rocks. The folded faults are cut by major normal faults, so I infer that the mélange had been formed and emplaced in pre-extension, probably Cretaceous, time.

What constraints can be placed on conditions and location of mélange formation? Mélange includes exotic blocks that are fragments of older sources that do not crop out in the region, and blocks that are fragments of footwall and hanging-wall rocks. Mélange must have formed, at least in part, after the metamorphic age of the youngest enclosed block (115–100 Ma blueschists). However, older block types may have been incorporated in the mélange earlier. Actinolite–chlorite schist rinds around many blocks attest to active chemical metasomatism between the blocks and the serpentinite and, together with the absence of rodingite from the mélange, show that the blocks were entrained after complete serpentinization of peridotite. Among the possible causes of mixing of blocks from diverse source areas are diapiric rise of serpentinite and pre-extension thrusting (see below).

Many components of the mélange are consistent with tectonic thinning of a mantle wedge and adjacent parts of the accretionary prism that once overlay the blueschists (Fig. 7). Ultramafic blocks and matrix may have such an origin, or alternatively may be derived from upper-plate ophiolitic terranes. Higher-grade blocks, such as amphibolite and eclogite, may be parts of a metamorphic sole that developed beneath the mantle wedge during the initiation of subduction in Jurassic time (Peacock 1988; Baldwin & Harrison 1992). Blueschist blocks that are much coarser grained (type 1) or higher temperature (type 2) than the coherent footwall blueschists may have been metamorphosed nearer the mantle wedge, where ambient temperatures were higher.

An alternative hypothesis is that many exotic blocks in mélange are remnants of cryptic terranes that overthrust the lower-plate coherent blueschists before a switch from thrust to normal sense of slip on the major fault zones (e.g. Rawling & Lister this volume). This hypothesis could account for greenschist and amphibolite-facies blocks with Early Cretaceous (138–120 Ma) metamorphic ages that are not simply explained by tectonic attenuation of the overburden of the coherent blueschists. Such a scenario also could explain the presence of eclogite, amphibolite, and coarse-grained blueschist blocks that have peak metamorphic pressures, temperatures, or both that are significantly higher than in current wall rocks. However, the major normal fault zones obliquely truncate exposed thrust faults, indicating a geometry that is more complex than straightforward reactivation of old thrust faults by younger normal faults.

Thermochronological events documented in many blocks were roughly coeval with the onset of exhumation at 100–95 Ma. Type 1 blueschists yield cooling ages of 115–94 Ma, eclogite and amphibolite-facies blocks have mid-Cretaceous blueschist overprints, and garnet amphibolite along an enigmatic fault within the ophiolite on Santa Margarita may have undergone a heating event at about 100 Ma (Baldwin & Harrison 1989, 1992; Bonini 1994; Bonini & Baldwin 1994; Baldwin 1996). These events and the onset of exhumation may reflect major changes in forearc dynamics because of subduction of older or younger plate, a change in the rate or obliquity of oblique convergence, or underplating of thick oceanic materials. This age also roughly coincides with that of intra-arc shortening in mainland Baja to the east (Griffith & Hoobs, 1993).

## Comparison with other high P/T rocks

Conclusions drawn from Mesozoic oceanic rocks in western Baja probably are more relevant to other occurrences of subduction-related high P/T rocks than to ultrahigh P/T terranes. An instructive case in light of the Baja results is California, where the Mesozoic geological framework closely resembles that of Baja. From east to west between 30°N and 24°N, Baja consists of a Cretaceous calc-alkalic magmatic arc (mostly buried south of 28°N), middle Cretaceous to Paleogene siliciclastic marine flysch, and Mesozoic oceanic terranes. Corresponding rocks in northern California include Jurassic–Cretaceous magmatic rocks of the Sierra Nevada, the upper Mesozoic Great Valley Group and associated Paleogene strata, and the Franciscan Complex and fragments of the Coast Range ophiolite. In both regions, these triads are interpreted as a subduction-related continental margin arc, a coeval forearc basin assemblage, and accreted oceanic terranes. Forearc basin strata in both regions overlap the inboard magmatic arc and outboard ophiolite and arc terranes.

Baja blueschists most closely resemble the eastern and coastal belts of the Franciscan Complex, where structural coherence is higher than in the central belt. There are several strong similarities: peak metamorphic conditions in those belts overlap those of the Baja blueschists, the Coast Range fault system that separates the Franciscan rocks from upper-plate ophiolite and forearc basin strata corresponds to the major normal fault zones in western Baja, and sparse outcrops of serpentinite-matrix mélange may occupy the same structural setting as in Baja. Previous workers have suggested that coherent Franciscan blueschists were exhumed primarily because of buoyancy and erosion (Ernst 1971, 1988), or extension and normal faulting (Jayko et al. 1987; Harms et al. 1992), with minimal contributions from ductile flow (Ring & Brandon 1994; Bohlar et al. 1997). However, Franciscan studies are hampered by discontinuous outcrop, urban development, and overprinting by complex tectonism associated with the birth and development of the San Andreas system. Superior exposure and comparatively minor overprinting in western Baja has allowed the documentation of syn-subduction extension and exhumation that have been inferred but are not easily demonstrable in California.

This work was partly supported by NSF grants EAR85–18871 and EAR91–04771. D. Larue collaborated on collection of fault slip data. I thank S. Baldwin and J. Bonini for access to unpublished data, discussions of tectonics and thermochronology, and accompaniment in the field. G. Gastil, J. Hagstrum, D. Kimbrough, T. Moore, and D. Smith have been sources of thought-provoking insights over the years. J. Selverstone, J. Wakabayashi, G. Lister, and S. Baldwin provided helpful reviews. Fault slip kinematics were calculated with the FAULTKIN program of R. Allmendinger, R. Marrett, and T. Cladouhos.

# References

AGUE, J. J. & BRANDON, M. T. 1992. Tilt and northward offset of Cordilleran batholiths resolved using igneous bathymetry. *Nature*, **360**, 146–149.

AVÉ LALLEMANT, H. G. & GUTH, L. R. 1990. Role of extensional tectonics in exhumation of eclogites and blueschists in an oblique subduction setting: northeastern Venezuela. *Geology*, **18**, 950–953.

AVIGAD, D. 1992. Exhumation of coesite-bearing rocks in the Dora Maira massif (western Alps, Italy). *Geology*, **20**, 947–950.

—— & GARFUNKEL, Z. 1991. Uplift and exhumation of high-pressure metamorphic terranes: the example of the Cycladic blueschists belt (Aegean Sea). *Tectonophysics*, **188**, 357–372.

BALDWIN, S. L. 1996. Contrasting P–T–t histories for blueschists from the Western Baja terrane and the Aegean: effects of synsubduction exhumation and backarc extension. *In*: BEBOUT, G. E., SCHOLL, D. W., KIRBY, S. H. & PLATT, J. P. (eds) *Subduction: Top to Bottom*. American Geophysical Union, Geophyiscal Monograph **96**, 135–141.

—— & HARRISON, T. M. 1989. Geochronology of blueschists from west–central Baja California and the timing of uplift of subduction complexes. *Journal of Geology*, **97**, 149–163.

—— & —— 1992. The P–T–t history of blocks in serpentinite-matrix melange, west-central Baja California. *Geological Society of America Bulletin*, **104**, 18–31.

BECK, M. E., JR 1991. Case for northward transport of Baja and coastal southern California: paleomagnetic data, analysis, and alternatives. *Geology*, **19**, 506–509.

BLAKE, M. C., JR, JAYKO, A. S., MOORE, T. E., CHAVEZ, V., SALEEBY, J. B. & SEEL, K. 1984. Tectonostratigraphic terranes of Magdalena Island, Baja California Sur. *In*: FRIZZELL, V. A. JR. (ed.) *Geology of the Baja California Peninsula*. Pacific Section, Society of Economic Paleontologists and Mineralogists, Book **39**, 183–191.

BOHLAR, R., RING, U. & BRANDON, M. T. 1997. Large-scale mass loss in the Franciscan subduction complex, California: implications for the exhumation of high-P rocks. *Geological Society of America Abstracts with Programs*, **29**(5), 5.

BOLES, J. R. & LANDIS, C. A. 1984. Jurassic sedimentary melange and associated facies, Baja California, México. *Geological Society of America Bulletin*, **95**, 513–521.

BONINI, J. A. 1994. $^{40}Ar/^{39}Ar$ geochronology of accreted terranes from southwestern Baja California Sur. MS thesis, University of Arizona, Tucson.

—— & BALDWIN, S. L. 1994. $^{40}Ar/^{39}Ar$ geochronology of accreted terranes from southwestern Baja California Sur. *US Geological Survey Circular*, **1107**, 34.

BOTTJER, D. J. & LINK, M. H. 1984. A synthesis of Late Cretaceous southern California and northern Baja California paleogeography. *In*: CROUCH, J. K. & BACHMAN, S. B. (eds) *Tectonics and Sedimentation along the California Margin*. Pacific Section, Society of Economic Paleontologists and Mineralogists, Book **38**, 171–188.

BUSBY-SPERA, C. J. 1988. Evolution of a Middle Jurassic back-arc basin, Cedros Island, Baja California: evidence from a marine volcaniclastic apron. *Geological Society of America Bulletin*, **100**, 218–233.

CARLSON, W. D. & ROSENFELD, J. L. 1981. Optical determination of topotactic aragonite–calcite growth kinetics: metamorphic implications. *Journal of Geology*, **89**, 615–638.

CLOOS, M. & SHREVE, R. L. 1988. Subduction-channel model of prism accretion, melange formation, sediment subduction, and subduction erosion at convergent plate margins: 2. Implications and discussion. *Pure and Applied Geophysics*, **128**, 501–545.

CLOWES, R. M., BRANDON, M. T., GREEN, A. G., YORATH, C. J., BROWN, A. S., KANASEWICH, E. R. & SPENCER, C. 1987. LITHOPROBE – southern Vancouver Island: Cenozoic subduction complex imaged by deep seismic reflections. *Canadian Journal of Earth Sciences*, **24**, 31–51.

COHEN, L. H., CONDIE, K. C., KUEST, L. J., JR, MACKENZIE, G. S., MEISTER, F. H., PUSHKAR, P. & STUEBER, A. M. 1963. Geology of the San Benito Islands, Baja California, México. *Geological Society of America Bulletin*, **74**, 1355–1370.

CONEY, P. J. 1980. Cordilleran metamorphic core complexes: an overview. *In*: CRITTENDEN, M. D., JR, CONEY, P. J. & DAVIS, G. H. (eds) *Cordilleran Metamorphic Core Complexes*. Geological Society of America, Memoir, **153**, 7–31.

—— & CAMPA-URANGA, M. F. 1987. Lithotectonic terrane map of Mexico (west of the 91st meridian). US Geological Survey Miscellaneous Field Studies Map **MF-1874-D**.

COUCH, R. W., NESS, G. E., SANCHEZ-ZAMORA, O., ET AL. 1991. Gravity anomalies and crustal structure of the Gulf and Peninsular Province of the Californias. *In*: DAUPHIN, J. P. & SIMONEIT, B. R. T. (eds) *The Gulf and Peninsular Province of the Californias*. American Association of Petroleum Geologists, Memoir, **47**, 25–45.

CRESPI, J. M., CHAN, Y.-C. & SWAIM, M. S. 1996. Synorogenic extension and exhumation of the Taiwan hinterland. *Geology*, **24**, 247–250.

DEWEY, J. F. 1988. Extensional collapse of orogens. *Tectonics*, **7**, 1123–1139.

ENGLAND, P. C. & HOLLAND, T. J. B. 1979. Archimedes and the Tauern eclogites: the role of buoyancy in the preservation of exotic tectonic blocks. *Earth and Planetary Science Letters*, **44**, 287–294.

ERNST, W. G. 1971. Metamorphic zonations on presumably subducted lithospheric plates from Japan, California, and the Alps. *Contributions to Mineralogy and Petrology*, **34**, 43–59.

—— 1988. Tectonic history of subduction zones inferred from retrograde blueschist P–T paths. *Geology*, **16**, 1081–1084.

EVANS, B. W. 1990. Phase relations of epidote-blueschists. *Lithos*, **25**, 3–23.

FEEHAN, J. G. & BRANDON, M. T. 1993. Quantitative estimate of the relative contributions of erosion and ductile flow to exhumation of high pressure metamorphic rocks, San Juan–Cascade nappes, NW Washington state. *Geological Society of America Abstracts with Programs*, **25**, A-282.

FENBY, S. S. & GASTIL, R. G. 1991. Geologic–tectonic map of the Gulf of California and surrounding areas. *In*: DAUPHIN, J. P. & SIMONEIT, B. R. T. (eds) *The Gulf and Peninsular Province of the Californias*. American Association of Petroleum Geologists, Memoir, **47**, 79–83.

FORMAN, J. A., BURKE, W. H., JR, MINCH, J. A. & YEATS, R. S. 1971. Age of the basement rocks at Magdalena Bay, Baja California, México. *Geological Society of America Abstracts with Programs*, **3**, 120.

FUIS, G. S. & PLAFKER, G. 1991. Evolution of deep structure along the Trans-Alaska Crustal Transect, Chugach Mountains and Copper River Basin, southern Alaska. *Journal of Geophysical Research*, **96**, 4229–4253.

GASTIL, G. 1991. Is there a Oaxaca–California megashear? Conflict between paleomagnetic data and other elements of geology. *Geology*, **19**, 502–505.

——, PHILLIPS, R. P. & ALLISON, E. C. 1975. *Reconnaissance Geology of the State of Baja California*. Geological Society of America, Memoir, **140**.

GRIFFITH, R., & HOOBS, J. 1993. Geology of the southern Sierra Calamajué, Baja California Norte, México. *In*: GASTIL, R. G. & MILLER, R.H. (eds) *The Prebatholithic Stratigraphy of Peninsular California*. Geological Society of America, Special Paper, **279**, 43–60.

GROVE, M., & BEBOUT, G. E. 1995. Cretaceous tectonic evolution of coastal southern California: insights from the Catalina Schist. *Tectonics*, **14**, 1290–1308.

HAGSTRUM, J. T. & SEDLOCK, R. L. 1990. Remagnetization and northward translation of Mesozoic red chert from Cedros Island and the San Benito Islands, Baja California, México. *Geological Society of America Bulletin*, **102**, 983–991.

HARMS, T. A., JAYKO, A. S. & BLAKE, M. C., JR 1992. Kinematic evidence for extensional unroofing of the Franciscan Complex along the Coast Range fault, northern Diablo Range, California. *Tectonics*, **11**, 228–241.

HSU, K. J. 1991. Exhumation of high-pressure metamorphic rocks. *Geology*, **19**, 107–110.

JAYKO, A. S., BLAKE, M. C., JR & HARMS, T. 1987. Attenuation of the Coast Range Ophiolite by extensional faulting, and nature of the Coast Range 'Thrust', California. *Tectonics*, **6**, 475–488.

JOHNSON, C., HARBURY, N. & HURFORD, A. J. 1997. The role of extension in the Miocene denudation of the Nevado–Filábride Complex, Betic Cordillera (SE Spain). *Tectonics*, **16**, 189–204.

JOLIVET, L., GOFFÉ, B., MONTÉ, P., TRUFFERT-LUXEY, C., PATRIAT, M. & BONNEAU, M. 1996. Miocene detachment in Crete and exhumation $P$–$T$–$t$ paths of high-pressure metamorphic rocks. *Tectonics*, **15**, 1129–1153.

JUN, G., GUOQI, H., MAOSONG, L., XUCHANG, X., YAOQING, T., JUN, W. & MIN, Z. 1995. The mineralogy, petrology, metamorphic $PTt$ trajectory and exhumation mechanism of blueschist, south Tianshan, northwestern China. *Tectonophysics*, **250**, 151–168.

KARIG, D. E. 1980. Material transport within accretionary prisms and the 'knocker' problem. *Journal of Geology*, **88**, 27–39.

KILMER, F. H. 1979. A geological sketch of Cedros Island, Baja California, México. *In*: ABBOTT, P. L. & GASTIL, R. G. (eds) *Baja California Geology*. Department of Geological Sciences, San Diego State University, San Diego, CA, 11–28.

KIMBROUGH, D. L. 1982. *Structure, petrology, and geochronology of Mesozoic paleooceanic terranes on Cedros Island and the Vizcaíno Peninsula, Baja California Sur, México*. PhD dissertation, University of California, Santa Barbara.

—— 1984. Paleogeographic significance of the Middle Jurassic Gran Cañon Formation, Cedros Island, Baja California Sur. *In*: FRIZZELL, V. A., JR (ed.) *Geology of the Baja California Peninsula*. Pacific Section, Society of Economic Paleontologists and Mineralogists, Book **39**, 107–117.

KIMURA, G. 1994. The latest Cretaceous–early Paleogene rapid growth of accretionary complex and exhumation of high pressure series metamorphic rocks in northwestern Pacific margin. *Journal of Geophysical Research*, **99**, 22147–22164.

LISTER, G. S., BANGA, G. & FEENSTRA, A. 1984. Metamorphic core complexes of Cordilleran type in the Cyclades, Aegean Sea, Greece. *Geology*, **12**, 221–225.

LITTLE, T. A., HOLCOMBE, R. J., GIBSON, G. M., OFFLER, R., GANS, P. B. & MCWILLIAMS, M. O. 1992. Exhumation of late Paleozoic blueschists in Queensland, Australia, by extensional faulting. *Geology*, **11**, 231–234.

LIU, M. & YUND, R. A. 1993. Transformational kinetics of polycrystalline aragonite to calcite: new experimental data, modelling, and implications. *Contributions to Mineralogy and Petrology*, **114**, 465–478.

LOCKWOOD, J. P. 1972. Possible mechanisms for the emplacement of Alpine-type serpentinite. *In*: Shagam, R. (ed.) *Studies in Earth and Space Sciences*. Geological Society of America, Memoir, **132**, 273–287.

LONSDALE, P. 1991. Structural patterns of the Pacific floor offshore of peninsular California. *In*: DAUPHIN, J. P. & SIMONEIT, B. R. T. (eds) *The Gulf and Peninsular Province of the Californias*. American Association of Petroleum Geologists, Memoir, **47**, 87–125.

LÓPEZ-RAMOS, E. 1985. *Geología de México*, Tomo II, edición 3a, primera reimpresión.

LUND, S. P. & BOTTJER, D. J. 1991. Paleomagnetic evidence for microplate tectonic development of southern and Baja California. *In*: DAUPHIN, J. P. & SIMONEIT, B. R. T. (eds) *The Gulf and Peninsular Province of the Californias*. American Association of Petroleum Geologists, Memoir, **47**, 231–248.

MANN, P. & GORDON, M. B. 1996. Tectonic uplift and exhumation of blueschist belts along transpressional strike-slip fault zones. *In*: BEBOUT, G. E., SCHOLL, D. W., KIRBY, S. H. & PLATT, J. P. (eds)

*Subduction: Top to Bottom.* American Geophysical Union, Geophysical Monograph, **96**, 143–154.

MARTÍNEZ-MARTÍNEZ, J. M. & AZAÑÓN, J. M. 1997. Mode of extensional tectonics in the southeastern Betics (SE Spain): implications for the tectonic evolution of the peri-Alborán orogenic system. *Tectonics*, **16**, 205–221.

MCNEILL, L. C., PIPER, K. A., GOLDFINGER, C., KULM, L. D. & YEATS, R. S. 1997. Listric normal faulting on the Cascadia continental margin. *Journal of Geophysical Research*, **102**, 12123–12138.

MICHARD, A., CHOPIN, C. & HENRY, C. 1993. Compression versus extension in the exhumation of the Dora–Maira coesite-bearing unit, Western Alps, Italy. *Tectonophysics*, **221**, 173–193.

MOORE, D. E. 1984. Metamorphic history of a high-grade blueschist exotic block from the Franciscan Complex, California. *Journal of Petrology*, **25**, 126–150.

MOORE, T. E. 1985. Stratigraphy and tectonic significance of the Mesozoic tectonostratigraphic terranes of the Vizcaíno Peninsula, Baja California Sur, México. *In*: HOWELL, D. G. (ed.) *Tectonostratigraphic Terranes of the Circum-Pacific Region*. Earth Science Series, **1**. Circum-Pacific Council for Energy and Mineral Resources, Houston, TX, 315–329.

—— 1986. Petrology and tectonic implications of the blueschist-bearing Puerto Nuevo melange complex, Vizcaíno Peninsula, Baja California Sur, México. *In*: EVANS, B. W. & BROWN, E. H. (eds) *Blueschists and Eclogites*. Geological Society of America, Memoir, **164**, 43–58.

NIE, S., YIN, A., ROWLEY, D. B. & JIN, Y. 1994. Exhumation of the Dabie Shan ultra-high-pressure rocks and accumulation of the Songpan–Ganzi flysch sequence, central China. *Geology*, **22**, 999–1002.

NORMARK, W. R., SPENCER, J. E. & INGLE, J. C., JR 1987. Geology and Neogene history of the Pacific continental margin of Baja California Sur, México. *In*: SCHOLL, D. W., GRANTZ, A. & VEDDER, J. G. (eds) *Geology and Resource Potential of the Continental Margin of Western North America and Adjacent Ocean Basins – Beaufort Sea to Baja California*. Circum-Pacific Council for Energy and Mineral Resources, Earth Science Series **6**, 449–472.

PEACOCK, S. M. 1988. Inverted metamorphic gradients in the westernmost Cordillera. *In*: Ernst, W. G. (ed.) *Metamorphism and Crustal Evolution of the Western United States (Rubey Volume VII)*. Prentice–Hall, Englewood Cliffs, NJ, 953–975.

PLATT, J. P. 1986. Dynamics of orogenic wedges and the uplift of high-pressure metamorphic rocks. *Geological Society of America Bulletin*, **97**, 1037–1053.

—— 1993. Exhumation of high-pressure rocks: a review of concepts and processes. *Terra Nova*, **5**, 119–133.

—— & VISSERS, R. L. M. 1989. Extensional collapse of thickened continental lithosphere: a working hypothesis for the Alboran Sea and Gibraltar arc. *Geology*, **17**, 540–543.

RANGIN, C. 1978. Speculative model of Mesozoic geodynamics, central Baja California to northeastern Sonora (México). *In*: HOWELL, D. G. & MCDOUGALL, K. A. (eds) *Mesozoic Paleogeography of the Western United States*. Pacific Section, Society of Economic Paleontologists and Mineralogists, Los Angeles, CA, 85–106.

RAWLING, T. J. & LISTER, G. S. 1999. Oscillating modes of orogeny in the Southwest Pacific and the tectonic evolution of New Caledonia. *This volume*.

REDFERN, S. A. T., SALJE, E. & NAVROTSKY, A. 1989. High-temperature enthalpy at the orientational order–disorder transition in calcite: implications for the calcite/aragonite phase equilibrium. *Contributions to Mineralogy and Petrology*, **101**, 479–484.

RING, U. & BRANDON, M. T. 1994. Ductile strain, mass loss, and exhumation of Franciscan rocks. *Geological Society of America Abstracts with Programs*, **26**, A73.

RUPPEL, C., ROYDEN, L. & HODGES, K. V. 1988. Thermal modelling of extensional tectonics: Application to P-T-t history of metamorphic rocks. *Tectonics*, **7**, 947–957.

RYAN, H. F. & SCHOLL, D. W. 1989. The evolution of forearc structures along an oblique convergent margin, central Aleutian arc. *Tectonics*, **8**, 497–516.

SAKAKIBARA, M. & KANISAWA, S. 1994. Metamorphic evolution of the Kamuikotan high-pressure and low-temperature metamorphic rocks in central Hokkaido, Japan. *Journal of Geophysical Research*, **99**, 22221–22235.

SEDLOCK, R. L. 1988a. Metamorphic petrology of a high-pressure, low-temperature subduction complex in west–central Baja California, México. *Journal of Metamorphic Geology*, **5**, 205–233.

—— 1988b. Tectonic setting of blueschist and island-arc terranes of west–central Baja California, México. *Geology*, **16**, 623–626.

—— 1993. Mesozoic geology and tectonics of blueschist and associated oceanic terranes in the Cedros–Vizcaíno–San Benito and Magdalena–Santa Margarita regions, Baja California, México. *In*: DUNNE, G. C. & MCDOUGALL, K. (eds) *Mesozoic Paleogeography of the Western United States – II*. Pacific Section, Society of Economic Paleontologists and Mineralogists, Book **71**, 113–125.

—— 1996. Syn-subduction forearc extension and blueschist exhumation in Baja California, México. *In*: BEBOUT, G. E., SCHOLL, D. W., KIRBY, S. H. & PLATT, J. P. (eds) *Dynamics of Subduction*. American Geophysical Union, Monograph, **96**, 155–162.

—— & ISOZAKI, Y. 1990. Lithology and biostratigraphy of Franciscan-like chert and associated rocks, west–central Baja California, México. *Geological Society of America Bulletin*, **102**, 852–864.

——, ORTEGA-GUTIÉRREZ, F. & SPEED, R. C. 1993. *Tectonostratigraphic Terranes and Tectonic Evolution of México*. Geological Society of America, Special Paper, **278**.

SELVERSTONE, J. 1988. Evidence for east–west crustal extension in the Eastern Alps: implications for the unroofing history of the Tauern Window. *Tectonics*, **7**, 87–105.

SIBSON, R. H. 1982. Fault zone models, heat flow, and depth distribution of earthquakes in the continental crust of the United States. *Seismological Society of America Bulletin*, **72**, 151–163.

SILVER, L. T. & CHAPPELL, B. W. 1988. The Peninsular Ranges Batholith: an insight into the evolution of the Cordilleran batholiths of southwestern North America. *Transactions of the Royal Society of Edinburgh: Earth Sciences*, **79**, 105–121.

SMITH, D. P. & BUSBY, C. J. 1993. Mid-Cretaceous crustal extension recorded in deep-marine half-graben fill, Cedros Island, Mexico. *Geological Society of America Bulletin*, **105**, 547–562.

SPENCER, J. E. & REYNOLDS, S. J. 1991. Tectonics of mid-Tertiary extension along a transect through west central Arizona. *Tectonics*, **10**, 1204–1221.

THOMPSON, A. B., SCHULMANN, K. & JEZEK, J. 1997. Extrusion tectonics and elevation of lower crustal metamorphic rocks in convergent orogens. *Geology*, **25**, 491–494.

WARD, P. L. 1991. On plate tectonics and the geologic evolution of southwestern North America. *Journal of Geophysical Research*, **96**, 12479–12496.

WESTBROOK, G. K. & SMITH, M. J. 1983. Long décollements and mud volcanoes: evidence from the Barbados Ridge complex for the role of high pore-fluid pressure in the development of an accretionary complex. *Geology*, **11**, 279–283.

WHEELER, J. & BUTLER, R. W. H. 1994. Criteria for identifying structures related to true crustal extension in orogens. *Journal of Structural Geology*, **16**, 1023–1027.

YEATS, R. S. & HAQ, B. U. 1981. Deep-sea drilling off the Californias: implications of Leg 63. *In*: YEATS, R. S. & HAQ, B. U. (eds) *Initial Reports of the Deep Sea Drilling Project 66*. US Government Printing Office, Washington, DC, 949–961.

# Ductile deformation and mass loss in the Franciscan Subduction Complex: implications for exhumation processes in accretionary wedges

UWE RING[1] & MARK T. BRANDON[2]

[1]*Institut für Geowissenschaften, Johannes Gutenberg-Universität, Becherweg 21, 55099 Mainz, Germany (e-mail: ring@mail.uni-mainz.de)*

[2]*Kline Geology Laboratory, Yale University, P.O. Box 208109, New Haven, CT 06520–8109, USA*

**Abstract:** Deformation measurements from 64 sandstone samples collected in three study areas from the Eastern Belt of the Franciscan Complex are used to evaluate how the high-pressure metamorphic interior of the Franciscan wedge was exhumed. Pressure estimates indicate 25–30 km of exhumation in this part of the Franciscan Complex. Much of the Eastern Belt has a semi-penetrative cleavage that formed by solution mass transfer (SMT) while the rocks were moving through the wedge. Individual samples have absolute principal stretches of $S_X = 1.00–1.52$, $S_Y = 0.60–1.21$, and $S_Z = 0.33–0.81$. Strain magnitudes and directions are quite variable at the local scale. The deformation at the regional scale is estimated by calculating tensor averages for groups of measurements. The three study areas, which are spaced over a distance of $c.$ 500 km along the Franciscan margin, give remarkably similar averages, which indicates that the deformation of the Eastern Belt is consistent at the regional scale. The tensor average for all data indicates a nearly vertical $Z$ direction with $S_X = 0.96$, $S_Y = 0.92$, and $S_Z = 0.70$. $S_X$ and $S_Y$ are near one because at the local scale, the $X$ and $Y$ directions vary considerably in orientation, which means that their stretch contributions are averaged out at the regional scale. This unusual strain type, consisting of both plane strain and uniaxial shortening, results from the fact that shortening in $Z$ was balanced by a pervasive mass-loss volume strain, averaging about 38%. The geometry of directed fibre overgrowths was used to measure internal rotations. These data indicate that in sandstones, SMT deformation was nearly coaxial (mean kinematic vorticity number is 0.05 at the regional scale and generally <0.4 for individual samples). A simple one-dimensional steady-state model indicates that ductile thinning accounted for only $c.$ 10% of the overall exhumation. Ductile shortening across the Franciscan wedge was very slow, at rates $<8 \times 10^{-17}$ s$^{-1}$ (<0.3% Ma$^{-1}$). Assuming that this strain was active in an across-strike zone <200 km wide, we estimate that horizontal ductile flow would have accounted for <0.25% of the total convergence across the Franciscan margin. We conclude that the SMT mechanism operated slowly as a background deformation process, and that the dislocation glide mechanism was completely inactive down to depths of 25–30 km. Thus, the stability of the Franciscan wedge was probably better defined by the Coulomb wedge criterion than by a viscous wedge criterion. No definitive normal faults have been found in or adjacent to the Eastern Belt. Therefore, we infer that wedge taper was mainly controlled by deep accretion and erosion of an emergent forearc high.

A fundamental problem in tectonics is the cause of deep exhumation of high-pressure metamorphic rocks commonly found in the interior of many convergent wedges. In recent years, extensional faulting has received considerable attention because it provides an elegant and possibly widely applicable mechanism for unroofing a wedge (e.g. Platt 1986, 1993; Dewey 1988). In addition to normal faulting, erosion (e.g. England & Richardson 1977; Rubie 1984; Brandon *et al.* 1998) and syn-convergent ductile flow (e.g. Selverstone 1985; Wallis 1992; Wallis *et al.* 1993; Feehan & Brandon 1999) can also contribute to exhumation of a convergent wedge.

Three different settings of deep exhumation have been recognized. One is in the internal zones of continent–continent collisions where exhumed high-pressure metamorphic rocks, including both high-temperature and low-temperature varieties, are found. This setting contains some of the deepest exhumed rocks, coesite- and diamond-bearing rocks coming from depths >100 km (e.g. Chopin 1984; Coleman & Wang 1995).

A second setting is found in ancient subduction complexes and is characterized by exotic blocks of blueschist and eclogite immersed in a matrix of highly deformed mudstone or serpentinite. These blocks typically come from depths of 30–60 km. A classic example of this setting is the knocker terrane of the Central Belt of the Franciscan Complex. The process of exhumation remains poorly understood. Platt (1986) favoured normal faulting, whereas Cloos & Shreve (1988) proposed upward transport by deep-seated channelized flow at the base of the wedge.

We focus here on a third setting, where large coherent tracts of low-temperature–high-pressure metamorphic rocks have been exhumed in subduction-related accretionary wedges. The metamorphic rocks involved are typically of prehnite–pumpellyite, lawsonite–albite, pumpellyite–actinolite, blueschist, or more rarely greenschist facies. The coherence of these terranes is indicated by the common preservation of metamorphic isograds within the terranes. The depth of exhumation is usually no greater than about 35 km, which is consistent with the maximum thickness of modern subduction-related wedges. Normal faulting is often cited as a likely exhumation process in this setting. The island of Crete, which marks the modern forearc high of the Hellenic convergent margin, provides an example of syn-convergent normal faulting (Fassoulas et al. 1994; Thomson et al. this volume). Normal faulting there seems to be controlled by roll-back of the Hellenic subduction zone. Platt (1986, 1987), Jayko et al. (1987), and Harms et al. (1992) have argued that coherent low-temperature–high-pressure metamorphic terranes in the Eastern Belt of the Franciscan Complex were unroofed by syn-convergent normal faulting. In contrast, erosion appears to be the primary exhumation process for the modern forearc high at the Cascadia margin (Brandon et al. 1998).

In this paper, we report new deformation measurements from the coherent metamorphic terranes of the Eastern Belt of the Franciscan Complex and use this information to resolve the deformation field that existed within the Franciscan wedge. The high-pressure metamorphic rocks of the Eastern Belt show extensive evidence of solution mass transfer (SMT) deformation. The clockwise $P$–$T$ loop for these rocks indicates a general displacement path involving subduction, then accretion at the base of the wedge, upward flow within the wedge, followed by exhumation and exposure at the Earth's surface. We maintain that these deeply exhumed rocks provide a path-integrated record of the deformation-rate field in the wedge. We utilize some new methods, which provide a full determination of the absolute finite deformation produced by the SMT mechanism, including volume strain and internal rotation. These methods are the projected dimension strain (PDS) method, the mode method and the semi-deformable antiaxial (SDA) fibre method. The contribution of vertical ductile thinning to exhumation is estimated using a simple one-dimensional steady-state model by Feehan & Brandon (1999). The data reported herein, combined with our work on the kinematic evolution of the Coast Range fault zone (Ring & Brandon 1994, 1997), allow us to approximate the relative contributions of different processes to the total exhumation of the Eastern Belt metamorphic rocks.

## Overview

### Geological and tectonic setting

The Franciscan Complex represents a long-lived accretionary wedge, Late Jurassic to Paleogene in age, which grew along the western edge of the North American Cordillera (Fig. 1a). This convergent margin is preserved in three northwest-trending tectonic zones: the Franciscan Complex was the subduction complex or accretionary wedge, the Sierra Nevada batholith was the magmatic arc, and the Great Valley basin formed between these as a broad forearc basin (e.g. Ingersoll 1978, 1979; Blake et al. 1988; Cowan & Bruhn 1992). The basement to the forearc is a Jurassic ophiolite, called the Coast Range ophiolite. At present, the Franciscan Complex is separated from the Coast Range ophiolite and overlying Upper Jurassic and Cretaceous strata of the Great Valley basin by the Coast Range fault zone. Platt (1986) called attention to the strong break in metamorphic grade across the Coast Range fault. The Coast Range ophiolite and the lowermost units of the Great Valley sequence show zeolite- to incipient prehnite–pumpellyite-facies metamorphism (Dickinson et al. 1969), whereas west of the fault, the Eastern Belt of the Franciscan Complex records blueschist-facies metamorphism (Fig. 1b).

The Franciscan Complex is dominated by clastic sedimentary rocks, interpreted as accreted trench sediments and superimposed trench-slope basins. It also includes subordinate mafic and keratophyric volcanic rocks and thick chert sequences, which represent, at least in part, accreted fragments of seamounts and oceanic plateaux, as well as imbricated slices from the overlying Coast Range ophiolite. In the northern Coast Range, the Franciscan Complex is commonly subdivided into three northwest-trending belts (Fig. 1a), which are, from west to east: the Coastal, Central and Eastern Belts

(Bailey et al. 1964; Berkland et al. 1972; Blake et al. 1988). Stratigraphic age, as well as the degree and age of metamorphism and deformation, generally increase to the east across these belts. South of San Francisco, this subdivision is not as apparent, probably because of disruption by young faulting associated with the modern San Andreas transform boundary.

We focus on the Eastern Belt, which constitutes the uppermost part of the Franciscan Complex (Fig. 1b). The term Eastern Belt refers to a gently east-dipping sequence of nappe-like units, each of which contains a relatively coherent internal stratigraphy (Suppe 1973; Cowan 1974; Worrall 1981; Blake et al. 1988; Jayko & Blake 1989). The faults that bound these units appear to be everywhere post-metamorphic (Suppe 1973; Cowan 1974; Platt 1975; Worrall 1981). Metamorphic grade is generally lawsonite–albite or blueschist facies as indicated by the presence of widespread lawsonite and aragonite, and more localized glaucophane and jadeite (e.g. Worrall 1981; Blake et al. 1988; Ernst 1993). The two main units within this structural succession are the Yolla Bolly terrane and the structurally higher Pickett Peak terrane (Blake et al. 1988; Fig. 1). The Pickett Peak terrane, which is restricted to the northern Coast Range, contains two structural units: the South Fork Mountain Schist and the underlying Valentine Springs Formation (Worrall 1981). Maximum temperatures were about 400°C in the South Fork Mountain Schist, and between 250°C and slightly greater than 310°C in the Valentine Springs Formation (Jayko et al. 1986; Blake et al. 1988; Tagami & Dumitru 1996). Maximum metamorphic pressures for both units were in the range 6–9 kbar (23–34 km depth assuming an average density of 2700 kg m$^{-3}$) (Blake et al. 1988). The Yolla Bolly terrane experienced lower metamorphic conditions, 125–200°C and 6–8 kbar (23–30 km) in the Yolla Bolly Mountains (Jayko et al. 1986; Bröcker & Day 1995), and 100–200°C and 7 to >8 kbar (26 to >30 km) in the Pacheco Pass area of the Diablo Range (Ernst 1993).

Isotopic ages from the Eastern Belt indicate a protracted history of high-pressure metamorphism, followed by slow cooling in Late Cretaceous and early Cenozoic time (Suppe & Armstrong 1972; Lanphere et al. 1978; McDowell et al. 1984; Mattinson 1988; Dumitru 1989; Wakabayashi 1992; Tagami & Dumitru 1996). The oldest metamorphic ages of 150–165 Ma are from high-grade blueschist-, eclogite- and amphibolite-facies blocks (K–Ar ages on phengite and hornblende, Rb–Sr mineral ages on aragonite/glaucophane and phengite–zoisite pairs, and a U–Pb isochron age for glaucophane, lawsonite and sphene; Coleman & Lanphere 1971; Nelson & DePaolo 1985; Mattinson 1988). As for the Pickett Peak terrane, its stratigraphic age remains poorly known, but regional high-pressure metamorphism is thought to have occurred at about 125–130 Ma (Ar–Ar whole-rock ages, Lanphere et al. 1978). The upper part of the Yolla Bolly terrane has yielded rare Berriasian to Valanginian fossils, indicating deposition sometime during 138–119 Ma. Regional metamorphism of this part of the Yolla Bolly terrane occurred at about 115–120 Ma (Rb–Sr whole-rock isochron, Peterman et al. 1967; U–Pb isochron on garnet, amphibole, and sphene, Mattinson 1986; Ar–Ar on hornblende, Weinrich et al. 1997). However, lawsonite- and aragonite-bearing shale and sandstone from the lower part of the Yolla Bolly terrane have yielded Cenomanian (97–91 Ma) *Inocerami* (Hull Mountain area in northern California, Blake et al. 1988). Metamorphism of this part of the Yolla Bolly terrane is younger as well, as indicated by U–Pb isotopic ages on sphene and plagioclase from a metamorphosed gabbro intrusion indicating peak metamorphism at 92 Ma (Ortigalita Peak gabbro in the Diablo Range, Mattinson & Echevirra 1980). We concur with others (e.g. Mattinson 1988; Wakabayashi 1992) that the range of metamorphic ages is probably related to protracted accretion and high-pressure metamorphism, but this interpretation remains difficult to prove.

A variety of evidence indicates that the Eastern Belt was mainly exhumed during Late Cretaceous and early Cenozoic time (Krueger & Jones 1989). In central California, Upper Cretaceous (Campanian?) trench-slope basins received blueschist and chert clasts derived from an uplifted Franciscan-like source (Cowan & Page 1975; Smith et al. 1979). Parts of the Great Valley basin itself became shallow and locally emergent during Late Cretaceous and early Tertiary time (Ingersoll 1978, 1979; Dickinson et al. 1982). Berkland (1973) recognized Paleocene (66–55 Ma) conglomerates in the northern Coast Ranges that contained abundant Franciscan detritus including clasts of lawsonite- and pumpellyite-bearing metagreywacke, which Berkland (1973) noted as 'indistinguishable in texture and mineralogy from the bedrock of the Eastern Belt of the Franciscan Complex'. In the central Coast Ranges, the first Franciscan detritus appears in the Lower Eocene deposits (c. 53 Ma, Domengine Formation; Moxon 1988). Page & Tabor (1967) and Pampeyan (1993) described a pre-Middle Eocene unconformity cut into the Franciscan Complex south of San Francisco.

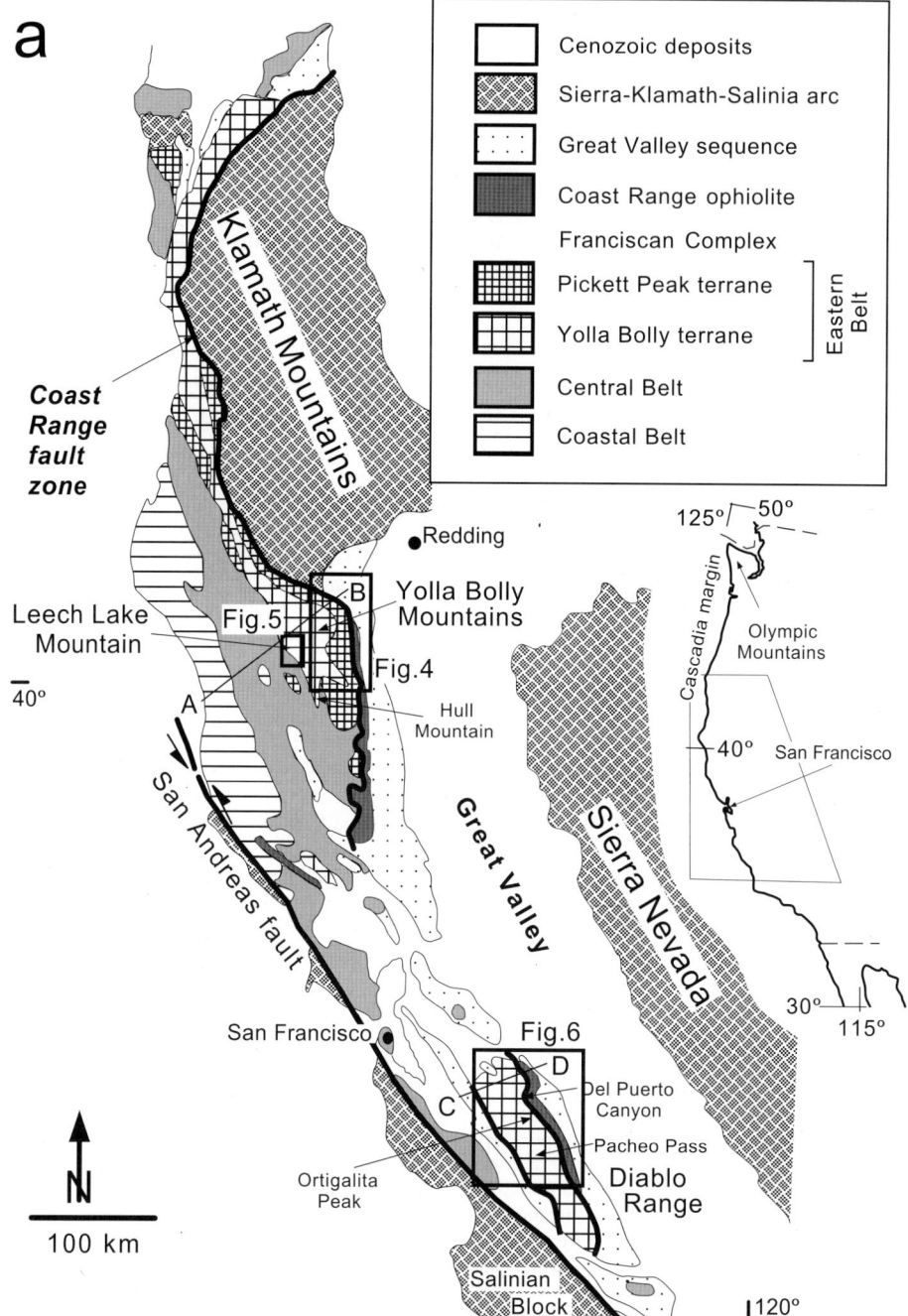

**Fig. 1.** (a) Geological map showing Mesozoic tectonic features of western California. The smaller inset map shows the location of the geological map relative to the coast of western North American. Our three study areas are located in the Yolla Bolly Mountains (Fig. 4) and at Leech Lake Mountain (Fig. 5) in northern California and in the Diablo Range (Fig. 6) of central California. Also shown are localities referred to in the text. (b) Generalized cross-section through the Yolla Bolly Mountains (A–B, after Cowan & Bruhn 1992) and the northern Diablo Range (C–D, after Bauder & Liou 1979).

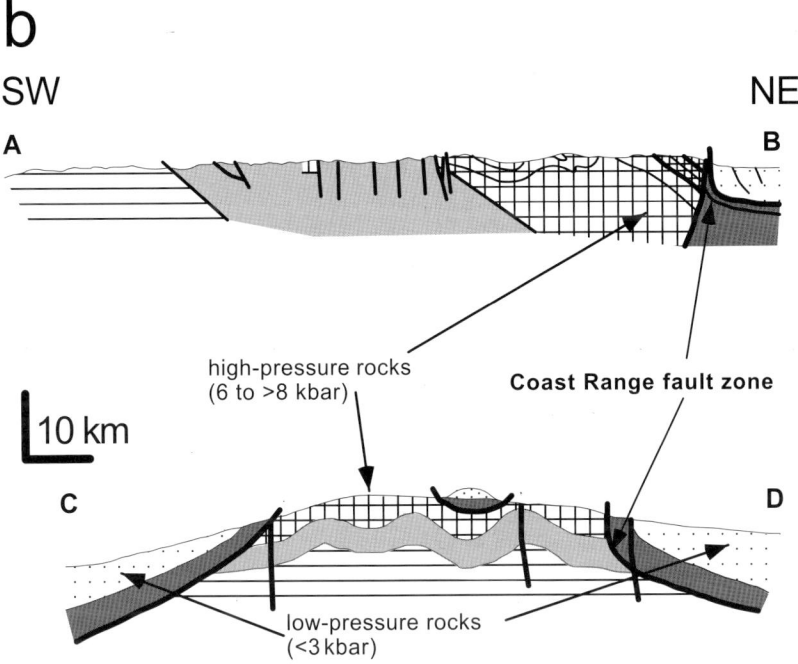

This sedimentological evidence is consistent with zircon and apatite fission-track ages by Dumitru (1989) and Tagami & Dumitru (1996) that indicate slow cooling during the Late Cretaceous and early Cenozoic time.

*Residence time and average exhumation rates*

The timing information given above is used here to estimate wedge residence times and exhumation rates for our study areas in the Eastern Belt (Table 1). First, we assume that the age of accretion is closely approximated by the age of peak metamorphism. To illustrate, let us consider an average accretionary wedge with a surface slope $c.$ 3° and basal décollement dipping at $c.$ 5° (Davis *et al.* 1983). Seven kilometres of convergence are needed to obtain 1 km of structural burial. Convergence rates at the Franciscan margin were $c.$ 100 km Ma$^{-1}$ in Late Cretaceous time (Engebretson *et al.* 1985). Thus, subduction to a depth of $c.25$–30 km would have taken less than $c.$ 2 Ma. Peak metamorphism should have followed rapidly after accretion to the base of the wedge. We infer from sedimentological data and apatite fission-track ages that Eastern Belt rocks exposed at present in our study areas were at or near the surface by $c.$ 60 Ma.

The wedge residence time for our northern

**Table 1.** *Orogenic parameters for Eastern Belt of the Franciscan Complex*

| Unit and study area | Depth of accretion (km) | Residence time (Ma) |
| --- | --- | --- |
| Valentine Springs Fm, Yolla Bolly Mtns | 23–34 | 65–70 |
| Yolla Bolly terrane, Yolla Bolly Mtns and Leech Lake Mtn | 23–30 | 55–60 |
| Yolla Bolly terrane, Diablo Range | 26–>30 | 32–37 |

California study areas in the Yolla Bolly Mountains and Leech Lake Mountain is estimated to about 65–70 Ma for the Valentine Springs Formation and 55–60 Ma for the upper part of the Yolla Bolly terrane. The Eastern Belt rocks in our Diablo Range study area are probably correlative with the lower part of the Yolla Bolly terrane. The 92 Ma metamorphic age for the Ortigalita Peak gabbro indicates a residence time of about 32–37 Ma for these rocks. These estimates are crude but should be accurate to within ±20%. Based on metamorphic depths and residence times, we estimate that exhumation rates were about 0.3–0.5 km Ma$^{-1}$ for the Valentine Springs Formation and Yolla Bolly terrane in the northern California study areas, and 0.7–0.9 km Ma$^{-1}$ for the Yolla Bolly terrane in the Diablo Range study area.

## Exhumation of Franciscan high-pressure rocks

Platt (1986) proposed the provocative idea that normal slip on the Coast Range fault zone was responsible for exhumation of the high-pressure metamorphic interior of the Franciscan wedge. He argued that underplating could drive a viscous convergent wedge into a supercritical taper where the upper part of the wedge would start to flow or fail by horizontal extension. Jayko et al. (1987) and Harms et al. (1992) provided additional evidence for normal slip on the Coast Range fault zone. Krueger & Jones (1989) and Wakabayashi & Unruh (1995) considered how tectonic exhumation might relate to the accretionary history of the Franciscan margin. We note that the Coast Range fault zone is the only structure to be considered for tectonic exhumation of the high-pressure Eastern Belt rocks. To our knowledge, no other exhumational structures have been recognized or proposed.

Most workers agree that the Franciscan and Coast Range Ophiolite were originally separated by a subduction-related thrust fault. This hypothetical structure would probably have been formed during the initiation of the Franciscan subduction zone, but such an early structure has never been definitively identified. Platt (1986), Jayko et al. (1987), and Harms et al. (1992) made the important point that much of the Coast Range fault zone was formed late relative to the initiation of the Franciscan subduction zone. Their arguments in favour of the normal fault interpretation were: (1) the Coast Range fault cuts out metamorphic section between high-pressure Franciscan metamorphic rocks and the very low-grade rocks of the Great Valley forearc basin and Coast Range ophiolite (Fig. 1b) (Platt 1986); (2) the Coast Range ophiolite, which lies above and within the Coast Range fault zone, appears to have been thinned by Coast Range faulting (Platt 1986; Jayko et al. 1987); (3) kinematic indicators from Franciscan rocks beneath the Del Puerto segment of the Coast Range ophiolite (Fig. 1) were used to infer top-east normal slip on the Coast Range fault zone (Harms et al. 1992).

We maintain that the first two arguments are inconclusive because out-of-sequence thrusting can also result in younger-over-older relationships and can attenuate metamorphic and stratigraphic sections. This will happen when contractional faults cut through a section that dips steeply in a direction opposite the hanging-wall transport direction (Ring & Brandon 1994; Wheeler & Butler 1994; Ring 1995). Kinematic data from the Coast Range fault zone (Ring & Brandon, 1994, 1997) indicate east-side-up motion, which is not compatible with the normal-slip interpretation. These data have been interpreted by Ring & Brandon (1994, 1997) to indicate that the Coast Range fault formed as a post-metamorphic, out-of-sequence thrust with a generally east-side-up or top-west sense of motion. We have already noted above that the Yolla Bolly and Pickett Peak terranes are cut by a series of widely spaced east-dipping faults, all of which appear to be post-metamorphic. Thus, these intra-Franciscan faults may have thickened the original metamorphic section by placing high-grade over low-grade rocks.

As noted above, Harms et al. (1992) used asymmetric fabrics in Eastern Belt sandstones to infer top-east shearing on the Coast Range fault zone at Del Puerto Canyon. The Del Puerto area is important because it may preserve a relative old segment of the fault zone, unaffected by Cenozoic tectonic wedging (Harms et al. 1992). Our measurements from Del Puerto indicate nearly coaxial deformation in the Eastern Belt, and thus are inconsistent with the ductile shearing interpretation of Harms et al. (1992). We return to this issue in the Discussion section.

We envision that many of the post-metamorphic faults in and adjacent to the Franciscan Complex are related to Cenozoic shortening across the Coast Range. The tectonic wedging models of Wentworth et al. (1984) and Unruh et al. (1995) have called attention to this young deformation but they do not address in any detail the possibility of young slip on the intra-Franciscan post-metamorphic faults. The point to be stressed is that, at present, there are no structures within the Great Valley–Franciscan boundary that can be equated to the normal faults predicted by Platt's (1986) model. The

Franciscan Complex might have been exhumed by normal faulting, but there is no longer any direct evidence for such structures.

Platt (1986) based his model for exhumation of the Eastern Belt on the critical-wedge concept as originally outlined by Chapple (1978) and Davis *et al.* (1983). Platt (1986) argued that the dominant ductile deformation mechanism within a subduction-related accretionary wedge is SMT, a linear viscous mechanism that operates by selective dissolution and precipitation along grain boundaries. In Platt's model, deep accretion and within-wedge ductile deformation would cause upward flow within the rear of the wedge, resulting in extensional failure in the upper part of the wedge (Platt 1986, 1987). An essential assumption is that within-wedge ductile flow was fast enough to influence the overall stability of the wedge. Platt's model does not provide specific quantitative predictions of strain rate, but he envisioned that ductile strain rates of $10^{-14}$ s$^{-1}$ (31% Ma$^{-1}$) or more would be required to destabilize the wedge (Platt, 1986, p. 1040). The model highlights the importance of within-wedge deformation for understanding wedge stability, but there have been surprisingly few quantitative studies of the magnitude, pattern, rate and nature of ductile strain in subduction complexes. Notable exceptions are the studies by Paterson & Sample (1988), Ring *et al.* (1989), Norris & Bishop (1990), Fisher & Byrne (1992), Wallis (1992), and Feehan & Brandon (1999).

## SMT deformation in the Franciscan wedge

Petrographic evidence clearly indicates that SMT was the dominant deformation mechanism operating in our sandstone samples. The sandstones are composed of first-cycle volcanic and plutonic detritus. Monocrystalline grains of volcanic quartz and plagioclase show little to no undulose extinction, deformation laminae, or deformation twinning. Polycrystalline quartz grains do show evidence of undulose extinction, but these may have been eroded from metamorphic source rocks. Quartz c-axis fabrics from six samples show no preferred crystallographic orientation (Ring 1996). Grain breakage is relatively uncommon. These collective observations are consistent with estimated metamorphic temperatures (see above) that were almost everywhere below the threshold needed to activate dislocation mechanisms in quartz (>300°C, e.g. Küster & Stöckhert 1997).

Quartz, and to a lesser degree feldspar, are truncated by thin selvages composed of insoluble minerals (Fig. 2). Microprobe work reveals that these selvages contain high concentrations of Fe-oxides, rutile, sphene, phengite, and chlorite (Ring 1996). The selvages can be regarded as planes of finite flattening that formed perpendicular to $Z$ (Ramsay & Huber 1983). $X$, $Y$ and $Z$ correspond to the principal strain directions for minimum, intermediate, and maximum shortening, respectively. Directed fibre overgrowths mantle those grain boundaries that lie at a high angle to the trace of cleavage (Fig. 2). Cathodoluminescence work by Ring (1996) indicates that all overgrowths on quartz grains have a directed fibre habit. These fibres are considered to record the entire extensional history of SMT deformation. The overgrowths are generally subparallel to the orientation of the selvages in $XZ$ sections (Fig. 2) indicating that the bulk deformation was close to coaxial.

As discussed by Feehan & Brandon (1999), the formation of a SMT fabric requires small offsets between grains on surfaces parallel to the cleavage selvages. The reason is that extension is accommodated in a heterogeneous fashion at the local scale because the sites of fibre growth are not coordinated with each other. The intergranular offsets must be small and discontinuous because the selvages typically are short in length (no more than a couple of grain dimensions) and anastomosing in form. This argument also indicates that during SMT deformation, intergranular slip typically does not occur on surfaces oblique to the selvages. If it did, we would see offsets and distortions of the selvages.

Textural evidence suggests that the sandstones had very little porosity at the start of SMT deformation. All of the space between grains is now occupied by selvages or directed fibre overgrowth. We have found no evidence for isotropic overgrowths. Displacement-controlled fibre overgrowths are thought to form only where the aperture for fibre growth was never greater than several microns (Urai *et al.* 1991; Fisher & Brantley 1992). Some pore space could have been removed where selvages have formed but there is no reason to suspect that the selvage sites had unusual large pore spaces given that the overgrowth sites show no evidence of significant original pore space. Mechanical compaction is characterized by slip on intergranular surfaces of all orientations to accommodate the motion of grains into intervening pore spaces. We have already noted that the parallelism of fibres and selvages indicates that oblique intergranular slip generally did not occur during SMT deformation. Therefore, we conclude that mechanical compaction had already removed much of the primary porosity before the onset of SMT deformation. This result is not unexpected

**Fig. 2.** Typical (**a**) *XZ* and (**b**) *XY* sections from Yolla Bolly sandstones. (**c**) Multidirectional fibres in a *XY* section indicating extension in both *X* and *Y*. (**d**) Overall parallelism between fibrous overgrowth and selvages in *XZ* section indicating little to no internal rotation. Long side of each photograph represents 1.4 mm for (**a**) and (**d**), and 0.8 mm for (**b**) and (**c**); all photomicrographs are in crossed-polarized light. *X* is horizontal in (**a**) and (**d**) and nearly vertical in (**b**).

for poorly sorted sands where grains of different sizes can move together into a tightly packed aggregate.

## Relative timing of SMT deformation and metamorphism

Ernst (1965, 1987, 1993) has published detailed textural observations for metamorphic minerals in Eastern Belt sandstones of the Diablo Range. He described host grains of plagioclase and quartz that were replaced by coarsely crystalline prisms of jadeite with concentric and oscillatory zoning. Tiny fibres of jadeite were observed extending from larger jadeite grains into the sandstone matrix. The fibres possessed the same chemical composition as the jadeite host (Ernst 1992). Euhedral jadeite and aragonite were found in quartzose veins as well. Sodic amphibole was observed in stringers and fracture-related patches.

Microprobe analyses (Ring 1996) of samples from Pacheco Pass indicate that the fibre overgrowths associated with SMT deformation are composed of quartz, muscovite, phengite, chlorite, and lawsonite. The Si content of the phengite is between 3.49 and 3.51 per formula unit. We also found veins of coarse dark green jadeite crosscut by a semi-penetrative SMT cleavage.

In the Yolla Bolly study area in northern California, fibre overgrowths in the sandstones are made up of quartz, lawsonite, aragonite, phengitic white mica, muscovite, and chlorite. The Si content of the fibrous phengite ranges between 3.50 and 3.58 per formula unit (Ring 1996). Crosscutting veins contain pumpellyite, aragonite, calcite, albite, and quartz. Sodic amphibole is present as a matrix phase. In metabasalt, aegirine–omphacitic pyroxene is found as rims or replacement of original igneous clinopyroxene (Blake *et al*. 1988), and sodic amphibole is present as a matrix phase or replacing igneous hornblende.

The high Si content of the phengite (Massone & Schreyer 1987) and the presence of lawsonite, aragonite, and jadeite in veins and fibres demonstrate that SMT deformation coincided with high-pressure metamorphism. Ductile strain may have continued to accumulate after peak metamorphism but for how much longer is not known.

## Methods

Our study employs some new methods for measuring deformation in sandstones deformed by the SMT mechanism. Traditional methods, such as the $R_f/\phi$ method, are not suitable because the grains did not deform as passive markers but rather by modification of their external shapes. The projected dimension strain (PDS) and mode methods (Brandon et al. 1994; Feehan & Brandon 1999) were used to measure contractional and extensional strains, respectively. The SDA-fibre method was used to estimate internal rotation. These methods are briefly summarized here. (All of the programs developed and used in this study are available from the Web in both compiled and source-code forms at http://love.geology.yale.edu/~brandon/brandon.html. They are written in Microsoft Professional Basic 7.0 and will run on a DOS or Windows computer using any standard printer.)

The first step is to determine the principal directions for SMT strain. The principal stretches themselves are designated as $S_X \geq S_Y \geq S_Z$, where $S$ is the final length/initial length. The cleavage plane is assumed to lie perpendicular to $Z$. To determine the $X$ direction, we measured the average orientation of directed fibre overgrowths in a cleavage-parallel ($XY$) thin section. $Y$ is then defined by its orthogonality with $X$ and $Z$. With this information, we are able to cut an $XZ$ thin section.

Most samples (36 of 64) have unidirectional fibres where both the $XY$ and $XZ$ sections showed well-organized fibre overgrowths oriented in the $X$ direction (Fig. 2a, b and d). This textural relationship indicates that only one of the principal stretches is extensional. For these samples, the strain type is constrictional or plane with $S_X \geq 1 \geq S_Y \geq S_Z$. The remaining samples (28 of 64) have multidirectional fibres defined by coplanar fibres radiating out in all directions in the $XY$ plane (Fig. 2c). This texture indicates a flattening-type strain with $S_X \geq S_Y > 1 \geq S_Z$.

### The mode method

The mode method is used to determine the extensional strain recorded by the fibre overgrowths. The modal abundance of fibres was determined by line traverses using a computer-driven micron-stepping petrographic microscope. Modes were measured in the $XZ$ section for unidirectional fibres and in the $XY$ section for multidirectional fibres. In the first case, the maximum stretch $S_X$ is related to the modal fraction of fibre $m$ by the relationship $S_X = 1/(1 - m)$. In the second case, extension has occurred in both the $X$ and $Y$ directions, so we can only determine the area dilatancy in the $XY$ section, defined by the product $S_X S_Y$ = final area/initial area = $1/(1 - m)$. Other information, introduced below, must be used to determine directions and magnitudes for $S_X$ and $S_Y$.

The more common and less time-consuming method for determining extension from fibre overgrowths involves measuring the length of the fibres relative to the section radius of the associated host grain (e.g. p. 271 in Ramsay & Huber 1983). We prefer our method because it provides a more direct measurement of the bulk extensional strain in the rock. Furthermore, our experience indicates that relative to the mode method, the fibre-length method commonly overestimates the extensional strain, perhaps because longer, better developed fibres are favoured when making length measurements.

A contractional principal direction, where $S \leq 1$, is characterized by an absence of fibre overgrowths parallel to that direction and by selvages and truncated grain boundaries lying perpendicular to that direction. Selvages can form in more than one orientation. Such a situation is expected where more than one principal direction is contractional. In this case, hand samples typically show a single well-developed cleavage, oriented perpendicular to $Z$, but observations with a petrographic microscope will reveal the presence of selvages in both the $XY$ and $XZ$ sections. Samples with $1 > S_Y \approx S_Z$ are characterized by the lack of a well-defined cleavage in hand sample. For these samples, thin section observations generally show many orientations of well-developed selvages, all of which are co-zonal with $X$.

### The PDS method

The PDS method is used to measure the amount of shortening produced by dissolution of grain boundaries. The method exploits the fact that for SMT deformation, the size of the detrital grains remains unchanged in length parallel to the $X$ direction. The projected dimension of a grain is equivalent to its caliper dimension. The main idea behind the PDS method is that a contractional stretch in a specific direction in the rock must be accompanied by a reduction of the average caliper dimension of the grains in that direction by a factor equal to the stretch. In comparison, the average initial caliper dimension of a detrital grain is preserved in the $X$ direction because the grains lack any significant internal deformation and because original grain boundaries are preserved in that direction beneath a mantle of fibre overgrowths. Therefore, a contractional principal stretch could be determined if we could measure the average caliper dimension of a large number of grains in a direction parallel to that principal direction and then divide that average by the average caliper dimension in the $X$ direction.

A practical problem is that our measurements are made in two-dimensional thin sections, and yet the necessary caliper dimensions should be made in three dimensions. This problem can be accurately accounted for using a simple correction based on the relationship between section measurements and volume measurements for truncated spheres (see Feehan & Brandon (1998) for further details). The correction works well as long as the aspect ratio of the initial grains is less than 3:1, which is commonly the case for sandstones (Paterson & Yu 1994; Ramthun et al. 1997). Our section measurements were made using a petrographic microscope with a drawing tube and digitizing tablet. Measurements are precise to better than ± 3 μm.

We have tested the PDS method using undeformed sandstone samples and found that it is relatively insensitive to primary grain fabrics that might be produced during deposition and compaction (Paterson & Yu 1994; Ramthun et al. 1997). Table 2 presents PDS measurements for seven samples of Great Valley sandstones unaffected by SMT deformation. For these samples, and others like them, we find that the average stretch is approximately one, as would be expected given the absence of SMT deformation. Deposition and compaction can produce a weak preferred orientation in grain shape. The PDS method, however, is based on the average projected dimension of the grains, and not on their orientations. Thus, the method is uniquely suited for measuring SMT strains in sandstones, but only as long as the assumption of no intragranular deformation holds (i.e. strains caused by dislocation glide are nil).

Next, we summarize uncertainties for our measurements. Orientation errors are estimated to be less than about ± 7°. To estimate uncertainties for the principal stretches, we accounted for the variance associated with the initial grain fabric as indicated by our PDS measurements from 'undeformed' sandstones (e.g. Table 2) and the variance associated with measurement errors as indicated by the bootstrap method of Efron (1982). These calculations indicate relative standard errors of c. 5%. In other words, a typical shortening stretch of 0.70 would have an absolute standard error of c. 0.035. The volume stretch, $S_V$, i.e. the final volume/initial volume, is defined by the product of the three principal stretches $S_V = S_X \cdot S_Y \cdot S_Z$. Propagation of standard errors for the principal stretches indicates that the relative standard error for $S_V$ is c. 9%. Thus, a typical $S_V = 0.66$ would have an absolute standard error of c. 0.06.

## The SDA-fibre method

The SDA-fibre method is used to measure the internal rotation associated with SMT deformation and the orientation and magnitude of $S_X$ and $S_Y$ for samples having multidirectional fibres in the $XY$ section. This method was developed by Brandon and a full description has yet to be published. The following brief summary should be sufficient for our purposes here. The two published methods for analysis of syntectonic fibre overgrowths are the rigid-fibre method (Durney & Ramsay 1973; Ramsay & Huber 1983) and the deformable-fibre method (Ramsay & Huber 1983; Ellis 1986). In the first method, the fibres and grains are assumed to remain undeformed. The only motion that is allowed is that associated with separation and growth of new fibres at the grain–fibre interface. The growth of fibres on pyrite framboids is commonly approximated by this model (p. 265 in Ramsay & Huber 1983). The bulk deformation causes the fibres to be translated and rotated relative to the sites of fibre accretion. Thus, the shape of the fibres is formed as the fibre grows, and not by deformation of pre-existing fibres. The deformable-fibre model also has fibres growing parallel to the current incremental extension direction, but after initial growth, the fibres deform in the same manner as the bulk rock. For this case, it is necessary to incrementally undeform the fibre to recover the history of extensional stretching and rotation (Ramsay & Huber 1983; Ellis 1986).

Our textural observations indicate that neither of these models correctly represents the geometry of fibres growing on detrital grains in SMT-deformed sandstones. Fibre bundles that lie between two detrital grains commonly show a 'bow tie' geometry with the central segment of the fibre bundle being much thinner than its ends. These fibre bundles commonly show continuity along their entire length, indicating that new material was added at the inference between fibre bundles and their host grains. The mineralogical contrast between fibre and grain ensures that this interface is crystallographically incoherent. This type of fibre growth was called 'pyrite-type' or antitaxial growth by Ramsay & Huber (1983). We interpret the thinned central region of the fibre bundle to have formed by dissolution and shortening between adjacent fibres within the bundle. This conclusion is supported by the observation that the degree of thinning is commonly much more pronounced in the $XZ$ section, where contractional strains are greater. Many

**Table 2.** *PDS ratios measured for undeformed Great Valley sandstones*

| Sample | PDS ratios | | |
| --- | --- | --- | --- |
| | AB | AC | BC |
| GV 93-1 | 0.958 | 1.101 | 1.096 |
| GV 93-2 | 1.034 | 1.057 | 1.147 |
| GV 93-5 | 0.996 | 1.197 | 1.333 |
| GV 93-6 | 1.113 | 0.991 | 1.060 |
| GV 93-45 | 1.012 | 0.993 | 1.078 |
| GV 93-52 | 0.978 | 1.011 | 0.989 |
| GV 93-55 | 1.067 | 0.981 | 0.998 |
| Log mean | 1.02 | 1.04 | 1.09 |

AB, AC, and BC define three orthogonal thin sections, with AB parallel to bedding.

of the *XY* sections, which typically have $S_Y$ close to unity, show relatively straight fibre bundles, with little to no thinning. These examples are similar to those shown by Ramsay & Huber (1983) to support their model of rigid-fibre overgrowths on pyrite grains.

We conclude that the SMT fibres in our sandstones do not fit the rigid-fibre model unless the contractional principal stretches are close to one. We also reject the deformable-fibre model because the fibres show no evidence that they were stretched or broken along their length. Extension appears to have been accommodated entirely by separation and growth of new fibre at the grain–fibre interface. In our view, the fibres were deformed only by dissolution and progressive thinning, which is most pronounced in the oldest parts of the fibre, located in the central segment. The deformable-fibre model specifies that, once formed, the fibres behaved as passive markers. At higher temperatures, this description might hold, but the temperatures associated with SMT deformation in our samples were too low for fibres to deform internally by intracrystalline mechanisms.

Brandon has developed two computer programs, similar in some respects to the one described by Ellis (1986), that analyse the growth of semi-deformable antitaxial fibres. The basic idea is that fibres are allowed to grow only at the grain–fibre interface, and once formed, they are allowed to shorten only perpendicular to the current incremental shortening direction. One of the programs (FBR-SIM) simulates the progressive growth of the fibres on a section through the centre of an 'average' three-dimensional detrital grain. The finite deformation is determined by integrating a constant velocity–gradient tensor. The evolving geometry of the fibres and the progressive orientation of the cleavage selvages are determined by calculating the displacement paths for particle points that originated at the grain boundary or within the cleavage selvage. The simulation can account for a general three-dimensional deformation including internal rotation. Some examples of simulated fibre overgrowths are shown in Fig. 3. We have found that the simulation program closely reproduces the size and form of unidirectional and multidirectional fibres in sections with different strains and degrees of internal rotation.

To analyse fibre overgrowths in our samples, Brandon designed a second program (FIBER), which includes the simulation routine and an inverse algorithm. The objective is to find the forward model that best represents the fibre trajectories in each thin section. A best-fit solution is found by using a non-linear least-squares algorithm based on the simplex method (Amoeba routine of Press *et al.* (1992, p. 402)). The required input data for each section includes the modal abundance of fibre, the average orientation of the selvages, and the digitized shapes of a random sample of about 30–40 non-truncated fibres. For sections with unidirectional fibres, the input must also include the contractional principal stretch for that section, as determined by the PDS method. The program calculates the best-fit solution for the internal rotation in that section. For sections with multidirectional fibres, it also gives the best estimate of the directions and magnitudes of the two extensional principal stretches. It should be noted that the program does not depend on the diameter of the host grains, but instead uses the mode and shape of the fibres to estimate the average section radius of the host grains.

To estimate the internal rotation in three dimensions, the components of the internal rotation vector are measured in the *XY* and *XZ* sections. (There is generally no record of internal rotation in the *YZ* section because fibres are typically small or absent in that section.) Each measured component is cast as a vector oriented normal to its section with a magnitude equal to the measured rotation angle. These components are then summed to give the internal rotation vector. The result is reported in Table 3 as a rotation axis and a finite rotation angle $\Omega_i$ with a positive angle indicating a right-handed rotation. Vector addition of finite rotations is appropriate as long as the rotation angles are less than c. 5° (p. 236 in Cox & Harte (1986)). Above this limit, the magnitude of the summed rotation remains fairly precise but the orientation of the rotation axis becomes increasingly imprecise. It should be noted that the FIBER program failed to find a stable solution for internal rotations for 15 samples (dashed entries in Tables 3 and 4). This problem typically occurs when fibres are very short or too variable in orientation.

More could be said about our SDA-fibre method, but for this study, the method provides quantitative support for a conclusion that is fairly obvious from the textures themselves. Namely, SMT deformation in the Eastern Belt involved only minor internal rotation and was approximately coaxial. The main evidence is that in *XZ* sections, the fibre overgrowths, which record the progressive orientation of the incremental extension direction, are, on average, parallel to the trace of cleavage (Fig. 2d), which marks the approximate orientation of the finite extension direction. This concordance indicates that there was little significant rotation of the incremental extension direction relative to the evolving finite extension direction. As a contrary example, the simulated simple-shear fabric illustrated in Fig. 3b shows a discordance of 21° between the orientation of fibres and the trace of cleavage in the *XZ* section (Fig. 3). Strongly asymmetric fabrics like this are not observed in our samples.

The finite strain and internal rotation data were used to calculate average kinematic numbers (p. 16 in Passchier & Trouw (1996)). Means *et al.* (1980) proposed using the kinematic vorticity number $W_k$ as a measure of the non-coaxiality of the deformation, where

$$W_k = \frac{w}{\sqrt{2}\, D_T}$$

In this equation, *w* is the magnitude of the internal vorticity vector, given by $w = 2\left|w_{32}^2 + w_{13}^2 + w_{21}^2\right|^{1/2}$, where $w_{ij}$ are the components of the internal vorticity tensor (equation (18) of Means *et al.* (1980)). $D_T$ is a scalar measure of the average strain rate given by $D_T = \left|D_1^2 + D_2^2 + D_3^2\right|^{1/2}$, where $D_1$, $D_2$, and $D_3$ are the principal values of the stretching tensor $D_{ij}$ (see pp. 145–150 in Malvern (1969), for more details about $w_{ij}$

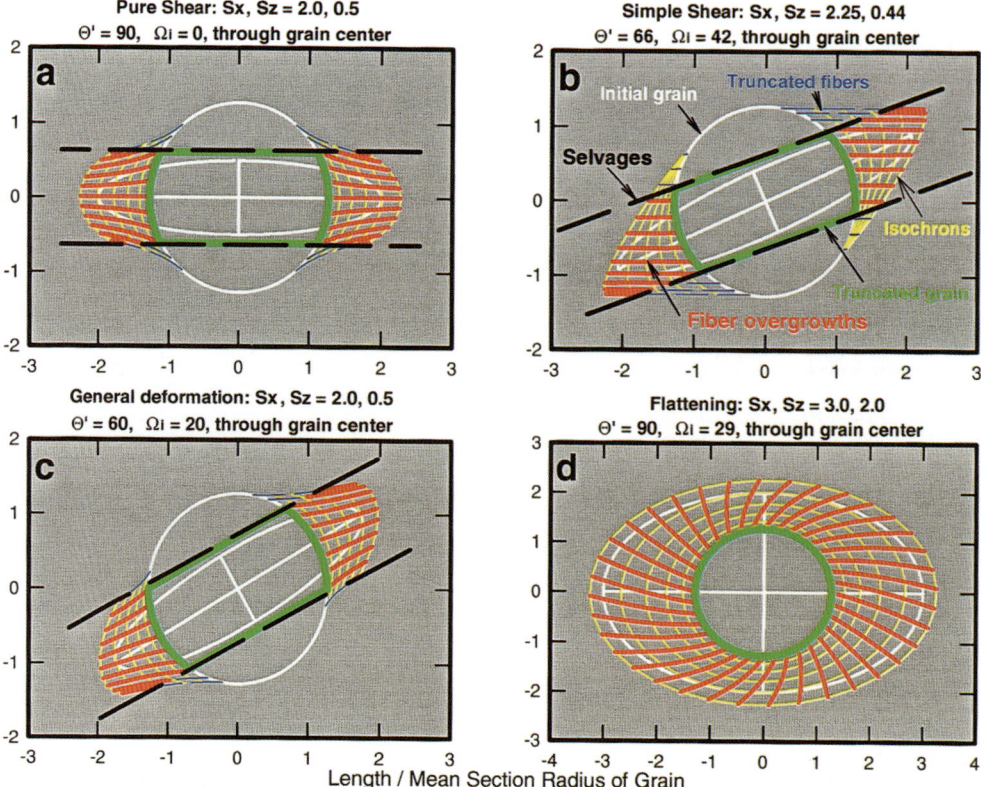

**Fig. 3.** Examples of SMT fabrics predicted by the semi-deformable antiaxial (SDA) fibre model. The figures are representative of the 'average' geometry of the deformation fabric through the centre of an average grain in the $XZ$ section. Dimensions are normalized to the mean section radius of the deforming grain. (Note that the true radius of a spherical grain, which is the outline shown in the diagrams, is 1.27 times the mean radius of all sections through the sphere.) $S_X$ and $S_Z$ are the principal stretches in the $X$ and $Z$ directions. The angle $\Theta'$ refers to the clockwise angle of $X$ relative to the vertical axis. The internal rotation axis is parallel to $Y$ and the internal rotation angle is reported as $\Omega_i$. An interesting result is that coaxial deformation can produce curved fibres (**a**) and non-coaxial deformation can produce straight fibres (**b**). The degree of non-coaxiality is best judged by the average angle between cleavage and the fibre overgrowths.

and $D_{ij}$). Passchier (1991) defined the kinematic dilatancy number $A_k$ as a relative measure of the volume strain rate with respect to the average strain rate. Generalizing his definition to three dimensions gives

$$A_k = \frac{D_V}{\sqrt{2}\, D_T}$$

where $D_V = D_1 + D_2 + D_3$ is the volume strain rate. When considering a general deformation that might included a volume strain, it is advantageous to replace $D_T$ with a scalar measure of the deviatoric strain rate defined by

$$D_D = \left[ (1/3)\left( (D_1 - D_2)^2 + (D_2 - D_3)^2 + (D_1 - D_3)^2 \right) \right]^{1/2}$$

(compare equation (6) in Brandon (1995)). $D_D$ describes the average rate of distortion caused by the deformation; it is proportional to the average shear strain rate and independent of the volume strain rate. It should be noted that $D_T^2 = D_D^2 + D_V^2$. The modified kinematic numbers, distinguished by asterisks, are

$$W_k^* = \frac{w}{\sqrt{2}\, D_D} \quad \text{and} \quad A_k^* = \frac{D_V}{\sqrt{2}\, D_D}$$

A coaxial deformation has $W_k = W_k^* = 0$ and a simple-shear deformation with no volume strain has $W_k = W_k^* = 1$. However, it should be noted that for simple shear with a volume strain, $W_k^*$ remains equal to one, but $W_k$ is no longer one and can take on any value. This example illustrates why we prefer the modified kinematic numbers. $A_k^* = 0$ indicates that the deformation is isochoric, $A_k^* > 0$ indicates a dilatant deformation, and $A_k^* < 0$ a compactive deformation. For example, a deformation involving uniaxial shortening and an

**Table 3.** Measurements of SMT deformation relative to present coordinates

| No. | Extension | | | Intermediate | | | Shortening | | | Volume | Internal rotation | | | Kinematic numbers | | |
|---|---|---|---|---|---|---|---|---|---|---|---|---|---|---|---|---|
| | tr. | pl. | $S_X$ | tr. | pl. | $S_Y$ | tr. | pl. | $S_Z$ | $S_V$ | tr. | pl. | $\Omega_i$ | $W_m$ | $W^*_m$ | $A^*_m$ |
| *Valentine Springs Formation, Yolla Bolly Mountains* | | | | | | | | | | | | | | | | |
| 1 | 5 | 14 | 1.49 | 102 | 27 | 0.82 | 251 | 59 | 0.47 | 0.57 | 103 | 31 | +7.0 | 0.21 | 0.23 | −0.48 |
| 2 | 348 | 26 | 1.35 | 104 | 42 | 0.70 | 239 | 38 | 0.58 | 0.55 | 255 | 9 | +8.5 | 0.31 | 0.36 | −0.68 |
| 3 | 69 | 13 | 1.17 | 174 | 49 | 0.94 | 328 | 38 | 0.33 | 0.36 | 326 | 42 | +2.4 | 0.05 | 0.06 | −0.75 |
| 4 | 202 | 17 | 1.13 | 80 | 60 | 1.06 | 300 | 24 | 0.60 | 0.72 | – | – | – | – | – | −0.47 |
| *Yolla Bolly terrane, Yolla Bolly Mountains* | | | | | | | | | | | | | | | | |
| 5 | 89 | 33 | 1.33 | 202 | 31 | 0.76 | 323 | 41 | 0.66 | 0.67 | 279 | 56 | −9.8 | 0.43 | 0.47 | −0.54 |
| 6 | 42 | 8 | 1.08 | 139 | 42 | 1.06 | 303 | 47 | 0.53 | 0.61 | – | – | – | – | – | −0.62 |
| 7 | 241 | 25 | 1.52 | 148 | 7 | 0.70 | 44 | 64 | 0.69 | 0.73 | 93 | 61 | −5.6 | 0.21 | 0.22 | −0.34 |
| 8 | 65 | 3 | 1.34 | 156 | 20 | 0.81 | 327 | 70 | 0.74 | 0.80 | 159 | 56 | +3.7 | 0.19 | 0.20 | −0.34 |
| 9 | 63 | 1 | 1.21 | 153 | 12 | 1.10 | 326 | 78 | 0.59 | 0.79 | 167 | 84 | −9.0 | 0.39 | 0.40 | −0.31 |
| 10 | 104 | 1 | 1.06 | 194 | 26 | 1.05 | 11 | 64 | 0.58 | 0.65 | – | – | – | – | – | −0.63 |
| 11 | 295 | 28 | 1.22 | 174 | 44 | 1.21 | 45 | 33 | 0.56 | 0.83 | 57 | 45 | −6.3 | 0.24 | 0.25 | −0.21 |
| 12 | 153 | 54 | 1.14 | 54 | 6 | 0.75 | 320 | 35 | 0.74 | 0.63 | 54 | 6 | +0.0 | 0.00 | 0.00 | −0.93 |
| 13 | 49 | 2 | 1.09 | 143 | 63 | 1.06 | 317 | 27 | 0.54 | 0.62 | 139 | 26 | −1.1 | 0.04 | 0.05 | −0.59 |
| 14 | 297 | 17 | 1.47 | 64 | 63 | 0.84 | 200 | 20 | 0.64 | 0.79 | 174 | 60 | −1.2 | 0.05 | 0.05 | −0.28 |
| 15 | 133 | 6 | 1.08 | 228 | 40 | 1.01 | 35 | 50 | 0.54 | 0.59 | – | – | – | – | – | −0.69 |
| 16 | 357 | 24 | 1.09 | 266 | 2 | 0.78 | 172 | 65 | 0.77 | 0.65 | – | – | – | – | – | −1.08 |
| 17 | 69 | 19 | 1.40 | 330 | 26 | 0.75 | 192 | 57 | 0.69 | 0.72 | 176 | 38 | +3.4 | 0.15 | 0.15 | −0.42 |
| 18 | 348 | 34 | 1.26 | 255 | 4 | 1.15 | 159 | 55 | 0.66 | 0.96 | 126 | 47 | −11.3 | 0.56 | 0.56 | −0.06 |
| 19 | 356 | 15 | 1.30 | 255 | 36 | 1.06 | 104 | 50 | 0.69 | 0.95 | 226 | 68 | −5.3 | 0.29 | 0.29 | −0.08 |
| 20 | 45 | 10 | 1.20 | 314 | 8 | 1.19 | 188 | 76 | 0.60 | 0.86 | 167 | 69 | +4.0 | 0.17 | 0.18 | −0.19 |
| *Yolla Bolly terrane, Leech Lake Mountain* | | | | | | | | | | | | | | | | |
| 21 | 83 | 8 | 1.14 | 345 | 45 | 1.09 | 181 | 44 | 0.61 | 0.76 | 184 | 53 | +4.9 | 0.23 | 0.25 | −0.40 |
| 22 | 284 | 2 | 1.36 | 14 | 4 | 0.75 | 167 | 85 | 0.72 | 0.73 | 190 | 59 | +8.9 | 0.43 | 0.46 | −0.43 |
| 23 | 270 | 6 | 1.14 | 3 | 23 | 1.08 | 167 | 66 | 0.67 | 0.82 | 158 | 75 | +2.4 | 0.14 | 0.14 | −0.33 |
| 24 | 101 | 26 | 1.07 | 350 | 37 | 1.06 | 217 | 42 | 0.64 | 0.73 | 212 | 36 | −1.6 | 0.09 | 0.09 | −0.54 |
| 25 | 137 | 2 | 1.10 | 45 | 30 | 1.08 | 230 | 60 | 0.61 | 0.72 | 258 | 86 | +1.3 | 0.06 | 0.07 | −0.48 |
| 26 | 91 | 19 | 1.32 | 183 | 6 | 1.06 | 290 | 70 | 0.61 | 0.85 | 336 | 50 | −8.0 | 0.35 | 0.35 | −0.20 |
| 27 | 284 | 3 | 1.17 | 193 | 25 | 0.66 | 20 | 65 | 0.63 | 0.49 | 19 | 61 | −18.4 | 0.72 | 0.94 | −1.03 |
| 28 | 329 | 11 | 1.36 | 92 | 71 | 0.67 | 236 | 15 | 0.61 | 0.56 | 61 | 10 | −7.0 | 0.24 | 0.28 | −0.67 |
| 29 | 106 | 13 | 1.33 | 5 | 37 | 0.70 | 212 | 50 | 0.54 | 0.50 | 294 | 76 | +6.7 | 0.23 | 0.27 | −0.74 |
| 30 | 15 | 22 | 1.10 | 109 | 11 | 1.01 | 225 | 65 | 0.65 | 0.72 | 139 | 55 | +0.9 | 0.05 | 0.06 | −0.58 |
| 31 | 59 | 8 | 1.10 | 325 | 27 | 0.83 | 164 | 62 | 0.64 | 0.58 | 168 | 67 | +1.1 | 0.06 | 0.07 | −0.99 |
| 32 | 6 | 24 | 1.20 | 98 | 6 | 1.00 | 202 | 65 | 0.53 | 0.64 | 108 | 26 | −1.1 | 0.04 | 0.04 | −0.53 |

**Table 3.** *continued*

| No. | Principal stretches | | | | | | | | | | | | | Internal rotation | | | Kinematic numbers | | |
|---|---|---|---|---|---|---|---|---|---|---|---|---|---|---|---|---|---|---|---|
| | Extension | | | Intermediate | | | Shortening | | | Volume | | | | | | | | |
| | tr. | pl. | $S_X$ | tr. | pl. | $S_Y$ | tr. | pl. | $S_Z$ | $S_V$ | tr. | pl. | $\Omega_i$ | $W_m$ | $W_m^*$ | $A_m^*$ |
| 33 | 319 | 21 | 1.03 | 64 | 35 | 1.03 | 204 | 48 | 0.55 | 0.58 | 119 | 68 | −0.1 | 0.00 | 0.00 | −0.74 |
| 34 | 312 | 10 | 1.13 | 51 | 43 | 1.07 | 212 | 45 | 0.50 | 0.60 | 211 | 49 | −1.5 | 0.05 | 0.06 | −0.55 |
| 35 | 296 | 11 | 1.31 | 38 | 47 | 0.66 | 196 | 41 | 0.62 | 0.54 | 202 | 24 | +5.8 | 0.21 | 0.24 | −0.75 |
| 36 | 225 | 2 | 1.01 | 134 | 26 | 1.00 | 320 | 64 | 0.54 | 0.55 | 316 | 19 | +0.0 | 0.00 | 0.00 | −0.85 |
| 37 | 328 | 43 | 1.00 | 110 | 41 | 0.72 | 218 | 20 | 0.63 | 0.45 | 70 | 13 | +0.0 | 0.00 | 0.00 | −1.66 |
| 38 | 3 | 43 | 1.12 | 131 | 34 | 1.11 | 243 | 29 | 0.62 | 0.77 | 263 | 10 | −0.2 | 0.01 | 0.01 | −0.38 |
| 39 | 337 | 12 | 1.00 | 98 | 67 | 0.87 | 243 | 19 | 0.81 | 0.70 | 223 | 62 | +0.0 | 0.00 | 0.00 | −1.63 |
| 40 | 287 | 22 | 1.28 | 28 | 25 | 0.82 | 161 | 56 | 0.58 | 0.61 | 42 | 46 | +3.5 | 0.14 | 0.15 | −0.63 |
| *Yolla Bolly terrane, Mount Hamilton, Diablo Range* | | | | | | | | | | | | | | | | |
| 41 | 21 | 54 | 1.27 | 142 | 20 | 0.71 | 243 | 28 | 0.54 | 0.49 | 221 | 34 | +3.3 | 0.11 | 0.13 | −0.82 |
| 42 | 33 | 14 | 1.06 | 300 | 14 | 1.06 | 168 | 70 | 0.49 | 0.55 | 167 | 70 | +0.5 | 0.02 | 0.02 | −0.67 |
| 43 | 168 | 50 | 1.06 | 318 | 36 | 0.78 | 59 | 15 | 0.50 | 0.41 | – | – | – | – | – | −1.17 |
| 44 | 323 | 46 | 1.04 | 202 | 26 | 1.04 | 94 | 32 | 0.63 | 0.68 | – | – | – | – | – | −0.66 |
| 45 | 315 | 34 | 1.22 | 201 | 31 | 0.77 | 80 | 40 | 0.64 | 0.60 | 47 | 3 | −0.8 | 0.04 | 0.04 | −0.77 |
| 46 | 111 | 54 | 1.19 | 249 | 28 | 0.66 | 350 | 20 | 0.62 | 0.49 | 234 | 21 | +2.8 | 0.11 | 0.14 | −1.00 |
| 47 | 317 | 36 | 1.04 | 213 | 19 | 1.04 | 100 | 48 | 0.62 | 0.67 | 206 | 27 | −0.9 | 0.05 | 0.05 | −0.67 |
| 48 | 7 | 61 | 1.09 | 253 | 13 | 0.84 | 157 | 25 | 0.56 | 0.51 | – | – | – | – | – | −1.00 |
| 49 | 158 | 35 | 1.02 | 56 | 17 | 0.79 | 304 | 50 | 0.50 | 0.40 | – | – | – | – | – | −1.26 |
| 50 | 219 | 61 | 1.10 | 1 | 24 | 1.09 | 98 | 16 | 0.70 | 0.84 | – | – | – | – | – | −0.34 |
| 51 | 187 | 19 | 1.22 | 74 | 49 | 0.83 | 291 | 35 | 0.60 | 0.61 | 8 | 71 | −2.7 | 0.12 | 0.13 | −0.70 |
| 52 | 232 | 60 | 1.30 | 119 | 12 | 0.76 | 23 | 27 | 0.59 | 0.58 | 130 | 6 | +7.1 | 0.29 | 0.33 | −0.67 |
| 53 | 232 | 30 | 1.06 | 140 | 4 | 0.96 | 43 | 60 | 0.66 | 0.67 | – | – | – | – | – | −0.80 |
| 54 | 192 | 17 | 1.07 | 101 | 4 | 0.75 | 357 | 72 | 0.46 | 0.37 | – | – | – | – | – | −1.18 |
| 55 | 233 | 27 | 1.01 | 134 | 17 | 1.01 | 15 | 57 | 0.45 | 0.46 | – | – | – | – | – | −0.83 |
| 56 | 235 | 26 | 1.05 | 61 | 64 | 1.04 | 325 | 3 | 0.61 | 0.67 | 33 | 63 | −4.7 | 0.23 | 0.26 | −0.65 |
| *Yolla Bolly terrane, Pacheco Pass, Diablo Range* | | | | | | | | | | | | | | | | |
| 57 | 24 | 5 | 1.14 | 292 | 17 | 1.03 | 130 | 72 | 0.66 | 0.77 | 128 | 70 | +1.7 | 0.10 | 0.10 | −0.44 |
| 58 | 36 | 24 | 1.02 | 164 | 54 | 0.66 | 294 | 25 | 0.50 | 0.34 | – | – | – | – | – | −1.52 |
| 59 | 248 | 35 | 1.04 | 148 | 14 | 1.01 | 40 | 51 | 0.57 | 0.60 | – | – | – | – | – | −0.76 |
| 60 | 245 | 2 | 1.25 | 155 | 7 | 1.05 | 355 | 83 | 0.51 | 0.67 | 342 | 72 | +4.6 | 0.16 | 0.17 | −0.42 |
| 61 | 18 | 15 | 1.15 | 270 | 51 | 0.60 | 119 | 35 | 0.49 | 0.34 | 119 | 35 | +9.1 | 0.27 | 0.38 | −1.22 |
| 62 | 103 | 32 | 1.17 | 10 | 5 | 0.89 | 272 | 58 | 0.62 | 0.65 | 271 | 58 | +1.9 | 0.09 | 0.10 | −0.69 |
| 63 | 280 | 8 | 1.14 | 184 | 35 | 0.97 | 21 | 54 | 0.53 | 0.59 | 21 | 54 | −4.0 | 0.15 | 0.17 | −0.66 |
| 64 | 191 | 7 | 1.21 | 98 | 23 | 0.69 | 297 | 66 | 0.66 | 0.55 | – | – | – | – | – | −0.88 |

tr. and pl. indicate trend and plunge. A right-handed internal rotation is indicated by a positive angle.

**Table 4.** Measurements of SMT deformation after unfolding of the effects of late-stage shallow folding

| No. | Extension | | | Intermediate | | | Shortening | | | Volume | Internal rotation | | | Kinematic numbers | | |
|---|---|---|---|---|---|---|---|---|---|---|---|---|---|---|---|---|
| | tr. | pl. | $S_X$ | tr. | pl. | $S_Y$ | tr. | pl. | $S_Z$ | $S_V$ | tr. | pl. | $\Omega_i$ | $W_m$ | $W^*_m$ | $A^*_m$ |

*Valentine Springs Formation, Yolla Bolly Mountains*

| 1 | 196 | 13 | 1.49 | 101 | 18 | 0.82 | 319 | 67 | 0.47 | 0.57 | 100 | 22 | +7.0 | 0.21 | 0.23 | −0.48 |
| 2 | 7 | 4 | 1.35 | 100 | 32 | 0.70 | 273 | 58 | 0.58 | 0.55 | 277 | 25 | +8.5 | 0.31 | 0.36 | −0.68 |
| 3 | 69 | 13 | 1.17 | 175 | 49 | 0.94 | 328 | 38 | 0.33 | 0.36 | 327 | 42 | +2.4 | 0.05 | 0.06 | −0.75 |
| 4 | 202 | 17 | 1.13 | 80 | 60 | 1.06 | 300 | 24 | 0.60 | 0.72 | — | — | — | — | — | −0.47 |

*Yolla Bolly terrane, Yolla Bolly Mountains*

| 5 | 100 | 46 | 1.33 | 192 | 3 | 0.76 | 286 | 43 | 0.66 | 0.67 | 243 | 37 | −9.8 | 0.43 | 0.48 | −0.54 |
| 6 | 42 | 8 | 1.08 | 139 | 41 | 1.06 | 303 | 47 | 0.53 | 0.61 | — | — | — | — | — | −0.62 |
| 7 | 241 | 25 | 1.52 | 148 | 6 | 0.70 | 44 | 64 | 0.69 | 0.73 | 94 | 61 | −5.6 | 0.21 | 0.22 | −0.34 |
| 8 | 65 | 3 | 1.34 | 156 | 20 | 0.81 | 327 | 70 | 0.74 | 0.80 | 159 | 56 | +3.7 | 0.19 | 0.20 | −0.34 |
| 9 | 63 | 1 | 1.21 | 153 | 12 | 1.10 | 326 | 78 | 0.59 | 0.79 | 167 | 84 | −9.0 | 0.39 | 0.40 | −0.31 |
| 10 | 104 | 1 | 1.06 | 195 | 26 | 1.05 | 11 | 64 | 0.58 | 0.65 | — | — | — | — | — | −0.63 |
| 11 | 295 | 28 | 1.22 | 177 | 44 | 1.21 | 45 | 33 | 0.56 | 0.83 | 58 | 46 | −6.2 | 0.24 | 0.24 | −0.21 |
| 12 | 161 | 48 | 1.14 | 54 | 15 | 0.75 | 311 | 39 | 0.74 | 0.63 | 54 | 15 | +0.0 | 0.00 | 0.00 | −0.93 |
| 13 | 48 | 11 | 1.09 | 156 | 58 | 1.06 | 310 | 29 | 0.54 | 0.62 | 142 | 23 | −1.1 | 0.04 | 0.05 | −0.59 |
| 14 | 302 | 17 | 1.47 | 57 | 55 | 0.84 | 203 | 29 | 0.64 | 0.79 | 166 | 68 | −1.2 | 0.05 | 0.05 | −0.28 |
| 15 | 134 | 8 | 1.08 | 233 | 49 | 1.01 | 36 | 40 | 0.54 | 0.59 | — | — | — | — | — | −0.69 |
| 16 | 185 | 2 | 1.09 | 276 | 17 | 0.78 | 92 | 73 | 0.77 | 0.65 | — | — | — | — | — | −1.08 |
| 17 | 252 | 4 | 1.40 | 342 | 9 | 0.75 | 140 | 81 | 0.69 | 0.72 | 156 | 59 | +3.4 | 0.15 | 0.15 | −0.42 |
| 18 | 0 | 10 | 1.26 | 266 | 23 | 1.15 | 113 | 64 | 0.66 | 0.96 | 101 | 43 | −11.3 | 0.56 | 0.56 | −0.06 |
| 19 | 352 | 23 | 1.30 | 249 | 29 | 1.06 | 114 | 52 | 0.69 | 0.95 | 220 | 59 | −5.3 | 0.29 | 0.29 | −0.08 |
| 20 | 231 | 18 | 1.20 | 141 | 1 | 1.19 | 47 | 72 | 0.60 | 0.86 | 21 | 69 | +4.0 | 0.17 | 0.18 | −0.19 |

*Yolla Bolly terrane, Leech Lake Mountain*

| 21 | 273 | 7 | 1.14 | 6 | 26 | 1.09 | 166 | 67 | 0.61 | 0.76 | 155 | 76 | +5.0 | 0.24 | 0.25 | −0.40 |
| 22 | 297 | 7 | 1.36 | 204 | 24 | 0.75 | 43 | 66 | 0.72 | 0.73 | 140 | 82 | +9.0 | 0.44 | 0.46 | −0.43 |
| 23 | 286 | 17 | 1.14 | 195 | 3 | 1.08 | 95 | 73 | 0.67 | 0.82 | 67 | 69 | +2.0 | 0.12 | 0.12 | −0.33 |
| 24 | 101 | 17 | 1.07 | 7 | 13 | 1.06 | 241 | 69 | 0.64 | 0.73 | 229 | 64 | −2.0 | 0.11 | 0.12 | −0.54 |
| 25 | 144 | 12 | 1.10 | 53 | 3 | 1.08 | 308 | 78 | 0.61 | 0.72 | 28 | 64 | +1.0 | 0.05 | 0.05 | −0.48 |
| 26 | 96 | 6 | 1.32 | 190 | 32 | 1.06 | 356 | 57 | 0.61 | 0.85 | 2 | 28 | −8.0 | 0.35 | 0.35 | −0.20 |
| 27 | 297 | 8 | 1.17 | 198 | 52 | 0.66 | 33 | 37 | 0.63 | 0.49 | 32 | 33 | −18.0 | 0.70 | 0.92 | −1.03 |
| 28 | 161 | 5 | 1.36 | 64 | 51 | 0.67 | 255 | 38 | 0.61 | 0.56 | 252 | 13 | −7.0 | 0.24 | 0.28 | −0.67 |
| 29 | 111 | 7 | 1.33 | 19 | 10 | 0.70 | 237 | 78 | 0.54 | 0.50 | 7 | 59 | +7.0 | 0.24 | 0.28 | −0.74 |
| 30 | 206 | 6 | 1.10 | 115 | 7 | 1.01 | 333 | 81 | 0.65 | 0.72 | 106 | 56 | +1.0 | 0.06 | 0.06 | −0.58 |
| 31 | 251 | 15 | 1.10 | 344 | 11 | 0.83 | 107 | 71 | 0.64 | 0.58 | 91 | 74 | +1.0 | 0.05 | 0.06 | −0.99 |

**Table 4.** *continued*

| No. | Extension | | | Principal stretches Intermediate | | | Shortening | | | Volume | Internal rotation | | | Kinematic numbers | | |
|---|---|---|---|---|---|---|---|---|---|---|---|---|---|---|---|---|
| | tr. | pl. | $S_X$ | tr. | pl. | $S_Y$ | tr. | pl. | $S_Z$ | $S_V$ | tr. | pl. | $\Omega_i$ | $W_m$ | $W^*_m$ | $A^*_m$ |
| 32 | 198 | 3 | 1.20 | 288 | 3 | 1.00 | 55 | 87 | 0.53 | 0.64 | 107 | 20 | −1.0 | 0.04 | 0.04 | −0.53 |
| 33 | 336 | 8 | 1.03 | 67 | 12 | 1.03 | 213 | 76 | 0.55 | 0.58 | 77 | 56 | +0.0 | 0.00 | 0.00 | −0.74 |
| 34 | 325 | 1 | 1.13 | 55 | 17 | 1.07 | 233 | 73 | 0.50 | 0.60 | 233 | 10 | −2.0 | 0.07 | 0.08 | −0.55 |
| 35 | 312 | 9 | 1.31 | 45 | 19 | 0.66 | 197 | 69 | 0.62 | 0.54 | 211 | 52 | +6.0 | 0.22 | 0.25 | −0.75 |
| 36 | 238 | 28 | 1.01 | 129 | 32 | 1.00 | 4 | 47 | 0.54 | 0.55 | 333 | 7 | +0.0 | 0.00 | 0.00 | −0.85 |
| 37 | 353 | 24 | 1.00 | 99 | 33 | 0.73 | 234 | 47 | 0.63 | 0.45 | 259 | 7 | +0.0 | 0.00 | 0.00 | −1.66 |
| 38 | 19 | 16 | 1.12 | 121 | 37 | 1.11 | 270 | 49 | 0.62 | 0.77 | 281 | 25 | +0.0 | 0.00 | 0.00 | −0.38 |
| 39 | 168 | 7 | 1.00 | 70 | 49 | 0.87 | 265 | 40 | 0.81 | 0.70 | 314 | 81 | +0.0 | 0.00 | 0.00 | −1.63 |
| 40 | 309 | 23 | 1.28 | 218 | 3 | 0.82 | 120 | 67 | 0.58 | 0.61 | 48 | 19 | +4.0 | 0.16 | 0.18 | −0.63 |

*Yolla Bolly terrane, Mount Hamilton, Diablo Range*

| No. | tr. | pl. | $S_X$ | tr. | pl. | $S_Y$ | tr. | pl. | $S_Z$ | $S_V$ | tr. | pl. | $\Omega_i$ | $W_m$ | $W^*_m$ | $A^*_m$ |
|---|---|---|---|---|---|---|---|---|---|---|---|---|---|---|---|---|
| 41 | 23 | 29 | 1.27 | 122 | 17 | 0.71 | 237 | 56 | 0.54 | 0.49 | 202 | 61 | +3.3 | 0.11 | 0.13 | −0.82 |
| 42 | 33 | 14 | 1.06 | 300 | 14 | 1.06 | 168 | 70 | 0.49 | 0.55 | 167 | 70 | +0.5 | 0.02 | 0.02 | −0.67 |
| 43 | 202 | 23 | 1.06 | 307 | 31 | 0.78 | 82 | 50 | 0.50 | 0.41 | – | – | – | – | – | −1.17 |
| 44 | 301 | 40 | 1.04 | 38 | 9 | 1.04 | 138 | 48 | 0.63 | 0.68 | 196 | 33 | −0.8 | 0.04 | 0.04 | −0.66 |
| 45 | 306 | 28 | 1.22 | 38 | 3 | 0.77 | 136 | 62 | 0.64 | 0.60 | 234 | 21 | +2.8 | 0.11 | 0.14 | −0.77 |
| 46 | 111 | 54 | 1.19 | 249 | 28 | 0.66 | 350 | 20 | 0.62 | 0.49 | 53 | 26 | −0.9 | 0.05 | 0.05 | −1.00 |
| 47 | 306 | 30 | 1.04 | 47 | 18 | 1.04 | 163 | 54 | 0.62 | 0.67 | – | – | – | – | – | −0.67 |
| 48 | 17 | 30 | 1.09 | 246 | 48 | 0.84 | 124 | 25 | 0.56 | 0.51 | – | – | – | – | – | −1.00 |
| 49 | 117 | 33 | 1.02 | 221 | 20 | 0.79 | 337 | 50 | 0.50 | 0.40 | – | – | – | – | – | −1.26 |
| 50 | 232 | 33 | 1.10 | 357 | 42 | 1.09 | 119 | 31 | 0.70 | 0.84 | – | – | – | – | – | −0.34 |
| 51 | 154 | 39 | 1.22 | 54 | 13 | 0.83 | 310 | 49 | 0.60 | 0.61 | 25 | 39 | −2.7 | 0.12 | 0.13 | −0.70 |
| 52 | 242 | 22 | 1.30 | 144 | 19 | 0.76 | 19 | 60 | 0.59 | 0.58 | 149 | 7 | +7.1 | 0.29 | 0.33 | −0.67 |
| 53 | 65 | 7 | 1.06 | 335 | 1 | 0.96 | 237 | 83 | 0.66 | 0.67 | – | – | – | – | – | −0.80 |
| 54 | 192 | 17 | 1.07 | 101 | 5 | 0.75 | 357 | 72 | 0.46 | 0.37 | – | – | – | – | – | −1.18 |
| 55 | 66 | 10 | 1.01 | 158 | 13 | 1.01 | 300 | 73 | 0.45 | 0.46 | – | – | – | – | – | −0.83 |
| 56 | 254 | 50 | 1.05 | 28 | 30 | 1.04 | 132 | 22 | 0.61 | 0.67 | 42 | 35 | −4.7 | 0.23 | 0.26 | −0.65 |

*Yolla Bolly terrane, Pacheco Pass, Diablo Range*

| No. | tr. | pl. | $S_X$ | tr. | pl. | $S_Y$ | tr. | pl. | $S_Z$ | $S_V$ | tr. | pl. | $\Omega_i$ | $W_m$ | $W^*_m$ | $A^*_m$ |
|---|---|---|---|---|---|---|---|---|---|---|---|---|---|---|---|---|
| 57 | 24 | 5 | 1.14 | 293 | 18 | 1.03 | 130 | 72 | 0.66 | 0.77 | 127 | 69 | +1.7 | 0.10 | 0.10 | −0.44 |
| 58 | 207 | 3 | 1.02 | 113 | 53 | 0.66 | 299 | 36 | 0.50 | 0.34 | – | – | – | – | – | −1.52 |
| 59 | 247 | 62 | 1.04 | 130 | 14 | 1.01 | 34 | 24 | 0.57 | 0.60 | – | – | – | – | – | −0.76 |
| 60 | 245 | 2 | 1.25 | 155 | 6 | 1.05 | 355 | 84 | 0.51 | 0.67 | 342 | 73 | +4.6 | 0.16 | 0.17 | −0.42 |
| 61 | 21 | 39 | 1.15 | 266 | 28 | 0.60 | 151 | 38 | 0.49 | 0.34 | 151 | 38 | +9.0 | 0.27 | 0.38 | −1.22 |
| 62 | 85 | 11 | 1.17 | 178 | 15 | 0.89 | 319 | 71 | 0.62 | 0.65 | 320 | 71 | +1.9 | 0.09 | 0.10 | −0.69 |
| 63 | 111 | 8 | 1.14 | 202 | 12 | 0.97 | 347 | 75 | 0.53 | 0.59 | 346 | 75 | −4.0 | 0.15 | 0.17 | −0.66 |
| 64 | 191 | 7 | 1.21 | 98 | 23 | 0.69 | 297 | 66 | 0.66 | 0.55 | – | – | – | – | – | −0.88 |

tr. and pl. indicate trend and plunge. A right-handed internal rotation is indicated by a positive angle.

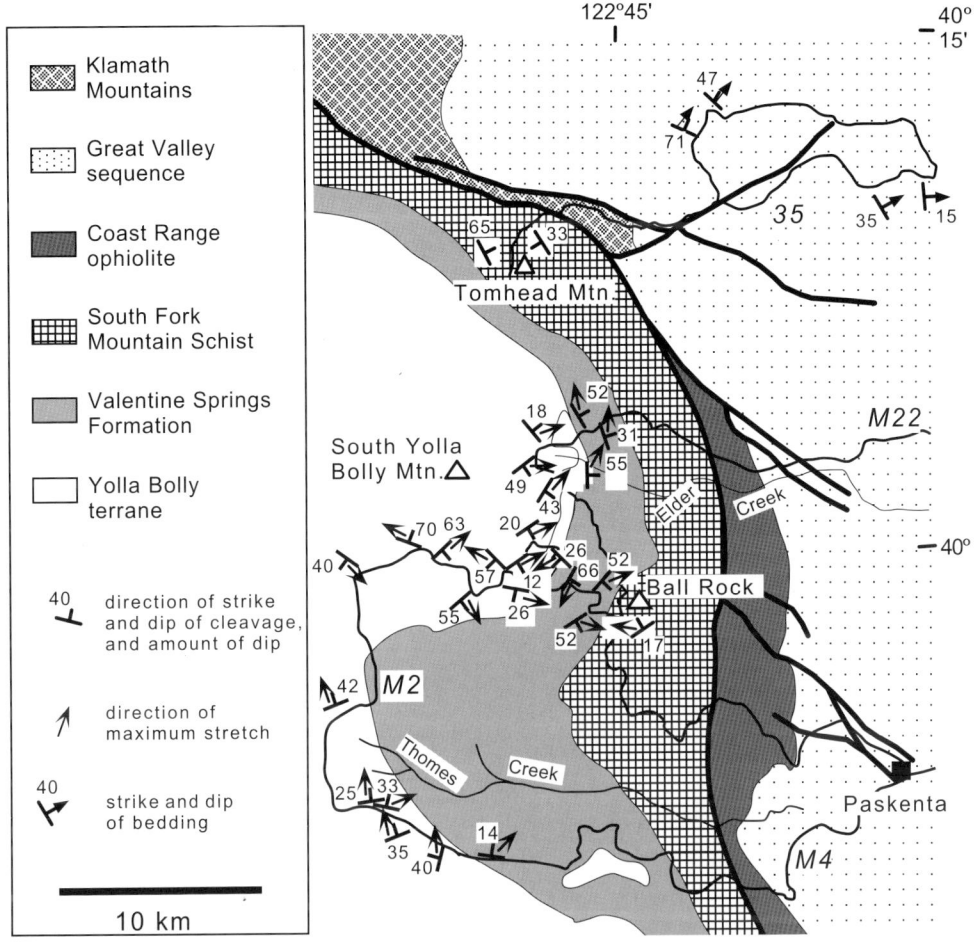

**Fig. 4.** Geological map of the Yolla Bolly Mountains study area showing principal stretch directions for SMT deformation in the Eastern Belt of the Franciscan Complex, which in this area includes the South Fork Mountain Schist, Valentine Springs Formation, and the Pickett Peak terrane. All directions are relative to present coordinates. Note the large degree of local variation in the principal directions. M2, M22, M4 and 35 refer to US Forest Service roads.

equal loss of volume, such as compaction in a sedimentary basin, would have $A^*_k = -1$ because the rates of volume strain and deviatoric strain would be equal in magnitude but opposite in sign.

The kinematic numbers are defined with respect to rates of deformation because they represent properties of the deformation at a point. They are, however, dimensionless, and therefore independent of the actual deformation rate. Application of these concepts to a finite deformation is complicated because the kinematic numbers associated with deformation at a material point can vary with time. None the less, one can define an average kinematic number by assuming a steady deformation and then finding an average set of rate parameters that when integrated give the observed finite deformation (Passchier 1988; Means 1994). (Note that as used here, steady means that the kinematic numbers at a material point remained constant with time, but it does not require that the rate of deformation remained constant.) The rate parameters are specified by the velocity gradient tensor ($L$ of Malvern (1969)). The kinematic numbers can be directly calculated from the average $L$. These average values are designated as $W_m$, $W^*_m$ and $A^*_m$. The values reported in Tables 3 and 4 were calculated numerically by searching for the unique average $L$ that when integrated gave the measured three-dimensional finite strain and internal rotation (see Feehan & Brandon (1999) for details about the numerical integration). The simple geometries associated with our fibre

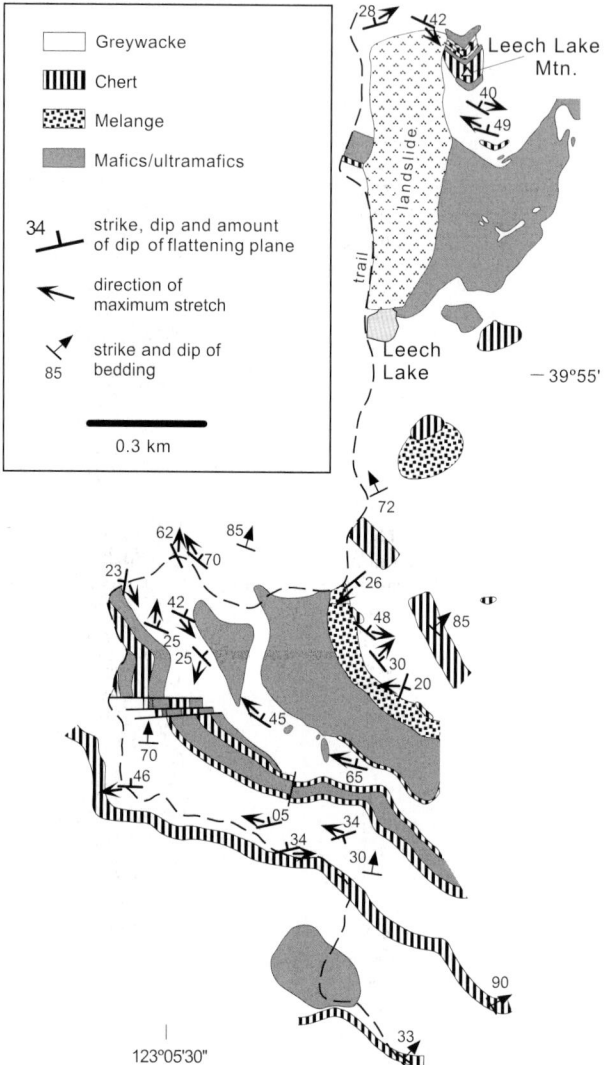

**Fig. 5.** Geological map of the Leech Lake Mountain study area showing principal stretch directions for SMT deformation. All directions are relative to present coordinates. Variability of principal directions is slightly less than that in the Yolla Bolly Mountains study area.

overgrowths indicate that SMT deformation in the Franciscan Complex probably did not depart strongly from the steady-deformation assumption used here.

## Results

### Sampling

To obtain reliable tensor averages for SMT deformation at a regional scale, we sampled randomly in regions where the rock types were homogeneous at the scale of sampling. We focused on two areas in northern California, located in the Yolla Bolly Mountains and at Leech Lake Mountain, and one in the Diablo Range of central California (Fig. 1a). All areas contain typical outcrops of the Eastern Belt. In the Yolla Bolly Mountains, we collected samples of metasandstones, metabasalt, shale and schist from the Pickett Peak terrane: one from the South Fork Mountain Schist and eight from the Valentine Springs Formation. Sixteen sandstone samples were collected from the Yolla Bolly terrane. Strain directions were

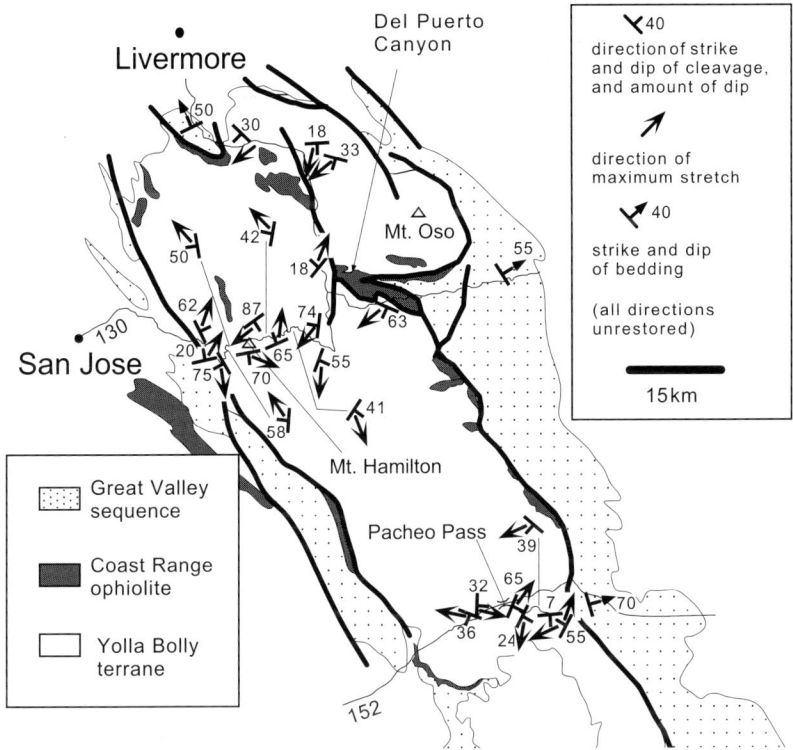

**Fig. 6.** Geological map of the Diablo Range with principal strain directions for SMT deformation. Mount Hamilton lies to the north and Pacheco Pass to the south. The variability in principal directions is similar to that in the Yolla Bolly Mountains. California State highways 130 and 152 are also shown.

determined for all and are shown in Fig. 4. Full deformation measurements were made for four of the Valentine Springs sandstones and all of the Yolly Bolly terrane samples. At Leech Lake Mountain (Fig. 5), we analysed 20 sandstones from the Yolla Bolly terrane. In the Diablo Range (Fig. 6), 16 sandstones for the Mt Hamilton–Mt Oso area and another eight sandstones from the Pacheco Pass area were analysed. Detailed descriptions of the geology of these study areas have been provided by Chesterman (1963), Worrall (1979, 1981), Blake et al. (1988), and Bolhar (1997) for the Yolla Bolly Mountains and Leech Lake Mountain, and Ernst (1965, 1987, 1993) and Cowan (1974) for the Diablo Range. Field measurements of cleavage, bedding, and stretching lineations were made for comparison with the deformation analysis.

*Analysis*

Our primary results are reported in Tables 3–5 and Fig. 7. The data are characterized in Fig. 8 according to strain symmetry (oblate v. prolate), strain type (constrictional v. flattening), and mean kinematic numbers.

An important initial observation is that individual principal strain directions and internal rotation axes are variable in orientation (e.g. Fig. 7). The observed variability is too large to be attributed to measurement error alone. Other studies (e.g. Feehan & Brandon 1999) suggest that SMT deformation might be typified by local-scale variability, especially in the orientation of $X$ and $Y$. Thus, we have found it useful to calculate averages for these data (Table 5). As discussed by Brandon (1995), it is not appropriate to average the principal stretches, principal directions, and internal rotation measurements separately. The data must be averaged in tensor form to ensure that the orthogonality of the axes is preserved and that the magnitudes and directions of the principal stretches are correctly associated. If the rotational component of the deformation is fairly small, then one can average the stretch tensor and the internal rotation tensor separately, without introducing significant errors (Brandon 1995). Average stretches and internal rotations in Table 5 were

**Table 5.** *Regional averages for SMT deformation in present (P) and unfolded (U) coordinates*

| | | | | Principal stretches | | | | | | | | | | | | | |
|---|---|---|---|---|---|---|---|---|---|---|---|---|---|---|---|---|---|
| | Extension | | | Intermediate | | | Shortening | | | Volume | Internal rotation | | | Kinematic numbers | | | |
| P/U | tr. | pl. | $S_X$ | tr. | pl. | $S_Y$ | tr. | pl. | $S_Z$ | $S_V$ | tr. | pl. | $\Omega_i$ | $W_m$ | $W^*_m$ | $A^*_m$ |

*Valentine Springs Formation, Yolla Bolly Mountains study area (N = 4)*
| P | 26 | 12 | 1.04 | 126 | 40 | 0.86 | 282 | 48 | 0.60 | 0.54 | 236 | 60 | +2.7 | 0.11 | 0.16 | −0.60 |
| U | 205 | 11 | 1.16 | 106 | 40 | 0.86 | 307 | 48 | 0.53 | 0.54 | 300 | 72 | +2.8 | 0.10 | 0.11 | −0.42 |

*Yolla Bolly terrane, Yolla Bolly Mountains study area (N = 16)*
| P | 255 | 2 | 1.02 | 164 | 14 | 0.92 | 354 | 76 | 0.77 | 0.73 | 80 | 80 | −2.7 | 0.25 | 0.30 | −0.54 |
| U | 240 | 13 | 1.02 | 146 | 15 | 0.96 | 10 | 70 | 0.75 | 0.73 | 139 | 76 | −2.4 | 0.22 | 0.26 | −0.48 |

*Yolla Bolly terrane, Leech Lake Mountain study area (N = 20)*
| P | 298 | 1 | 1.06 | 28 | 31 | 0.91 | 206 | 59 | 0.66 | 0.64 | 199 | 6 | +1.4 | 0.08 | 0.09 | −1.04 |
| U | 128 | 0 | 1.06 | 38 | 3 | 0.91 | 224 | 87 | 0.66 | 0.64 | 165 | 43 | +1.0 | 0.05 | 0.06 | −1.04 |

*Grand mean for Yolla Bolly terrane, Yolla Bolly and Leech Lake Mountain study areas (N = 36)*
| P | 105 | 3 | 1.02 | 14 | 17 | 0.90 | 206 | 72 | 0.73 | 0.68 | 28 | 45 | −1.3 | 0.09 | 0.12 | −1.47 |
| U | 118 | 2 | 1.01 | 208 | 6 | 0.94 | 12 | 84 | 0.71 | 0.68 | 4 | 59 | −0.5 | 0.04 | 0.04 | −1.31 |

*Yolla Bolly terrane, Mount Hamilton, Diablo Range study area (N = 17)*
| P | 196 | 23 | 0.90 | 303 | 34 | 0.82 | 80 | 47 | 0.76 | 0.56 | 356 | 7 | −1.3 | 0.10 | 0.37 | −1.58 |
| U | 37 | 0 | 0.95 | 307 | 10 | 0.84 | 130 | 80 | 0.70 | 0.56 | 189 | 6 | +1.6 | 0.11 | 0.21 | −0.85 |

*Yolla Bolly terrane, Pacheco Pass, Diablo Range study area (N = 7)*
| P | 202 | 6 | 0.97 | 111 | 9 | 0.82 | 326 | 79 | 0.64 | 0.51 | 135 | 52 | +2.6 | 0.13 | 0.19 | −1.42 |
| U | 212 | 2 | 0.94 | 121 | 10 | 0.84 | 315 | 79 | 0.65 | 0.51 | 152 | 52 | +2.6 | 0.13 | 0.17 | −1.52 |

*Grand mean for Yolla Bolly terrane, Diablo Range study area (N = 24)*
| P | 196 | 14 | 0.91 | 288 | 9 | 0.81 | 50 | 74 | 0.74 | 0.55 | 161 | 22 | +1.4 | 0.10 | 0.28 | −2.83 |
| U | 216 | 0 | 0.95 | 306 | 2 | 0.83 | 119 | 88 | 0.69 | 0.55 | 178 | 26 | +1.7 | 0.11 | 0.18 | −1.88 |

*Grand mean, all study areas (N = 64)*
| P | 103 | 5 | 0.92 | 13 | 5 | 0.90 | 236 | 83 | 0.75 | 0.62 | 16 | 18 | −1.0 | 0.08 | 0.12 | −2.14 |
| U | 221 | 3 | 0.96 | 129 | 5 | 0.92 | 345 | 84 | 0.70 | 0.62 | 184 | 6 | +0.6 | 0.04 | 0.05 | −1.45 |

tr. and pl. indicate trend and plunge. A right-handed internal rotation is indicated by a positive angle.

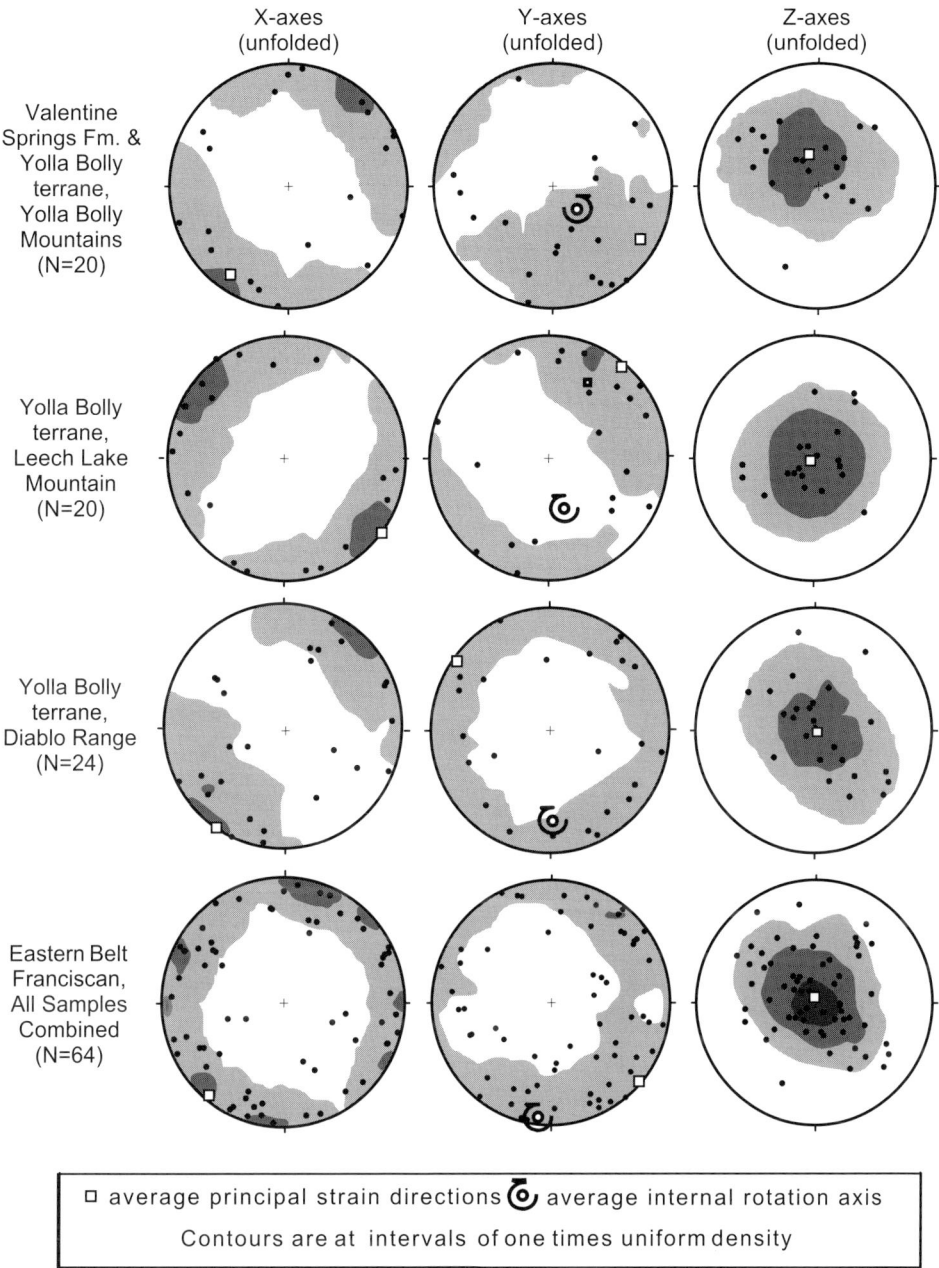

**Fig. 7.** Lower-hemisphere equal-area stereograms for 'unfolded' principal stretch directions in the Yolla Bolly Mountains, at Leech Lake Mountain, in the Diablo Range, and for all data combined. Also shown are axes and relative shear sense for the tensor-averaged internal rotations. Contours were determined using the method of Kamb (1959) and are shown at intervals of one times uniform density.

determined separately using the Hencky and velocity-gradient methods, respectively (see Appendix B of Brandon (1995) for details about methods).

The measurements in Table 4 and Fig. 7 have been adjusted for the effects of broad folding, probably late Cenozoic in age. Apatite fission-track cooling ages (Dumitru 1989) indicate that

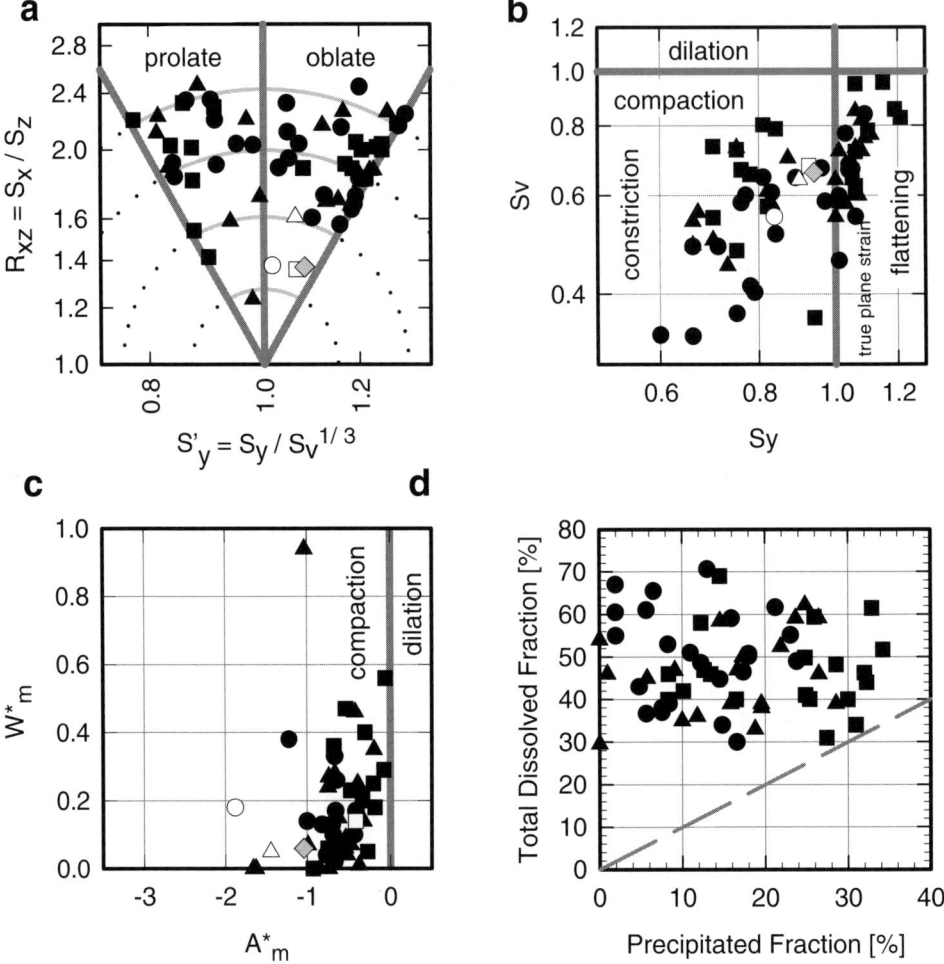

**Fig. 8.** Graphical summary of deformation data. Plots symbols are: squares for Yolla Bolly Mountain study area, triangles for Leech Lake Mountain, and circles for Diablo Range. Open symbols show averages for each study area. Grey diamond shows grand mean for all data. (**a**) Nadai plot showing strain symmetry and deviatoric strain magnitude (Brandon 1995). Circular contours mark 0.2 increments of octahedral shear strain, a measure of deviatoric strain magnitude. The data indicate both oblate and prolate symmetries. (**b**) Strain type, as indicated by the $S_V - S_Y$ plot of Brandon (1995). Note that the individual measurements fall mainly in the constrictional field, but the tensor averages plot near the plane strain boundary. (**c**) Plot of the kinematic vorticity number versus the kinematic dilatancy number. (**d**) Total percentage of dissolved grain mass versus percentage precipitated as fibre overgrowths. The difference between these two measures indicates the amount of grain mass removed from the rock.

the sampled rocks were at less than c. 3 km depth during much of Cenozoic time. We infer that, at shallow depths, Cenozoic folding would have been accommodated by flexural slip on discrete surfaces. Thus, sandstone units would be expected to have rotated in a rigid fashion as limb dip increased. To restore the effects of this young and relatively modest deformation, we have estimated the geometry and limb angles associated with regional-scale folds in our study areas. The measured principal directions for each sample were then rotated about the fold axis and by the amount of local limb dip. For the Yolla Bolly and Leech Lake areas, the average fold axis of broad regional antiforms and synforms (Fig. 4) has a trend and plunge of 120° and 20°. Restoration of limb dips required between 10° and 30° of rotation around this fold axis. The overall structure of the Diablo Range is more complicated, with a number of variably oriented antiforms and synforms which mainly trend 140–160° with a plunge of 10–20°. In rare cases, the folds trend 100–110° with plunges of 10–20°. Restoration of limb dips involve 10–40° rotation about these fold axes. The corrected data are designated as 'unfolded' in Tables 4 and 5, whereas the original, uncorrected data are designated as 'present coordinates' (Tables 3 and 5). In most cases, the 'unfolding' correction produces only a modest change in the orientation of the individual measurements. Tensor averages for each study area (Table 5) are also little affected by the 'unfolding' corrections.

*Finite strain*

*Yolla Bolly Mountains.* The field orientations of the measured finite-strain axes show scattered orientations (Fig. 4, Table 3). In the Yolla Bolly terrane, cleavage typically has a moderate dip and a variable strike. $X$ generally plunges gently within the cleavage plane. Stereograms of 'unfolded' principal directions (Fig. 7) reveal a well-defined subvertical maximum for Z and a subhorizontal girdle for $X$ with maxima plunging gently to the NE and SW. The maxima in the stereograms have similar orientations to the principal directions of the average strain tensor (Table 5), which are shown as open squares in Fig. 7.

Modes indicate that the sandstones are made up of 3–34% fibre by volume. $S_X$ shows a relatively wide scatter, ranging from 1.06 to 1.52. $S_Z$ ranges from 0.53 to 0.77, and $S_Y$ ranges from 0.70 to 1.19 with the data split evenly between constrictional ($S_Y < 1$) and flattening ($S_Y > 1$) strain types (Fig. 8b). The tensor average ($S_X:S_Y:S_Z = 1.02:0.96:0.75$, Table 5) indicates little shortening in $X$ and $Y$ at the regional scale. What may seem odd about this result is that the tensor average for $S_X$ falls outside the range of individual $S_X$ values. The reason is that the individual strain tensors do not share the same principal directions. In particular, because $X$ and $Y$ are variable at the local scale, local variations in $S_X$ and $S_Y$ tend to be averaged out, which accounts for why the tensor averages for $S_X$ and $S_Y$ are c. 1.

The principal directions for sandstones from the Valentine Springs Formation of the Pickett Peak terrane are similar to those from the Yolla Bolly terrane (Fig. 4, Tables 3 and 4), but the tensor average ($S_X:S_Y:S_Z = 1.16:0.86:0.53$, Table 5) indicates smaller $S_Y$ and $S_Z$ values and a slightly larger $S_X$ value.

*Leech Lake Mountain.* The field orientations of the finite-strain axes and the orientation of cleavage are similar to those in the Yolla Bolly Mountains (Fig. 5, Table 3). The only difference is that the average $X$ is horizontal in a NW–SE direction, as indicated by the tensor average and the stereograms (Table 5, Fig. 7). Fibre modes for individual samples are between 0% and 29%, with three samples lacking fibres at all. The absolute stretches range from 1.00 to 1.36 for $S_X$, 0.66 to 1.09 for $S_Y$, and 0.50 to 0.81 for $S_Z$. The tensor average is $S_X:S_Y:S_Z = 1.06:0.91:0.66$ (Table 5).

*Diablo Range.* The field orientations of the finite-strain axes and the cleavage attitudes resemble those in the Yolla Bolly Mountains and at Leech Lake Mountain (Fig. 6, Table 3). This area is similar to the Yolla Bolly Mountains in that the average $X$ is horizontal in a NE–SW direction (Table 5, Fig. 7). Fibre modes for individual samples are between 2% and 23%. Absolute stretches for $S_X$ range from 1.01 to 1.30. $S_Y$ is once again both constrictional and flattening (Fig. 8b) with values ranging from 0.60 to 1.09. $S_Z$ ranges from 0.45 to 0.70. The tensor average gives principal stretches of $S_X:S_Y:S_Z = 0.95:0.83:0.69$.

*All areas combined.* The tensor averages indicate that SMT deformation at the regional scale was similar in all three study areas. This result is remarkable given that the study areas are spaced over a distance of c. 500 km along the Franciscan margin. The grand tensor average for all strain measurements gives $S_X:S_Y:S_Z = 0.96:0.92:0.70$ (Table 5), which indicates contraction in Z but almost no strain in $X$ and $Y$. This deformation is best viewed as plane strain uniaxial shortening, with the shortening in Z balanced by a comparable mass-loss volume strain. The average Z direction is subvertical and indicates moderate vertical thinning of the wedge.

*Volume strain*

All Eastern Belt sandstones show significant compactional volume strains, ranging from 4% to 66% with an average of 38% (Table 5). We reiterate that these measurements represent mass-loss volume strains, because they are defined by the difference between the loss of mass from the detrital grains and the precipitation of new mass in the fibre overgrowths (Fig. 8d). As noted above, other sources of volume strain, such as changes in porosity or average mineral density, are considered to have been unimportant during SMT deformation and to have had no significant influence on our PDS and mode measurements. Our modal measurements provide a complete inventory of the volume of grains, selvage, and fibre in each sample. Thus, we contend that the missing mass has moved beyond the scale of an individual rock sample.

At the outcrop scale, there is no evidence of a sink for this missing mass. Modal measurements performed at Leech Lake Mountain along line traverses across outcrops show that veins make up no more than 3% by volume of a typical outcrop with the average about 1%. We conclude that SMT deformation was accomplished by a relatively large flux of fluid that was able to dissolve the grains and transport the dissolved load over great distances (several kilometres and perhaps more). Work in progress in the Central and Coastal Belts of the Franciscan indicates that the missing mass was not precipitated there. This leaves us with two options: either the missing mass was precipitated at a deeper level within the wedge, or it left the wedge entirely, perhaps carried by fluids that were vented on the sea floor or subducted into the mantle.

SMT deformation included both closed and open exchange involving local precipitation of fibre overgrowth and wholesale loss of mass from the rock (Fig. 8d). The open-system exchange was apparently controlled by dissolution and removal of the more soluble components of the rock. In contrast, the closed-system exchange was controlled by grain-scale transport of the relatively insoluble components of the rock. The presence of relatively Al-rich phases in the fibre overgrowths, such as phengite, chlorite, and lawsonite, is consistent with the very low solubility of Al species in a normal metamorphic fluid. Thus, the growth of the extensional fibres is probably well approximated by the closed-system Coble creep mechanism (Elliott 1973), whereas the mass-loss volume strain must be related to the dissolution and transport properties of an advecting fluid.

*Internal rotation*

For individual samples, the measured internal rotation axis is commonly subparallel to $Y$ (Table 3). The rotation axes vary considerably at the local scale and show no systematic patterns in map view. Their distribution is similar to that observed for the $Y$ directions. Internal rotation angles are generally small, with $\Omega_i < 10°$ in all but two samples. Individual measurements of $W^*_m$ are generally <0.4 (Table 3; Fig. 8c).

The tensor-averaged internal rotation axes (Table 5; centre stereograms in Fig. 7) show considerable variation between the study areas. Furthermore, there is no longer any parallelism between the rotation axes and $Y$. These results might be due to the fact that the average internal rotation angles and kinematic vorticity numbers for the study areas are very small, with $\Omega_i < 3°$ and $W^*_m < 0.26$ (Table 5). The main conclusion is that SMT deformation in the Eastern Belt sandstones was close to coaxial, both at the local and regional scale.

## Discussion

Our study provides only a preliminary view of deformation in the Franciscan wedge. None the less, it challenges many first-order predictions about deformation in accretionary wedges. Cloos & Shreve (1988) argued that subduction-related accretionary wedges are dominated by a strongly shearing flow, but our results show only low-magnitude coaxial deformation. Almost all wedge models assume isochoric deformation, which forces a balance between shortening and extension, but our results indicate that mass-loss volume strains are an integral part of the deformation. It is commonly inferred that dislocation glide becomes important in wedges that are thicker than about 15 km (Davis *et al.* 1983; Pavlis & Bruhn 1983) but our observations indicate that the SMT mechanism remained the dominant ductile deformation mechanism to depths of 25–30 km in the Franciscan wedge. Wakabayashi (1992) has argued that margin-parallel linear fabrics in the Franciscan Complex might have been produced by oblique convergence but our results indicate that when averaged at the regional scale, margin-parallel ductile strains are close to zero.

Issues that remain unresolved in our study include the role of shear partitioning, either as spin of beds or blocks or as enhanced noncoaxial deformation in mudstone-rich units. The sandstone units that we have sampled show no evidence of large spin-induced rotations (i.e. rigid rotations). We have no quantitative information about deformation in mudstone lithologies, which occur as interbeds in turbidite sequences and in local mélange units. We note, however, that mudstones typically make up less than 40% of the Eastern Belt, and are even less common in the Yolla Bolly terrane, which is dominated in many places by massive sandstones. Mudstones of the Eastern Belt do not show any unusually strong fabrics, which suggests that the degree of deformation is not greatly different from that measured in the sandstones. Furthermore, the mudstones do not appear to be unusual in composition, which might be the case if the mass dissolved from the sandstones was precipitated in adjacent mudstone units. These issues are being addressed by work in progress, but nothing has been found yet to suggest that our preliminary inferences here are incorrect.

## Ductile fabrics at Del Puerto Canyon

We have already outlined the proposal of Platt (1986) that the Eastern Belt was exhumed by normal slip on the Coast Range fault zone. Subsequent work has indicated that much of what we now see as the Coast Range fault zone formed during Cenozoic time above an eastward-moving tectonic wedge or triangle zone (Wentworth et al. 1984; Glen 1990; Unruh et al. 1995; Wakabayashi & Unruh 1995).

Harms et al. (1992) proposed that the Del Puerto area (Fig. 6) might preserve an older record of Coast Range faulting because the gentle dip of the fault zone at Del Puerto indicated that it was little affected by Cenozoic tectonic wedging. Harms et al. (1992) reported the presence of a 'mylonitic lineation' in Yolla Bolly terrane sandstones exposed directly below the Coast Range fault zone at Del Puerto Canyon (Fig. 6). They argued that an anastomosing SMT cleavage present in the sandstone was analogous to S–C fabrics typically found in high-temperature metamorphic rocks (Berthé et al. 1979). The direction and sense of shear inferred from this fabric were variable (fig. 3 in Harms et al. 1992): out of a total of 22 measurements, 13 were top-east, seven top-west, and two top-north. The predominance of top-east indicators was taken as evidence for a normal sense of shear within the easternmost Franciscan Complex and, by inference, across the Coast Range fault zone as well. This interpretation predicts strong non-coaxial ductile deformation within the easternmost Franciscan Complex.

We were not able to directly sample the S–C fabrics of Harms et al. (1992) because of private property restrictions. None the less, some of our samples are from Del Puerto Canyon (Fig. 6) and at least one is adjacent to or within the proposed ductile shear zone. We did not see any mylonitic rocks in the field, and low strain magnitudes indicate that none of our samples from Del Puerto Canyon and adjacent areas have a mylonitic fabric (see Nos 41–56 in Table 3). The strain results do indicate large local variations in $X$ in the Del Puerto area (Fig. 6). Internal rotation axes are also variable in orientation, but ductile deformation appears to have been approximately coaxial ($W^*_m < 0.35$). Linear fabrics are present at the local scale, but our tensor-averaged results indicate that there was little to no extension associated with regional-scale ductile deformation (see Yolla Bolly terrane, Mount Hamilton area, in Table 5). Deformation measurements for rocks containing the S–C fabrics of Harms et al. (1992) are needed before this problem can be fully resolved. However, at present, our deformation measurements are difficult to reconcile with the interpretation of a normal-sense ductile shear zone within the easternmost Franciscan Complex.

## Viscous versus Coulomb wedges

Our results do not provide a definitive test of Platt's (1986) viscous wedge model but they do help to characterize the nature and magnitude of ductile flow within the Franciscan wedge. We have shown that strain magnitudes were low, despite the relatively deep accretion and long residence time of the Eastern Belt rocks. The largest strains are in the $Z$ direction and indicate an average of c. 30% shortening. Assuming a coaxial deformation and residence times of 60–30 Ma, these shortening strains would require average strain rates of $(2 \text{ to } 4) \times 10^{-16} \text{ s}^{-1}$, which is 2–25 times slower than the $10^{-14} \text{ s}^{-1}$ strain rates anticipated by Platt (1986). This comparison is crude, and ignores the fact that strain rates are clearly not uniform throughout the wedge. None the less, the point we want to emphasize is that the magnitudes of SMT strains in the Franciscan wedge are fairly small. From this we infer that the stability of the Franciscan wedge is probably better represented by the Coulomb wedge criterion than by a viscous wedge criterion as advocated by Platt (1986). We infer that the SMT mechanism operated as a background deformation process. Ductile strains were fast enough and residence times long enough for the rocks to develop a fabric, but they were probably too slow to significantly influence the stability of the wedge.

Platt (1986) argued that underplating could drive the upper part of a viscous Franciscan wedge into extensional failure which would allow syn-convergent tectonic exhumation. This interpretation, however, is probably only tenable for a viscous wedge. Dahlen (1984) showed that accretion alone was not capable of driving a convergent Coulomb wedge from a thrust-dominated taper (Region I in the taper stability plot of Dahlen (1984)) to a taper where normal faults are active (Regions II and III of Dahlen (1984)). Such a transition would require a significant decrease in dip and/or strength of the décollement. Krueger & Jones (1989) and Harms et al. (1992) have argued that the transition to a shallowly subducting slab during the Laramide orogeny might have triggered extensional failure of the Franciscan wedge. The décollement dip for modern accretionary wedges is generally <8° (Davis et al. 1983). Thus, the amount of rotation of the wedge caused by shallowing of the Laramide slab was probably <8°. This amount of rotation would not be sufficient to cause extensional failure in a Coulomb wedge unless the décollement was very weak (Dahlen 1984). A weak décollement might result from high pore fluids at the base of the wedge or rheological changes, such as the thermal activation of dislocation-controlled ductile flow (Pavlis & Bruhn 1983).

## Exhumation by ductile thinning

Ductile deformation may not have been an important factor in controlling the stability of the Franciscan wedge, but ductile thinning did contribute to exhumation. Feehan & Brandon (1999) introduced a one-dimensional steady-state model to estimate the contribution of ductile deformation to exhumation of a generic accretionary wedge (Fig. 9). The model tracks the vertical progress of a material point through a steady-state wedge from its site of initial accretion at depth $z_b$ to its exhumation at the surface of the wedge at $z = 0$. Ductile thinning is defined as the change in thickness of the overburden above a material point because of ductile deformation within the overburden. To estimate the contribution of ductile thinning to exhumation, it is necessary to specify $L(z)$, which is the velocity-gradient tensor as a function of depth. Ductile SMT deformation is considered to be the primary deformation mechanism that operated within the Franciscan wedge. Based on the simple coaxial fabrics observed in our samples, we infer that the principal strain rates for SMT deformation remained constant in direction and polarity throughout the wedge. To mimic this condition, $L(z)$ is scaled so that its components increase linearly with depth. As discussed by Feehan & Brandon (1999), a rheologically based rate law would be preferable but one is not available. A depth-proportional relationship provides a reasonable first-order approximation given that silica solubility, mean effective stress, and deviatoric stress are all expected to increase with depth.

The kinematic model assumes a steady-state wedge, where the rates of basal accretion $\dot{a}$, ductile deformation $L(z)$, and near-surface exhumation caused by erosion $\dot{\varepsilon}$ and shallow normal faulting $\dot{\eta}$ all remain constant with time (Fig. 9). If the steady-state assumption were correct, then all material points within the overburden would have experienced the same deformation history as the material point being tracked. We can solve for specific values for $L(z)$ given the fact that integration of $L(z)$ along the material path through the wedge should give the finite strain and internal rotation measured in exhumed rocks. The EXHUME program (Feehan & Brandon 1999) was used to calculate results reported here.

It is difficult to evaluate if and when the Franciscan wedge achieved a steady-state configuration, defined by an accretionary flux into the wedge equal to the erosional flux out of the

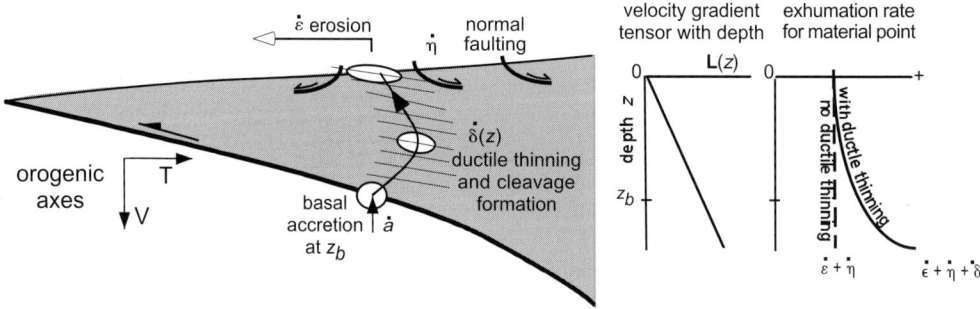

**Fig. 9.** Exhumation model of Feehan & Brandon (1999) illustrating the relationship between basal accretion, ductile flow, and exhumation in a steady-state wedge. A particle enters the base of the wedge at $z_b$, where it starts to deform in a ductile manner as it moves upward through the wedge. The rate of basal accretion is indicated by $\dot{a}$, and the deformation rate along the exhumation path is represented by the velocity gradient tensor $L(z)$, which is assumed to increase linearly with $z$. The rate of exhumation, which changes along the exhumation path, is equal to the sum of the rate of ductile thinning of the overlying cover $\dot{\delta}(z)$, and rate of thinning by shallow normal faulting $\dot{\eta}$, and the erosion rate $\dot{\varepsilon}$. The model assumes a steady-state wedge, where accretion, ductile flow, erosion and normal faulting remain constant with time. Orogenic axes $T$ and $V$ indicate the across-strike and vertical directions, respectively.

forearc high. The general inference is that an accretionary wedge will gradually approach a steady-state configuration as long as the rate of accretion and the internal state of the wedge remain fairly constant with time. As an example, Brandon et al. (1998) estimated that about 20–50 Ma are needed to achieve a steady-state configuration at the Cascadia accretionary wedge.

Our strain results indicate that the horizontal strain rate through the Franciscan wedge was probably contractional throughout the entire wedge. In this case, the only way that the wedge could have reached a steady state would have been by emergence and erosion of the forearc high. Sedimentological data indicate strong uplift on the west side of the Great Valley basin starting at 84–75 Ma (Moxon & Graham 1987; Moxon 1988). We have already summarized evidence that uplift and exhumation in the Franciscan forearc high had brought the Eastern Belt to the surface by Paleocene time (66–55 Ma) (e.g. Berkland 1973). Therefore, we infer that the Franciscan forearc high became emergent sometime in latest Cretaceous time, but no earlier than c. 84 Ma. By then, the accretionary wedge had been growing for some 75 Ma. If the wedge was exhumed mainly by erosion, then it seems likely that it reached a steady state during latest Cretaceous and early Tertiary time. To be consistent with our steady-state assumption, we limit our analysis here to the Yolla Bolly terrane, which was accreted and exhumed during this period of time.

To solve for the best-fit parameters, the EXHUME program iterates through a series of guesses to find parameters that give the observed amount of exhumation and finite strain (see Feehan & Brandon (1999) for details of best-fit calculation). Residence time and depth of accretion are taken from Table 1. The tensor averages in Table 5 are used to represent the ductile deformation. The results of the calculation (Table 6; Fig. 10) indicate that ductile strain contributed about 8–13% to the overall exhumation. This low value reflects the fact that ductile strains are fastest at depth as prescribed by the depth-proportional relationship for $L(z)$. The contribution would be even less if $L(z)$ increased nonlinearly with depth. Conversely, a constant $L(z)$ would give the largest contribution, about 12–19%, but such a constant deformation rate with depth seems unlikely.

Thus, we conclude that ductile thinning accounts for only about c. 3 km of the exhumation of the Eastern Belt. The remaining c. 24 km must have been accommodated by some other process such as normal faulting or erosion. We have already noted that there is no direct evidence for normal faulting in the Franciscan Complex. Thus, we suggest that erosion might have been the dominant exhumation process. The predicted rates, c. 0.4–0.8 km Ma$^{-1}$, are not unusual for erosion of an emergent forearc high. For instance, Brandon & Vance (1992) and Brandon et al. (1998) estimated that the long-term erosion rate for the Olympic Mountains segment of the Cascadia forearc high ranged from 0.3 to 0.75 km Ma$^{-1}$. By analogy to the Olympic Mountains, we envision that most of the sediments eroded from the Franciscan forearc high were transported to the west and redeposited in trench-slope basins and in the trench itself (Ring & Brandon 1994). This would account for why there is so little evidence in the Great Valley basin for deep erosion of the forearc high (Wakabayashi & Unruh 1995).

The model calculation also provides vertically averaged ductile strain rates parallel to the principal directions of the Franciscan wedge ($T$, $P$, and $V$ in Table 7). We can use these values to make a rough estimate of how much plate

**Table 6.** *Results of exhumation calculations for the Yolla Bolly terrane using a depth-dependent rate for ductile deformation*

| Study areas | Rates (km Ma$^{-1}$) | | | Integrated values | | |
|---|---|---|---|---|---|---|
| | $\dot{a}$ | $\dot{\delta}(z_b)$ | $\dot{\varepsilon} + \dot{\eta}$ | $\delta$ (km) | $\varepsilon + \eta$ (km) | $\dfrac{\delta}{(\delta + \varepsilon + \eta)}$ (%) |
| Yolla Bolly Mountains | 0.54 | 0.12 | 0.42 | 2.2 | 23.8 | 8 |
| Leech Lake Mountain | 0.59 | 0.20 | 0.39 | 3.5 | 22.5 | 13 |
| Diablo Range | 1.10 | 0.33 | 0.73 | 3.3 | 24.7 | 12 |

$\dot{a}$, basal accretion rate; $\dot{\delta}(z_b)$, vertical ductile thinning rate for a point at depth $z_b$, the base of the wedge; $\dot{\varepsilon} + \dot{\eta}$, combined exhumtion rates for erosion $\dot{\varepsilon}$ and normal faulting $\dot{\eta}$; $\delta$, amount of exhumation caused by vertical ductile thinning; $\varepsilon + \eta$, amount of exhumation caused by erosion $\varepsilon$ and shallow normal faulting $\eta$; $\delta + \varepsilon + \eta$, total exhumation; $\delta/(\delta + \varepsilon + \eta)$, relative contribution of ductile thinning to exhumation.

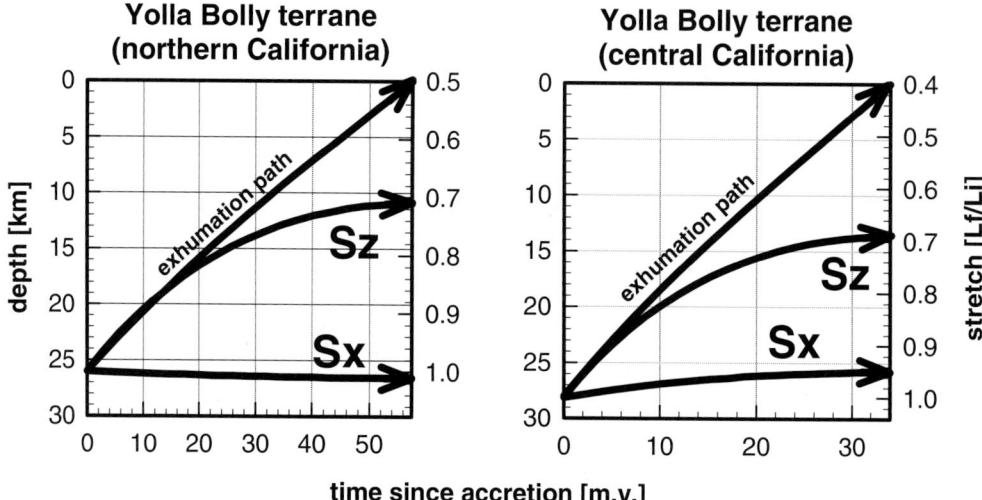

**Fig. 10.** Strain and exhumation paths estimated for the Yolla Bolly terrane. For this calculation, the rate of ductile deformation was assumed to increase linearly with depth. See Tables 6 and 7 for numerical results.

**Table 7.** *Strains and vertically averaged strains rates for SMT deformation in the Yolla Bolly terrane*

| Unit and study area | Across-strike ($T$) | Parallel-to-strike ($P$) | Vertical ($V$) |
|---|---|---|---|
| Yolla Bolly terrane, Yolla Bolly Mtns | +0% <br> +4 × 10$^{-18}$ s$^{-1}$ | −6% <br> −4 × 10$^{-17}$ s$^{-1}$ | −23% <br> −2 × 10$^{-16}$ s$^{-1}$ |
| Yolla Bolly terrane, Leech Lake Mtn | −6% <br> −4 × 10$^{-17}$ s$^{-1}$ | +2% <br> +2 × 10$^{-17}$ s$^{-1}$ | −34% <br> −2 × 10$^{-16}$ s$^{-1}$ |
| Yolla Bolly terrane, Diablo Range | −8% <br> −8 × 10$^{-17}$ s$^{-1}$ | −15% <br> −2 × 10$^{-16}$ s$^{-1}$ | −31% <br> −4 × 10$^{-16}$ s$^{-1}$ |

Strains and strain rates were calculated using averages from Table 5 for 'unfolded' data and assuming a depth-dependent deformation rate. Maximum strain rates would be twice the vertically averaged rates given here. Average strike of Franciscan wedge is about 155°.

convergence was accommodated by ductile deformation within the Franciscan wedge. The average across-strike strain rate indicates a horizontal shortening rate <0.25% Ma$^{-1}$ ($T$ in Table 7). This upper bound probably holds for the more outboard part of the wedge, which was thinner and cooler, and therefore should have had even lower rates of ductile shortening. The width of the actively deforming accretionary wedge was probably <200 km. Thus, the rate of convergence accommodated by within-wedge ductile flow was probably <0.5 km Ma$^{-1}$. Convergence rates at the Franciscan margin were about 100 km Ma$^{-1}$ (Engebretson *et al.* 1985), which means that *c.* 6000 km of oceanic crust would have been subducted in 60 Ma. Our estimates suggest that <30 km of this convergence was accommodated by within-wedge ductile deformation. From this we infer that the wedge remained largely decoupled from the subducting plate. The fact that rocks within the wedge were able to form significant ductile fabrics is due to their long residence time in the wedge.

## Conclusions

This paper presents the first comprehensive study of ductile deformation in the Franciscan wedge. The Franciscan Complex is often considered to be the prototypical sediment-rich accretionary wedge and therefore our results should have broad implications for other subduction-related wedges. Our major conclusions are as follows:

(1) Ductile deformation was slow within the Franciscan wedge, probably because of the low thermal gradients that typify this setting. SMT was the most active ductile mechanism and

appears to have remained so to depths of 25 and 30 km. Even so, the slow rates for SMT deformation indicate that it operated as a background deformation process and probably had little influence on the stability of the wedge.

(2) At the regional scale, SMT deformation was characterized by uniaxial vertical shortening with little to no strain in the horizontal. Vertical shortening was accommodated by a pervasive mass-loss volume strain, averaging about 38%. There is no evidence of where the dissolved mass went, so we are forced to consider large-scale transport by an advecting fluid. Unlike many other strain studies, our methods have allowed us to measure absolute strains associated with SMT deformation. As a result, our estimates of mass-loss volume strains are based on direct measurements and cannot be easily dismissed. The reliability of our results are supported by their consistency over such a large region of the Eastern Belt, and by the fact that tensor averages for each study area indicate a deformation that is consistent with plane strain across the Franciscan wedge.

(3) Within-wedge ductile deformation was nearly coaxial at the regional scale. This result poses a serious challenge for geodynamic models that postulate strongly shearing flows within the wedge or for tectonic models that advocate normal-sense ductile shear zones along the eastern side of the Franciscan wedge.

(4) Ductile thinning can account for about 10% of the exhumation of the Eastern Belt rocks. The remaining $c.$ 24 km of exhumation is attributed to erosion of an emergent forearc high. We show that the inferred rates of erosion, between 0.4 and 0.8 km/ $Ma^{-1}$, are reasonable when compared with modern erosion rates for the Cascadia forearc high.

This work was supported by National Science Foundation grants EAR9005777, EAR9305367, and INT9513911 to M.T.B., and a Feodor-Lynen fellowship from the Alexander von Humboldt-Stiftung and grant Ri538/8–1 from the Deutsche Forschungsgemeinschaft to U.W. We thank R. Bolhar in particular for letting us use data on Leech Lake Mountain from his Diploma thesis. J. Wakabayashi kindly provided invaluable information about various details of Franciscan geology. We thank the two anonymous referees and associate editor S. Willett for thoughtful reviews.

## References

BAILEY, E. H., IRWIN, W. P. & JONES, D. L. 1964. Franciscan and related rocks, and their significance in the geology of western California. *California Division of Mines and Geology Bulletin*, **183**, 1–177.

BAUDER, J. M. & LIOU, J. G. 1979. Tectonic outlier of Great Valley sequence in Franciscan terrane, Diablo Range, California. *Geological Society of America Bulletin*, **90**, 561–568.

BERKLAND, J. O. 1973. Rice Valley Outlier, new sequence of Cretaceous–Paleocene strata in northern Coast Ranges, California. *Geological Society of America Bulletin*, **84**, 2389–2405.

——, RAYMOND, L. A., KRAMER, J. C., MOORES, E. M. & O'DAY, M. 1972. What is Franciscan? *Bulletin, American Association of Petroleum Geologists*, **56**, 2295–2302.

BERTHÉ, D., CHOUKROUNE, P. & JEGUZO, P. 1979. Orthogneiss, mylonite and non-coaxial deformation of granites. *Journal of Structural Geology*, **1**, 31–42.

BLAKE, M. C., JR, JAYKO, A. S., MCLAUGHLIN, R. J. & UNDERWOOD, M. B. 1988. Metamorphic and tectonic evolution of the Franciscan Complex, northern California. In: ERNST, W. G. (ed.) *Metamorphism and Crustal Evolution of the Western United States*. Prentice–Hall, Englewood Cliffs, NJ, 1036–1059.

BOLHAR, R. 1997. *Absolute strain analysis of metasediments from the Franciscan subduction complex, Californian Coast Ranges*. Diploma thesis, Universität Mainz.

BRANDON, M. T. 1995. Analysis of geologic strain data in strain–magnitude space. *Journal of Structural Geology*, **17**, 1375–1385.

—— & VANCE, J. A. 1992. Tectonic evolution of the Cenozoic Olympic subduction complex, Washington State, as deduced from fission track ages for detrital zircons. *American Journal of Science*, **292**, 565–636.

——, COWAN, D. S. & FEEHAN, J. G. 1994. Fault-zone structures and solution-mass-transfer cleavage in Late Cretaceous nappes, San Juan Islands, Washington. In: SWANSON, D. A. & HAUGERUD, R. A. (eds) *Geologic Field Trips in the Pacific Northwest*. Geological Society of America Annual Meeting, Seattle, WA, 1994, 2L, 1–19.

——, RODEN-TICE, M. K. & GARVER, J. I. 1998. Late Cenozoic exhumation of the Cascadia accretionary wedge in the Olympic Mountains, NW Washington State. *Geological Society of America Bulletin*, **110**, 985–1009.

BRÖCKER, M. & DAY, H. W. 1995. Low-grade blueschist facies metamorphism of metagreywackes, Franciscan Complex, northern California. *Journal of Metamorphic Geology*, **13**, 61–78.

CHAPPLE, W. M. 1978. Mechanics of thin-skinned fold-and-thrust belts. *Geological Society of America Bulletin*, **89**, 1189–1198.

CHESTERMAN, C. W. 1963. Intrusive ultrabasic rocks and their metamorphic relationships at Leech Lake Mountain: Mendocino County, California. *California Division of Mines and Geology Bulletin*, **82**, 5–10.

CHOPIN, C. 1984. Coesite and pure pyrope in high-grade blueschists of the Western Alps; a first record and some consequences. *Contributions to Mineralogy and Petrology*, **86**, 107–118.

CLOOS, M. & SHREVE, R. 1988. Subduction-channel model of prism accretion, melange formation, sediment subduction, and subduction erosion at convergent plate margins: 2. Implications and discussion. *Pageoph*, **128**, 501–545.

COLEMAN, R. G. & LANPHERE, M. A. 1971. Distribution and age of high-grade blueschists, associated eclogites, and amphibolites from Oregon and California. *Geological Society of America Bulletin*, **82**, 2397–2412.

—— & WANG, X. 1995. *Ultrahigh Pressure Metamorphism*. Cambridge University Press, Cambridge.

COWAN, D. S. 1974. Deformation and metamorphism of the Franciscan subduction zone complex northwest of Pacheco Pass, California. *Geological Society of America Bulletin*, **85**, 1623–1634.

—— & BRUHN, R. L. 1992. Late Jurassic to early Late Cretaceous geology of the U.S. Cordillera. *In*: BURCHFIEL, B. C. et al. (eds) *The Cordilleran Orogen: Conterminous US*. Geological Society of America, Geology of North America, Boulder, CO, **G-3**, 169–203.

—— & PAGE, B. M. 1975. Recycled material in Franciscan melange west of Paso Robles, California. *Geological Society of America Bulletin*, **86**, 1089–1095.

COX, A. & HARTE, R. B. 1986. *Plate Tectonics*. Blackwell, Oxford.

DAHLEN, F. A. 1984. Noncohesive critical Coulomb wedges: an exact solution. *Journal of Geophysical Research*, **89**, 10125–10133.

DAVIS, D., SUPPE, J. & DAHLEN, F. A. 1983. Mechanics of fold-and-thrust belts and accretionary wedges. *Journal of Geophysical Research*, **88**, 1153–1172.

DEWEY, J. F. 1988. Extensional collapse of orogens. *Tectonics*, **7**, 1123–1140.

DICKINSON, W. R., INGERSOLL, R. V., COWAN, D. S., HELMOLD, K. P. & SUCZEKI, C. A. 1982. Provenance of Franciscan greywackes in coastal California. *Geological Society of America Bulletin*, **93**, 95–107.

——, OJAKANGAS, R. W. & STEWART, R. J. 1969. Burial metamorphism of the late Mesozoic Great Valley sequence, Cache Creek, California. *Geological Society of America Bulletin*, **80**, 519–526.

DUMITRU, T. A. 1989. Constraints on uplift in the Franciscan Subduction Complex from apatite fission track analysis. *Tectonics*, **8**, 197–220.

DURNEY, D. W. & RAMSEY, J. G. 1973. Incremental strains measured by syntectonic crystal growths. *In*: DEJONG, K. A. & SCHOLTEN, R. (eds) *Gravity and Tectonics*. Wiley, New York, 67–96.

EFRON, B. 1982. *The Jackknife, the Bootstrap, and other Resampling Methods*. CBMS–NSF Regional Conference Series in Applied Mathematics, Society for Industrial and Applied Mathematics, Philadelphia, PA.

ELLIOTT, D. 1973. Diffusion flow laws in metamorphic rocks. *Geological Society of America Bulletin*, **84**, 2645–2664.

ELLIS, M. A. 1986. The determination of progressive deformation histories from antitaxial syntectonic crystal fibres. *Journal of Structural Geology*, **8**, 701–710.

ENGEBRETSON, D. G., COX, A. & GORDON, R. C. 1985. Relative motions between oceanic and continental plates in the Pacific basin. Geological Society of America, Special Papers, **206**.

ENGLAND, P. C. & RICHARDSON, S. W. 1977. The influence of erosion upon the mineral facies of rock from different metamorphic environments. *Journal of the Geological Society, London*, **134**, 201–213.

ERNST, W. G. 1965. Mineral parageneses in Franciscan metamorphic rocks, Panoche Pass, California. *Geological Society of America Bulletin*, **76**, 879–914.

—— 1987. Jadeitized Franciscan metamorphic rocks of the Pacheco Pass–San Luis Reservoir area, central California Coast Ranges. *Geological Society of America Centennial Field Guide – Cordilleran Section*, **57**, 245–250.

—— 1992. Response to R.H. Grapes. *New Zealand Journal of Geology and Geophysics*, **35**, 385–387.

—— 1993. Metamorphism of Franciscan tectonostratigraphic assemblage, Pacheco Pass area, east-central Diablo Range, California Coast Ranges. *Geological Society of America Bulletin*, **105**, 618–636.

FASSOULAS, C., KILIAS, A. & MOUNTRAKIS, D. 1994. Postnappe stacking extension and exhumation of high-pressure/low-temperature rocks in the island of Crete, Greece. *Tectonics*, **13**, 127–138.

FEEHAN, J. G. & BRANDON, M. T. 1999. Contribution of ductile flow to exhumation of low $T$ – high $P$ metamorphic rocks: San Juan – Cascade Nappes, NW Washington State. *Journal of Geophysical Research*, in press.

FISHER, D. M. & BRANTLEY, S. L. 1992. Models of quartz overgrowth and vein formation: deformation and episodic fluid flow in an ancient subduction zone. *Journal of Geophysical Research*, **97**, 20043–20061.

—— & BYRNE, T. 1992. Strain variations in an ancient accretionary complex: implications for forearc evolution. *Tectonics*, **11**, 330–347.

GLEN, R. A. 1990. Formation and thrusting in some Great Valley rocks near the Franciscan Complex, California, and implications for the tectonic wedging hypothesis. *Tectonics*, **9**, 1451–1477.

HARMS, T., JAYKO, A. S. & BLAKE, M. C., JR 1992. Kinematic evidence for extensional unroofing of the Franciscan Complex along the Coast Range fault zone, northern Diablo Range, California. *Tectonics*, **11**, 228–241.

INGERSOLL, R. V. 1978. Paleogeography and paleotectonics of the late Mesozoic forearc basin of northern and central California. *In*: HOWELL, D. G. & MCDOUGALL, K. A. (eds) *Mesozoic Paleogeography of the Western United States*. Pacific Coast Paleogeography Symposium 2, Pacific Section, Society of Economic Paleontologists and Mineralogists, 471–482.

—— 1979. Evolution of the Late Cretaceous forearc basin, northern and central California. *Geological Society of America Bulletin*, **90**, 813–826.

JAYKO, A. S. & BLAKE, M. C., JR 1989. Deformation of the eastern Franciscan Belt, Northern California. *Journal of Structural Geology*, **11**, 375–390.

——, —— & BROTHERS, R. N. 1986. Blueschist metamorphism of the Eastern Franciscan belt, northern California. *In*: EVANS, B. W. & BROWN, E.

H. (eds) *Blueschists and Eclogites.* Geological Society of America, Memoirs, **164**, 107–124.

——, —— & HARMS, T. 1987. Attenuation of the Coast Range ophiolite by extensional faulting, and nature of the Coast Range 'thrust', California. *Tectonics*, **6**, 475–488.

KAMB, W. B. 1959. Ice petrofabric observations from Blue Glacier, Washington, in relation to theory and experiment. *Journal of Geophysical Research*, **64**, 1891–1909.

KRUEGER, S. & JONES, D. 1989. Extensional fault uplift of regional Franciscan blueschists due to subduction shallowing during the Laramide orogeny. *Geology*, **17**, 1157–1159.

KÜSTER, M. & STÖCKHERT, B. 1997. Density changes of fluid inclusions in high-pressure/low-temperature metamorphic rocks from Crete: a thermobarometric approach to the creep strength of the host minerals. *Lithos*, **41**, 151–167.

LANPHERE, M. B., BLAKE, M. C. & IRWIN, W. P. 1978. Early Cretaceous metamorphic age of the South Fork Mountain schist in the northern Coast Ranges of California. *American Journal of Science*, **278**, 798–815.

MALVERN, L. E. 1969. *Introduction to the Mechanics of a Continuum Medium.* Prentice–Hall, Englewood Cliffs, NJ.

MASSONE, H. J. & SCHREYER, W. 1987. Phengite geobarometry based on the limiting assemblage with K-feldspar, phlogopite and quartz. *Contributions to Mineralogy and Petrology*, **96**, 212–224.

MATTINSON, J. M. 1986. Geochronology of high pressure–low temperature Franciscan metabasites – a new approach using the U–Pb system. *In:* EVANS, B. W. & BROWN, E. H. (eds) *Blueschists and Eclogites.* Geological Society of America Memoir, **164**, 95–105.

—— 1988. Constraints on the timing of Franciscan metamorphism: geochronological approaches and their limitations. *In:* ERNST, W. G. (ed.) *Metamorphism and Crustal Evolution of the Western United States. Rubey Volume VII.* Prentice–Hall, Englewood Cliffs, NJ, 1023–1034.

—— & ECHEVIRRA, L. M. 1980. Ortigalita Peak gabbro, Franciscan complex – U–Pb ages of intrusion and high pressure–low temperature metamorphism. *Geology*, **8**, 589–593.

MCDOWELL, F., LEHMANN, D. H., GUCWA, P. R., FRITZ, D. & MAXWELL, J. C. 1984. Glaucophane schists and ophiolites of the northern California Coast Ranges: isotopic ages and their tectonic implications. *Geological Society of America Bulletin*, **95**, 1373–1382.

MEANS, W. D. 1994. Rotational quantities in homogeneous flow and the development of small-scale structure. *Journal of Structural Geology*, **16**, 437–445.

——, HOBBS, B. E., LISTER, G. S. & WILLIAMS, P. F. 1980. Vorticity and non-coaxiality in progressive deformation. *Journal of Structural Geology*, **5**, 279–286.

MOXON, I. W. 1988. Sequence stratigraphy of the Great Valley basin in the context of convergent margin tectonics. *In:* GRAHAM, S. A. & OLSON, H. C. (eds) *Studies of the geology of the San Joaquin Basin.* Field Trip Guidebook, Pacific Section, Society of Economic Paleontologists and Mineralogists, **60**, 3–28.

—— & GRAHAM, S. A. 1987. History and controls of subsidence in the Late Cretaceous–Tertiary Great Valley forearc basin, California. *Geological Society of America Bulletin*, **15**, 626–629.

NELSON, B. K. & DEPAOLO, D. J. 1985. Isotopic investigation of metasomatism in subduction zones – The Franciscan complex, California. *Geological Society of America Abstracts with Programs*, **17**(7), 674–675.

NORRIS, R. J. & BISHOP, D. G. 1990. Deformed conglomerates and textural zones in the Otago Schists, South Island, New Zealand. *Tectonophysics*, **174**, 331–349.

PAGE, B. M. & TABOR, L. T. 1967. Chaotic structure and décollement in Cenozoic rocks near Stanford University, California. *Geological Society of America Bulletin*, **78**, 1–12.

PAMPEYAN, E. 1993. *Geologic map of the Palo Alto and part of the Redwood Point 7–1/2' quadrangles, San Mateo and Santa Clara Counties, California.* US Geological Survey, Map **I-2371**, scale 1:24 000.

PASSCHIER, C. W. 1988. The use of Mohr circles to describe non-coaxial progressive deformation. *Tectonophysics*, **149**, 323–338.

—— 1991. The classification of dilatant flow types. *Journal of Structural Geology*, **13**, 101–104.

—— & TROUW, R. A. J. 1996. *Microtectonics.* Springer, Berlin.

PATERSON, S. R. & SAMPLE, J. C. 1988. The development of folds and cleavages in slate belts by underplating in accretionary complexes; a comparison of the Kodiak Formation, Alaska and the Calaveras Complex, California. *Tectonics*, **7**, 859–874.

—— & YU, H. 1994. Primary fabric ellipsoids in sandstones: implications for depositional processes and strain analysis. *Journal of Structural Geology*, **16**, 505–517.

PAVLIS, T. L. & BRUHN, R. L. 1983. Deep-seated flow as a mechanism for the uplift of broad forearc ridges and its role in the exposure of high $P/T$ metamorphic terranes. *Tectonics*, **2**, 473–497.

PETERMAN, Z. E., HEDGE, C. E., COLEMAN, R. G. & SNAVELY, P. D., JR 1967. $^{87}Sr/^{86}Sr$ ratios in some eugeosynclinal sedimentary rocks and their bearing on the origin of granitic magma in orogenic belts. *Earth and Planetary Science Letters*, **2**, 433–439.

PLATT, J. P. 1975. Metamorphic and deformational processes in the Franciscan Complex, California: some insights from the Catalina schist terrane. *Geological Society of America Bulletin*, **86**, 1337–1347.

—— 1986. Dynamics of orogenic wedges and the uplift of high-pressure metamorphic rocks. *Geological Society of America Bulletin*, **97**, 1037–1053.

—— 1987. The uplift of high-pressure–low-temperature metamorphic rocks. *Philosophical Transactions of the Royal Society of London, Series A*, **321**, 87–103.

—— 1993. Exhumation of high-pressure rocks: a review of concepts and processes. *Terra Nova*, **5**, 119–133.

PRESS, W. H., TEUKOLSKY, S. A., VETTERLING, W. T. & FLANNERY, B. P. 1992. *Numerical Recipes in Fortran: The Art of Scientific Computing*. Cambridge University Press, Cambridge.

RAMSAY, J. G. & HUBER, M. I. 1983. *The Techniques of Modern Structural Geology, Volume 1: Strain Analysis*. Academic Press, London.

RAMTHUN, A., BRANDON, M. T. & RING, U. 1997. Fabric analysis in the Ukelayet Flysch in the footwall of the Vatyna thrust zone, Kamchatka, Russia: sedimentary or tectonic fabrics? *Terra Nova*, **9**, 377.

RING, U. 1995. Horizontal contraction or horizontal extension: heterogeneous Late Eocene and Early Oligocene general shearing during blueschist- and greenschist-facies metamorphism at the Pennine–Austroalpine boundary zone in the Western Alps. *Geologische Rundschau*, **84**, 843–859.

—— 1996. *Kinematic analysis of heterogeneous brittle deformation at the Coast Range fault zone and ductile strain and mass loss in the Eastern Franciscan belt (Franciscan subduction complex, U.S.A.): implications for the exhumation of high-pressure/low-temperature metamorphic rocks*. Habilitation thesis, Universität Mainz.

—— & BRANDON, M. T. 1994. Kinematic data for the Coast Range fault zone and implications for the exhumation of the Franciscan Subduction Complex. *Geology*, **22**, 735–738.

—— & —— 1997. Kinematics of the Coast Range fault zone, California: implications for the evolution of the Franciscan subduction complex. *Geological Society of America Abstracts with Programs*, **29**(5), 60.

——, RATSCHBACHER, L., FRISCH, W., BIEHLER, D. & KRALIK, M. 1989. Kinematics of the Alpine plate margin: structural styles, strain and motion along the Penninic–Austroalpine boundary in the Swiss–Austrian Alps. *Journal of the Geological Society, London*, **146**, 835–849.

RUBIE, D, C. 1984. A thermal–tectonic model for high-pressure metamorphism and deformation in the Sesia zone, western Alps. *Journal of Geology*, **92**, 21–36.

SELVERSTONE, J. 1985. Petrologic constraints on imbrication, metamorphism, and uplift in the SW Tauern window, Eastern Alps. *Tectonics*, **4**, 687–704.

SMITH, G. W., HOWELL, D. G. & INGERSOLL, R. V. 1979. Late Cretaceous trench-slope basins of central California. *Geology*, **7**, 303–306.

SUPPE, J. 1973. *Geology of the Leech Lake Mountain–Ball Mountain Region, California: a Cross-section of the Northeastern Franciscan Belt and its Tectonic Implications*. University of California Publications in Geological Sciences, **107**.

—— & ARMSTRONG, R. L. 1972. Potassium–argon dating of Franciscan metamorphic rocks. *American Journal of Science*, **272**, 217–233.

TAGAMI, T. & DUMITRU, T. A. 1996. Provenance and thermal history of the Franciscan accretionary complex: constraints from zircon fission track thermochronology. *Journal of Geophysical Research*, **101**, 11353–11364.

THOMSON, S. N., STÖCKHERT, B. & BRIX, M. A. 1999. Miocene high-pressure metamorphic rocks of Crete, Greece: rapid exhumation by buoyancy escape. This volume.

UNRUH, J. R., LOEWEN, B. A. & MOORES, E. M. 1995. Progressive arcward contraction of a Mesozoic–Tertiary fore-arc basin, southwestern Sacramento Valley, California. *Geological Society of America Bulletin*, **107**, 38–53.

URAI, J., WILLIAMS, P. & VAN ROERMUND, H. 1991. Kinematics of crystal growth in syntectonic fibrous veins. *Journal of Structural Geology*, **13**, 823–836.

WAKABAYASHI, J. 1992. Nappes, tectonics of oblique plate convergence, and metamorphic evolution related to 140 million years of continuous subduction, Franciscan Complex, California. *Journal of Geology*, **100**, 19–40.

—— & UNRUH, J. 1995. Tectonic wedging, blueschist metamorphism, and exposure of blueschists: are they compatible? *Geology*, **23**, 85–88.

WALLIS, S., PLATT, J. & KNOTT, S. 1993. Recognition of syn-convergence extension in accretionary wedges with examples from the Calabrian Arc and the Eastern Alps. *American Journal of Science*, **293**, 463–494.

WALLIS, S. R. 1992. Vorticity analysis in a metachert from the Sanbagawa Belt, SW Japan. *Journal of Structural Geology*, **14**, 271–280.

WEINRICH, A., SHARP, W. D., RENNE, P. R. & MERTZ, D. F. 1997. Alkaline diabases in a near trench environment, Franciscan complex, California: implications from geochemical and $^{40}Ar/^{39}Ar$ data. *Terra Nova*, **9**, 352.

WENTWORTH, C. W., BLAKE, M. C., JONES, D. L., WALTER, A. W. & ZOBACK, M. D. 1984. Tectonic wedging associated with emplacement of the Franciscan assemblage, California Coast Ranges. *In*: BLAKE, M. C. (ed.) *Franciscan Geology of northern California*. Society of Economic Paleontologists and Mineralogists, Pacific Section, 163–173.

WHEELER, J. & BUTLER, R. W. H. 1994. Criteria for identifying structures related to true crustal extension in orogens. *Journal of Structural Geology*, **16**, 1023–1027.

WORRALL, D. M. 1979. *Geology of the South Yolla Bolly area, northern California and its tectonic implications*. PhD thesis, University of Texas, Austin.

—— 1981. Imbricate low-angle faulting in uppermost Franciscan rocks, South Yolla Bolly area, northern California. *Geological Society of America Bulletin*, **92**, 703–729

# Miocene high-pressure metamorphic rocks of Crete, Greece: rapid exhumation by buoyant escape

STUART N. THOMSON, BERNHARD STÖCKHERT & MANFRED R. BRIX

*Institut für Geologie, Ruhr-Universität Bochum, D-44780 Bochum, Germany*

**Abstract:** The pre-Neogene thrust sheets of Crete, Greece, accreted during Oligocene and Early Miocene time, can be divided into two main groups juxtaposed by a Miocene extensional detachment. Oligo-Miocene high pressure–low temperature (HP–LT) metamorphic rocks crop out in the lower plate to the detachment, and rocks that show no evidence of Tertiary metamorphism in the upper plate. Detailed pressure, temperature and structural information from the HP–LT metamorphic rocks combined with new fission-track data from the upper plate reveal that the lower plate rocks were subducted, then pervasively deformed and metamorphosed at their maximum depth of burial (30–35 km, 300–400°C) between c. 24 and 19 Ma and then exhumed to less than 10 km depth at rates >4 km Ma$^{-1}$ before c. 17 Ma. Microstructural studies reveal that during exhumation the lower plate acted as a coherent block, with deformation and retrograde metamorphism localized along the extensional detachment. The rocks of the upper plate can be shown to have remained in the upper 4–7 km of the crust since at least Eocene time. After cessation of movement along the main extensional detachment, both the upper and lower plates were subjected to brittle extension and increased erosion initiated at c. 16–17 Ma. These boundary conditions imply continuous subduction retreat since at least Eocene time along the Aegean segment of the Hellenic convergent plate boundary. A tectonic model is presented where exhumation is driven by positive buoyancy of the subducted continental crust following lithospheric delamination. We propose that the subducted microcontinent was exhumed by a process of 'oblique buoyant escape' and entered the space created by the retreating subducting oceanic slab.

The processes involved in the burial, formation, and the subsequent exhumation of high pressure–low temperature (HP–LT) metamorphic rocks, particularly at active convergent plate boundaries and in collision zones, are still relatively poorly understood. This study concerns itself mainly with the recognition of exhumation by extensional unroofing. This is now seen as an important contributor to the overall exhumation of many HP–LT rocks and is recognized in a number of orogenic belts. This paper presents a detailed and integrated study from the island of Crete, Greece, that combines thermochronological, structural, and metamorphic information from rocks both below and above a major extensional detachment. By combining all of these aspects, a detailed assessment of the relative role of tectonic, as opposed to erosional, exhumation processes in the overall exhumation of the HP–LT rocks is made possible. Our results provide important new constraints on the tectonic development of the Hellenic convergent plate boundary during Tertiary time, and provide an opportunity to present a new tectonic model to explain the exhumation of the HP–LT rocks of Crete during continuing subduction roll-back.

## Cretan geology

The island of Crete is situated above the present-day Hellenic Subduction Zone of the eastern Mediterranean Sea and is one of the few places in the world where very young (Miocene) HP–LT metamorphic rocks are recognized as having been juxtaposed by extensional detachment (Fassoulas *et al.* 1994; Kilias *et al.* 1994; Jolivet *et al.* 1996) against nonmetamorphic rocks within a convergence zone dominated by oceanic subduction.

From simple stratigraphic relationships, Crete has been shown to be composed of a series of thrust sheets or tectonic slices (Bonneau 1984), which were formed by collision of the northern margin of the Adria microcontinent with the southern margin of the Eurasian Plate during Oligocene and Miocene time at a convergent plate boundary dominated by subduction (Robertson *et al.* 1991). Traditionally, the rocks of Crete, as with much of Greece, the Balkans, and Turkey, have been subdivided into a series of tectonic units or 'isopic zones' largely according to rock type, stratigraphy, metamorphism, style of deformation, and probable pre-collisional palaeogeography (see Robertson *et al.* 1991). Unfortunately, these names vary considerably in the literature. To avoid ambiguity

THOMSON, S. N., STÖCKHERT, B. & BRIX, M. R. 1999. Miocene high-pressure metamorphic rocks of Crete, Greece: rapid exhumation by buoyant escape. *In:* RING, U., BRANDON, M. T., LISTER, G. S. & WILLETT, S. D. (eds) *Exhumation Processes: Normal Faulting, Ductile Flow and Erosion.* Geological Society, London, Special Publications, **154**, 87–107.

we use the most widely accepted nomenclature of Bonneau (1984). The relationship between these units is outlined in Fig. 1, along with a geological map showing their distribution.

On Crete two main rock groups can be identified: a lower group comprising rocks that show pervasive Oligocene–Miocene HP–LT metamorphism, and an upper group that shows no evidence of Tertiary metamorphism. This major metamorphic break in the sequence has long been difficult to explain (see Robertson & Dixon 1984, p. 47). However, Lister et al. (1984) proposed a model where the HP–LT metamorphic rocks of Crete were exhumed along a shallow dipping extensional detachment from under a stretching and fracturing upper plate composed mainly of unmetamorphosed sedimentary rocks and ophiolite suite rocks. More recent studies by Fassoulas et al. (1994), Kilias et al. (1994), Jolivet et al. (1996) and Thomson et al. (1998a) also concluded, albeit with an opposite sense of motion along the extensional detachment, that the two rock groups must have been juxtaposed by a major low-angle extensional detachment sometime during the Miocene. It should be pointed out, however, that convincing exposures of a major detachment are difficult to find on Crete, as it is obliterated and cut by numerous Neogene cataclastic faults, landslips and slope screes. The proposals for major crustal extension on Crete rely mainly on the combination of indirect metamorphic, geochronological and structural criteria, as outlined by Wheeler & Butler (1994), to prove the occurrence of true crustal extension.

Three main thrust sheets occur in the 'upper plate' (above the detachment fault; see Lister & Davis (1989)). These are labelled, from top to bottom, the Uppermost, Pindos, and Tripolitza units. These units represent a shallow accretionary sequence formed before and during the Oligocene to Miocene collision of the Adria microcontinent with Eurasia. The Uppermost unit is generally considered to represent a mélange or olistostrome (see Seidel et al. (1981), Hall et al. (1984) and Bonneau (1984) for variations on this theme) formed at the southern leading edge of the Eurasian Plate during northward directed subduction of the Pindos branch of the Neotethys Ocean (Robertson et al. 1991) and before the collision of Adria. It contains elements derived from both the Jurassic–Cretaceous Pindos Ocean, such as oceanic pillow basalts, gabbros, deep-water carbonates and cherts, and serpentinized mantle fragments, and the overriding Eurasian Plate, including Jurassic HP–LT metamorphic, and Late Cretaceous high temperature–low pressure (HT–LP) metamorphic and plutonic rocks. The rocks of the Pindos unit represent the accreted distal parts of the northern continental margin of the Adria microcontinent, and comprise largely deep-water Mesozoic to early Tertiary carbonates and cherts. The Tripolitza unit contains rocks of similar age that were accreted from the more proximal shallow-water platform carbonates of the colliding Adria microplate. From Palaeocene to Eocene time onwards, both the Pindos and Tripolitza units are dominated by terrigenous flysch deposits derived from the overriding Eurasian Plate (Hall et al. 1984).

Directly below the main extensional detachment HP–LT metamorphic rocks of the Phyllite–Quartzite (PQ) unit are found, and below this, representing the deepest tectonostratigraphic unit seen in Crete, is the Plattenkalk (PK) unit, which also shows evidence of an HP–LT metamorphic overprint (Seidel et al. 1982). The PQ unit contains largely Carboniferous to mid-Triassic clastic sedimentary rocks, with some local carbonates and evaporites (Krahl et al. 1983). We believe that this unit represents the older stratigraphic continuation of the carbonate Tripolitza unit, as originally proposed by Bonneau (1976). Whereas the Tripolitza unit underwent accretion at shallow depth, the PQ unit would then have continued its downward subduction, probably underthrust along a weak mid-Triassic evaporite horizon, and subsequently accreted at much greater depth. It is also likely that these evaporites, now largely seen as large gypsum bodies within the upper parts of the PQ unit, formed a weak zone allowing the development of the main extensional detachment during the subsequent exhumation of the HP–LT lower plate of Crete. The PK unit is dominated by Permian to Eocene, initially shallow-water, but mainly deep-water carbonates and cherts, with a minor Eocene to Oligocene calcareous flysch in its uppermost part (Seidel et al. 1982). These rocks are thought to represent the sedimentary cover of the southern continental margin of the Adria Microplate, and have been correlated with the carbonates of the Ionian Zone of western mainland Greece by Bonneau (1984). Carbonate breccias within the PK unit (the Tripali unit of Creutzburg & Seidel 1975) probably represent submarine fault scarp breccias formed either during an Early Jurassic phase of collapse of the PK carbonate platform (Hall et al. 1984), or during flexure of the foreland, shortly before Eocene to Oligocene collision with Eurasia. The crystalline continental basement, which must have underlain many of these accreted sedimentary units or thrust sheets, is generally not

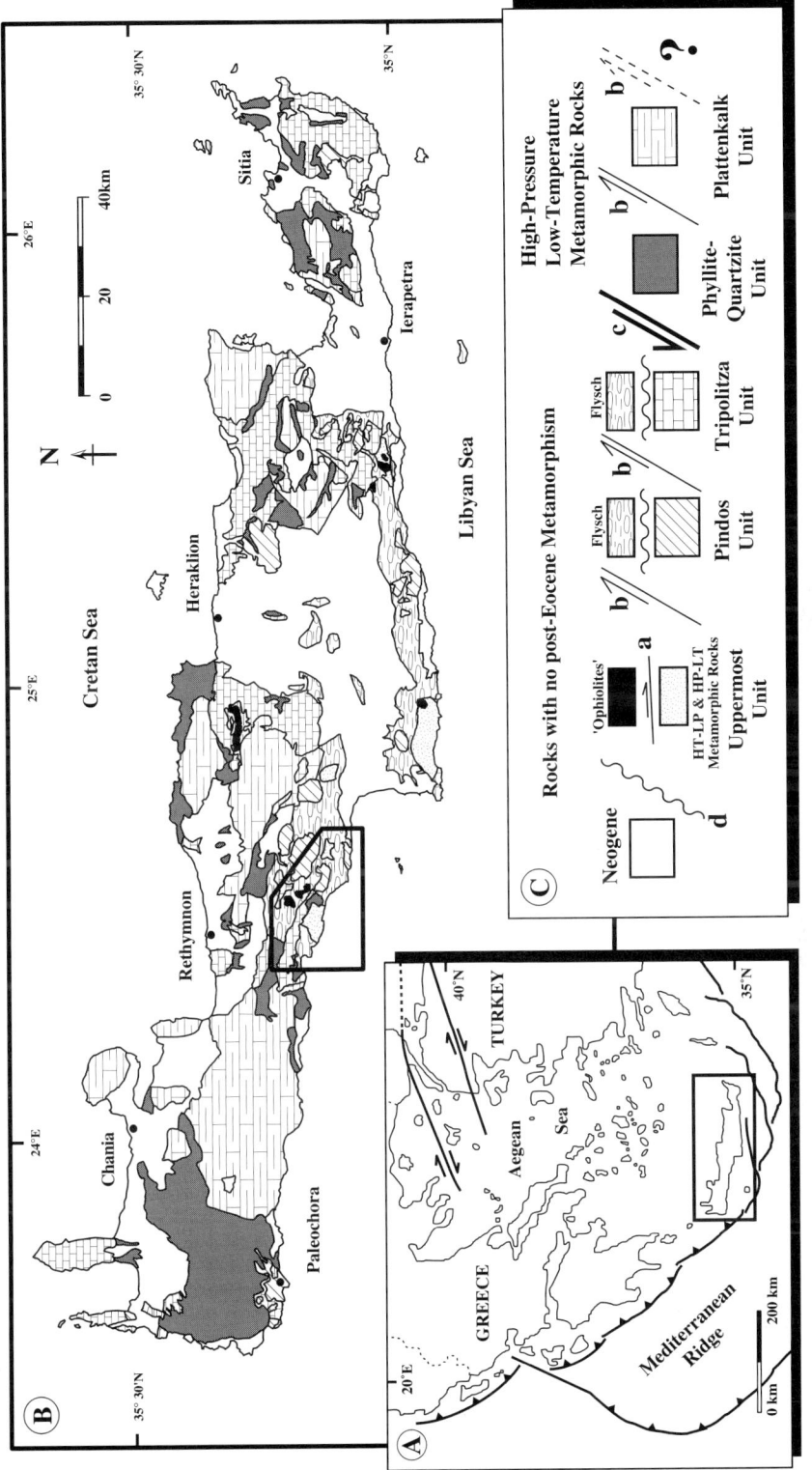

Fig. 1. (A) Location of Crete within the eastern Mediterranean. (B) Geological map of Crete, modified from Creutzburg & Seidel (1975), showing juxtaposition of rock units with and without Oligocene HP–LT metamorphism. The location of Fig. 3 (the Plakias area) is indicated by the box. (C) Key and simplified tectonostratigraphy. a, thrust of post-Cretaceous age; b, major thrust contacts of Oligocene–Miocene age; c, extensional detachment of Miocene age, recognized by Jolivet *et al.* (1996); d, Neogene sedimentary rocks of Tortonian age unconformably overlie all the tectonic units of Crete. (After Thomson *et al.* 1998*a*.)

seen in Crete, although a few small outcrops of Hercynian metamorphic rocks are exposed associated with PQ unit rocks in eastern Crete (Wachendorf *et al.* 1974; Seidel *et al.* 1982).

## Thermochronology

Wheeler & Butler (1994) pointed out that to identify exhumation of HP–LT rocks by true crustal extension using solely structural and metamorphic arguments is of limited value unless detailed information is also available on the thermal history of rocks from both above and below any extensional detachment. Therefore, in this section we present a detailed look at the differences between the thermal histories of the HP–LT lower plate and the upper plate rocks during orogenesis. This was done both to constrain the speed and timing of exhumation, and additionally to assess the relative roles of tectonic and erosional processes during exhumation. New apatite fission-track data from the upper plate of Crete are presented, along with a review and re-evaluation of previously collected isotopic age data (Seidel *et al.* 1982; Jolivet *et al.* 1996) and fission-track data (Thomson *et al.* 1998*a, b*) from both the upper and lower plates.

### Thermal history of the HP–LT lower plate

The thermal history of the HP–LT lower plate rocks of Crete has been dealt with in detail by Thomson *et al.* (1998*a*), and is summarized here and in Fig. 2. The first constraint on this plot is the age of the youngest rocks of the PK unit. These have been dated to mid-Oligocene age (*Globigerina ampliapetura* zone) by Bizon *et al.* (1976) (29.3–28.3 Ma according to the time scale of Harland *et al.* (1990)). Therefore, these rocks must have entered the subduction or collision zone more recently than *c.* 29 Ma. The PQ unit, however, from which most of the thermochronological data originate, lay, according to the palaeogeography of Bonneau (1984), at least 150 km to the north of the PK unit before collision. It thus follows that the PQ unit entered the collision zone before deposition of the youngest flysch upon the PK unit sometime before *c.* 29 Ma. Dewey *et al.* (1989) estimated a plate convergence rate of *c.* 1.7 cm/year at this time. However, Wortel *et al.* (1993), by modelling seismic tomography data of the subducted slab, revealed that because of subduction retreat or roll-back, the true subduction rate at *c.* 29 Ma was between 2 and 4 cm/year. The PQ unit rocks would, therefore, have entered the subduction or collision system some 4–7 Ma before the youngest parts of the PK unit.

The post-metamorphic cooling history of the lower plate is constrained by K–Ar (Seidel *et al.* 1982) and Ar/Ar (Jolivet *et al.* 1996) ages in white mica, and zircon and apatite fission-track data (Thomson *et al.* 1998*a*). In Fig. 2, we use a closure temperature range of 350–430°C for $^{40}$Ar in white mica, as calculated for muscovite between 0 and 10 kbar and cooling rates of 10–100°C Ma$^{-1}$ by Lister & Baldwin (1996). For the seven zircon fission-track ages (mean age of 19 ± 2 Ma; Thomson *et al.* 1998*a*) a range of 260 ± 30°C is used to represent the closure temperature for fission tracks in zircon. This temperature range encompasses values calculated for geologically realistic cooling rates (e.g. between 5 and 500°C Ma$^{-1}$) by Brandon *et al.* (1998) from experimental data. It also encompasses values estimated from calibrations using Ar/Ar studies in K-feldspar and biotite (Foster *et al.* 1996), and values higher than 255°C required by zircon fission-track data from the bottom of the KTB deep drillhole (Coyle & Wagner 1996). The *T–t* histories derived from the apatite fission-track data (Thomson *et al.* 1998*a*), which indicate rapid cooling from *c.* 120°C to 60°C at 15 ± 1 Ma, were derived using the modelling approach of Gallagher (1995), based on the apatite annealing model of Laslett *et al.* (1987).

In summary, these data reveal that the HP–LT rocks of Crete were subducted between 36 and 29 Ma, heated to peak temperatures in excess of 350°C, then rapidly cooled to below *c.* 290°C before *c.* 19 Ma and to below 60°C at *c.* 15 Ma. This implies that the PQ unit rocks were subducted, metamorphosed at high pressure, and subsequently cooled to less than 60°C within a maximum time span of *c.* 20 Ma, and more likely within *c.* 15 Ma. The PK unit rocks must have undergone the same sequence of events within an even shorter time period as low as 10 Ma.

### Thermal history of the upper plate

Most of the Tripolitza and Pindos units above the postulated extensional detachment are generally unsuitable for thermochronological studies as they are largely made up of carbonate rocks. Thus, to constrain the thermal history of the upper plate to the detachment, we have concentrated on the more lithologically suitable rocks of the Uppermost (UM) tectonic unit and parts of the Tertiary flysch that stratigraphically overlie the carbonate rocks of the Pindos unit.

Previous isotopic age data from the UM unit are limited to a number of K–Ar ages (Seidel *et al.* 1976, 1977, 1981). HT–LP metamorphic rocks and associated granitic intrusions yield K–Ar ages of between 75 and 66 Ma from both

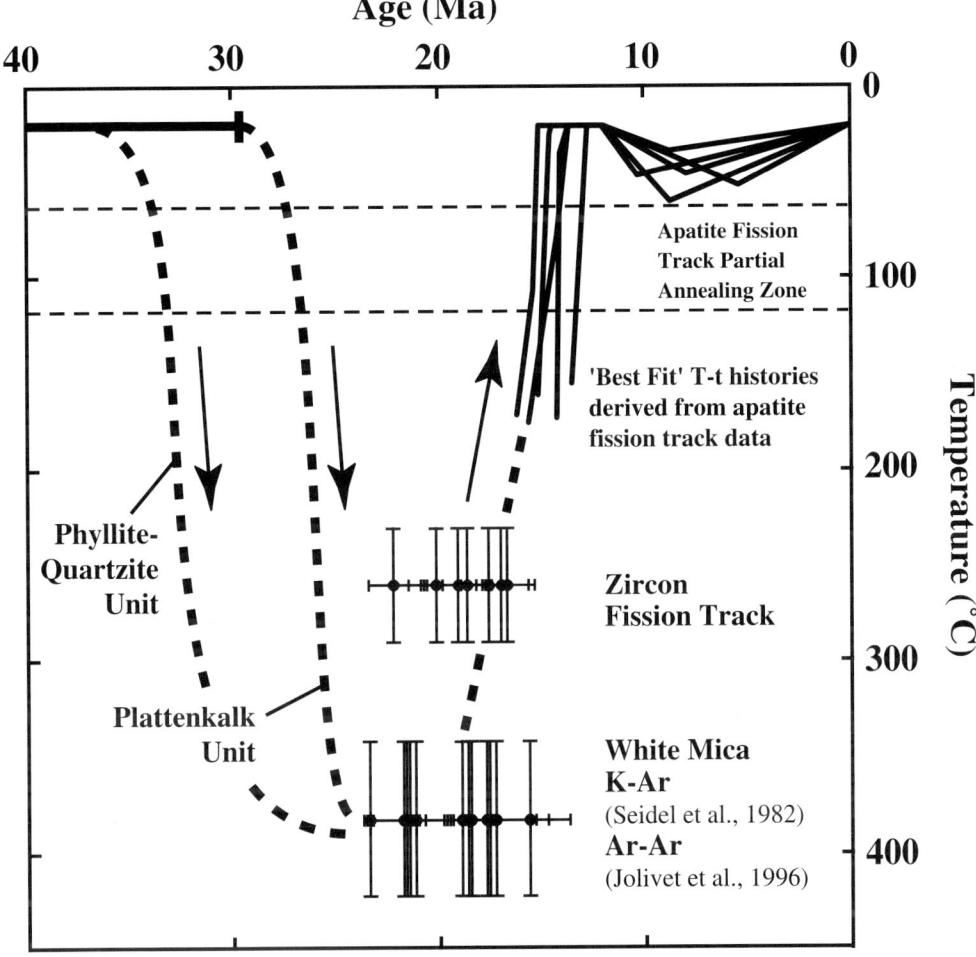

**Fig. 2.** *T–t* history of the Plattenkalk and Phyllite–Quartzite units of the lower plate of Crete (after Thomson *et al.* 1998*a*).

hornblende and biotite. The respective effective closure temperatures for Ar diffusion in these minerals are c. 500–560°C in hornblende and c. 310–360°C in biotite at pressures below 5 kbar and cooling rates of between 10°C Ma$^{-1}$ and 100°C Ma$^{-1}$ (Lister & Baldwin, 1996). This indicates a rapid cooling of these rocks during Late Cretaceous time. Seidel *et al.* (1976) also concluded that maximum conditions of metamorphism (700°C, 3–4 kbar) occurred in these rocks immediately before this cooling at c. 75 Ma. Hornblende K–Ar ages from the ophiolite bodies that form the upper part of the UM unit give, in contrast, ages between 135 and 156 Ma (Seidel *et al.* 1977). This provides a Mid-Jurassic minimum age for the formation of these ophiolites and indicates that they are genetically unrelated to the HT–LP metamorphic rocks which they now overlie. One other group of rocks from the UM unit, the HP–LT rocks of the island of Gavdos, has also been dated (Seidel *et al.* 1977). These rocks give roughly concordant K–Ar amphibole and phengite ages of between 135 and 150 Ma.

Apatite fission-track data have been collected from various components of the UM unit in the Lendas region of southern Crete (Thomson *et al.* 1998*b*). These data reveal that the UM unit rocks were emplaced upon Eocene flysch above the Pindos unit during the initial phase of collision at c. 35 Ma, then remained within the apatite fission-track partial annealing zone

(c. 120°C to 60°C; i.e. in the upper 4–7 km of the crust, dependent on the geothermal gradient) until a phase of accelerated cooling related to increased erosion occurred at c. 16–17 Ma. These data are complemented in this paper by 11 new apatite fission-track ages and seven new apatite confined track length measurements from the UM unit in the Plakias area of southern Crete (see Table 1 and map in Fig. 3). One apatite fission-track age (TH 208) was also obtained from a tuff within Tripolitza unit carbonates of the upper plate. This gives an age of 17.5 ± 2.5 Ma, revealing that although these rocks show no evidence of Tertiary metamorphism, they have experienced temperatures of at least c. 120°C since late Oligocene time.

To interpret such a mixed regional dataset, it is common practice to plot the apparent apatite fission-track age against the mean track length and standard deviation of the apatite track length distribution (Green 1986). Figure 4 shows these relationships for the data obtained from the UM unit of Crete. Gallagher & Brown (1997) have shown that the distinctive curved distribution of data seen in Fig. 4a and b, commonly referred to as a 'boomerang', can be compared to synthetic curves produced by applying an apatite annealing model to a simple cooling history. The synthetic curves shown in Fig. 4 are the data that would be expected from a simulated cooling episode shown in Fig. 5. This latter plot simulates the simple situation where samples reside at similar temperatures in the crust between 40 and 16 Ma, and are subjected to a cooling episode, where all samples cool to 20°C at the same rate, which in our case is initiated at 16 Ma, with a maximum duration of 1 Ma. The annealing model of Laslett et al. (1987) is then used to predict the fission-track age and length parameters produced by the $T$–$t$ history of each sample represented by the synthetic curves shown in Fig. 4. The predicted and observed data for length against age (Fig. 4a) fit very well. However, in the other two plots (Fig. 4b and c), the observed standard deviations, especially of the 'bimodal' or 'mixed' (Gleadow et al. 1986) samples with mean track lengths less than c. 14 μm, appear to underestimate by c. 0.5 μm those predicted by the Laslett et al. (1987) model for the cooling episode outlined in Fig. 5. The cause of this misfit is uncertain. Two possibilities are (1) that it is due to an additional bias in the measurement of short tracks larger than that built into the Laslett et al. (1987) annealing

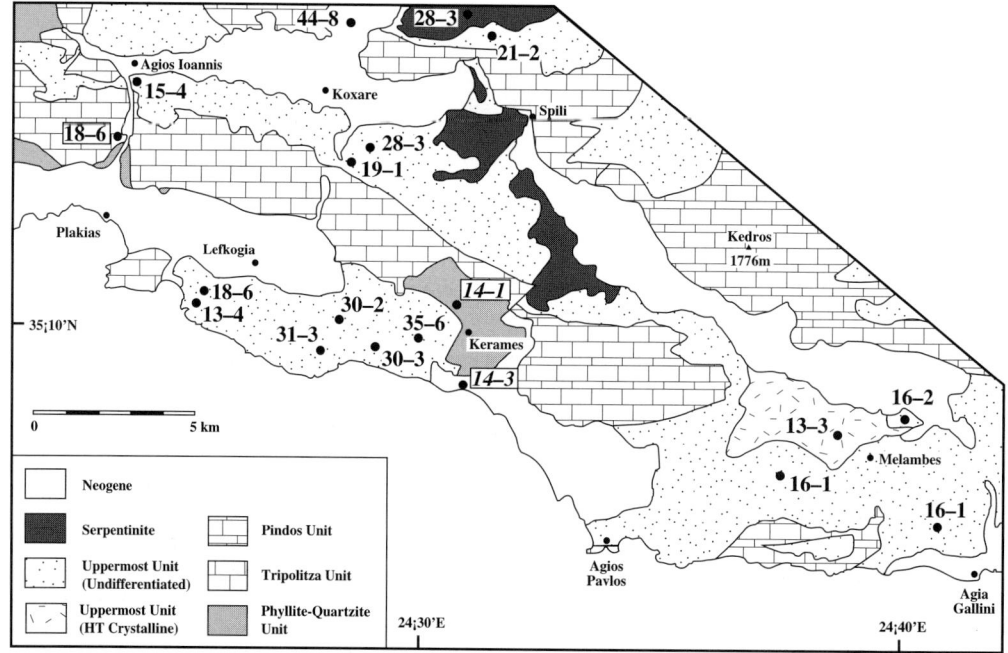

**Fig. 3.** Geological map of Plakias area of southern Crete (see Fig. 1 for location) with apatite fission-track ages presented in this study from the Uppermost unit of the upper plate. Also shown are two ages from the Phyllite–Quartzite unit of the lower plate (italicized within box) and one age from the Tripolitza unit of the upper plate (non-italicized within box).

Table 1.

| Sample number Rock type (Geological unit) | Location latitude, longitude | Height | No. of crystals | Track density ($\times 10^6$ tr cm$^{-2}$) | | | Age dispersion $(P\chi^2)$* | Apatite FT central age ($\pm 1\sigma$) | Apatite mean track length ($\mu$m $\pm$1 s.e.) (no. of tracks) | Standard deviation ($\mu$m) |
|---|---|---|---|---|---|---|---|---|---|---|
| | | | | Spontaneous $\rho_s(N_s)$ | Induced $\rho_i(N_i)$ | Dosimeter $\rho_d(N_d)$ | | | | |
| **Plakias Area** | | | | | | | | | | |
| TH 69 Quartzite (UM) | Koxare 35°14'59"N, 24°28'39"E | 330 m | 9 | 0.0433 (35) | 0.2261 (183) | 1.259 (8695) | 0% (64%) | 44.3 ± 8.2 | 13.86 ± 0.16 (44) | 1.07 |
| TH 71 Conglom. (UM) | Melambes 35°08'30"N, 24°40'22"E | 190 m | 40 | 0.0956 (91) | 1.298 (1239) | 1.160 (8011) | 0% (83%) | 15.7 ± 1.7† | 14.50 ± 0.41† (6) | 0.91† |
| TH 72 Greywacke (UM) | Melambes 35°06'27"N, 24°41'11"E | 210 m | 20 | 0.3358 (269) | 4.601 (3685) | 1.160 (8009) | 0% (98%) | 15.6 ± 1.0† | 14.05 ± 0.25† (16) | 0.97† |
| TH 73 Gneiss (UM) | Melambes 35°07'56"N, 24°39'38"E | 510 m | 9 | 0.1448 (19) | 2.470 (324) | 1.161 (8018) | 0% (n/a) | 12.6 ± 3.0† | – | – |
| TH 74 Greywacke (UM) | Saktouria 35°07'29"N, 24°37'03"E | 500 m | 20 | 0.4063 (335) | 5.526 (4556) | 1.159 (8005) | 0% (99%) | 15.7 ± 0.9† | 14.37 ± 0.13† (51) | 0.93† |
| TH 75 Quartzite (PQ) | Kerames 35°09'53"N, 24°30'43"E | 370 m | 20 | 0.2825 (103) | 4.391 (1601) | 1.159 (8002) | 0% (93%) | 13.7 ± 1.4 | – | – |
| TH 76 Greywacke (UM) | Kanevos 35°13'55"N, 24°24'06"E | 480 m | 4 | 0.1413 (16) | 1.978 (224) | 1.160 (8012) | 0% (65%) | 15.3 ± 4.0 | – | – |
| TH 121 Phyllite (UM) | Moni Preveli 35°09'26"N, 24°27'45"E | 200 m | 20 | 0.3434 (175) | 2.632 (1341) | 1.283 (8860) | 0% (97%) | 30.8 ± 2.5† | 12.31 ± 0.49† (17) | 1.97† |
| TH 122 Phyllite (UM) | Kato Moni Preveli 35°10'00"N, 24°28'09"E | 90 m | 20 | 0.3103 (207) | 2.445 (1631) | 1.275 (8806) | 0% (78%) | 29.8 ± 2.2† | 12.44 ± 0.39† (50) | 2.70† |
| TH 198 Conglom. (UM) | Karines 35°14'34"N, 24°31'16"E | 600 m | 20 | 0.1483 (95) | 1.322 (847) | 1.010 (6976) | 0% (99%) | 20.9 ± 2.3 | – | – |
| TH 208 Tuff (Tripolitza) | Kotsifou Gorge 35°13'13"N, 24°24'22"E | 350 m | 20 | 0.0409 (55) | 0.4354 (585) | 1.012 (6989) | 0% (47%) | 17.5 ± 2.5 | – | – |
| TH 209 Quartzite (UM) | Frati 35°12'39"N, 24°28'53"E | 310 m | 20 | 0.4648 (224) | 4.538 (2187) | 1.014 (7002) | 0% (94%) | 19.1 ± 1.4 | 13.54 ± 0.20 (71) | 1.70 |
| TH 210 Sandstone (UM) | Skinaria 35°10'06"N, 24°25'42"E | 30 m | 6 | 0.1337 (12) | 1.983 (178) | 1.016 (7015) | 0% (n/a) | 12.6 ± 3.8 | 14.93 ± 0.55 (6) | 1.24 |

**Table 1.** continued

| Sample number Rock type (Geological unit) | Location latitude, longitude | Height | No. of crystals | Track density ($\times 10^6$ tr cm$^{-2}$) | | | Age dispersion $(P\chi^2)$* | Apatite FT central age ($\pm 1\sigma$) | Apatite mean track length ($\mu$m $\pm 1$ s.e.) (no. of tracks) | Standard deviation ($\mu$m) |
|---|---|---|---|---|---|---|---|---|---|---|
| | | | | Spontaneous $\rho_s(N_s)$ | Induced $\rho_i(N_i)$ | Dosimeter $\rho_d(N_d)$ | | | | |
| TH 211 Quartzite (UM) | Skinaria 35°10'35"N, 24°25'92"E | 100 m | 7 | 0.0431 (11) | 0.4423 (113) | 1.018 (7029) | 0% (n/a) | 18.3 ± 5.8 | – | – |
| TH 212 Sandstone (UM) | Frati 35°12'72"N, 24°29'19"E | 380 m | 20 | 0.4687 (145) | 3.200 (990) | 1.020 (7042) | 0% (97%) | 27.5 ± 2.5 | 12.97 ± 0.30 (57) | 2.26 |
| TH 213 Gneiss (UM) | Karines 35°15'00"N, 24°30'95"E | 520 m | 20 | 0.5488 (144) | 3.739 (981) | 1.022 (7055) | 0% (94%) | 27.6 ± 2.5 | 12.11 ± 0.32 (51) | 2.24 |
| TH 214 Quartzite (PQ) | Agia Fotini 35°08'83"N, 24°30'58"E | 10 m | 5 | 0.1253 (18) | 1.720 (247) | 1.024 (7068) | 0% (87%) | 13.7 ± 3.4 | – | – |
| TH 215 Hornblendite (UM) | Korifi (Kerames) 35°39'65"N, 24°29'84"E | 260 m | 20 | 0.0874 (43) | 0.4699 (231) | 1.026 (7081) | 0% (n/a) | 35.1 ± 5.8 | 13.72 ± 0.50 (8) | 1.33 |
| TH 216 Breccia (UM) | Preveli 35°09'65"N, 24°28'82"E | 140 m | 9 | 0.5720 (94) | 3.651 (600) | 1.028 (7095) | 0% (55%) | 29.6 ± 3.3 | 13.25 ± 0.48 (19) | 2.04 |

Analyses by external detector method using 0.5 for the $4\pi/2\pi$ geometry correction factor.
Ages calculated using dosimeter glasses: CN5 with $\zeta_{CNS} = 368.9 \pm 8$.
$P\chi^2$ is the probability of obtaining a $\chi^2$ value for $\nu$ degrees of freedom where $\nu$ = no. of crystals –1.
Abbreviations for geological units: UM, Uppermost Unit; PQ, Phyllite–Quartzite Unit.
* The $\chi^2$ probability was only calculated using crystal combinations where $N_s$ and $N_i \geq 5$. This was done because the results of $\chi^2$-Test tend to become distorted when low grain counts are included (see Galbraith & Laslett 1993).
† Previously presented in Thomson et al. (1998a).
n/a Not applicable (≤1 relevant grain).

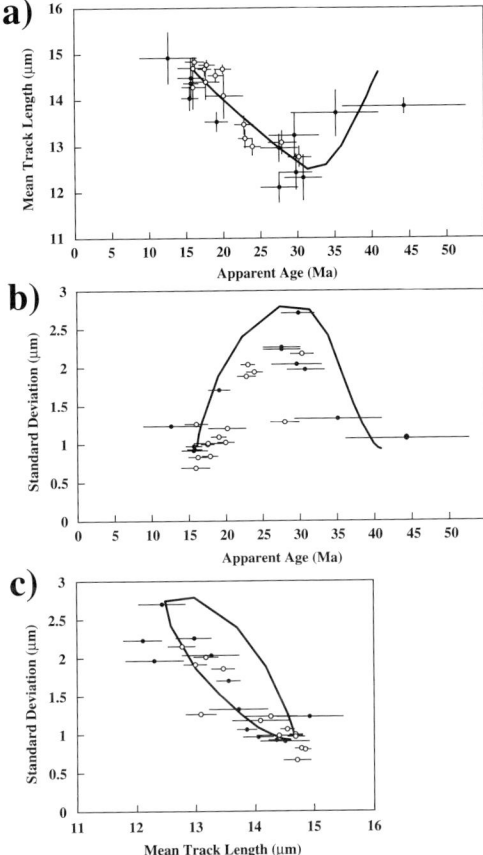

**Fig. 4.** Apparent age v. mean track length and standard deviation, and mean track length v. standard deviation plotted from the apatite fission-track data from the Uppermost tectonic unit of Crete. Data presented in this study are shown with filled symbols; data from Thomson et al. (1998b) are shown with open symbols. The synthetic curves or 'boomerangs' were obtained by modelling the thermal histories in Fig. 5 (see text for details).

apatite fission-track analysis from the UM unit also require a period of rapid cooling at c. 40 Ma. This may be related to the very early stages of collision of the leading edge of the Adria Microcontinent with Eurasia. The UM unit rocks must then have remained within similar levels of the crust, some within the apatite partial annealing zone between c. 120°C and 60°C, until a second stage of cooling at c. 16 Ma, when all of the samples of the UM unit cooled to below c. 60°C, where they remained until the present day.

The results of modelling the $T-t$ histories in Fig. 5 can also be used to produce apparent apatite age and length v. palaeotemperature or palaeodepth profiles, as outlined by Brown et al. (1994). Such plots are commonly used to interpret apatite fission-track data from vertical or near-vertical sampling profiles to detect the presence of an exhumed apatite partial annealing zone (see Fitzgerald et al. 1995). The profiles predicted by the cooling episode outlined in Fig. 5 are shown in Fig. 6. The data obtained from the UM unit rocks of the Plakias area are plotted, as close as possible, to where both the age and track length intersect the predicted profiles. An excellent fit can be seen, with almost every data point able to be placed on these profiles. It is thus possible to use the position of the data on the profile to estimate the palaeotemperature of each sample at 40 Ma. The palaeodepth of each sample and hence the amount of overburden removed from above this sample since 40 Ma is more difficult to estimate, as this relies on an assumed geothermal gradient of c. 30°C km$^{-1}$ at 40 Ma. However, if this is taken as an assumption then the maximum amount of overburden removed from the UM unit since c. 40 Ma is c. 4.5 km. If a geothermal gradient of 20°C km$^{-1}$ is assumed, perhaps more realistic in this subduction zone setting, then the removal 6.5 km of overburden is required since 40 Ma.

## Metamorphic history of the HP–LT rocks

Assessing the maximum $P-T$ conditions of the lower plate Plattenkalk unit is made difficult by the lack of suitable lithologies. Nevertheless, metabauxites found in central Crete have been shown by Theye et al. (1992) to bear assemblages containing lawsonite, magnesio-carpholite and pyrophyllite plus diaspore, indicating maximum $P-T$ conditions of c. 10 kbar and c. 350°C.

In contrast, the favourable pelitic lithologies of the PQ unit allow a detailed assessment of its $P-T-t$ history to be made. Theye & Seidel (1991, 1993) and Theye et al. (1992), using mineral parageneses in metapelites and aragonite marbles,

model (the failure to measure just one or two short track lengths can significantly reduce the observed standard deviation), or (2) that is due to the composition of the apatites measured differing from the standard Durango composition used in the Laslett et al. (1987) model. With reference to composition it was also pointed out by Thomson et al. (1998b) that the 'mixed' age samples from the Lendas area of Crete, as with those presented here, show a very narrow spread of single grain ages. This is indicative of very small compositional differences between individual grains within each sample.

According to our data analysis, the results of

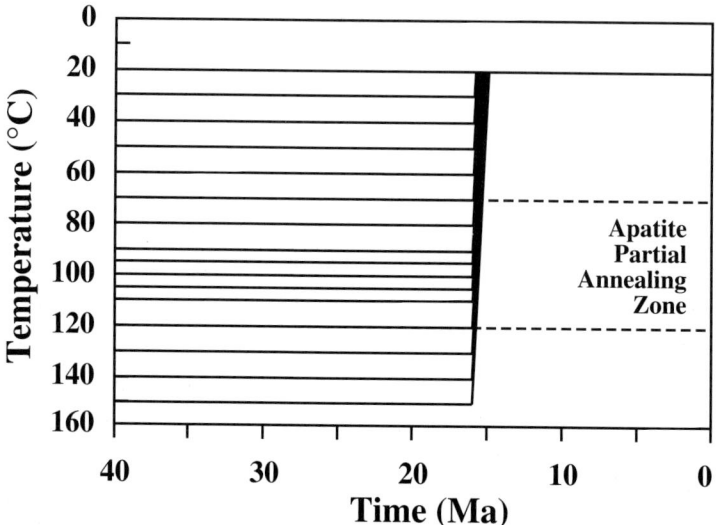

**Fig. 5.** $T$–$t$ diagram showing the simple thermal histories used to produce the synthetic curves in Fig. 4. This plot represents a series of samples that resided before 40 Ma at different temperatures (defining an apatite fission-track 'stratigraphy'; Brown et al. 1994) before a hypothetical cooling event, initiated at 16 Ma. This event cools all the samples to 20°C at the same rate within a maximum time period of 1 Ma, after which all samples remain at or near the surface until the present.

showed that the maximum $P$–$T$ conditions increase from $300 \pm 50$°C and $8 \pm 3$ kbar in the east to $400 \pm 50$°C and $10 \pm 2$ kbar in the west of Crete. Seidel et al. (1982) and Jolivet et al. (1996) both argued that K–Ar and Ar/Ar ages of between 24 and 19 Ma from metamorphic white micas from the PQ unit of central and western Crete (see Fig. 2) represent an age close to that of HP–LT metamorphism. Seidel et al. (1982) also proposed that because of excess radiogenic argon within actinolite fibres in some metamorphic phengite, the younger ages of 18–20 Ma are more likely to represent the true age of HP–LT metamorphism. The least disturbed phengite Ar/Ar step heating profiles of Jolivet et al. (1996) also give younger plateau ages of 19–21 Ma.

The exhumation path of the PQ unit rocks of western Crete is constrained by quartz fluid inclusion data (Küster & Stöckhert 1997). These demonstrate that near-isothermal decompression from $10 \pm 2$ kbar to 3–4 kbar occurred before the rocks had cooled to below c. 300°C, and thus before cooling through the closure temperature for fission tracks in zircon at the cooling rates implied. The mean zircon fission-track age of the PQ unit rocks of western Crete ($19 \pm 2$ Ma) implies exhumation from c. 35 km to c. 10 km occurred in the time period between 24–19 Ma (constrained by the K–Ar and Ar/Ar ages) and, at the latest, 17 Ma (constrained by the mean zircon fission-track age). This corresponds to a minimum exhumation rate of 4 km $Ma^{-1}$. Final cooling to below 60°C took place at c. 15 Ma, as has already been demonstrated in Fig. 2. The simple clockwise $P$–$T$–$t$ path of the PQ unit rocks of western Crete (after Thomson et al. 1998a) is summarized in Fig. 7.

## Deformation history of the HP–LT rocks

The rocks of the lower plate, in particular the siliciclastic and carbonate parts of the PQ unit, show intense ductile deformation. Within the Plattenkalk unit rocks of central Crete, ductile deformation is largely limited to two sets of large- and small-scale recumbent folds (Krahl et al. 1988). The majority of these folds have E–W to ESE–WNW fold axes and are south-vergent. The axes of the other set of folds trend approximately N–S.

The bulk of the PQ unit rocks also show two sets of recumbent nearly isoclinal folds (Greiling 1982; Hall & Audley Charles 1983; Schwarz & Stöckhert 1996). Our measurements in western Crete (Fig. 8) show that the first generation of folds have approximately NNE–SSW-trending axes and an associated strong, almost horizontal $S1$ cleavage and a N–NNE-trending stretching

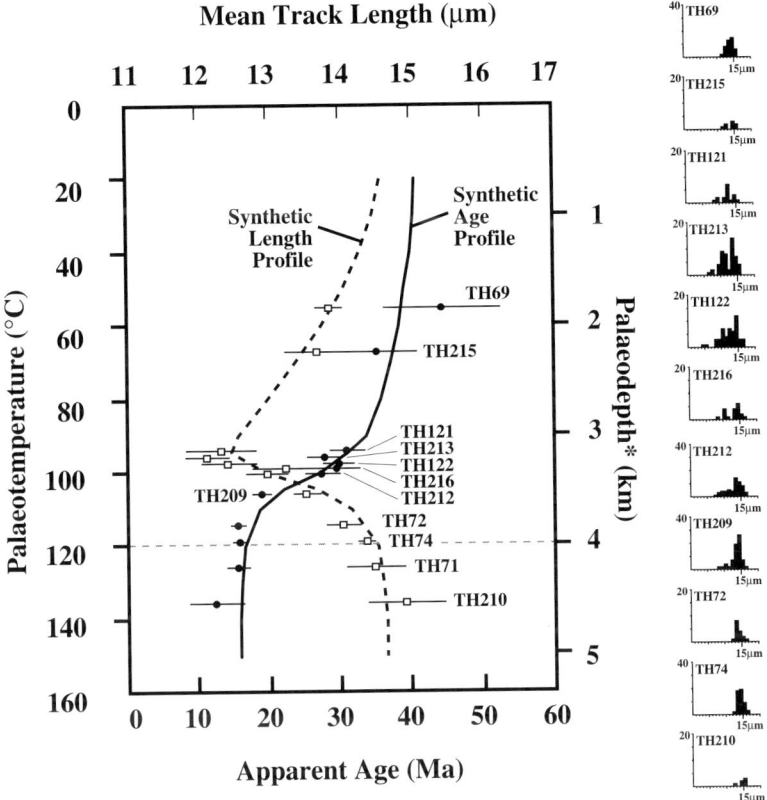

**Fig. 6.** Synthetic apatite apparent age and length profiles, produced by modelling the thermal histories of Fig. 5, plotted relative to palaeotemperature (i.e. the temperature at which the samples would have resided before the 16 Ma cooling event). The dashed horizontal line represents the depth at which the apparent apatite age would have been zero before the cooling event or the base of a fossil apatite partial annealing zone (see Brown et al. (1994) and Fitzgerald et al. (1995)). The palaeodepth is obtained by converting the palaeotemperature using a geothermal gradient of 30°C km$^{-1}$ (see text). The real data obtained from the Uppermost unit of the Plakias Area are shown alongside the synthetic profiles for comparison. A remarkably good fit can be achieved; this means that the temperatures at which these samples resided between 40 and 16 Ma can be estimated without the need for detailed thermal modelling. The amount of overburden removed from above each sample since 40 Ma can be estimated from the palaeodepth axis.

lineation (Fig. 9a). The second generation of folds (Fig. 9b) have approximately E–W-trending fold axes which in places are associated with an *S2* crenulation cleavage.

Microstructural studies from various lithologies within the PQ unit (Wachmann 1997; Stöckhert et al. 1998) show that a large portion of the contractional ductile deformation of these rocks took place under HP–LT conditions. Between this HP–LT pervasive ductile deformation at c. 35 km and the start of brittle deformation for quartz-rich rocks at c. 10 km (Küster & Stöckhert, 1997), no pervasive deformation and retrograde metamorphism related to exhumation occurred within the bulk of PQ unit rocks, even though the rocks passed through greenschist-facies metamorphic conditions (see Fig. 7). Stöckhert et al. (1998) have shown that many of the microstructures developed during HP–LT metamorphism are still well preserved, with, for example, static replacement of aragonite by calcite. Such observations imply that the bulk of the PQ unit was exhumed as a coherent block.

Some localized pervasive deformation and greenschist retrograde metamorphism does occur within parts of the PQ unit, and has been described by Greiling (1982), Fassoulas et al. (1994) and Jolivet et al. (1996). However, this appears to be largely restricted to localized extensional shear zones related to the main

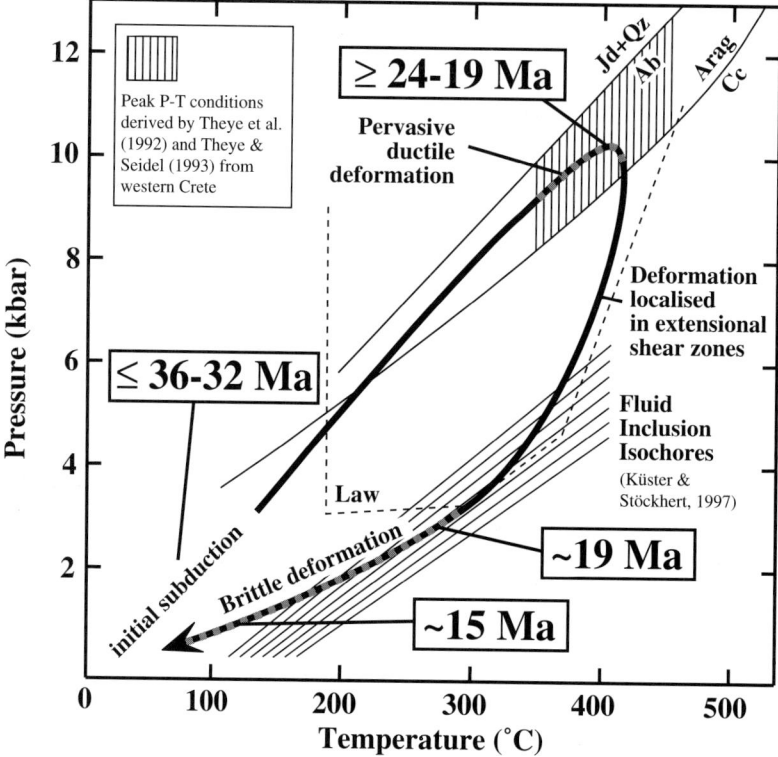

**Fig. 7.** P–T–t history of the Phyllite–Quartzite unit rocks of western Crete, also including references to the type of deformation experienced by these rocks during burial, metamorphism and subsequent exhumation (after Thomson et al. 1998a).

extensional detachment that exhumed the HP–LT lower plate rocks. According to our observations, intense horizontal extension within the PQ unit itself was confined to the brittle field and localized in numerous low- to moderate-angle normal faults (Fig. 9e and f) controlled by earlier boudins and folds and reactivation of earlier structures. Jolivet et al. (1996) suggested that this late-stage extension was distinctly asymmetric, and related to a north-dipping extensional shear zone. However, our observations (Fig. 10b and c) contradict these findings, and we believe that it is not possible to observe any asymmetry of these late-stage brittle extensional faults.

This late-stage extensional deformation within the brittle field must have occurred at temperatures below c. 300°C, and from the constraints supplied in this paper, to pressures of less than 3–4 kbar corresponding to depths of less than c. 10 km. Thus, these faults could only have played a late-stage minor role in the exhumation of the HP–LT PQ unit rocks. It is likely that the apatite fission-track data from the PQ unit rocks, which record rapid cooling at around 15 Ma, date the approximate timing of this late-stage extensional event. Finally, once the HP–LT lower plate rocks of Crete were within the brittle uppermost crust, and juxtaposed against the upper plate thrust sheets of present-day Crete, all the main tectonic units and some overlying late Miocene sedimentary rocks underwent minor approximately N–S shortening, with the development of open folds, kink bands (Figs 8d and 9d) and minor thrusting (Postma et al. 1993; Fassoulas et al. 1994).

## Discussion and tectonic model

Before any tectonic model to describe the post-Eocene evolution of Crete can be applied, the boundary conditions that need to be placed on any such model need to be defined. The results

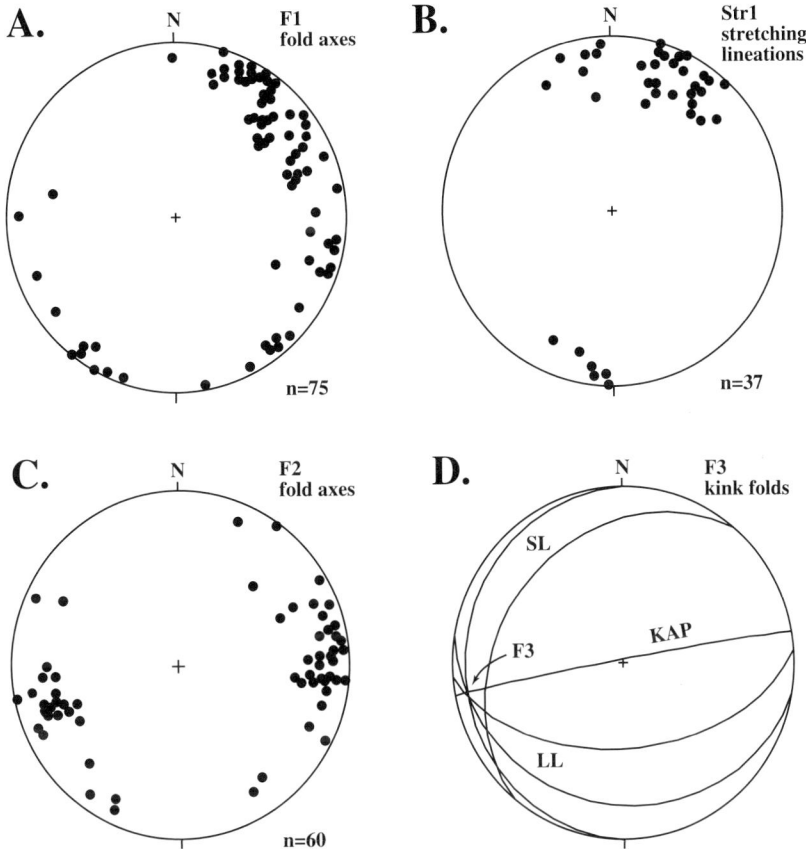

**Fig. 8.** Stereographic projections (equal area, lower hemisphere) showing the orientation of structural features of the Phyllite–Quartzite unit in western Crete. (**a**) First generation F1 fold axes. (**b**) Mineral stretching lineation Str1 on near-horizontal S1 schistosity planes. In many places Str1 is roughly parallel to F1. (**c**) Second generation F2 fold axes. (**d**) Post-exhumation late-stage compressional kink folds from near Agios Theodori, western Crete, showing E–W kink axes approximately orthogonal to the Miocene to present plate convergence motion vector (F3, kink axes; KAP, kink fold axial plane; LL, long limb; SL, short limb).

presented in this paper show that HP–LT rocks of Crete entered the subduction zone between 36 and 29 Ma. They were then subducted, metamorphosed and pervasively deformed at high pressure (10 ± 2 kbar or 35 ± 7 km) before 19–24 Ma. Immediately after metamorphism, these rocks then followed a nearly isothermal decompression path to 3–4 kbar before c. 19 Ma, during which no further pervasive deformation occurred. All deformation and any retrograde metamorphism during this exhumation phase was restricted to the vicinity of the main extensional detachment responsible for the majority of their exhumation. During the time between c. 40 Ma and 16–17 Ma, Thomson et al. (1998b), using apatite fission-track thermochronology, have shown that the rocks in the uppermost part of the accretionary complex above the detachment remained within or below the apatite annealing zone, which is confined to the upper c. 4–7 km of the crust. Significant increased erosion becomes evident only after c. 16–17 Ma. This implies that only minor erosion could have occurred in the upper plate to the detachment at the same time as the main exhumation of the lower plate rocks occurred. Such a lack of erosion, as well as being indicative of a low topographic relief within the upper plate rocks at this time, implies that extensional tectonic unroofing must have contributed c. 80–90% of the overall c. 25 km exhumation of the HP–LT lower plate rocks of Crete before c. 19 Ma. Once all the

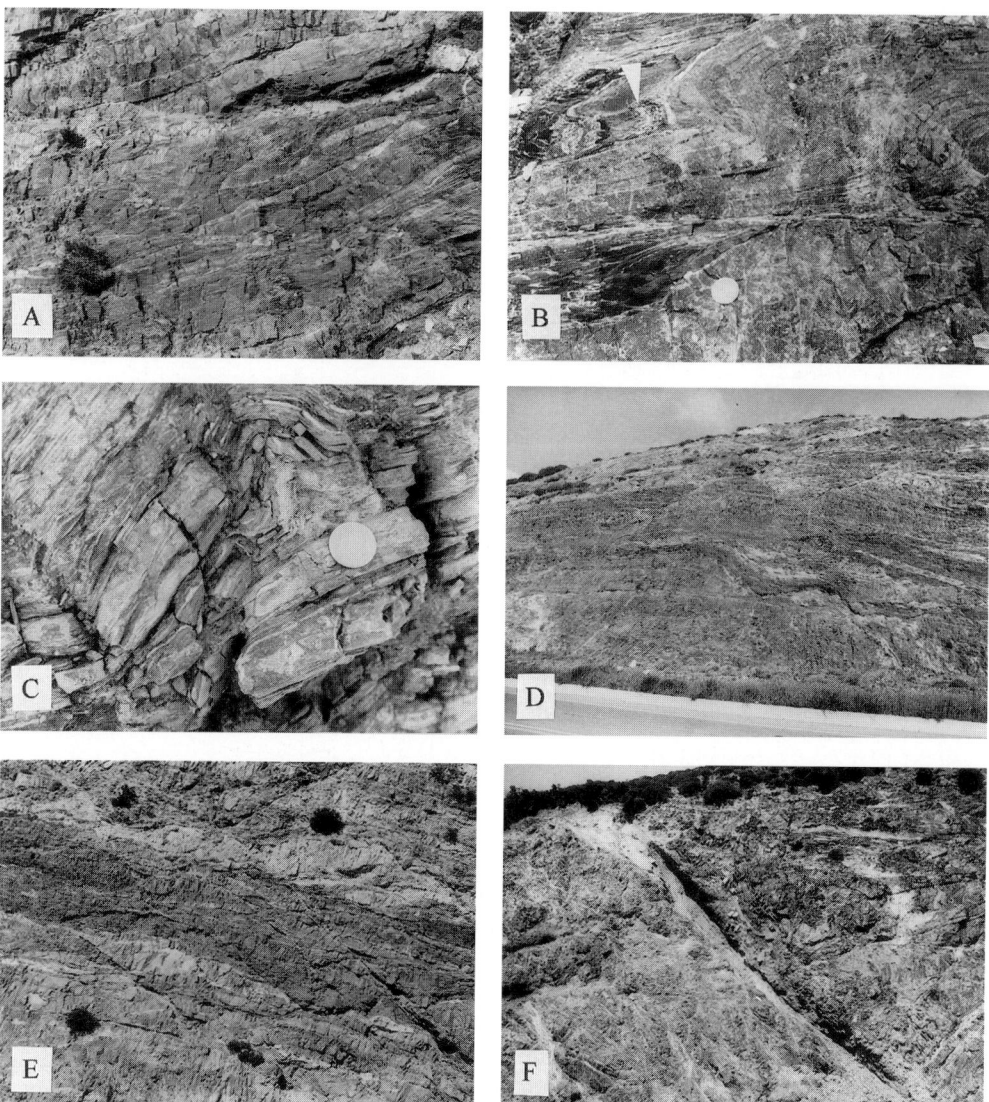

**Fig. 9.** Mesoscopic structures in the HP–LT metamorphic Phyllite–Quartzite unit in western Crete. (**a**) First generation (F1) isoclinal folds of quartzite within a phyllitic matrix. (**b**) Second generation asymmetric fold (F2) in quartzite layers. A crenulation cleavage is developed within adjacent phyllites (white arrow). (**c**) Late kink folds and a small-scale thrust in stiff carbonate layers intercalated with phyllites. (**d**) Low-angle normal faults developed in the brittle regime, following pre-existing discontinuities. (**e**) Close-up view of brittle low-angle normal faults. (**f**) Major Neogene normal fault cross-cutting earlier structures. The thickness of the fault gouge is between 0.5 and 1 m.

currently exposed rocks of Crete were confined to the brittle upper crust, a period of intense horizontal extension occurred with the formation of low to moderate angle normal faults, particularly in the PQ unit. This coincides with a period of rapid cooling of the PQ unit between about 120°C and 60°C at c. 15 Ma (Thomson et al. 1998a). Surface exposure of the HP–LT rocks by c. 10 Ma at the latest is indicated by detritus in Tortonian sedimentary rocks (Postma et al. 1993). The subsequent development of Crete is dominated by normal and block faulting and the

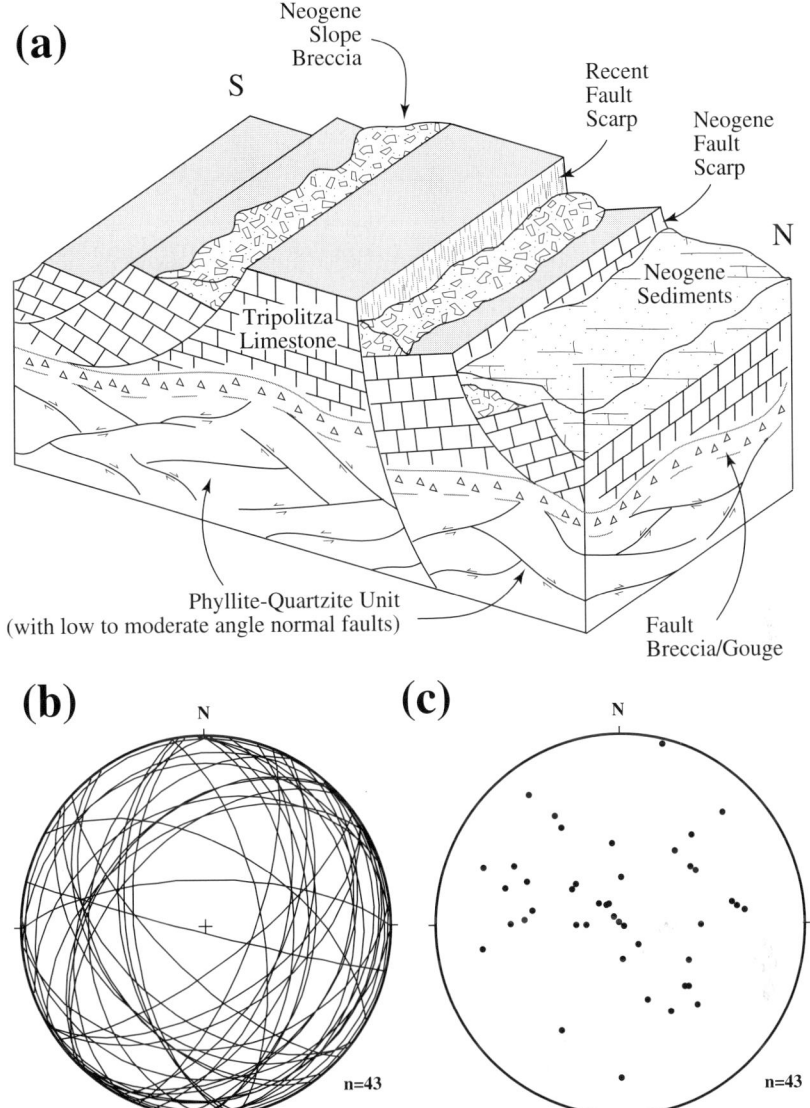

**Fig. 10.** (**a**) Interpretative and highly simplified sketch to show the relationship between the PQ unit below (with abundant late-stage low- to moderate-angle brittle normal faults) and the Tripolitza unit above, separated by a brittle fault breccia. A later Neogene high-angle normal fault is also shown. (**b** and **c**) Orientations of low-angle normal faults in some outcrops of the PQ along the western coast of Crete, near Sfinari (see also Fig. 9d–f). (Note that the faults show no distinct pattern of extension.)

formation of Neogene sedimentary basins, disrupted by occasional small compressional events (Postma *et al.* 1993).

As mentioned by Thomson *et al.* (1998*b*), the Late Eocene to Miocene evolution of the HP–LT rocks and the overlying upper plate accretionary complex of Crete is entirely consistent with the features described by Royden (1993), which typify continental accretion at a retreating subduction boundary dominated by slab pull forces. The apatite fission-track data from the Uppermost unit require that little or no erosion occurred during the collision of Adria with Eurasia between c. 40 Ma and 16–17 Ma. By definition, retreating subduction boundaries exhibit extension within the back-arc region. The consequence is that no significant mountain belt, with related topographic relief to allow widespread erosion, can occur. In addition, very little crystalline basement is found within the Cretan thrust sheets. Royden (1993) stated that this is also diagnostic of collision at a retreating subduction boundary, where horizontal compressional stresses must be sufficiently small to allow regional extension in the overriding plate, and thus, generally too small to allow penetrative deformation within strongly crystalline metamorphic basement rocks. On Crete, and throughout much of Greece and the Aegean, sedimentation from Paleocene to Oligocene time was dominated by deep-water flysch sedimentation (Hall et al. 1984; Mascle et al. 1986). Again, this is typical of accretion at a retreating subduction boundary (Royden 1993), as during pre- and syn-collision, anomalously deep foredeep basins tend to be developed because of downbending related to the gravitational pull exerted by the dense downgoing slab. Such a setting, with only minor erosion of the overriding plate because of its low topographic relief, will then be conducive to the deposition of flysch-type deep-water submarine fan type deposits.

Before we discuss our preferred tectonic model, it should be pointed out that the boundary conditions outlined above are not consistent with the recently popular tectonic models applied to the Aegean segment of the Hellenic convergent plate boundary, which propose the development of a thickened crust followed by the tectonic exhumation of HP–LT rocks by gravitational collapse (e.g. Jolivet et al. 1994, 1996; Avigad et al. 1997). Such models would require significant erosion of the upper plate rocks between 40 Ma and 16–17 Ma, the probable involvement of basement in thrusting, and widespread shallow marine or continental molasse style sedimentation. As we have demonstrated, no evidence for such diagnostic criteria are seen in Crete or, indeed, in much of Greece.

Continuous subduction retreat during collision of the Adria microplate with Eurasia is further supported by the following tectonic constraints. If one takes the overall convergence rate calculated by Dewey et al. (1989) of c. 1.7 cm/year between Africa and Eurasia between c. 38 Ma and 9 Ma, then this would require that subduction and accretion of the c. 300 km wide (as estimated by Bonneau (1984)) Adria microcontinent would take at least 18 Ma. However, as we have shown, the PQ unit rocks were not only subducted, but also metamorphosed at high pressure, and subsequently exhumed to less than 60°C within c. 15 Ma, whereas the PK unit underwent the same sequence of events within a time period as short as 10 Ma. For subduction to proceed faster than the true overall plate convergence between Africa and Eurasia requires that significant rollback or subduction retreat must have been active during the collision, accretion and exhumation process.

Any model developed for the post-Eocene tectonic evolution of the Crete, or the Aegean segment of the Hellenic plate boundary, needs to take into account all the above-mentioned boundary conditions. Our preferred model, which does account for these conditions, is outlined in Fig. 11.

The late Eocene to early Oligocene stage of the model (Fig. 11a) is based on Bonneau (1984). At this time, old dense Pindos oceanic lithosphere was being subducted beneath the continental lithosphere of the Pelagonian or Eurasian Plate. Today metamorphic rocks of the UM unit of Crete, originating from the Pelagonian or Eurasian Plate (Robertson et al. 1991), are found thrust upon Eocene flysch (above the deep-water Pindos carbonate rocks) with only minor amounts of oceanic material between. This indicates that at this convergent margin, active since Late Cretaceous time, the majority of incoming sediment was not accreted at the front of the wedge, but rather subducted and accreted at depth. This is a typical feature of a retreating subduction boundary dominated by roll-back (e.g. Cloos & Shreeve 1988). The approach of the Adria microcontinent towards this convergent margin is recorded by the southward migration of flysch sedimentation, derived from the Eurasian Plate, upon the carbonate rocks of its northern continental margin (Hall et al. 1984).

Collision of the Adria microcontinent occurred between Oligocene and mid-Miocene time (Fig. 11b). Because of continued subduction retreat or roll-back, this resulted in shallow accretion of the rocks of the northern continental margin of Adria, now represented by the largely non-metamorphic thrust sheets of the Pindos and Tripolitza units on Crete. The present-day occurrence of fragments of the uppermost tectonic unit of Crete, derived from the Pelagonian or Eurasian Plate, also means that parts of the overriding plate must have remained at the leading edge of the accretionary wedge throughout the collision process. The PQ

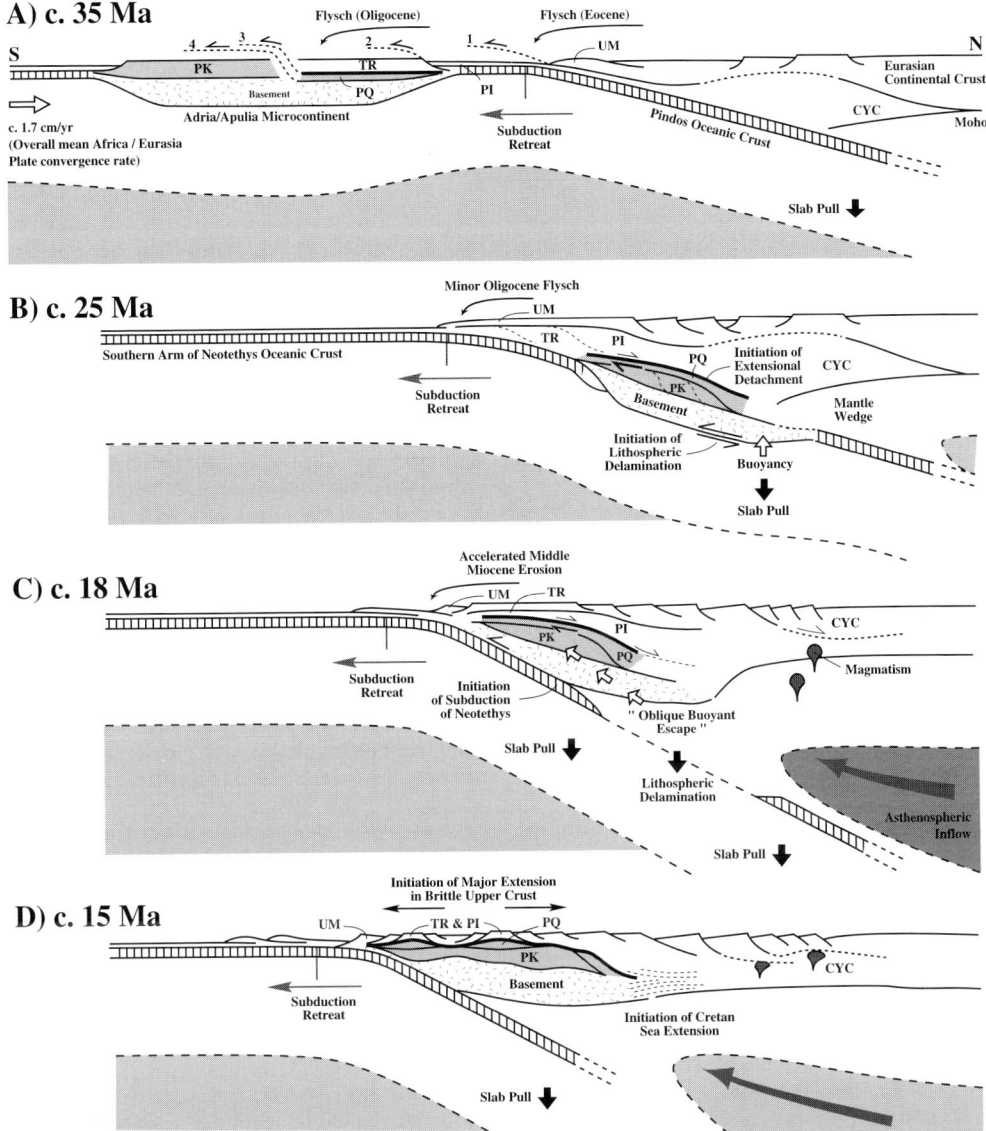

**Fig. 11.** Schematic N–S cross-sections that illustrate our preferred model for the late Eocene to mid-Miocene tectonic development of the Cretan segment of the Hellenic convergent boundary. The details are described in the text. Vertical exaggeration is c. 3×, with the crustal layers further exaggerated for clarity. The age sequence of accretion of the different tectonic units of Crete is numbered in (**a**). The HP–LT lower plate rocks exposed at present on Crete are shaded throughout for clarity. PK, Plattenkalk unit; PQ, Phyllite–Quartzite unit; TR, Tripolitza unit; PI, Pindos unit; UM, Uppermost unit; CYC, present-day rocks of the Cyclades.

unit rocks, in contrast, could not have been accreted in the shallow parts of the crust, but must have been detached from the overlying Tripolitza platform carbonates, probably along a weak décollement in mid-Triassic evaporites, and accreted at depths of c. 35 km where they underwent pervasive ductile deformation and metamorphism. These rocks must then have been detached from their crystalline basement and underthrust by the PK unit carbonates and related basement of the southern continental margin of the Adria microcontinent. Again, it must be emphasized that no large-scale crustal thickening and creation of large topographic relief during collision is required by this model, and such models are not supported by any of the evidence we have presented.

According to the boundary conditions the HP–LT rocks of the PQ and PK units were then rapidly exhumed soon after cessation of metamorphism and deformation at their depth of maximum burial between 24 and 19 Ma. This exhumation was largely achieved by tectonic unroofing along a discrete extensional detachment at rates of over 4 km Ma$^{-1}$, with erosion playing only a minor role during this period. The speed of the exhumation, and the observation that the bulk of HP–LT rocks were exhumed as a coherent block, unaffected by any pervasive deformation or metamorphism during their exhumation, effectively rules out any models that propose extension, and hence exhumation, driven by the gravitational collapse of overthickened crust, such as in accretionary wedges thickened by underplating (e.g. Platt 1986). The regional nature of the Cretan HP–LT rocks also rules out models that invoke 'corner flow' (see discussion by Platt (1993)). To fit our observations, we prefer a model, similar to that proposed by Wijbrans et al. (1993), where exhumation is driven by buoyancy forces. It was pointed out by Cloos (1993) that when lithosphere capped by continental crust is subducted, the positive buoyancy of continental crust will resist the gravitational pull of the subducting lithospheric slab. This causes underthrusting to stop. Additionally, at weak zones within the lithosphere, where the buoyancy force exceeds the lithospheric strength, detachment can occur (Wijbrans et al. 1993). This results in 'collisional or lithospheric delamination' with either separation of the lithologically different lower and upper continental crust, or the separation of the whole of continental crust from its mantle root. The likely thin nature of the continental crust of the Adria microplate (probably around 30 km) probably favours the development of a mantle–crust detachment (Fig. 11c). Any decoupled body of continental crust can have a density contrast of up to 500 kg m$^{-3}$ with the adjacent and/or overlying mantle wedge (Cloos 1993). This can then allow exhumation driven by buoyancy of the previously subducted microcontinental domain, whereas the denser mantle lithosphere continues to subduct (Wijbrans et al. 1993). The continued availability of the gravitational pull of this dense lithospheric slab can then propagate into the oceanic lithosphere of the southern arm of Neotethys. This, along with continued overall plate convergence (Dewey et al. 1989), then allows the initiation of subduction, and continued subduction retreat of this oceanic segment between c. 15 Ma and the present day (Meulenkamp et al. 1994). One other consequence of this lithospheric delamination between 24 and 19 Ma is that a sudden influx of asthenospheric mantle in the back-arc region could occur (Fig. 11c), to induce melting and the emplacement of granites and related greenschist-facies metamorphism that occurred after c. 22 Ma in the overlying crust that underlies the present-day Cyclades region. A similar model, to explain the post-Eocene volcanism seen in the Aegean, was proposed by Zeilinga de Boer (1989).

A model proposing the complete 'slab break-off' of the downgoing oceanic lithosphere of the Pindos Ocean in the Aegean has been presented by Davies & von Blanckenburg (1995). This model explains many of the geological features of the Aegean region. However, models of the subducting lithospheric slab produced by Wortel et al. (1993) by modelling seismic tomographic data, propose that continuous subduction has occurred at the Hellenic convergent plate boundary since at least 26 Ma. This suggests that complete slab break-off could not have occurred between 24 and 19 Ma. Slab break-off would also remove the slab pull forces causing rollback or subduction retreat, and is thus incompatible with the evidence of continuous slab retreat during the whole of Late Tertiary time presented in this paper.

Because of isostatic adjustment after unloading, an increase in topography would then result owing to the thickened crust, and any exhumation would be accommodated by extension related to gravitational collapse. However, as we have already stated, such a model is not consistent with our observations. Instead, our preferred proposal is that the presence of a structurally weak horizon directly above the PQ unit rocks, in the form of Triassic evaporites, allowed the development of a discrete extensional detachment. This then allowed the positively buoyant subducted Adria continental basement (as a form of 'keel' surrounded by mantle lithosphere) along with the HP–LT rocks of the PQ and PK units to rise obliquely, and as a largely coherent block, into the space created by the roll-back of the Neotethyan oceanic lithosphere. We label this process 'oblique buoyant escape'.

In contrast, east of Crete and Rhodes, the initiation of renewed subduction retreat of the southern arm of Neotethys has not occurred.

Thus, here buoyant escape of the subducted and now positively buoyant basement of the Adria microplate has not resulted. Instead, and consistent with our model, a thickened crustal profile developed resulting in the Miocene symmetric gravitational collapse of the Menderes region of central western Turkey to the north (Hetzel et al. 1995).

The accelerated erosion of the UM unit at c. 16–17 Ma was probably due to a minor phase of uplift caused by 'excess buoyancy' or isostatic adjustment within the crustal section that remained after cessation of movement on the extensional detachment between the HP rocks and unmetamorphosed accretionary sequence of Crete. Another possibility is that this late stage of accelerated erosion was related to the initiation of 'Anatolian Push' related to the tectonic escape of Turkey initiated in Late Miocene time (Meijer & Wortel 1997). The subsequent pervasive brittle extension between c. 19 Ma and 15 Ma was probably caused by the transfer of crustal extension related to continued subduction retreat to the whole crustal section (Fig. 11d). The continued influx of warm back-arc mantle asthenosphere in the back-arc region could then cause the weakening of the overriding lithosphere further arcward in the system, resulting in final transfer of crustal extension in the overriding plate to the region of the present-day Cretan Sea at c. 12 Ma (Meulenkamp et al. 1994). The subsequent Neogene tectonism on Crete is dominated by the migration of Crete as a coherent block to the south and its fragmentation into several horsts and basins.

This work has been supported by the Deutsche Forschungsgemeinschaft grants STO 196/6 and STO 196/8. Fission-track sample preparation was carried out under the supervision of F. Hansen. Thanks are due to the IGME for permission to collect samples from Crete, and to N. Fytrolakis and many other Greek colleagues for their interest and continuous support. E. Seidel and H. Rauche provided interesting discussions in the field, and K. Gallagher is thanked for providing his modelling program MONTETRAX. Finally this paper benefited greatly from reviews by G. Batt, M. Brandon, G. Lister and S. Willet.

## References

AVIGAD, D., GARFUNKEL, Z., JOLIVET, L. & AZAÑÓN, J. M. 1997. Back-arc extension and denudation of Mediterranean eclogites. *Tectonics*, **16**, 924–941.

BIZON, G., BONNEAU, M., LEBOULENGER, P., MATESCO, S. & THIÉBAULT, F. 1976. Sur la signification et l'extension des «massifs cristallins externes» en Péloponnèse méridional et dans l'Arc égéen. *Bulletin de la Société géologique de France*, **18**, 337–345.

BONNEAU, M. 1976. Esquisse structurale de la Crète alpine. *Bulletin de la Société géologique de France*, **18**, 351–353.

—— 1984. Correlation of the Hellenic Nappes in the south-east Aegean and their tectonic reconstruction. *In*: DIXON, J. E. & ROBERTSON, A. H. F. (eds) *The Geological Evolution of the Eastern Mediterranean*. Geological Society, London, Special Publications, **17**, 517–527.

BRANDON, M. T., RODEN-TICE, M. K. & GARVER, J. I. 1998. Late Cenozoic exhumation of the Cascadia accretionary wedge in the Olympic Mountains, NW Washington State. *Geological Society of America Bulletin*, **110**, 985–1009.

BROWN, R. W., SUMMERFIELD, M. A. & GLEADOW, A. J. W. 1994. Apatite fission track analysis: Its potential for the estimation of denudation rates and implications for models of long-term landscape development. *In*: KIRKBY, M. J. (ed.) *Process Models and Theoretical Geomorphology*. Wiley, Chichester, 23–53.

CLOOS, M. 1993. Lithospheric buoyancy and collisional orogenesis: subduction of oceanic plateaus, continental margins, island arcs, spreading ridges, and seamounts. *Geological Society of America Bulletin*, **105**, 715–737.

—— & SHREEVE, R. L. 1988. Subduction-channel model of prism accretion, melange formation, sediment subduction, and subduction erosion at convergent plate margins: 1. Background and description. *Pure and Applied Geophysics*, **128**, 455–500.

COYLE, D. A. & WAGNER, G. A. 1996. Fission-track dating of zircon and titanite from the 9101 m deep KTB: observed fundamentals of track stability and thermal history reconstruction. *International Workshop on Fission-Track Dating, Gent 1996, Abstracts*, 22.

CREUTZBURG, N. & SEIDEL, E. 1975. Zum Stand der Geologie des Präneogens auf Kreta. *Neues Jahrbuch für Geologie und Paläontologie, Abhandlungen*, **141**, 259–285.

DAVIES, J. H. & VON BLANCKENBURG, F. 1995. Slab breakoff: a model of lithosphere detachment and its test in the magmatism and deformation of collisional orogens. *Earth and Planetary Science Letters*, **129**, 85–102.

DEWEY, J. F., HELMAN, M. L., TURCO, E., HUTTON, D. W. H. & KNOTT, S. D. 1989. Kinematics of the western Mediterranean. In: COWARD, M. P. & DIETRICH, D. (eds) *Alpine Tectonics*. Geological Society, London, Special Publications, **45**, 265–283.

FASSOULAS, C., KILIAS, A. & MOUNTRAKIS, D. 1994. Postnappe stacking extension and exhumation of high-pressure/low-temperature rocks in the island of Crete, Greece. *Tectonics*, **13**, 127–138.

FITZGERALD, P. G., SORKHABI, R. B., REDFIELD, T. F. & STUMP, E. 1995. Uplift and denudation of the central Alaska Range: a case study in the use of apatite fission track thermochronology to determine absolute uplift parameters. *Journal of Geophysical Research*, **100**, 20175–20191.

FOSTER, D. A., KOHN, B. P. & GLEADOW, A. J. W. 1996.

Sphene and zircon fission track closure temperatures revisited: empirical calibrations from $^{40}Ar/^{39}Ar$ diffusion studies of K-feldspar and biotite. *International Workshop on Fission-Track Dating, Gent 1996, Abstracts*, 37.

GALBRAITH, R. F. & LASLETT, G. M. 1993. Statistical models for mixed fission track ages. *Nuclear Tracks and Radiation Measurements*, **21**, 459–470.

GALLAGHER, K. 1995. Evolving temperature histories from apatite fission-track data. *Earth and Planetary Science Letters*, **126**, 421–435.

—— & BROWN, R. 1997. The onshore record of passive margin evolution. *Journal of the Geological Society, London*, **154**, 451–457.

GLEADOW, A. J. W., DUDDY, I. R., GREEN, P. F. & LOVERING, J. F. 1986. Confined fission track lengths in apatite: a diagnostic tool for thermal history analysis. *Contributions to Mineralogy and Petrology*, **94**, 405–415.

GREEN, P. F. 1986. On the thermotectonic evolution of northern England: evidence from fission track analysis. *Geological Magazine*, **123**, 493–506.

GREILING, R. 1982. The metamorphic and structural evolution of the Phyllite–Quartzite Nappe of western Crete. *Journal of Structural Geology*, **4**, 291–297.

HALL, R. & AUDLEY-CHARLES, M. G. 1983. The structure and regional significance of the Talea Ori, Crete. *Journal of Structural Geology*, **5**, 167–179.

——, AUDLEY-CHARLES, M. G. & CARTER, D. J. 1984. The significance of Crete for the evolution of the Eastern Mediterranean. *In*: DIXON, J. E. & ROBERTSON, A. H. F. (eds) *The Geological Evolution of the Eastern Mediterranean*. Geological Society, London, Special Publications, **17**, 499–516.

HARLAND, W. B., ARMSTRONG, R. L., COX, A. L., CRAIG, L. E., SMITH, A. G. & SMITH, D. G. 1990. *A Geological Time Scale, 1989*. Cambridge University Press, Cambridge.

HETZEL, R. PASSCHIER, C. W., RING, U. & DORA, Ö. O. 1995. Bivergent extension in orogenic belts: the Menderes massif (southwestern Turkey). *Geology*, **23**, 455–458.

HURFORD, A. J. 1990. Standardization of fission track dating calibration: recommended by the Fission Track Working Group of the I.U.G.S. Subcommission on Geochronology. *Chemical Geology (Isotope Geosciences Section)*, **80**, 171–178.

——, HUNZIKER, J. C. & STÖCKHERT, B. 1991. Constraints on the late thermotectonic evolution of the western Alps: evidence for episodic rapid uplift. *Tectonics*, **10**, 758–769.

JOLIVET, L., DANIEL, J. M., TRUFFERT, C. & GOFFÉ, B. 1994. Exhumation of deep crustal metamorphic rocks and crustal extension in arc and back-arc regions. *Lithos*, **33**, 3–30.

——, GOFFÉ, B., MONIÉ, P., TRUFFERT-LUXEY, C., PATRIAT, M. & BONNEAU, M. 1996. Miocene detachment in Crete and exhumation $P$–$T$–$t$ paths of high-pressure metamorphic rocks. *Tectonics*, **15**, 1129–1153.

KILIAS, A., FASSOULAS, C. & MOUNTRAKIS, D. 1994. Tertiary extension of continental crust and uplift of Psiloritis metamorphic core complex in the central part of the Hellenic Arc (Crete, Greece). *Geologische Rundschau*, **83**, 417–430.

KRAHL, J., KAUFFMANN, G., KOZUR, H., RICHTER, D., FÖRSTER, O. & HEINRITZ, F. 1983. Neue Daten zur Biostratigraphie und zur tektonischen Lagerung der Phyllit-Gruppe und der Trypali-Gruppe auf der Insel Kreta (Griechenland). *Geologische Rundschau*, **72**, 1147–1166.

——, RICHTER, D., FÖRSTER, O., KOZUR, H. & HALL, R. 1988. Zur Stellung der Talea Ori im Bau des kretischen Deckenstapels (Griechenland). *Zeitschrift der deutschen geologischen Gesellschaft*, **139**, 191–227.

KÜSTER, M. & STÖCKHERT, B. 1997. Density changes of fluid inclusions in high-pressure low-temperature metamorphic rocks from Crete: a thermobarometric approach based on the creep strength of the host minerals. *Lithos*, **41**, 151–167.

LASLETT, G. M., GREEN, P. F., DUDDY, I. R. & GLEADOW, A. J. W. 1987. Thermal annealing of fission tracks in apatite: 2. A quantitative analysis. *Chemical Geology (Isotope Geosciences Section)*, **65**, 1–13.

LISTER, G. S. & BALDWIN, S. L. 1996. Modelling the effect of arbitrary $P$–$T$–$t$ histories on argon diffusion in minerals using the MacArgon program for the Apple Macintosh. *Tectonophysics*, **253**, 83–109.

—— & DAVIS, G. A. 1989. The origin of metamorphic core complexes and detachment faults formed during Tertiary continental extension in the northern Colorado River region, U.S.A. *Journal of Structural Geology*, **11**, 65–94.

——, BANGA, G. & FEENSTRA, A. 1984. Metamorphic core complexes of Cordilleran type in the Cyclades, Aegean Sea, Greece. *Geology*, **12**, 221–225.

MASCLE, J., LE CLEAC'H, A. & JONGSMA, D. 1986. The Eastern Hellenic Margin from Crete to Rhodes: example of progressive collision. *Marine Geology*, **73**, 145–168.

MEIJER, P. T. & WORTEL, M. J. R. 1997. Present-day dynamics of the Aegean region: a model analysis of the horizontal pattern of stress and deformation. *Tectonics*, **16**, 879–895.

MEULENKAMP, J. E., VAN DER ZWAAN, G. J. & VAN WAMEL, W. A. 1994. On Late Miocene to recent vertical motions in the Cretan segment of the Hellenic arc. *Tectonophysics*, **234**, 53–72.

PLATT, J. P. 1986. Dynamics of orogenic wedges and the uplift of high-pressure metamorphic rocks. *Geological Society of America Bulletin*, **97**, 1037–1053.

—— 1993. Exhumation of high-pressure rocks: a review of concepts and processes. *Terra Nova*, **5**, 119–133.

POSTMA, G., FORTUIN, A. R. & VAN WAMEL, W. A. 1993. Basin-fill patterns controlled by tectonics and climate: the Neogene 'fore-arc' basins of eastern Crete as a case history. *In*: FROSTICK, L. E. & STEEL, R. J. (eds) *Tectonic Controls and Signatures in Sedimentary Successions*. International Association of Sedimentologists, Special Publications, **20**, 335–362.

ROBERTSON, A. H. F. & DIXON, J. E. 1984. Introduction: aspects of the geological evolution of the Eastern Mediterranean. *In*: DIXON, J. E. & ROBERTSON, A. H. F. (eds) *The Geological Evolution of the Eastern Mediterranean*. Geological Society, London, Special Publication, **17**, 1–74.

——, CLIFT, P., DEGNAN, P. J. & JONES, G. 1991. Palaeogeographic and palaeotectonic evolution of the Eastern Mediterranean Neotethys. *Palaeogeography, Palaeoclimatology, Palaeoecology*, **87**, 289–343.

ROYDEN, L. H. 1993. The tectonic expression of slab pull at continental convergent boundaries. *Tectonics*, **12**, 303–325.

SCHWARZ, S. & STÖCKHERT, B. 1996. Pressure solution in siliciclastic HP–LT metamorphic rocks – constraints on the state of stress in deep levels of accretionary complexes. *Tectonophysics*, **255**, 203–209.

SEIDEL, E., KREUZER, H. & HARRE, W. 1982. A Late Oligocene/Early Miocene high pressure belt in the External Hellenides. *Geologisches Jahrbuch*, **E23**, 165–206.

——, OKRUSCH, M., KREUZER, H., RASCHKA, H. & HARRE, W. 1976. Eo-alpine metamorphism in the uppermost unit of the Cretan nappe system – petrology and geochronology – Part 1. The Léndas area (Asteroussia Mountains). *Contributions to Mineralogy and Petrology*, **57**, 259–275.

——, ——, ——, & —— 1981. Eo-alpine metamorphism in the uppermost unit of the Cretan nappe system – petrology and geochronology – Part 2. Synopsis of high-temperature metamorphics and associated ophiolites. *Contributions to Mineralogy and Petrology*, **76**, 351–361.

——, SCHLIESTEDT, M., KREUZER, H. & HARRE, W. 1977. Metamorphic rocks of Late Jurassic age as components of the ophiolitic mélange on Gavdos and Crete, Greece. *Geologisches Jahrbuch*, **B28**, 3–21.

STÖCKHERT, B., WACHMANN, M., KÜSTER, M. & BIMMERMAN, S. 1998. Low effective viscosity during high-pressure metamorphism due to dissolution precipitation creep: the record of HP–LT metamorphic carbonates and siliciclastic rocks from Crete. *Tectonophysics*, in press.

THEYE, T. & SEIDEL, E. 1991. Petrology of low-grade high-pressure metapelites from the External Hellenides (Crete, Peloponnese). A case study with attention to sodic minerals. *European Journal of Mineralogy*, **3**, 343–366.

—— & —— 1993. Uplift-related retrogression history of aragonite marbles in Western Crete. *Contributions to Mineralogy and Petrology*, **114**, 349–356.

——, —— & VIDAL, O. 1992. Carpholite, sudoite, and chloritoid in low-grade high-pressure metapelites from Crete and the Peloponnese. *European Journal of Mineralogy*, **4**, 487–507.

THOMSON, S. N., STÖCKHERT, B. & BRIX, M. R. 1998a. Thermochronology of the high-pressure metamorphic rocks of Crete, Greece: implications for the speed of tectonic processes. *Geology*, **26**, 259–262.

——, ——, RAUCHE, H. & BRIX, M. R. 1998b. Apatite fission-track thermochronology of the uppermost tectonic unit of Crete, Greece: implications for the post-Eocene tectonic evolution of the Hellenic subduction system. *In*: VAN DEN HAUTE, P. & DE CORTE, F. (eds) *Advances in Fission-Track Geochronology*. Kluwer Academic, Dordrecht, 187–205.

WACHENDORF, H., BAUMANN, A., GWOSDZ, W. & SCHNEIDER, W. 1974. Die „Phyllit-Serie" Ostkretas – eine Mélange. *Zeitschrift der deutschen geologischen Gesellschaft*, **125**, 237–251.

WACHMANN, M. 1997. *Die strukturelle Entwicklung hochdruckmetamorpher Karbonatgesteine bei Agios Theodori, SW Kreta*. Bochumer geologische und geotechnische arbeiten, **48**.

WHEELER, J. & BUTLER, R. W. H. 1994. Criteria for identifying structures related to true crustal extension in orogens. *Journal of Structural Geology*, **16**, 1023–1027.

WIJBRANS, J. R., VAN WEES, J. D., STEPHENSON, R. A., & CLOETINGH, S. A. P. L. 1993. Pressure–temperature–time evolution of the high-pressure metamorphic complex of Sifnos, Greece. *Geology*, **21**, 443–446.

WORTEL, M. J. L., GOES, S. D. B. & SPAKMAN, W. 1993. Structure and seismicity of the Aegean subduction zone. *Terra Nova*, **2**, 554–562.

ZEILINGA DE BOER, J. 1989. The Greek Enigma: is development of the Aegean Orogene dominated by forces related to subduction or obduction? *Marine Geology*, **87**, 31–54.

# Oscillating modes of orogeny in the Southwest Pacific and the tectonic evolution of New Caledonia

T. J. RAWLING & G. S. LISTER

*Australian Geodynamics Cooperative Research Centre, VIEPS, Department of Earth Sciences, Monash University, Melbourne, Vic. 3168, Australia (e-mail: timr@earth.monash.edu.au)*

**Abstract:** This paper presents a new interpretation of the tectonic evolution of New Caledonia, based on the extrapolation of detailed structural analysis at several different scales. The coherent high-pressure schist belt of northern New Caledonia contains clear evidence for cyclicity in the orogenic process. Two switches from large-scale crustal shortening to extensional tectonism can be recognized. We propose that orogenesis initially resulted in significant crustal thickening, obduction, and high-pressure (15–20 kbar) metamorphism. There is a close temporal link between ophiolite emplacement and high-pressure metamorphism. The high-pressure rocks were exhumed during the first period of extensional tectonism (c. 40–36 Ma) as a large coherent terrane (>1500 km$^2$). During the second period of crustal shortening the New Caledonia orogen was folded into upright megafolds. Basin and Range style normal faulting then took place, bringing the folded extensional terrane and the high-pressure rocks to their present crustal level. Metamorphic grade changes in the high-pressure belt are structurally controlled by these late-stage normal faults. However, movement on these faults is responsible for only the very last stages of exhumation.

This paper considers the role of normal faulting and ductile shear zones in the exhumation of a coherent high-pressure terrane in the Southwest Pacific, during the post-Eocene evolution of this part of the Alpine–Himalayan collision belt. We argue that the rocks of New Caledonia record a history of 'oscillating orogeny', and that this sheds new light on the complex puzzle of how high-pressure metamorphic rocks are formed, and how they are exhumed. Unlike the proponents of alternative theories for the origin of such terranes (e.g. Cloos 1984; Platt 1986; Platt & England 1994) we argue that their evolution can be the result of processes that occur solely in the overriding plate, in front of a major subduction system. The rocks of the high-pressure orogenic belt of New Caledonia need never have been part of a downgoing (subducting) slab.

The Mesozoic tectonic evolution of the Southwest Pacific involved extensional break-up of the active eastern margin of Gondwana (Weissel & Hayes 1977; Cande et al. 1989; Veevers et al. 1990; Scheibner et al. 1991; Veevers & Li 1991; Yan & Kroenke 1993). The pattern of extension appears to have resulted in the formation of large slivers of continental crust (Lister & Etheridge 1989; Lister et al. 1991) separated from the coast of Gondwana–Australia by a number of deep basins (Fig. 1). These extensional basins matured and new oceanic crust began to form, in the New Caledonia Basin (Laird 1993), in the Tasman Sea (Hayes & Ringis 1973; Shaw 1990), and in the Coral Sea (Kroenke 1984). At the same time, there have been complex interactions between the Indo-Australian and Pacific plates resulting in the formation of numerous subduction and collision zones (Kroenke 1984; Seno & Peterson 1984; Daniel et al. 1986; Ferriere & Chanier 1993; Yan & Kroenke 1993).

The tectonic evolution of the Southwest Pacific (from c. 110 Ma to the present) involved eastward migration of the collisional boundary between the Indo-Australian and the Pacific plates. During Late Jurassic and Early Cretaceous time (c. 160–110 Ma) the Lord Howe Rise and the Norfolk Ridge system (Fig. 1) lay adjacent to the active eastern margin of the Gondwanan supercontinent (Blake et al. 1977; Kroenke 1984; Lister et al. 1991). Rifting of this continental margin began in Late Cretaceous time and by the beginning of Tertiary time (75 Ma) the New Caledonia Basin was a relatively wide oceanic basin (Yan & Kroenke 1993). New oceanic crust had begun to form in the southern Tasman Basin (Hayes & Ringis 1973; Shaw 1990) as well as between Antarctica and Australia (Veevers et al. 1990).

By mid-Eocene time, spreading had stopped in the New Caledonia Basin, but was continuing in the Tasman Sea (Yan & Kroenke 1993). A series of northeast-dipping subduction zones

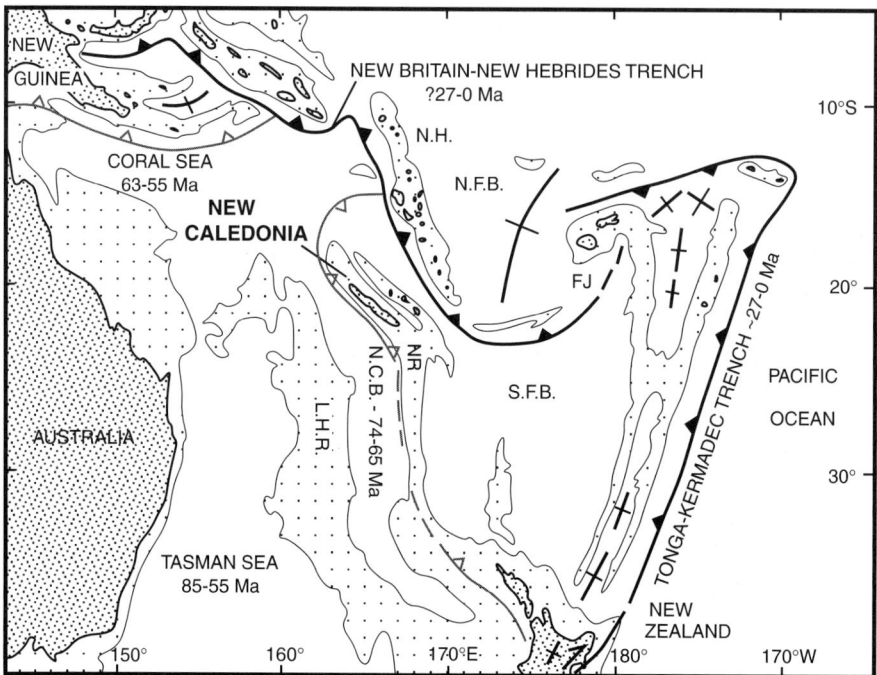

**Fig. 1.** Tectonic map of the Southwest Pacific, showing the present location of New Caledonia on the Norfolk Ridge, and the ribbons of continental crust and intervening basins produced by dismemberment of Gondwana. Currently active subduction zones are indicated in black. Possible trace of the Eocene subduction system is indicated in grey. FJ, Fiji; LHR, Lord Howe Rise; NCB, New Caledonia Basin; NFB, North Fiji Basin; NH, New Hebrides; NR, Norfolk Rise; SFB, South Fiji Basin.

formed beneath various parts of the Norfolk Ridge. Subduction of the relatively young oceanic crust of the New Caledonia Basin beneath New Caledonia had begun, resulting in extension and spreading in the Loyalty Basin, the back-arc to this system (Fig. 1). Kroenke (1984) suggested that the subducting segments at New Caledonia, northern New Zealand and the Rennell Arc were joined by long transverse faults, involving strike-slip displacements, but subduction may have been more extensive along this zone.

At the end of Eocene epoch, subduction had ceased in the New Caledonia Basin and the locus of convergence had shifted to a new southwest-dipping subduction system in the Manus–Solomon–Vitiaz region (Yan & Kroenke 1993). Early Eocene back-arc ophiolites were overthrust in Papua New Guinea (Davies & Jaques 1984; Crook 1989), New Caledonia (Paris 1981; Brothers & Lillie 1988) and northern New Zealand (Malpas et al. 1992). However, timing of ophiolite emplacement in many of these terranes is not well constrained (see below).

Convergence ceased along the Manus–Solomon–Vitiaz trench as a result of the arrival of the Ontong Java Plateau at the subduction zone (sometime before the end of the Oligocene). Subduction was transferred to a south-dipping subduction zone north of Papua New Guinea and a west-northwest-dipping subduction zone at the proto-Kermadec–Tonga trench. Spreading had also begun in the Caroline and South Fiji Basins. During the late Miocene the extensive northeast-dipping Solomon–New Hebrides subduction system was initiated. 'Rollback' of the subducting slab, possibly in response to a more east–west directed convergence vector in the Pacific, resulted in migration of this system toward the southwest (Yan & Kroenke 1993; Scholl 1995). Several, transient, east-dipping subduction zones may also have formed beneath the Three Kings Arc and potentially within the Loyalty Basin at this time (Kroenke 1984).

Although the overall tectonic pattern is evident, there are many uncertainties that need to be resolved before the tectonic evolution of the Southwest Pacific is understood. New Caledonia is situated in the centre of this complex region, and is thus strategically located in terms of its ability to provide critical information to constrain different hypotheses. Ever since plate tectonic theory was first applied to the Southwest Pacific (Carey 1958; Coleman 1967), it has been recognized that a detailed understanding of the evolution of the high-pressure schist belt on New Caledonia was critical to any tectonic interpretation or reconstruction of the region.

## Regional geology of New Caledonia

The regional geology of New Caledonia has been described in detail by Paris (1981) and Brothers & Lillie (1988) and more recently by Cluzel et al. (1994) and Aitchison et al. (1995). Brothers & Lillie (1988) divided the regional geology into five major structural units and classified these by their regional distribution. These are the eastern and western coastal belts, the central chain, the ultramafic nappe and the northeastern region (Fig. 2a). This classification was expanded on by Cluzel et al. (1994), who defined eight distinct terranes. The northern region contains all the variably metamorphosed metasedimentary and metavolcanic sequences of the Eocene high-pressure schist belt (Fig. 2b). This region has been divided into western and eastern sectors by Paris (1981) and Brothers & Lillie (1988), and into a Koumac Terrane, Diahot Terrane and Pouébo Terrane by Cluzel et al. (1994).

The Pouébo terrane (Cluzel et al. 1994) is defined as 'the core of a foliation antiform' in the NE part of the island, made up of a mélange containing large mafic eclogitic boulders enclosed in an argillaceous or serpentinitic matrix (Maurizot et al. 1989). These rocks yield the highest pressure estimates obtained from the New Caledonia orogen, as described by Clarke et al. (1997). These workers stated that this terrane occupies the deepest structural levels, because they believed these rocks occupy the core of a large antiform. However, detailed mapping shows that this is not the case (Occhipinti 1994; Streets 1995; Rawling & Lister 1997; and see below). The Pouébo terrane may exist as a discrete entity, but parts of it occupy a structurally high position, above bounding shear zones, on the NE flank of the eclogite–blueschist terrane (see below).

The Diahot terrane (Cluzel et al. 1994) comprises mainly lawsonite-glaucophane-bearing schists, but this is fault bounded to the east and west by relatively young normal faults (Fig. 2b; Rawling & Lister 1997; and see below). The Koumac terrane of Cluzel et al. (1994) lies to the west of the Diahot terrane. The Koumac terrane represents the lowest grade metasediments of the high-pressure belt and the rocks within this region exhibit the highest pressure/temperature ratios of any rocks in the belt (Brothers 1970). The Koumac terrane is bounded to the west by a large strike-slip fault system known as the 'sillon' (Brothers & Blake 1973; Paris 1981). There is very little reason to separate the fault-bounded slivers of the Koumac and western Diahot terranes. Detailed mapping (Rawling & Lister 1997; and see below) shows that the high-pressure schist belt is divided up into a number of fault-bounded blocks and that the demarcation between the Koumac and Diahot terranes is probably defined by several southwest-dipping normal faults that separate different crustal levels of the same metamorphic pile (see below).

## The origin of the blueschists of New Caledonia

Aitchison et al. (1995), Cluzel et al. (1995a,b) and Clarke et al. (1997) suggested that ocean–ocean collision has taken place resulting in east to northeast directed subduction beneath the Loyalty Basin. They suggested that subduction was interrupted by the arrival of a sliver of continental crust from the west. This material was partially subducted, resulting in stacking of continental thrust slices at depth and crustal thickening. Buoyancy of this material resulted in failure of the subduction zone and emplacement of the ophiolite from the northeast via obduction. Exhumation was accomplished by erosion and tectonic unroofing. Aitchison et al. (1995) argued that late-stage extensional tectonism resulted in the formation of a metamorphic core complex in which the blueschists and eclogites are exposed. They described 'a metamorphic core that was shedding lower-grade rocks including the ophiolite nappe and overlying blueschists'.

Concurrent work by Streets (Streets et al. 1993; Streets 1995), Occhipinti (1994) and Rawling & Lister (1997) resulted in the identification of a number of large-scale ductile shear zones along the northeast coast, similar to those described by Clarke et al. (1997). Individual shear zones range in thickness from 10 to 500 m and can often be traced laterally for several kilometres. Based on detailed structural mapping, microstructural analysis and petrology, as well as determination of the pressure–temperature conditions from both the upper and lower plates of

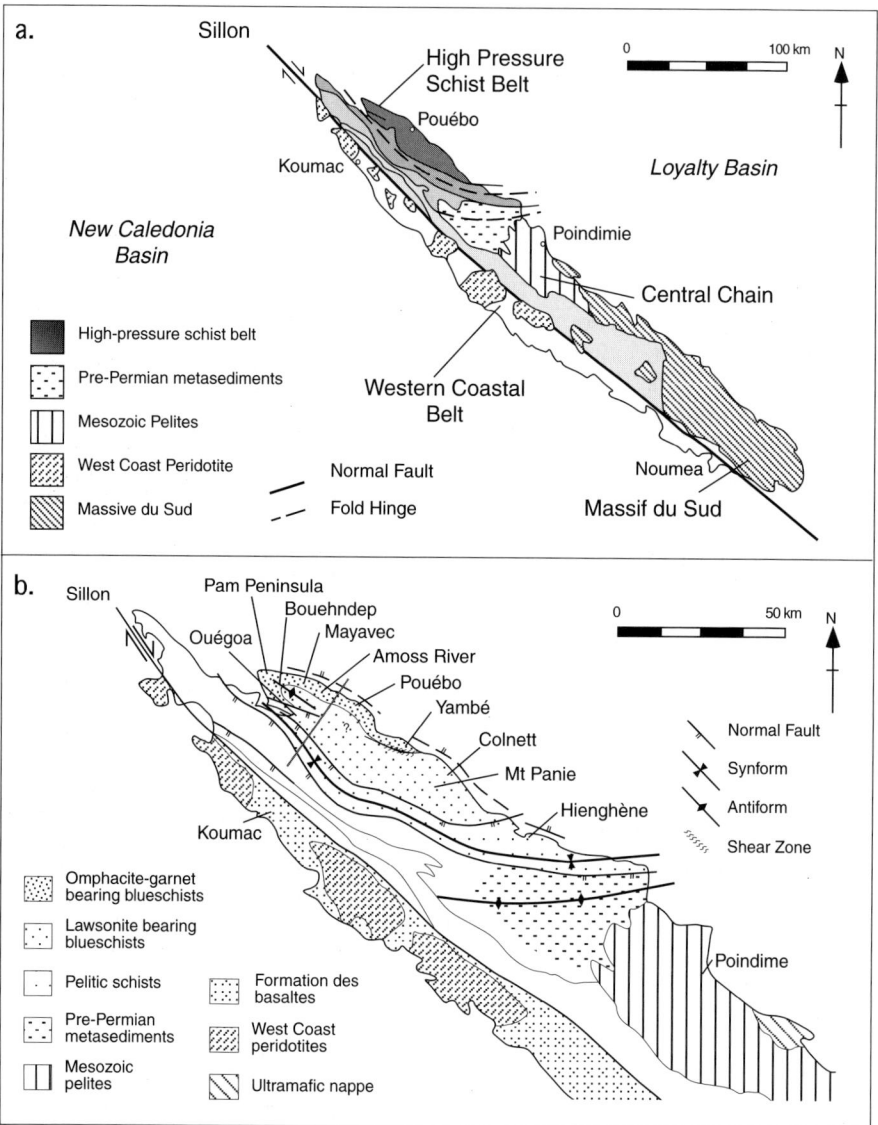

**Fig. 2.** (a) Geological map of New Caledonia outlining the extent of the high-pressure schist belt and its relationship to other major rock units and regions as well as to the 'sillon' or Trans-Caledonian Fault. (b) Geological map of the high-pressure schist belt of northern New Caledonia showing significant jumps in metamorphic grade associated with orogen-parallel late-normal faulting. The normal faults dip southwest. Grade increases towards the northeast until finally a zone of omphacite-bearing eclogite and blueschist facies rocks is reached along the coastal mountain range (modified after Brothers 1985). Grey line indicates position of cross-section presented in Fig. 5.

the system, these shear zones were interpreted to be extensional in nature (Streets 1995). Based on these data, it was suggested that these were major zones of mylonitization that represent a carapace shear zone to a large metamorphic core complex (Streets *et al.* 1993; Streets 1995). However, detailed regional mapping (Rawling & Lister 1997; and see below) showed that the carapace shear zone is tightly folded, and dissected by brittle normal faults.

## Approach

Given the structural complexity of the region, the relatively poor outcrop because of the tropical climate and the potential for controversy, an approach involving detailed structural analysis was undertaken. A network of well-studied key localities was established, concurrent with regional mapping and structural traverses. This work allowed the correlation of fabrics and microstructures over a wide region, allowing links to be made between the timing of the processes of deformation and metamorphism and the development of large-scale structures. The purpose of this paper is not to provide exhaustive detail of this analysis, but to provide an overview of the main results, and an attempt at synthesis.

We approach the problem by first establishing the details of the geometry (at a variety of scales), and then removing the effects of each major deformation event in turn. In this way, we reconstruct the configuration of the high-pressure schist belt of New Caledonia, as it was, at an early stage of its genesis. Then we speculate on the geodynamic processes that led to high-pressure metamorphism, and on the denudation and exhumation of this coherent metamorphic terrane.

## Orogen-parallel late normal faults

Previous workers have identified a series of roughly northwest-striking metamorphic isograds within the northern high-pressure schist belt (Bell & Brothers 1985; Yokoyama *et al.* 1986). These isograds define a normal sequence of high-pressure, low-temperature metamorphic facies. These are characterized, in turn, by the first appearance of lawsonite, Mn-garnet, epidote, and omphacite (Yokoyama *et al.* 1986). However, when detailed mapping was carried out at localities where grade changes occur, no evidence for progressive metamorphism was obtained. Rather, evidence for large-scale faulting was revealed (Fig. 3).

The faults are difficult to identify in the field because of poor outcrop surrounding the fault zones. However, within *c.* 15 m of the inferred position of the faults, older penetrative fabrics are locally sheared and reoriented into parallelism with a steeply southwest-dipping spaced cleavage (Fig. 3c). This new fabric is developed within only *c.* 5 m of the inferred position of the faults and it results in a marked increase in fissility of the rocks. The orientation of this spaced fabric may approximate that of the adjacent fault, in which case these late-stage faults dip between *c.* 30° and *c.* 70° towards the southwest. The intensity of these fault-related fabrics increases from southwest to northeast.

Several generations of structural fabrics can be recognized (Fig. 3). The youngest are a series of kinks and warps which maintain a relatively consistent shallow northwesterly plunge across the map area (and through much of the high-pressure schist belt). These structures overprint all other fabrics.

There are many older mesoscopic upright folds that reorient schistosity throughout the low- to medium-grade schist belt. The orientation of the fold axes varies considerably between different fault-bounded slices. In the low-grade schist domain (Fig. 3b) they typically plunge shallowly to the northwest. In the lawsonite-bearing schist domain they plunge steeply to the southwest, and in the garnet–glaucophane schist domain, the folds plunge gently to the south-southwest (Fig. 3b). The younger fabrics maintain a consistent orientation between individual fault-bounded blocks, but each fault-bounded block contains differently oriented upright folds of the same generation (Fig. 3). This suggests that fault activity post-dates upright folding but pre-dates late kinking and warping.

Some rotation must have occurred along these faults as well as dip-slip movement, resulting in the reorientation of the older structures and fabrics, and juxtaposition of different crustal levels that display different metamorphic grades. The magnitude of displacement along these faults is difficult to estimate because of the lack of easily identifiable offset marker horizons. However, based on drops in pressure from footwall to hanging wall, we estimate that the displacement must have been of the order of several kilometres on each of the faults.

The late faults consistently dip to the southwest and separate lower-pressure material in the hanging wall from higher-pressure material in the footwalls. In isolation this relationship may be indicative of either normal faulting or out of sequence thrusting. However, given the consistency of this relationship across several of these faults, as well as the lack of evidence for typical thrust faulting relationships in this region, it seems more probable that these faults represent large-scale normal faults.

Regional mapping in the Diahot region shows that, at least on the scale of tens of kilometres, these faults are laterally continuous (Fig. 3a). Based on regional correlations we interpret that the same faults may extend as far south as Hienghène. These faults are thus developed on the same scale as that of the New Caledonia

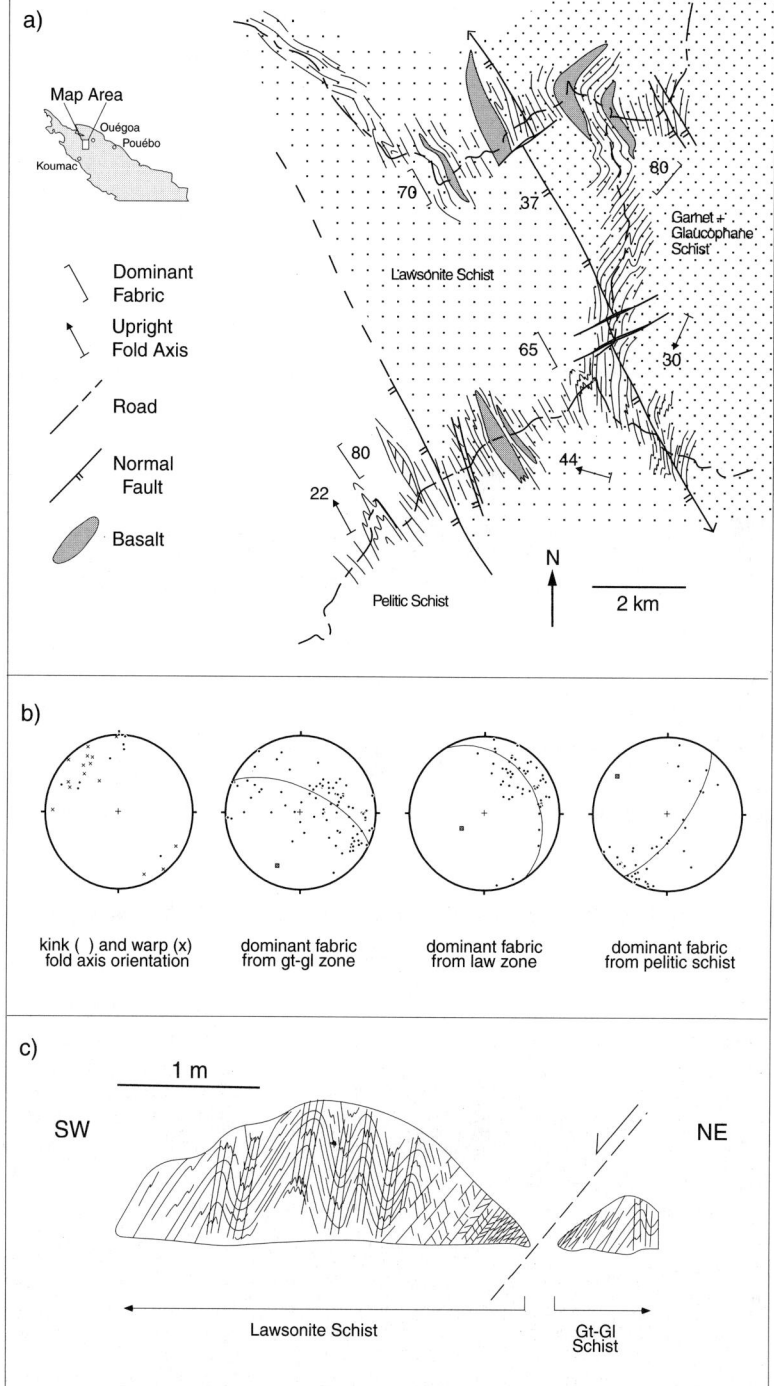

**Fig. 3.** (a) Map of structural relationships across several late brittle normal faults in the Ouégoa region. (b) Stereographic projections, from each fault-bounded block, showing the orientation of upright folds as well as orientation of kinks and warps across the entire map area. (c) Field sketch of roadside outcrops on either side of an inferred normal fault showing fabric development and reorientation close to these faults.

schist belt. They are arcuate in form and run parallel to the grain of the orogenic belt (Fig. 2b).

These orogen-scale structures are overprinted by one or more generations of younger faults. The northeastern margin of the mountains in the region between Pouébo and Hienghène have a flat slope that dips seaward at around 15–25° (Fig. 4a). The slope has consistent orientation over a large area, except where incised by steep river systems. This slope is interpreted as a fault scarp that has retreated to a variable degree owing to high erosion rates. No kinematic indicators have been preserved as the fault surface itself has been removed. However, the recrystallized karst limestones exposed off the coast near Hienghène appear to be similar to those that occur in the lower-pressure metamorphic assemblages from higher in the tectonostratigraphy

**Fig. 4.** (a) Photograph of the northeastern coast of New Caledonia showing the effect of late seaward-dipping normal faulting on the geomorphology of the high-pressure schist belt. (b) Photograph of strongly folded mylonitic schists of the Col d'Amoss shear zone. Pencil is 14 cm in length.

east of Koumac (Carroué 1971). This suggests that the hanging wall is downthrown and that these faults have a normal movement sense. Submerged normal fault traces have been identified that are parallel to these faults within the South Loyalty Basin (Lafoy et al. 1996), and these are probably of the same age.

Late-stage normal faulting (both southwest- and northeast-dipping) must contribute in part to the exhumation of the high-pressure belt exposed on the northeast coast of New Caledonia.

## Orogen-scale tight upright megafolds

The dominant fabric in the low- to medium-pressure schist belt is generally steep, but it is folded by several younger generations of upright folds. There is systematic variation of vergence of these folds across the belt, indicating that the dominant fabric was originally gently dipping, and was steepened during upright folding.

The most common and regionally significant generation of folding is common at both microscopic and mesoscopic scales. It is defined by upright angular folds that are open in the western, low-grade part of the schist belt, and are tighter and locally isoclinal in the east near Ouégoa and Col d'Amoss. Although the orientation of the axial plane of these folds varies between different fault-bounded slices, systematic analysis of overprinting relations allows folds of the same generation to be identified in each slice. In the garnet–glaucophane and lawsonite-bearing rocks (Fig. 3) the folds plunge variably to the southwest whereas in the lower-grade rocks to the west the fold axes plunge gently to the northwest.

The amplitude of the regional-scale upright folds can be estimated by constructing enveloping surfaces, smoothing out the effect of mesoscopic-scale parasitic folds. The asymmetry of the enveloping surfaces suggests that macro-scale vergence changes several times across the high-pressure schist belt, defining moderate to tight, crustal-scale upright folds, with amplitudes of the order of 10–20 km. These are schematically illustrated in Fig. 5a.

It is unusual to identify structures of this magnitude, as it implies geometry on a crustal scale. These very large scale folds were obviously considerably disrupted by the late normal faults described above. Although the actual synformal closure cannot be identified, vergence changes suggest that the southwest limb of the synform occurs in the low-grade western zone, and the northeast limb in the moderate- to high-grade Ouégoa zone (see Figs 2 and 5a). The hinge zone has been faulted out by late normal faults.

A possible reconstruction of the crustal geometry before movement on late brittle normal faults is presented in Fig. 5b. Orogen-scale tight upright folding reoriented a gently dipping fabric into more steeply dipping attitudes. There is a significant difference in the amount of exhumation, from the western part of the fold to the northeast, suggesting that the now exposed limbs of the synform come from markedly different crustal levels, possibly because of the effect of earlier extension (see below).

This hypothesis, that a crustal-scale mega-syncline exists within the schist belt, was tested by searching for adjacent orogen-scale folds, which should in this case be tight upright antiforms. The core of a large antiform outcrops to the southwest of the orogen (in the Poindimié area; see Maurizot 1984). There is an overall arcuate trend to the orogenic belt, as can be noted in the changing trend of structures in the high-pressure schist belt. The Poindimié antiform follows this trend, and its axial zone curves towards a northwest–southeast trend. However, this regional antiform terminates at the younger strike-slip 'sillon' fault zone and thus is not exposed on the approximately east–west Koumac–Ouégoa traverse.

The core of the adjacent orogen-scale antiform to the northeast may be defined by the outcrop of the high-grade metamorphic rocks, in the eclogite–blueschist domain along the northeast coast. The highest grade block of the high-pressure schist belt (on the Pam Peninsula) has been referred to as domal or antiformal by numerous previous workers (e.g. Clarke et al. 1997). A traverse across the Bouehndep region revealed the presence of a large upright antiform, and this has folded all previously formed fabrics. However, further to the south, the crest of this regional structure is truncated by the relatively young normal fault system that defines the coastal geomorphology.

## Middle-stage extensional shear zones mantling the eclogite–blueschist domain

A series of discrete mesoscopic shear zones can be identified within many exposures of blueschist and eclogite along the northeast coast (e.g. in the Mayavéc or Yambé valleys). These shear zones typically transpose the older peak high-pressure fabrics within discrete zones of intense S–C fabric development. They range in width from tens of centimetres to tens of metres. The orientation of this generation of shear fabric

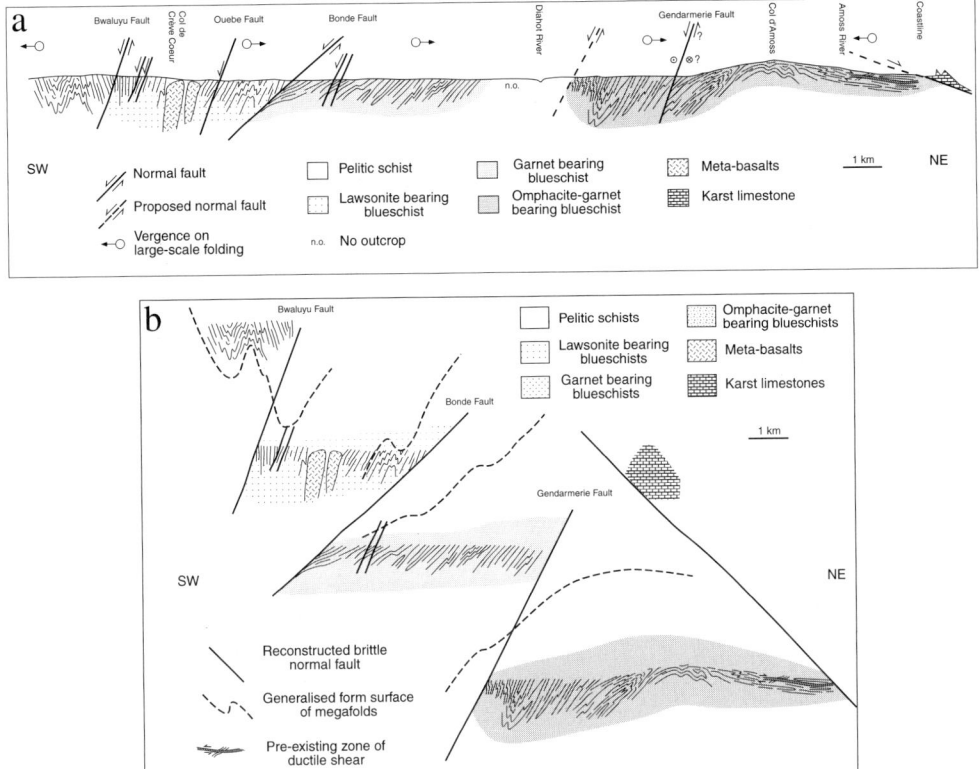

**Fig. 5.** (a) Schematic cross-section across the high-pressure schist belt between the Amoss River and Vallee Tomy in northern New Caledonia. In the northeast, there are young fault scarps marking structures that dip northeast and which may have down-dropped limestones in the Hienghène area against the high-pressure metamorphic rocks. Also shown are several major retrograde ductile shear zones formed during metamorphic core complex development which juxtapose eclogite-bearing remnants of the ultramafic nappe against deformed blueschist facies metasediments in the lower plate. Towards the southwest, a number of brittle normal faults are shown and vergence changes indicated which suggest the existence of orogen-scale tight upright folds of an earlier more gently dipping fabric. (b) A schematic restored cross-section of the same region of the high-pressure schist belt as is presented in (a). This representation removes effects of the different generations of late normal faults and shows how the hinge of the synformal megafold has been removed by faulting.

is typically sub-horizontal, although locally the fabrics are reoriented by later upright folding.

These middle-stage ductile shear zones can be correlated between outcrops based on their geometry and timing relations. Throughout the entire schist belt of northeast New Caledonia the development of these shear zones was immediately preceded by an episode of widespread recumbent folding, which was contemporaneous with peak metamorphic conditions. Temperature outlasted deformation in these folds, because previously formed (high-pressure) foliations are decussately recrystallized in the fold hinges. Small-scale shear zones, in which deformation outlasts temperature, then developed. These often coincide with fold limbs, which become highly attenuated as a result. However, the shear zones also crosscut limbs and less commonly the hinge zones of these folds, and thus overprint the recumbent folds. Therefore they form a distinct generation of fabric.

Deformation in these shear zones occurred while significant exhumation was accomplished. Foliation boudinage of older high-pressure fabrics is common immediately adjacent to these zones. The dilational space in the necks of these boudins is often filled with glaucophane and quartz. This suggests that middle-stage shearing commenced under high-pressure conditions. In the adjacent fabric glaucophane needles (1–5 mm in length) and zoisite laths are aligned along shear planes, generally with a northeast–southwest trend. Deformation clearly continued after high-pressure metamorphic minerals stopped growing, leading to the development of localized, intensely foliated and lineated metamorphic tectonites. However, metamorphic conditions continued to diminish while deformation took place. The shear planes in S–C fabrics are defined by aligned zones of shredded phengitic white mica. Shear fabrics overprint peak high-pressure assemblages, indicating that deformation continued until greenschist facies conditions were prevalent. The shear zones are typically retrogressed to greenschist facies and are strongly chloritized, indicating fluid infiltration synchronous with shearing.

The microstructures in the small-scale shear zones described above are indistinguishable from those in the major ductile shear zones that flank the eclogite–blueschist core of the high-pressure belt.

*Shear zones on the northeast coast*

Major middle-stage shear zones have been recognized at various localities along the northeast coast. These systems typically juxtapose an upper plate of serpentine matrix mélange containing mafic boulders of relatively fresh eclogite against a lower plate dominated by blueschist–greenschist facies schists and gneisses. The schists of the lower plate are derived from a sedimentary precursor. The 'knockers' in the upper plate have an unknown provenance. They have much in common with other mélange occurrences in which 'knockers' are reported in serpentinitic, argillaceous, or altered ultramafic matrices (e.g. the Franciscan Complex, Cloos 1986; Syros, Ridley 1984).

The matrix material of the upper plate is strongly altered, to chlorite, actinolite and talc. However, the mafic pods (which range in scale from metres to tens of metres) are relatively unaffected by this alteration, and they contain fresh undeformed eclogitic assemblages. The siliceous schists and gneisses of the lower plate have been strongly retrogressed from eclogite to greenschist or blueschist facies.

*Mylonitic schist of the Col d'Amoss*

A second larger-scale shear zone has also been identified at Col d'Amoss in the northeastern high-pressure schist belt. This shear zone is exposed as a sequence of folded mylonitic schists of the order of 1 km thick (Fig. 4b).

Intense deformation in the shear zone at Col d'Amoss resulted in extreme transposition of older fabrics. The intensity of deformation, in association with subsequent retrogression and the effects of fluid movement, obliterated almost all of the original mineralogy in the shear zone. Only quartz and phengite remain, except locally within narrow microlithons where relics of the older metamorphic mineral parageneses are preserved.

Mylonitic shear fabrics associated with this shear zone have identical overprinting relationships to those described from the smaller-scale shear zones described above. In addition, however, the mylonitic schists are overprinted by several generations of upright folds, kinks, and a number of steeply dipping brittle faults. The shear zone of the Col d'Amoss has been tightly folded subsequent to its formation.

## Tectonic evolution of the high-pressure metamorphic belt of New Caledonia

The structural relations discussed above allow the relative chronology of various deformation and thermal events to be described in detail (Fig. 6). As a result, we can begin to restore the large-scale geometry of the dominant fabrics, at least in schematic form, as they were before regional upright folding and the subsequent late-stage normal faulting (Fig. 7). The significance of the middle-stage ductile shear zones that mantle the eclogite–blueschist domain now becomes apparent.

The sequence of discrete mesoscopic shear zones in the Mayavéc, Amoss and Yambé streams juxtapose ultramafic rocks against blueschist facies siliceous continental metasediments. These shear zones may be part of a single large shear zone on the northeast flank of the core zone, with a high-pressure ultramafic nappe as its upper plate. Similarly, mylonitic schists at a variety of locations along the southwest flank, including the Col d'Amoss, may also be part of a single large shear zone, with lower-grade (accretionary prism?) metasediments as its upper plate. It is the essence of this geometry that we need to explain if we are to reach an understanding of the tectonic evolution of New Caledonia.

The large-scale geometry of these carapace shear zones that bound the flanks of the high-pressure eclogite–blueschist core is illustrated in Fig. 7, as it would have been before the second cycle of crustal shortening and extension. There may be additional complexities that need to be added to this picture; for example, there may be several interleaved shear zones on the northeast

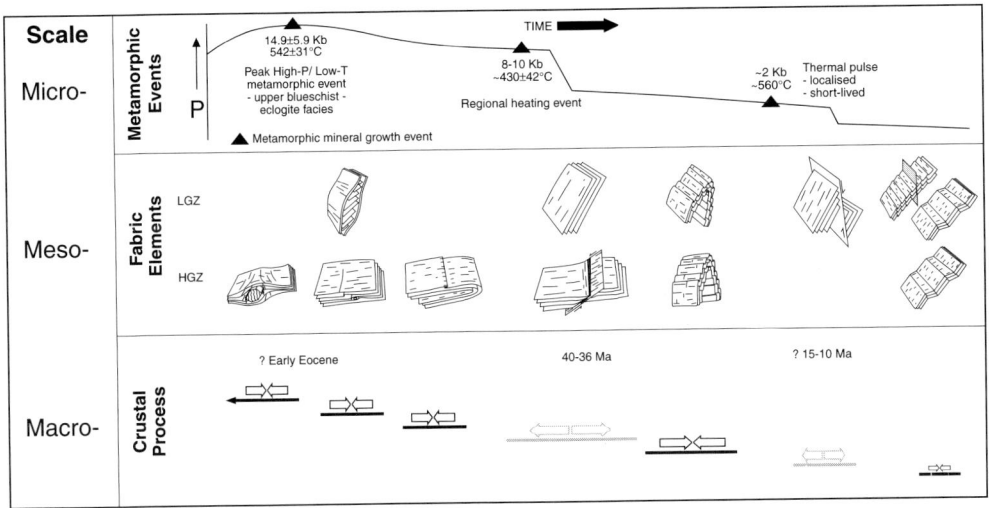

**Fig. 6.** Schematic time-line showing relative ages of pressure prograde metamorphic events, pressure retrograde thermal events, periods of deformation, resultant fabrics and interpreted tectonic conditions. LGZ, lower-grade zone (region between Ouégoa and Koumac); HGZ, higher-grade zone (region of the Pam Peninsula).

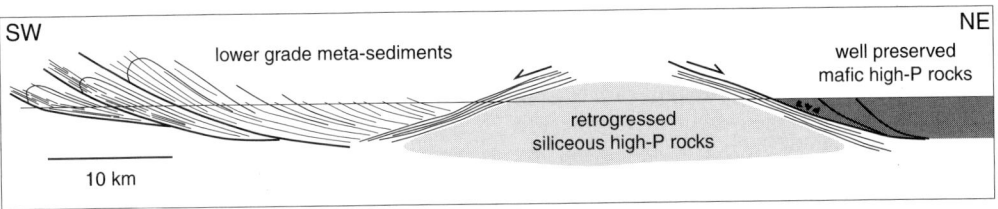

**Fig. 7.** Schematic restoration of the large-scale geometry of the dominant fabrics, as they were before regional upright folding and the subsequent late-stage normal faulting.

flank. However, these do not influence the overall conclusion. These mylonitic high-strain zones are analogous to the domed mylonite and detachment zones of the metamorphic core complexes of the US Cordillera (e.g. Spencer & Reynolds 1989). Although more than one crustal-scale shear zone is involved, this is not dissimilar to what is reported from the US core complexes (Davis 1983; Coney & Harms 1984; Davis *et al.* 1987). However, the structure differs from that of the classic core complexes in that the orogen has been subsequently subject to large-scale crustal shortening, and then been dismembered by normal faults.

The shear zones beneath the high-pressure ultramafic nappe form the northeastern carapace to the high-pressure eclogite–blueschist core zone. The mylonitic schists of the Col d'Amoss form the southwestern carapace.

However, the present outcrop pattern is the result of a combination of different events. The folded Col d'Amoss shear zone is cut by late-stage normal faults. No detachment fault related to the earlier core complex forming event can be recognized.

## Discussion

The upper plates of the shear zones on the northeast flank typically contain pristine knockers of undeformed mafic eclogite entrained in a deformed serpentinized matrix. The lower plates contain retrogressed siliceous blueschists. These blueschists do not contain any serpentinized material but locally relict peak eclogite mineralogies have been identified that are strongly overprinted by blueschist and greenschist assemblages. This is consistent with a history of thrusting followed by extension, as observed in several other locations along the Alpine–Himalayan collisional system (e.g. Forster & Lister this volume), as well as other high-pressure belts (e.g. Baja peninsula: Sedlock, this volume).

The preservation of pristine eclogitic assemblages within undeformed mafic knockers implies that the upper plate cooled after metamorphism, and that it remained cool; for example, as the result of exhumation. The pervasive overprinting of similar assemblages in the lower plate suggests that these rocks remained at deeper crustal levels, where temperatures remained elevated, allowing the effects of subsequent metamorphic events and retrogression to take place. The subsequent juxtaposition of the already cooled eclogites of the upper plate against retrograded equivalents in the lower plate suggests that the middle-stage shear zones are extensional in their nature, and that the eclogites of the upper plate were already at relatively high crustal levels when these shear zones began to operate.

The lower plate to the mylonitic shear zone in the Col d'Amoss (i.e. the schists and gneisses of the eclogite–blueschist domain) has preserved microstructures and metamorphic mineral assemblages that formed later than the microstructures and the assemblages preserved in the mafic blocks of the upper plate (i.e. in the ultramafic nappe). These shear zones are therefore interpreted to be the result of large-scale crustal extension resulting in exhumation and cooling. This cooling may be responsible for the c. 38 Ma cooling age (Blake et al. 1977; Ghent et al. 1994) reported for many white micas from the high-pressure schist belt using the $^{40}Ar/^{39}Ar$ thermochronology technique. As described above, the Amoss–Mayavéc shear zone preserves very similar microstructures and overprinting relationships and can thus be interpreted to have formed in response to the same major crustal extension event. The mesoscopic shear zones described above were critical to the exhumation of high-pressure metamorphic rocks in New Caledonia.

## Tectonic evolution of New Caledonia

Numerous tectonic models have been presented over the last 20 years that attempt to explain the formation and exhumation of the New Caledonia blueschist–eclogite assemblages within the known tectonic framework of the Southwest Pacific region. More recently published tectonic models include northeast-directed obduction (Brothers 1974; Briggs et al. 1978; Paris 1981; Brothers & Lillie 1988), southwest-directed obduction–subduction and core complex development (Aitchison et al. 1995), northeast-directed subduction (Brothers 1974), and south-directed subduction (Brothers & Blake 1973). Many are intuitive and novel explanations of a very complex problem. There are a number of problems involved in interpreting the tectonic history of this region in terms of a classical subduction or obduction setting.

There are no exposed Cenozoic volcanic arcs in the region and little of the regional plutonism that is usually indicative of a subduction zone setting (Brothers & Lillie 1988). In addition, there is a well-preserved obducted ophiolite exposed in the southern and western parts of New Caledonia that appears to have been emplaced from the northeast at the same time as high-pressure metamorphism was occurring in the north (Avias 1967; Collot et al. 1987; Brothers & Lillie 1988). Yet, where exposed, this ophiolite is relatively thin and the rocks beneath it are only weakly metamorphosed. There is no similar obducted material preserved in the region of the high-pressure metamorphic belt.

Finally, although coherent high-pressure metamorphic terranes are relatively common around the world, it is difficult to explain the exhumation of large volumes of rock as coherent units to the surface. It is currently believed that most high-pressure rocks that form near active margins are a result of material being forced down a subduction zone to great depth and then returned to the surface via complex mechanisms of underplating, corner flow, buoyancy, extrusion or dynamic wedge tectonics (e.g. Cloos 1984; Platt 1986; Michard et al. 1993; Hacker et al. 1995; Mancktelow 1995; Chemenda et al. 1996; Thompson et al. 1997). Although it is clear that in many places there is evidence to suggest that some or all of these processes are important, often they do not adequately describe the frequently very rapid exhumation of large coherent belts of rock, pristine examples of which are exposed in New Caledonia. The following model was developed to explain the geology of northern New Caledonia. It considers the currently exposed geometry of the belt, the inferred chronology of tectonic events, and modern and ancient regional tectonic settings (Kroenke 1984; Malpas et al. 1992; Kroenke & Yan 1993; Laird 1993; Yan & Kroenke 1993). The model, which can be divided into six tectonic stages, is presented as a series of lithospheric crustal sections (Fig. 8) and is described in detail below.

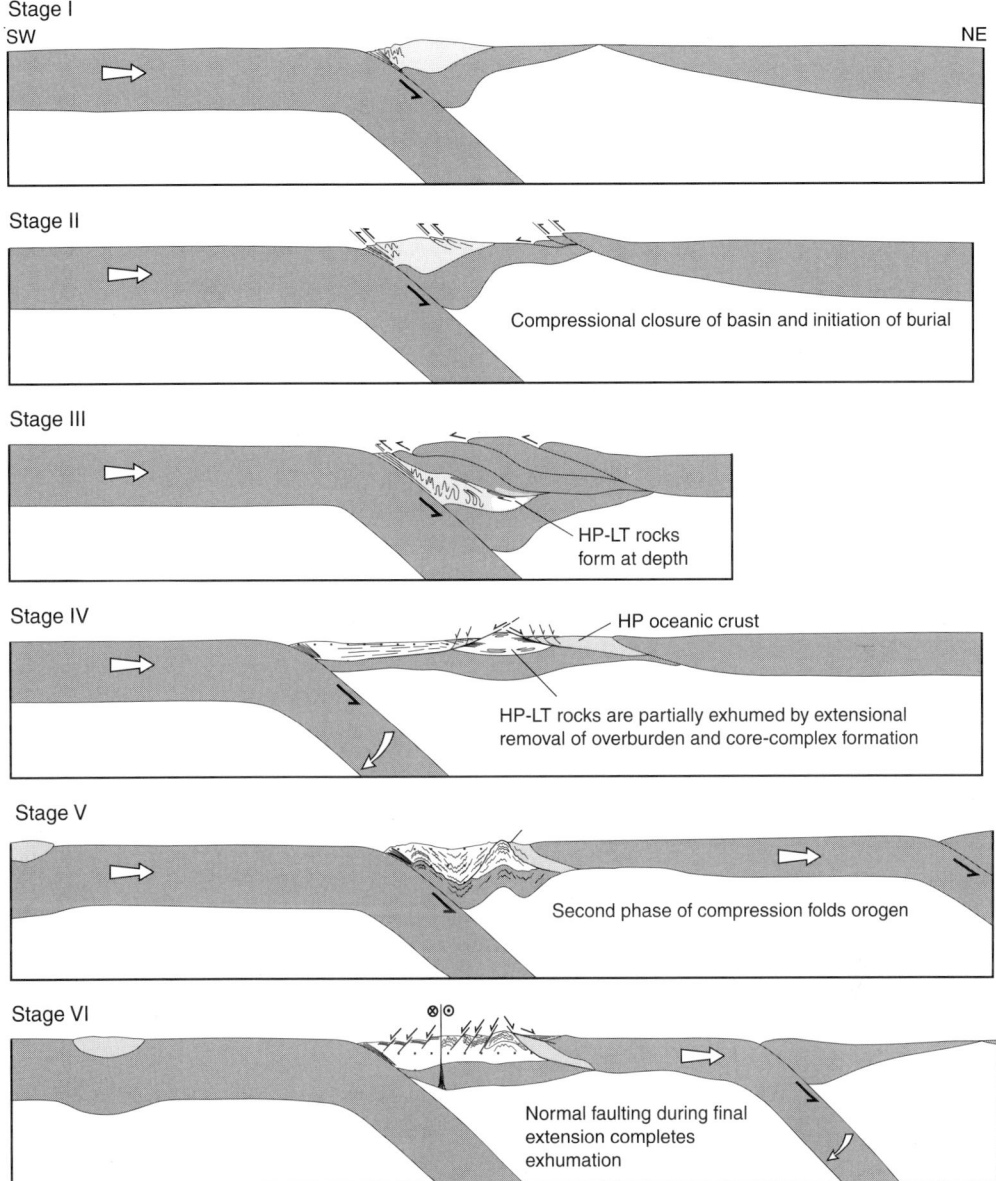

**Fig. 8.** Lithospheric cross-sections representing various times during the cyclic process of mountain belt formation and destruction (see detailed descriptions in text).

## A new model for the tectonic evolution of New Caledonia

Northeast-directed subduction was initiated to the west of the present island system at about 55 Ma in response to a major change in the direction of Indo-Australian plate motion (Stage I; Yan & Kroenke 1993). This subduction system may have extended as far north as New Guinea and has been referred to as the Papuan–New Caledonia–Norfolk trench system by Yan & Kroenke (1993). Subduction progressed beneath a sliver of continental crust rifted from the eastern margin of Gondwana during Cretaceous time (Lister & Etheridge 1989; Lister *et al.* 1991). As subduction continued, roll-back and associated back-arc spreading created new crust behind the sliver of continental crust (Fig. 8a). Subsequently, back-arc basin inversion was initiated (Stage II) in response to a change in Pacific and Indo-Australian Plate convergence direction at or before 43 Ma. This resulted in compressional orogeny along the northern margin of the Australian Plate.

Back-arc inversion progressed as a result of continued compression during Stage III and resulted in the closure of the newly formed back-arc basin and emplacement of a series of thrust slices onto the wedge of continental crust (Fig. 8c). This resulted in deep burial (>70 km) of the mixed continental and oceanic material at the base of the thrust pile and consequently high-pressure metamorphism of these rocks. This event was probably the same as that which drove the large-scale obduction in the Owen Stanley belt of Papua New Guinea (Davies & Jaques 1984; Crook 1989). New Caledonia thus represents a fragment of the root of a major mountain belt that extended over much of the Pacific. This mountain belt has subsequently been destroyed as the result of subsequent large-scale extension.

Another change in convergence direction at about 40 Ma caused slab roll-back beneath New Caledonia resulting in regional extension. At this time Melanesian arc volcanism was also initiated (Yan & Kroenke 1993). During Stage IV the overthickened mountain belt, to the northeast of the New Caledonia trench, underwent crustal-scale extension resulting in the removal of much of the ophiolitic material emplaced during Stage III. A number of crustal-scale extensional ductile shear zones formed at this time resulting in significant exhumation of the high-pressure rocks (Fig. 8d). Two of these ductile shear zones are exposed today as the carapace to the core complex in the Amoss–Mayavéc region, which also developed at this time (Streets 1995).

This model explains the problematic existence of a metamorphic core complex with a high-pressure core composed of dominantly continental crust, which separates very high pressure (deformed and metamorphosed) oceanic rocks (to the northeast) from lower-pressure (deformed and metamorphosed) continental rocks (in the southwest).

During Stage V the orogen was once again undergoing compressional orogenesis. This period of contraction resulted in the development of major (orogen-scale) megafolds. These folds strongly deform older structures at all scales, including the Amoss–Mayavéc metamorphic core complex which is at present exposed in the core of a major Stage V fold (Fig. 8e).

Stage VI was the final epoch of extension resulting in the present exposure of the rocks in the region. During this period brittle normal faults dissected the orogen and juxtaposed blocks from different crustal levels (Fig. 8f). Cessation of subduction as well as major strike-slip movements on the sillon fault probably occurred in response to another major change in plate motions.

This model for the tectonic evolution of the New Caledonia high-pressure schist belt represents a departure from previous interpretations of the development of this region as well as from many accepted models for the generation of high-pressure rocks in general.

The model requires that subduction beneath the rifted continental sliver resulted in extension in the overriding plate and back-arc spreading. This back-arc spreading generated new, relatively buoyant, oceanic crustal material that was later obducted as the ophiolitic nappe and involved in considerable crustal thickening in northern New Caledonia. This is not the first time that obduction and blueschist generation in New Caledonia have been linked (see Aitchison et al. 1995). However, we suggest that collision and subduction resulted in the formation of the ophiolitic material in the back arc and that it was a result of the overthrusting of this material that crustal thickening and high-pressure metamorphism occurred in New Caledonia.

Several workers have suggested that ophiolites were generated as a result of syn-collisional back-arc extension, based on close temporal and spatial relationships (e.g. Oman, Coleman 1981; Lippard et al. 1986; Bay of Islands, Cawood & Suhr 1992; Banda Arc and Brooks Range, Harris 1992; Ballantrae Complex, Smellie & Stone 1992). Harris (1992) has been able to show that small extensional basins (back-arc) formed during collision at both the Banda Arc in Indonesia and the Brooks Range orogens in Alaska, and that these basins were closed and obducted, probably during the same collisional events. In both cases, complex multi-plate boundary interaction resulted in the progressive style of convergence.

Lippard et al. (1986) described a model for the generation and emplacement of the Oman ophiolite that is similar to the model presented, in this paper, for New Caledonia. They described new ocean crust generation in the back-arc of an active subduction system and how this subduction system is chocked when the continental margin arrives and begins to subduct. A large thrust then forms at the edge of the new basin and much of the oceanic material is overthrust onto the now frozen collisional zone.

Lippard *et al.* attributed high-pressure metamorphism in the region to classical style subduction tectonics, whereby a downthrust slab (or lower plate of the subduction system) is the source for the high-pressure metamorphic material. They suggested that the entire subduction system is overthrust by the ophiolite and that later erosion and normal faulting are responsible for the exhumation and denudation of the high-pressure material. Our model differs in that we suggest crustal thickening associated with stacking of thrust slices of ophiolitic material is sufficient to allow incipient high-pressure metamorphism.

The model presented by Smellie & Stone (1992) for the Ballantrae Complex of southwest Scotland is the most similar to the model presented in this paper. On the basis of a detailed geochemical study they described ocean–ocean collision, outboard of a passive continental margin, in which the active arc splits and a marginal sea forms as a result of roll-back. In a process they described as 'backarc imbrication' (compare with back-arc inversion described above) the marginal basin is closed and a series of imbricate slices of new oceanic crust are emplaced onto the active(?) subduction zone. The major difference between the two models is that the arc was volcanic and no continental material was involved until the point when the continental margin entered the active subduction zone, resulting in subduction polarity reversal.

Working on the Bay of Islands Ophiolite, Cawood & Suhr (1992) linked the formation of ophiolite and its subsequent obduction to arc–continent collision, involving the North American continental margin and the Dunnage Zone island arc. This model differs from the model presented here in that: (1) the ophiolitic material was hot enough at the time of emplacement to produce high-temperature, low-pressure metamorphism in a shear zone now emplaced in the footwall sequences; (2) the thickness of ophiolitic material emplaced was not large enough to cause significant crustal thickening (high-pressure metamorphism); (3) the associated collisional zone involved arc–continent collision and the continental margin involved was irregular in form (promontories and re-entrants).

However, there are a number of important similarities between the models that are significant with respect to formation of ophiolites and mountain building in collisional environments. Cawood & Suhr (1992) stated that the Bay of Islands Ophiolite was formed in an extensional environment in the back-arc of an active collision zone. We believe the same is true for the New Caledonia ophiolite. The Bay of Islands Ophiolite was emplaced via obduction shortly after its generation. This was also true for the ophiolite in New Caledonia. Obduction was initiated in the back-arc in response to cessation of subduction or a change in the rate of subduction. Again, we believe this to be the case in New Caledonia. Finally, the deformations recorded by the rocks of both the upper and lower plates (of the obduction system) attest to episodicity between extension and contraction during their tectonic evolution.

The significance of these models is that they link the processes of obduction and high-pressure metamorphism to processes that occur continually in the hanging wall adjacent to a major subduction zone. If circumstances alter so that convergence accommodated by subduction cannot continue, the adjacent areas of thinned lithosphere (in the back-arc basin) will be subject to crustal shortening, and an episode of consequent crustal thickening can be the cause of an episode of high-pressure metamorphism. However, once retreat of the hinge of the adjacent subduction zone recommences, the overriding plate will be subject to a period of extensional tectonism, during which time remnants of the upthrust sheets will be stranded, and the underlying zone of high-pressure metamorphism will be exhumed as a coherent terrane.

The essential episodicity of processes in the hanging wall to a major subduction zone has been highlighted by the present study. We expect that future work will lead to the recognition of several other examples of oscillating orogeny along the length of the Alpine–Himalayan collision belt.

## Conclusions

The major ductile shear zones and the late brittle normal faults described above formed in response to two distinct phases of continental extension. The first of these appears to have resulted in exhumation of the coherent high-pressure eclogite–blueschist domain from beneath an overriding ultramafic nappe (to the northeast), and from beneath metasediments (derived from an accretionary prism?) to the southwest. This period of extension was associated with core complex development. The late-stage brittle normal faults were associated with relatively narrow ductile shear zones, and these have a geometry more akin to Basin and Range style tilt block development, with an array of arcuate brittle normal faults developed on the scale of the entire orogen.

The core complex in the Amoss–Mayavéc region of New Caledonia is not a 'classical' core complex as described by Coney (1980). Subsequent to the initiation of core complex development the entire region was folded into regional-scale megafolds (as described above). This included both upper and lower plates of the metamorphic core complex as well as the carapace shear zone. This shortening event may have occurred in response to the dramatic thinning, attenuation and weakening of the crust that resulted from core complex development. This now thinned, weakened and relatively hot crust would have been more susceptible to changes in the stress regime resulting from variations in the regional tectonic setting (i.e. changes in local spreading or subduction regimes and rates).

Perhaps the most important observation to come from this study is that the nature of deformation in New Caledonia was highly episodic. We have described structures and fabrics formed during compressional orogenesis associated with collision early in the tectonic cycle. These structures appear to have been overprinted by fabrics formed during a major period of extension, associated with core complex formation. The crust was then buckled and folded during an orogen-scale upright folding event. Finally the orogen was once again extended, resulting in its dissection by brittle normal faults. The orogen thus appears to have evolved in an accordion-like manner, involving two cycles of contractional orogenesis followed by a major period of extensional tectonism.

The late-stage orogen-scale normal faults caused significant extension. Large jumps in metamorphic grade occur across each of several southwest-dipping normal faults. Traversing to the northeast, the late faults mark abrupt increases in metamorphic grade until finally the strip of eclogite–blueschist facies rocks is reached, along the northeast coast. However, displacement along these late-stage normal faults is not solely responsible for the exposure of the high-pressure metamorphic belt. This is more probably a relict effect of the earlier episode of extension. The majority of retrogression and decompression occurred before the formation of these late brittle features.

Most of the exhumation of the high-pressure rocks took place during a major extensional tectonic event before c. 38 Ma (Blake et al. 1977; Ghent et al. 1994; Verts & Baldwin 1996). This event produced numerous shallowly dipping shear zones that juxtapose eclogitic assemblages (contained in retrogressed ultramafic bodies) against structurally lower, highly deformed (blueschist facies) metamorphic tectonites that may also once have included eclogite facies assemblages.

The dominant high-pressure fabrics formed early in this tectonic history because continental crust was overridden by oceanic lithosphere. This is the ultimate cause of the high-pressure metamorphism in New Caledonia and is recorded in the complex fabrics produced during oscillating orogeny.

This research was supported by an Australian Research Council to study 'Continental Extension Tectonics'. S. Baldwin, C. Venn and S. Occhipinti are thanked for stimulating discussions and their efforts in this collaborative project between the Monash group and the University of Arizona, Tucson, USA. We gratefully acknowledge the support of the Bureau de Recherches Géologique et Minières and thank in particular J. J. Espirat, P. Maurizot and J. Leguere for their help, suggestions and interest in the project. This manuscript was greatly improved thanks to the thorough reviews by R. J. Holcombe, R. A. Harris, J. Aitchison, M. T. Brandon and P. Betts. We would also like to sincerely thank H. Hounda, the petit chief of Tribu de Colnett, for his unsurpassed hospitality. Work reported here was conducted as part of the Australian Geodynamic Cooperative Research Centre and this paper is published with the permission of the Director, AGCRC.

## References

AITCHISON, J., CLARKE, G. L., MEFFRE, S. & CLUZEL, D. 1995. Eocene arc–continent collision in New Caledonia and implications for regional Southwest Pacific tectonic evolution. *Geology*, **23**, 161–164.

AVIAS, J. 1967. Overthrust structure of the main ultrabasic New Caledonian massifs. *Tectonophysics*, **4**, 531–541.

BELL, T. H. & BROTHERS, R. N. 1985. Development of P–T prograde and P-retrograde, T-prograde isogradic surfaces during blueschist to eclogite regional deformation/metamorphism in New Caledonia, as indicated by progressively developed porphyroblast microstructures. *Journal of Metamorphic Geology*, **3**, 59–78.

BLAKE, M. C., BROTHERS, R. N. & LANPHERE, M. A. 1977. Radiometric ages of blueschists in New Caledonia. In: *International Symposium on Geodynamics in South-West Pacific*. Technipublication, Paris, 279–281.

BRIGGS, R. M., LILLIE, A. R. & BROTHERS, R. N. 1978. Structure and high-pressure metamorphism in the Diahot region, northern New Caledonia. *Bulletin du Bureau de Recherches Géologiques et Minières, Section 4: Géologie Générale*, **3**, 171–189.

BROTHERS, R. N. 1970. Lawsonite–albite schists from northernmost New Caledonia. *Contributions to Mineralogy and Petrology*, **25**, 185–202.

—— 1974. High-pressure schists in northern New Caledonia. *Contributions to Mineralogy and Petrology*, **46**, 109–127.

―― 1985. Regional mid-Tertiary blueschist–eclogite metamorphism in northern New Caledonia. *Géologie de la France*, **1**, 37–44.

―― & BLAKE, M. C. 1973. Tertiary plate tectonics and high-pressure metamorphism in New Caledonia. *Tectonophysics*, **17**, 337–358.

―― & LILLIE, A. R. 1988. Regional geology of New Caledonia. *In*: NAIRN, A. E. M., STEHLI, F. G. & UYEDA, S. (eds) *The Ocean Basins and Margins*. Plenum, New York, 325–374.

CANDE, S. C., LABRECQUE, J. L., LARSON, R. L., PITTMAN, W. C., III, GOLOVCHENKO, X. & HAXBY, W. F. 1989. *Magnetic Lineations of the World's Ocean Basins (Map Sheet)*. American Association of Petroleum Geologists, Tulsa, OK, **13**.

CAREY, S. W. 1958. The tectonic approach to continental drift. *In*: *Continental Drift; a Symposium*. Tasmania University Geology Department, 177–355.

CARROUÉ, J. P. 1971. *Carte géologique à l'échelle du 1/50 000; Hienghène*. Bureau de Recherches Géologiques et Minières, Orleans.

CAWOOD, P. A. & SUHR, G. 1992. Generation and obduction of ophiolites; constraints for the Bay of Islands Complex, western Newfoundland. *Tectonics*, **11**, 884–897.

CHEMENDA, A. I., MATTAUER, M. & ALEXANDER, N. B. 1996. Continental subduction and a mechanism for exhumation of high-pressure metamorphic rocks; new modelling and field data from Oman. *Earth Planetary Science Letters*, **143**, 173–182.

CLARKE, G. L., AITCHISON, J. & CLUZEL, D. 1997. Eclogites and blueschists of the Pam Peninsula, NE New Caledonia: a reappraisal. *Journal of Petrology*, **38**, 843–876.

CLOOS, M. 1984. Flow melanges and the structural evolution of accretionary wedges. *In*: RAYMOND, L. A. (ed) *Melanges; their Nature, Origin and Significance*. Geological Society of America, Memoir, **164**, 77–93.

―― 1986. Blueschists in the Franciscan Complex of California; petrotectonic constraints on uplift mechanisms. *In*: EVANS, B. W. & BROWN, E. H. (eds) *Blueschists and Eclogites*. Geological Society of America, Special Paper, **198**, 71–79.

CLUZEL, D., AITCHISON, J., CLARKE, G., MEFFRE, S. & PICARD, C. 1994. Point de vue sur l'évolution tectonique et géodynamique de la Nouvelle Calédonie. *Compte Rendus de l'Académie des Sciences*, **319**, 683–690.

――, ――, ――, & ―― 1995a. Dénudation tectonique du complexe à noyau métamorphique de haute pression d'âge tertiaire (Nord de la Nouvelle Calédonie, Pacifique, France). Données cinématiques. *Compte Rendus de l'Académie des Sciences*, **321**, 57–64.

――, CLARKE, G. & AITCHISON, J. 1995b. Northern New Caledonia high-pressure metamorphic core complex. From continental subduction to extensional exhumation. *1995 Pacrim Conference Proceedings*, 129–134.

COLEMAN, R. G. 1967. Glaucophane schists from California and New Caledonia. *Tectonophysics*, **4**, 479–498.

―― 1981. Tectonic setting for ophiolite obduction in Oman. *Journal of Geophysical Research, B*, **86**, 2497–2508.

COLLOT, J. Y., MALAHOFF, A., RECY, J., LATHAM, G. & MISSEGUE, F. 1987. Overthrust emplacement of New Caledonia ophiolite; geophysical evidence. *Tectonics*, **6**, 215–232.

CONEY, P. J. 1980. Cordilleran metamorphic core complexes; an overview. *In*: CRITTENDEN, M. D., JR, CONEY, P. J. & DAVIS, G. H. (eds) *Cordilleran Metamorphic Core Complexes*. Geological Society of America, Memoir, **153**, 7–31.

―― & HARMS, T. A. 1984. Cordilleran metamorphic core complexes; Cenozoic extensional relics of Mesozoic compression. *Geology*, **12**, 550–554.

CROOK, K. A. W. 1989. Implication of the Finisterre terrain collision for development of the New Guinea Orogen. Australasian tectonics. *Geological Society of Australia, Abstracts*, **24**, 31–32.

DANIEL, J., PONTOISE, B. & RECY, J. 1986. Arc–ridge collisions and back-arc rifting in the South-West Pacific. *In*: *IOC Symposium on Marine Science in the Western Pacific; the Indo-Pacific Convergence; Abstracts*. Workshop Report. UNESCO, Intergovernmental Oceanographic Commission, **47**, 30.

DAVIES, H. L. & JAQUES, A. L. 1984. Emplacement of ophiolite in Papua New Guinea. *In*: GASS, I. G., LIPPARD, S. J. & SHELTON, A. W. (eds) *Ophiolites and Oceanic Lithosphere*. Geological Society, London, Special Publications, **13**, 341–349.

DAVIS, G. H. 1983. Shear-zone model for the origin of metamorphic core complexes. *Geology*, **11**, 342–347.

――, GARDULSKI, A. F. & LISTER, G. S. 1987. Shear zone origin of quartzite mylonite and mylonitic pegmatite in the Coyote Mountains metamorphic core complex, Arizona. *Journal of Structural Geology*, **9**, 289–297.

FERRIERE, J. & CHANIER, F. 1993. La tectonique des plaques à l'épreuve de la réalité; SW Pacifique et Nouvelle-Zélande. (Plate tectonics faces reality; the Southwest Pacific and New Zealand.) *Géochronique*, **45**, 14–20.

FORSTER, M. A. & LISTER, G. S. 1999. Detachment faults in the Aegean core complex of Ios, Cyclades, Greece. This volume.

GHENT, E. D., RODDICK, J. C. & BLACK, P. M. 1994. $^{40}$Ar/$^{39}$Ar dating of white micas from the epidote to omphacite zones, northern New Caledonia: tectonic implications. *Canadian Journal of Earth Sciences*, **31**, 995–1001.

HACKER, B. R., RATSCHBACHER, L., WEBB, L. & DONG, S. 1995. What brought them up? Exhumation of the Dabie Shan ultrahigh-pressure rocks. *Geology*, **23**, 743–746.

HARRIS, R. A. 1992. Peri-collisional extension and the formation of Oman-type ophiolites in the Banda Arc and Brooks Range. *In*: PARSON, L. M., MURTON, B. J. & BROWNING, P. (eds) *Ophiolites and their Modern Oceanic Analogues*. Geological Society, London, Special Publications, **60**, 301–325.

HAYES, D. E. & RINGIS, J. 1973. Seafloor spreading in the Tasman Sea. *Nature*, **244**, 454–458.

KROENKE, L. W. 1984. New Caledonia: the Norfolk and Loyalty Ridges: the New Caledonia and Loyalty Basins. In: Cenozoic Tectonic Development of the Southwest Pacific. Technical Bulletin for the Committee for the Co-Ordination of Joint Prospecting for Mineral Resources in South Pacific Offshore Areas, 15–28.

—— & YAN, C. Y. 1993. An animated plate tectonic reconstruction of the Southwest Pacific, 0–100 Ma, based on the hotspot frame of references. Eos, Transactions, American Geophysical Union, **74**(Supplement), 286.

LAFOY, Y., MISSEGUE, F., CLUZEL, D. & LE SUAVE, R. 1996. The Loyalty–New Hebrides arc collision: effects on the Loyalty Ridge and Basin system, southwest Pacific (first results of the ZoNéCo Programme). Marine Geophysical Researches, **18**, 337–356.

LAIRD, M. G. 1993. Cretaceous continental rifts; New Zealand region. In: BALLANCE, P. F. (ed.) South Pacific Sedimentary Basins. Sedimentary Basins of the World, **2**. Elsevier, Amsterdam, 37–49.

LIPPARD, S. J., SHELTON, A. W. & GASS, I. G. 1986. The ophiolite of northern Oman. Geological Society, London, Memoir, **11**.

LISTER, G. S. & ETHERIDGE, M. A. 1989. Detachment models for uplift and volcanism in the Eastern Highlands, and their application to the origin of passive margin mountains. In: JOHNSON, R. W. (ed.) Intraplate Volcanism in Eastern Australia and New Zealand. Cambridge University Press, Cambridge, 297–313.

——, —— & SYMONDS, P. A. 1991. Detachment models for the formation of passive continental margins. Tectonics, **10**, 1038–1064.

MALPAS, J., SPOERLI, K. B., BLACK, P. M. & SMITH, I. E. M. 1992. Northland Ophiolite, New Zealand, and implications for plate-tectonic evolution of the Southwest Pacific. Geology, **20**, 149–152.

MANCKTELOW, N. S. 1995. Nonlithostatic pressure during sediment subduction and the development and exhumation of high-pressure metamorphic rocks. Journal of Geophysical Research, B, Solid Earth and Planets, **100**1, 571–583.

MAURIZOT, P. 1984. Carte géologique à l'échelle du 1/50 000 et Notice explicative su la feuille – Touho-Poindimié. Bureau de Recherches Géologiques et Minières, Orleans.

——, EBERLE, J.-M., HABAULT, C. & TESSAROLLO, C. 1989. Carte géologique à l'échelle du 1/50 000 et Notice explicative su la feuille – Pam-Ouégoa. Bureau de Recherches Géologiques et Minières, Orleans.

MICHARD, A., CHOPIN, C. & HENRY, C. 1993. Compression versus extension in the exhumation of the Dora-Maira coesite-bearing unit, Western Alps, Italy. Tectonophysics, **221**, 173–193.

OCCHIPINTI, S. A. 1994. The tectonic and metamorphic evolution of the high-pressure schists of the Yambé–Colnett region, northeastern New Caledonia. MSc thesis, Monash University, Melbourne, Vic.

PARIS, J. P. 1981. Géologie de la Nouvelle Calédonie: un essai de synthese. Mémoires du Bureau de Recherches Géologiques et Minières, **279**.

PLATT, J. P. 1986. Dynamics of orogenic wedges and the uplift of high-pressure metamorphic rocks. Geological Society of America Bulletin, **97**, 1037–1053.

—— & ENGLAND, P. C. 1994. Convective removal of lithosphere beneath mountain belts; thermal and mechanical consequences. American Journal of Science, **294**, 307–336.

RAWLING, T. J. & LISTER, G. S. 1997. The structural evolution of New Caledonia; a lost fragment of the Australian Plate? In: Geodynamics and Ore Deposits Conference, Australian Geodynamics Cooperative Research Centre, Ballarat, Victoria, 62–64.

RIDLEY, J. 1984. The significance of deformation associated with blueschist facies metamorphism on the Aegean island of Syros. In: DIXON, J. E. & ROBERTSON, A. H. F. (eds) The Geological Evolution of the Eastern Mediterranean. Geological Society, London, Special Publications, **17**, 545–550.

SCHEIBNER, E., SATO, T. & CRADDOCK, C. 1991. Tectonic Map of the Circum-Pacific Region, Southwest Quadrant. Circum-Pacific Map Series, US Geological Survey.

SCHOLL, D. W. 1995. Initiation of new arc–backarc systems of crustal generation in response to ridge–trench collision; example of the Lau–Tonga region by activation of rapid trench rollback. AGU 1995 fall meeting. Eos, Transactions, American Geophysical Union, **76** (Supplement), 614.

SEDLOCK, R. L. 1999. Evaluation of exhumation mechanisms for coherent blueschists in western Baja California, Mexico. This volume.

SENO, T. & PETERSON, E. T. 1984. Fluctuated motions of the Pacific Plate and collision episodes in the southwestern Pacific. AGU 1984 fall meeting. Eos, Transactions, American Geophysical Union, **65**, 1100.

SHAW, R. D. 1990. Development of the Tasman Sea and easternmost Australian continental margin; a review. In: FINLAYSON, D. M. (ed.) The Eromanga–Brisbane Geoscience Transect; a Guide to Basin Development across Phanerozoic Australia in Southern Queensland. Bulletin, Australian Bureau of Mineral Resources, Geology and Geophysics, **232**, 53–66.

SMELLIE, J. L. & STONE, P. 1992. Geochemical control on the evolutionary history of the Ballantrae Complex, SW Scotland, from comparisons with recent analogues. In: PARSON, L. M., MURTON, B. J. & BROWNING, P. (eds) Ophiolites and their Modern Oceanic Analogue. Geological Society, London, Special Publications, **60**, 171–178.

SPENCER, J. E. & REYNOLDS, S. J. 1989. Tertiary structure, stratigraphy, and tectonics of the Buckskin Mountains. In: SPENCER, J. E. & REYNOLDS, S. J. (eds) Geology and Mineral Resources of the Buckskin and Rawhide Mountains, West–Central Arizona. Bulletin, State of Arizona, Bureau of

Geology and Mineral Technology, Geological Survey Branch, **198**, 103–167.

STREETS, C. J. 1995. *The tectonic evolution of New Caledonia: implications of a detailed study of the high-pressure metamorphic tectonites of the Amoss, Mayavéc and Oue Bato river valleys*. MSc thesis, Monash University, Melbourne, Vic.

——, RAWLING, T. J . & LISTER, G. S. 1993. Structure and tectonic evolution of high-pressure metamorphic rocks of northeastern New Caledonia. *In*: *SGTSG Conference, Jindabyne, Geological Society of Australia Abstracts*, **36**, 160–161.

THOMPSON, A. B., SCHULMAN, K. & JEZEK, J. 1997. Extrusion tectonics and elevation of lower crustal metamorphic rocks in convergent orogens. *Geology*, **25**, 491–494.

VEEVERS, J. J. & LI, Z. X. 1991. Review of seafloor spreading around Australia; II, Marine magnetic anomaly modelling. *Australian Journal of Earth Sciences*, **38**, 391–408.

——, STAGG, H. M. J., WILLCOX, J. B. & DAVIES, H. L. 1990. Pattern of slow seafloor spreading (<4 mm/year) from breakup (96 Ma) to A20 (44.5 Ma) off the southern margin of Australia. *Bureau of Mineral Resources Journal of Australian Geology and Geophysics*, **11**, 499–507.

VERTS, L. A. & BALDWIN, S. L. 1996. Late Eocene rapid exhumation of New Caledonian eclogites: insights from thermochronology and numerical modeling. *Eos Supplement, American Geophysical Union*, **77**, 144.

WEISSEL, J. K. & HAYES, D. E. 1977. Evolution of the Tasman Sea reappraised. *Earth and Planetary Science Letters*, **36**, 77–84.

YAN, C. Y. & KROENKE, L. W. 1993. A plate tectonic reconstruction of the Southwest Pacific, 0–100 Ma. *Proceedings of the Ocean Drilling Program, Scientific Results*, **130**. Ocean Drilling Program, College Station, TX, 697–709.

YOKOYAMA, K., BROTHERS, R. N. & BLACK, P. M. 1986. Regional eclogite facies in the high-pressure metamorphic belt of New Caledonia. *In*: EVANS, B. W. & BROWN, E. H. (eds) *Blueschists and Eclogites*. Geological Society of America, Special Paper, **164**, 407–423.

# Exhumation of the Sanbagawa blueschist belt, SW Japan, by lateral flow and extrusion: evidence from structural kinematics and retrograde $P-T-t$ paths

R. P. WINTSCH[1], T. BYRNE[2] & M. TORIUMI[3]

[1]*Department of Geological Sciences, Indiana University, Bloomington, IN 47405, USA*
[2]*Department Geology and Geophysics, University of Connecticut, Storrs, CT 06269, USA*
[3]*Geological Institute, Faculty of Sciences, University of Tokyo, Hongo 7–3–1, Tokyo 113, Japan*

**Abstract:** The correlation in time of a pervasive, nearly strike-parallel, retrograde stretching lineation in the Sanbagawa blueschists with a nearly down-dip, prograde lineation in the structurally underlying Shimanto belt suggests that the oblique accretion and underplating of this belt was at least in part responsible for the longitudinal exhumation of the blueschists. A series of sub-horizontal collinear overgrowths of alkali amphiboles on cores of riebeckite in radiolarian cherts of the Sanbagawa belt, western Shikoku, Japan, indicate essentially continuous retrograde crystallization and extension during nearly strike-parallel exhumation. Needles of magnesio-riebeckite are progressively overgrown by beards or infillings of magnesio-riebeckite, winchite, actinolite, and finally quartz. The Na/(Na + Ca) compositions of overgrowths over many grains show a continuous range from >0.95 to <0.2. We interpret this to reflect continuously decreasing metamorphic conditions from an inferred $c$. 10 to $c$. 4 kbar ($c$. 35–15 km), and from $c$. 550°C to <300°C during continuous deformation of these metacherts. Published thermochronological data in related and correlative rocks show that this cooling occurred from mid-Cretaceous to Early Eocene time. Structural and geochronological studies of rocks of the northern Shimanto Belt show that they were accreted to southwest Japan by north-directed underthrusting contemporaneously with exhumation of the Sanbagawa belt. We propose that the retrograde metamorphic fabrics of the Sanbagawa belt developed as these initially high-pressure rocks were exhumed and laterally extruded to the east (present coordinates) during oblique, north-directed underthrusting of the Izanagi (or an equivalent) plate and its thick sedimentary cover (i.e. the Shimanto belt).

Thermal inversions associated with subduction zones are well known to be capable of generating the low-temperature – high-pressure conditions necessary for the development of blueschist facies mineral assemblages (e.g. Ernst 1977; Peacock 1993; and many others). Explaining the exhumation, preservation and, in some cases, the continued growth of the high-pressure minerals as the rocks are exhumed, however, has remained a challenge to earth scientists, and numerous kinematic models have been proposed (e.g. Cowan & Silling 1978; Ernst 1988; Shreve & Cloos 1986). These models range from the diapiric rise of a less dense matrix that supports the blueschists (Shreve & Cloos 1986) to underplating and normal faulting (Platt 1986). In addition, several blueschist facies belts display a stretching lineation that is nearly parallel to the inferred plate boundary, suggesting along-strike stretching during, or after, peak metamorphism (e.g. Sanbagawa, Alps, Canadian Cordillera, southern Alaska) and requiring 3D exhumation models (e.g. Avé Lallemant & Guth 1990; Toriumi 1990; Hansen 1992; Wallis & Platt 1995).

The Sanbagawa belt of southwest Japan (Fig. 1 inset) is similar to other high-pressure metamorphic belts for which numerous 2D and 3D kinematic models have been proposed (e.g. Toriumi, 1990; Hara *et al.* 1992; Ernst & Peacock, 1996). The Sanbagawa belt, however, appears to be exceptional in terms of the substantial amount of structural and thermochronological data that are available due, at least in part, to the early work of Miyashiro, who first proposed that the Sanbagawa belt represented part of his now classic 'paired metamorphic belts' (Miyashiro 1961). Since the work of Miyashiro, numerous studies have documented the complex thermal and structural histories of the belt (e.g. see Hara *et al.* (1992) for one of the most thorough reviews), and, most recently, Wallis *et al.* (1992) and Takasu *et al.* (1994) have proposed that the dominant fabric present in the Sanbagawa belt formed during retrograde (rather than prograde) metamorphism. Wallis *et al.* (1992) also

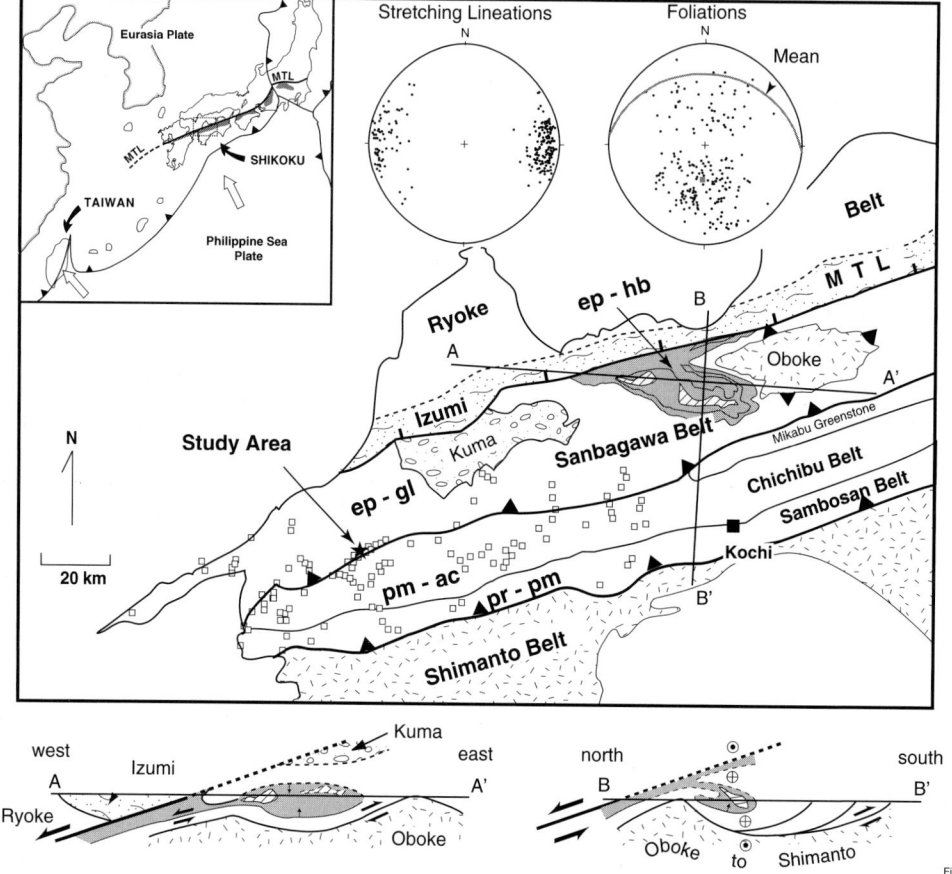

**Fig. 1.** Structural data, geological map and cross-sections of northwestern Shikoku; map is modified from Toriumi & Noda (1986), Hara *et al.* (1992) and Wallis (1998). Regional-scale fold structure shown in cross-sections is based on the northward fold closure proposed by Wallis (1992). Sanbagawa belt is subdivided into Oboke (stippled) and Besshi (remaining part of Sanbagawa belt) thrust sheets. Shaded area represents higher grades of metamorphism within the Besshi thrust sheet; fine lines differentiate different lithologies and different metamorphic zones, with diagonal lines marking oligoclase–biotite metamorphic zone (after Wallis 1998). Small arrows in these high-grade zones show the trends of increasing metamorphic grade. Tectonic blocks of metagabbro that preserve evidence of high-pressure eclogite facies metamorphism are locally present in the oligoclase–biotite zone. Boxes show locations of samples for our earlier study (Toriumi & Noda 1986) and the star shows the location of rocks examined in detail for this study. Unmetamorphosed sedimentary rocks of the Kuma Group unconformably overlie the Sanbagawa belt and are of Eocene age, whereas unmetamorphosed sedimentary rocks of the Izumi Group are of late Cretaceous age and contemporaneous with exhumation of the Sanbagawa belt. Kinematic data from high structural levels are from Wallis (1990, 1995), whereas data from lower structural levels are from Faure (1985) and Wallis (1995). MTL, Median Tectonic Line; pr-pm, prehnite–pumpellyite; pm-ac, pumpellyite–actinolite; ep-gl, epidote–glaucophane; ep-hb, epidote–hornblende. Moderately dipping MTL with Izumi and Ryoke units in the hanging wall is based on geophysical data obtained about 20 km east of area shown (Ito *et al.* 1996). It should be noted that some workers consider the Chichibu and Sambosan belts to be lower-grade parts of the Sanbagawa belt.

proposed that many of the multiple deformations recognized previously were part of a single, continuous deformation related to lateral flow of the blueschist facies rocks.

In this paper, we build on these earlier studies and propose that the lateral flow of the Sanbagawa belt represents the lateral extrusion and exhumation of the blueschist facies rocks, and that exhumation was driven by the accretion and underplating of younger, more seaward units (represented today by the northern Shimanto belt). In addition, we propose that exhumation was nearly orthogonal to the late Cretaceous plate convergence vector. Evidence for this interpretation comes from the direct correlation of retrograde metamorphic assemblages with the oblique stretching lineation in the Sanbagawa belt, the temporal correlation of peak heating in the northern Shimanto belt with exhumation of the Sanbagawa belt, and kinematic data from the northern Shimanto belt that suggest oblique plate convergence in late Cretaceous time. Finally, we suggest that a modern analogue for oblique convergence and lateral extrusion and exhumation may be represented by the medium- to high-grade metamorphic rocks in the Taiwan arc–continent collision.

## Sanbagawa belt

As discussed by Miyashiro (1961, 1973) and many subsequent workers (e.g. Banno 1986; Banno & Sakai 1989; Takasu 1989) most of the Sanbagawa belt records high-pressure–low-temperature metamorphism whereas several exotic tectonic blocks composed dominantly of mafic igneous rocks also preserve evidence of an earlier, eclogite facies metamorphism. Although peak metamorphic conditions have been well documented and inverted metamorphic gradients are generally accepted (Fig. 1) (e.g. Banno 1964; Hara et al. 1990; Higashino 1990), only recently have workers recognized the importance of retrograde conditions in the development of the dominant structural fabric (e.g. Wallis & Banno 1990; Hara et al. 1992; Takasu et al. 1994). For example, in central Shikoku, where the highest-grade rocks are exposed, Wallis et al. (1992) have recently proposed that many of the pervasive and relatively early-formed structures, including west-trending sheath folds, stretching lineations and a crenulation cleavage (D1 and D2 of Wallis et al. (1992)), formed during a single, progressive event associated with decreasing $P-T$ conditions. That is, the dominant structural fabrics in the Sanbagawa belt, including the stretching lineations we examine in more detail below, reflect exhumation rather than accretion.

*Metamorphic history and geochronology.* More mildly metamorphosed sections of the Sanbagawa belt suggest the presence of two dominant litho-tectonic units that were accreted to the Japanese continental margin in late Mesozoic time: the Besshi unit and the structurally underlying Oboke unit. The Besshi unit is composed largely of pelitic, mafic and siliceous schists derived from deep marine protoliths (including volcanic rocks and chert). Regional metamorphic grade reaches epidote–amphibolite facies (c. 550°C and 10 kbar) with exotic tectonic blocks preserving eclogite facies (c. 18 kbar and (c. 800°C) conditions (Fig. 1) (Takasu et al. 1994). Our detailed studies of needles of metamorphic amphiboles come primarily from the radiolarian-rich cherts in rocks equivalent to the lower structural levels of the Besshi unit (see Toriumi & Noda 1986). Microfossils retrieved from the Besshi unit range from Carboniferous to late Jurassic in age (Suyari et al. 1980; Iwasaki et al. 1984; Faure et al. 1991), and Isozaki & Itaya (1990) suggested that the unit accumulated in a trench environment in latest Jurassic time, probably 120–130 Ma. The Oboke unit is composed of psammitic and pelitic schists that are interpreted to have originated in a deep-sea fan environment (Kenzan 1984). Metamorphic grade lies in the middle pumpellyite–actinolite zone, and $^{40}Ar/^{39}Ar$ whole-rock ages of 70–77 Ma are interpreted to record the time of peak metamorphism (Takasu & Dallmeyer 1990b). Zircon fission-track ages of 56–69 Ma from the Oboke unit are consistent with this interpretation (Shinjoe & Tagami 1994). The Besshi and Oboke units both appear to be involved in an east-trending, regional-scale, recumbent fold that deforms the high-grade metamorphic isograds preserved in the Besshi unit (Fig. 1) (Wallis 1998; Wallis et al. 1992) although the detailed geometry of this structure is debated (Faure 1985; Banno & Sakai 1989; Hara et al. 1992).

Interpretations of the $P-T$ paths followed by the Sanbagawa rocks are complex and some appear contradictory. Studies focusing on metamorphic field gradients, for example, have led to the interpretation of clockwise $P-T$ paths (e.g. Banno 1984; Banno 1989; Enami 1994; Nakamura 1994 (Enami 1998)). However, the petrology is complicated because many of the minerals that define the metamorphic zones are themselves chemically zoned or contain inclusions that show systematic changes in chemical composition from core to rim (e.g. Banno et al. 1984; Otsuki & Banno 1990). The ecologite-grade blocks that occur in amphibolite facies rocks also contain mineral assemblages that record both prograde and retrograde clockwise and counter-clockwise $P-T$ paths (Takasu 1989; Takasu et al. 1994). In addition, the compositions of plagioclase, garnet, and clinopyroxene throughout the Besshi unit show prograde zoning (Otsuki & Banno 1990; Enami et al. 1994) whereas textural and chemical relations within and among amphibole grains reflect both prograde and retrograde amphibole growth (e.g. Otsuki 1980; Banno et al. 1984; Otsuki & Banno 1990; Takasu et al. 1994; this study). Detailed studies of the amphibole zoning patterns, however, show that counter-clockwise $P$–$T$ paths are common (Hara et al. 1990; Wallis et al. 1992; Enami et al. 1994; Nakamura & Enami, 1994; Wintsch et al. 1994). Combined structural and metamorphic studies have also shown the dominance of retrograde mineral assemblages in defining the penetrative structural fabrics (Takasu et al. 1994).

Available thermochronological data suggest a 60 Ma cooling history for the Sanbagawa metamorphic belt. Mineral ages published before 1990 (summarized by Isozaki & Itaya (1990)) and subsequent data (Takasu & Dallmeyer 1990; Dallmeyer & Takasu 1991*b*) suggest peak metamorphism occurred about 110 Ma, followed by late Cretaceous cooling. The Jurassic and early Cretaceous whole-rock ages of Suzuki et al. (1990) and Kawato et al. (1991) are interpreted here to reflect mixed ages of detrital and authigenic micas. Argon ages of metamorphic hornblendes from the Sanbagawa yield 90 Ma ages and are interpreted to record the initial cooling of the Sanbagawa belt. Detrital clasts of Sanbagawa rocks occur in the Middle Eocene Kuma Group, indicating exhumation of the belt by about 50 Ma (Takasu & Dallmeyer 1990a).

*Structural geology.* Structural and geophysical studies of the Sanbagawa belt show that the belt is a generally north-dipping package of regional-scale nappes or folds sandwiched between the Cretaceous Ryoke and Shimanto belts (Fig. 1). Field-based structural studies of the Sanbagawa belt have documented a well-developed, east-trending stretching lineation (Fig. 1), asymmetric fabrics, outcrop-scale sheath folds as well as folds that verge both north and south, and a regional-scale inversion of the metamorphic isograds within the Besshi unit (Hara et al. 1977, 1990; Faure 1985; Toriumi 1985; Wallis & Banno 1990). As a result of this inversion, the highest-grade rocks in the Sanbagawa belt, including the ecologitic blocks, occur near the middle of the present structural pile of the Besshi unit. Both regional-scale nappes and recumbent folding have been used to explain the metamorphic inversion (Banno et al. 1978; Faure 1985; Hara et al. 1990; Wallis et al. 1992). Most recently, however, Wallis (1998) has proposed that the only regional-scale, structural discontinuity in the Sanbagawa belt occurs between the Besshi unit and the underlying Oboke unit. Wallis (1998), therefore, argued that the metamorphic inversion is the result of a regional-scale recumbent fold with northward closure; the fold is considered to have formed contemporaneously with the pervasive, east-trending stretching lineation.

Strain and textural studies of radiolarian-rich cherts in the Besshi unit suggest a constant volume deformation that resulted in a range of strain states. Strain ellipsoids (based on the aspect ratios of deformed radiolaria) at low metamorphic grades have $X/Y$ ratios (where $X$, $Y$ and $Z$ represent the long, intermediate and short axes of the strain ellipsoid) of less than 1.5 and are oblate in shape (Toriumi & Noda 1986). At higher grades (e.g. epidote–amphibolite), $X/Y$ ratios range up to 24 and the strain ellipsoid has a prolate shape (Toriumi & Noda 1986). In essentially all cases, the long axis of the strain ellipsoid ($X$) plunges gently east or west and parallels a conspicuous metamorphic lineation defined, in part, by elongate amphibole grains (Fig. 1). The general eastward trend of the stretching lineation, therefore, is slightly oblique to the ENE structural grain of southwest Japan. Constant volume during deformation is indicated, at local scales, by studies of deformed vein arrays (Wallis 1992) and, at more regional scales, by the dominance of crystal plastic deformation mechanisms (Toriumi et al. 1986).

Previous workers have interpreted the penetrative, retrograde fabrics in the Sanbagawa belt to reflect lateral flow of the belt, although the dominant flow direction (relative to the inferred hanging wall, the Ryoke magmatic arc) has been debated. For example, Faure (1985) suggested that a top-to-east sense of shear dominates, whereas Wallis (1990) and Hara et al. (1992) suggested that top-to-the west dominates. Most recently, Wallis (1995) has found that top-to-the-west shear indicators are dominant at higher present structural levels (and generally higher grades of metamorphism) whereas top-to-the-east shear indicators are dominant at lower structural levels (and lower grades). These data therefore suggest that the Sanbagawa belt flowed from west to east as it was exhumed (Fig. 1). In addition, Wallis (1995) used a combination of strain analysis techniques to argue that the Besshi unit of the Sanbagawa belt represents a regional-scale shear zone with mean vorticity numbers generally between 0.5 and 0.9 and strain ratios of between 3.5 and 6. These results suggest tectonic thinning of the Sanbagawa belt of about 50% (Wallis 1995) and are consistent with metamorphic pressure gradients that are steeper than the geobaric gradient (Wallis & Banno 1990; Hara et al. 1992).

## Amphibole needle microstructures and compositions

Building on a reconnaissance field and petrographic survey of >100 radiolarian-rich samples from western Shikoku (Fig. 1), we selected a few samples for detailed microchemical analysis, including sample Ik1 of Toriumi & Noda (1986). All of the results come from an area in the northwestern corner of region A–A' of Toriumi & Noda (1986), which is near the southern boundary of the epidote–glaucophane metamorphic zone (Toriumi & Noda 1986). The metacherts in this area contain primarily quartz (50–95%), with lesser white mica (5–35%), and minor chlorite, calcite, amphibole and epidote, depending on the metamorphic grade. Much of the quartz occurs within and as matrix to deformed and recrystallized radiolaria, although numerous, micron-scale inclusions of quartz are also present in the amphibole needles described below.

Almost all of the amphiboles occur as strongly lineated needles that range from 20 to 200 μm in length and up to c. 30 μm in width (Figs 2 and 3). Although the needles are dominantly riebeckite in composition, detailed petrographic and microprobe analyses of the longer grains (>50 μm) show that many of the needles are stretched or necked, and amphiboles of different compositions (e.g. winchite or actinolite) correlate with the regions of necking and boudinage (Fig. 4). Locally, the necked regions may also be filled with quartz (e.g. Fig. 3). The relation of these various amphiboles to each other and to the deformation and metamorphic histories of the rocks of the Sanbagawa belt is discussed below.

*Riebeckite.* Riebeckite grains occur in the cores of amphibole needles that are up to 30 μm wide and 200 μm long. These grains are commonly fractured and pulled apart, with necks and ends overgrown by winchite (Figs 4 and 5). The riebeckite grains have a relatively high $SiO_2$

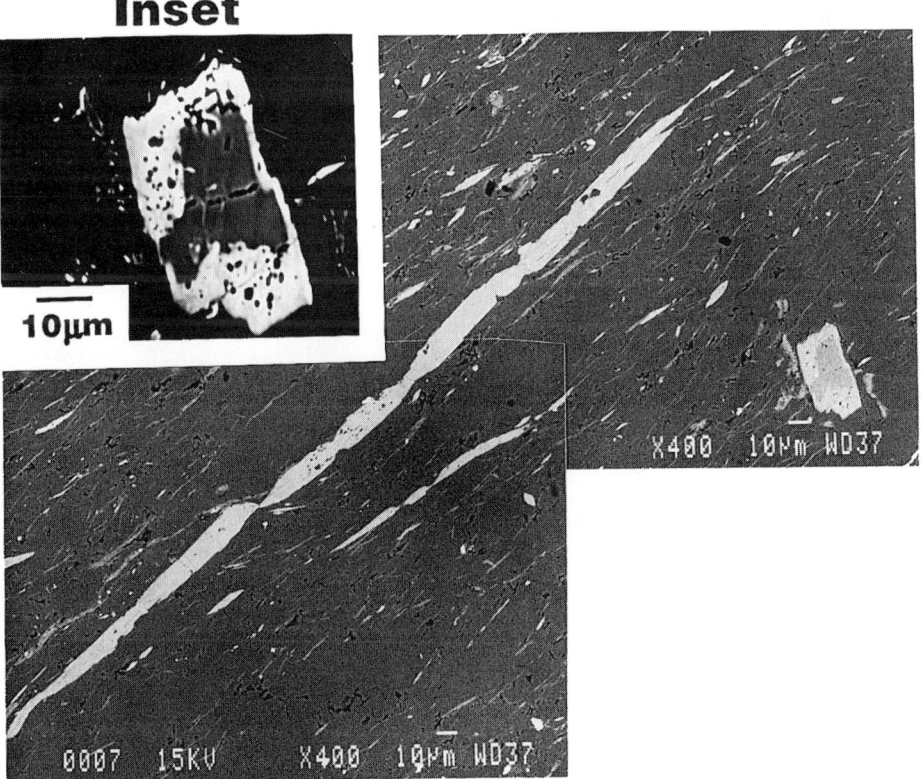

**Fig. 2.** Back-scattered electron (BSE) image of a typical >200 mm long compound amphibole needle with smaller needles in a matrix of quartz (medium grey). Individual quartz grains are outlined by <5 mm grains of chlorite and amphibole (light grey). Necked amphibole needles are connected with winchite, actinolite and quartz. Inset shows small magnesio-hornblende included in riebeckite.

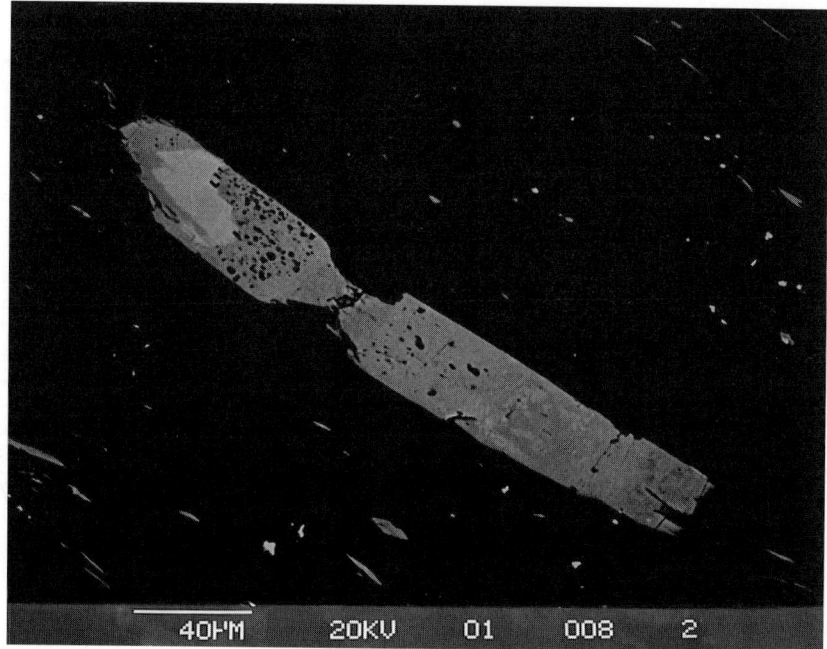

**Fig. 3.** BSE image of a relatively wide amphibole needle, minor needles of riebeckite and plates of chlorite in a matrix of quartz. (Note hornblende inclusion at upper left (light grey) with necks and overgrowths of winchite (see also Fig. 5).)

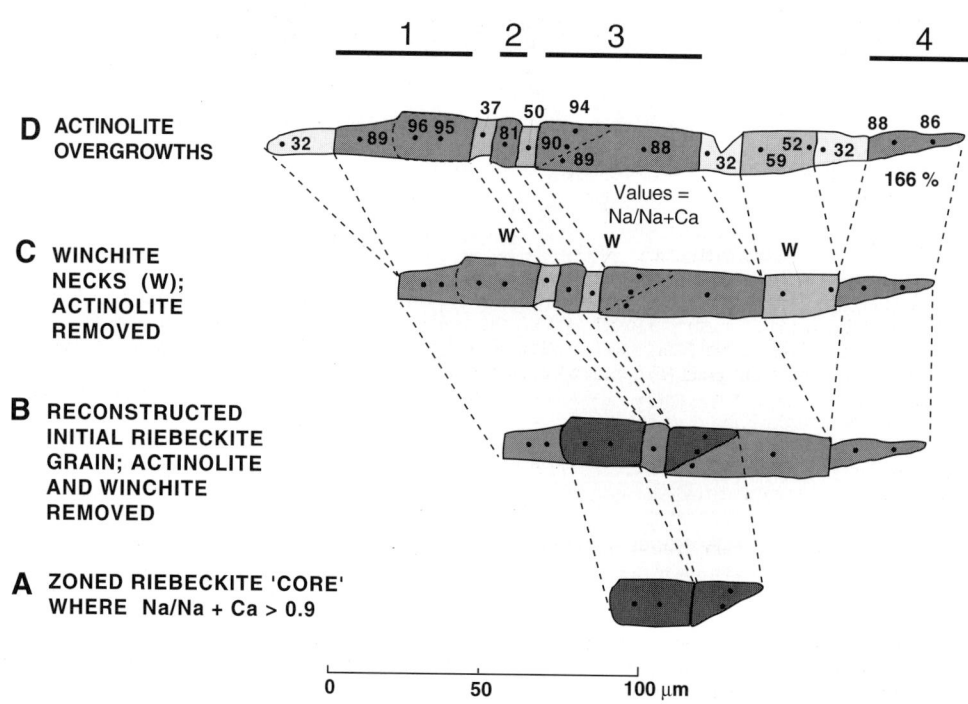

content, commonly >55 wt % and most formulae show greater than 8.0 Si atoms per formula unit (p.f.u.) when iron is calculated as FeO. However, the formulae reported (Table 1, Figs 6–8) have been calculated by oxidizing $Fe^{2+}$ to $Fe^{3+}$ until Si atoms reach 8.0 (Table 1). These grains are very sodic with a mean value of Na/(Na + Ca) of about 0.93 (Fig. 8) and the calculated ferric iron content is >66 % of total Fe (Table 1). This calculation also increases the total wt % to >98 wt %. Compositional zoning within individual riebeckite needles or needle segments commonly varies by >10% Na/(Na + Ca), and zoning patterns within some grain segments can be chaotic (Fig. 5). However, the composition of one riebeckite grain can differ from the next by 5–10% Na/(Na + Ca), such that the total range of Na/(Na + Ca) among many grains includes the entire compositional range of riebeckite.

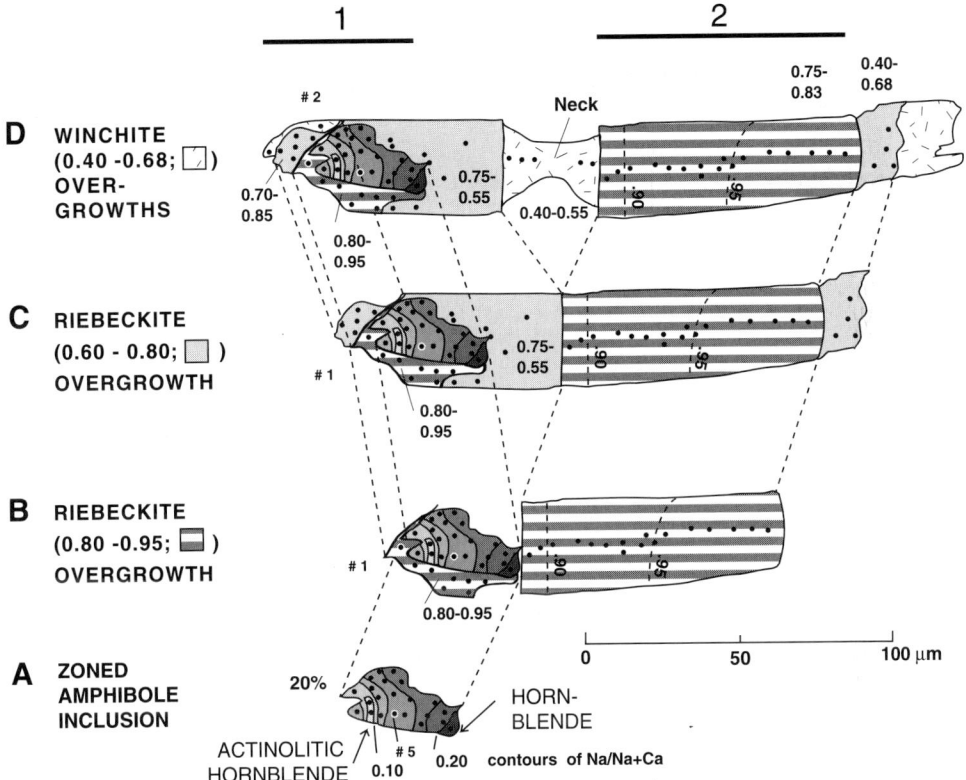

**Fig. 5.** Tracing of back-scattered electron image (d) shown in Fig. 3. Microprobe analyses (•) show values of Na/(Na + Ca). Progressive removal of winchite from d (Na/(Na + Ca) = 0.40–0.68) and two riebeckite segments from c and b (Na/(Na + Ca) = 0.60–0.80 and 0.80–0.95) results in an irregularly shaped zoned grain of hornblende. The core of this grain (a) is zoned from actinolitic hornblende to hornblende (Na/(Na + Ca) zoning indicated). This grain was apparently incompletely resorbed and subsequently overgrown by riebeckite (0.8–0.95 Na/(Na + Ca); striped pattern). This composite grain was then broken, extended, and overgrown by riebeckite (c. 0.6–0.8 Na/(Na + Ca); shaded grey; c), which was further broken, extended, and overgrown by winchite (c.0.4–0.6 Na/(Na + Ca); hatched). A stretch of c. 1.9 (following Ferguson 1981) is recorded by the two marked segments (from b to d).

**Fig. 4.** Tracing of a typical composite alkali amphibole grain from a BSE image. Microprobe analyses (•) show that values of Na/(Na + Ca) typically range from >0.90 to <0.35. Chemical and back-scattered electron images (not shown) document abrupt compositional boundaries between grain segments. Progressive removal of actinolite (Na/(Na + Ca) < 0.33; b) and winchite segments results in a reconstructed riebeckite grain. Discontinuous and continuous zoning within riebeckite 'cores' further shows growth stages of earlier riebeckite needles (a). A stretch of approximately 1.9 (following Ferguson 1981) is recorded by the four marked segments (from b to d).

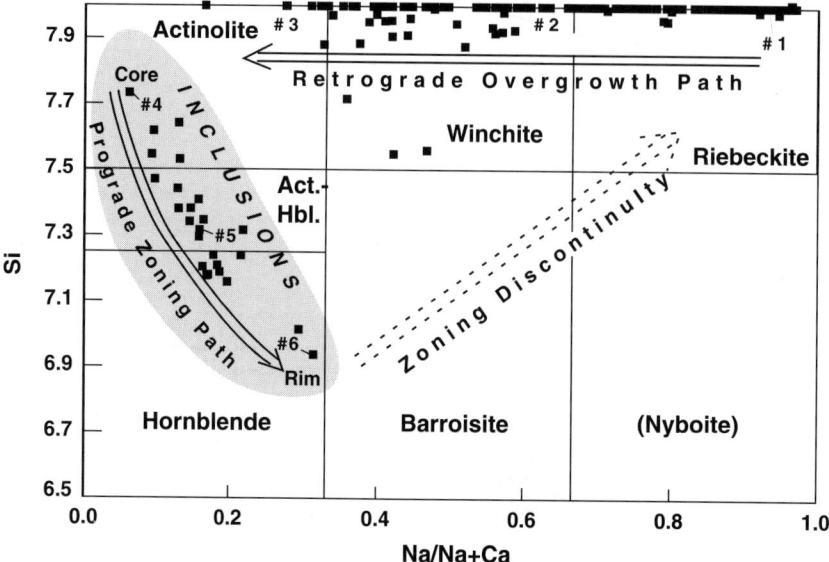

**Fig. 6.** The range of amphibole compositions analysed in this study in terms of Si and Na/(Na + Ca) atoms in an amphibole formula with 23 oxygen atoms. The diagram shows the continuous range in chemical compositions of zoned hornblende inclusions (see Nos 4–6, Table 1), and the continuous range in compositions of alkali amphiboles from nearly theoretical riebeckite (No. 1) to sodic actinolite (No. 3; see Table 1). A counter-clockwise metamorphic $P$–$T$ path is indicated (Fig. 10).

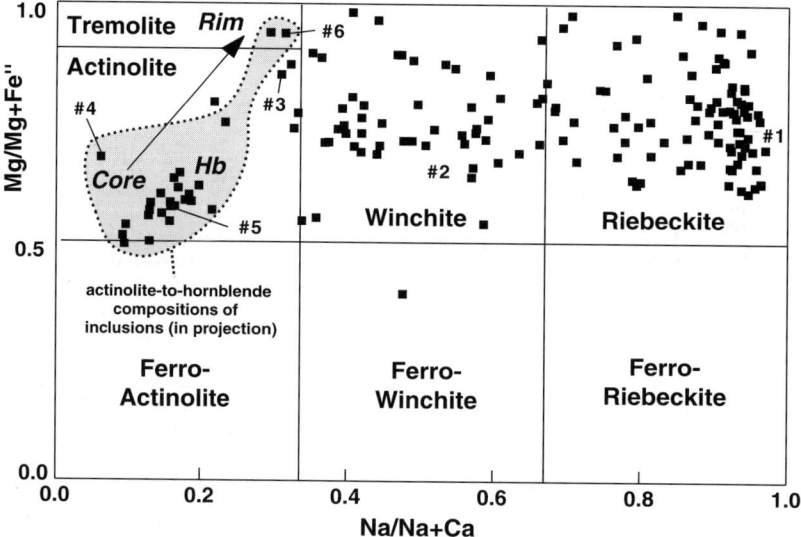

**Fig. 7.** Compositions of actinolite–winchite–riebeckite solid solutions in terms of calculated $Mg/(Mg + Fe^{2+})$ v. Na/(Na + Ca). Compositions of actinolite to actinolitic hornblende to hornblende solid solutions are projected onto the diagram for comparison. (See Table 1 for compositions of amphiboles Nos 1–6.

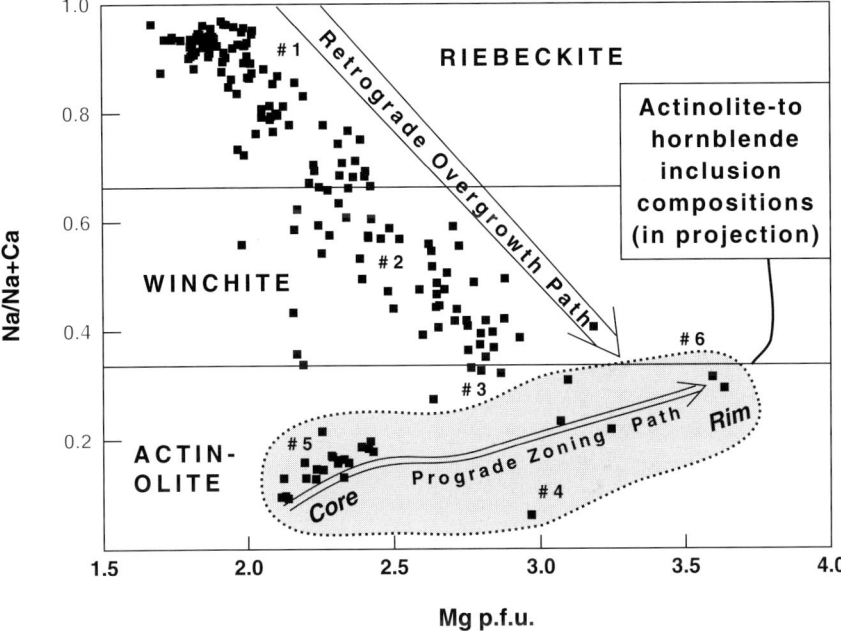

**Fig. 8.** Compositions of alkali amphiboles in terms of Na/(Na+Ca) and Mg per calculated formula unit (see Table 1 for methods and analyses 1–6). Compositions of hornblende in inclusions are projected onto the diagram for comparison.

*Winchite.* Winchite occurs both as overgrowths or 'beards' on riebeckite grains, and as segments between riebeckite grains; it never occurs as discrete, isolated grains. The boundaries between these two sodic amphiboles is always sharp, with differences in Na/(Na + Ca) of >20%. The compositional range in Na/(Na + Ca) within individual crystal segments can be as much as 15% Na/(Na + Ca), but compilations of many grain segments suggest a complete range in compositions (Figs 7 and 8).

*Actinolite.* Actinolite is less common than the other sodic amphiboles, and typically occurs as overgrowths on riebeckite grains, and as segments, that are sometimes necked, between riebeckite grains or riebeckite and winchite grains (Fig. 4). Actinolite has not been observed as discrete, individual grains. The actinolite is richer in magnesium than the other amphiboles, but it still has a relatively high calculated ferric iron content, much higher than that in the cores of the hornblende inclusions (Table 1).

*Miscibility gap.* In total, the solid solution between riebeckite and actinolite appears to be complete with Na/(Na + Ca) ratios varying from about 0.2 to >0.9 (Figs 7 and 8). Calculated Mg/(Mg + Fe$^{2+}$) ratios are consistently greater than 0.5 (Fig. 7), and actinolite is much more common than tremolite. A strong correlation also exists between Ca and Mg within riebeckite–winchite solid solutions (Figs 7 and 8). This continuous evolution in composition suggests that a solvus within the alkali amphiboles (Toriumi 1975; Spear 1982) may not exist, at least at these *P–T* conditions.

*Hornblende inclusions.* Relatively small (10–20 μm) zoned grains of amphibole occur as rare inclusions within a few riebeckite needles (Fig. 3). These inclusions have irregular anhedral shapes and are themselves zoned, with actinolite cores and more aluminous and more sodic, but less magnesian, hornblende rims (Fig. 5). These zoned grains occur exclusively within riebeckite grains; they never share grain boundaries with quartz or other matrix silicate grains. Chemical zoning within these inclusions shows compositions continuously evolving from actinolite to hornblende with a minimum Si content of about 6.9 atoms p.f.u. (Fig. 6; Table 1). There is a

**Table 1.** *Representative chemical analyses of amphiboles*

|  | 1* | 2 | 3 | 4 | 5 | 6 |
|---|---|---|---|---|---|---|
| $SiO_2$ | 55.666 | 54.056 | 57.495 | 53.309 | 49.466 | 48.947 |
| $TiO_2$ | 0.098 | 0.086 | 0.035 | 0.125 | 0.172 | 0.296 |
| $Al_2O_3$ | 1.548 | 1.618 | 0.800 | 2.762 | 4.181 | 7.767 |
| FeO | 21.167 | 17.761 | 13.762 | 14.557 | 21.932 | 9.224 |
| MnO | 0.636 | 1.245 | 1.216 | 0.383 | 0.234 | 0.403 |
| MgO | 9.415 | 11.25 | 13.816 | 13.728 | 9.954 | 17.000 |
| CaO | 0.720 | 5.888 | 8.631 | 11.376 | 10.152 | 10.383 |
| $Na_2O$ | 6.811 | 4.301 | 2.267 | 0.412 | 1.058 | 2.631 |
| $K_2O$ | 0.006 | 0.030 | 0.000 | 0.191 | 0.011 | 0.173 |
| Total | 96.153 | 98.224 | 98.022 | 96.843 | 97.160 | 96.824 |

*Number of cations per formula unit†*

|  | 1* | 2 | 3 | 4 | 5 | 6 |
|---|---|---|---|---|---|---|
| $Si^{IV}$ | 8.000 | 7.922 | 8.000 | 7.733 | 7.316 | 6.938 |
| $Al^{IV}$ | 0.000 | 0.078 | 0.000 | 0.267 | 0.684 | 1.062 |
| $Al^{VI}$ | 0.263 | 0.201 | 0.132 | 0.205 | 0.044 | 0.235 |
| Ti | 0.011 | 0.010 | 0.004 | 0.014 | 0.02 | 0.031 |
| $Fe^{+3}$ | 1.775 | 0.780 | 1.184 | 0.347 | 1.079 | 0.856 |
| Mg | 2.017 | 2.458 | 2.866 | 2.969 | 2.194 | 3.592 |
| $Fe^{+2}$ | 0.769 | 1.396 | 0.418 | 1.419 | 1.634 | 0.238 |
| Mn | 0.077 | 0.154 | 0.143 | 0.047 | 0.029 | 0.048 |
| Na | 1.888 | 1.070 | 0.611 | 0.116 | .308 | 0.391 |
| Ca | 0.111 | 0.924 | 1.287 | 1.768 | 1.608 | 1.577 |
| K | 0.001 | 0.006 | 0.000 | 0.036 | 0.002 | 0.032 |
| $Na^A$ | 0.011 | 0.153 | 0.000 | 0.000 | 0.000 | 0.333 |
|  | riebeckite | winchite | actinolite | actinolite | act. horn. | hornblende |

†Number refers to numbered (e.g. No. 6) analysis in Figs 6–8.
Numbers 2 and 4 reduced by setting $Fe^{3+}$ to Si cations value of 8.0. Others were reduced to total cations of 13, exclusive of Ca, Na, and K as recommended by Cosca *et al.* (1991).
$^{IV}$Tetrahedral site occupancy; $^{IV}$octahedral site occupancy; $^A$A-site occupancy.

relatively large, abrupt compositional gap between these hornblende inclusions and the riebeckite overgrowths (Figs 6 and 8).

The possibility might be raised that the hornblende inclusions could be detrital. We cannot prove that this is not the case, but we believe this unlikely for several reasons. First, this type of prograde zoning is common in many other Sanbagawa rocks (Fig. 9 and Hara *et al.* (1992)). Second, the occurrence of these metamorphic grains in a pelagic radiolarian sediment rules out volcanic air-borne clasts, and requires deposition by deep-water currents. The absence of other detrital grains and of sedimentary structures consistent with such deposition makes this unlikely. Finally, the interpretation that these grains are detrital requires a pre-Mid-Jurassic middle amphibolite facies metamorphic event for which we know of no other evidence. We thus make the simplest interpretation that the zoning of these grains preserves the prograde portion of a Cretaceous metamorphic event.

*Comparison with other data.* The spectrum of amphibole compositions described above appears to be relatively common in Sanbagawa rocks (Fig. 9) (see, e.g. Houseman & England 1986; Hara *et al.* 1990; Otsuki & Banno 1990; Nakamura & Enami 1994). For example, the actinolite-to-hornblende zoning of the cores of grains described here, and shown as line segment C1 of Fig. 9 (from Fig. 6) has been described by Takagi & Hara (1979), Takasu & Dallmeyer (1992), and Nakamura & Enami (1994), but has more sodic intermediate compositions (C3–R3 and C7, Fig. 9). Riebeckite overgrowths like those we describe (shown as C1 and R1, Fig. 9) are type IV of Hara *et al.* (1990, 1992), with both discontinuous (fig. 4 of Nakamura & Enami (1994)) and continuous patterns (C4–R4, Fig. 9). The high-pressure, retrograde path of the riebeckite to winchite to actinolite is also common (e.g. C4–R4; Hara *et al.* 1990; Otsuki & Banno 1990; Takasu *et al.* 1994).

The counter-clockwise chemical path defined by the above patterns is in contrast with the less common reverse zoning patterns represented by paths C2 (type III of Hara *et al.* (1990, 1992), C6, and possibly C5 of Fig. 9 (Banno *et al.* 1984; Hara *et al.* 1990, 1992; Wallis *et al.* 1992). In spite of the

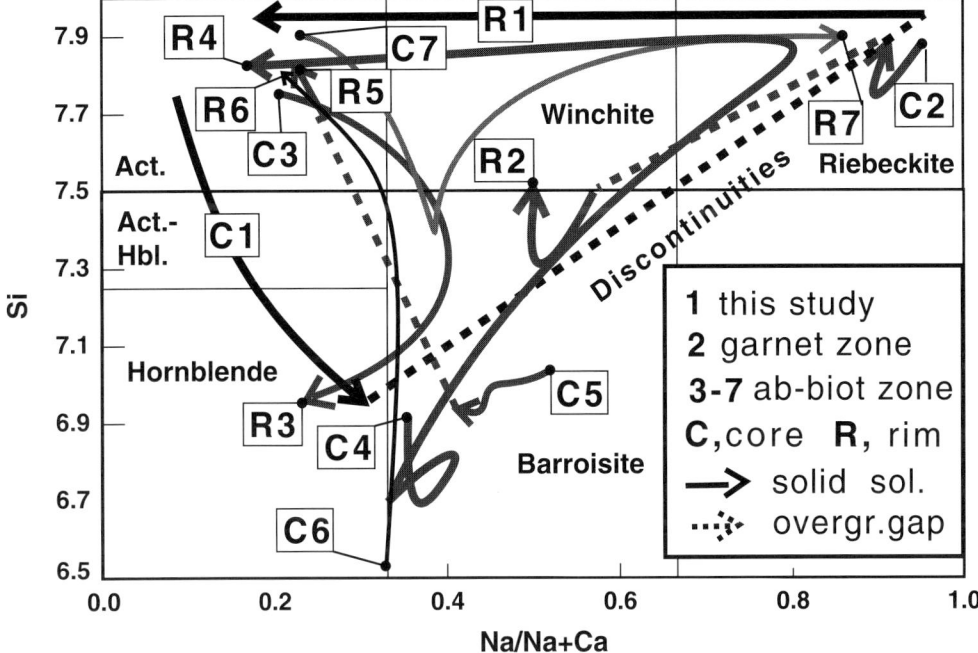

**Fig. 9.** Representative zoning patterns from core (C) to rim (R) defined by the compositions of calcic and alkali amphiboles. Continuous zoning is represented by the continuous lines, and discontinuities between cores and overgrowths are given by the dashed lines. Continuous line-segment C1 reflects the actinolite to hornblende prograde zoning of the cores of the amphibole presented here, and continuous line-segment R1 from riebeckite to actinolite summarizes the compositions of retrograde alkali amphibole overgrowths and rims. Curves 2–5 are calculated from Wallis *et al.* (1992) and Nakamura & Enami (1994). Sample numbers are: 2, AS62; 3, HU1005; 4, ST2204; 5, TN1503; 6, Wallis *et al.* (1992, fig. 19c); 7, Wallis *et al.* (1992, fig. 17). Collectively these data show that a counter-clockwise chemical path is common.

complexities that these reverse patterns must reflect, the larger volume of the data is consistent with a three-stage event. The first early, low-pressure, actinolite to hornblende growth event is followed by a second, higher-pressure, lower-temperature riebeckite overgrowth event, which in turn is followed by a third, higher-pressure, low-temperature, retrograde riebeckite to winchite to actinolite growth event. The pressure of the stage 2 event is at a minimum, consistent with the growth of barroisite or winchite, and riebeckite or even glaucophane is common in some rocks (e.g. Hara *et al.* 1990).

*Pressure–temperature path*

We are unable to quantitatively estimate the P–T conditions for the assemblages described here because chemical equilibrium cannot be established for the present assemblage, the earlier-formed amphiboles have been at least partially removed by dissolution, and buffering assemblages (e.g. Triboulet 1992) are not present. However, the relative history of growth of the various compositions of amphibole, from zoned hornblende to riebeckite to winchite to actinolite, can be deduced from the cross-cutting relations described above. Zoned hornblende inclusions preserve cores of actinolite and overgrowths of magnesian hornblende (Fig. 6) with c. 6.9 Si atoms p.f.u. (Table 1). This compositional gradation is typical of the changes in compositions of prograde amphiboles in regional metamorphic rocks (e.g. Laird 1982), and are interpreted here to reflect amphibole growth during a prograde, low- to moderate-pressure metamorphic event. The local inclusion of these zoned hornblendes in relatively large riebeckite needles and the presence of a large, compositional jump across the inclusion boundary require the riebeckite to have grown after the growth of, and after the partial dissolution of these zoned hornblendes. Winchite and actinolite must have followed riebeckite growth because neither have been observed to include the zoned hornblendes, and both grow on or between riebeckite segments. The dominance of actinolite in necks and as terminal overgrowths (e.g. Fig. 4) and the continuous gradient in composition from winchite to actinolite (Figs 6–8) suggest that actinolite precipitation followed winchite, and actinolite was the last amphibole to grow.

Although quantitative estimates of $P$ and $T$ are not possible, qualitative estimates can be made, because the varying temperature and pressure conditions of growth of these amphibole compositions are known (e.g. Holland & Richardson 1979; Laird 1982; Wakabayashi 1990; Triboulet 1992). Moreover, these $P$–$T$ conditions have been empirically calibrated in the Sanbagawa belt (e.g. Banno et al. 1984; Banno & Sakai 1989). Consequently, the relative metamorphic conditions present at the time of actinolite, hornblende, riebeckite and winchite growth can be estimated by correlation with other studies (Fig. 10). This path is characterized by a prograde, lower-pressure, actinolite-to-hornblende growth event preserved in the few inclusions. This event was followed by a major dissolution event that destroyed most of the prograde hornblende grains. These components reprecipitated as a sodic riebeckite overgrowth event that is higher in pressure and lower in temperature than the hornblende growth event. The growth of riebeckite begins a retrograde metamorphic growth path that passes continuously through the winchite and actinolite fields (Figs 6–8). An abrupt discontinuity between the hornblende and riebeckite compositions makes it impossible to infer the maximum temperatures and pressures experienced by these rocks, and that part of the $P$–$T$ path in Fig. 10 is queried. The retrograde $P$–$T$ path reflects a shallow geothermal gradient (10–15°C km$^{-1}$), consistent with geotherms calculated from other $P$–$T$ estimates for the development of riebeckite in these rocks (Otsuki & Banno 1990; Triboulet 1992; Enami et al. 1994).

Similar amphibole compositions and zoning patterns appear to be relatively widespread in the Sanbagawa belt; thus, the counter-clockwise path defined by our samples must be as pervasive as is the zoning pattern (e.g. Hara et al. 1990). In particular, the high-pressure retrograde metamorphic path (Fig. 10) suggests that the Sanbagawa belt stayed in a tectonic setting with a relatively low geothermal gradient as it was exhumed. Apparently, the belt was not underplated beneath or exhumed through the higher-temperature regime of the hanging wall as suggested by Enami et al. (1994) for the higher-grade rocks in the Sanbagawa belt. These pervasive, relatively cool retrograde histories

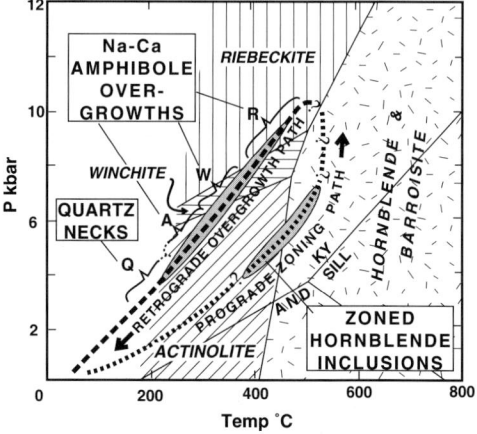

**Fig. 10.** Pressure–temperature diagram showing the stability fields of the alkali amphiboles (from Banno & Sakai 1989) and aluminosilicates (from Holdaway & Mukhopadhyay 1993) for comparison. Metamorphic $P$–$T$ path followed by the zoned hornblende inclusions and composite alkali amphibole overgrowths (shaded lens-shaped regions) are shown. The approximate $P$–$T$ regions of riebeckite (R) grains, and overgrowths of winchite (W), actinolite (A) and quartz (Q) define a continuous retrograde metamorphic path in the blueschist facies. KY, kyanite; AND, andalusite; SILL, sillimanite.

are also inconsistent with the subduction of an active spreading centre in late Cretaceous time (e.g. Maruyama 1997). We return to this problem of a cool retrograde path below.

## Deformation history and metamorphism

The necking of the riebeckite grains with infilling and overgrowth of winchite and actinolite is interpreted to reflect a microboudinage process (Misch 1969) where the amphiboles behaved in a brittle manner in their surrounding ductile quartz matrix. The chemical data suggest, however, that the deformation was heterogeneous at the scale of individual crystals, but more homogeneous at the scale of multiple crystals (e.g. the scale of a thin-section). Individual amphibole needles show sharp discontinuities in compositions of riebeckite, winchite and actinolite, and the compositions of amphiboles in a single neck or in the beard of an individual needle are relatively limited (Figs 4 and 5). It thus appears that individual needles were broken and extended over a relatively narrow range of $P$–$T$–$X$ conditions, such that the instantaneous strain and strain rate around a single crystal may have been relatively high. At the scale of a thin-section, however, amphibole compositions within different crystals vary, such that over many crystals a continuous variation among all the alkali amphiboles is present. Thus, on the scale of a single amphibole needle, the microboudinage process was heterogeneous and discontinuous, whereas at the scale of millimetres or centimetres, it was homogeneous and continuous. A similar heterogeneous distribution of slip among textural domains was proposed by Knipe & Wintsch (1985) for strongly deformed, quartz-rich mylonites.

We interpret the complicated shapes and patterns of zoning of the amphibole grains (Figs 3–6) to reflect a cyclic dissolution–precipitation process that was linked to deformation during waning metamorphic conditions. Early formed riebeckite grains appear to have grown over still earlier formed zoned hornblende grains. The hornblende grains are included within the riebeckites, and their zoning is universally and abruptly truncated by the overgrowing riebeckite grains (Fig. 5). Thus, these hornblende grains were largely dissolved before the precipitation of sodic riebeckite, and their dissolution must have contributed critical components to the fluid from which the riebeckite precipitated (see also Otsuki & Banno (1990)). The relatively mafic-poor assemblage of minerals in these metacherts also suggests chemical recycling of riebeckite to form winchite and actinolite. Finally, because these retrograde products occur only as elongate overgrowths (necks, fibres and beards) that parallel the principal stretching axis, and never as concentric overgrowths enveloping the entire crystal, we infer that dissolution and precipitation were deformation induced.

Recognizing the continuous change in compositions from riebeckite to actinolite over several crystals is important to understanding the deformation history because it provides a reference frame for calibrating the stretch values resulting from microboudinage. That is, the compositional bands serve as a proxy for elongation; larger bands suggest larger elongation. Using individual needles that preserve the largest range in amphibole compositions (e.g. Fig. 4) and following Ferguson (1981) we calculate minimum elongation values $(1+e_1)$ in sample IK1 (Toriumi et al. 1986) of 1.9. Deformed radiolaria in the quartz-rich portions of this sample (i.e., in the area where the amphibole needles are located) indicate $X/Z$ ratios of 4.4/1 (reformulated from Toriumi et al. (1986)), suggesting $1+e_1$ values of 2.1, if volume was constant and plane strain applies. The boudinaged amphiboles therefore record about 90% of the elongation recorded by the radiolaria.

Apparently, if microboudinage of the riebeckite needles reflects only the retrograde, syn-exhumation part of the metamorphic–deformation path and the quartz-filled radiolaria record the total ductile strain, then very little, if any, of the ductile strain in these rocks is related to the prograde metamorphic history. One possibility is that the Sanbagawa protoliths were isolated from the deformation associated with accretion because they were carried on the downgoing plate beneath the décollement. That is, the Sanbagawa protoliths may not have been penetratively deformed until they were underplated at substantial depths; penetrative deformation began only as the rocks were transferred from the footwall and exhumed.

## Temperature–time path

Geochronological data provide an additional important constraint on the cooling and exhumation histories of Sanbagawa rocks (Fig. 11). No thermochronological data are available for the immediate vicinity of our samples of the Besshi unit, but data are available for similar, low-grade Besshi rocks along strike. Although apparent age ranges overlap and more detailed interpretations may be possible (e.g. Dallmeyer & Takasu 1991b), average $^{40}Ar/^{39}Ar$ cooling ages of about 90 Ma and 75 Ma for hornblende and muscovite, respectively, indicate regional cooling into the greenschist facies in mid- to late Cretaceous time (Isozaki & Itaya 1990; Dallmeyer & Takasu 1991a, b). Thermochronological data from the Oboke unit define a cooling curve indistinguishable from the younger cooling stages of the Besshi unit (Fig. 11). Muscovite and whole-rock $^{40}Ar/^{39}Ar$ ages of 70–77 Ma (Takasu et al. 1994) are interpreted to be crystallization ages, and shown in Fig. 11 at a temperature of about 300°C. The zircon fission-track ages of 56–69 Ma (Shinjoe & Tagami 1994) support this early Tertiary cooling history. Final exhumation of the Sanbagawa belt (including both the Besshi and Oboke units) is defined by the occurrence of clasts derived from this belt in the Kuma Group conglomerates of Eocene age (c. 50 Ma) (Takasu & Dallmeyer 1992). Taken together, these data suggest a linear cooling history from about 500°C to 25°C during a time period of about 40 Ma with a cooling rate of c. 12°C Ma$^{-1}$ (Fig. 10). When combined with the pressure data obtained from the $P$–$T$ path, the 40 Ma exhumation history of rocks from at least a depth of 35 km suggests a maximum exhumation rate of about 0.9 mm a$^{-1}$, which is relatively slow.

In summary, our petrological and structural data show that sub-horizontal stretching

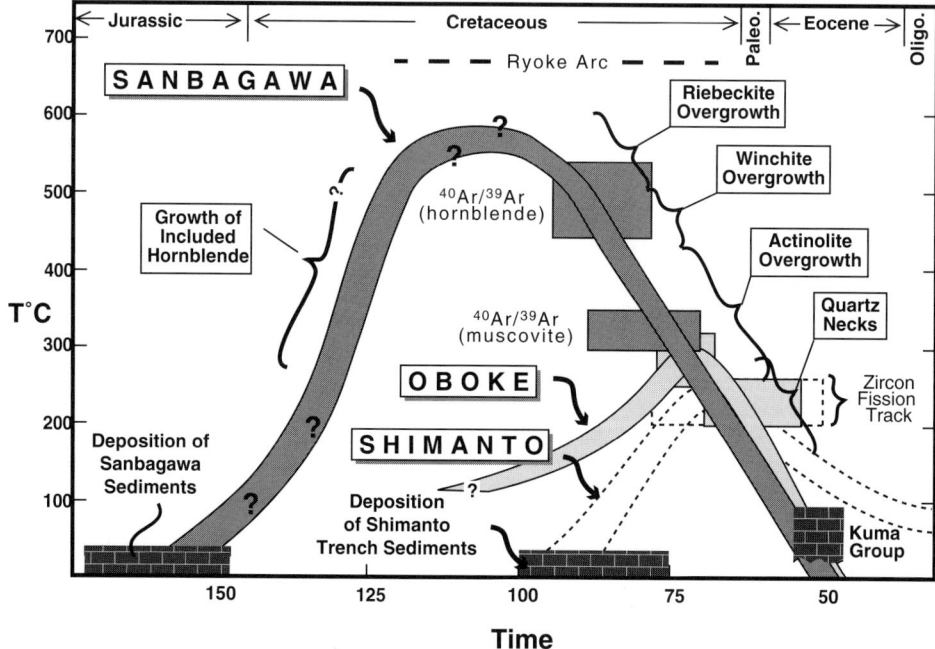

**Fig. 11.** Temperature–time path reconstructed for the Sanbagawa and Shimanto belts. The loading history of the Sanbagawa belt is constrained to be between early Cretaceous deposition (Isozaki & Itaya 1990) and late Cretaceous hornblende, and muscovite cooling ages. Exhumation occurred between latest Cretaceous muscovite ages and the unconformably overlying Eocene sediments of the Kuma Group. Inferred times of growth of zoned hornblende inclusions and alkali amphibole needles are indicated. The retrograde path of the Oboke unit within the Sanbagawa belt is based on whole-rock $^{40}Ar/^{39}Ar$ ages (Takasu *et al.* 1994) and zircon cooling ages (Shinjoe & Tagami 1994) (shaded boxes). The prograde and retrograde paths of Shimanto belt rocks are constrained to be younger than mid-Cretaceous deposition and early Tertiary zircon ages. (Note that exhumation and extrusion of the Sanbagawa blueschists correlates with the accretion and loading (underplating?) of Shimanto and Oboke rocks.)

occurred essentially continuously from high- to low-pressure metamorphic conditions or from about 35 to <15 km from late Cretaceous to earliest Tertiary time. Although we have calculated exhumation and cooling rates of 0.9 mm $a^{-1}$ and 12°C $Ma^{-1}$, respectively, the exact particle path cannot be established because the trend and plunge of the particle trajectory in the plane of the subduction zone (or channel) are not known. At present, the mean foliation dips moderately northward, the mean lineation is nearly strike-parallel, and both have trends that are oblique to the present structural grain of southwest Japan (070°) (Fig. 1). If we assume a similar geometry for late Cretaceous to earliest Tertiary time, then one possibility is that the rocks of the Sanbagawa belt may have been exhumed from beneath a north-dipping hanging wall as it flowed from west to east. The general north dip of most of the accreted units in southwest Japan and the present north dip of the northern boundary of the Sanbagawa belt (the Median Tectonic Line or MTL) are consistent with this interpretation. In the following sections, we propose that this lateral, west to east exhumation resulted from north-directed underthrusting of the Northern Shimanto belt in latest Cretaceous and early Tertiary time.

## The Shimanto Complex

The Shimanto Complex crops out seaward of the Sanbagawa belt (Fig. 11) and previous studies have recognized two regional-scale belts: the northern and southern Shimanto belts. The northern Shimanto belt ranges from latest early Cretaceous to latest Cretaceous age (Taira 1985) and is characterized by alternating sequences of relatively thick and coherent turbidite deposits (dominantly sandstone and shale) and more complexly deformed belts of shale-rich, tectonic mélange, including blocks of ribbon cherts and pillow lava. Taira *et al.* (1988) and Kimura & Mukai (1991) have proposed that the mélanges occur along a décollement zone that separated dominantly oceanic sediments from thick, trench-fill clastic deposits. Kimura (1994) has also recently proposed that the relatively thick trench-fill records the rapid accumulation of collision-related sediments derived from northeast Asia. The southern Shimanto belt ranges from Paleocene to latest Miocene age and is also characterized by alternating sequences of turbidite deposits and tectonic mélanges (Taira 1985). The Tertiary mélanges, however, tend to be less regular along- and across-strike and often grade into the sandstone and shale turbidite deposits. The Tertiary mélanges also lack blocks of ribbon cherts.

Thermochronological data from the northern Shimanto belt and the northern part of the southern Shimanto belt indicate a relatively low grade of metamorphism with ages of peak metamorphism overlapping, in part, with the time of exhumation of the Sanbagawa belt (Fig. 11) (Hasebe *et al.* 1993, 1997; Tagami *et al.* 1995). For example, illite crystallinity, vitrinite reflectance and zircon fission-track studies show that the sediments of the northern Shimanto belt were not heated to above about 225°C (DiTullio & Hada 1993; Hasebe *et al.* 1993). The zircon fission-track data also indicate that peak metamorphism occurred about 60–75 Ma (Hasebe *et al.* 1993). This range is similar to K–Ar ages of cleavage-forming micas (grain size <0.5 μm) from these rocks (Agar *et al.* 1989; MacKenzie *et al.* 1990). Similar data from the northern part of the southern Shimanto belt indicate slightly higher temperatures of metamorphism (peak temperatures of 250–300°C) and early Tertiary ages for peak metamorphism (Tagami *et al.* 1995). Taken together, the thermal data indicate that these belts were being accreted and metamorphosed from about 100–80 Ma to 50 Ma, which overlaps in time with the development of the retrograde overgrowths of riebeckite, winchite, and actinolite in the Sanbagawa rocks (Fig. 11).

Structural and kinematic data from the Cretaceous and earliest Tertiary Shimanto belts suggest north-directed underthrusting and are consistent with slightly oblique plate convergence in latest Cretaceous to early Tertiary time. Although we have only limited kinematic data from the Shimanto belt directly south of our study area, a compilation of kinematic data from several along-strike localities shows a consistent pattern for over 600 km (Fig. 12). In the Akaishi Mountains near Tokyo asymmetric mélange fabrics from two areas of the Cretaceous Shimanto belt show well-defined, north-directed underthrusting (Kano *et al.* 1991). On Kii Peninsula, relatively early formed conjugate fractures in Paleocene sediments indicate north–south shortening during fold development (Nakamura 1986; Byrne & DiTullio 1992). Further west, in the eastern part of Shikoku, detailed studies of extensional shear bands in mélanges in both the Cretaceous and early Tertiary Shimanto belts indicate north-directed underthrusting (Kimura & Mukai 1991; Lewis & Byrne unpublished data). In western Shikoku, in the Cretaceous Shimanto belt south of the study area (Fig. 12), outcrop-scale folds and duplex structures suggest north-directed underthrusting (Taira *et al.* 1988), although detailed kinematic data are not available. In the Paleocene to early Eocene Shimanto belt, however, Byrne & DiTullio (1992) documented a north-trending stretching lineation and local asymmetric structures indicate north-directed underthrusting. Finally, further west, on Kyushu, Fabbri (1989) and MacKenzie *et al.* (1990) have documented north-plunging stretching lineations in, respectively, the Cretaceous and middle Eocene Shimanto belts, and both workers argued for north-directed underthrusting. Overall, therefore, kinematic data from several localities that together span nearly the total length of the Cretaceous and early Tertiary Shimanto belts in southwest Japan (over 600 km along-strike) indicate north–south shortening, and several of the localities also indicate north-directed underthrusting. This direction of shortening and underthrusting is slightly oblique to the inferred plate boundary (the structural grain of southwest Japan, 070°), suggesting oblique plate convergence in late Cretaceous and early Tertiary time.

## Discussion: possible kinematic regimes for exhumation during oblique plate convergence

The petrological, geochemical and thermochronological data from the Sanbagawa and Shimanto belts presented above suggest a temporal correlation between the exhumation of the Sanbagawa belt and peak heating in the Cretaceous Shimanto belt. Specifically, our study of the oriented amphibole needles shows that the dominant lineation in the Sanbagawa belt formed during retrograde metamorphism, which, by correlation with published thermochronological data, occurred between about 110 Ma and

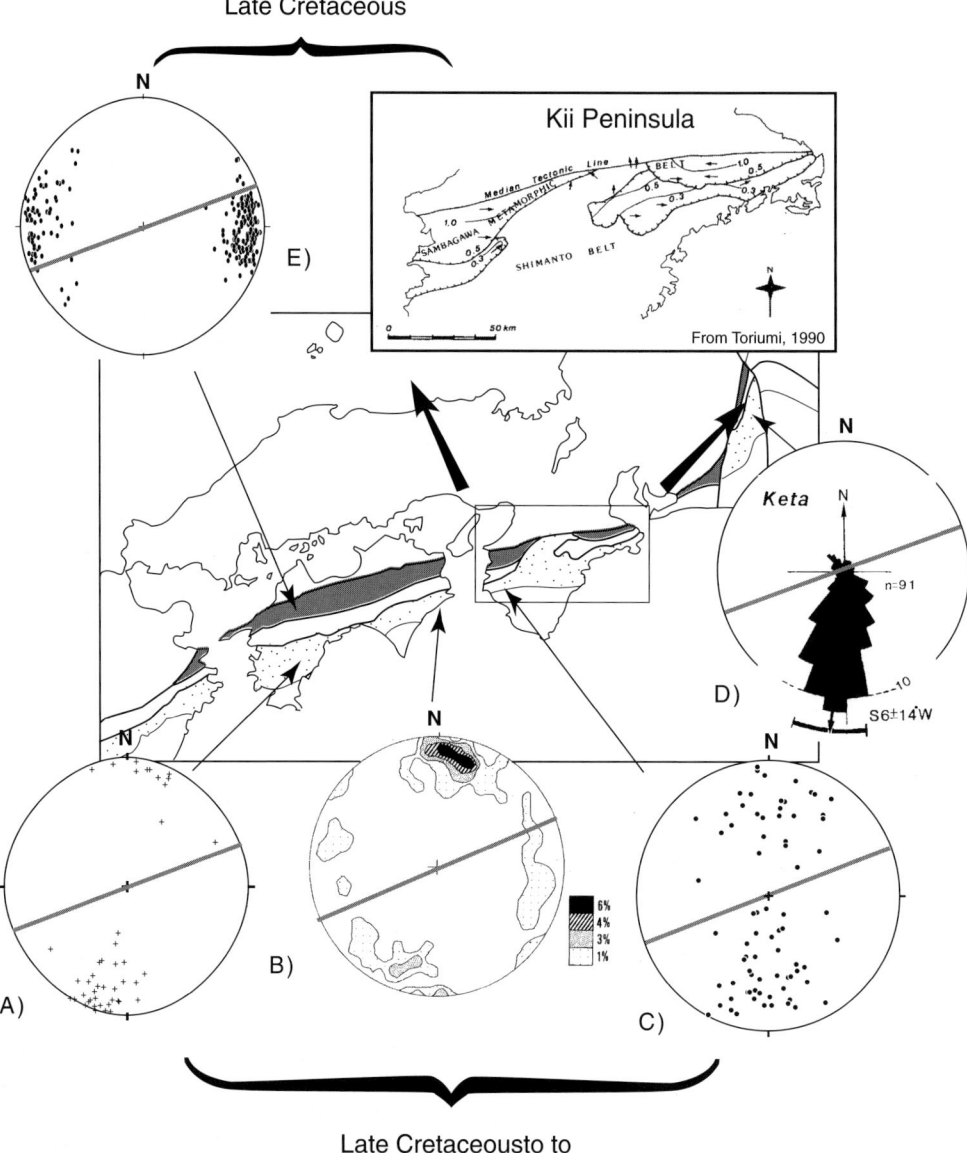

**Fig. 12.** Structural data from the Sanbagawa belt (upper stereonet, (**e**) and inset map) and late Cretaceous to early Tertiary Shimanto belt (lower stereonets). Data from Sanbagawa are from Toriumi & Noda (1986) and Torimui (1990). Shimanto belt data are: (**a**) slickenlines in mélanges (Byrne & DiTullio 1992); (**b**) slickenlines and extensional shear bands in mélanges (Kimura & Mukai 1991); (**c**) $\sigma_1$ directions in coherent sandstone–shale sequences (Nakamura 1986; Byrne & DiTullio 1992); and (**d**) extensional shear bands in mélange (Kano *et al.* 1991). In all stereonets, a 070° reference line (thick grey line), representing the present structural grain of southwest Japan, shows that all of the lineations are oblique to the structural grain. Younger rotations (mostly landward tilting) have been removed from some of the data (see original references). (Note that lineations from the Sanbagawa and Shimanto belts are nearly perpendicular.) Dark stippled area is Sanbagawa belt, light stippled area is northern Shimanto belt, and clear area between represents Chichibu, Sambosan and Kuroseqawa belts.

50 Ma. In addition, kinematic studies summarized above suggest that the Sanbagawa belt was extruded from west to east. This period of exhumation correlates with the age of deposition, accretion and metamorphism of the currently exposed northern (Cretaceous) Shimanto belt. We now propose that rocks equivalent to the Cretaceous Shimanto belt were accreted and underplated beneath the Sanbagawa belt in latest Cretaceous time. Tagami *et al.* (1995) and Kimura (1994) have also recently proposed a similar interpretation based on thermochronological data and palaeogeographical reconstructions, and Hara *et al.* (1992) has suggested that the Oboke unit represents the underplated rocks equivalent to the Shimanto belt.

Compiled structural data from the Sanbagawa and northern Shimanto belts, however, raise an important conundrum about the kinematics of the accretion and exhumation of these two belts. Specifically, although both of these belts yield kinematic axes that are oblique to the structural grain of southwest Japan, the observed stretching axes, and inferred transport directions for each belt are essentially orthogonal to each other. In the Himalaya, where contemporaneous accretion and exhumation was first well documented (Burg *et al.* 1984; Burchfiel & Royden 1985; Hodges *et al.* 1996), the accretion-related and exhumation-related stretching lineations are essentially parallel. That is, the retrograde particle path is inferred to have been similar to the prograde particle path. If our interpretation of underplating and exhumation for the northern Shimanto and Sanbagawa belts is correct, the kinematic regime must have been different from the Himalayan model. To resolve this problem, we propose that oblique plate convergence (proposed previously by several workers including Engebretson *et al.* (1985) and Jolivet *et al.* (1989)) was responsible for the oblique subduction and exhumation. Oblique convergence is also suggested by plate reconstructions for late Cretaceous time and by the occurrence of late Cretaceous left-lateral strike-slip faulting in southwest Japan. In the following paragraphs, we summarize the relevant aspects of this evidence and elaborate on possible accretion and exhumation models for late Cretaceous time.

*Plate tectonic setting: evidence for oblique convergence*

Plate reconstructions and kinematics indicate that an oceanic plate (probably the 'Izanagi' plate) was being subducted to the north-northwest beneath southwest Japan in Late Cretaceous time, suggesting left-lateral oblique convergence (Fig. 13) (Engebretson *et al.* 1985; Jolivet *et al.* 1989). Although there is substantial debate concerning the details of these reconstructions (e.g. what plate was being subducted and was an active spreading centre also subducted), two aspects of the reconstructions appear to be relatively well accepted in the literature. First, most researchers agree that the Japan Sea opened as a back-arc basin in late Tertiary time, and that before this time Japan formed part of an Andean-like convergent plate boundary along the southeast China continental margin (Jolivet *et al.* 1989; Taira *et al.* 1992; and many others). This plate boundary generally trended to the northeast (*c.* 030° ± 10°; see Jolivet (1994) for discussion; Jolivet *et al.* 1989; Otofugi 1996) and separated Eurasia from the oceanic plates of the Pacific Basin (e.g. Morgan 1972; Engebretson *et al.* 1985; Jolivet *et al.* 1989). Second, most plate reconstructions show north to northwest plate convergence along this plate boundary in late Cretaceous time (e.g. Engebretson *et al.* 1985; Jolivet *et al.* 1989).

Although the uncertainties are relatively large for the late Cretaceous reconstructions, north to northwest convergence is consistent with several aspects of the geology of southwest Japan. First

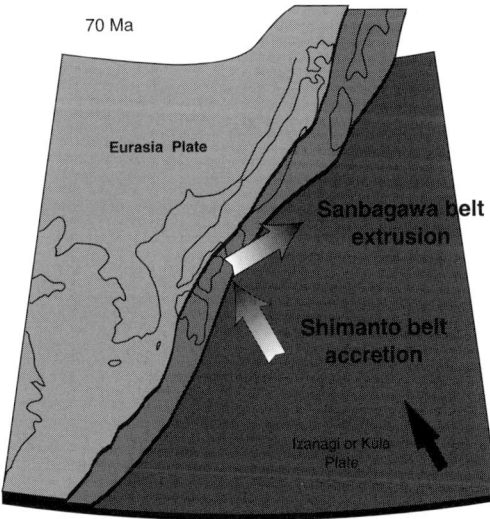

**Fig. 13.** Reconstruction of southwest Japan in late Cretaceous time with the Japan Sea closed and southwest Japan rotated counter-clockwise 40° (from Taira *et al.* 1992). Motion of oceanic plate is relative to Eurasia (from Engebretson *et al.* 1985). Accretion and extrusion directions of the Shimanto and Sanbagawa belts, respectively, are based on structural data presented in this study.

are the kinematic data from the Cretaceous and early Tertiary Shimanto belts presented above. Second, mildly deformed and unmetamorphosed late Cretaceous sedimentary rocks of the Izumi Group fill a strike-slip fault-related pull-apart basin along the MTL (Fig. 1) marking the present fault boundary between the Sanbagawa and Ryoke belts in southwest Japan (Taira *et al.* 1980; Miyata 1990). Kinematic data from the associated fault zones also indicate left-lateral displacements (Takagi 1986). Thus, the geometry of the inferred basin, the kinematic data from the fault zones and the kinematic data from the Shimanto belt are all consistent with the plate reconstructions that suggest oblique, north-directed underthrusting in late Cretaceous time (Fig. 12).

## Exhumation of the Sanbagawa belt: extrusion or deformation partitioning?

The oblique, exhumation-related amphibole lineations in the Sanbagawa belt and the obliquely converging plate boundary suggest two possible regional-scale kinematic interpretations (Fig. 14a and b): (1) the oblique lineations may reflect strike-slip faulting as the oblique plate convergence vector was partitioned into strike-parallel and plate boundary normal components (e.g. Fitch 1972) or (2) the lineations and retrograde P–T conditions may reflect lateral extrusion and exhumation of the blueschist facies rocks as the Shimanto belt was underplated. In the following paragraphs, we argue that the second model (lateral extrusion) is more consistent with our observations and the available geological data from southwest Japan. We present three general arguments for extrusion, and discuss in more detail the application of this interpretation to the Sanbagawa belt.

First, the orientations of the lineations in the Sanbagawa rocks favour an extrusion model. The lineations in the Sanbagawa belt are not parallel to the 070° structural grain of southwest Japan as defined by the trends of the lithotectonic belts (Fig. 1). We consider these belts to have been parallel to the Cretaceous plate boundary, and the 090° orientation of the mineral lineations reflects extension 20° clockwise from this trend. The oblique lineations, therefore, record a gentle, up-to-the-east extrusion direction (Fig. 14b). There is also a conspicuous lack of evidence for regional-scale strike-slip faulting (e.g. vertical foliations and

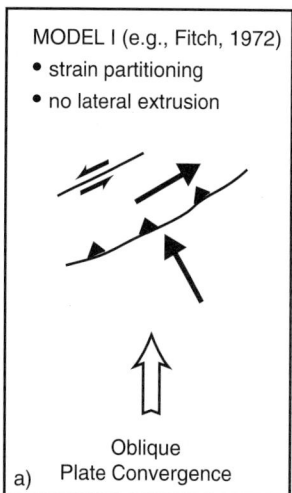
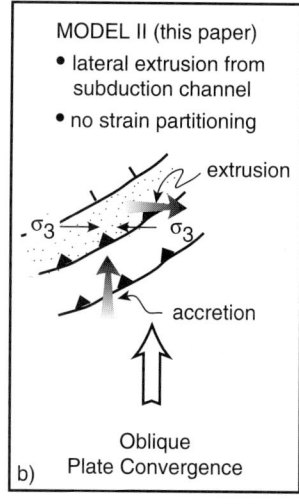
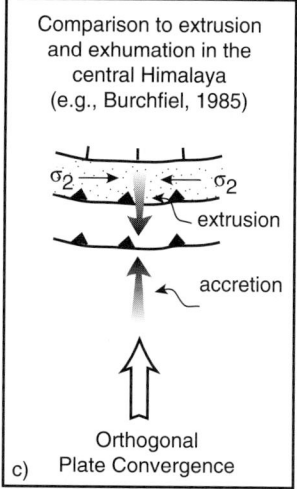

**Fig. 14.** Possible kinematic consequences of oblique plate convergence and comparison with interpretations for extrusion during orthogonal plate convergence. In Model I (**a**), the forearc is decoupled from both the downgoing and overriding plates, and shortening related to oblique convergence is partitioned into plate-boundary normal and plate-boundary parallel motions (based on Fitch 1972). In Model II (**b**), supported here, the forearc remains coupled to the overriding plate and plate convergence is not partitioned. Instead, previously underplated high-pressure rocks are extruded laterally parallel to the least principal stress ($\sigma_3$) and towards the 'free face' provided by the subducting oceanic plate. For comparison, in the central Himalaya, where late Tertiary extrusion is well documented and plate convergence has been nearly orthogonal (**c**), lateral flow is constrained by strike-parallel stresses and extrusion is nearly parallel to the plate convergence (Royden & Burchfiel 1987).

flower structures) expected for deformation partitioning, and simple models of strike-slip faulting would not explain the 30–40 km of relatively continuous, penetrative exhumation required by the petrological data. More complex models of deformation resulting from oblique convergence also suggest strike-parallel foliations (e.g. Thompson *et al.* 1999), strike-parallel lineations (e.g. Platt 1993), a gradual change in lineations across strike (e.g. Wallis & Platt 1995) or a complex set of along-strike extensional structures associated with along-strike transport of the forearc region (e.g. Avé Lallemant & Guth 1990). None of these structures have been documented in the Sanbagawa belt. The structural relations in the Sanbagawa belt rocks, therefore, support extrusion over a deformation partitioning.

Second, recent compilations of earthquake slip vectors, trench geometries and relative plate motions suggest that relatively large obliquities (defined as the angle between the plate convergent vector and the normal to the trench) may be necessary before plate convergent shortening is partitioned into trench-parallel and trench-normal components. For example, in a study of the Sumatran and Aleutian subduction zones, McCaffrey (1992) found that obliquities of 25–40° were necessary before partitioning occurred. A study of 40 great earthquakes ($M > 8.0$) from around the Pacific (McCaffrey 1993) also shows that the earthquake slip vectors consistently record the plate convergence vector (within *c.* 10°) even when obliquities are as high as 60°. Although small obliquities may not preclude strain partitioning (McCaffrey 1993) they do seem to subdue them. In southwest Japan, the north-trending lineations in the Shimanto belt indicate minimum obliquities of 20° and, based on observations from modern subduction zones, do not require deformation partitioning. Although larger obliquities are possible, the absence of well-defined trench-parallel lineations or transport in the Sanbagawa or Shimanto belt argues against deformation partitioning associated with larger angles of oblique convergence.

Finally, the active Taiwan orogenic belt may provide a possible modern analogue for the southwest Japan kinematic regime in late Cretaceous time. In Taiwan, the plate convergence vector trends 310°, which is 20° to the plate boundary normal (Fig. 15), and kinematic data (e.g. stretching lineations and finite strain markers) from moderate structural levels (e.g. lower greenschist facies) consistently record the plate convergence vector (Fig. 15) (Tillman & Byrne 1995). At deeper structural levels (e.g.

upper greenschist to amphibolite facies), however, the dominant foliation dips gently east beneath the accreted volcanic arc and a well-developed stretching lineation trends slightly oblique to the plate boundary (the Longitudinal Valley). The lineation, therefore, is nearly perpendicular to the plate convergence vector (Fig. 15). Sense-of-shear indicators record a top-to-the-northeast sense of shear with movement of the high-grade rocks to the southwest relative to the inferred hanging wall (the rocks of the accreted arc). Apparently, lateral flow and exhumation are facilitated by the oblique convergence between the Chinese continental margin and the Philippine Sea plate, and the unconstrained boundary condition provided by the South China Sea southwest of Taiwan (Fig. 15). A recently completed global positioning system survey across southern Taiwan (Yu & Lee 1986) suggested that lateral extrusion (to the southwest) is continuing today.

The geometries and kinematics recognized in the moderate and deep structural levels of the modern Taiwan orogenic belt are nearly identical to the geometries and kinematics inferred for

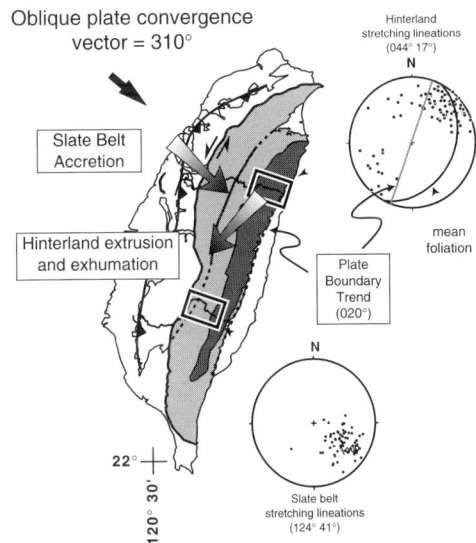

**Fig. 15.** Generalized geological map of Taiwan and structural data from the slate belt (light shading) and the pre-Tertiary metamorphic basement (dark shading). Both stereonets show mineral stretching lineations and large arrows show motion of the footwall rocks. It should be noted that plate convergence is slightly oblique to the plate boundary (*c.* 20°) and that both lineations are oblique to the plate boundary. This obliquity and the nearly orthogonal relation between the two lineations is similar to the lineation patterns in southwest Japan.

the late Cretaceous Sanbagawa and Shimanto belts of southwest Japan (compare Figs 12 and 15). Although the plate convergence vector is undefined in Japan for Cretaceous time, we propose that it was nearly parallel to the lineations in the Shimanto belt, and that underthrusting and underplating of rocks equivalent to the Shimanto belt during late Cretaceous time caused the lateral extrusion, exhumation and uplift of the Sanbagawa belt.

*Proposed kinematic interpretation for extrusion and exhumation.* We envision southwest Japan as forming part of the NNE-trending southeast Asian continental margin (030° after the Japan Sea is closed) in late Cretaceous time to early Tertiary time (Fig. 13). Plate convergence is inferred to have been to the northwest and approximately parallel to the structural lineations in the Shimanto belt (*c.* 320° after the Japan Sea is closed) and oblique to the inferred plate boundary. Oblique convergence results in a relatively unconstrained lateral boundary (in this case, to the northeast) and we show material moving to the northeast, perpendicular to the plate convergence vector (Fig. 13). Material flowing parallel to the stretching lineation towards this unconstrained boundary would therefore have had a slight up-dip component, resulting in gradual exhumation.

In our interpretation, movement of the Sanbagawa rocks from west to east (present coordinates), slightly up the dip of the subducting slab (Figs 14b and 16) involved a combination of underplating beneath an arcward-dipping backstop, tectonic thinning of the Sanbagawa metasediments, partial closure of the subduction channel and oblique plate convergence. Underplating of the Shimanto–Oboke unit is inferred to have thickened the accretionary wedge, which was accommodated, at least in part, by tectonic thinning of previously accreted Sanbagawa belt rocks (Wallis 1995). If volume remained constant (see, e.g. Toriumi *et al.* (1986) and Wallis (1992)) as the Sanbagawa belt was deformed by a combination of simple and pure shear (with thinning of *c.* 50%; Wallis 1995) and the subduction channel was constricted (e.g. by the accretion of the Shimanto belt rocks; Kimura 1997) then thinning would have resulted in the upward

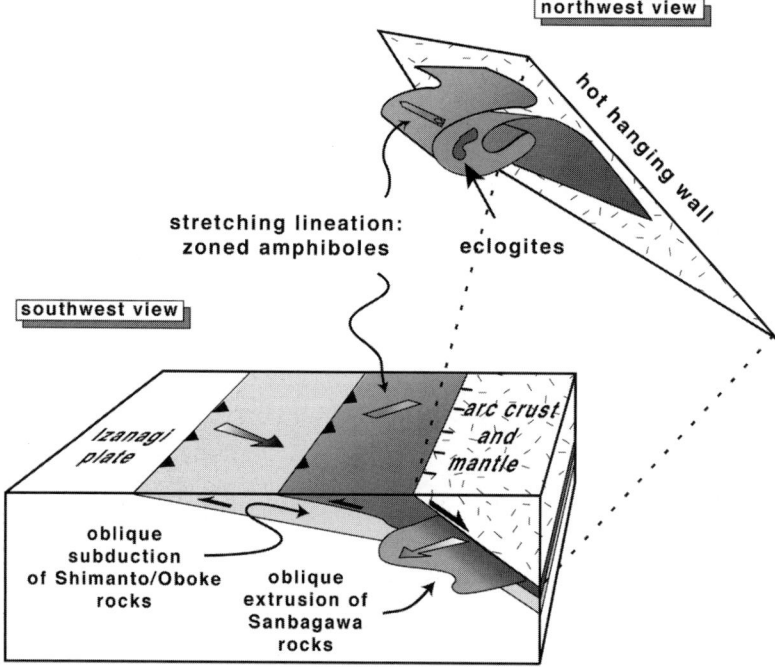

**Fig. 16.** 3D block diagram and 3D perspective view (note different view directions) that show our kinematic interpretation for the extrusion and exhumation of the Sanbagawa belt. Movement of the Sanbagawa from west to east (present coordinates), slightly up the dip of the subducting slab to shallower structural levels, is inferred to have been driven by a combination of oblique plate convergence, underplating, tectonic thinning, and partial closure of the subduction channel. Regional-scale recumbent fold deforms peak metamorphic isograds and is based, in part, on Wallis (1992, 1998).

extrusion of the Sanbagawa belt. Extrusion would have been at a low angle to the plate boundary (Figs 13, 14b and 16), however, rather than nearly orthogonal to it (Fig. 14c) (e.g. the Himalaya; Burchfiel & Royden 1985), because plate convergence was oblique, which reoriented $\sigma_3$ from vertical to sub-horizontal (compare Fig. 14b and c).

Quantitative estimates of the exhumation related to extrusion can be made if we assume a relatively simple closed subduction channel and a 30–40° dip for the subducting plate. For example, if the Sanbagawa belt had an initial width of 50 km parallel to the retrograde stretching lineation, thinning of the belt by 50% would have resulted in 50 km of extrusion parallel to the lineation. The vertical component of extrusion parallel to the lineation and obliquely up the dipping slab would be to 10–15 km. This estimate is less than half of the total exhumation required by the metamorphic history (c. 30 km), suggesting that other exhumation mechanisms are necessary.

Wallis (1995) has recently proposed that thinning alone could account for as much as 15 of the 30 km needed to bring the Sanbagawa rocks to the surface. This interpretation requires, however, that the thinning documented in the Sanbagawa belt is representative of the entire hanging wall, and this may not have been the case. Thinning of the high-pressure rocks in the Alps, for example, was apparently limited to the material within the subduction channel and was accommodated by the underthrusting of progressively thicker continental crust (Escher & Beaumont 1997); the overlying mantle and crust were not thinned. The frontal nose of thinned nappe, however, rose c. 20 km vertically (from 45 to 25 km) and was forced to return up the subduction zone as it was constrained by the closed subduction channel (Escher & Beaumont 1997). Thus, without independent evidence for thinning of the hanging wall, tectonic thinning and extrusion of the Sanbagawa belt rocks can account for no more than half of the required exhumation.

A third exhumation mechanism is return flow caused by pressure gradients within the accretionary wedge or subduction channel (Cowan & Silling 1978; Cloos 1982). For example, both Cowan & Silling (1978) and Cloos (1982) assumed a relatively rigid, arcward-dipping backstop that acts as a buttress which diverts material from the downgoing channel upward and towards the surface. During orthogonal convergence, strike-parallel stresses inhibit lateral flow, and exhumation occurs through erosion and tectonic extrusion with normal faulting at high structural levels and thrust faulting at moderate to low structural levels (Fig. 14c). During oblique convergence, the return flow may be more complex (Toriumi 1990; Wallis & Platt 1995). The first-order change in boundary conditions resulting from oblique convergence, however, is probably the reorientation of $\sigma_2$ and $\sigma_3$ (e.g. compare Fig. 14b and c), which would produce conditions favourable for the lateral escape, or flow, of material (Fig. 14b).

These different mechanisms of exhumation are not necessary mutually exclusive and all three may have been acting in concert. In fact, the 'underplating and extension' model of Platt (1986, 1987) incorporates many of these concepts and is similar to the model we envision for the exhumation of the Sanbagawa belt. In our model, however, oblique extension occurs dominantly on a regional-scale, arcward-dipping normal fault (e.g. the MTL of Fig. 1) rather than on a system of seaward- and arcward-dipping faults. This difference probably reflects the anisotropy provided by the initial arcward-dipping plate boundary beneath which the Sanbagawa belt was underthrust. In addition, structural studies show that the Sanbagawa belt itself was thinned substantially as it was exhumed (Wallis 1995) which, if the subduction channel was constricted, suggests that thinning would have been accommodated by extrusion. Our interpretation also considers the 3D aspects for exhumation.

The palaeo-pressure and -temperature data presented above may also be used to constrain the accretion and exhumation history of the Sanbagawa belt. For example, the inverted metamorphic gradients preserved in the Sanbagawa belt may reflect subduction initiation during late Cretaceous time because inverted thermal gradients are known to decay relatively rapidly (Oxburgh & Turcotte 1974), even when relatively cool oceanic crust and sediments are progressively underthrust (Peacock 1987). The counter-clockwise P–T path and progressive cooling of the Sanbagawa belt rocks are also consistent with this interpretation (e.g. Wakabayashi 1990; Hacker 1991) as are geological data which suggest the rejuvenation of subduction in late Cretaceous time (e.g. Otsuki 1992). Alternatively, the progressive cooling of the Sanbagawa belt rocks may reflect their progressive exhumation (e.g. Rubie 1984), and collapsed metamorphic field gradients in the Sanbagawa belt (Wallis et al. 1992) are consistent with this interpretation. The relatively cool metamorphic history appears to be inconsistent, however, with recent suggestions that an active spreading centre was subducted in late Cretaceous time (e.g. Maruyama 1997).

Finally, the lateral extrusion interpretation raises a fundamental question about oblique convergence and accretion: how was lateral extrusion accommodated along-strike? Unfortunately, we have few constraints because very little is known about the initial thickness of the Sanbagawa belt or how its present thickness varies along-strike. Wallis (1992) has argued for structural thinning of the Sanbagawa belt as it was exhumed, whereas Toriumi & Noda (1986) showed evidence of thickening of the Sanbagawa belt after the ductile (exhumation-related) fabrics formed. Presumably, as the Sanbagawa belt was extruded (to the east in the present reference frame), it (or rocks laterally equivalent to it) was also imbricated and thickened, at least locally. Without more data, however, we are unable to speculate on details of these processes.

## Comparison with previous interpretations of the Sanbagawa belt

The underplating–extrusion interpretation presented above shares several similarities with previous models for the exhumation of the Sanbagawa belt, and with blueschist belts in general. The interpretation, however, is also fundamentally different. Major differences centre on: (1) the retrograde path suggested by the segmented amphiboles, as discussed above, which is counter-clockwise and relatively cool; (2) structural data which suggest a 3D model with an arcward-dipping backstop rather than a 2D exhumation model; (3) the relative importance of lateral extrusion versus deformation partitioning during oblique convergence. We discuss and elaborate on the last two differences below.

*Three-dimensional models.* A 3D exhumation model is required for the Sanbagawa belt because the structural and metamorphic data indicate lateral exhumation (slightly oblique to the plate boundary) (Fig. 16). Most previous studies that have related underplating of the Shimanto belt to exhumation of the Sanbagawa belt, however, have assumed or implied 2D models (e.g. Kimura 1994; Tagami et al. 1995). These models have also implied that underplating and exhumation were parallel to the plate boundary normal. Although plate normal (or 'frontal') exhumation appears to be common in other orogenic belts (e.g. the central Himalaya, Burchfiel & Royden 1985; the Franciscan complex, Jayko & Blake 1987), the structural data from the Sanbagawa belt require lateral motions during exhumation.

Faure (1985), Toriumi (1985), Toriumi & Noda (1986), and Wallis & Platt (1995) recognized the well-developed, nearly strike-parallel lineation and argued that it reflected strike-parallel transport during oblique plate convergence. Toriumi & Kohsaka (1995) also documented cyclic, high-pressure events in eclogitic blocks within the Sanbagawa belt and proposed that they reflected an along-strike 'spiralling' motion. In their model, the spiralling motion resulted from the corner flow caused by oblique convergence (Toriumi & Noda 1986). Wallis & Platt (1995) have suggested that the change from north-directed underthrusting in the northern Shimanto belt to lateral flow in the Sanbagawa belt records a progressive increase in the strike-slip component of deformation during oblique convergence, and that this progressive change suggests a bulk viscous rheology for the accretionary prism. As they pointed out, however, their modelling results (see also Platt (1983)] indicate that the increase in strike-slip component should be gradational across the prism. In contrast, we prefer a model in which the Sanbagawa belt is decoupled from its hanging wall and footwall. This model is preferred primarily because the lineation data appear to be clustered into two groups (east- and north-trending, Fig. 11) and because the sense of shear appears to change from top-to-the west at high structural levels to top-to-the east at low structural levels (Wallis et al. 1992).

Finally, an important characteristic of all of the 3D models is that they assume a steep backstop with some degree of deformation partitioning and strike-parallel motions (spiral or otherwise). Recent geophysical data (e.g. refraction, reflection, and gravity; Ito et al. 1996) indicate, however, that the northern boundary of the Sanbagawa belt dips north 30–40°, which is consistent with lineation and foliation data from the belt (Fig. 1). Although the geometry of this boundary may have changed since late Cretaceous time, structural and thermal studies of younger, more seaward accretionary packages indicate only progressive arcward tilting, which would have steepened the older fabrics. We therefore consider the dip of the northern boundary to be a maximum and envision a gently arc-dipping backstop or 'lid'.

*Extrusion versus deformation partitioning.* A common conception of the consequences of oblique convergence is that plate convergence-related shortening is typically partitioned into plate-boundary normal and plate-boundary parallel components (Fitch 1972; Beck 1983; McCaffrey 1992). We have argued in the previous section, however, that this concept does not apply to the Sanbagawa belt. Instead, we propose that the belt was extruded at a low angle to the plate boundary. One way to reconcile this fundamental difference in understanding the consequences of oblique convergence is to consider that the upper and lower crusts are decoupled. That is, the upper, relatively brittle, crust may follow the Fitch model with the development of plate-boundary parallel strike-slip faults and plate-boundary normal shortening. At deeper structural levels (e.g. greenschist facies and higher), however, the flowing crust may be decoupled from the kinematics of the upper crust. Presumably, the decoupling occurs along zones that are much weaker (e.g. composed of weaker materials or overpressured) than the hanging-wall and footwall rocks. Decoupling of the upper and lower crust during exhumation has also been proposed to explain lateral flow in the lower crust of the Himalaya (Royden 1996).

In summary, we envision a model of exhumation and extrusion for the Sanbagawa belt that is similar to the frontal extrusion model for the Himalaya (Burchfiel & Royden 1985; Royden & Burchfiel 1987), the Andes (Dalmayrac & Molnar 1981), locally in the Alps (Escher & Beaumont 1997) and the Franciscan complex (Jayko & Blake 1987) in which the rearward boundary, or backstop, dips arcward. In the case of the Sanbagawa belt, however, the extruding rocks moved laterally at a low angle to the plate boundary rather than towards the deformation front. Lateral extrusion (relative to the inferred plate convergence vector) has also been proposed for high-pressure rocks in the Aegean Sea (Avigad & Garfunkel 1991) and in the northwest Italian Alps (Inger & Ramsbotham 1997). Ultimately, lateral extrusion in the Sanbagawa belt is thought to have been driven by oblique convergence, which resulted in a less constrained lateral boundary; tectonic thinning, which resulted in substantial stretching; and a partially closed subduction channel, which forced the stretched material to move up-dip.

## Conclusions

In this study we have shown that the Sanbagawa blueschists were exhumed along a gently rising eastward trajectory out of the late Cretaceous subduction zone that dipped under Japan. The synchronous subduction and metamorphism of rocks of the Shimanto belt (and equivalent units) probably helped drive the higher-pressure rocks to the surface. This mechanism requires that the blueschist facies rocks were not mechanically bound to either the hanging wall or footwall (Fig. 16). Such a mechanism holds much promise for explaining how blueschists return to the surface without becoming retrograded to greenschist or amphibolite facies assemblages. In explaining the exhumation of blueschists, the underplating–extrusion model also explains the continuous retrograde metamorphic reactions, the collapse of metamorphic zones, the preservation of low geothermal gradients, and the apparently slow rates of exhumation of some blueschists terranes.

This work was initiated while T.B. and R.P.W. were supported by separate Monbusho Fellowships from the Government of Japan. Subsequent work has been supported by NSF grants EAR-9418203 (R.P.W.) and EAR-9418344 (T.B.). Numerous colleagues in Japan have generously provided their time and assistance both in the field and the laboratory throughout the length of our studies. We are very grateful for their support. T.B. is also especially indebted to T. Tokunaga and A. Nomizo, who introduced him to the challenges of the Sanbagawa belt. S. Wallis, M. Faure, D. Cowan, and R. McCaffrey provided very helpful and constructive reviews of an earlier draft of this paper.

## References

AGAR, S., CLIFF, R., DUDDY, I. & REX, D. 1989. Accretion and uplift in the Shimanto Belt, SW Japan. *Journal of the Geological Society, London*, **146**, 893–896.

AVÉ LALLEMANT, H. G. & GUTH, L. R. 1990. Role of extensional tectonics in exhumation of eclogites and blueschists in an oblique subduction setting: northwestern Venezuela. *Geology*, **18**, 950–953.

AVIGAD, D. & GARFUNKEL, Z. 1991. Uplift and exhumation of high-pressure metamorphic terrains: the example of the Cycladic blueschist belt (Aegean Sea). *Tectonophysics*, **188**, 357–372.

BANNO, M. 1964. Petrologic studies on Sambagawa crystalline schists in the Bessi-Ino district, central Shikoku, Japan. *Journal, Faculty of Science, University of Tokyo, Section 2*, **15**, 203–319.

—— 1986. The high-pressure metamorphic belts of Japan: a review. *In*: EVANS, B. W. & BROWN, E. H. (eds) *Blueschists and Eclogites*. Geological Society of America, Memoir, **164**, 365–374.

—— & SAKAI, C. 1989. Geology and metamorphic evolution of the Sanbagawa metamorphic belt, Japan. *In*: DALY, J. S., CLIFF, R. A. & YARDLEY, B. W. D. (eds) *Evolution of Metamorphic Belts*. Geological Society, London, Special Publications, **43**, 519–531.

——, HIGASHINO, T., OTSUKI, M., ITALYA, T. & NAKAJIMA, T. 1978. Thermal structure of the Sanbagawa metamorphic belt in central Shikoku. *Journal of Physics of the Earth*, **26**, 345–356.

——, SAKAI, C. & OTSUKI, M. 1984. Thermal history of the Sambagwa metamorphic rocks. *Chigaku Zasshi*, 515–527.

BECK, M. 1983. On the mechanism of tectonic transport in zones of oblique subduction. *Tectonophysics*, **93**, 1–11.

BURCHFIEL, B. C. & ROYDEN, L. H. 1985. North–south extension within the convergent Himalayan region. *Geology*, **13**, 679–682.

BURG, J. P., BRUNEL, M., GAPAIS, D., CHEN, G. M. & LIU, G. H. 1984. Deformation of leucogranites of the crystalline Main Central Sheet in southern Tibet (China). *Journal of Structural Geology*, **6**(5), 535–542.

BYRNE, T. & DITULLIO, L. 1992. Evidence for changing plate motions in southwest Japan and reconstructions of the Philippine Sea plate. *Island Arc*, **1**, 148–165.

CLOOS, M. 1982. Flow melanges: numerical modelling and geologic constraints on their origin in the Franciscan subduction complex. *Bulletin, Seismological Society of America*, **93**, 330–345.

COSCA, M. A., ESSENE, E. T. & BOWMAN, J. R. 1991. Complete chemical analyses of metamorphic hornblendes; implications for normalizations, calculated $H_2O$ activites, and thermobarometry. *Contributions to Mineralogy and Petrology*, **108**, 472–484.

COWAN, D. S. & SILLING, R. M. 1978. A dynamic, scaled model of accretion at trenches and its implications for the tectonic evolution of subduction complexes. *Journal of Geophysical Research*, **83**, 5389–5396.

DALLMEYER, R. D. & TAKASU, A. 1991a. Middle Paleocene terrane juxtaposition along the median tectonic line, southwest Japan: evidence from $^{40}Ar/^{39}Ar$ mineral ages. *Tectonophysics*, **200**, 281–297.

—— & —— 1991b. Tectonometamorphic evolution of the Sebadani eclogitic metagabbro and the Sambagawa schists, central Shikoku, Japan: $^{40}Ar/^{39}Ar$ mineral age constraints. *Journal of Metamorphic Geology*, **9**, 605–618.

DALMAYRAC, B. & MOLNAR, P. 1981. Parallel thrust and normal faulting in Peru and constraints on the state of stress. *Earth and Planetary Science Letters*, **55**, 473–481.

DITULLIO, L. & HADA, S. 1993. Regional and local variations in the thermal history of the Shimanto Belt, southwest Japan. *In*: UNDERWOOD, M. (ed.) *Thermal Evolution of the Tertiary Shimanto Belt, Southwest Japan: An example of Ridge–Trench interaction*. Geological Society of America, Special Paper, **172**, 103–114.

ENAMI, M. 1998. Pressure–temperature path of Sanbagawa prograde metamorphism deduced from grossular zoning of garnet. *Journal of Metamorphic Geology*, **16**, 97–106.

——, WALLIS, S. R. & BANNO, Y. 1994. Paragenesis of sodic pyroxene-bearing quartz schists: implications for the P–T history of the Sambagawa belt. *Contributions to Mineralogy and Petrology*, **116**, 182–198.

ENGEBRETSON, D. C., COX, A. & GORDON, R. G. Ieds) 1985. *Relative Motions between Oceanic and Continental Plates in the Pacific Basin*. Geological Society of America, Special Paper, **59**.

ERNST, W. G. 1977. Mineral parageneses and plate tectonic settings of relatively high-pressure metamorphic belts. *Fortschritte der Mineralogie*, **54**, 192–222.

—— 1988. Tectonic history of subduction zones inferred from retrograde blueschist P–T paths. *Geology*, **16**, 1081–1084.

—— & PEACOCK, S. M. 1996. A thermotectonic model for preservation of ultrahigh-pressure phases in metamorphosed continental crust. *In*: BEBOUT, G. E., SCHOLL, D. W., KIRBY, S. H. & PLATT, J. P. (eds) *Subduction: top to bottom*. American Geophysical Union, Geophysical Monographs, **96**, 171–178.

ESCHER, A. & BEAUMONT, C. 1997. Formation, burial and exhumation of basement nappes at crustal scale: geometric model based on the Western Swiss–Italian Alps. *Journal of Structural Geology*, **19**, 955–974.

FABBRI, O. 1989. *Contribution à l'étude de la marge Asiatique: la chaîne Tertiare de Shimanto (Japon sw)*. PhD thesis, University of Orleans.

FAURE, M. 1985. Microtectonic evidence for eastward ductile shear in the Jurassic orogen of SW Japan. *Journal of Structural Geology*, **7**(2), 175–186.

——, IWASAKI, M., ICHIKAWA, K. & YAO, A. 1991. The significance of Upper Jurassic radiolarians in high pressure metamorphic rocks of SW Japan. *Journal of Southeast Asian Earth Science*, **6**(2), 131–136.

FERGUSON, C. 1981. A strain reversal method for estimating extension from fragmented rigid inclusions. *Tectonophysics*, **79**, T43–T52.

FITCH, T. J. 1972. Plate convergence, transcurrent faults, and internal deformation adjacent to Southeast Asia and the western Pacific. *Journal of Geophysical Research*, **77**(23), 4432–4460.

HACKER, B. 1991. The role of deformation in the formation of metamorphic gradients: ridge subduction beneath the Oman ophiolite. *Tectonics*, **10**(2), 455–473.

HANSEN, V. 1992. Backflow and margin-parallel shear within an ancient subduction complex. *Geology*, **20**, 71–74.

HARA, I., HIKE, K., TAKEDA, K., TSUDUTA, E., TOKUDA, M. & SHIOTA, T. 1977. Tectonic movement of the Sambagawa belt. *In*: HIDE, K. (ed.) *The Sambagawa Belt*. Hiroshima University Press, 307–390.

——, SHIOTA, T., HIDE, K., *et al.* 1992. Tectonic evolution of the Sambagawa schists and its implications in convergent margin processes. *Journal of Hiroshima University, Series C9*, 495–595.

——, ——, ——, OKAMOTO, K., TAKEDA, K., HAYASAKA, Y. & SAKURAI, Y. 1990. Nappe structure of the Sambagawa belt. *Journal of Metamorphic Geology*, **8**, 441–456.

HASEBE, N., TAGAMI, T. & NISHIMURA, S. 1997. Melange-forming processes in the development of an accretionary prism: evidence from fission-track thermochronology. *Journal of Geophysical Research*, **102**(B4), 7659–7672.

——, —— & —— 1993. Evolution of the Shimanto accretionary complex: a fission-track thermochronologic study. *In*: UNDERWOOD, M. (ed.) *Thermal Evolution of the Tertiary Shimanto Belt, Southwest Japan: An example of Ridge–Trench Interaction*. Geological Society of America Special Papers, **273**, 121–136.

HIGASHINO, T. 1990. The higher grade metamorphic zonation of the Sambagawa metamorphic belt in

central Shikoku, Japan. *Journal of Geological Society of Japan*, **96**, 703–718.
HODGES, K. V., PARISH, R. R. & SEARLE, M. P. 1996. Tectonic evolution of the central Annapurna Range, Nepalese Himalayas. *Tectonics*, **15**, 1264–1291.
HOLDAWAY, M. J. & MUKHOPADHYAY, B. 1993. A reevaluation of the stability relations of andalusite; thermochemical data and phase diagram for the aluminum silicates. *American Mineralogist*, **78**(3–4), 298–315.
HOLLAND, T. J. B. & RICHARDSON, S. W. 1979. Amphibole zonation in metabasites as a guide to the evolution of metamorphic conditions. *Contributions to Mineralogy and Petrology*, **70**, 143–148.
HOUSEMAN, G. A. & ENGLAND, P. C. 1986. Finite strain calculations of continental deformation, 2, method and general results for convergent zones. *Journal of Geophysical Research*, **91**, 3651–3663.
INGER, S. & RAMSBOTHAM, W. 1997. Syn-convergent exhumation implied by progressive deformation and metamorphism in the Valle dell'Orco transect, NW Italian Alps. *Journal of the Geological Society, London*, **154**, 667–677.
ISOZAKI, Y. & ITAYA, T. 1990. Chronology of Sambagawa metamorphism. *Journal of Metamorphic Geology*, **8**, 401–411.
ITO, T., IKAWA, T., ADACHI, I., et al. 1996. Geophysical exploration of the subsurface structure of the Median Tectonic Line, East Shikoku, Japan. *Journal of Geological Society of Japan*, **102**(4), 346–360.
IWASAKI, M., ICHIKAWA, K., YAO, A. & FAURE, M. 1984. On the age of the Mikabu green rocks, eastern Shikoku, Japan. *Newsletter, Kansai Branch*, Geological Society of Japan, **97**, 21.
JAYKO, A. S. & BLAKE, M. C. 1987. Attenuation of the Coast Range ophiolite by extensional faulting and the nature of the Coast Range 'thrust'. *Tectonics*, **6**, 475–488.
JOLIVET, L. 1994. Clockwise tectonic rotation of Tertiary sedimentary basins in central Hokkaido, northern Japan: Comment and Reply. *Geology*, **22**, 94–95.
——, HUCHON, P. & RANGIN, C. 1989. Tectonic setting of Western Pacific marginal basins. *Tectonophysics*, **160**, 23–47.
KANO, K.-I., NAKAJI, M. & TAKEUCHI, S. 1991. Asymmetrical melange fabrics as possible indicators of the convergent direction of plates: a case study from the Shimanto Belt of the Akaishi Mountains, central Japan. *Tectonophysics*, **185**, 375–388.
KAWATO, K., ISOZAKI, Y. & ITAYA, T. 1991. Geotectonic boundary between the Sambagawa and Chichibu belts in central Shikoku, Southwest Japan. *Journal of Geological Society of Japan*, **97**, 959–975.
KENZAN, R. G. 1984. Stratigraphy and geological structure of the Sanbagawa metamorphic belt in the Oboke area, central Shikoku, Japan. *Earth Sciences*, **39**, 53–63.
KIMURA, G. 1994. Rapid growth of the Latest Cretaceous accretionary complex and exhumation of high pressure type metamorphic rocks in the NW Pacific region. *Journal of Geophysical Research*, **99**, 22148–22164.
—— 1997. Cretaceous episodic growth of the Japanese Islands. *Island Arc*, **6**, 52–68.
—— & MUKAI, A. 1991. Underplated units in an accretionary complex: melange of the Shimanto Belt of Eastern Shikoku, Southwest Japan. *Tectonics*, **10**(1), 31–50.
KNIPE, R. J. & WINTSCH, R. P. 1985. Heterogeneous deformation, foliation development and metamorphic processes in a polyphase mylonite. *In*: THOMPSON, A. B. & RUBIE, D. C. (eds) *Advances in Physical Geochemistry*. Springer, New York, 180–210.
LAIRD, J. 1982. Phase relations of metamorphic amphiboles; natural occurrence and theory; amphiboles in metamorphosed basaltic rocks; blueschist–greenschist–eclogite relations. *In*: VEBLEN, D. R. & RIBBE, P. H. (eds) *Amphiboles; Petrology and Experimental Phase Relations*. Mineralogical Society of America, Reviews in Mineralogy, 73–122.
MACKENZIE, J. S., TAGUCHI, S. & ITAYA, T. 1990. Cleavage dating by K–Ar isotopic analysis in the Paleogene Shimanto Belt of eastern Kyushu, S.W. Japan. *Journal of Mineralogy, Petrology and Economic Geology*, **85**, 161–167.
MARUYAMA, S. 1997. Pacific-type orogeny revisited: Miyashiro-type orogeny proposed. *Island Arc*, **6**, 91–120.
MCCAFFREY, R. 1992. Oblique plate convergence, slip vectors, and forearc deformation. *Journal of Geophysical Research*, **97**(B6), 8905–8915.
—— 1993. On the role of the upper plate in great subduction zone earthquakes. *Journal of Geophysical Research*, **98**, 11953–11966.
MISCH, P. 1969. Paracrystalline microboudinage of zoned grains and other criteria for synkinematic growth of metamorphic minerals. *American Journal of Science*, **267**, 43–63.
MIYASHIRO, A. 1961. Evolution of metamorphic belts. *Journal of Petrology*, **2**, 277–311.
—— 1973. *Metamorphism and Metamorphic Belts*. Halsted, New York.
MIYATA, T. 1990. Slump strain indicative of paleoslope in Cretaceous Izumi sedimentary basin among Median Tectonic Line, Southwest Japan. *Geology*, **18**, 392–394.
MORGAN, J. 1972. Plate motions and deep mantle convection. *In: Studies in Earth and Space Science*. Geological Society of America, Memoir, **132**, 7–22.
NAKAMURA, C. & ENAMI, M. 1994. Prograde amphiboles in hematite-bearing basic and quartz schists in the Sanbagawa belt, central Shikoku: evolution of metamorphic field gradient and individual $P$–$T$ path. *Journal of Metamorphic Geology*, **12**, 841–852.
NAKAMURA, K. 1986. *Deformation process and condition of the Otonashigawa Group in the Shimanto Terrain, southwest Japan*. Science Report of Niigata University, Series E, Geology and Mineralogy, 25–87.
OTOFUGI, Y. 1996. Large tectonic movement of the Japan Arc in late Cenozoic times inferred from paleomagnetism: review and synthesis. *Island Arc*, **5**, 229–249.

OTSUKI, K. 1992. Oblique subduction, collision of microcontinents and subduction of oceanic ridge: their implications on the Cretaceous tectonics of Japan. *Island Arc*, **1**, 51–63.

OTSUKI, M. 1980. *Petrological studies of the basic Sambagawa metamorphic rocks in central Shikoku, Japan*. PhD thesis, University of Tokyo.

—— & BANNO, S. 1990. Prograde and retrograde metamorphism of hematite bearing basic schists in the Sambagawa belt in central Shikoku. *Journal of Metamorphic Geology*, **8**, 425–439.

OXBURGH, E. R. & TURCOTTE, D. L. 1974. Thermal gradients and regional metamorphism in overthrust terrains with special reference to the eastern Alps. *Schweizerische Mineralogische und Petrographische Mitteilungen*, **54**, 641–662.

PEACOCK, S. 1987. Creation and preservation of subduction-related inverted metamorphic gradients. *Journal of Geophysical Research*, **92**(B12), 12763–12781.

PEACOCK, S. M. 1993. The importance of blueschist–eclogite dehydration reactions in subducting oceanic crust. *Geological Society of America Bulletin*, **105**, 684–694.

PLATT, J. 1993. Mechanics of oblique convergence. *Journal of Geophysical Research*, **98**(B9), 16239–16256.

PLATT, J. P. 1983. Progressive refolding in ductile shear zones. *Journal of Structural Geology*, **5**(6), 619–622.

—— 1986. Dynamics of orogenic wedges and the uplift of high-pressure metamorphic rocks. *Geological Society of America Bulletin*, **97**, 1037–1053.

—— 1987. The uplift of high-pressure low-temperature metamorphic rocks. *Philosophical Transactions of the Royal Society of London, Series A*, **321**, 87–103.

ROYDEN, L. 1996. Coupling and decoupling of crust and mantle in convergent orogens: implications for strain partitioning in the crust. *Journal of Geophysical Research*, **101**, 17679–17705.

ROYDEN, L. H. & BURCHFIEL, B. C. 1987. Thin-skinned N–S extension within the convergent Himalayan region: gravitational collapse of a Miocene topographic front. *In*: DEWEY, J. F. & HANCOCK, P. L. (eds) *Continental Extensional Tectonics*. Geological Society, London, Special Publications, **28**, 611–619.

RUBIE, O. C. 1984. A thermal–tectonic model for high-pressure metamorphism and deformation in the Sesia zone, Western Alps. *Journal of Geology*, **92**, 21–36.

SHINJOE, H. & TAGAMI, T. 1994. Cooling history of the Sambagawa metamorphic belt inferred from fission track zircon ages. *Tectonophysics*, **239**, 73–79.

SHREVE, R. L. & CLOOS, M. 1986. Dynamics of sediment subduction, melange formation, and prism accretion. *Journal of Geophysical Research*, **91**, 10229–10245.

SPEAR, F. 1982. Phase equilibria of amphiboles from the Post Pond Volcanics, Mt. Cube Quadrangle, Vermont. *Journal of Petrology*, **23**, 383–426.

SUYARI, K., KUWANO, U. & ISHIDA, K. 1980. Discovery of late Triassic conodonts from the Sanbagawa metamorphic belt in western Shikoku. *Journal of Geological Society of Japan*, **86**, 827–828.

SUZUKI, H., ISOZAKI, Y. & ITAYA, T. 1990. Tectonic superposition of the Kurosegawa Terrane upon the Sambagawa metamorphic belt in eastern Shikoku, southwest Japan: K–Ar ages of weakly metamorphosed rocks in northwestern Kamikatsu town, Tokushima prefecture. *Journal of Geological Society of Japan*, **96**, 143–153.

TAGAMI, T., HASEBE, N. & SHIMADA, C. 1995. Episodic exhumation of accretionary complexes: fission-track thermochronologic evidence from the Shimanto Belt and its vicinities southwest Japan. *Island Arc*, **4**, 209–230.

TAIRA, A. 1985. Sedimentary evolution of Shikoku subduction zone: the Shimanto Belt and Nankai Trough. *In*: NASU, N. (ed.) *Formation of Active Ocean Margins*. Terra Scientific, Tokyo, 835–851.

——, BYRNE, T. & ASHI, J. 1992. *Photographic Atlas of an Accretionary Prism, Geologic Structures of the Shimanto Belt, Japan*. University of Tokyo Press.

——, KATTO, J., TASHIRO, M., OKAMURA, M. & KODAMA, K. 1988. The Shimanto Belt in Shikoku, Japan – evolution of Cretaceous to Miocene accretionary prism. *Modern Geology*, **12**(1–4), 5–46.

——, TASHIRO, M., OCAMURA, M. & KATTO, J. 1980. The geology of the Shimanto Belt in Kochi Prefecture, Shikoku, Japan. *In*: TAIRA, A. & TASHIRO, M. (eds) *Geology and Paleontology of the Shimanto Belt, Selected Papers Honoring Prof. J. Katto*. Rinyakosaikai Press, Kochi, 319–389.

TAKAGI, H. 1986. Implications of mylonitic microstructures for the geotectonic evolution of the Median Tectonic Line, Central Japan. *Journal of Structural Geology*, **8**, 3–14.

—— & HARA, I. 1979. Relationship between growth of albite porphyroblasts and deformation in a Sambagawa schist, central Shikoku, Japan. *Tectonophysics*, **58**, 113–125.

TAKASU, A. 1989. *P–T* histories of peridotite and amphibolite tectonic blocks in the Sambagawa metamorphic belt, Japan. *In*: DALY, J. S., CLIFF, R. A. & YARDLEY, B. W. D. (eds) *Evolution of Metamorphic Belts*. Geological Society, London, Special Publications, **43**, 538–553.

—— & DALLMEYER, R. D. 1990. $^{40}$Ar/$^{39}$Ar mineral age constraints for the tectonothermal evolution of the Sambagawa metamorphic belt, central Shikoku, Japan: a Cretaceous accretionary prism. *Tectonophysics*, **185**, 111–139.

—— & —— 1992. $^{40}$Ar/$^{39}$Ar mineral ages of metamorphic clasts from the Kuma group (Eocene), Central Shikoku, Japan; implications for the tectonic development of the Sambagawa accretionary prism. *Lithos*, **28**, 69–84.

——, WALLIS, S. A., BANNO, S. & DALLMEYER, R. D. 1994. Evolution of the Sambagawa metamorphic belt, Japan. *Lithos*, **33**, 119–133.

THOMPSON, A. B., SCHULMANN, K. & JEZEK, J. 1999. Thermal evolution and exhumation in obliquely convergent (transpressive) orogens. *Tectonophysics*, in press.

TILLMAN, K. S. & BYRNE, T. 1995. Kinematic analysis of the Taiwan Slate Belt. *Tectonics*, **14**(2), 322–341.

TORIUMI, M. 1975. *Petrological Study of the Sambagawa Metamorphic Rocks: the Kanto Mountains, central Japan*. University Museum, University of Tokyo.

—— 1985. Two types of ductile deformation/regional metamorphic belt. *Tectonophysics*, **113**, 307–326.

—— 1990. The transition from brittle to ductile deformation in the Sambagawa metamorphic belt, Japan. *Journal of Metamorphic Geology*, **8**, 457–466.

—— & KOHSAKA, Y. 1995. Cyclic *P–T* path and plastic deformation of eclogite mass in the Sambagawa metamorphic belt. *Journal of Faculty of Science*, **22**, 211–231.

—— & NODA, H. 1986. The origin of strain patterns resulting from contemporaneous deformation and metamorphism in the Sambagawa metamorphic belt. *Journal of Metamorphic Geology*, **4**, 409–420.

——, TERUYA, J., MASUI, M. & KUWAHARA, H. 1986. Microstructures and flow mechanisms in regional metamorphic rocks of Japan. *Contributions to Mineralogy and Petrology*, **94**, 54–62.

TRIBOULET, C. 1992. The (Na–Ca) amphibole–albite–chlorite–epidote–quartz geothermobarometer in the system S–A–F–M–C–N–H$_2$O: an empirical calibration. *Journal of Metamorphic Geology*, **10**, 545–556.

WAKABAYASHI, J. 1990. Counterclockwise *P–T–t* paths from amphibolites, Franciscan Complex, California: relics from the early stages of subduction zone metamorphism. *Journal of Geology*, **98**(5), 657–680.

WALLIS, S. 1990. The timing of folding and stretching in the Sambagawa Belt; the Asemigawa region, central Shikoku. *Journal of the Geological Society of Japan*, **96**, 345–357.

—— 1998. Exhuming the Sanbagawa metamorphic belt: the importance of tectonic discontinuities. *Journal of Metamorphic Geology*, **16**, 83–95.

—— & BANNO, S. 1990. The Sambagawa Belt – trends in research. *Journal of Metamorphic Geology*, **8**, 393–399.

—— & PLATT, J. 1995. Oblique subduction and kinematics of deformation in southwest Japan: implications for the bulk rheology of convergent margins. *Journal of the Czech Geological Society*, **40**(3), C-128.

WALLIS, S. R. 1992. Vorticity analysis in a metachert from the Sanbagawa Belt, SW Japan. *Journal of Structural Geology*, **14**(3), 271–280.

—— 1995. Vorticity analysis and recognition of ductile extension in the Sanbagawa belt, SW Japan. *Journal of Structural Geology*, **17**(8), 1077–1093.

——, BANNO, S. & RADVANEC, M. 1992. Kinematics, structure and relationship to metamorphism of the east–west flow in the Sambagawa belt, southwest Japan. *Island Arc*, **1**, 175–185.

WINTSCH, R. P., TORIUMI, M. & BYRNE, T. 1994. Syntectonic retrograde amphibole overgrowth as evidence for late Cretaceous oblique plate convergence in the Sanbagawa metamorphic belt, SW Japan. *Geological Society of America Abstracts with Programs*, A212.

YU, S. B. & LEE, C. 1986. Geodetic measurement of horizontal crustal deformation in eastern Taiwan. *Tectonophysics*, **125**, 73–85.

# Spatial and temporal variations in exhumation of the central Swiss Alps and implications for exhumation mechanisms

FRITZ SCHLUNEGGER[1,2] & SEAN WILLETT[1]

[1] *Penn State University, Department of Earth Sciences, Deike Building, University Park, PA 16802, USA*

[2] *Present address: Institut für Geowissenschaften, Friedrich-Schiller-Universität Jena, Burgweg 11, 07749 Jena, Germany (e-mail: fritz@geo.uni-jena.de)*

**Abstract:** Information about the structural and thermal evolution of the Alps, interpreted through thermal and mechanical models, provides an improved understanding of the processes that led to the exhumation of the Alps. We synthesize published thermochronometric data, analyse these data in terms of cooling rates and interpret the spatial and temporal patterns of cooling. Cooling rates are interpreted in terms of exhumation rates, aided by the use of a one-dimensional thermal model. Our study reveals that rapid exhumation of the Lepontine core occurred during the interval of 35–20 Ma. Existing data cannot determine whether this was by rapid erosion at rates exceeding 1 km Ma$^{-1}$ or by a relatively brief period of tectonic exhumation, although the correlation between extensional fault motion and high cooling rates supports the tectonic exhumation hypothesis. Peripheral regions of the Alps cooled at rates consistent with low to moderate exhumation rates of 400–500 m Ma$^{-1}$, initiating later than cooling of the Lepontine core, consistent with outward growth of the orogen. Outward growth of the orogen is potentially the result of either lower exhumation rates or higher rates of crustal accretion as demonstrated by a two-dimensional, coupled erosion–deformation model. In particular, the growth of the Southern Alps after c. 20 Ma is evidence for a decrease in the exhumation rate relative to the crustal accretion rate. This could represent a decrease in exhumation rate after cessation of normal faulting, or it could reflect initiation of accretion of a larger fraction of the European crust.

Orogens are the result of multiple lithospheric and surface processes that interact simultaneously and with feedback. Possible controls on the growth of orogens have been intensively explored by theoretical models and numerous case studies (Davis *et al.* 1983; Allmendinger *et al.* 1990; Pfiffner 1992; Willett *et al.* 1993; Beaumont *et al.* 1996a; Schmid *et al.* 1996; Meigs & Burbank 1997). Similarly, the importance of erosion to the orogen structure and the rate and patterns of exhumation of deeply buried rocks has been shown by theoretical models (Willett *et al.* 1993; Avouac & Burov 1996; Batt & Braun 1997), and has been illustrated by case histories (Hoffman & Grotzinger 1993; Okaya *et al.* 1996a,b; Beaumont *et al.* 1996b). The Oligocene–Miocene Alps of central Switzerland serve as a good example in which the exhumation history provides important constraints on theories of coupled deformation, heat transfer and erosion.

In this paper, we present a detailed analysis and tectonic synthesis of the thermochronological data that have been collected in the last 30 years by workers at the universities of Heidelberg, Lausanne, Bern and Zürich (e.g. Schaer *et al.* 1975; Hurford 1986; Michalski & Soom 1990) and that have been compiled by Hunziker *et al.* (1992, 1997). These data have been traditionally used to calculate average rates of exhumation assuming constant geothermal gradients (Hurford 1986; Giger 1991; Pfiffner *et al.* 1997a), or to estimate average production rates of sediment supplied to neighbouring foreland basins (Sinclair & Allen 1992; Schlunegger *et al.* 1997a). However, the relationship between cooling rates, rock exhumation rates and rates of sediment supply to basins is complicated by other processes including tectonic exhumation by normal faulting (Mancktelow 1985; Graseman & Mancktelow 1993; Steck & Hunziker 1994; Nievergelt *et al.* 1996) and by advection of heat by exhumation (Mancktelow & Grasemann 1997).

The goal of this paper is to provide a synthesis of thermochronometric data from the Alps, compare these data with thermal and mechanical models, and to interpret the results in terms of exhumation mechanisms. The thermochronometric data synthesis relies primarily on the compilation of Hunziker *et al.* (1992, 1997) and is used to characterize the cooling and exhumation of Alpine rocks over the last 30–40 Ma. The chronology of crustal extension is also

SCHLUNEGGER, F. & WILLETT, S. 1999. Spatial and temporal variations in exhumation of the central Swiss Alps and implications for exhumation mechanisms. *In:* RING, U., BRANDON, M. T., LISTER, G. S. & WILLETT, S. D. (eds) *Exhumation Processes: Normal Faulting, Ductile Flow and Erosion.* Geological Society, London, Special Publications, **154**, 157–179.

considered because of the proposal that much of the exhumation of the high-grade Lepontine metamorphic dome was by this mechanism of tectonic unroofing (Mancktelow 1992). By considering the spatial and temporal patterns in cooling and by comparing these patterns with those produced by tectonic and erosional exhumation we attempt to evaluate the relative importance of these mechanisms.

## Geological setting

The Alps (Fig. 1a) comprise a doubly vergent orogen which consists of a highly metamorphosed upper-crustal crystalline core and mostly unmetamorphosed thick-skinned and thin-skinned thrust sheets on the external flanks (Fig. 1b). In the north, the present-day Alps comprise the external massifs (e.g. the Aar massif) and the Helvetic thrust nappes (Fig. 1b). They are structurally overlain by the piggy-back stack of north-vergent Penninic and Austroalpine nappes that were emplaced before 35 Ma (Fig. 2; Schmid et al. 1996). Thrusting of the Penninic and Austroalpine nappes above the Helvetic nappes (Fig. 2) resulted in low-grade metamorphism of the latter units sometime between 30 and 35 Ma (Frey et al. 1980; Hunziker et al. 1992). Subsequent emplacement of the Helvetic nappes above the Aar massif along the basal Helvetic thrust (BH, Fig. 1b) occurred before 30 Ma in the east, between c. 30 and 25 Ma in the centre, and between 30 and 20 Ma in the west (Fig. 3), resulting in low-grade metamorphism of about 350°C of the footwall units by 25 Ma at the latest (Burkhard 1988).

The Penninic and Austroalpine nappes are separated from the south-vergent Southern Alps by the E–W striking Insubric Line (Fig. 1b). This fault is interpreted to have accommodated the late stages of collision between the Central Alps and the Adriatic promontory (Southern Alps) by steep S-directed backthrusting and right-lateral strike-slip movement (Fig. 2; Schmid et al. 1989, 1996). The phase of enhanced rates of backthrusting is considered to be diachronous, starting before 30 Ma in the east and at c. 25 Ma in the west (Burkhard 1988; Schmid et al. 1989, 1996).

The central Alps comprise a Barrow-type facies series of Tertiary metamorphism that has been referred to as the Lepontine dome in the classical Alpine literature (Fig. 1a) (see discussion by Engi et al. (1995)). Surface rocks exposed in the Lepontine dome are interpreted to have experienced maximum temperatures between c. 35–30 Ma (Steck & Hunziker 1994) and form an asymmetric dome of isograd surfaces which has been exposed by late Oligocene to present-day exhumation. Alternative studies, however, suggest that the peak of thermal metamorphism in this part of the Alps occurred between c. 27 and 21 Ma (Engi et al. 1995). Metamorphic grade increases from diagenetic conditions in the north to amphibolite facies in the south near the Insubric Line and to granulite facies in the Bergell area. The southern limit of the Lepontine dome is formed by the Insubric Line (Fig. 1a; Frey et al. 1980). The Lepontine dome is laterally flanked by low-angle normal faults and shear zones along which orogen-parallel crustal extension was interpreted to have occurred (Fig. 1a) (Mancktelow 1985, 1992; Steck & Hunziker 1994). Because cooling ages from basement rocks exposed in the Lepontine dome provide information only about the exhumation path of those rocks at present at the Earth's surface, they provide limited information about the erosional mechanisms of this part of the Alps. Therefore, as pointed out by Grasemann & Mancktelow (1993) and Mancktelow & Grasemann (1997), a convincing reconstruction of the erosional mechanisms requires an integrated discussion of cooling ages, tectonic structures and the thermal structure of the lithosphere.

### The detachment systems

The shear zones bordering the Lepontine dome have been interpreted as having accommodated normal faulting (Mancktelow 1985, 1992; Steck & Hunziker 1994; Nievergelt et al. 1996) possibly leading to significant tectonic exhumation of the Lepontine core (Mancktelow 1992; Grasemann & Mancktelow 1993; Steck & Hunziker 1994). A detailed analysis of their structure and inferred time of motion is crucial to our analysis.

The Simplon Fault Zone (Mancktelow 1992; Grasemann & Mancktelow 1993) borders the Lepontine dome to the west (Fig. 1a) and consists of two components (Mancktelow 1992): a c. 8 km broad and c. 10–15 km thick shear zone of strongly foliated and variably lineated mylonite displaying SW–NE-oriented stretching lineations within the footwall (Simplon Shear Zone, Fig. 1a, Steck & Hunziker 1994). This shear zone is separated from the hanging wall by a narrow cataclastic discontinuity (Rhône–Simplon Line, Fig. 1a, Mancktelow 1985). The cataclasites that are associated with the Rhône–Simplon Line are clearly younger than the mylonites and directly overprint them (Mancktelow 1985, 1992).

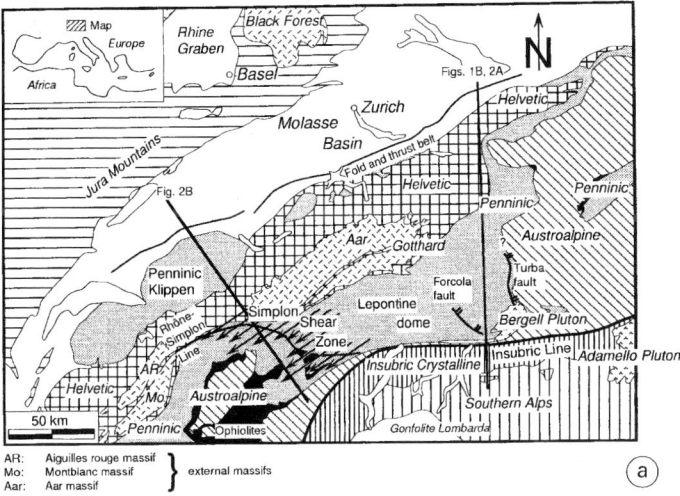

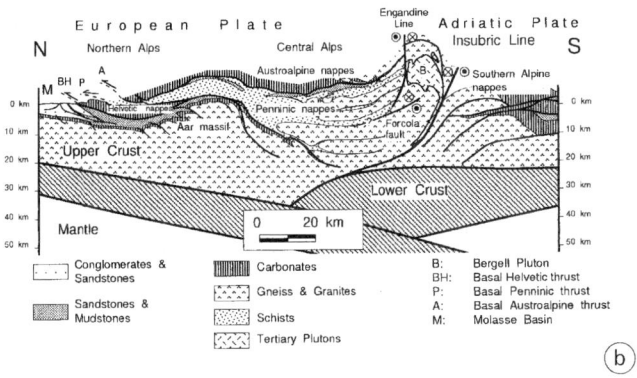

**Fig. 1.** (**a**) Geological map of the Oligo-Miocene Swiss Molasse Basin and the adjacent Alpine orogen. The major tectonic units are labelled. The map also shows the traces of the major low-angle normal faults and locations of the cross-sections of Fig. 2. It should be noted that in the eastern Alps, the Penninic and Austroalpine nappes are part of the northern Alps. In the analysed cross-section, however, they crop out in the central Alps south of the Aar massif. The arrows indicate the sense of motion recorded by the stretching lineations in the Simplon Shear Zone (modified after Steck & Hunziker 1994). (**b**) Synthetic cross-section through the present-day Alpine hinterland. The eroded Penninic and Austroalpine nappes are projected into the cross-section to illustrate the architecture of the Alpine edifice before erosion. (Modified after Schmid *et al.* 1996.)

The Simplon Shear Zone (Steck & Hunziker 1994) separates amphibolite facies rocks in the footwall from greenschist facies rocks in the hanging wall. Similarly, the metamorphic grade during mylonitization increases with distance into the footwall, perpendicular to strike (Mancktelow 1992). In the south, the shear zone crosses the zone of high metamorphic grade. Farther to the northwest, it follows the southern rim of the Aar massif and the southernmost Helvetic nappes at an oblique angle, gradually changing into a dextral shear zone. Sense of shear criteria suggest that displacement of the hanging wall occurred towards the SW (Mancktelow 1992), and the amount of relative displacement along this zone was estimated to

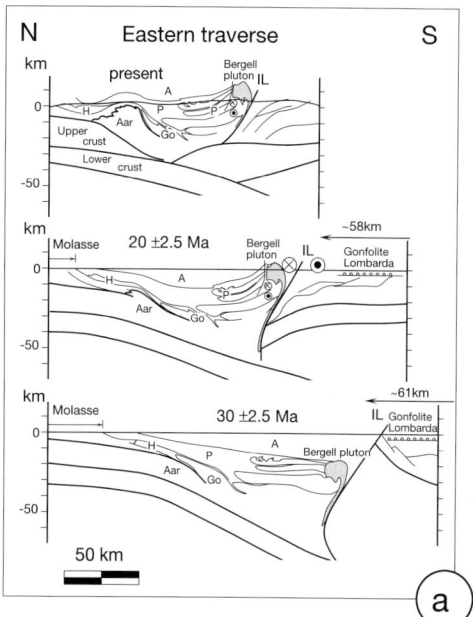
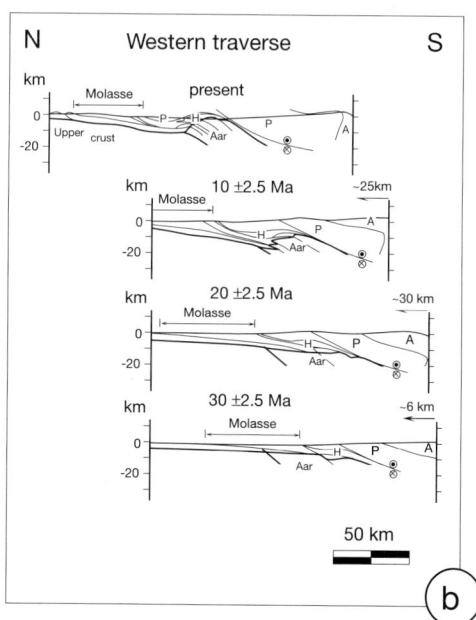

**Fig. 2.** Structural evolution of the Alps in (**a**) the east and (**b**) the west of the study area. The sequential palinspastic restorations are modified after Schmid et al. (1996) and Burkhard (1988) for the east and the west, respectively. Aar, Aar massif; Go, Gotthard 'massif'; A, Austroalpine; P, Penninic; H, Helvetic; IL, Insubric Line.

be c. 80 ± 40 km (Steck 1990). The initiation and duration of slip along this shear has been assessed using (1) cross-cutting relationships between temporally calibrated metamorphic fabrics and the shear zone (Steck & Hunziker 1994; Trommsdorff & Evans 1974), and (2) two-dimensional finite-difference thermal modelling of normal faulting (Grasemann & Mancktelow 1993). According to the former workers, slip movement along this zone occurred between c. 30 and 15 Ma. However, because Alpine geologists strongly disagree about the age of the peak thermal metamorphism in the Lepontine dome (e.g. 25 Ma v. 35 Ma) (Deutsch & Steiger 1985), assessments of the chronology of movement along the Simplon Shear Zone are controversial.

In this paper, we follow the finding of Jäger (1973) that the peak of thermal metamorphism of the Lepontine dome occurred between 35 and 30 Ma (see below for justification). This implies that extension along the Simplon detachment fault commenced later than c. 30 Ma. Furthermore, based on the numerical simulation of the cooling history of the Lepontine dome, Grasemann & Mancktelow (1993) suggested that the rate of slip along the Simplon Shear Zone was probably highest between 18 and 15 Ma, reaching maximum values of c. 1 km Ma$^{-1}$. However, because these workers did not take into account the large temporal variations of cooling ages into their model (Hunziker et al. 1992, 1997), their estimates of a 3 Ma duration of enhanced rates of cooling might be too low.

The present-day trace of the discrete Rhône–Simplon Line follows the regional trend of the Simplon Shear Zone (Fig. 1a). The Rhône–Simplon Line is interpreted to have accommodated SW–NE-oriented synorogenic extension by SW-directed brittle cataclastic movement of the hanging wall for the last 10–15 Ma (Mancktelow 1985; Steck & Hunziker 1994). The Simplon Shear Zone appears to have evolved continuously into the Rhône–Simplon

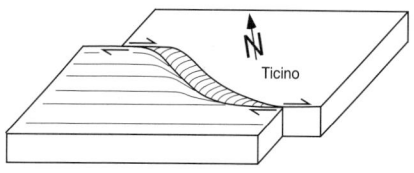

**Fig. 3.** Sketch figure showing the geometrical configuration of the Simplon Shear Zone.

Line. According to Grasemann & Mancktelow (1993), exhumation and cooling of the footwall led to a progressive concentration of deformation into a narrower zone and the transition to cataclastic reworking of the earlier mylonites. The three-dimensional reconstruction of the present-day geometry of the Rhône–Simplon Line reveals that the plane along which normal faulting occurred dips at c. 25° to the SW, whereas the plane that accommodates strike-slip movement dips at c. 18° to the SE (Mancktelow 1985, 1992; Steck & Hunziker 1994). According to Hunziker (1969) and Mancktelow (1985), the amount of post 10–15 Ma displacement along the Simplon fault was c. 12 km.

The geometry of the Simplon Fault Zone as outlined above results in a characteristic pattern of tectonic exhumation. Along the SE-dipping part of the detachment, only right-lateral strike-slip movement occurred, which resulted in zero tectonic exhumation (Fig. 3). However, major tectonic exhumation was accomplished by SW-directed normal faulting along the SW-dipping part of the fault (Steck & Hunziker 1994).

The Forcola and Turba normal faults border the Lepontine dome to the east (Fig. 1a). These faults were interpreted by Nievergelt et al. (1996) to have accommodated strike-parallel synorogenic extension by E-directed slip movement of the hanging wall. At present, the Turba low-angle normal fault separates low- to medium-grade sedimentary and crystalline rocks in the hanging wall from medium- to high-grade gneiss and schists in the footwall (Liniger & Nievergelt 1990). Mapping of the trace of the Turba fault has not been completed (Schmid et al. 1997), but present-day geological maps suggest that it extends >25 km in a N–S direction (Nievergelt et al. 1996) (Fig. 1a). Using cross-cutting relationships between the fault plane and temporally calibrated metamorphic fabrics (von Blankenburg 1990, 1992; Schreurs 1993), Nievergelt et al. (1996) speculated that slip along the Turba fault occurred sometime between 45 and 30 Ma. New studies in the area surrounding the Bergell pluton suggest that extension along the Turba fault might have occurred between c. 33 and 30 Ma (Berger et al. 1996). However, the amount of displacement along this fault is not known.

Situated at deeper structural levels and c. 20 km farther west, NNW–SSE-striking steeply inclined E-dipping normal faults exhibit a systematic downthrow towards the ENE, documented by stretching lineations (Baudin et al. 1993). The largest of these faults, the Forcola normal fault, has not yet been systematically investigated. However, the trace of this fault has been mapped between the area of the Bergell pluton and the tip of the Penninic crystalline nappes, by Schmid et al. (1997) (Fig. 1a). These workers suggested that E-directed slip movement of the hanging wall occurred between 25 and 20 Ma, which, however, is poorly constrained. The amount of displacement along this fault is not known.

## Cooling rates and patterns in the Alpine orogen

As a first step in our analysis, we summarize existing thermochronometric data bearing on the cooling history of the Alps based largely on the compilation of Hunziker et al. (1992, 1997) and similar to the synthesis given by Ménard (1988). Subsequently, we use these data to infer cooling rates for the last 30 Ma of the orogen history. This reconstruction is performed in three steps. First, we determine the spatial distribution of the maximum temperatures, or the peak of thermal metamorphic conditions, for the present-day exposed rocks. Second, by assuming closure temperatures for individual isotopic systems, we map the age of cooling through specific closure isotherms. These age–temperature data are subsequently converted to maps of temperature at specific ages. Finally, by subtracting these temperature fields, we obtain maps of cooling rates which are compared with models and specific Alpine structures. Details of this analysis and the resulting patterns of cooling are presented in the following three sections.

### The thermochronometric data of the Alps

The history of cooling of Alpine rocks begins with the peak of thermal metamorphic conditions. These temperatures are interpreted to have occurred at c. 35 Ma south of the Gotthard 'massif', and between 30 and 25 Ma in the area of the external massifs (Hunziker et al. 1992). We used (i) the thermal calibration of reaction isograds (Niggli & Niggli 1965; Tromsdorff 1966; Frey 1987; Hunziker et al. 1992), and (ii) the thermal calibration of mineral parageneses that were interpreted to have formed in a thermodynamic equilibrium during metamorphism (Frey et al. 1980; Ayrton et al. 1982; Klaper 1982; Hunziker et al. 1986; Burkhard 1988; Bowtell 1991; Engi et al. 1995). The locations of the low-temperature isograds (200–300°C) are taken from Frey (1974, 1986) and Burkhard (1988).

The assumption that maximum temperatures were achieved in the Lepontine dome at c. 35 Ma (Jäger 1973) is not universally accepted. Major controversies arise from the <30 Ma K–Ar ages for amphibole (Deutsch & Steiger 1985), U–Pb ages for monazite and zircon, and the Sm–Nd ages for garnet that were measured for the highly metamorphosed core ($T_{max}$ >500°C) of the Lepontine dome (Köppel & Grünenfelder 1975; Köppel et al. 1981; Vance & O'Nions 1992). These ages have been interpreted either as cooling ages with unknown or disputable closure temperatures (e.g. Hunziker et al. 1992; Steck & Hunziker 1994), or as formation ages of high-grade mineral parageneses, which would require a post 30 Ma thermal event in the core of the Lepontine dome (Deutsch & Steiger 1985; Engi et al. 1995). Indeed, using a thermobarometric recalibration of the monazite, zircon and garnet ages, Engi et al. suggested that the 35 Ma mineral parageneses that formed below the staurolite isograd (<500°C, Fig. 4a) were formed during south-dipping continental subduction, whereas the post 30 Ma $T_{max}$ in the central and southern parts of the Lepontine dome ($T_{max}$ >500°C, Fig. 4a) document later heating. These workers suggested that at c. 30 Ma hot thrust slices (Cima Lunga tectonic mélange) were expelled from deep crustal positions (>100 km) onto the lower Pennine stack at c. 20 km depth. According to their model, thermal relaxation of the initially inverted thermal gradient provided heat to the footwall underneath the hot lid (i.e. the present-day Lepontine dome). The model of Engi et al. appears to be confirmed by the presence of 40-Ma-old eclogites in the Cima Lunga tectonic mélange (Central Alps) (Becker 1993) and by the occurrence of ultrahigh-pressure metamorphic rocks in the western Alps (Van der Klauw et al. 1997). However, we doubt that the heat advected by this relatively small tectonic sliver (Spicher 1980) is large enough to cause the Barrovian-type facies series of the Lepontine dome. Therefore, as long as the model of Engi et al. (1995) is not confirmed by structural data nor by thermokinematic forward models (e.g. Okaya et al. 1996a; Schmid et al. 1996), we tentatively follow the findings of Jäger (1973), that (i) the main thermal metamorphism of the Lepontine dome (formation of the Barrovian-type facies series) occurred at c. 35 Ma, (ii) this phase of thermal overprint was followed by a phase of continuous cooling, and (iii) the hornblende, monazite and garnet ages of <30 Ma represent cooling ages (e.g. Steck & Hunziker 1994). Similarly, structural and chronological data from the Bergell area (Fig. 1a) suggest that heat was advected by intrusion of magma to mid-crustal levels between c. 33 and 28 Ma (Berger et al. 1996). Nevertheless, we do not think that this thermal event affected the cooling processes of the Lepontine dome significantly because of the large distance between the Bergell pluton and the Lepontine dome (Fig. 1a), and because the restricted volume of the pluton appears to be too small to carry enough heat to cause a regional metamorphic event. The latter statement, however, needs to be tested by thermal modelling.

Subsequent to the peak of the thermal metamorphic conditions, cooling histories are recorded by thermochronometric data. We chose closure temperatures of $110 \pm 20$°C for apatite (Fig. 4b), and $240 \pm 30$°C for zircon fission-track systems (Fig. 4c), appropriate for the average cooling rates of c. 15°C Ma$^{-1}$ (Hurford 1986). Further cooling ages are taken from Rb–Sr and K–Ar systems for biotite (Fig. 4d) ($300 \pm 50$°C, Amstrong et al. 1966), and the system of Rb–Sr for K-white mica (Fig. 4e) ($500 \pm 50$°C, Jäger et al. 1967). The complete list of papers that presented and discussed the thermochronometric data has been published by Hunziker et al. (1992) and reviewed by Hunziker et al. (1997). Additional apatite and zircon fission-track data are taken from Seward & Mancktelow (1994), Lihou et al. (1995) and Rahn et al. (1997). The closure temperatures for biotite and white mica were established based on the fact that rejuvenated and/or newly grown biotite and white mica overlap with the stilpnomelane–biotite and staurolite–chloritoid isograds, respectively (Jäger et al. 1967; Dempster 1986). However, because the closure temperatures depend on various factors such as strain rates, grain size, duration of metamorphism, rates of cooling and the chemical composition of minerals, this calibration has been a matter of debate (e.g. Deutsch & Steiger 1985). Nevertheless, as cooling ages are regionally consistent and independent of lithological parameters (e.g. Soom 1990), we consider the concept of closure temperature of Jäger et al. (1967) and Jäger (1973) as reasonable.

Because the Dora Maira Massif (western Alps) and the Cima Luna tectonic mélange (central Alps) experienced maximum pressure conditions (eclogite facies) between 40 and 35 Ma (Becker 1993; Gebauer et al. 1995), the published K–Ar cooling ages of biotite especially in these part of the Alps might be too old, because of the retention of 'unsupported' Ar (Soom 1990). Therefore, for these areas, we only considered ages that were presumably not affected by the high-pressure metamorphism in Tertiary time (Rb–Sr systems), or that are considered to detect the retention or inheritance of 'unsupported' Ar (biotite $^{40}$Ar–$^{39}$Ar ages) (see Hunziker et al. (1992), for detailed discussion). The ages were then contoured in 5 Ma time intervals.

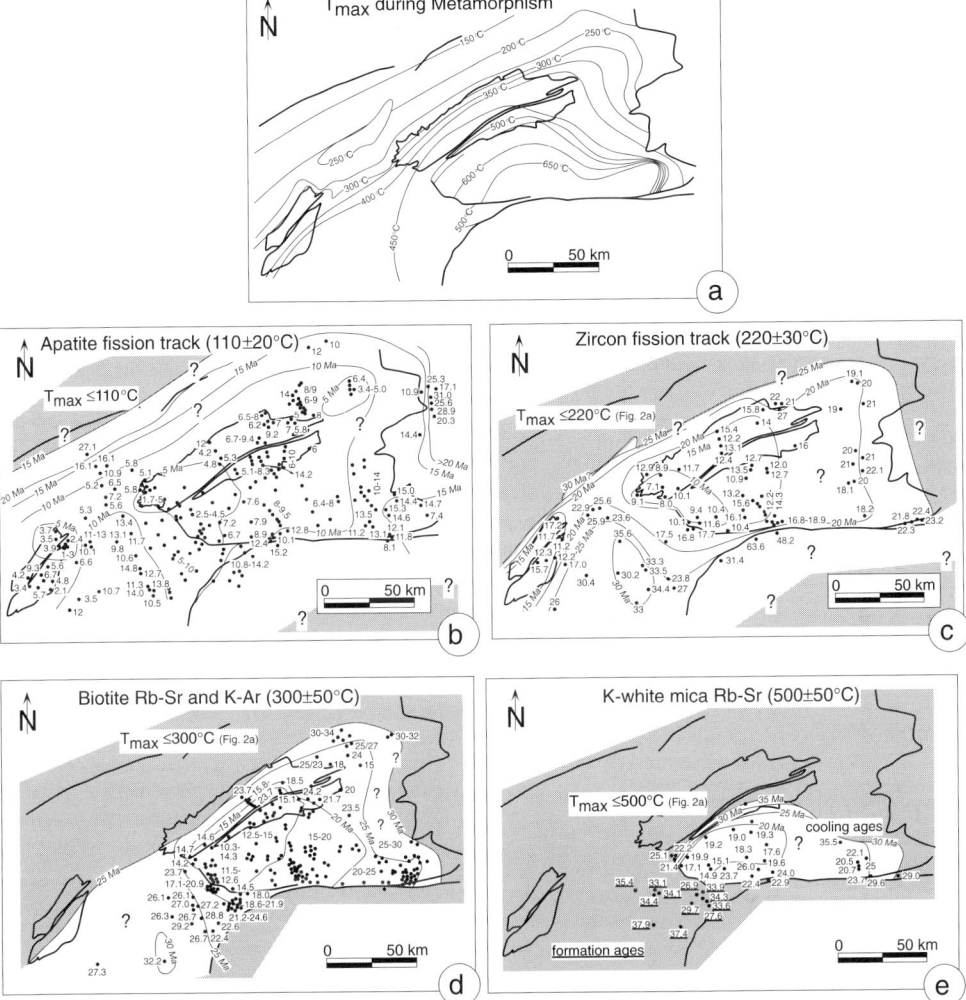

**Fig. 4.** Maps showing (**a**) the spatial distribution of the maximum temperatures in Oligocene and early Miocene time of the present-day exposed rocks, and compilations of (**b**) apatite and (**c**) zircon fission-track ages, (**d**) Rb–Sr and K–Ar ages measured on biotite, and (**e**) Rb–Sr ages determined on K-white mica. (See text for data sources.) The grey shaded areas represent regions where $T_{max}$ did not reach the closure temperatures of the various systems. The apatite isochron of 15 Ma at the northern border of the map is guided by the fact that thrusting in the Molasse Basin started at c. 20 Ma (Schlunegger et al. 1998a).

It should be noted that the Rb–Sr ages of muscovite and the Rb–Sr and K–Ar ages of biotite sampled for areas where $T_{max}$ is ≤500 and ≤300°C (Fig. 4d and e), respectively, most probably represent formation ages (Hunziker et al. 1992).

## Analysis of thermochronometric data

A map representing the spatial distribution of maximum thermal overprint of present-day exposed rocks is presented in Fig. 4a. The 250°C, 300°C and 350°C isograds are all deflected especially around the western Aar massif, indicating that thrusting of this part of the Alps post-dates the peak of thermal overprint (Pfiffner et al. 1997a). Furthermore, the >35-Ma-old metamorphism of the base of the Penninic Klippen belt (Fig. 3a) as well as the c. 25-Ma-old 250°C isograd beneath the Helvetic thrust nappes represent transported metamorphisms with offsets of c. 15 and 5–10 km, respectively (see discussion by Burkhard (1988) and Rahn et al. (1995)). It should be noted that in the east, a major discontinuity of metamorphism exists between the Penninic nappes and the orogenic lid (Austroalpine nappes). The age at which $T_{max}$ (Fig. 4a) was achieved is interpreted to have been at 25–20 Ma for the northern border of the external massifs (Burkhard 1988; Rahn et al. 1995; Pfiffner et al. 1997a), between 25 and 30 Ma for the central part of the external massifs (e.g. Hunziker et al. 1986; Burkhard 1988) and the area surrounding the Bergell intrusion (Berger et al. 1996), and at 35 Ma for the northern border of the Lepontine area (Jäger 1973) and the Klippen Belt (Burkhard 1988). It should be noted that except for the Bergell intrusion, the ages for $T_{max}$ are rather badly constrained and may vary within 2–5 Ma (Burkhard 1988). Because of the presence of a discontinuity in the metamorphism at the base of the Penninic Klippen (Fig. 1a), a similar discontinuity might be expected in the apatite and zircon isochrons. However, this is not observed, suggesting that apatite and zircon fission-track ages postdate the emplacement of the Penninic Klippen.

Rates of temperature change, i.e. cooling rates, potentially contain a decipherable record of tectonic and erosional events. We have therefore converted the age data of Fig. 4 to cooling rate data through a two-step process. First, isotherms are calculated in 100°C steps for specific ages (50°C steps for 10 Ma) by linearly interpolating the isochrons of 100°C, 220°C, 300°C and 500°C that are presented in Fig. 4b–e (Fig. 5). Because no isochrons are available for temperatures ≥500°C (Fig. 3), we assume that the rate of cooling was constant between 35 Ma (age of $T_{max}$ in the Lepontine dome) and 30–25 Ma (ages for which muscovite and biotite cooling ages are available). The maps shown in Fig. 5 represent an estimate of the temperature of the rocks currently at the surface at the ages of 30 Ma (Fig. 5a) to 10 Ma (Fig. 5e). Next, these temperature maps were differenced to give cooling rates over the 5 Ma intervals represented in Fig. 6. These cooling rates exhibit distinctive patterns that are characterized below. Except for the emplacement of the Helvetic thrust nappes above the Aar massif and slip along the Simplon Fault Zone, the post 30 Ma tectonic movements did not significantly modify the architecture of the Alps (Schmid et al. 1996). Specifically, the relative position of the Aar massif and the Helvetic thrust nappes with respect to the Penninic nappes did not change. Therefore, the current distribution of cooling appears to be a reasonable approximation of the original form of cooling, with the exception of the hanging wall of the Simplon Fault Zone (see discussion below).

## Patterns of cooling

The thermal evolution of the present-day exposed rocks of the Alps is presented in Fig. 6. It shows that average cooling was <20°C Ma$^{-1}$ (see also Hurford (1986)), and that areas where cooling was ≥20°C Ma$^{-1}$ are spatially limited. Therefore, these areas are interpreted as cooling anomalies.

Between 30 and 25 Ma, average cooling rates <20°C Ma$^{-1}$ are observed for the Penninic crystalline nappes and the Austroalpine nappes between the Gotthard 'massif' and the Insubric Line (Fig. 6a). Cooling anomalies of ≥20°C Ma$^{-1}$ are recorded in the area of the Bergell pluton (60–80°C Ma$^{-1}$) and in a 10–20 km wide zone north of the Insubric Line between the Bergell pluton and the present-day Rhône–Simplon Line (20–40°C Ma$^{-1}$). Because of the high density (Fig. 4) and the good quality of the thermal data for this area, the 30–25 Ma cooling anomalies of this part of the Alps are interpreted to be of geological significance. However, as ages older than 20 Ma are rare and commonly of questionable quality for the southeastern border of the Gotthard 'massif' and for the eastern part of the orogenic lid (Fig. 4), we do not attribute significant geological meaning to the 20–40°C Ma$^{-1}$ thermal anomalies in these areas (Fig. 6a). North of the present-day Gotthard 'massif' in the area of the external massifs and the overlying Helvetic thrust nappes and the Penninic sedimentary nappes, as well as south of the Insubric Line, either no cooling (Helvetic and Penninic thrust nappes, Southern Alpine nappes) or continuing burial and heating because of forward propagation and thickening of the wedge (external massifs) occurred (Fig. 2; Hunziker et al. 1992; Rahn et al. 1994, 1995; Schumacher et al. 1997).

Between 25 and 20 Ma, the area of the orogen that underwent synorogenic cooling of <20°C Ma$^{-1}$ expanded farther north to the area of the external massifs (Aar, Aiguilles Rouges and Mont Blanc) (Fig. 6b). The lithotectonic units between the external massifs and the present-day Alpine thrust front, i.e. the Helvetic thrust nappes in the east and the Penninic sedimentary nappes in the west (Fig. 1), as well as the crystalline and sedimentary rocks south of the Insubric Line, underwent no cooling (Burkhard 1988; Schumacher et al. 1997). During this time interval, cooling anomalies of ≥20°C Ma$^{-1}$ occurred in the core of the Lepontine dome adjacent to the Forcola normal fault and the

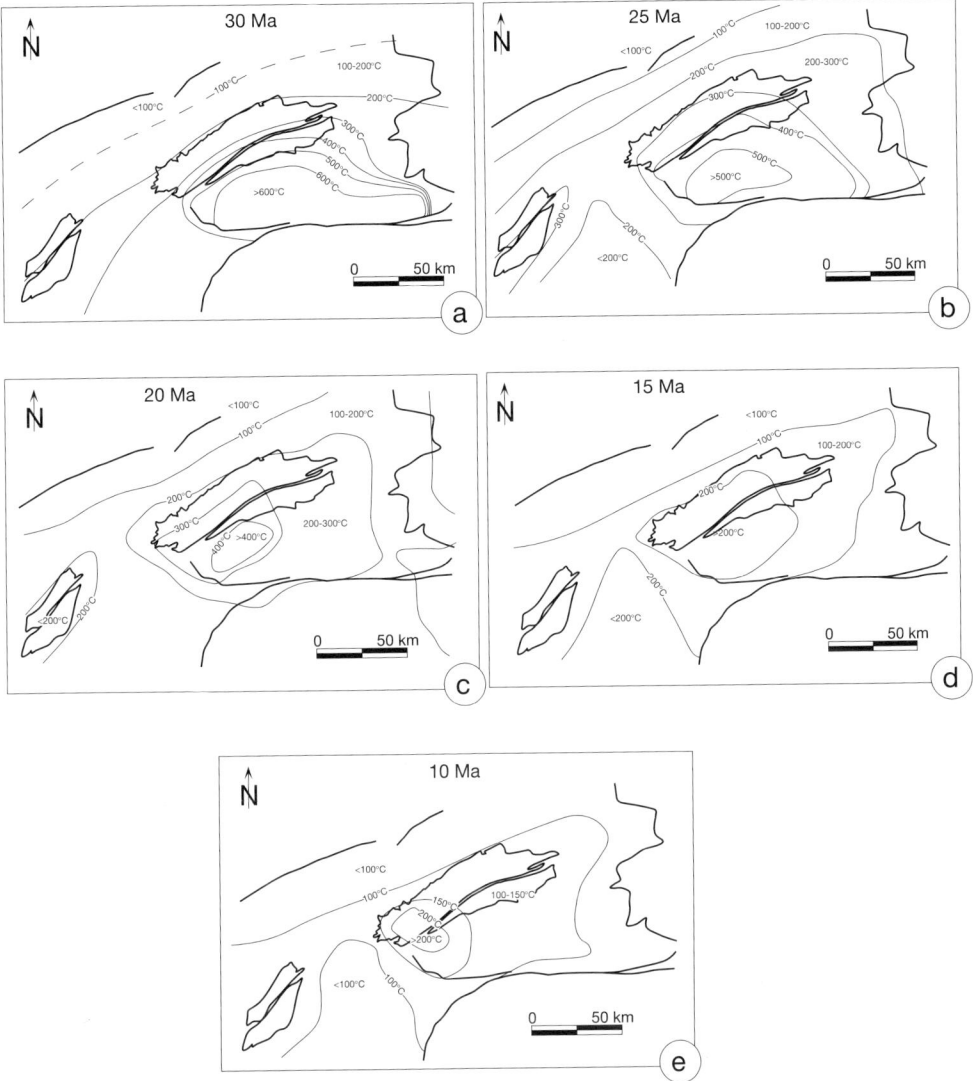

**Fig. 5.** Temperature of the present-day exposed rocks at (**a**) 30 Ma, (**b**) 25 Ma, (**c**) 20 Ma, (**d**) 15 Ma, and (**e**) 10 Ma. These maps were drawn using the isotherms of Fig. 4b–e, and the maximum temperatures (Fig. 4a).

eastern part of the Simplon Shear Zone. The anomaly covers an area of >2500 km², and extends from the present-day culmination of the Gotthard 'massif' to the area between the eastern termination of the Simplon Shear Zone and the Bergell pluton. The amount of cooling increased from the margin of the anomaly to its centre located in the core of the Lepontine dome c. 10 km north of the Insubric Line. A second cooling anomaly (≥60°C Ma$^{-1}$) is located near the southeastern termination of the Simplon Shear Zone.

Between 20 and 15 Ma, the whole Alpine orogen including the Helvetic thrust nappes and the Penninic sedimentary nappes, as well as the Southern Alpine nappes (Figs 1 and 6c) underwent cooling. Because of scarcity of thermochronological data northwest of the Aiguilles Rouges massif (Fig. 4), the apparent heating of this area is probably not of geological significance. During this time interval, the area of maximum cooling shifted c. 25 km to the west. As in the 25–20 Ma interval, the cooling anomaly defines a triangular region with the maximum

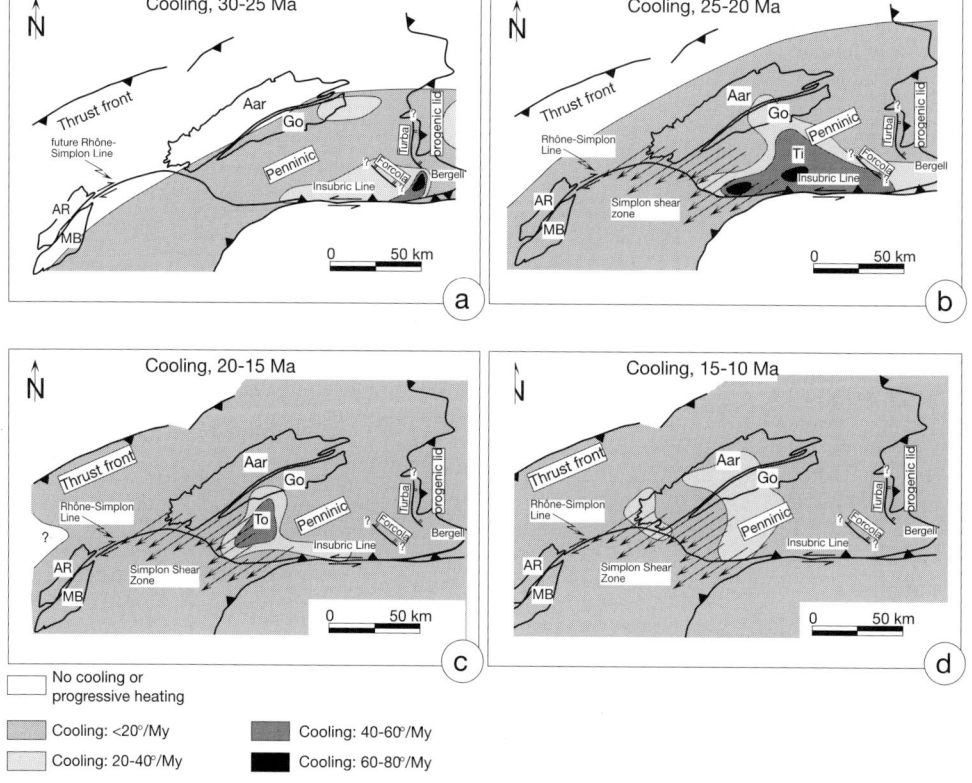

**Fig. 6.** Cooling rate of the present-day exposed rocks for time intervals (**a**) 30–25 Ma, (**b**) 25–20 Ma, (**c**) 20–15 Ma, and (**d**) 15–10 Ma. Average cooling rate of active orogen is $c.$ 20°C Ma$^{-1}$. (See text for further explanation.)

cooling rate (40–60°C Ma$^{-1}$) located in its centre. The size of the area, however, is $c.$ 900 km$^2$, which is $c.$ 30% smaller than the 25–20 Ma cooling anomaly.

The time interval between 15 and 10 Ma (Fig. 6d) is characterized by rather uniform cooling rates of <20°C Ma$^{-1}$ which are observed for the whole Alpine orogen. Enhanced cooling rates of 20–40°C km$^{-1}$, however, occurred in the core of the Aar and Gotthard massifs as well as near the northwestern termination of the Simplon Shear Zone.

## Modelling of Alpine exhumation

### Exhumation rates: 1D thermal model

The cooling rates presented in Fig. 6 potentially contain information about the rates and the pattern of exhumation and erosion. However, the temporal relationship between cooling rate and exhumation rate is complicated by the thermal processes associated with erosional and/or tectonic exhumation. In particular, we need to consider the effects of relative upward advection of heat. To assess this and other thermal effects, we have constructed a 1D model of advective–diffusive heat transfer which we use to model cooling rates for specific regions of the Alps. One-dimensional, time-dependent heat transfer is described by

$$\frac{\partial T}{\partial T} = \frac{K}{\rho c} \frac{\partial^2 T}{\partial z^2} + \frac{A}{\rho c} + v \frac{\partial T}{\partial z}$$

where $T$ is the temperature, $t$ is the time, $\rho$ is the density, $c$ is the specific heat capacity, $K$ is the thermal conductivity, $A$ is the radiogenic heat production, and $v$ is the erosion rate. This equation was solved using a finite-element method in space and an implicit finite-difference method in time. Boundary conditions for all models include constant upper surface temperature (0°C) and constant heat flux at the base of the lithosphere. Although possible spatial variations of the surface temperature (e.g. 0–10°C) influence low-temperature thermochronometers such as apatite fission-track dating (see also Mancktelow & Grasemann (1997)), the use of a constant upper surface temperature of 0°C is justified given the relatively large errors in the geochronological data.

The three regions that we modelled include the Penninic crystalline nappes of eastern Switzerland, the Aar massif and the Lepontine core. We chose these regions because these areas underwent significantly different tectonic histories in Oligocene and Miocene time (Schmid et al. 1996), and because the thermochronometric data of these regions are of good quality. We took average temperature and time data for each of these regions from Fig. 4 for modelling purposes. Initial conditions were obtained by calculating steady-state geotherms with a 40 km thick crust and a constant heat production within the crust calibrated against peak metamorphic conditions (Fig. 7). The assumption of a steady-state initial condition is not well justified, but by calibrating our initial geotherm to peak metamorphic conditions, we minimize differences between this steady-state condition and any possible transient geotherm that must also satisfy the peak metamorphic conditions. Geotherms for a range of heat production and surface heat flow are shown in Fig. 7 along with the peak metamorphic conditions for four subdomains of the Alps. The western Alps appear to have anomalously low peak metamorphic temperatures, but the other regions are fitted well using geotherms with surface heat flow between 80 and 110 mW m$^{-2}$. In all models, we used intermediate values for heat production of 1.5 μW m$^{-3}$ and surface heat flow of 90 mW m$^{-2}$. The average geothermal gradient for this geotherm is $c$. 30°C km$^{-1}$. Although the thermal effects associated with motion on individual faults require 2D treatment, 1D models are sufficient for analysis of the regional patterns of exhumation and cooling.

*Exhumation of the Aar massif and the peripheral Penninic nappes.* The Aar massif and Penninic nappes show cooling at a near-constant rate (within the resolution of the data), although cooling in the Penninic nappes probably began considerably earlier than in the Aar massif (Michalski & Soom 1990). We modelled exhumation of these regions by forward simulations of the time–temperature histories using the initial conditions specified by Fig. 7. A range of surface exhumation rates were investigated; those that give an adequate fit to the closure ages

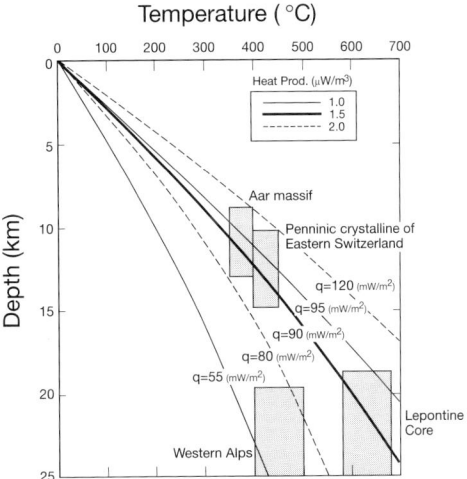

**Fig. 7.** Steady-state geotherms for the Alps calculated with a crustal heat production of 1.0 μW m$^{-3}$ and surface heat flow of 55 and 95 mW m$^{-2}$ (fine lines), a crustal heat production of 2.0 μW m$^{-3}$ and surface heat flow of 80 and 120 mW m$^{-2}$ (dashed lines), and a crustal heat production of 1.5 μW m$^{-3}$ and surface heat flow of 90 mW m$^{-2}$ (bold line). This last model was used for all 1D thermal models. Peak thermal metamorphism *P–T* data used for calibration (shaded regions) are taken from Klaper (1982) and Engi *et al.* (1995) for the Lepontine dome, Dempster (1986) for the Aar massif, and Bowtell (1991) for the Western Alps. The *P–T* conditions of the Penninic crystalline nappes of eastern Switzerland are based on the calibration of the stilpnomelan-in isograd (e.g. Hunziker *et al.* 1992).

are shown in Fig. 8. We used a constant rate of erosion initiating at 25 Ma for the Aar massif and 40 Ma for the Penninic nappes (Fig. 8). Although the data lack resolution, the Aar massif cooling data are best fitted by models with erosion rates of 400–500 m Ma$^{-1}$ and the data from the Penninic nappes by models with erosion rates between 250 and 350 m Ma$^{-1}$ (Fig. 8).

*Erosional exhumation of the Lepontine core (western part).* The cooling data from the Lepontine core exhibit considerably more complexity than those of the Aar massif and the Penninic crystalline nappes of eastern Switzerland (Fig. 9). In particular, the rapid cooling between the closure temperature of white mica and that of biotite implies variable rates of exhumation. Because the rate of cooling was higher between 40 and 20 Ma than between 20 Ma and the present, we model the erosion with two phases at different rates. The cooling data were best fitted by early, high erosion rates of between

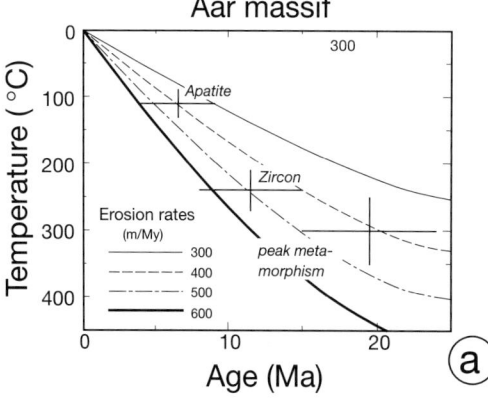

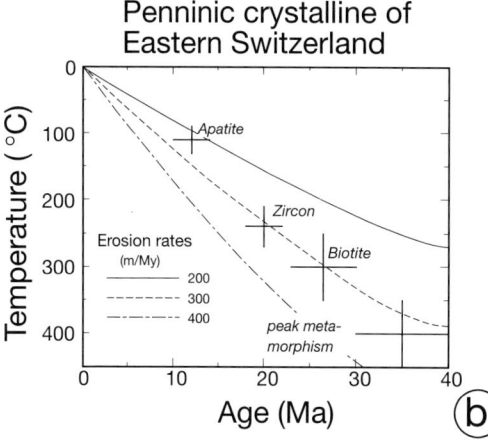

**Fig. 8.** Models of thermal history of surface rocks of (**a**) the Aar massif and (**b**) the Penninic crystalline nappes of eastern Switzerland. Models were calculated for constant erosion rates of 200–600 m Ma$^{-1}$ from initial conditions of Fig. 7. The thermochronometric data are taken from Fig. 4.

about 800 and 1000 m Ma$^{-1}$ with the subsequent rates decreasing to between 300 and 500 m Ma$^{-1}$ (Fig. 9a and b).

*Tectonic exhumation of the Lepontine core (western part).* The variable rates of cooling in the Lepontine core can also be interpreted as evidence for tectonic exhumation. We tested this hypothesis by modelling the cooling data with a short period of very rapid exhumation. It is impractical to attempt to simulate specifics of the extensional process in a 1D thermal model, so we simply applied an erosion rate of 5000 m Ma$^{-1}$ over a 2 Ma period from 22 to 20 Ma, thereby simulating the rapid removal of a 10 km thick upper plate. This rate is sufficiently fast that thermal differences between a model of instantaneous removal of a hanging wall block and this rapid erosion model will be significant only in the upper few kilometres of the footwall. Results in Fig. 10 show that we obtain a good fit to the cooling data if erosion rates before and after the simulated extension vary between 200 and 600 m Ma$^{-1}$. Even with the tectonic exhumation the data suggest a higher erosion rate of 400–600 m Ma$^{-1}$ before 22 Ma and lower rates of 200–400 m Ma$^{-1}$ after 20 Ma (Fig. 10a and b).

*Exhumation patterns: 2D exhumation model*

The first-order feature of the patterns of thermal exposure (Fig. 5) and cooling rate (Fig. 6) is the high inferred exhumation over the Lepontine core. Although this exposure probably reflects in part tectonic exhumation, the significant cooling rates before and after the estimated time of motion on the bounding shear zones suggest that surface erosional processes are still important to the observed pattern of exhumation. Furthermore, the progressive broadening of the zone of exhumation to include the Aar massif (Fig. 6b) and other peripheral regions of the Alps (Fig. 6c–6d) reflects important erosional and tectonic processes given the accompanying outward progression of deformation shown in Fig. 2.

*Mechanical–erosion model.* To investigate this pattern of broadening of the zone of exhumation and deformation, we have developed a 2D, coupled erosion–mechanical model for the evolution of the Alps on a crustal scale. The deformation component of the model is based on a mantle-subduction, crustal-accretion model (Willett *et al.* 1993; Beaumont *et al.* 1996*a*) and uses a finite element technique to solve for large-strain, Coulomb plastic and temperature-dependent viscous deformation (Willett 1992; Fullsack 1995). The initial and boundary conditions are shown in Fig. 11. The domain is a single layer of thickness $H$ that represents the deformable crust. Relative convergence velocity $V_c$ between two rigid mantle plates induces shear tractions on the crust and drives deformation, which is initially focused at the contact point, S (Fig. 11), where the mantle plate on the left detaches from the crust and descends beneath the mantle plate to the right. The problem can be scaled with $H$ and $V_c$ and expressed in non-dimensional form. Temperature is assumed to be steady state and at the diffusive limit so that there is no thermal time lag between crustal thickening and heating. With progressive deformation, crustal thickening

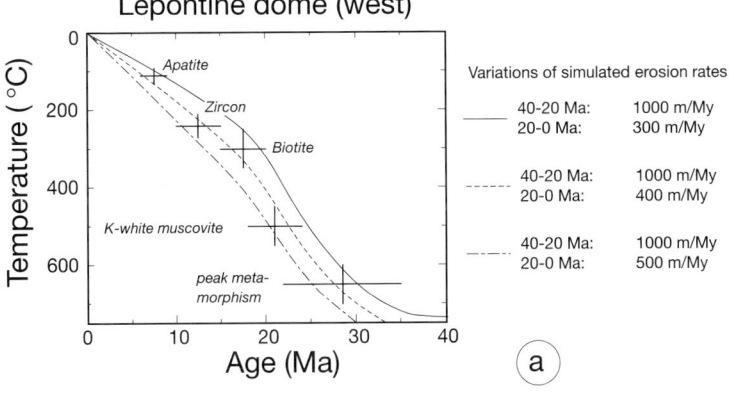

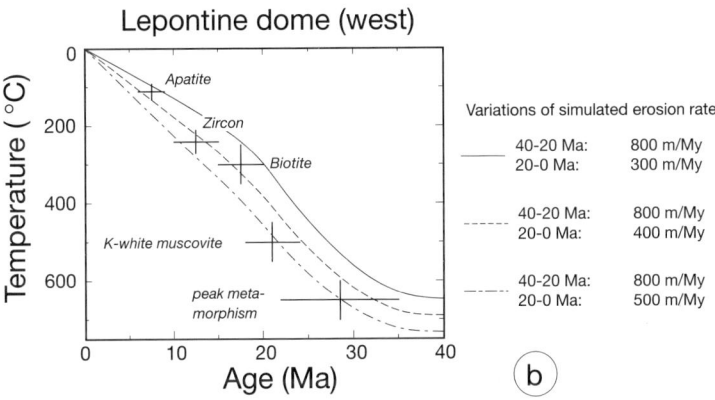

**Fig. 9.** Models of thermal history of the surface rocks of the Lepontine dome (western part) assuming exhumation is by surface processes. Exhumation history consists of an erosion rate of 1000 m Ma$^{-1}$ (**a**) or 800 m Ma$^{-1}$ (**b**) from 40 to 20 Ma and erosion rates of 300, 400 and 500 m Ma$^{-1}$ from 20 Ma to the present (continuous and dashed curves). The thermochronological data are taken from Fig. 4.

leads to the formation of a doubly vergent orogenic wedge that grows outward in both directions (Willett et al. 1993). As the crust thickens, the base of the crust becomes hotter and hence less viscous so that the wedge geometry which depends on this viscosity changes with time. The deformation fronts to the orogen are zones of high deformation which we refer to as the pro- and retro-shear zones; pro-referring to the subduction side of the orogenic system (Fig. 11).

Beaumont et al. (1996a) presented a similar model to explain the formation of the Insubric Line as the preserved record of the transition from a subduction to a collisional boundary. In their model an increasing fraction of the crust was accreted causing the retro-deformation front to propagate outward, leaving the Insubric Line preserved in the orogen interior. However, Beaumont et al. (1996a) did not explicitly include erosion. As erosion plays an important role in the large-scale development of orogens (Avouac & Burov 1996), our model is designed to investigate this process. Our intent in the modelling exercise is not a complete simulation of the formation of the Alps, but rather is to demonstrate the effects of

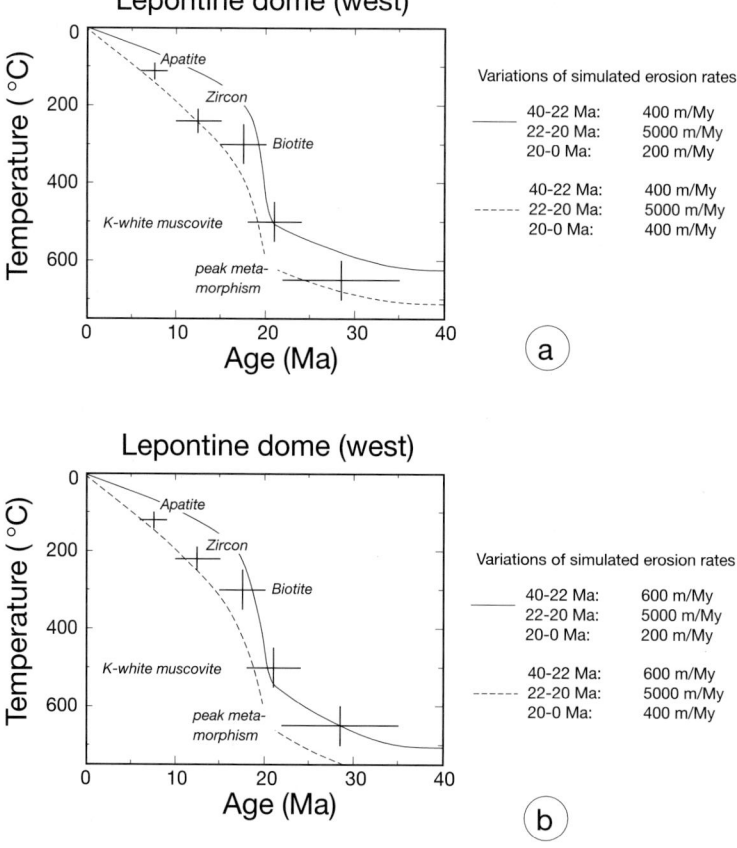

**Fig. 10.** Models of thermal history of the surface rocks of the Lepontine dome (western part) assuming that exhumation includes a period of tectonic unroofing. Tectonic unroofing is simulated by removal of 10 km of rock from 22 to 20 Ma. Erosion rates before tectonic unroofing vary between 400 m Ma$^{-1}$ (**a**) and 600 m Ma$^{-1}$ (**b**). Erosion rates following tectonic unroofing vary from 200 m Ma$^{-1}$ (continuous lines) to 400 m Ma$^{-1}$ (dashed lines). The thermochronometric data are taken from Fig. 4.

erosion on exhumation patterns in space and time. Furthermore, as we concentrate mainly on the large-scale exhumation patterns, we do not attempt to resolve tectonic features on the scale of individual thrusts or nappes.

Erosion of the orogenic wedge affects the rate of growth and the distribution of deformation, but, most importantly for this paper, leads to a distinctive pattern of exhumation. Erosion rates are enhanced in mountain belts by surface uplift, which leads to increased relief, and through climatic feedback such as orographically enhanced precipitation. This feedback can be modelled directly by including a model of precipitation, runoff collection and fluvial incision (Beaumont et al. 1996b), or, more simply, by linking erosion rates to elevation or relief characteristics. We have pursued this latter alternative by coupling the erosion rate to the surface slope. We do not advocate any specific erosional mechanism which might behave in this manner, but simply adopt regional slope as a proxy for local relief which is related to erosion rate (Ahnert 1984). We could also use elevation as a proxy for relief and would obtain a similar result. In the model, eroded sediment is removed from the system, rather than being deposited elsewhere so that there is a mass flux out of the model. Results from this simple experiment are shown in Fig. 12. After convergence of $4H$, a small crustal orogen has developed with sufficient elevation to induce significant erosion as illustrated

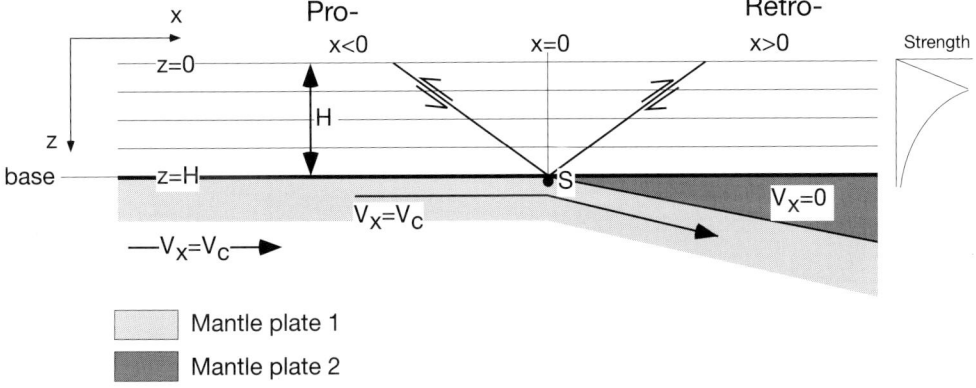

**Fig. 11.** Initial and boundary conditions for 2D, plane-strain thermomechanical model. Crustal layer of thickness $H$ has material properties of constant Coulomb angle of friction and temperature-dependent viscosity. Boundary conditions are a stress-free upper surface and a specified velocity simulating sub-crustal subduction on the base of the layer. Horizontal velocity is constant on the base for $x < 0$ and zero for $x > 0$. Vertical motion of the base is calculated by response of an elastic plate to surface loading. (Modified after Willett *et al.* 1993). It should be noted that the layer could represent the entire continental crust or merely the accreted portion if the lower crust is also subducted. Pro- and retro-prefixes are applied to structures above the subducting and overriding plates, respectively.

graphically by the portion of the mesh above the ground surface (Fig. 12a). Deformation is nearly symmetric about the contact point S, and erosion rates are approximately constant across the uplifted region, but the surface pattern of exhumation is strongly asymmetric in response to the polarity of subduction. The exhumation of material currently exposed at the surface in Fig. 12a increases progressively from the pro-deformation front towards the orogen interior. At the retro-deformation front there is a strong gradient in depth of exhumation, decreasing to no exhumation right of the deformation front. This is approximately the pattern observed in the Alps, particularly in the early phases of development, with depth of exhumation increasing progressively towards the Insubric Line at the southern limit of the Lepontine core.

However, this model fails to explain some important features of the Alps. The model of Fig. 12a has reached a mass-flux steady state such that the addition of mass into the orogen by convergence is balanced by the erosional mass flux out of the orogen. Continued convergence (Fig. 12b) leads to increased exhumation, but not to outward growth of the orogen. From a perspective fixed to the pro-plate (northern Europe), there is still relative advance of deformation as the undeformed crust is advected southward into the orogen to replace eroded material, but the retro-deformation front, analogous to the Insubric Line, remains fixed regardless of the total convergence (Fig. 12b). This has not been the case for the recent history of the Alps, as the Lepontine core and the Insubric Line have moved southward with respect to the stable Adriatic plate concurrent with shortening of the Southern Alps (Schmid *et al.* 1996) (Fig. 2).

The southern edge of the Lepontine core, i.e. the Insubric Line, is an important boundary between highly exhumed rocks to the north and less deeply exhumed rocks to the south. In the context of the asymmetric doubly vergent orogen model (Fig. 12), this boundary represents deformation in a retro-shear zone, localized by erosion. This interpretation combined with the observation of late contraction of the Southern Alps requires a change in erosion conditions, a situation we consider in the model of Fig. 13. In this model there are two phases of erosion. In the first phase, erosion occurs at rates equivalent to those in the model of Fig. 12. This continues until total convergence reaches $4H$ (Fig. 12a). At this point, the system is in steady state, the erosional flux is equal to the convergence flux, implying an average erosion rate of $0.2V_c$ as the orogen is approximately five times wider than the crustal thickness. In the second phase, the relative erosion rate is decreased by a factor of eight giving an average erosion rate of $0.025V_c$. This is not sufficient to balance the convergent mass flux and the orogen goes from steady state to a constructive state of growth.

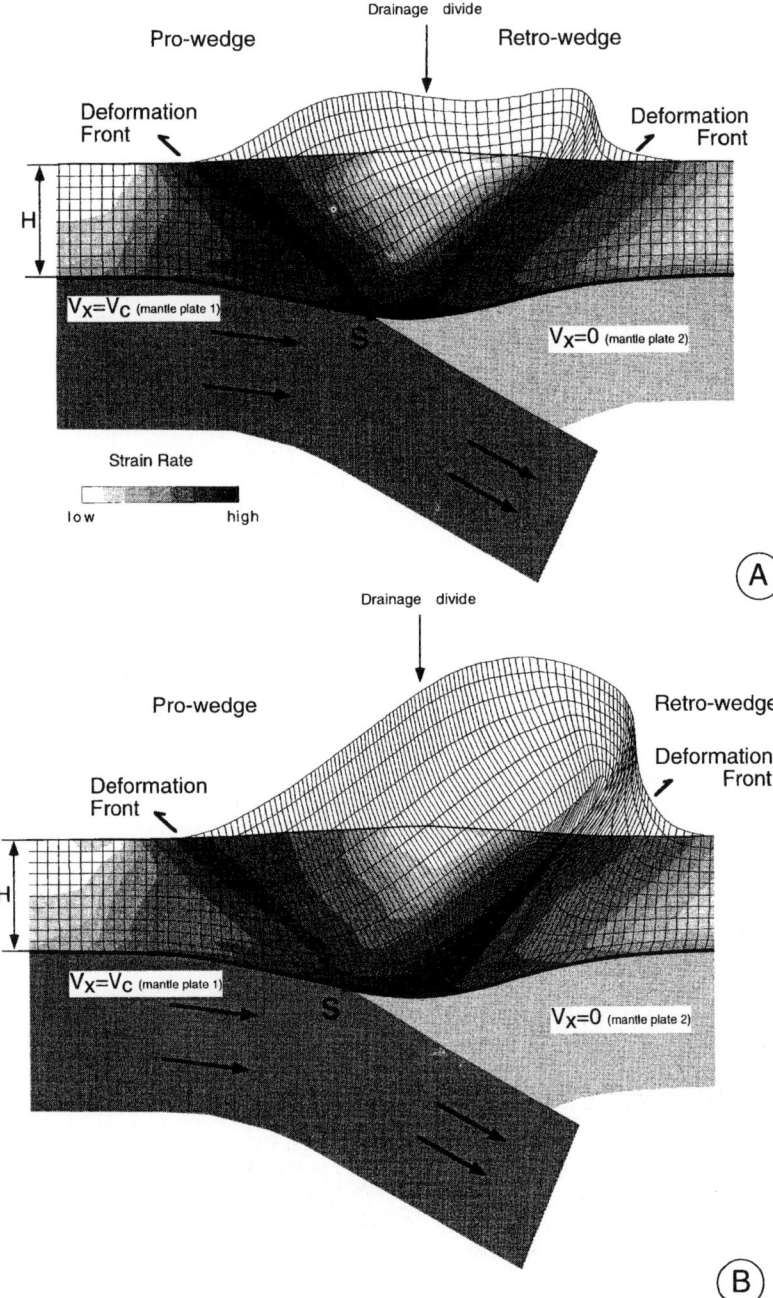

**Fig. 12.** Predicted deformation and erosion in thermomechanical model after convergence of $4H$ (**a**) and $8H$ (**b**). Erosion rate is proportional to surface slope. Eroded mass is shown schematically by unfilled mesh above the ground surface. Active deformation is given by the non-dimensional strain rate, which is shaded according to value. Orogen has reached a steady state as indicated by the lack of outward propagation of deformation fronts.

After a period of growth representing an additional convergence of $4H$, we obtain the configuration shown in Fig. 13. In this model, the initial core of the orogen is highly exhumed during the first, high erosion rate, phase. This central core is preserved in the interior of a much wider orogenic belt that is the result of the second phase of constructive growth with lower erosion rates. The peripheral regions of the orogen are only moderately exhumed and represent the youngest deformation. It should be noted that this model orogen will continue to grow outward in both directions until the erosional flux increases to the point that a new steady state is achieved.

This pattern of exhumation and deformation is reasonably consistent with observations from the Alps in which the Lepontine core was exhumed by backthrusting on the Insubric Line with high rates of surface exhumation, but was subsequently translated southward concomitant with shortening in the southern Alps. It is important to note that the implication of a change in erosion rate is non-unique in that the crucial model parameter giving rise to the structure of Fig. 13 is erosion rate relative to convergence rate. The same result would be obtained by an increase in convergence rate rather than a decrease in erosion rate. More specifically, it is the rate of mass accretion that must be compared with an erosional mass flux to determine an orogen state of growth, decay or stability (Jamieson & Beaumont 1988).

We also note that the mechanism of exhumation is not resolved (Figs 9 and 10); the exhumation of the Lepontine core could be due to either extension or surface erosional processes and would yield a similar result. The primary factor in controlling the tectonic evolution of the Alps is the relative rate between accretion into the orogen and mass removal by any or all mechanisms. This means that although tectonic exhumation is governed by a different relationship from that of surface erosion, it is not really important

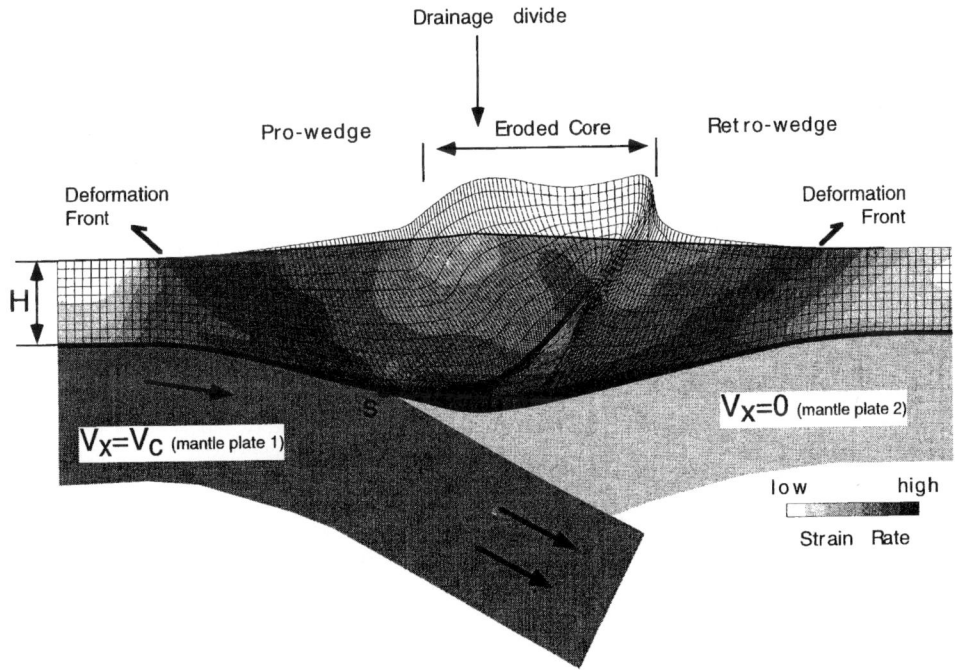

**Fig. 13.** Predicted deformation and erosion in thermomechanical model of an Alpine orogen which experiences a decrease in erosion rate. Erosion rate is proportional to surface slope and is identical to the erosion function in the model of Fig. 12 up to total convergence of $4H$ (Fig. 12a). Subsequently, the proportionality factor is decreased by a factor of eight, resulting in a decrease in average erosion rates. Deformation and exhumation are shown at a total convergence of $8H$. Eroded mass is shown schematically by unfilled mesh above ground surface. Active deformation is given by the non-dimensional strain rate. (Note outward propagation of deformation fronts (compare with Fig. 12b), indicating that the orogen is in a constructive state.) Previously eroded core with initial retro-deformation front, analogous to the Insubric Line, is preserved.

for the thermal and mechanical models whether exhumation occurred by tectonic processes or by surface erosion. The important and resolvable factor for the mechanical model is the rate of surface lowering and the relative balance between this rate and the rate of crustal accretion.

## Discussion

If both temporally calibrated thermal and barometric data were available, the processes that resulted in exhumation of deeply buried rocks could be reconstructed, as hanging-wall and footwall blocks of low-angle normal faults undergo significantly different P–T–t paths during synorogenic extension. Although any rapid near-surface exhumation process would result in an initial phase of isothermal decompression, followed by isobaric cooling, P–T–t differences between footwall and hanging wall can resolve whether this process is tectonic or erosional. Examples are described from the Devonian Hornelen Basin detachment system of western Norway (Wilks & Cuthbert 1994), and from the Variscan Tormes Gneiss Dome of western Spain (Viruete et al. 1994).

In the case of the Alps, where the resolution of the thermal data is rather low and barochronological data are limited, the reconstruction of the exhumation history, especially for the Lepontine dome, is not unique. Indeed, our 1D thermal model reveals that the evolution of the Lepontine dome can be adequately simulated by either erosional or tectonic exhumation. Nevertheless, we evaluate the likelihood of crustal extension as a control on the exhumation of the Lepontine core complex by relating the rates and the 3D pattern of cooling to (i) the structural evolution of the Alps, (ii) the chronology of slip movement along normal faults, and (iii) the geometry of exhumation caused by crustal extension. Although these arguments provide only circumstantial evidence for the occurrence of tectonic exhumation, they enable us to reconstruct a most likely scenario for exhumational processes (see also Mancktelow & Grasemann (1997)). In a further step, we compare the average erosion rates of the orogenic front (Aar massif, Fig. 8a) with those of the rear of the wedge (Penninic crystalline nappes of eastern Switzerland, Fig. 8b) that were determined by the 1D thermal model. Finally, we evaluate the controls of erosion on the outward growth of the orogen by combining the results of the 2D plane-strain forward model with the chronology of tectonic events in the Alps. All three aspects of the discussion require knowledge about the crustal dynamics of the Alps, which are presented in Fig. 2 and which will be integrated in our discussion.

## Cooling pattern and tectonic processes

In this section we relate the cooling pattern of the Alps to the tectonic processes of the evolving orogen and to the chronology of movement along the detachment faults. Correlation between extensional fault motion and high footwall cooling rates provides support for the importance of tectonic unroofing to cooling. Below, we interpret the pattern of Alpine cooling between 30 and 10 Ma in four time slices (Fig. 6a–d).

*30–25 Ma*. The 30–25 Ma cooling pattern of the Alpine orogen (Fig. 6a) is interpreted to be primarily the result of erosional exhumation, although cooling in the vicinity of the Bergell pluton might be due to the local thermal perturbation associated with intrusion (Köppel & Grünenfelder 1975). Comparative petrographic and chronological studies carried out on high-grade rocks north of the Insubric Line, in the area of the Bergell pluton and on pebbles and boulders deposited in the 30–c. 14 Ma southern Alpine foreland basin (Gonfolite Lombarda, Fig. 1a) indicate rapid erosion in the rear of the wedge (Giger 1991). Although the 30–25 Ma cooling anomaly at the northeastern border of the Gotthard 'massif' (20–40°C Ma$^{-1}$) might be an artifact of insufficient thermochronometric data (see above), enhanced exhumation in this part of the orogen is supported by sedimentological data from the Molasse Basin. Using a 3D reconstruction of the 30–25 Ma architecture of the basin fill, Schlunegger et al. (1998a) showed that the supply rate of sediment of the Alpine rivers that crossed the eastern flank of the Gotthard 'massif' increased between 30 and 25 Ma. Because there is no correlation between the cooling pattern and the trace of the low-angle normal faults (Fig. 6a), we interpret that the 30–25 Ma Alpine cooling was not affected by extension along western or eastern detachment faults. Furthermore, it appears that tectonic exhumation caused by faulting along the Turba fault did not significantly disturb the thermal structure of the Eastern Alps.

*25–15 Ma*. Whereas some cooling at rates ≥20°C Ma$^{-1}$ occurred in the area of the 30-Ma-old Bergell pluton (Fig. 6b), the location of maximum cooling shifted to the area that is referred to as the 'Ticino culmination', and then to the area called the 'Toce culmination' some

25 km to the west (Fig. 6b and c). There, the geometry of the cooling anomaly suggests control by extension-related crustal exhumation (see Fig. 3). Slip along the normal faults bordering the Lepontine dome was initiated no earlier than 30 Ma. The cooling model of the Lepontine dome suggests that if crustal extension did cause significant exhumation of this area, it probably occurred sometime around 20 Ma. These findings are supported by the 2D thermal model of Grasemann & Mancktelow (1993), who suggested that enhanced rates of slip along the fault occurred between 18 and 15 Ma. Finally, the 25 km westward shift of the location of maximum cooling from the 'Ticino' to the 'Toce culmination' (Fig. 6b and c) is in agreement with the slip direction of the hanging wall of the Simplon detachment fault.

*15–10 Ma.* The >20°C $Ma^{-1}$ cooling anomaly that is observed in the area of the Aar massif is likely to have been caused by erosional rather than by tectonic exhumation. Indeed, stratigraphic information from the Molasse Basin suggests that the litho-tectonic units overlying the central Aar massif at that time underwent enhanced erosional exhumation (Schlunegger *et al.* 1997b). However, the >20°C $Ma^{-1}$ cooling anomaly adjacent to the Simplon detachment fault was possibly controlled by movement along this fault (Steck & Hunziker 1994).

## Rates of surface processes in the frontal and central Alps

The numerical simulation of the thermochronometric evolution of the Aar massif and the Penninic crystalline nappes of eastern Switzerland reveals, as outlined above, constant erosion rates in Oligocene and Miocene time. These areas, as well as the Western Alps, are characterized by constant cooling rates of <20°C $Ma^{-1}$ (Fig. 6). Because in these parts of the Alps, no low-angle normal faults and ductile shear zones have been mapped so far, it appears most likely that the Aar massif as well as the Penninic crystalline nappes of eastern and western Switzerland were exhumed by erosion. These areas underwent average erosion rates between 0.25 and 0.5 km $Ma^{-1}$ (Fig. 8). It appears that the average erosion rates of the frontal Alpine nappes (0.4–0.5 km $Ma^{-1}$) were higher than those in the rear of the wedge (0.25–0.35 km $Ma^{-1}$). This might reflect high rates of crustal uplift of the Aar massif that started in Miocene time (Pfiffner *et al.* 1997a). Petrographic data from the Molasse Basin suggest that the frontal Alpine units underwent enhanced rates of erosion after 20 Ma (Schlunegger *et al.* 1997b).

## Erosion rates and orogen growth

The results of the 2D mechanical models illustrate how erosion is linked to the outward growth of the orogen (Figs 12 and 13). As suggested by the thermochronometric data (Fig. 6) and the results of the 1D thermal forward models (Figs 8–10), average erosion rates, with or without tectonic unroofing, decreased significantly after 20 Ma. This result is supported by the 2D thermomechanical modelling, which suggests that extensive backthrusting north of the Insubric Line, followed by southward growth of the Alps, is possible only with a relative decrease in erosion rate. In support of this scenario, observations include the fact that before 20 Ma, the northern tip of the Alpine wedge was located at c. 30 km distance from the present-day Aar massif, whereas the southern tip was located immediately south of the Insubric Line (Schmid *et al.* 1996) (e.g. Fig. 12a). Cross-cutting relationships between thrusts and temporally calibrated sediments reveal that deformation of the Jura Mountains and the Southern Alps was initiated after 20 Ma (Schmid *et al.* 1996; Pfiffner & Heitzmann 1997; Pfiffner *et al.* 1997b). The tips of these structural units are located at >50 km distance from the thrust fronts before 20 Ma. This simultaneous outward growth in both directions suggests a constructive phase of orogen growth consistent with low exhumation rate relative to the rate of crustal accretion.

We recognize three potential mechanisms for inducing this response. First, the decrease in relative erosion rate could reflect a decrease in the efficiency of erosional surface processes. These could include a climatic change, i.e. lower precipitation rates, or changes in the fluvial network leading to less efficient exhumation.

Second, exhumation of the Lepontine dome may have been dominated by crustal extension and tectonic exhumation. Cooling data are consistent with short periods of rapid exhumation consistent with removal of hanging-wall blocks (Figs 6c and 10) (Grasemann & Mancktelow 1993). The southwestward shift of the area of high rates of cooling between 25 and 15 Ma (Fig. 6) would thus reflect progressive motion along the Simplon Fault Zone. Later exhumation would be the result of surface erosional process. In this scenario, erosion rates have operated at c. 300–500 m $Ma^{-1}$ (Fig. 10) for the last 40 Ma. Tectonic exhumation at high rates kept the system in near steady state for only a short period of time, but was sufficient to exhume the Lepontine dome.

Third, the constructive outward growth of the orogen could reflect an increase in the crustal accretion rate, potentially in response to a decrease in the proportion of the crust subducted with the continental mantle (Beaumont et al. 1996a). In the model proposed by Beaumont et al. (1996a) the crustal accretion rate increased in Oligocene time, erosion was not able to keep up with accretion, the orogen entered a phase of construction, the deformation fronts propagated outward and the Southern Alps formed south of the Insubric Line. This explanation cannot explain an increase in cooling rate or exhumation rate in the Lepontine core. However, if the accelerated orogen growth were accompanied by extensional unroofing as suggested by the coincidence between periods of enhanced rates of back-thrusting and phases of orogen-parallel extension, this would explain both the outward growth and the higher rate of exhumation.

These mechanisms are not independent and the possibility of a combined mechanism exists. Indeed, we consider this to be the most likely case. For example, the Eo-Oligocene Alps may have developed into a steady state in which accretion rate was balanced by erosion. At one or more points in time tectonic exhumation may have accelerated exhumation. Near the end of Oligocene time, balance in the system was perturbed either by an increase in accretion rate or a climatically driven decrease in erosional efficiency leading to a constructional period of orogenic growth. At this point, thermochronometric data cannot distinguish between these models.

## Conclusions

Information about the structural and the thermal evolution of the Alps in combination with thermal and mechanical models allows an improved understanding of the rates and mechanisms of exhumation of the Alps. Our study reveals that cooling of the high-grade metamorphic (Lepontine) core of the Alps occurred at variable rates with higher rates of exhumation before about 20 Ma. Although our study is consistent with very rapid tectonic exhumation of the Lepontine core at about 20 Ma, we cannot rule out higher rates of erosional exhumation over a longer period of time. Nevertheless, the coincidence of regions of high exhumation rate adjacent to extensional structures suggests that tectonic exhumation was an important mechanism.

Thermochronometric data suggest that exhumation by erosion was significant in the Alps from Eocene time to the present, although rates may have been slower since early Miocene time. A decrease in exhumation rate is also supported by models of the structural evolution of the Alps as demonstrated by thermomechanical models simulating critical orogenic wedge growth. The formation of the Insubric Line as a long-lived structural boundary between the highly exhumed Lepontine core and unexhumed, undeformed rocks to the south implies a balance between erosion and convergence mass fluxes during the early history of the Alps. This balance was broken at the end of Oligocene or in early Miocene time with exhumation rates failing to keep up with mass accretion by convergence. This led to a period of orogen construction in which the Insubric Line ceased to serve as the deformation front and the Southern Alps formed. Whether this change reflects a decrease in surface erosion rate or an increase in accretion rate cannot be resolved.

This project was funded by the sedimentological research group of the University of Berne, Switzerland, by National Science Foundation Grants EAR 94-17766 and EAR 95-26954. We would like to thank R. Slingerland and K. Furlong, Penn State University, for stimulating discussions. Special thanks go to A. Pfiffner, University of Bern, for continuing scientific exchange.

## References

AHNERT, F., 1984. Local relief and the height limits of mountain ranges. *American Journal of Science*, **284**, 1035–1055.

ALLMENDINGER, R. W., FIGUEROA, D., SNYDER, D., BEER, J., MPODOZIS, C. & ISACKS, B. L. 1990. Foreland shortening and crustal balancing in the Andes at 30°S latitude. *Tectonics*, **9**, 789–809.

AMSTRONG, R. L., JÄGER, E. & EBERHARDT, P. 1966. A comparison of K–Ar and Rb–Sr ages on Alpine biotite. *Earth and Planetary Science Letters*, **1**, 13–19.

AVOUAC, J. P. & BUROV, E. B. 1996. Erosion as driving mechanism of intracontinental mountain growth. *Journal of Geophysical Research*, **101**, 17747–17769.

AYRTON, S., BUGNON, C., HAARPAINTNER, T., WEIDMANN, M. & FRANK, E. 1982. Géologie du front de la nappe de la Dent-Blanche dans la région des Monts-Dolins, Valais. *Eclogae Geologicae Helvetiae*, **75**, 269–286.

BATT, G. E. & BRAUN, J. 1997. On the thermomechanical evolution of compressional orogens. *Geophysical Journal International*, **128**, 364–382.

BAUDIN, T., MARQUER, D. & PERSOZ, F. 1993. Basement–cover relationships in the Tambo nappe (Central Alps, Switzerland): geometry, structure and kinematics. *Journal of Structural Geology*, **15**, 543–553.

BEAUMONT, C., ELLIS, S. HAMILTON, J. & FULLSACK, P. 1996a. Mechanical model for subduction–collision tectonics of Alpine-type compressional orogens. *Geology*, **24**, 675–678.

——, KAMP, P. J. J., HAMILTON, J. & FULLSACK, P. 1996b. The continental collision zone, South Island, New Zealand; comparison of geodynamic models and observations, *Journal of Geophysical Research*, **101**, 3333–3359.

BECKER, H. 1993. Garnet peridotite and eclogite Sm–Nd mineral ages from the Lepontine dome (Swiss Alps): new evidence for Eocene high-pressure metamorphism in the central Alps. *Geology*, **21**, 599–602.

BERGER, A., ROSENBERG, C. & SCHMID, S. 1996. Ascent, emplacement and exhumation of the Bergell pluton within the Southern Steep Belt of the Central Alps. *Schweizerische Mineralogische und Petrographische Mitteilungen*, **76**, 357–382.

BOWTELL, S. A. 1991. *Geochronological and geochemical studies of the Zermatt-Saas Fee Ophiolite, Western Alps*. PhD thesis, University of Leeds, UK.

BURKHARD, M. 1988. L'Hélvetique de la bordure occidentale du massif de l'Aar (évolution tectonique et métamorphique). *Eclogae Geologicae Helvetiae*, **81**, 63–114.

DAVIS, D. M., SUPPE, J. & DAHLEN, F. A. 1983. Mechanics of fold-and-thrust belts and accretionary wedges. *Journal of Geophysical Research*, **88**, 1153–1172.

DEMPSTER, T. J. 1986. Isotope systematics in minerals: biotite rejuvenation and exchange during Alpine metamorphism. *Earth and Planetary Science Letters*, **78**, 355–367.

DEUTSCH, A. & STEIGER, H. 1985. Hornblende K–Ar ages and the climax of Tertiary metamorphism in the Lepontine Alps (south–central Switzerland): an old problem reassessed. *Earth and Planetary Science Letters*, **72**, 175–189.

ENGI, M., TODD, S. C. & SCHMATZ, D. R. 1995. Tertiary metamorphic conditions in the eastern Lepontine Alps. *Schweizerische Mineralogische und Petrographische Mitteilungen*, **75**, 347–396.

FREY, M. 1974. Alpine metamorphism of pelitic and marly rocks of the Central Alps. *Schweizerische Mineralogische und Petrographische Mitteilungen*, **54**, 489–506.

—— 1986. Very low-grade metamorphism of the Alps – an introduction. *Schweizerische Mineralogische und Petrographische Mitteilungen*, **66**, 13–27.

—— 1987. The reaction-isograd kaolinite + quartz = pyrophyllite + $H_2O$, Helvetic Alps, Switzerland. *Schweizerische Mineralogische und Petrographische Mitteilungen*, **67**, 1–11.

——, BUCHER, K., FRANK, E. & MULLIS, J. 1980. Alpine metamorphism along the geotraverse Basel–Chiasso: a review. *Schweizerische Mineralogische und Petrographische Mitteilungen*, **73**, 527–546.

FULLSACK, P. 1995. An arbitrary Lagrangian–Eulerian formulation for creeping flows and its application in tectonic models. *Geophysical Journal International*, **120**, 1–23.

GEBAUER, D., SCHERTL, H.-P. & SCHREYER, W. 1995. A 35 Ma old ultrahigh-pressure-metamorphism in the Dora Maira massif and its geodynamic implications for the Pennine zone of the central and the western Alps. *Bochumer Geologische und Geotechnische Arbeiten*, **44**, 49–52.

GIGER, M. 1991. *Geochronologische und petrographische Studien an Geröllen und Sedimenten der Gonfolite Lambarda Gruppe (Südschweiz und Norditalien) und ihr Vergleich mit dem alpinen Hinterland*. PhD thesis, University of Berne.

GRASEMANN, B. & MANCKTELOW, N. S. 1993. Two-dimensional thermal modelling of normal faulting; the Simplon fault zone, Central Alps, Switzerland. *Tectonophysics*, **225**, 155–165.

HOFFMAN, P. F. & GROTZINGER, J. P. 1993. Orographic precipitation, erosional unloading, and tectonic style, *Geology*, **21**, 195–198.

HUNZIKER, J. C. 1969. Rb–Sr Altersbestimmungen aus den Walliser Alpen – Hellglimmer- und Gesamtgesteinsalterswerte. *Eclogae Geologicae Helvetiae*, **62**, 527–542.

——, DESMONS, J. & HURFORD, A. J. 1992. Thirty-two years of geochronological work in the Central and Western Alps: a review on seven maps. *Mémoires de Géologie (Lausanne)*, **13**.

——, FREY, M., CLAUER, N., *et al.* 1986. The evolution of illite to muscovite: mineralogical and isotopic data from the Glarus Alps, Switzerland. *Contributions to Mineralogy and Petrology*, **92**, 157–180.

——, HURFORD, A. J. & CALMBACH, L. 1996. Alpine cooling and uplift. *In*: PFIFFNER, O. A., LEHNER, P., HEITZMANN, P., MÜLLER, ST. & STECK, A. (eds) *Results of the National Research Program 20 (NRP 20)*. Birkhäuser, Basel, 260–264.

HURFORD, A. J. 1986. Cooling and uplift patterns in the Lepontine Alps South Central Switzerland and an age of vertical movement on the Insubric fault line. *Contributions to Mineralogy and Petrology*, **93**, 413–427.

JÄGER, 1973. Die alpine Orogenese im Lichte der radiometrischen Altersbestimmung. *Eclogae Geologicae Helvetiae*, **66**, 11–21.

——, NIGGLI, E. & WENK, E. 1967. Rb–Sr Altersbestimmungen an Glimmern der Zentralalpen. *Beiträge zur Geologischen Karte der Schweiz*, **N.F. 134**.

JAMIESON, R. A. & BEAUMONT, C. 1988. Temperature-time paths with applications to the Central and Southern Appalachians. *Tectonics*, **7**, 417–446.

KLAPER, E. M. 1982. Deformation und Metamorphose in der nördlichen Maggia-Zone. *Schweizerische Mineralogische und Petrographische Mitteilungen*, **62**, 47–76.

KÖPPEL, V. & GRÜNENFELDER, M. 1975. Concordant U–Pb ages of monazite and xenotime from the Central Alps and the timing of the high temperature Alpine metamorphism, a preliminary report. *Schweizerische Mineralogische und Petrographische Mitteilungen*, **55**, 129–132.

——, GÜNTHERT, A. & GRÜNENFELDER, M. 1981. Patterns of U–Pb zircon and monazite ages in polymetamorphic units of the Swiss Central Alps. *Schweizerische Mineralogische und Petrographische Mitteilungen*, **61**, 97–119.

LIHOU, J., HURFORD, A. & CARTER, A. 1995. Preliminary fission-track ages on zircon and apatites from the Sardona unit, Glarus Alps, eastern Switzerland: late Miocene–Pliocene exhumation rates. *Schweizerische Mineralogische und Petrographische Mitteilungen*, **75**, 177–186.

LINIGER, M. & NIEVERGELT, P. 1990. Stockwerk-Tektonik im südlichen Graubünden. *Schweizerische Mineralogische und Petrographische Mitteilungen*, **70**, 95–101.

MANCKTELOW, N. S. 1985. The Simplon Line: a major displacement zone in the western Lepontine Alps. *Eclogae Geologicae Helvetiae*, **78**, 73–96.

—— 1992. Neogene lateral extension during convergence in the Central Alps: evidence from interrelated faulting and backfolding around the Simplonpass (Switzerland). *Tectonophysics*, **215**, 295–317.

—— & GRASEMANN, B. 1997. Time-dependent effects of heat advection and topography on cooling histories during erosion. *Tectonophysics*, **270**, 167–195.

MEIGS, A. J. & BURBANK, D. W. 1997. Growth of the South Pyrenean orogenic wedge. *Tectonics*, **16**, 239–258.

MÉNARD, G. 1988. *Structure et cinématique d'une chaîne de collision: Les Alpes occidentales et Centrales*. Thèse de doctorat, Observatoire de Grenoble et IRIGM.

MICHALSKI, I. & SOOM, M. 1990. The Alpine thermotectonic evolution of the Aar and Gotthard massifs, Central Switzerland: fission track ages on zircon and apatite and K–Ar mica ages. *Schweizerische Mineralogische und Petrographische Mitteilungen*, **70**, 373–387.

NIEVERGELT, P., LINIGER, M., FROITZHEIM, N. & MÄHLMANN, R. F. 1996. Early to mid Tertiary crustal extension in the Central Alps: the Turba Mylonite Zone (Eastern Switzerland). *Tectonics*, **15**, 329–340.

NIGGLI, E. & NIGGLI, C. R. 1965. Karten der Verbreitung einiger Mineralien der alpidischen Metamorphose in den Schweizer Alpen. *Eclogae Geologicae Helvetiae*, **58**, 335–368.

OKAYA, N., CLOETINGH, S. & MÜLLER, St. 1996a. A lithospheric cross-section through the Swiss Alps – II. Constraints on the mechanical structure of a continent–continent collision zone. *Geophysical Journal International*, **127**, 399–414.

——, FREEMAN, R., KISSLING, E. & MÜLLER, St. 1996b. A lithospheric cross-section through the Swiss Alps – I. Thermokinematic modelling of the Neoalpine orogeny. *Geophysical Journal International*, **125**, 504–518.

PFIFFNER, O. A. 1992. Alpine orogeny. *In*: BLUNDELL, D., FREEMAN, R. & MÜLLER, S. (eds) *The European Geotraverse*. Cambridge University Press, Cambridge, 180–189.

—— & HEITZMANN, P. 1996. Geologic interpretation of the seismic profiles of the Central Traverse (lines C1, C2, C3-north). *In*: PFIFFNER, O. A., LEHNER, P., HEITZMANN, P., MÜLLER, ST. & STECK, A. (eds) *Results of the National Research Program 20 (NRP 20)*. Birkhäuser, Basel, 115–122.

——, ERARD, P.-F. & STÄUBLE, M. 1997b. Two cross-sections through the Swiss Molasse Basin (lines E4–E6, W1, W7–W10). *In*: PFIFFNER, O. A., LEHNER, P., HEITZMANN, P., MÜLLER, ST. & STECK, A. (eds) *Results of the National Research Program 20 (NRP 20)*. Birkhäuser, Basel, 64–72.

——, SAHLI, S. & STÄUBLE, M. 1997a. Stucture and evolution of the external basement uplifts. *In*: PFIFFNER, O. A., LEHNER, P., HEITZMANN, P., MÜLLER, ST. & STECK, A. (eds) *Results of the National Research Program 20 (NRP 20)*. Birkhäuser, Basel, 139–153.

RAHN, M., HURFORD, A. J. & FREY, M. 1997. Rotation and exhumation of a thrust plane: apatite fission-track data from the Glarus thrust, Switzerland. *Geology*, **25**, 599–602.

——, MULLIS, J., ERDELBROCK, K. & FREY, M. 1994. Very low-grade metamorphism of the Taveyanne greywacke, Glarus Alps, Switzerland. *Journal of Metamorphic Geology*, **12**, 625–641.

——, ——, ——, & —— 1995. Alpine metamorphism of the Taveyanne greywacke, Glarus Alps, Switzerland. *Eclogae Geologicae Helvetiae*, **88**, 125–178.

SCHAER, J. P., REIMER, G. M. & WAGNER, G. A. 1975. Actual and ancient uplift rate in the Gotthard region, Swiss Alps; a comparison between precise levelling and fission-track apatite age. *Tectonophysics*, **29**, 293–300.

SCHLUNEGGER, F., JORDAN, T. E. & KLAPER, E. M. 1997a. Controls of erosional denudation in the orogen on foreland basin evolution: the Oligocene central Swiss Molasse Basin as an example. *Tectonics*, 823–840.

——, LEU, W. & MATTER, A. 1997b. Sedimentary sequences, seismofacies, subsidence analysis, and evolution of the Burdigalian Upper Marine Molasse Group (OMM) of central Switzerland. *Bulletin, American Association of Petroleum Geologists*, 1185–1207.

SCHMID, S. M., AEBERLI, H. R., HELLER, F. & ZINGG, A. 1989. The role of the Periadriatic Line in the tectonic evolution of the Alps. *In*: COWARD, M., DIETRICH, D. & PARK, R. (eds) *Alpine Tectonics*. Geological Society, London Special Publications, **45**, 153–171.

——, PFIFFNER, O. A., FROITZHEIM, N., SCHÖNBORN, G. & Kissling, E. 1996. Geophysical–geological transect and tectonic evolution of the Swiss–Italian Alps. *Tectonics*, **15**, 1036–1064.

——, —— & SCHREURS, G. 1997. Rifting and collision in the Penninic Zone of eastern Switzerland. *In*: PFIFFNER, O. A., LEHNER, P., HEITZMANN, P., MÜLLER, ST. & STECK, A. (eds) *Results of the National Research Program 20 (NRP 20)*. Birkhäuser, Basel, 160–185.

SCHREURS, G. 1993. Structural analysis of the Schams nappes and adjacent tectonic units: implications for the orogenic evolution of the Penninic zone in eastern Switzerland. *Bulletin de la Société Géologique de France*, **164**, 415–435.

SCHUMACHER, M. E., SCHÖNBORN, G., BERNOULLI, D. & LAUBSCHER, H. P. 1997. Rifting and collision in the Southern Alps. *In*: PFIFFNER, O. A., LEHNER, P., HEITZMANN, P., MÜLLER, ST. & STECK, A. (eds)

*Results of the National Research Program 20 (NRP 20).* Birkhäuser, Basel, 186–204.
SEWARD, D. & MANCKTELOW, N. S. 1994. Neogene kinematics of the central and western Alps: evidence from fission-track dating. *Geology*, **22**, 803–806.
SINCLAIR, H. P. & ALLEN, P. A. 1992. Vertical versus horizontal motions in the Alpine orogenic wedge: stratigraphic response in the foreland basin. *Basin Research*, **4**, 215–232.
SOOM, M. 1990. *Abkühlungs- und Hebungsgeschichte der Externmassive und der penninischen Decken beidseits der Simplon–Rhone Line seit dem Oligozän: Spaltspurdatierungen and Apatit/Zirkon und K–Ar-Datierungen an Biotit/Muskovit (Westliche Zentralalpen)*. PhD thesis, University of Berne.
SPICHER, A. 1980. *Tektonische Karte der Schweiz 1:500 000*. Schweizerische Geologische Kommission, Basel.
STECK, A. 1990. Une carte des zones de cisaillement ductile des Alpes Centrales. *Eclogae Geologicae Helvetiae*, **83**, 603–627.
—— & HUNZIKER, J. 1994. The Tertiary structural and thermal evolution of the Central Alps – compressional and extensional structures in an orogenic belt. *Tectonophysics*, **238**, 229–254.
TROMSDORFF, V. 1966. Progressive Metamorphose kieseliger Karbonatgesteine in den Zentralalpen zwischen Bernina und Simplon. *Schweizerische Mineralogische und Petrographische Mitteilungen*, **46**, 431–460.
—— & EVANS, B. W. 1974. Alpine metamorphism of peridotitic rocks. *Schweizerische Mineralogische une Petrographische Mitteilungen*, **54**, 333–352.
VAN DER KLAUW, S. N. G. C., REINECKE, T. & STÖCKHERT, B. 1997. Exhumation of ultrahigh-pressure metamorphic oceanic crsut from Lago di Cignana, Piemontese zone, western Alps: the structural record in metabasites. *Lithos*, **41**, 79–102.
VANCE, D. & O'NIONS, R. K. 1992. Prograde and retrograde thermal histories from the central Swiss Alps. *Earth and Planetary Science Letters*, **114**, 113–129.
VIRUETE, J. E., ARENAS, R. & MARTÍNEZ CATALÁN, J. R. 1994. Tectonothermal evolution associated with Variscan crustal extension in the Tormes Gneiss Dome (NW Salamanca, Iberian Massif, Spain). *Tectonophysics*, **238**, 117–138.
VON BLANCKENBURG, F. 1990. *Isotope geochemical and geochronolgcial case studies of Alpine magmatism amd metamorphism: the Bergell intrusion and Tauern Window*. PhD thesis, ETH Zürich.
—— 1992. Combined high-precision chronometry and geochemical tracing using accessory minerals: applied to the Central-Alpine Bergell intrusion (central Europe). *Chemical Geology*, **100**, 19–40.
WILKS, W. J. & CUTHBERT, S. J. 1994. The evolution of the Hornelen Basin detachment system, western Norway: implications for the style of late orogenic extension in the southern Scandinavian Caledonides. *Tectonophysics*, **238**, 1–30.
WILLETT, S. D. 1992. Dynamic and kinematic growth and change of a Coulomb wedge. *In*: MCCLAY, K. R. (ed.) *Thrust Tectonics*. Chapman and Hall, London, 19–31.
WILLETT, S., BEAUMONT, C. & FULLSACK, P. 1993. Mechanical model for the tectonics of doubly vergent compressional orogens. *Geology*, **21**, 371–374.

# Exhumation of migmatites in two collapsed orogens: Canadian Cordillera and French Variscides

OLIVIER VANDERHAEGHE[1,3], JEAN-PIERRE BURG[2] & CHRISTIAN TEYSSIER[1]

[1]*Department of Geology and Geophysics, University of Minnesota, 310 Pillsbury Drive, Minneapolis MN 55455, USA*
[2]*Geologisches Institut, ETH Zentrum, Sonneggstrasse 5, CH-8092 Zurich, Switzerland*
[3]*Present address: Earth Sciences and Oceanography Department, Dalhousie University, Halifax, NS B3H 4J1, Canada*

**Abstract:** The Shuswap Metamorphic Core Complex in the Canadian Cordillera and the Velay Dome in the French Massif Central are examples of metamorphic core complexes formed in the hinterland of collapsed orogens. They display structural sections which allow the assessment of the efficiency of various exhumation mechanisms at different crustal levels. Erosion affected the relief generated during crustal thickening and redistributed the upper level of the mountain belt into nearby extensional basins or larger foreland basins. Faulting accommodated horizontal extension and lateral sliding of the upper crust. A detachment zone separates the upper brittle crust from ductilely deformed lower crust and achieved mechanical decoupling between these two major crustal layers. Below the detachment zone, the major fabric of high-grade rocks developed in the presence of melt and indicates vertical thinning along with horizontal extension. The lowest structural level exposed consists of migmatites which rose during crustal-scale boudinage of the overlying units and appear in the cores of dome-shaped culminations. The combination of these various exhumation mechanisms within a period of a few million years brought into close proximity migmatites which formed at depth of about 20–25 km and sediments deposited at the surface.

The plate tectonic theory has been applied to explain structural and metamorphic features of modern and ancient orogenic belts (Dewey & Bird 1970; Dewey & Burke 1973). In particular, high-pressure–low-temperature metamorphic belts associated in time and space with active convergent plate margins such as the circum-Pacific regions have been recognized to represent material subducted and exhumed owing to a combination of thrusting, erosion and tectonic denudation at the front of orogenic belts (Ernst 1973; Miyashiro 1973; Andersen *et al.* 1991; Platt 1993; Chemenda *et al.* 1995). Two major mechanisms have been invoked to explain the formation of high-temperature–low-pressure migmatite–granite terranes in eroded mountain belts. (1) The first mechanism ascribes a major role to the generation and rise of magmas related to the descending slab in a subduction zone. This mechanism accounts for the generation of calc-alkaline or I-type granitoids that form island arcs and active continental margins. (2) The second mechanism, whose importance was recognized more recently, ascribes a major role to thermal equilibration and subsequent partial melting of thickened continental crust (England & Thompson 1984; Dewey 1988). In the latter case, the melt produced is peraluminous and enriched in potassium (S-type). Both mechanisms explain the formation of migmatite–granite terranes. However, processes involved in the exhumation of these terranes are debated. Two competing models are (1) a 'simple shear model' in which extension is accommodated along a lithospheric shear zone (Wernicke 1981; Lister & Davis 1989); and (2) a two-layer crust model where localized extension of the upper brittle crust is accommodated by pervasive flow of the underlying ductile crust (Gans 1987; Block & Royden 1990; Buck 1991; Brun *et al.* 1994).

This study focuses on the exhumation (i.e. the relative vertical component of motion of rocks with respect to the surface) of migmatite terranes in orogens. We investigate various mechanisms that were operating at different structural levels during late-orogenic extension on the basis of similarities noticed during field investigation of two examples of collapsed orogens: the Mesozoic Canadian Cordillera (Fig. 1) and the Palaeozoic French Variscides (Fig. 2). In both orogens, migmatites were brought into contact with sub-contemporaneous volcaniclastic sedimentary rocks. In this paper,

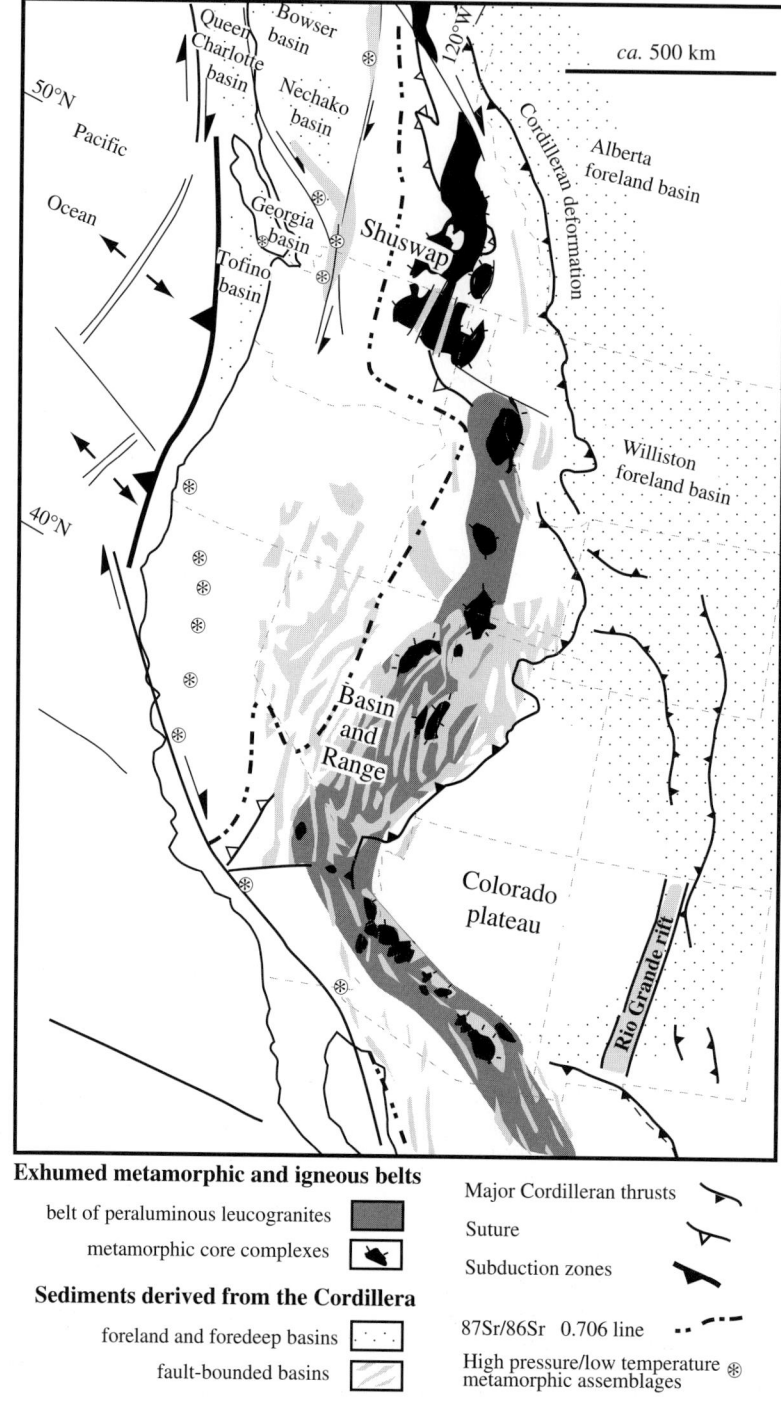

**Fig. 1.** Schematic tectonic map of western North America illustrating the relationship between major Cordilleran tectonic units and metamorphic core complexes formed from late Mesozoic to late Cenozoic time (modified after Miller & Bradfish 1980; Armstrong 1982; Armstrong & Ward 1991).

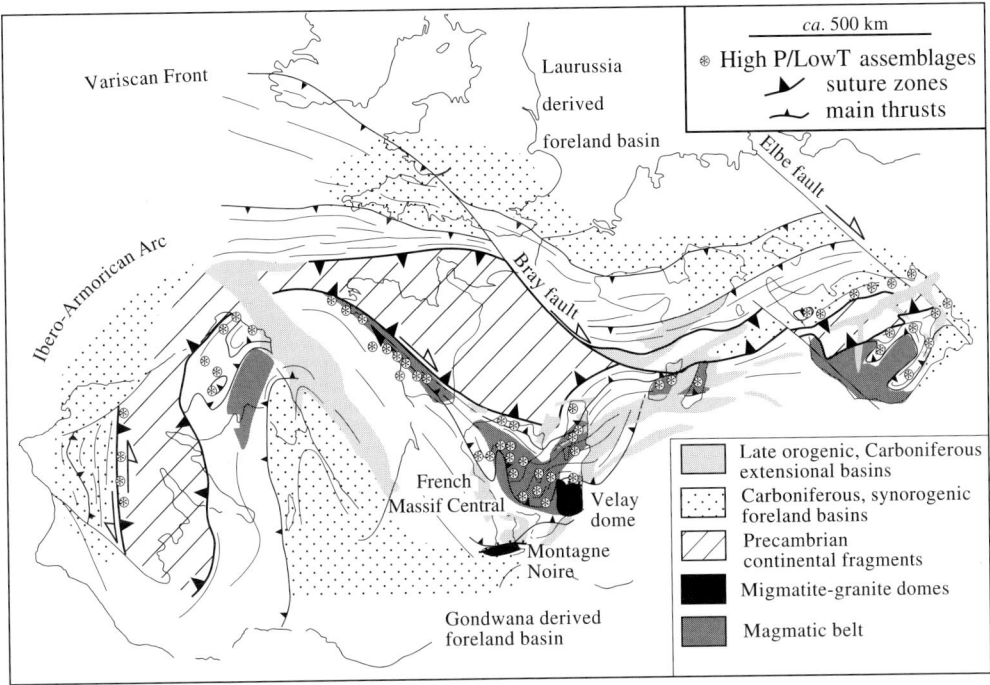

**Fig. 2.** Schematic tectonic map of the Variscan belt of eastern Europe (modified after Burg & Matte 1978; Franke & Engel 1986; Ledru *et al.* 1989; Burg *et al.* 1994).

we review the main features of these orogenic segments and propose a conceptual model that hints at a link between the thermo-mechanical evolution of the orogenic crust and exhumation mechanisms specific to late-orogenic collapse. While erosion and sedimentation redistributed the material at the surface, partially molten rocks rose in necked zones between crustal-scale boudins of the brittlely extended upper crust.

## Characteristics of collapsed orogens

In this part we refer to the sections across the collapsed Southern Canadian Cordillera and French Massif Central. The geological information for each of these orogenic segments is summarized in maps, cross-sections and diagrams depicting the pressure and temperature path defined by metamorphic petrology, and the thermal history constrained by thermochronological data. Even though they show variations in the details, the tectonic histories and the current structural sections of both areas present similar first-order features: (1) both orogens resulted from collision during and after subduction; (2) continental collision produced crustal thickening while deformation migrated from the hinterland to the foreland; (3) loading of the underthrust margin resulted in the formation of a foreland flexural basin; (4) several tens of million years after crustal thickening, metamorphic core complexes with high-temperature migmatites were exhumed during extensional collapse of the orogenic belt.

The Shuswap Metamorphic Core Complex (MCC) in the Canadian Cordillera (Figs 3–5), and the Velay Dome in the French Massif Central (Figs 6 and 7) comprise three lithologically and stratigraphically complex units that are distinguished on the basis of their contrasting rheological behaviour during the formation of the core complexes (Vanderhaeghe 1997; Vanderhaeghe & Teyssier 1997). A low-angle detachment zone separates remnants of a dismembered upper unit from the exhumed metamorphic cores. Below the detachment zone, the exhumed and ductilely deformed high-grade rocks consist of an amphibolite-facies middle unit that forms domes cored by migmatites and granites (lower unit). Although we focus on these two examples, other metamorphic core complexes in both orogens (Figs 1 and 2) display similar rock units and geological histories.

### Canadian Cordillera

*Upper unit and detachment.* The upper unit of the Shuswap MCC (Fig. 5) is composed dominantly

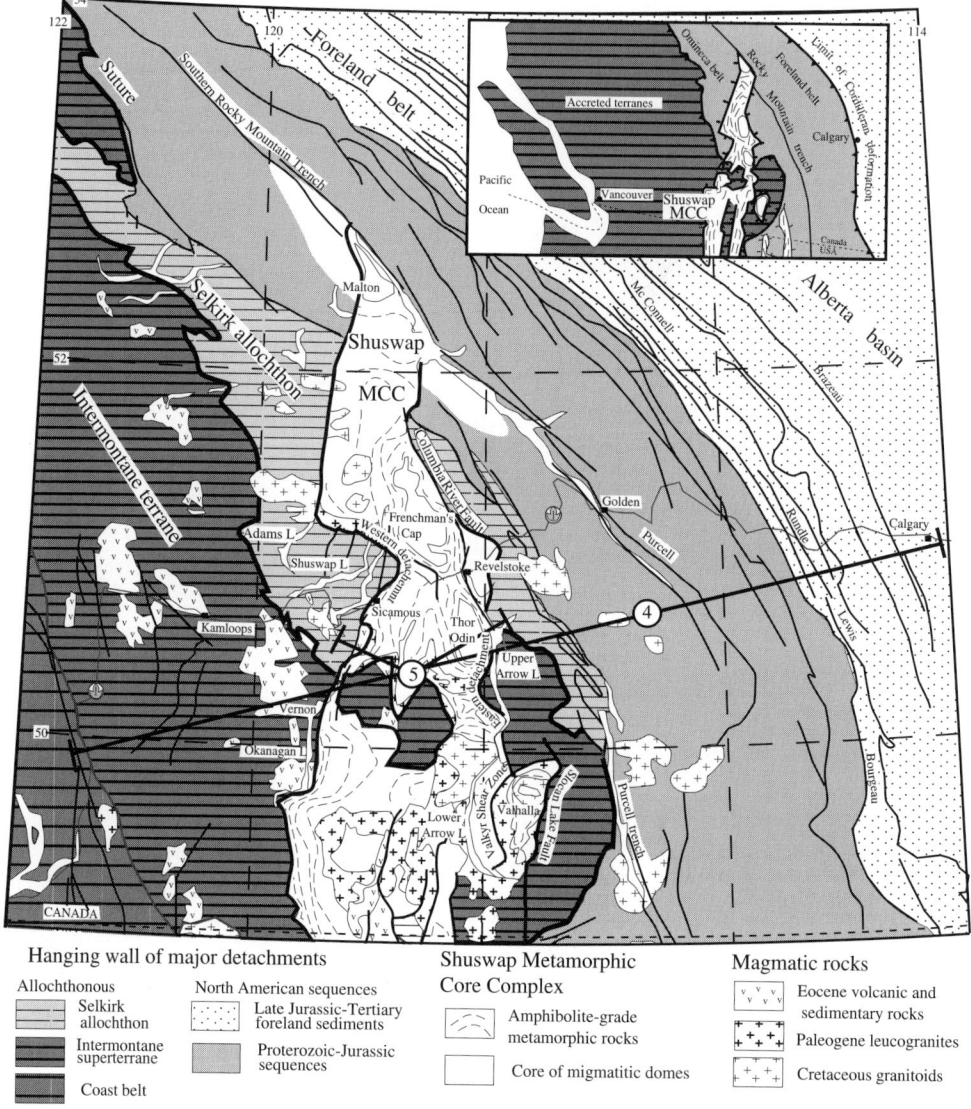

**Fig. 3.** Geological map of the Shuswap Metamorphic Core Complex. Cross-sections represented in Figs 4 and 5 are indicated (modified after Wheeler & McFeely 1991).

of rocks from the allochthonous Intermontane superterrane which comprises the Quesnel arc underlain to the east by the Slide Mountain terrane with oceanic affinities and to the west by the Cache Creek oceanic terrane that contains high-pressure–low temperature mineral assemblages. The arc was active until $c.162$ Ma (Parrish & Armstrong 1987) and was thrust eastward over the sedimentary cover of the North American continental margin to form the Rocky Mountains fold and thrust belt and the Alberta foreland basin (Monger et al., 1982; Brown et al. 1986; Price 1986). Crustal thickening was responsible for Barrovian metamorphism in the Selkirk allochthon and in the Kootenay terrane to the west of the Shuswap MCC, and was followed by a period of cooling during Mesozoic time as recorded by whole-rock and mineral Rb–Sr and K–Ar ages (Mathews 1981; Archibald et al. 1983; Colpron et al. 1996).

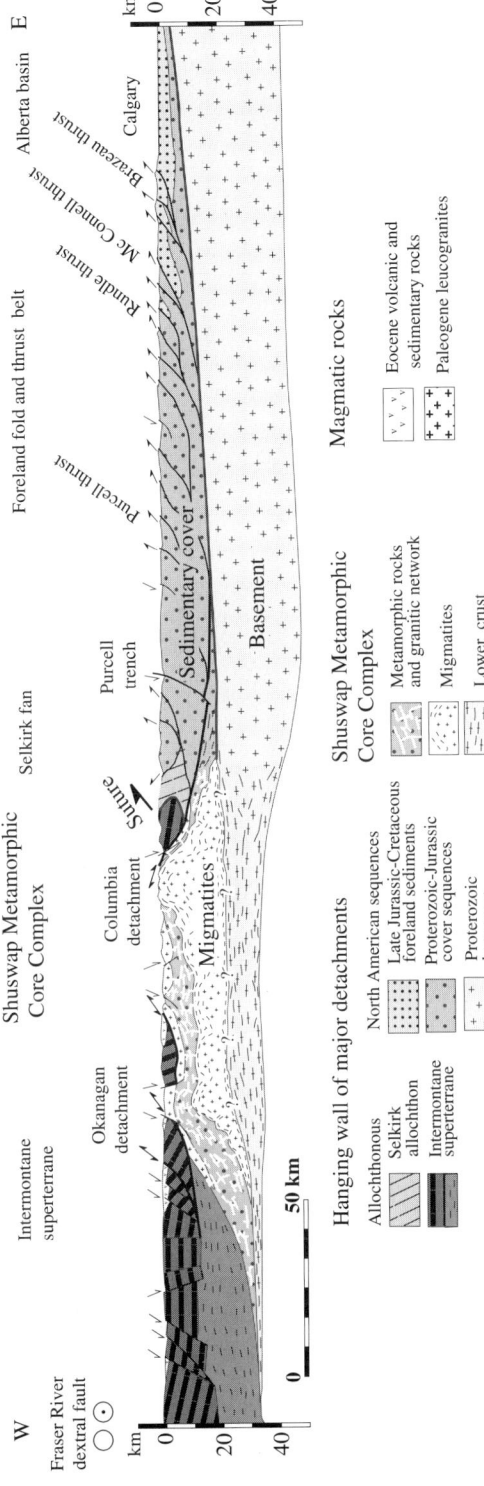

**Fig. 4.** Cross-section through the Canadian Cordillera (modified after Brown & Journeay 1987; Price & Mountjoy 1970). The lower-crustal part is interpreted on the basis of the Lithoprobe profile (Cook *et al.* 1988, 1992).

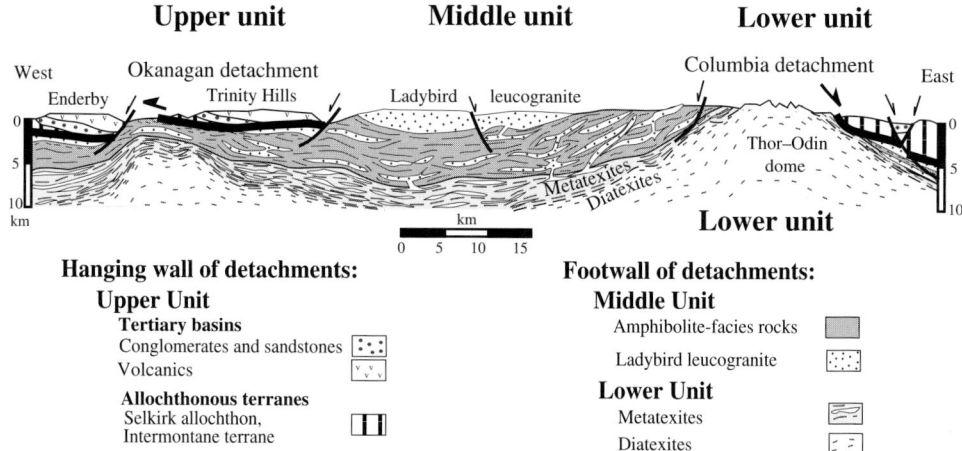

**Fig. 5.** Section across the Shuswap MCC at the latitude of the Thor–Odin dome (modified after Vanderhaeghe & Teyssier 1997).

In our structural section (Figs 4 and 5), the upper unit occurs as thin rafts on top of detachment zones bounding the MCC. These detachment zones have been defined by various names in the past, Okanagan-Eagle River-Adams detachment to the west and and Columbia river detachment to the east (Read & Brown 1981; Lane 1984; Tempelman-Kluit & Parkinson 1986; Johnson & Brown 1996), but in this paper are referred to as the western and eastern detachments. The detachment zones are characterized by progressive overprinting of an amphibolite-facies fabric by a greenschist-facies mylonitic fabric that grades into cataclasite, breccia, and pseudotachylite zones toward the top, and into amphibolite-facies fabrics downward, towards the core of the dome. In the hanging wall, the upper unit is affected by an array of steeply dipping brittle strike-slip and normal faults that accommodated E–W extension (Johnson & Brown 1996). High angle normal faults affecting the upper unit root into and/or terminate at the detachment zone and bound the early Cenozoic wedge-shaped Enderby, Trinity Hills, and Vernon basins. The entire area is also dissected by a number of late high-angle normal faults. In order to distinguish these various generations of structures we define a western and an eastern detachment, which are restricted to mylonitic zones marking the boundary of the MCC on each sides. In the fault-bounded basins, sedimentary infilling begins with alternating sandstones and coal-bearing layers overlain by fanglomerates containing large angular clasts (Mathews 1981). The clast population includes greenschists and granodiorites similar to the rocks of the upper unit, but also leucogranites and quartzites recognized in the footwall of the detachment. Clasts of the footwall appear in increasing proportions upward through the sequence, reflecting the progressive exhumation and erosion of footwall high-grade rocks. Elevated coefficients of vitrinite reflectance in the basal coal-bearing sediments indicate that they experienced temperatures of 175–200°C for a few million years (Mathews 1981), indicating either a high vertical geothermal gradient at the time of deposition or lateral heat advection from the adjacent high-grade rocks. The sedimentary sequences are intruded and capped by rhyolitic to basaltic volcanites dated at 47–49 Ma (K–Ar ages, Mathews 1981).

*Middle unit.* The middle unit (Fig. 5) is composed of intensely deformed Palaeozoic to Mesozoic metasedimentary rocks of the North American passive margin (Read 1980; Simony *et al.* 1980; Price 1986; Scammell & Brown 1990). Widespread amphibolite-facies metamorphism locally reached conditions of partial melting (Fig. 8), especially in fertile lithologies such as metapelites. Temperatures of 720–820°C at 7.5–9 kbar imply minimum burial depths of 20–25 km (Nyman *et al.* 1995).

The major fabric is generally concordant with the flat-lying detachment zones except around the dome-shaped culminations (Vanderhaeghe & Teyssier 1997). This main fabric carries an E–W trending mineral and stretching lineation. The major foliation is boudinaged in several

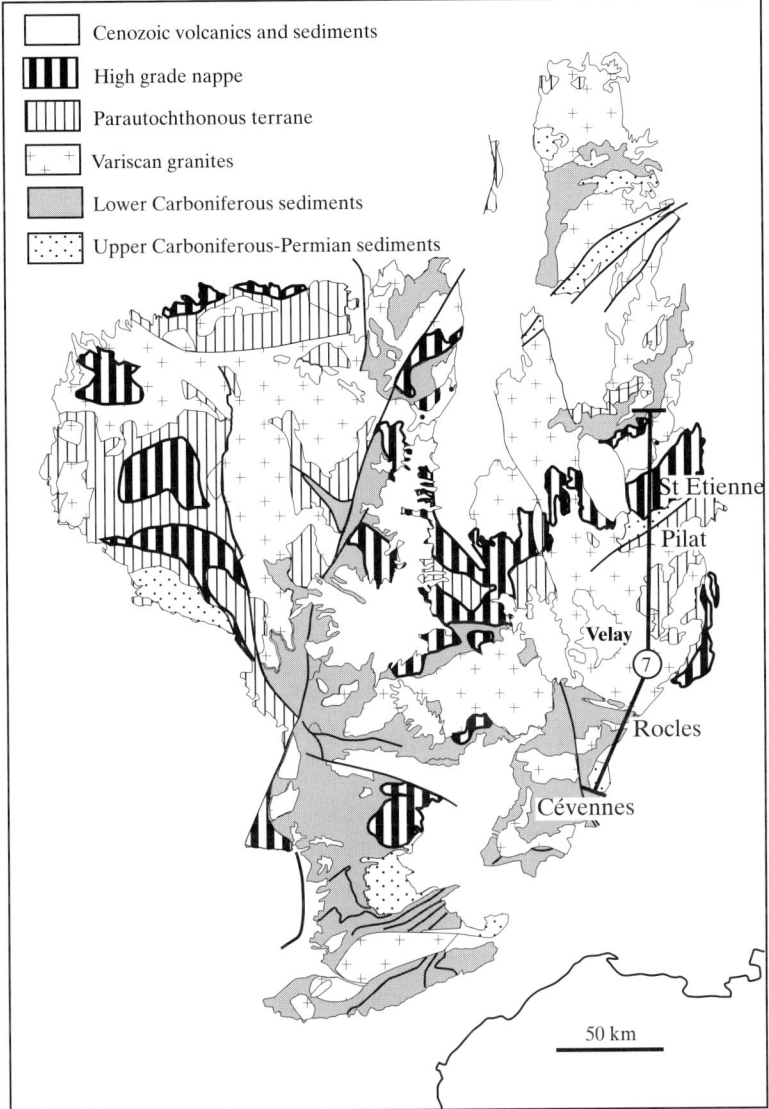

**Fig. 6.** Geological map of the French Massif Central (modified after Burg & Leyreloup, unpublished data, 1988; Ledru *et al.* 1989). The section represented in Fig. 7 is indicated.

directions resulting in chocolate-tablet structures. The middle unit is permeated by a network of sills (concordant with the foliation) and dykes of pegmatitic and aplitic granite structurally connected to the larger laccoliths of Ladybird leucogranite intruded within the detachment zone (Carr 1992). A greenschist facies mylonitic fabric overprints the magmatic texture and carries an E–W trending stretching lineation. Associated kinematic criteria indicate outward motion of the hanging wall with respect to the core of the MCC. Granitic veins and pods occupy several structural sites such as layers concordant with the main foliation, shear zones, pressure fringes around garnets, boudin necks and fractures breaking garnets, indicating that some granites were coeval with the major deformation. U–Pb ages of zircon from leucogranites range from 100 Ma to 50 Ma with a dominant population at 60–55 Ma confirmed by more

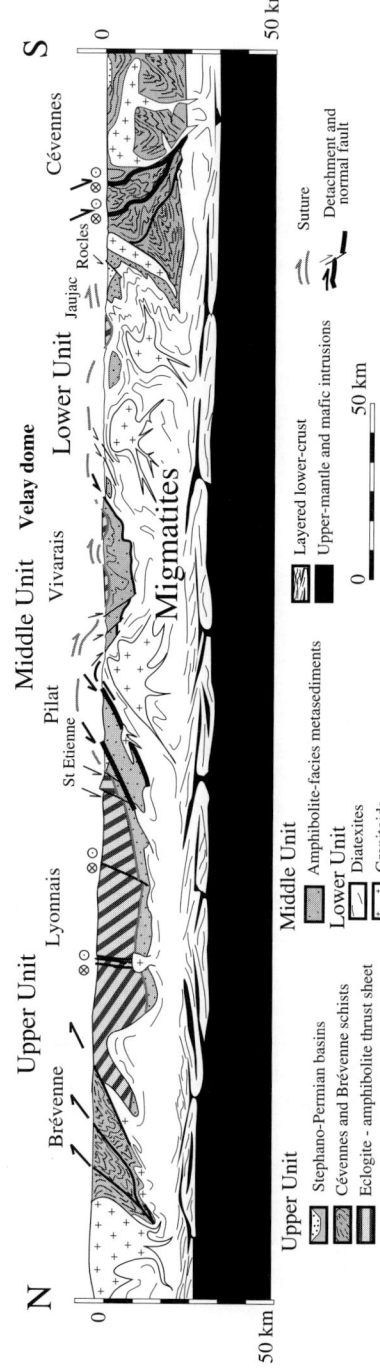

Fig. 7. Section across the French Massif Central through the Velay Dome (modified after Burg & Vanderhaeghe 1993). The interpretation of the lower crust is based on extrapolation of the ECORS seismic profiles and on xenoliths brought to the surface by Neogene volcanoes.

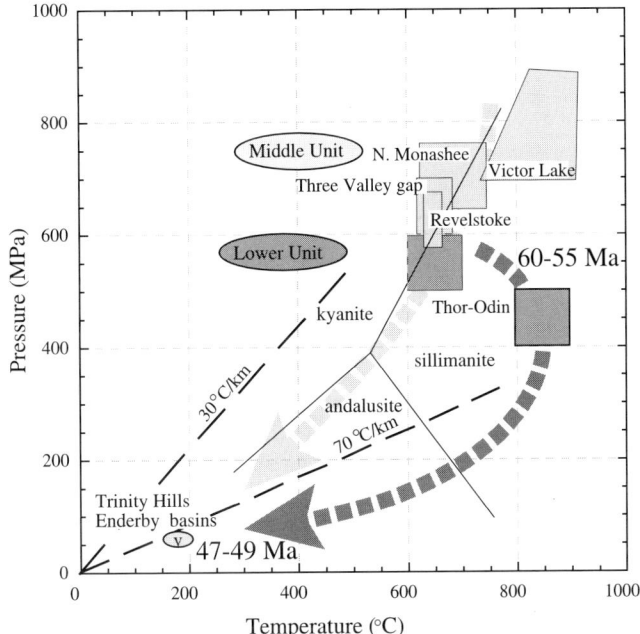
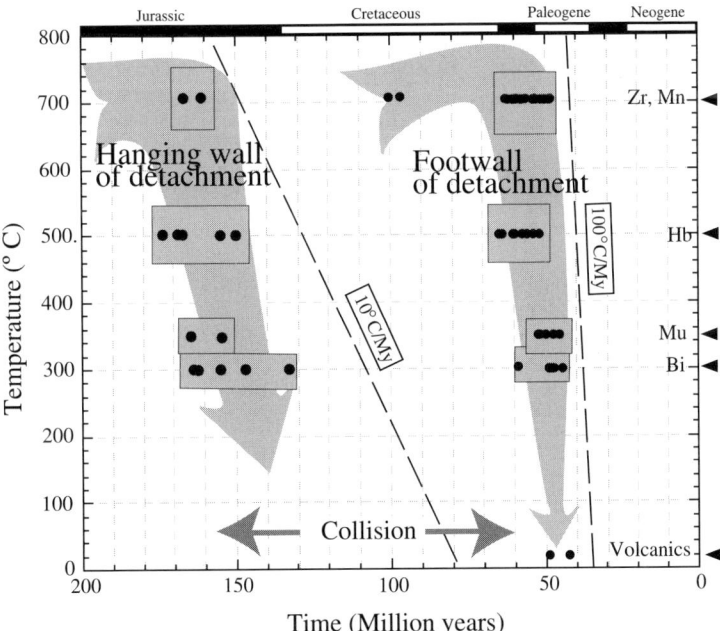

**Fig. 8.** Synthesis of thermobarometric and thermochronological data from the Shuswap MCC (Mathews 1981; Archibald *et al.* 1983; Journeay 1986; Sevigny *et al.* 1989, 1990; Carr 1991, 1992; Parkinson 1991, 1992; Colpron *et al.* 1996).

constant monazite ages (Parrish et al. 1988; Sevigny et al. 1989; Carr 1992; Parkinson 1992; Parrish 1995) and by U–Pb SHRIMP ages from zircons of the Ladybird leucogranite (Vanderhaeghe 1997). Ages as young as c.50 Ma have been obtained on zircons from pegmatites intruded within the detachment zone (Parkinson 1992). The thermal history of the middle unit is also constrained by K–Ar, Rb–Sr, and Ar–Ar ages indicating rapid cooling between 55 and 50 Ma (Fig. 8). Results from a recent thermochronological study in the area of the Thor–Odin dome refine the cooling history of the footwall of the detachments (Vanderhaeghe 1997). Hornblende yields $^{40}Ar/^{39}Ar$ ages ranging from 59 to 54 Ma, and muscovite and biotite ages are clustered between 47 and 49.5 Ma, and are consistent with closure of K-feldspars typically ranging from 50 to 43 Ma, except for samples in the immediate footwall of the Columbia river fault (a brittle normal fault that offsets the eastern detachment zone) which yield ages as young as 26 Ma. These data constrain a first period of rapid cooling through the closure temperature of hornblende (c. 500°C) to the closure temperature of muscovite and biotite (c. 350°C and c. 300°C, respectively), followed by a period of thermal stability and a second period of rapid cooling at c. 45 Ma constrained by thermal modelling of the K-feldspar.

*Lower unit and metatexite/diatexite transition.* The lower unit (Fig. 5) consists of migmatites exposed in the cores of dome-shaped culminations aligned along the strike of the belt (from north to south, Malton, Frenchman's Cap, Thor–Odin, Valhalla, Fig. 3). We distinguish two types of migmatites. (1) Metatexites are characterized by a continuous gneissic layering. They correspond to the 'layered gneisses' of Reesor & Moore (1971). (2) Diatexites are composed of more or less heterogeneous granitoids. They correspond to the monzogranitic to granodioritic core gneisses of Reesor & Moore (1971). The lithological heterogeneity of this lower unit reflects the composite nature of its protolith. Paleoproterozoic North American basement is revealed by discordant U–Pb ages on zircons with upper intercepts of 2.2 Ga in paragneisses (Parkinson 1991), to 1.87–2.10 Ga in orthogneisses (Wanless & Reesor 1975; Armstrong et al. 1991; Parkinson 1991). However, in the Thor–Odin dome (Fig. 5), diatexites show outcrop and map-scale cross-cutting contacts with lower Cambrian quartzites (Reesor & Moore 1971; Vanderhaeghe & Teyssier 1997), which suggests that anatexis may have affected part of the Palaeozoic sedimentary sequence.

Metamorphic assemblages (Fig. 8) show that partial melting took place at a temperature of c. 700°C and at 5–6 kbar followed by isothermal decompression to 4 kbar (Duncan 1984; Journeay 1986). The pressure recorded by diatexites is lower by 2–3 kbar than the highest pressures recorded by metapelites of the middle unit, although the migmatites are structurally below the pelites.

In the Thor–Odin dome, the transition from the core diatexites to the surrounding metatexites occurs over a few hundred metres (Vanderhaeghe 1997; Vanderhaeghe & Teyssier 1997). On the southern limb of the Thor–Odin dome the magmatic foliation carries a down-dip biotite and sillimanite lineation perpendicular to the horizontal E–W trending mineral and stretching lineation of metatexites. Microscopic observation of the diatexites shows that they preserve a magmatic texture marked by interfingered quartz and feldspar grains, slightly overprinted by high-temperature plastic deformation. Rare and small chlorite- or muscovite-bearing shear zones with a normal sense of shear are consistent with the diatexites moving upward with respect to metatexites. Synmigmatitic way-up criteria indicate outward tilting of the migmatitic foliation around the diatexites (Vanderhaeghe & Teyssier 1997).

## French Variscides

*Upper unit and detachment.* The nappe pile preserved in the upper unit of the French Massif Central (Figs 6 and 7) comprises from top to bottom (Burg & Matte 1978; Ledru et al. 1989):

(1) A nappe of migmatitic metasedimentary rocks that have undergone an early high-pressure granulite-facies metamorphism. At the base of the nappe, the so-called *leptynite–amphibolite gneiss complex* may represent a strongly sheared active margin dated at 500–480 Ma (Duthou et al. 1981). The leptynite–amphibolite gneiss includes eclogites related to subduction at 450–400 Ma (Burg & Matte 1978; Duthou et al. 1981; Pin & Lancelot 1982; Gardien et al. 1988; Lardeaux et al. 1989).

(2) Parautochthonous Late Proterozoic to Early Palaeozoic metasediments and granitic orthogneisses which were progressively involved in the Variscan collision. Burial of the parautochthonous rocks resulted in widespread Barrovian amphibolite-facies metamorphism (Gardien et al. 1997). Retrogression and rapid cooling ($>50°C\,Ma^{-1}$) was dated by Ar-Ar thermochronology on micas and amphiboles at 340 Ma (Costa et al. 1993). Thrusting of the allochthonous nappe over the continental margin was

associated with subsidence of a flexural basin to the south of the French Massif Central, which was then progressively involved in the collision between 350 Ma and 320 Ma (Maluski *et al.* 1991; Arnaud & Burg 1993). An array of brittle strike-slip and normal faults that bound pull-apart and extensional basins filled by late Carboniferous lava flows and continental sediments (Ménard & Molnar 1988; Burg *et al.* 1990) is related to N–S extension of the upper unit. For example, the Saint Etienne basin, which rests directly on the Pilat detachment, to the north of the Velay Dome (Fig. 3b), is filled by sedimentary sequences typically composed of proximal coarse clastic and volcaniclastic deposits alternating with coal horizons (Malavieille *et al.* 1990). The degree of maturity of organic matter implies a steep geothermal gradient of about 50°C km$^{-1}$ and periglacial deposits suggest mountain altitudes of 5000 m (Becq-Giraudon & Van den Driessche 1994). To the south of the Velay Dome, the Jaujac and Cévennes basins present similar stratigraphic characteristics and have a fan geometry consistent with syndepositional northward tilting (Allemand *et al.* 1997).

The base of the upper unit is delineated by several detachment zones. The Velay Dome is bounded to the north by the Pilat detachment zone (Vitel 1988; Malavieille *et al.* 1990). The mylonitic fabric carries a N–S stretching lineation and kinematic criteria indicate a top-to-the-north, normal sense of shear initiated under amphibolite conditions and finishing at greenschist-facies conditions (Malavieille *et al.* 1990). The south limb of the Velay Dome is subvertical (Burg & Vanderhaeghe 1993) and local chlorite-bearing shear zones indicate south-side down, normal shear.

*Middle unit.* The middle unit of the Velay Dome is 2–3 km thick (Fig. 7). It consists of paragneisses intruded by Late Proterozoic to Early Palaeozoic granitoids (Pin & Lancelot 1982) and is preserved as pendants within the migmatitic lower unit. Metamorphic assemblages in metapelites indicate amphibolite-facies metamorphism (Fig. 9) with a first stage at 8–9 kbar, 650°C, followed by a second at 3–5 kbar, 700–780°C, responsible for partial melting of metapelites (Caen-Vachette *et al.* 1984; Gardien *et al.* 1997). The middle unit is permeated by abundant leucosomes contained by, and flattened within, the composite foliation, and by a network of leucogranitic dykes and sills. Leucogranitic pods yielded a whole-rock Rb/Sr isochron at 322 ± 9 Ma (Caen-Vachette *et al.* 1984). Monazites from leucosomes yielded a U–Pb age of 314 ± 5 Ma (Mougeot *et al.* 1997). The pattern of the major foliation delineates the shape of the Velay Dome and is concordant with the detachment. In contrast, to the south the vertical contact between the middle unit and the Cévennes schists of the upper unit is intruded by the Rocles leucogranite laccolith whose magmatic fabric, overprinted by a solid-state fabric, is oriented parallel to the detachment and the regional major foliation. The Rocles leucogranite has been dated by Rb/Sr whole-rock at 302 ± 4 Ma (Caen-Vachette *et al.* 1981). The pluton floor is to the north and the roof to the south, which suggests that the detachment zone was active during leucogranite intrusion. Both have been later tilted to their present orientation (Burg & Vanderhaeghe 1993).

*Lower unit and metatexite–diatexite transition.* Diatexites and granites of the Massif Central lower unit constitute most of the detachment footwall in the Velay Dome (Dupraz & Didier 1988; Burg & Vanderhaeghe 1993). Scattered U–Pb zircon and monazite ages from migmatites (Fig. 9) suggest a long anatexis history from 320 to 290 Ma (Mougeot *et al.* 1997) but also reflect the heterogeneity of the protolith. A water-present stage of migmatization at <750°C, >5 kbar, was followed by water-absent melting at 760–850°C, 4–5 kbar (Montel *et al.* 1992).

On the southern limb of the Velay Dome, the transition from the Cévennes schists of the upper unit to the diatexites occurs over a few kilometres (Burg & Vanderhaeghe 1993). Although diatexites cross-cut gneisses of the middle unit at the outcrop scale, lithological boundaries at a map scale are parallel to the main foliation and delineate the grossly circular shape of the dome. Complex disharmonic and non-cylindrical folding of the migmatitic foliation, which carries a N–S trending biotite lineation, reflects flow of the migmatitic Velay Dome during its emplacement (Burg & Vanderhaeghe 1993; Lagarde *et al.* 1994). Synmigmatitic way-up criteria indicate outward tilting and overturning of the migmatitic foliation around the diatexites (Burg & Vanderhaeghe 1993). Conjugate and post-foliation shear bands suggest N–S oriented coaxial shortening. Accordingly, the asymmetric shape of the Velay Dome is related to its southward lateral expansion while the Pilat detachment was active to the north (Burg & Vanderhaeghe 1993; Lagarde *et al.* 1994).

*Lower crust.* Knowledge of the lower crust beneath the French Variscides comes from the

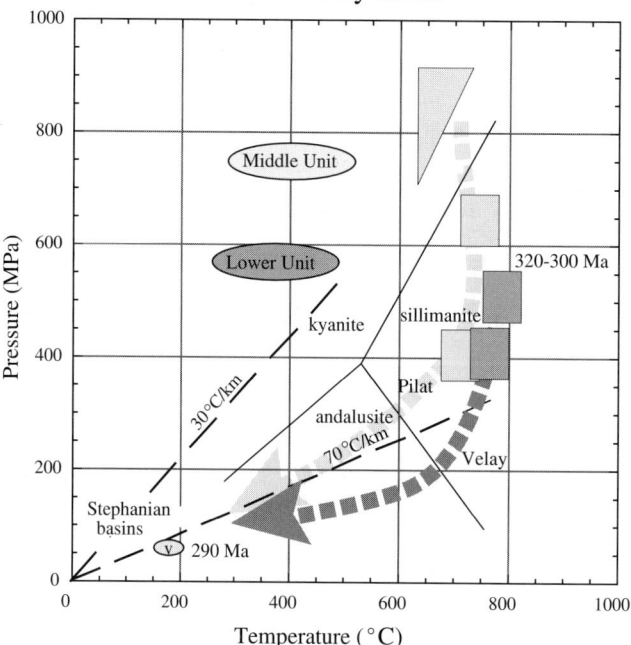

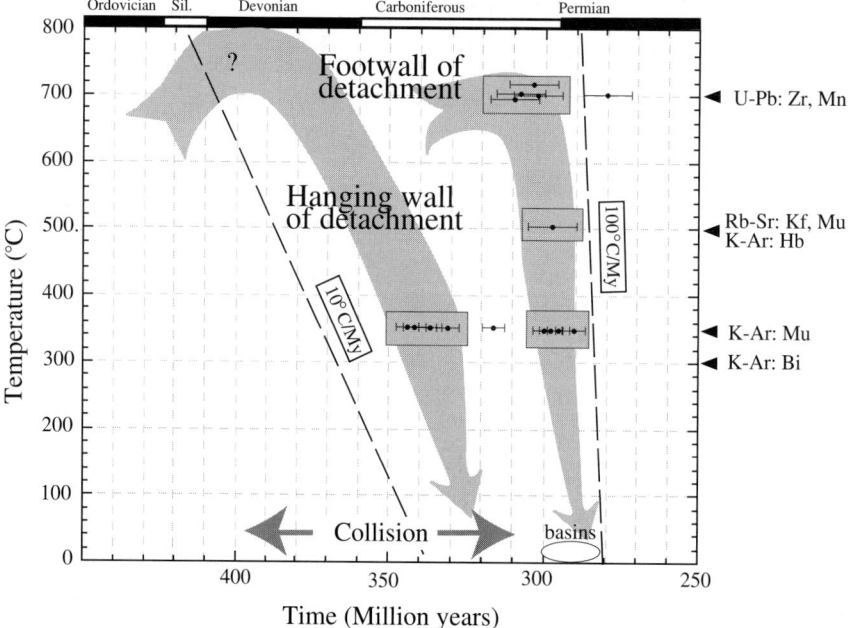

**Fig. 9.** Synthesis of thermobarometric and thermochronological data from the Velay Dome (Gardien 1990; Maluski *et al.* 1991; Costa 1991–1992; Montel *et al.* 1992; Costa *et al.* 1993; Costa & Rey 1995; Gardien *et al.* 1997).

interpretation of geophysical data, in particular the European ECORS deep seismic profiles (Bois & ECORS Scientific Party 1990). The profiles show a strongly reflective lower crust underlain by the Moho. Direct information on the lower crust of the Velay Dome is provided by xenoliths found in Neogene volcanic pipes (Downes & Leyreloup 1986). Lithological, tectonic and geochronological characteristics of these lower-crustal xenoliths favour formation of layering by horizontal flow and tectonic transposition of a lithologically heterogeneous lower crust (Rey 1993; Costa & Rey 1995), along with the intrusion of underplated charnockitic plutons (Leyreloup et al. 1977) during late-orogenic collapse of the Variscan belt.

## Exhumation of the migmatites

We will now try to provide a crude estimate of the amount of material removed from above the migmatites along the studied sections. The amount that can be accounted for by erosion is derived from the estimation of the amount of sediments preserved along the line of the sections. Tectonic denudation during late-orogenic collapse implies that the upper unit slides on the sides of the metamorphic core complex. The amount of exhumation attributable to tectonic denudation and associated upwelling of the ductile crust is not directly estimated but is discussed after reconstruction of crustal-scale sections before and after late-orogenic extension (Fig. 10). We are aware that this approach relies heavily on the assumption that material moved only within the plane of section and is far from precise, but we claim that this exercise allows us to bracket the relative importance of each mechanism.

### Canadian Cordillera

*Timing and amount of exhumation.* Exhumation of the Canadian Cordillera hinterland started in

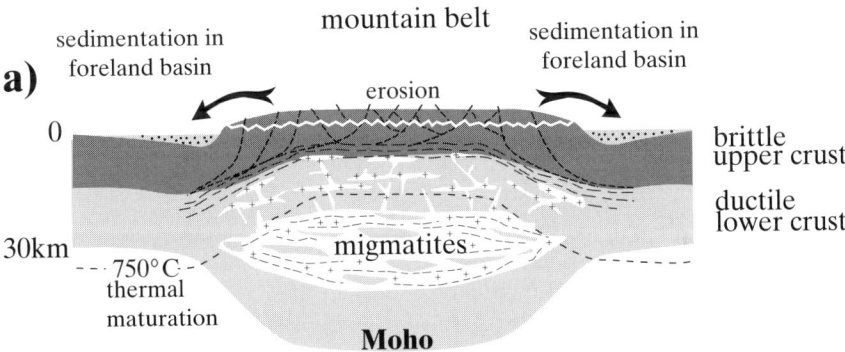

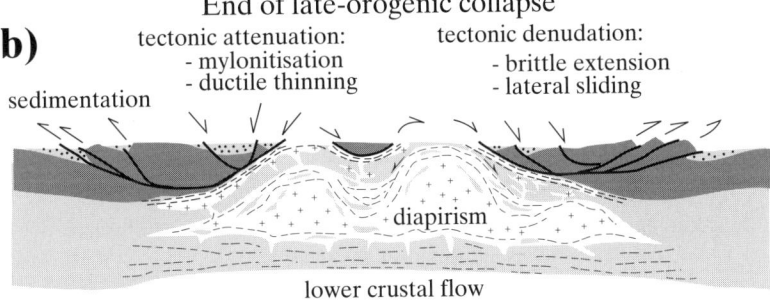

Fig. 10. Mechanisms of exhumations. (a) During the period of crustal thickening, erosion is redistributing material from the mountain belt to the foreland and foredeep basins. Thermal maturation related to thermal relaxation and radioactive heat production causes a rise of the isotherms through the crust. (b) During late-orogenic collapse, exhumation of migmatitic terranes is potentially achieved by a combination of mechanisms such as erosion, tectonic denudation, lower-crustal flow and diapirism.

Late Jurassic time during crustal thickening and ended with the formation of the Shuswap MCC during Cenozoic time (Fig. 11). Reconstruction shows that the present-day 150 ± 20 km wide Canadian Cordillera hinterland was about 100 ± 20 km wide before extension (Coney & Harms 1984; Parrish et al. 1988; Johnson & Brown 1996). During Late Jurassic–Cretaceous crustal thickening, cooling is recorded by thermochronology in the upper unit of the MCC. The cooling is interpreted to represent $c.10$ km of exhumation during the growth of the orogen (Mathews 1981; Archibald et al. 1983; Colpron et al. 1996), which implies the removal of 1000 ± 200 km$^2$ of upper crust from the pre-extension section. This amount of exhumation is consistent with metamorphic decompression in the middle unit from a depth of 20–25 km (maximum burial recorded by Barrovian metamorphism) to 10–15 km (Brown & Journeay 1987).

The second period of exhumation corresponds to the early Cenozoic formation of the Shuswap MCC, whose current width on the Thor–Odin section is $c.50$ km. The high-grade middle and lower units were buried at depths of 15–20 km before their Cenozoic exhumation, which yields removal of 750–1000 km$^2$ of upper unit material between $c.60$ and $c.56$ Ma (U–Pb crystallization age of zircons in migmatite leucosomes) (Vanderhaeghe 1997).

Accordingly, an orogenic crustal thickness of 60 ± 5 km can be restored by adding to the present $c.$ 33 km thickness underneath the Shuswap MCC (Sweeney et al. 1991), (1) 15–20 km of exhumation during late-orogenic collapse, and (2) >10 km of erosion during crustal thickening. These values are consistent with previous estimates (Coney & Harms 1984; Johnson & Brown 1996).

*Erosion.* The material removed from the growing mountains during the period of crustal thickening was mostly transported into two distinct foreland basins, the Alberta basin to the east, and the Bowser–Nechako basin to the west (Fig. 1). The Bowser–Nechako basin, which is currently north of the Shuswap MCC, may not have arrived in this position until late Cretaceous time (Irving & Wynne 1991). Most of the sediments filling this basin came from the west or from the nearby Intermontane superterrane, except at its top, where metamorphic minerals attributed to the Omineca belt are recognized (Yorath 1991). Consequently, the Bowser–Nechako basin did not absorb a significant amount of erosion from the Canadian Cordillera hinterland. The 300–500 km wide Alberta basin (Figs. 1–3) is a typical flexural basin (Price & Mountjoy 1970; Price 1973; Beaumont 1981). It is deepest on its western side at the front of the major thrust, with 3–4 km of Jurassic–Cretaceous sedimentary rock overlain by 1 km of Paleogene sedimentary rock (Yorath 1991). This represents about 600–1250 km$^2$ of sediments (500–1000 km$^2$ of Jurassic–Cretaceous sediments plus 150–250 km$^2$ of Paleogene sediments). Comparatively, the amount of Jurassic–Cretaceous sediments derived from the Canadian Cordillera hinterland and deposited in the pull-apart basins opened along strike-slip faults is not significant (20–50 km$^2$, if we take a depth of 4–5 km and a width of 5–10 km). The volume of Jurassic–Cretaceous sediments deposited in the foreland basin is of the same order of magnitude as the amount of material removed from the Canadian Cordillera hinterland at the same time. Accordingly, we suggest that most of the relief formed during crustal thickening was eroded and transported to the foreland basin (Fig. 11a). However, we appreciate that sediments were also transported out of the plane of the section and that some sediments bypassed the basin to be transported as far as the Gulf of Mexico (Beaumont 1981). On the other hand, a significant portion of the sediments deposited in the Alberta basin originated from the fold and thrust belt, which results in exhumation not of the Canadian Cordillera hinterland but of the Rockies in the foreland (Price & Mountjoy 1970; Yorath 1991).

Cenozoic extension and formation of the Shuswap MCC is associated with reactivation of the foreland basin and the formation of basins (Trinity Hills, Enderby and Vernon basins in the Shuswap MCC) along normal faults caused by brittle extension of the upper crust. These basins are deep (2–3 km) but limited in extent (2–20 km wide by 5–40 km long). Half of the Cenozoic fill consists of volcanites, the other half of sediments derived from both the upper unit and the detachment footwall. Thus, sediments deposited in these basins account for only a few per cent of the amount of material removed from the top of the Shuswap MCC (Fig. 11b).

*Tectonic denudation and upwelling of the ductile crust.* As discussed above, before Cenozoic late-orogenic extension the hinterland of the Canadian Cordillera was about 100 km wide and 50 km thick (Fig. 11b). The current crustal thickness underneath the Shuswap MCC is about 33 km, and seismic profiles indicate that for at least 50 km on each sides of the 50 km wide Shuswap MCC, the crust has a similar thickness and a flat Moho (Cook et al. 1988, 1992). Accordingly, the pre-extension crustal section represents about

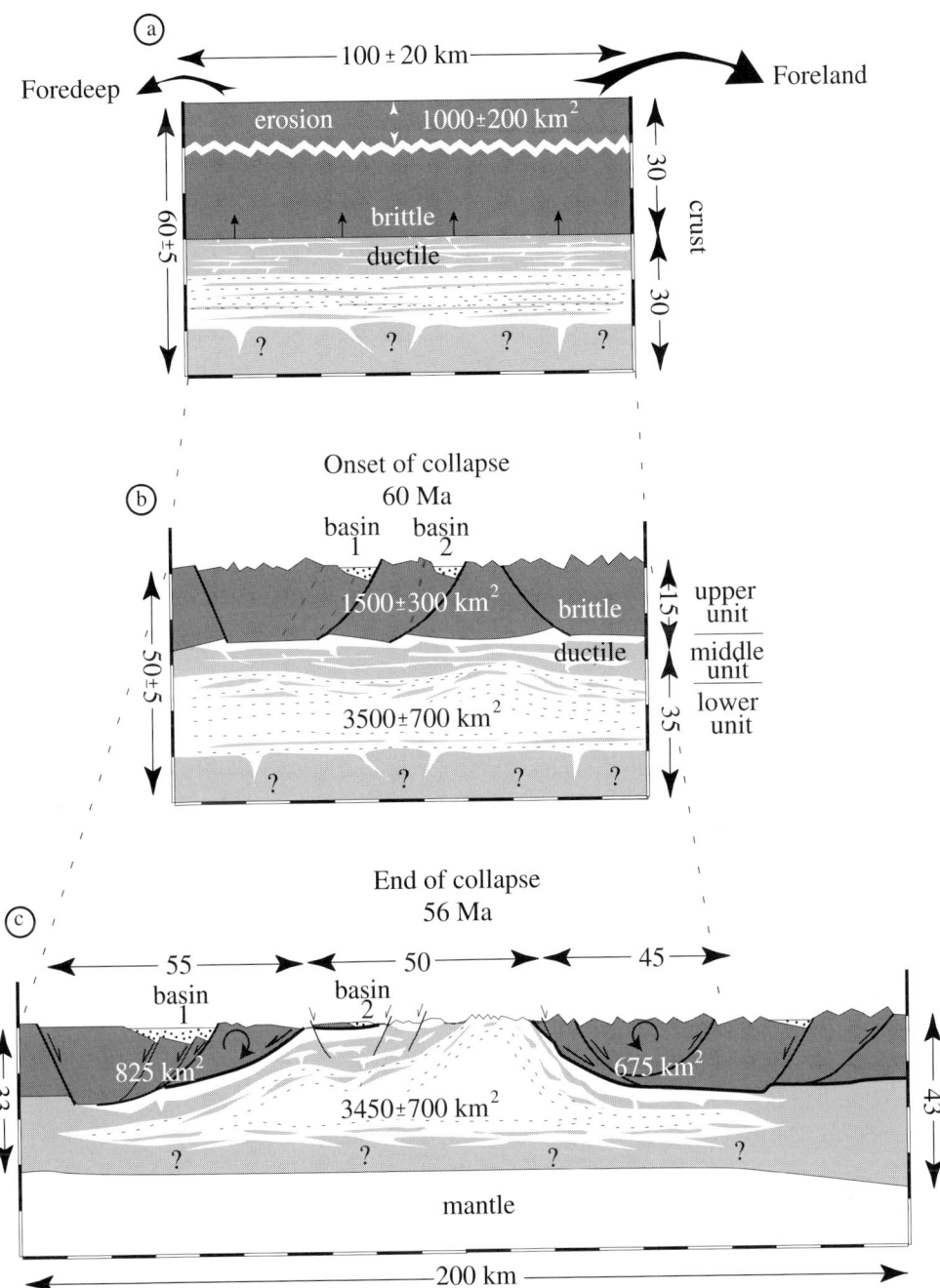

**Fig. 11.** Model of exhumation of the Shuswap Metamorphic Core Complex. The section passes through the Thor–Odin dome and is approximately the same as that in Fig. 5. In the crust, surfaces in dark grey correspond to the brittle layer of the model (upper unit of the geological section), and those in white and light grey correspond to the ductile layer of the model (middle and lower unit of the geological section, plus the lower crust).

500 km² which is possible to balance in two dimensions with the post-extension c.150 × 33 km² zone characterized by a flat-Moho. Furthermore, the simple calculation proposed in Fig. 11 shows that it is possible to balance separately a brittle upper crust and a ductile lower crust if we consider that the brittle upper crust is about 15 km thick; this is approximately the amount of exhumation required at the top of the middle unit of the Shuswap MCC and is also the depth at which the main detachments occur on each side of the complex root in the crust according to the Lithoprobe seismic profiles (Cook *et al.* 1988, 1992). Reconstruction of the section before and after extension shows that the surface areas of upper brittle and lower ductile layers are balanced from one stage to the other, assuming that a 150 km wide belt has been extended which is consistent with the estimates of Johnson & Brown (1996). The kinematic criteria described above suggest that one detachment was activated on each side of the MCC. The western (Okanagan) detachment was responsible for eastward tilting of upper unit blocks. The eastern (Columbia) detachment caused westward tilting of crustal blocks. The structural link observed on the Lithoprobe profile between the detachment and the basal décollement of the fold-and-thrust belt (Brown *et al.* 1992) and early Cenozoic K–Ar ages on clay minerals from gouge collected in the major thrusts of the Rocky Mountain foreland (Covey *et al.* 1994) suggest that part of the lateral sliding of the upper unit was accommodated by reactivation of thrusts in the Rocky Mountain to the east and probably also in the Selkirk fan. Ductile flow of the middle and lower units filled the gap between disrupted upper-crustal segments and accommodates thinning of the crust during collapse. In addition to tectonic denudation and ductile thinning, diapiric rise of the partially molten rocks probably contributed to some extent to the exhumation of the migmatites of the lower unit. However, according to our reconstructions and to the metamorphic histories recorded in the middle and upper units, this mechanism does not appear to play the major role. Therefore, we relate exhumation of the Shuswap MCC mainly to tectonic denudation, the upper brittle unit being disrupted on the scale of the core complex accommodated by ductile flow of the mid- to lower crust.

## French Massif Central

*Timing and amount of exhumation.* In the French Massif Central, a similar sequence of events is suggested by the data presented above. Reconstruction accounting for late-orogenic extension (Fig. 12) shows that the current *c.* 240 km wide hinterland of the Variscan belt was 150–180 km wide before extension (Burg *et al.* 1994). The first exhumation period corresponds to denudation of the upper unit during crustal thickening, which is associated with cooling at *c.* 340 Ma of rocks that were 20–25 km deep at the time of the Barrovian metamorphism (Costa *et al.* 1993; Gardien *et al.* 1997). Metamorphic decompresson implies that about 10 km of material was removed during this period, corresponding to 1500–1800 km² on the section.

The second period of exhumation was associated with the formation of large migmatitic domes during Carboniferous time. The Velay Dome is *c.* 90 km wide and pressures calculated from thermobarometry indicate that the migmatites exhumed in its core were at depths of 13–17 km before exhumation at *c.* 300 Ma (Montel *et al.* 1992), corresponding to the removal of 1300–1700 km² of upper crust in a period of *c.* 10 Ma. An original crustal thickness of 48–52 km is obtained by adding (1) 13–17 km of exhumation during late-orogenic collapse, and (2) 10 km of exhumation during crustal thickening, to the *c.* 25 km of present-day thickness, which is a minimum because the crust here has undergone a few per cent of Paleogene and Neogene extension.

*Erosion.* Large Viséan foreland basins are known on both sides of the Variscan belt (Franke & Engel 1986) but, owing to Mesozoic and Cenozoic tectonics, the preserved record is probably not representative of their original extent and location. Therefore, we cannot derive a meaningful amount of sediments eroded from the growing Variscan belt.

Stephanian basins, coeval with extension of the upper crust during late-orogenic collapse, are deep but of limited extent. Along the cross-section of the Velay Dome, the St Etienne basin is *c.* 5 km wide and 3–4 km deep, which represents at most 15–20 km². To the south of the Velay Dome, the Cévennes basin has a similar size and the Jaujac basin is smaller. These basins together do not account for the amount of material removed from the top of the Velay Dome at the time they formed.

*Tectonic denudation and upwelling of the ductile crust.* The metamorphic record of the Velay Dome migmatites indicates that they were exhumed from a depth of 15–20 km. The dome is *c.* 90 km wide, filling a space between two large segments of the upper unit. Reconstruction of the Massif Central section through the Velay Dome before and after extension

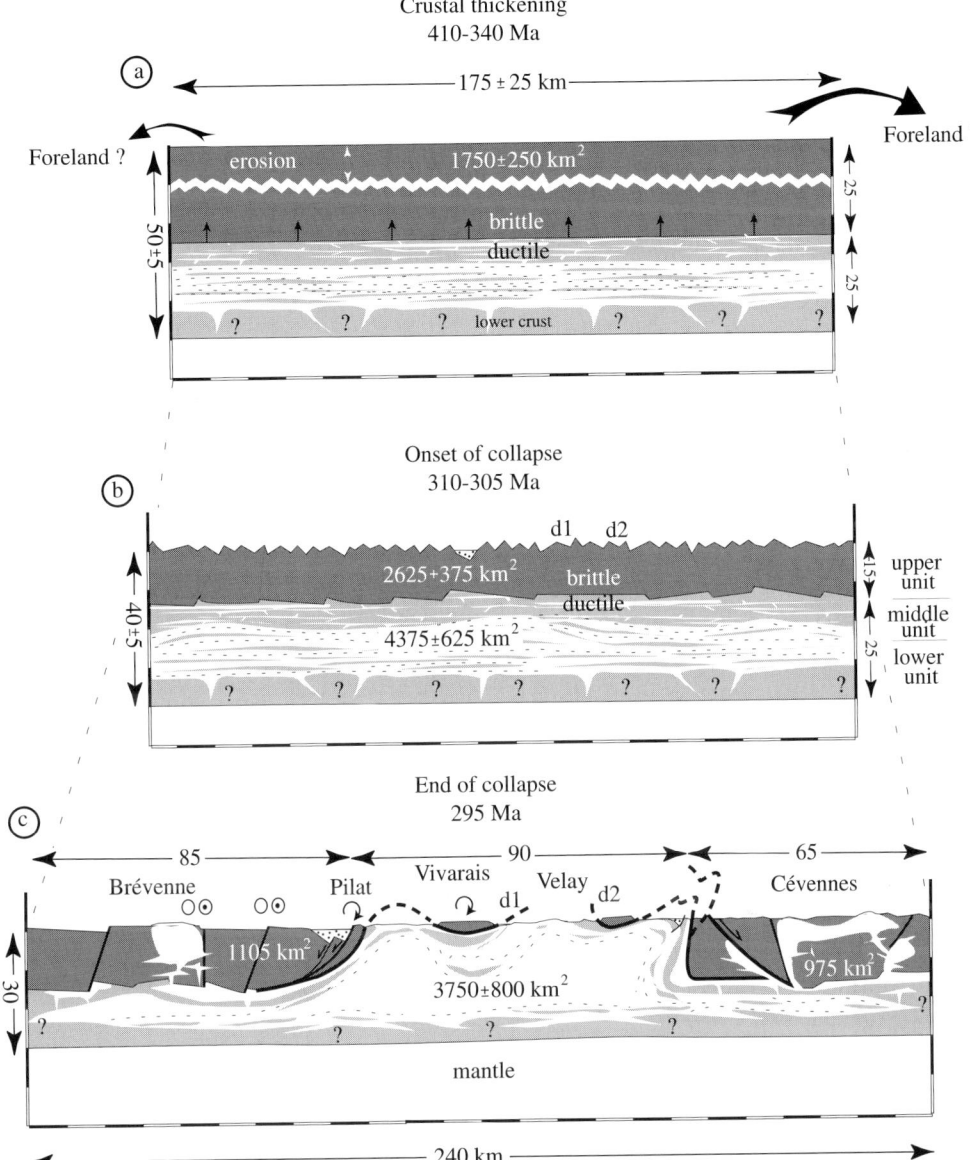

**Fig. 12.** Model of exhumation of the Velay Dome. The section is the same as the one shown in Fig. 7. In the crust, surfaces in dark gray correspond to the brittle layer of the model (upper unit of the geological section), and those in white and light grey corresponds to the ductile layer of the model (middle and lower unit of the geological section, plus the lower crust).

(Fig. 12b and c) shows that surfaces of the brittle upper crust and ductile lower crust are almost balanced if we consider a width of 240 km, which is consistent with the amount of extension estimated along other sections of the eastern French Massif Central (Burg et al. 1994). Therefore, it is possible to attribute exhumation of the migmatites mainly to tectonic denudation during late-orogenic collapse accommodated by ductile flow of the mid- to lower crust.

## Discussion

### Thermal and mechanical implications

In summary, intra-continental convergence was largely accommodated by crustal thickening during several tens of million years and culminated in high-temperature ($T > 750°C$) metamorphism in both the Canadian Cordillera and the French Variscides. Late-orogenic collapse lasted a few million years and coincided with the presence of a partially molten middle crust now seen in the metamorphic domes. Brittle extension of the upper unit was combined with ductile flow of the mid- and lower crust, which resulted in crustal-scale boudinage ductile thinning and upwelling of high-temperature migmatitic terranes forming domes between the boudins of upper crust. The three-dimensional dome shape of the migmatitic terranes also suggest that upwelling was controlled in part by the development of gravitational instablities related to inverted density gradients.

The orogenic evolution summarized above suggests a genetic link between the formation of a layer of partially molten rocks and late-orogenic collapse. The mode of extension leading to exhumation of migmatites is probably controlled by the rheology of the crust, which, in turn, is probably weakened by partial melting. In the Canadian Cordillera and the French Variscides U–Pb ages of leucogranite intrusions indicate that magmatism occurred over a period of about 10 Ma and pre-dated late-orogenic collapse and exhumation dated by crystallization of the migmatites and thermochronology. Therefore crustal anatexis was a potential trigger to gravitational collapse and exhumation of the deep crust. Note that in the active and still overthickened southern Tibetan plateau, the INDEPTH traverse across southern Tibet is characterized by 'bright spots' at depths of 15–20 km interpreted to represent magmatic bodies (Nelson *et al.* 1996). Below the bright spots the crust is marked by high electrical conductivity and low velocity of seismic waves, features interpreted as a partially molten layer (Chen *et al.* 1996; Nelson *et al.* 1996).

Two major heat sources can produce temperatures high enough for partial melting at 15–20 km depth in the crust: (1) asthenospheric upwelling and/or intrusion of mantle-derived magmas in the lower crust (Wickham & Oxburgh 1985); and (2) after a characteristic time of 20–30 Ma, internal heat production associated with radiogenic decay of elements such as U or Th concentrated in metapelites (England & Thompson 1984; Chamberlain & Sonder 1990; Huerta *et al.* 1996). In the Velay and Shuswap examples, collapse was c. 60–80 Ma younger than the onset of collision, which provides the necessary time gap for thermal maturation of the thickened crust, without an external heat source. Moreover, the partially molten layer described above was mainly derived from fertile metapelites, accreted during continental collision, suggesting that their distribution within a thickened crust may control the localization and geometry of partially molten levels (Patiño Douce *et al.* 1990). Experimental and theoretical work shows that melting drastically weakens rocks (Arzi 1978; Van der Molen & Paterson 1979). The effect of a partially molten layer on the bulk strength of the crust is not easy to quantify. The strength of the lithosphere is usually approximated as its integral over depth (Brace & Kohlstedt 1980; Kusznir & Park 1987; Sonder *et al.* 1987). On the other hand, if we consider the strength of the crust alone, the generation of a weak partially molten layer at middle-crustal depth within a thickened crust could potentially decouple the upper crust from the underlying stronger mantle. In this case, the strength of the crust is controlled by its strongest layer, the upper brittle crust and, when the upper crust yield strength is reached, the partially molten middle crust may flow laterally to accommodate the excess of potential energy accumulated during crustal thickening (Artyushkov 1973).

### Models of metamorphic core complex formation

Several models have been proposed to explain the exhumation of high-grade rocks below low-angle detachments. The 'simple shear model' is based on the inference that extension is accommodated along one lithospheric shear zone (Wernicke 1981; Lister & Davis 1989). This model has been successfully applied to explain exhumation of mantle rocks along extremely extended continental margins (Boillot *et al.* 1987). However, it implies that deformation is localized within one major shear zone separating two essentially rigid blocks, which is not consistent with the field data described in exhumed high-grade rocks in the hinterland of collapsed orogens such as the one presented in this paper. Consequently, we contend that models with a brittle crust overlying a ductile crust (Gans 1987; Block & Royden 1990; Buck 1991; Brun *et al.* 1994) are more relevant to examples of thickened and thermally weakened crust. In such models, brittle extension of the upper layer is accommodated by pervasive ductile flow in the

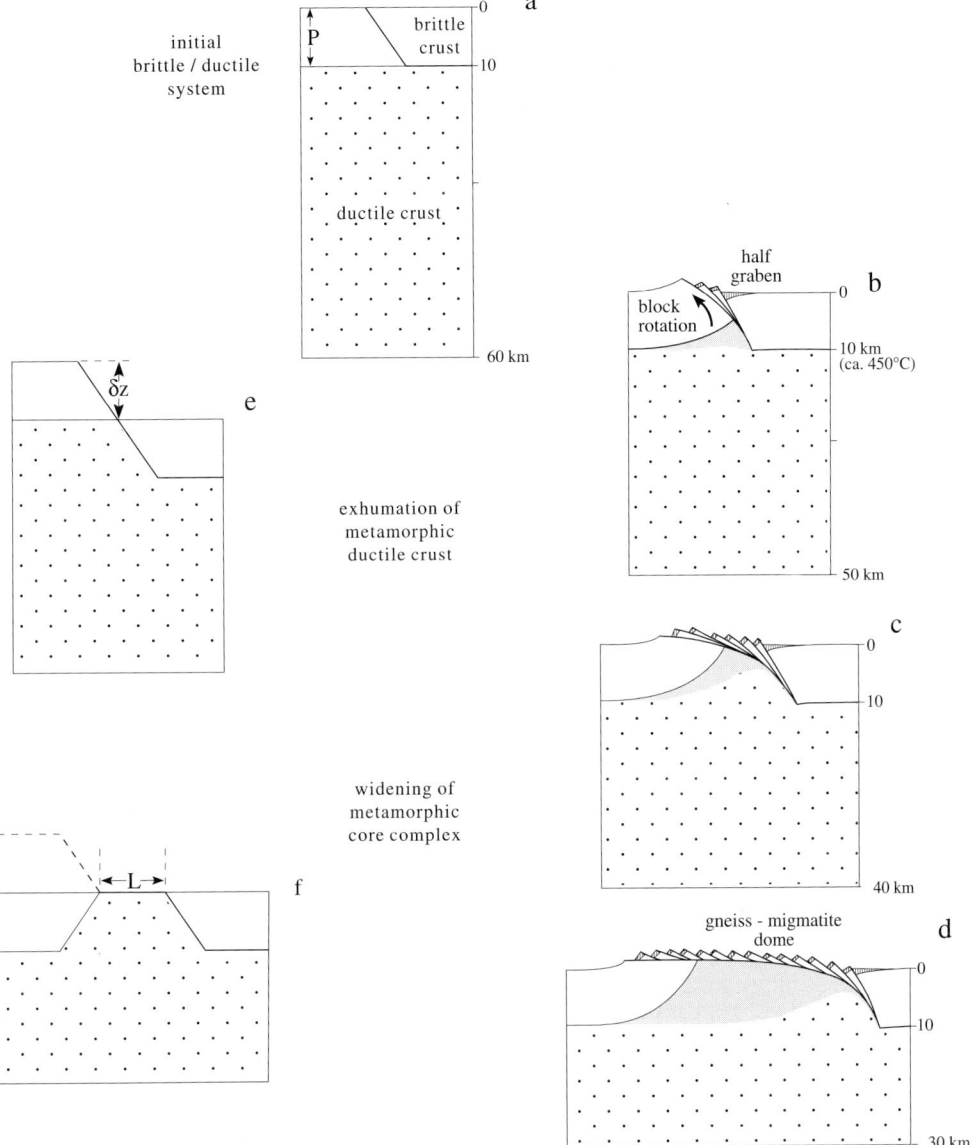

**Fig. 13.** Metamorphic core complex formation (modified after Brun *et al.* 1994). (**a**) Initial stage after crustal thickening and thermal relaxation at the onset of late-orogenic collapse. On the right side the diagrams (**b**), (**c**), and (**d**) illustrate progressive thinning of the crust with activation of normal faults affecting the brittle upper crust and homogeneous flow of the ductile crust. On the left side, diagrams (**e**) and (**f**) illustrate the amount of exhumation achieved by removal of the brittle upper crust ($\delta z$) and the amount of widening of the metamorphic core complex ($L$).

lower layer (Fig. 13). The style of extension of the upper brittle crust (symmetrical v. asymmetrical) controls the final geometry of the metamorphic core complex.

In the asymmetrical case (Buck 1991), extension starts with the activation of a single normal fault (Fig. 13a and b). Due to subsequent isostatic rebound of the ductile crust, the normal fault is rotated to shallower dips and is then abandoned. Further extension of the upper unit is accommodated by activation of new normal faults, which are in turn rotated and inactivated, slicing up the upper crust (Fig. 13c and d). During widening of the core complex, the inactive trailing part of the detachment is passively rotated into a subhorizontal attitude, thus explaining flat-lying thin rafts of upper units above the detachment (Fig. 13d). Once all the upper crust has been tectonically removed from above the lower crust, further extension results in widening the area of exhumed lower crust (Burg *et al.* 1990, 1994; Brun *et al.* 1994). New normal faults may migrate forward, dissecting the half-graben formed in the hanging wall. This is consistent with the presence of tilted early Cenozoic sediments resting on the Shuswap MCC detachment (Fig. 12c). Alternatively, symmetrical boudinage of the upper crust results in more symmetrical and spatially periodical upwelling of the ductile lower crust.

According to this model, the maximum amount of exhumation that can be achieved by tectonic denudation is the removal of the entire upper crust from above the ductile unit (Fig. 13e and f). Therefore, the maximum tectonic denudation is the depth of the brittle–ductile transition when extension began, which is equated with the 450°C isotherm within the continental crust (Brace & Kohlstedt, 1980). The depth of this isotherm depends on the geothermal gradient. Crustal thickening causes vertical stretching of the isotherms, which depresses the brittle–ductile transition. However, thermal relaxation and heat generation by radioactive decay should overcome this effect after a characteristic time of *c.*20–30 Ma, raising back the brittle–ductile transition to shallow levels (England & Thompson 1984). In present-day situations, it is located at the depth where seismic activity ceases (<20 km in general and as low as 10 km in southern Tibet). In old orogenic systems, the palaeo brittle–ductile transition is defined by combining structural analysis, thermochronology and thermobarometry. Following our interpretation of the seismic layering in the Cordilleran crust, the brittle-ductile transition correlates with the top of the zone of dense seismic reflections. These analyses yield a 10–15 km deep brittle–ductile transition in thermally relaxed orogens.

Both the Velay Dome and the Shuswap MCC are asymmetrical and are bounded by kinematically conjugate detachments. Accordingly, extension in these examples was intermediate between the symmetrical and asymmetrical end-members presented above. In both cases, the low-angle detachment corresponds to a zone of mechanical decoupling between the brittle upper and the ductile middle crust. The brittle upper layer correlates with the upper unit, and the ductile layer comprises both the middle unit and the migmatitic lower unit.

## Conclusion

The comparison of migmatitic core complexes formed in the crystalline axes of the collapsed Canadian Cordillera and French Variscides shows a 15–20 km exhumation of high-temperature migmatites. The efficiency of various exhumation mechanisms active at different structural levels is assessed based on a compilation of thermobarometric, gechronologic, sedimentologic and structural data. This compilation serves as a basis to constrain simple crustal-scale two-dimensional reconstruction of the tectonic evolution of these two orogens, which suggests the following points:

- Erosional exhumation was mainly achieved during crustal thickening that also produced low geothermal gradients. Perhaps more than 10 km of erosion would account for decompression of units that have recorded Barrovian metamorphic conditions. This erosion-related denudation supplied sediment to foreland basins.
- After a few tens of million years, thermal maturation of the thick crust generated a layer of partially molten rocks at 15–20 km depth. The weakened crust was rheologically characterised by a relatively shallow brittle-ductile transition. The mid- to lower molten crust may have allowed decoupling from the underlying lithosphere and thus triggered late-orogenic collapse even though lithospheric plates were still converging.
- Collapse was accommodated by normal faulting and extension of the upper brittle crust, while the buoyant and low-viscosity partially molten layer flowed to fill gaps between large upper crustal boudins, forming elongate domes. This tectonic process accounts for *c.*75–80 % of the exhumation of high-temperature migmatites exposed in metamorphic core complexes.
- In comparison, erosion recorded in syn-extensional fault-bounded basins and additional sedimentation in foreland basins does not play a significant role during the late orogenic exhumation of HT–LP migmatites in the hinterland of collapsed orogens.

O.V. gratefully acknowledges summer support and a Gruner–Emmons fellowship from the Department of Geology and Geophysics, University of Minnesota, Sigma Xi Grant-in-Aid of Research, and the Geological Society of America Research Grants; J.P.B. is supported by the ETH. This work was supported by

National Science Foundation grant EAR 9509750. C.T. gratefully acknowledges support from the Bush sabbatical grant, University of Minnesota, as well as support from ETH. This paper benefited from constructive reviews by R. A. Jamieson, J. M. Bartley, and R. A. Price, although the reviewers might not necessarily share the views presented in this paper.

## References

ALLEMAND, P., LARDEAUX, J.-M., DROMART, G. & ADER, M. 1997. Extension tardi-orogénique et formation des bassins intracontinentaux: le bassin stephanien des Cévennes. *Geodinamica Acta*, **10**(2), 60–70.

ANDERSEN, T. B., JAMTVEIT, B., DEWEY, J. F. & SWENSSON E. 1991. Subduction and eduction of continental crust: major mechanisms during continent–continent collision and orogenic extensional collapse, a model based on the south Norwegian Caledonides. *Terra Nova*, **3**(3), 303–310.

ARCHIBALD, D. A., GLOVER, J. K., PRICE, R. A., FARRAR, E. & CARMICHAEL, D. M. 1983. Geochronology and tectonic implications of magmatism and metamorphism, southern Kootenay arc and neighboring regions, southeastern British Columbia. Part I: Jurassic to mid-Cretaceous. *Canadian Journal of Earth Sciences*, **20**, 1891–1913.

ARMSTRONG, R. L. 1982. Cordilleran metamorphic core complexes – from Arizona to southern Canada. *Annual Review of Earth and Planetary Science*, **10**, 129–154.

—— & WARD, P. 1991. Evolving geographic patterns of Cenozoic magmatism in the North American Cordillera: the temporal and spatial association of magmatism and metamorphic core complexes. *Journal of Geophysical Research*, **96**(B8), 13201–13224.

——, PARRISH, R. R., VAN DER HEYDEN, P., SCOTT, K., RUNKLE, D. & BROWN, R. L. 1991. Early Proterozoic basement exposures in the southern Canadian Cordillera: core gneiss of Frenchman Cap, Unit of Grand Forks Gneiss and the Vasseaux Formation. *Canadian Journal of Earth Sciences*, **28**, 1169–1201.

ARNAUD, F. A. & BURG, J.-P. 1993. Microstructures des mylonites schisteuses: cartographie des chevauchements varisques dans les Cévennes et détermination de leur cinématique. *Comptes Rendus de l'Académie des Sciences*, **317**(II), 1441–1447.

ARTYUSHKOV, E. V. 1973. Stresses in the lithosphere caused by crustal thickness inhomogeneities. *Journal of Geophysical Research*, **78**, 7675–7708.

ARZI, A. A. 1978. Critical phenomena in the rheology of partially melted rocks. *Tectonophysics*, **44**, 173–184.

BEAUMONT, C. 1981. Foreland basins. *Geophysical Journal of the Royal Astronomical Society*, **55**, 471–498.

BECQ-GIRAUDON, J.-F. & VAN DEN DRIESSCHE, J. 1994. Dépôts périglaciaires dans le Stéphano-Autunien du Massif Central: témoin de l'effondrement gravitaire d'un haut plateau hercynien. *Comptes Rendus de l'Académie des Sciences*, **318**(II), 675–682.

BLOCK, L. & ROYDEN, L. H. 1990. Core complex geometries and regional scale flow in the lower crust. *Tectonics*, **9**(4), 557–567.

BOILLOT, G., RECQ, M., WINTERER, E. L., ET AL. 1987. Tectonic denudation of the upper mantle along passive margins: a model based on drilling results (ODP 103, western Galicia margin, Spain). *Tectonophysics*, **132**, 334–342.

BOIS, C. & ECORS SCIENTIFIC PARTY 1990. Major geodynamic processes studied from the ECORS deep seismic profiles in France and adjacent areas. *Tectonophysics*, **173**, 397–410.

BRACE, W. F. & KOHLSTEDT, D. T. 1980. Limits on lithospheric stress imposed by laboratory experiments. *Journal of Geophysical Research*, **85**, 6248–6252.

BROWN, R. L., CARR, S. D., JOHNSON, B. J., COLEMAN, V. J., COOK, F. A. & VARSEK, J. L. 1992. The Monashee décollement of the southern Canadian Cordillera: a crustal scale shear zone linking the Rocky Mountain Foreland belt to lower crust beneath accreted terranes. *In*: MCCLAY, K. R. (eds.), *Thrust Tectonics*. Chapman & Hall, London 353–364.

—— & JOURNEAY, J. M. 1987. Tectonic denudation of the Shuswap metamorphic terrane of Southeastern British Columbia. *Geology*, **15**, 142–146.

——, ——, LANE, L. S., MURPHY, D. C. & REES, C. J. 1986. Obduction, backfolding and piggyback thursting in the metamorphic hinterland of the Southeastern Canadian Cordillera. *Journal of Structural Geology*, **8**(3/4), 255–268.

BRUN, J.-P., SOKOUTIS, D. & VAN DEN DRIESSCHE, J. 1994. Analogue modeling of detachment fault systems and core complexes. *Geology*, **22**(4), 319–322.

BUCK, W. R. 1991. Modes of continental lithospheric extension. *Journal of Geophysical Research*, **96**(B12), 20161–20178.

BURG, J.-P. & MATTE, P. 1978. A cross section through the French Massif Central and the scope of its Variscan geodynamic evolution. *Zeitschrift der Deutschen Geologischen Gessellschaft*, **129**, 429–460.

—— & VANDERHAEGHE, O. 1993. Structures and way-up critera in migmatites, with application to the Velay dome (French Massif Central). *Journal of Structural Geology*, **15**(11), 1293–1301.

——, BRUN, J.-P. & VAN DEN DRIESSCHE, J. 1990. Le Sillon Houiller du Massif Central français: faille de transfert pendant l'amincissement crustal de la chaîne varisque? *Comptes Rendus de l'Académie des Sciences*, **311**(II), 147–152.

——, VAN DEN DRIESSCHE, J. & BRUN, J.-P. 1994. Syn- to post-thickening extension in the Variscan Belt of Western Europe: mode and structural consequences. *Géologie de la France*, **3**, 33–51.

CAEN-VACHETTE, M., COUTURIÉ, J.-P. & DIDIER, J. 1981. Age Westphalien du granite de Rocles (Cévennes, Massif Central français). *Comptes Rendus de l'Académie des Sciences*, **293**, 957–960.

——, GAY, M., PETERLONGO, J.-M., PITIOT, P. & VITEL, G. 1984. Age radiométrique du granite syntectonique du gouffre d'Enfer et du metamorphisme hercynien dans la série de basse pression du Pilat (Massif Central français). *Comptes Rendus de l'Académie des Sciences*, **299**, 1201–1204.

CARR, S. D. 1991. U–Pb zircon and titanite ages of three Mesozoic igneous rocks south of the Thor–Odin–Pinnacles area, southern Omineca Belt, British Columbia. *Canadian Journal of Earth Sciences*, **28**, 1877–1882.

—— 1992. Tectonic setting and U–Pb geochronology of the early Tertiary Ladybird leucogranite suite, Thor–Odin–Pinnacles area, southern Omineca belt, British Columbia. *Tectonics*, **11**(2), 258–278.

CHAMBERLAIN, C. P. & SONDER, L. J. 1990. Heat-producing elements and the thermal and baric patterns of metamorphic belts. *Science*, **250**, 763–769.

CHEMENDA,, A. I., MATTAUER M., MALAVIEILLE, J. & BOKUN, A. N. 1995. A mechanism for syn-collisional rock exhumation and associated normal faulting: results from physical modelling. *Earth and Planetary Science Letters*, **132**, 225–232.

CHEN, L., BOOKER, J. R., JONES, A. G., WU, N., UNSWORTH, M. J., WEI, W. & TAN, T. 1996. Electrically conductive crust in southern Tibet from INDEPTH magnetotelluric surveying. *Science*, **274**, 1694–1697.

COLPRON, M., PRICE, R. A., ARCHIBALD, D. A. & CARMICHAEL, D. M. 1996. Middle Jurassic exhumation along the western flank of the Selkirk fan structure: thermobarometric and thermochronometric constraints from the Illecillewaet synclinorium, southeastern British Columbia. *Geological Society of America Bulletin*, **108**(11), 1372–1392.

CONEY, P. J. & HARMS, T. A. 1984. Cordilleran metamorphic core complexes: Cenozoic extensional relics of Mesozoic compression. *Geology*, **12**, 550–554.

COOK, F. A., GREEN, A. G., SIMONY, P. S. ET AL. 1988. Lithoprobe seismic reflection structure of the Southern Canadian Cordillera: initial results. *Tectonics*, **7**, 157–180.

——, VARSEK, J. L., CLOWES, R. M. ET AL. 1992. Lithoprobe crustal reflection cross section of the southern Canadian Cordillera, 1. Foreland and Thust and Fold Belt to Fraser River Fault. *Tectonics*, **11**, 12–35.

COSTA, S. 1991–1992. East–west diachronism of the collisional stage in the French Massif Central: implications for the European Variscan Orogen. *Geodinamica Acta*, **5**(1–2), 51–68.

—— & REY, P. 1995. Lower crustal rejuvenation and crustal growth during post-thickening collapse: insights from a crustal cross section through a metamorphic core complex. *Geology*, **23**, 905–908.

——, MALUSKI, H. & LARDEAUX, J.-M. 1993. $^{40}$Ar–$^{39}$Ar chronology of Variscan tectono-metamorphic events in an exhumed crustal nappe: the Monts du Lyonnais complex (Massif Central, France). *Chemical Geology*, **105**, 339–359.

COVEY, M. C., VROLIJK, P. J. & PEVEAR, D. R. 1994. Direct dating of fault movement in the Rocky Mountain front ranges of southern Alberta.

*Geological Society of America, Annual Meeting, Seattle. Abstracts with Programs*, A-467.

DEWEY, J. F. 1988. Extensional collapse of orogens. *Tectonics*, **7**(6), 1123–1139.

—— & BIRD, J. M. 1970. Mountain belts and the new global tectonics. *Journal of Geophysical Research*, **75**, 2625–2647.

—— & BURKE, K. 1973. Tibetan, Variscan and Precambrian basement reactivation: products of continental collision. *Journal of Geology*, **81**, 683–692.

DOWNES, H. & LEYRELOUP, A. 1986. Granulitic xenoliths from the French Massif Central: petrology, Sr and Nd isotope systematics and model ages estimates. *In*: DAWSON, J. B. (ed.) *The Nature of the Lower Continental Crust*. Geological Society, London, Special Publications, **24**, 319–330.

DUNCAN, I. J. 1984. Structural evolution of the Thor–Odin gneiss dome. *Tectonophysics*, **101**, 87–130.

DUPRAZ, J. & DIDIER, J. 1988. Le complexe anatectique du Velay (Massif Central francais): structure d'ensemble et évolution géologique. *Bulletin du, Série 'Géologie de la France' Bureau de Recherches Géologiques et Minières* **4**, 73–88.

DUTHOU, J.-L., PIBOULE, M., GAY, M. & DUFOUR, E. 1981. Datations radiométriques Rb–Sr sur les ortho-granulites des Monts du Lyonnais (Massif Central français). *Comptes Rendus de l'Académie des Sciences*, **292**(II), 749–7526.

ENGLAND, P. C. & THOMPSON, A. B. 1984. Pressure–Temperature–time paths of regional metamorphism I. Heat transfer during the evolution of regions of thickened continental crust. *Journal of Petrology*, **25**(4), 894–928.

ERNST, W. G. 1973. Blueschists metamorphism and *P–T* regimes in active subduction zones. *Tectonophysics*, **17**, 255–272.

FRANKE, W. & ENGEL, W. 1986. Synorogenic sedimentation in the Variscan Belt of Europe. *Bulletin de la Société Géologique de France*, **8/2**(1), 25–33.

GANS, P. B. 1987. An open system, two-layer crustal stretching model for the eastern Great Basin. *Tectonics*, **6**(1), 1–12.

GARDIEN, V. 1990. Reliques de grenat et de staurotide dans la série métamorphique de basse pression du Mont Pilat (Massif Central français): témoins d'une évolution tectonométamorphique polyphasée. *Comptes Rendus de l'Académie des Sciences*, **310**(II), 233–240.

——, LARDEAUX, J.-M., LEDRU, P., ALLEMAND, P. & GUILLOT, S. 1997. Metamorphism during late-orogenic extension: insights from the French Variscan belt. *Bulletin de la Société Géologique de France*, **168**(3), 271–286.

——, —— & MISSERI, M. 1988. Les péridotites des Monts du Lyonnais (M.C.F.): témoins privilégiés d'une subduction de lithosphère paléozoïque. *Comptes Rendus de l'Académie des Sciences*, **307**, 1967–1972.

HUERTA, A. D., ROYDEN, L. H. & HODGES, K. V. 1996. The interdependence of deformational and thermal processes in mountain belts. *Science*, **273**, 637–639.

IRVING, E. & WYNNE, P. J. 1991. Paleomagnetism:

review and tectonic implications. *In*: GABRIELSE, H. & YORATH, C. J. (eds) *Geology of the Cordilleran Orogen in Canada*. Geological Society of America Geology of North America, **G2** and Geological Survey of Canada, **4**, 61–86.

JOHNSON, B. J. & BROWN, R. L. 1996. Crustal structure and early Tertiary extensional tectonics of the Omineca belt at 51°N latitude, southern Canadian Cordillera. *Canadian Journal of Earth Sciences*, **33**, 1596–1611.

JOURNEAY, J. M. 1986. *Stratigraphy, internal strain and tectono-metamorphic evolution of northern Frenchman Cap dome: an exhumed duplex structure, Omineca hinterland, S.E. Canada, Cordillera*. PhD thesis, Queen's University, Kingston, Ont.

KUSZNIR, N. J. & PARK, R. G. 1987. The extensional strength of the continental lithosphere: its dependence on geothermal gradient, crustal composition and thickness. *In*: COWARD, M. P., DEWEY, J. F. & HANCOCK, P. L. (eds) *Continental Extensional Tectonics*. Geological Society, London, Special Publications, **28**, 35–52.

LAGARDE, J.-L., DALLAIN, C., LEDRU, P. & COURRIOUX, G. 1994. Strain pattern within the Variscan granite dome of Velay, French Massif Central. *Journal of Structural Geology*, **16**, 839–852.

LANE, L. S. 1984. Brittle deformation in the Columbia River fault zone near Revelstoke, southeastern British Columbia. *Canadian Journal of Earth Sciences*, **21**, 584–598.

LARDEAUX, J. M., REYNARD, B. & DUFOUR, E. 1989. Granulites à kornérupine et décompression post-orogénique des Monts du Lyonnais. *Comptes Rendus de l'Académie des Sciences*, **308**, 1443–1449.

LEDRU, P., LARDEAUX, J. M., SANTALLIER, D., ET AL. 1989. Où sont les nappes dans le Massif central français? *Bulletin de la Société Géologique de France*, **8**(3), 605–618.

LEYRELOUP, A., DUPUY, C. & ANDRIAMBOLONA, R. 1977. Catazonal xenoliths in French Neogene volcanic rocks, constitution of the lower crust. *Contributions to Mineralogy and Petrology*, **62**, 283–300.

LISTER, G. S. & DAVIS, G. A. 1989. The origin of metamorphic core complexes and detachment faults formed during Tertiary continental extension in the northern Colorado River region, U.S.A. *Journal of Structural Geology*, **11**(1/2), 65–94.

MALAVIEILLE, J., GUIHOT, P., COSTA, S., LARDEAUX, J.-M. & GARDIEN, V. 1990. Collapse of a thickened Variscan crust in the French Massif Central: Mont-Pilat extensional shear zone and Saint-Etienne Upper Carboniferous basin. *Tectonophysics*, **177**, 139–149.

MALUSKI, H., COSTA, S. & ECHTLER, H. 1991. Late Variscan tectonic evolution by thinning of earlier thickened crust. An $^{40}Ar-^{39}Ar$ study of the Montagne Noire, southern Massif Central, France. *Lithos*, **26**, 287–304.

MATHEWS, W. H. 1981. Early Cenozoic resetting of potassium–argon dates and geothermal history of north Okanagan area, British Columbia.

*Canadian Journal of Earth Sciences*, **18**, 1310–1319.

MÉNARD, G. & MOLNAR, P. 1988. Collapse of a Hercynian Tibetan Plateau into a Late Palaeozoic European Basin and Range Province. *Nature*, **334**, 235–237.

MILLER, C. F. & BRADFISH, L. J. 1980. An inner Cordilleran belt of muscovite-bearing plutons. *Geology*, **8**, 412–416.

MIYASHIRO, A. 1973. Paired and unpaired metamorphic belts. *Tectonophysics*, **17**, 241–254.

MONGER, J. W. H., PRICE, R. A. & TEMPELMAN-KLUIT, D. J. 1982. Tectonic accretion and plutonic welts in the Canadian Cordillera. *Geology*, **10**, 70–75.

MONTEL, J. M., MARIGNAC, C., BARBEY, P. & PICHAVANT, M. 1992. Thermobarometry and granite genesis: the Hercynian low-$P$ high-$T$ Velay anatectic dome (French Massif Central). *Journal of Metamorphic Geology*, **10**(1), 1–15.

MOUGEOT, R., RESPAUT, J.-P., LEDRU, P. & MARIGNAC, C. 1997. U–Pb chronology on accessory minerals of the Velay anatectic dome (French Massif Central). *European Journal of Mineralogy*, **9**, 141–156.

NELSON, K. D., WENJIN, Z., BROWN, L. D., ET AL. 1996. Partially molten middle crust beneath southern Tibet: synthesis of project INDEPTH results. *Science*, **274**, 1684–1688.

NYMAN, M. W., PATTISON, D. R. M. & GHENT, E. D. 1995. Melt extraction during formation of K-feldspar–sillimanite migmatites, west of Revelstoke, British Columbia. *Journal of Petrology*, **36**(2), 351–372.

PARKINSON, D. L. 1991. Age and isotopic character of Early Proterozoic basement gneisses in the southern Monashee Complex, southeastern British Columbia. *Canadian Journal of Earth Sciences*, **28**, 1159–1168.

—— 1992. *Age and tectonic evolution of the southern Monashee complex, southeastern British Columbia: a window into the deep crust*. PhD thesis, University of California at Santa Barbara.

PARRISH, R. R. 1995. Thermal evolution of the southeastern Cordillera. *Canadian Journal of Earth Sciences*, **32**, 1618–1642.

—— & ARMSTRONG, R. L. 1987. The *ca.* 162 Ma Galena Bay stock and its relationship to the Columbia River fault zone, southeast British Columbia. *In: Radiogenic age and isotopic studies: Report 1.* Geological Survey of Canada, Papers, **87.2**, 25–32.

——, CARR, S. D. & PARKINSON, D. L. 1988. Eocene extensional tectonics and geochronology of the Southern Omineca belt, British Columbia and Washington. *Tectonics*, **7**(2), 181–212.

PATIÑO DOUCE, A. E., HUMPHREYS, E. D. & JOHNSTON, A. D. 1990. Anatexis and metamorphism in tectonically thickened continental crust exemplified by the Sevier hinterland, western North America. *Earth and Planetary Science Letters*, **97**(3/4), 290–315.

PIN, C. & LANCELOT, J. 1982. U/Pb dating of an early Paleozoic bimodal magmatism in the French Massif Central and of its further metamorphic evolution. *Contributions to Mineralogy and Petrology*, **79**, 1–12.

PLATT, J. P. 1993. Exhumation of high-pressure rocks: a review of concepts and processes. *Terra Nova*, **5**(2), 119–133.

PRICE, R. A. 1973. Large scale gravitational flow of supra-crustal rocks, southern Canadian Rockies. *In*: DE JONG, K. A. & SCHOLTEN, R. (eds) *Gravity and tectonics*. Wiley, New York, 491–502.

—— 1986. The Canadian Cordillera: Thrust faulting, tectonic wedging, and delamination of the lithosphere. *Journal of Structural Geology*, **8**(3/4), 238–254.

—— & MOUNTJOY, E. W. 1970. Geologic structures of the Canadian Rocky Mountains between the Bow and Athabasca Rivers – A Progress Report. *In*: WHEELER, J. D. (ed.) *Structure of the Southern Canadian Cordillera*. Geological Association of Canada, Special Papers, **6**, 7–25.

READ, P. B. 1980. Stratigraphy and structure: Thor–Odin to Frenchman Cap domes, Vernon east-half southern British Columbia. *Current Research, Part A*, **80**(1A), 19–25.

—— & BROWN, R. L. 1981. Columbia River fault zone: southeastern margin of the Shuswap and Monashee complexes, southern British Columbia. *Canadian Journal of Earth Sciences*, **18**, 1127–1145.

REESOR, J. E. & MOORE, J. M. J. 1971. *Thor–Odin Dome. Shuswap Metamorphic Complex, British Columbia*. Geological Survey of Canada Bulletin, **195**,

REY, P. 1993. Seismic and tectonometamorphic characters of the lower continental crust in Phanerozoic areas: a consequence of post-thickening extension. *Tectonics*, **12**, 580–590.

SCAMMELL, R. J. & BROWN, R. L. 1990. Cover gneisses of the Monashee terrane: a record of synsedimentary rifting in the North American Cordillera. *Canadian Journal of Earth Sciences*, **27**, 712–726.

SEVIGNY, J. H., PARRISH, R. R., DONELCIK, R. A. & GHENT, E. D. 1990. Northern Monashee mountains, Omineca crystalline Belt, British Columbia: Timing of metamorphism, anatexis, and tectonic denudation. *Geology*, **18**, 103–106.

——, PARRISH, R. R. & GHENT, E. D. 1989. Petrogenesis of peraluminous granites, Monashee Moutains, southeastern Canadian Cordillera. *Journal of Petrology*, **30**(3), 557–581.

SIMONY, P. S., GHENT, E. D., CRAW, D., MITCHELL, W. & ROBBINS, D. B. 1980. Structural and metamorphic evolution of the northeast flank of the Shuswap complex in southern Canoe River area, British Columbia. *In*: CRITTENDEN, M. D., CONEY, P. J. & DAVIS, G. H. (eds) *Cordilleran Metamorphic Core Complexes*. Geological Society of America, Memoir **153**, 445–462.

SONDER, L. J., ENGLAND, P. C., WERNICKE, B. P. & CHRISTIANSEN, R. L. 1987. A physical model for Cenozoic extension of western North America. *In*: COWARD, M. P., DEWEY, J. F. & HANCOCK, P. L. (eds) *Continental Extensional Tectonics*. Geological Society, London, Special Publication, **28**, 187–201.

SWEENEY, J. F., STEPHENSON, R. A., CURRIE, R. G. & DE LAURIER, J. M. 1991. Part C. Crustal geophysics. *In*: GABRIELSE, H. & YORATH, C. J. (eds) *Geology of the Cordilleran Orogen in Canada*. Geological Society of America Geology of North America, **G2** and Geological Survey of Canada, **4**, 39–59.

TEMPELMAN-KLUIT, D. & PARKINSON, D. 1986. Extension across the Eocene Okanagan crustal shear in southern British Colombia. *Geology*, **14**, 318–321.

VAN DER MOLEN, I. & PATERSON, M. S. 1979. Experimental deformation of partially-melted granite. *Contributions to Mineralogy and Petrology*, **70**, 299–318.

VANDERHAEGHE, O. 1997. Role of partial melting during late orogenic collapse. PhD thesis, University of Minnesota.

—— & TEYSSIER, C. 1997. Formation of the Shuswap metamorphic core complex during late-orogenic collapse of the Canadian Cordillera: role of ductile thinning and partial melting of the mid- to lower crust. *Geodinamica Acta*, **10**(2), 41–58.

VITEL, G. 1988. Le granite du gouffre d'enfer (Massif Central français). Petrologie d'un marqueur tectonique varisque. *Bulletin de la Société Géologique de France*, **8**(II), 907–915.

WANLESS, R. K. & REESOR, J. E. 1975. Precambrian zircon age of orthogneiss in the Shuswap Metamorphic complex, British Columbia. *Canadian Journal of Earth Sciences*, **12**, 326–332.

WERNICKE, B. 1981. Low-angle normal faults in the Basin and Range Province: nappe tectonics in an extending orogen. *Nature*, **291**(5817), 645–648.

WHEELER, J. O. & MCFEELY, P. 1991. *Tectonic assemblage map of the Canadian Cordillera and adjacent parts of the United States of America*, scale 1:2 000 000. Geological Survey of Canada, Map **1712A**.

WICKHAM, S. M. & OXBURGH, E. R. 1985. Continental rifts as a setting for regional metamorphism. *Nature*, **318**, 330–333.

YORATH, C. J. 1991. Upper Jurassic to Paleogene assemblages. *In*: GABRIELSE, H. & YORATH, C. J. (eds) *Geology of the Cordilleran Orogen in Canada*. Geological Society of America Geology of North America, **G2** and Geological Survey of Canada, **4**, 329–371.

# Diapiric ascent and cooling of a sillimanite gneiss dome revealed by $^{40}Ar/^{39}Ar$ thermochronology: the Kigluaik Mountains, Seward Peninsula, Alaska

ANDREW T. CALVERT[1], PHILLIP B. GANS[1] & JEFFREY M. AMATO[2]

[1] *Department of Geological Sciences, University of California, Santa Barbara, CA 93106, USA*

[2] *Department of Geology and Geophysics, University of Wisconsin–Madison, 1215 West Dayton Street, Madison, WI 53706, USA*

**Abstract:** The upper amphibolite to granulite facies Kigluaik gneiss dome cooled rapidly in late Cretaceous time as it rose through the crust and was emplaced against the older blueschist–greenschist facies Nome Group. $^{40}Ar/^{39}Ar$ data from metamorphic rocks reveal prolonged moderately rapid cooling histories that vary somewhat with structural depth and geographical position. Hornblende ages ($T_c$ c. 535°C) range from 86 to 82 Ma, mica ages ($T_c$ c. 300–400°C) range from 85 to 83 Ma, and K-feldspar spectra yield age gradients that record cooling from c. 300° to c. 150°C between 82 and 65 Ma. These high-precision cooling ages on hornblende, white mica and biotite, and detailed temperature–time curves obtained from diffusion modelling of K-feldspar age spectra allow us to reconstruct the 3D geometry of isotherms during cooling and local reheating of the gneiss dome. Isotherms were parallel to lithological contacts during high-temperature cooling, then became sub-horizontal in present-day coordinates as layering was domed by c. 84 Ma. Low-temperature cooling (300–150°C) of the gneiss dome is asymmetrical, with the north side cooling several million years before the south side. The contact between high-grade gneisses of the Kigluaik Mountains and surrounding lower-grade rocks is not a fault, but rather a steep metamorphic field gradient where closely spaced Barrovian isograds and partially reset mica ages document progressive thermal overprint of the older blueschist–greenschist facies Nome Group rocks. These combined geological and thermochronological data are most compatible with a model wherein the Kigluaik gneiss dome rose diapirically from mid-crustal levels between 91 Ma and c. 82 Ma and cooled through low temperatures differentially as it was tilted gently southward between 83 and 65 Ma.

The Kigluaik Mountains are one of three sillimanite-grade gneiss domes of Cretaceous age that are flanked by older structurally overlying blueschist–greenschist facies metamorphic rocks on the Seward Peninsula, Alaska (Figs 1 and 2). These metamorphic culminations are common in the hinterlands of other orogenic belts including the Alps, Himalayas and the North American Cordillera. Several models have been put forward to explain the exhumation of these high-grade gneiss complexes. These include: (1) buoyancy-driven rise of mid-crustal material, either heated sufficiently to reduce its density or carried upward by a rising magma diapir (Berner *et al.* 1972; Ramberg 1980); (2) exhumation of mid-crustal rocks during crustal shortening and thickening (by duplexing coupled with rapid erosion at the surface) (Burg *et al.* 1984); (3) extensional unroofing of mid-crustal rocks in the footwalls of major normal faults as is widely postulated for the Cordilleran metamorphic core complexes (Crittenden *et al.* 1980). Each of these models makes predictions about the nature of the boundary between the gneiss dome and surrounding lower-grade rocks, as well as on the spatial variation in cooling histories that would be expected within the gneiss dome.

Here we present data from a detailed $^{40}Ar/^{39}Ar$ thermochronological investigation of the Kigluaik Mountains gneiss dome using a suite of minerals with different closure temperatures ($T_c$) to reconstruct the cooling history as a proxy for the exhumation history of the complex. Although cooling rates alone cannot prove an exhumation mechanism, analysing structural relationships and relative field constraints with detailed cooling histories can yield an understanding of exhumation of these mid-crustal complexes (e.g. Hill *et al.* 1992; John & Foster 1993). The thermochronological study discussed here was carried out with several goals: (1) to document the high- to low-temperature cooling history of the high-grade rocks using hornblende, white mica, biotite and K-feldspar

CALVERT, ANDREW T., GANS, PHILLIP, B. & AMATO, JEFFREY M. 1999. Diapiric ascent and cooling of a sillimanite gneiss dome revealed by $^{40}Ar/^{39}Ar$ thermochronology: the Kigluaik Mountains, Seward Peninsula, Alaska. *In:* RING, U., BRANDON, M. T., LISTER, G. S. & WILLETT, S. D. (eds) *Exhumation Processes: Normal Faulting, Ductile Flow and Erosion.* Geological Society, London, Special Publications, **154**, 205–232.

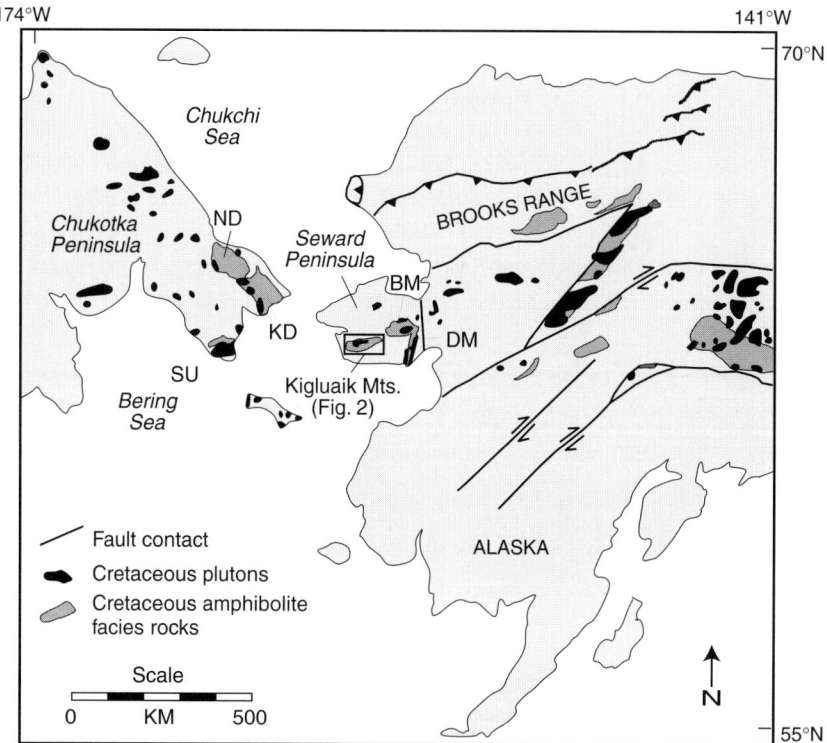

**Fig. 1.** Index map of Alaska and eastern Russia with Bering Strait region gneiss domes: Bendeleben Mountains (BM), Darby Mountains (DM) and Kigluaik Mountains on the Seward Peninsula and the Neshkan Dome (ND), Kool'en Dome (KD) and Senyavin Uplift (SU) on the Chukotka Peninsula. Cretaceous amphibolite-facies metamorphic rocks in the region are shaded, major fault systems and delineated and Cretaceous plutons are black (after Dusel-Bacon et al. 1989; Amato et al. 1994).

thermochronology. Specifically, we set out to examine spatial variations in that cooling history, so as to understand the timing, rate and mechanism(s) of exhumation of these amphibolite to granulite facies rocks; (2) to determine whether intrusive activity in the gneiss dome was coeval with metamorphism and assess whether the intrusive and metamorphic rocks share a similar cooling history; (3) to attempt to directly date the peak metamorphism and assess the thermal history of surrounding lower-grade rocks by analysing white mica separates from closely spaced samples across the greenschist to amphibolite facies transition on the flanks of the gneiss dome. In total, we present $^{40}Ar/^{39}Ar$ data for detailed incremental heating experiments on 33 samples (932 individual steps) of hornblende, muscovite, biotite and K-feldspar from 26 rocks collected across the range. These samples span >6 km of structural section as measured perpendicular to the dominant foliation defining the dome and allow us to examine along- and across-strike variations in the cooling histories as well as variations with rock type and structural depth. Estimated closure temperatures ($T_c$) for the minerals we analysed permit us to evaluate and quantify the temperature–time history for metamorphic and plutonic rocks from >500°C to ≤150°C. These new thermochronological data place important new constraints on how these mid-crustal rocks were exhumed and on the relationship between metamorphism and plutonism in a classic gneiss dome.

## Regional geological setting

The Bering Sea region is a vast expanse of mostly submerged continental crust connecting western Alaska and the Russian Far East (Fig. 1). Mesozoic and older gneisses of the Chukotka and Seward Peninsulas are part of the Arctic Alaska Terrane, which also includes the Brooks Range

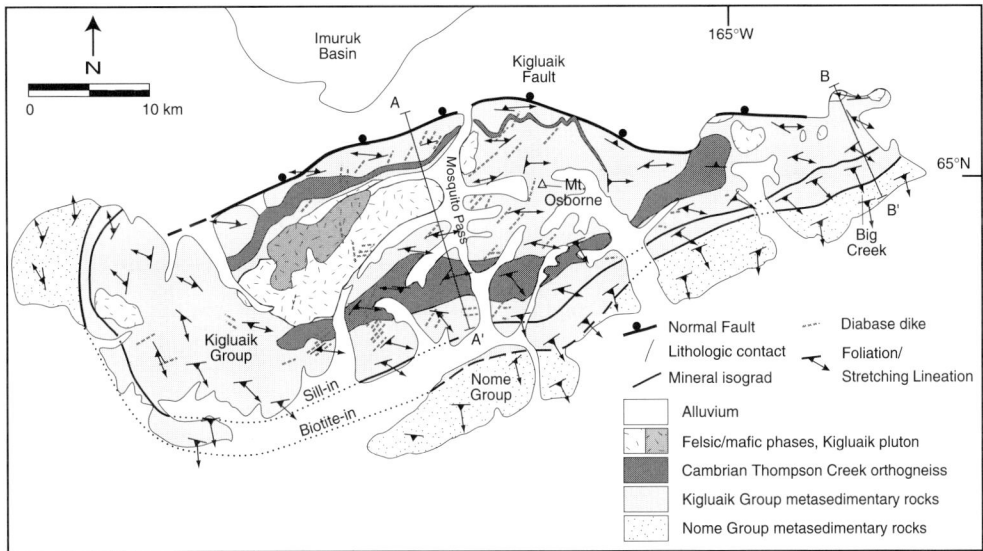

**Fig. 2.** Simplified geological map of the Kigluaik Mountains sillimanite gneiss dome with the biotite-in and sillimanite-in isograds. The edge of the gneiss dome is defined as the area lower than biotite grade, which coincides well with gentler topography of the Nome Group.

of northern Alaska. Isolated exposures of metamorphic rocks preserve evidence that the area was subjected to Mesozoic deformation, magmatism and metamorphism (Armstrong et al. 1986; Patrick & Lieberman 1988; Dusel-Bacon et al. 1989; Amato et al. 1994). Late Cretaceous magmatic rocks are prevalent in the Bering Sea region, volcanic and plutonic rocks are common on the Chukotka Peninsula, but no Cretaceous volcanic rocks are exposed on the Seward Peninsula. Cretaceous(?)–Eocene to Present transtensional basins occur offshore and may indicate late Mesozoic to Tertiary extension (Worrall 1991; Dumitru et al. 1995). The Chukotka and Seward Peninsula amphibolite facies rocks are mainly restricted to discrete structural culminations or 'gneiss domes' surrounded by lower-grade (blueschist–greenschist facies) to unmetamorphosed country rocks (Fig. 1). The main focus of this paper is the origin and exhumation history of one of these gneiss domes.

Previous workers have outlined three principal metasedimentary units on the Seward Peninsula: the Slate of the York Region (Sainsbury et al. 1972; Till et al. 1986), a weakly metamorphosed unit of silty limestone and slate in fault contact with overlying incipiently metamorphosed Ordovician to Mississippian limestones (Hannula et al. 1995); the underlying blueschist–greenschist facies metamorphosed schists, marbles and metabasites of the Nome Group (Till et al. 1986; Patrick 1988; Miller et al. 1992; Hannula et al. 1995) widely exposed in the Seward Peninsula lowlands; and the Kigluaik Group, the upper amphibolite to granulite facies metamorphosed schists, quartzites, marbles and calcsilicate rocks of the Kigluaik Mountains (Sainsbury et al. 1972; Till et al. 1986; Miller et al. 1992; Amato et al. 1994). Protoliths for these metamorphic rocks are of Proterozoic (Till et al. 1986; Amato et al. 1994; Patrick & McClelland 1995) to lower Palaeozoic age (Till et al. 1986).

All of these metasedimentary rocks are inferred to have been deposited on a south-facing continental slope (Mayfield et al. 1988; Moore et al. 1992). Telescoping and thrust imbrication during the Jurassic–Cretaceous Brookian orogeny is inferred to be responsible for the locally preserved high-pressure–low-temperature assemblages in some of these Nome Group rocks (Evans & Patrick 1987; Patrick & Evans 1989). Exhumation of the Nome Group high-pressure rocks has been variously attributed to crustal extension (Dumitru et al. 1995), or continued thrusting and erosion (Patrick & Lieberman 1988). Previous K/Ar dating shows Jurassic and Early Cretaceous ages (Armstrong et al. 1986) for Nome Group rocks and Late Cretaceous hornblende and biotite (Turner & Swanson 1981; Sturnick 1984) ages for Kigluaik Mountains metamorphic rocks, but ages on these samples were too variable in the Nome Group and errors too large (>2.5 Ma) in the Kigluaik Group to quantify cooling rates or discern spatial variations in cooling histories. Recent work shows that Nome Group plutonic rocks have lower U/Pb zircon intercepts of 124 ± 6 and 117 ± 17 Ma (Patrick & McClelland 1995) and metasedimentary rocks cooled through white mica closure temperatures (c. 350°C) at c. 120 Ma (Hannula & McWilliams 1995), and were at or slightly above apatite fission-track annealing temperatures (60–120°C) during c. 91 Ma Kigluaik Group peak metamorphism (Dumitru et al. 1995).

## Kigluaik Mountains field relations

The amphibolite to granulite facies metasedimentary and metaplutonic rocks (Kigluaik Group) structurally underlie the Nome Group (Figs 2 and 3). The rock types, fabrics, metamorphic assemblages and general structural framework of the Kigluaik Mountains have been previously described by Till et al. (1986), Lieberman (1988), Patrick & Lieberman (1988), Miller et al. (1992) and Amato et al. (1994). Compositional layering and foliations in L–S tectonites define a simple doubly plunging antiform or dome that is 60 km (E–W) by 20 km (N–S). The transitional boundary between the Kigluaik gneiss dome and the enveloping Nome Group rocks is defined by a narrow zone of steeper foliations and closely spaced biotite, garnet, staurolite, sillimanite and white mica-out isograds. These isograds are concentric around the dome, except where cut by the active Kigluaik normal fault, and appear to be everywhere parallel to foliations and lithological contacts. Occasional variation in map width of isograds (Fig. 2) is due to topography, exposure and bulk composition.

Significant strain accompanied the high-grade metamorphism, as evidenced by widespread synkinematic sillimanite, and the transposition of older Nome Group fabrics in sillimanite- and higher-grade rocks within the gneiss dome. This high-temperature strain has been previously attributed to an episode of regional crustal thinning (Miller et al. 1992; Dumitru et al. 1995). Foliations and lineations appear to vary continuously (Fig. 2) from high structural levels in the blueschist–greenschist facies Nome Group (subhorizontal foliations and N–S stretching lineations) to the isograd transition (moderately dipping foliations and lineation azimuths of N–S to WNW–ESE) and to the high-grade core of the complex (subhorizontal foliations and E–W stretching lineations). However, relative timing constraints from field relations and absolute constraints from thermochronology (described below) suggest that the fabrics present at these three structural levels formed at distinct times.

(1) At the highest structural levels, the dominant Nome Group fabrics (D2 of Patrick 1988; Hannula et al. 1995) adjacent to the Kigluaik Mountains were developed at greenschist facies conditions and are defined by white mica and chlorite growth axial-planar to a transposed older foliation, and stretched quartz lineations and elongate pressure shadows around porphyroblasts. There is significant deformation that post-dates the peak of blueschist–greenschist metamorphism (i.e. ribbon quartz, chlorite pressure shadows, foliation that wraps around albite porphyroblasts) and the sense of shear is consistently top-to-the-north (Patrick 1988; Hannula et al. 1995). The age of the Nome Group fabric is not well constrained, but is thought to mainly post-date white mica $^{40}Ar/^{39}Ar$ ages of c. 120 Ma (Hannula & McWilliams 1995) and pre-date Late Cretaceous (75–106 Ma) apatite fission-track ages (Dumitru et al. 1995). These apatite fission-track analyses were problematic because of low uranium contents, but suggest that these samples were at low temperatures (<120°C) during at least part of this interval. Low-temperature $^{40}Ar/^{39}Ar$ thermochronology on biotite and K-feldspar from the Cape Nome Orthogneiss yields a biotite age of 108 Ma and an unmodelled K-feldspar spectrum suggesting that the orthogneiss was at <200°C by 90–95 Ma (Hannula 1993). The available data are compatible with Nome Group rocks having been penetratively deformed after 120 Ma, but largely cooled to ≤100°C by 80–90 Ma.

(2) At the deepest structural levels, gneissic foliations in the core of the Kigluaik Mountains are granoblastic, generally dip shallowly and contain a strong E–W stretching lineation defined by high-temperature metamorphic minerals and elongate pressure shadows around porphyroblasts. There is no consistent sense of shear in these rocks. It is important to emphasize that the rocks in the core document very high strain at peak metamorphic conditions, and almost no low-temperature strain or retrogression. In a few places we identified thin, chlorite-grade mylonitic shear zones within the core gneissic rocks that have well-developed N–S lineations and S–C fabrics indicating down-dip transport. The high-temperature, high-strain event that produced the gneissic foliation and E–W stretch is apparently tightly bracketed by metamorphic monazite at 91 ± 1 Ma (Amato & Wright 1998) and post-tectonic intrusion of the undeformed 90 ± 1 Ma Kigluaik Pluton (see discussion below).

(3) Fabrics in the intervening isograd transition are interpreted to be younger than either

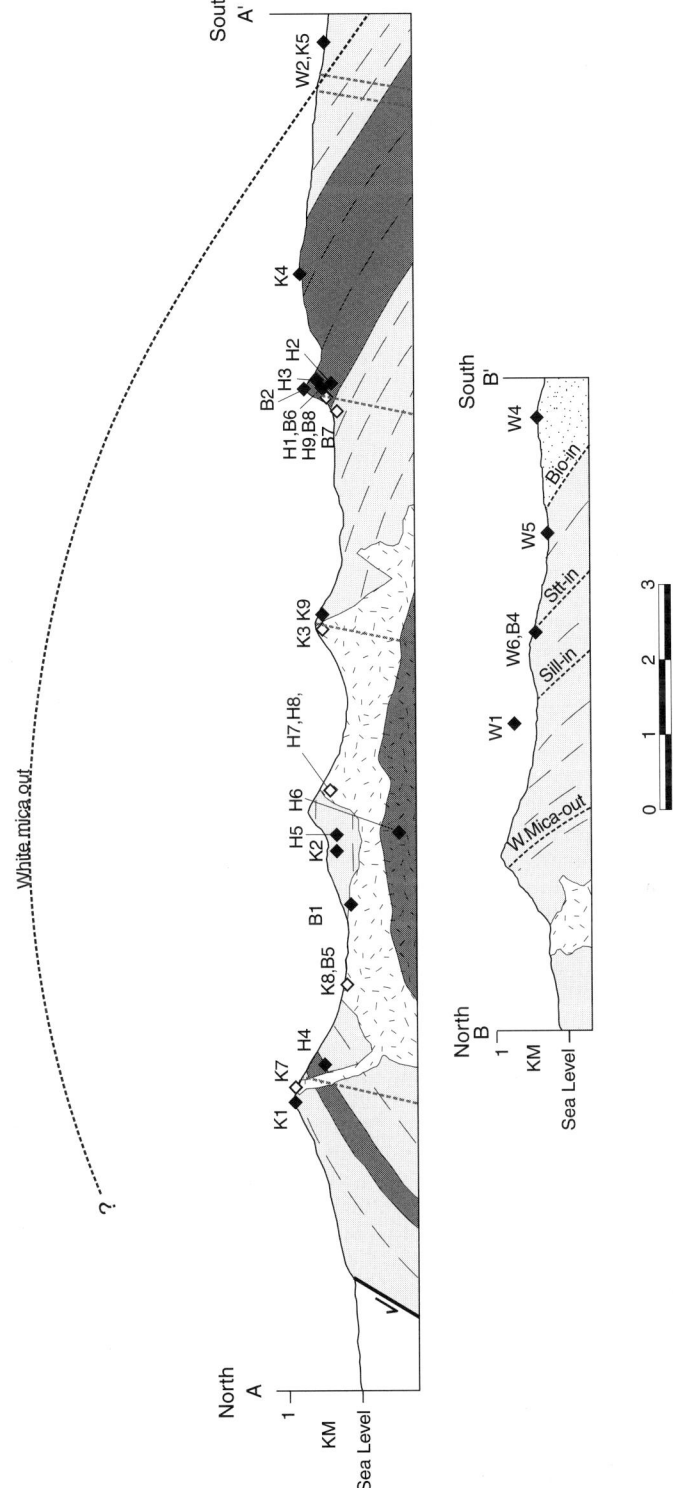

**Fig. 3.** Simplified cross-sections of the central and eastern Kigluaik Mountains. Geochronology samples are projected into the line of cross-section according to their structural level; ◆, metamorphic samples; ◇, samples from the Kigluaik pluton and associated dykes. Shading of units is the same as in Fig. 2.

the Nome Group fabric or the gneissic core fabrics (see discussion below). Foliations steepen to c. 45° in this interval and at first inspection, appear to grade into the high-strain, synmetamorphic fabric at deep levels and the axial planar foliation near the Nome Group. Lineations are defined by elongate metamorphic minerals, quartz streaks and elongate pressure shadows around porphyroblasts, and are generally oblique to the lineations above and below. Here the strain clearly post-dates the peak Barrovian metamorphic conditions because porphyroblasts are broken apart and filled with chlorite, and quartz–chlorite-filled tension gashes are common. Sense of shear is down-dip in the isograd region and is defined by asymmetric pressure shadows, oblique grain-shape foliations in quartz and mesoscopic late-stage asymmetric folds. These folds, largely restricted to incompetent carbonate and quartz-rich intervals, fold the dominant foliation and lineation and are consistently overturned down the present dip of the gneiss dome. The termination of penetrative strain in the isograd region is c. 83 Ma and is best dated by rapid cooling of this interval from >350°C to <200°C between 84 and c. 82 Ma (see below).

The Kigluaik Group is believed to have experienced the high-$P$–low-$T$ metamorphism preserved in the Nome Group rocks, followed by a lower-$P$–higher-$T$ overprint (Patrick & Lieberman 1988; Hannula et al. 1995). Metamorphic petrological work has focused on peak conditions (Lieberman 1988; Patrick and Lieberman, 1988) or pluton emplacement depths (Amato 1995) rather than exhumation paths, but these studies show that little retrograde information is preserved in these rocks. Peak assemblages are very well preserved and available metamorphic mineral zonation in garnets generally shows only prograde or no zonation (Lieberman 1988). Sillimanite + K-feldspar zone garnets have flat compositional profiles with only slight retrograde zoning at the rims (Lieberman 1988). From this we can glean little information about the retrograde path.

Lieberman (1988) and Patrick & Evans (1989) presented thermobarometry data from different levels of the complex and pointed out that pressure differences across mapped isograds significantly exceed the c. 4 km of structural thickness between blueschist–greenschist facies rocks and sillimanite + K-feldspar gneisses. Blueschist facies rocks across the area yield pressures and temperatures of 1200 MPa at 450°C (Patrick & Evans 1989), but they had already been exhumed to shallow crustal levels during Kigluaik metamorphism (Hannula & McWilliams 1995); staurolite zone rocks record 350–450 MPa at 525–575°C; sillimanite zone rocks record 400–500 MPa at 625–700°C; sillimanite + K-feldspar zone rocks record 500–600 MPa at 700–750°C and garnet–spinel lherzolite at the deepest structural levels in the dome yields pressures from 800–1400 MPa at 800–850°C. This pressure variation in the isograd and deepest levels suggests that rocks at different levels reached their peak temperatures at different times during exhumation (Lieberman 1988; Patrick & Evans 1989).

Late Cretaceous I-type mafic to silicic intrusions core the western half of the Kigluaik gneiss dome and have been interpreted as synmetamorphic and the heat source for the amphibolite facies metamorphism (Amato et al. 1994; Amato & Wright 1997). It is important to emphasize that the intrusions are entirely undeformed, with hypidiomorphic granular textures typical of epizonal plutons. Most plutonic contacts are sharp and the associated silicic to intermediate dykes generally cut across the dominant foliation in the metamorphic rocks. Except for several dykes with a weak alignment of biotite parallel to the regional foliation, they are undeformed. Though high-temperature metamorphic minerals in the gneisses are generally synkinematic to the dominant foliation and lineation, locally near intrusions high-temperature minerals have static textures (Amato et al. 1994). Thus, field relationships suggest that the pluton was emplaced and crystallized after peak metamorphism and certainly post-dates the development of the gneissic foliation. We mapped the intrusive complex in the west–central part of the range as a single large pluton because of spatial and compositional continuity. Subsequent geochronology suggests at least two ages of intrusions – a permissible result given relatively poor exposure. U/Pb zircon from an undeformed granodiorite in the mingling zone of the Kigluaik pluton yields 90 ± 1 Ma (Amato & Wright 1998), but thermochronology on silicic intrusive rocks which we mapped as part of the Kigluaik pluton suggest younger ages of c. 83 Ma. This also appears to be the age of the diabase dyke swarm with an average orientation of c. N40°E that cuts all metamorphic units and the pluton (Fig. 2) (Amato & Wright 1998).

Recent models for the development of the gneiss dome are diverse. Patrick & Lieberman (1988) postulated that the Kigluaik Mountains and adjacent Nome Group blueschist facies rocks represented an intact crustal column that equilibrated initially at 0.8–1.2 GPa pressures and was statically overprinted at higher temperatures during exhumation. Miller et al. (1992)

cited significant strain along the boundary between the Nome and Kigluaik Groups and suggested that the areas represented very different levels of the crust, juxtaposed by an extensional shear zone. Amato et al. (1994) argued that the heat source for Kigluaik metamorphism in the Kigluaik Mountains was magmatic and that exhumation was in part extensional and in part diapiric during intrusion of the large dioritic to granitic Kigluaik pluton. A key test of these ideas is a detailed knowledge of the internal cooling history of the gneiss dome and how it varies spatially.

## $^{40}$Ar/$^{39}$Ar thermochronology

### Analytical methods

Geochronology samples were collected from metasedimentary rocks, orthogneisses, plutonic rocks and diabase dykes from throughout the high-grade core and from a transect across the isograd region (Fig. 2). All but the two western samples are projected onto two N-S cross-sections, one across the central part of the range (Fig. 3a) and one across the Barrovian isograd region in the eastern part of the range (Fig. 3b). From these samples, a total of 33 mineral separates of hornblende, muscovite, biotite and K-feldspar were analysed, and these results are summarized in Tables 1 and 2 and Fig. 4. Complete data tables, age spectra, isochron plots and a description of analytical procedures for all samples are available upon request from the first author or on the internet (http://www.geol.ucsb.edu/~calvert/kigluaik.html) and have been deposited with the Society Library and British Library at Boston Spa, West Yorkshire, UK, as Supplementary Publication No. SUP18125 (29 pp.).

All $^{40}$Ar/$^{39}$Ar dates were obtained by conventional step heating using a Staudacher-type resistance furnace. The data were obtained over a period of several years in the argon geochronology laboratories at UCSB and Stanford University. Equipment is virtually identical at the two facilities and numerous replicate analyses have yielded indistinguishable results. Flux monitors used were Taylor Creek Sanidine (US Geological Survey (USGS) standard 85G003, Dalrymple & Duffield 1988) with an assigned age of 27.92 Ma (G. B. Dalrymple, pers. comm.), and Charcoal Ovens Tuff, an internal standard with an assigned age of 35.88 Ma. Samples were crushed and appropriate size fractions (the 106-125 μm fraction for hornblende and 250-425 μm fraction for mica and alkali feldspar) were concentrated using heavy liquids, paper shaking and magnetic separation techniques. Samples were cleaned in an ultrasonic bath and remaining contaminant minerals were hand-picked under reflected light. Separates were then packaged in copper or aluminium foil and loaded into a quartz vial with packaged flux monitors. Vials were irradiated at the TRIGA reactors at the USGS-Denver and Oregon State University for 30 h over a period of 5 days. Samples and monitors were heated in the resistance furnace and gas was purified continuously during extraction by two SAES ST-172 porous getters and analysed on a MAP 216 mass spectrometer fitted with a Baur-Signer source and a Johnston MM1 multiplier with a sensitivity of $2.0 \times 10^{-14}$ mol V$^{-1}$. Our Staudacher-type resistance furnace uses a coiled tungsten filament, a tantalum crucible with no liner and tungsten-rhenium thermocouple. Analyses are blank-corrected and our typical system blanks on m/e 40 vary from $2.0 \times 10^{-16}$ moles at low temperatures (<1000°C) and climb to $6.0 \times 10^{-16}$ mol at 1300°C for 15 min heating steps. All parts of the step-heating experiments are automated with pneumatically actuated, all-metal valves, a Macintosh computer and custom software.

Two sigma errors are reported on plateau and inverse isotope correlation diagram (isochron) ages. Plateau ages are reported where more than 50% of the $^{39}$Ar is released within 2σ error and for these samples isochron ages are concordant. For disturbed spectra interpreted ages are reported and are either isochron ages or discontinuous plateau ages (Table 1). In general, we quote age spectra rather than isochron ages, because of the variation in ages possible with isochron analysis. We find that the lowest MSWD (mean square weighted deviation, a goodness of fit index; Roddick 1978) of a sample will commonly be from the 10-50% portion of $^{39}$Ar released during a step heating experiment. Typically, the first several steps have $^{40}$Ar/$^{36}$Ar$_i$ intercepts near atmospheric ratios and this is interpreted as argon adsorbed during sample preparation. Subsequent steps generally have decreasing ages and increasing radiogenic yields, which often provide a good linear fit. High-temperature steps (50-100% of the $^{39}$Ar) often do not lie on this chord, and when this is true we reject the isochron data and instead choose the age spectrum minimum as a maximum sample age. In presenting the isochron data we plot a linear fit with the selected steps providing the lowest MSWD even though it is sometimes less than 50% of the gas. Hornblende data are complicated (see below for a discussion of spectra and isochron interpretations). Mica data were also complicated; in three cases, two steps with

**Table 1.** *Analytical data and sample information for hornblende (HBL), white mica (WM) and biotite (BIO)*

| Sample | Field name | Latitude | Longitude | Mineral | Total steps | Age (Ma) ± 2σ | Temp. (°C) steps used | %$^{39}$Ar (spectrum) | Isochron age (MA) ± 2σ | Isochron MSWD | Total fusion age (Ma) |
|---|---|---|---|---|---|---|---|---|---|---|---|
| H1 | 90P12-4b | 60°8'4" | 165°31'49" | HBL | 17 | **86.4 ± 0.7** | 1075–1165 | 72 | 86.9 ± 0.9 | 1.68 | 85.9 ± 4.6 |
| H2 | 90K12-7a | 64°55'18" | 165°33'3" | HBL | 12 | **86.1 ± 0.4** | 990–1070 | 48 | 86.7 ± 0.6 | 3.00 | 86.5 ± 0.5 |
| H3 | 90P12-5b | 64°55'43" | 165°32'0" | HBL | 20 | **85.6 ± 0.4** | 1030–1080 | 57 | 84.3 ± 1.1 | 2.14 | 87.9 ± 0.6 |
| H4 | 90K10-3 | 65°1'0" | 165°32'0" | HBL | 12 | 84.5 ± 0.3 | 1030–1260 | 95 | 84.8 ± 0.7 | 3.97 | 84.3 ± 0.5 |
| H5 | 90J11-7 | 64°59'58" | 165°25'16" | HBL | 12 | 85.4 ± 0.5 | 1070–1090 | 15 | 84.4 ± 0.7 | 1.60 | 89.1 ± 0.4 |
| H6 | 91-C-5 | 65°1'0" | 165°20'10" | HBL | 12 | 84.0 ± 0.3 | 1050–1090 | 42 | 82.7 ± 0.6 | 0.76 | 84.6 ± 0.3 |
| H7 | 924A-120 | 64°56'11" | 165°47'27" | HBL | 12 | **82.0 ± 0.3** | 950–1070 | 81 | 81.9 ± 0.5 | 1.53 | 81.6 ± 0.4 |
| H8 | 924A-119 | 64°57'12" | 165°46'6" | HBL | 14 | **82.2 ± 0.3** | 1050–1190 | 52 | 82.1 ± 0.6 | 2.61 | 82.0 ± 0.7 |
| H9 | 922A-40 | 64°54'58" | 165°37'8" | HBL | 14 | **82.8 ± 0.4** | 900–1000 | 81 | 82.8 ± 1.1 | 5.60 | 82.1 ± 0.4 |
| W1 | 89S-MC-54 | 65°0'37" | 164°48'26" | WM | 9 | **83.0 ± 0.8** | 830–1350 | 80 | 83.2 ± 0.8 | 0.71 | 82.1 ± 1.8 |
| W2 | 90P8-13b | 64°53'34" | 165°27'49" | WM | 17 | **83.7 ± 0.5** | 830–1450 | 91 | 83.6 ± 0.8 | 0.48 | 83.6 ± 1.0 |
| W3 | 89S-LMC-47 | 64°50'35" | 166°2'49" | WM | 13 | **83.1 ± 0.5** | 650–1125 | 95 | 83.2 ± 0.5 | 1.15 | 83.1 ± 0.7 |
| W4 | 90S-EK-2 | 64°57'41" | 164°48'34" | WM | 16 | Disturbed | | | | | 139.9 ± 0.9 |
| W5 | 90S-EK-5 | 64°58'45" | 164°49'23" | WM | 12 | Disturbed | | | | | 126.5 ± 0.7 |
| W6 | 90S-EK-9 | 64°59'19" | 164°50'0" | WM | 13 | Disturbed | | | | | 121.8 ± 0.9 |
| B1 | 90J11-5 | 64°59'38" | 165°24'48" | BIO | 29 | **83.9 ± 0.4** | 840–1250 | 95 | 83.9 ± 0.5 | 4.41 | 83.3 ± 0.5 |
| B2 | 90P8-4A | 64°55'11" | 165°31'49" | BIO | 26 | **83.9 ± 0.4** | 775–970, 1090–1210 | 83 | 84.0 ± 0.6 | 5.04 | 83.7 ± 0.6 |
| B3 | 90-Hens | 65°8'4" | 164°51'4" | BIO | 25 | **84.7 ± 0.2** | 790–980, 1050–1400 | 82 | 85.0 ± 0.5 | 7.84 | 84.5 ± 0.3 |
| B4 | 90S-EK-9 | 64°59'19" | 164°50'0" | BIO | 17 | **83.8 ± 1.1** | 775–1250 | 81 | 83.8 ± 1.1 | 0.65 | 83.4 ± 1.1 |
| B5 | 90P1-1 | 65°0'34" | 165°31'1" | BIO | 29 | *83.5 ± 0.4* | 800–930, 1030–1100 | 79 | 83.4 ± 0.8 | 5.90 | 83.0 ± 0.5 |
| B6 | 90P12-4A | 64°8'4" | 165°31'49" | BIO | 18 | 83.6 ± 0.3 | 700–1260 | 95 | 83.5 ± 0.6 | 3.33 | 82.6 ± 0.7 |
| B7 | 92.3A-101 | 64°55'10" | 165°46'0" | BIO | 8 | **82.5 ± 0.3** | 1070–1155 | 48 | 82.5 ± 0.4 | 2.22 | 82.1 ± 0.3 |
| B8 | 92.2A-40 | 64°54'58" | 165°37'8" | BIO | 8 | **82.1 ± 0.3** | 1125–1205 | 47 | 75.6 ± 2.64 | 27.86 | 80.2 ± 0.3 |

Plateau ages are shown in bold, and are defined by 50% or more of the gas within 2σ error. Interpreted ages are given when this criterion is not met and are shown in italics. Goodness of fit index (MSWD) is from Roddick (1978).

**Table 2.** *Analytical data and sample information for K-feldspar samples*

| Sample name | Field name | Latitude | Longitude | Total steps | Activation energy (kcal mol$^{-1}$) | Frequency factor (s$^{-1}$) | No. of domains |
|---|---|---|---|---|---|---|---|
| K1 | 90P1-7 | 65°1′11S″ | 165°32′15″ | 56 | 32.44 | 2.251 | 6 |
| K2 | 90J11-5 | 64°59′38″ | 165°24′48″ | 89 | 35.43 | 2.944 | 6 |
| K3 | 90P5-7 | 64°57′38″ | 165°30′86″ | 48 | 47.31 | 7.119 | 8 |
| K4 | 90P8-4A | 64°55′11″ | 165°31′49″ | 104 | 45.97 | 6.825 | 7 |
| K5 | 90P8-13B | 64°53′34″ | 165°27′49″ | 47 | 54.38 | 9.507 | 8 |
| K6 | 89S-LMC-51 | 64°51′37″ | 166°1′53″ | 20 | | | |
| K7 | 90P1-13 | 65°0′38″ | 165°36′45″ | 57 | 36.35 | 3.302 | 7 |
| K8 | 90P1-1 | 65°0′34″ | 165°31′1″ | 54 | 46.31 | 6.92 | 6 |
| K9 | 90P5-8 | 64°57′38″ | 165°30′47″ | 50 | 47.13 | 4.747 | 6 |
| K10 | 90P12-4A | 64°8′4″ | 165°31′49″ | 34 | | | |

Activation energy, frequency factors and domain calculations are obtained using the technique of Lovera (1992).

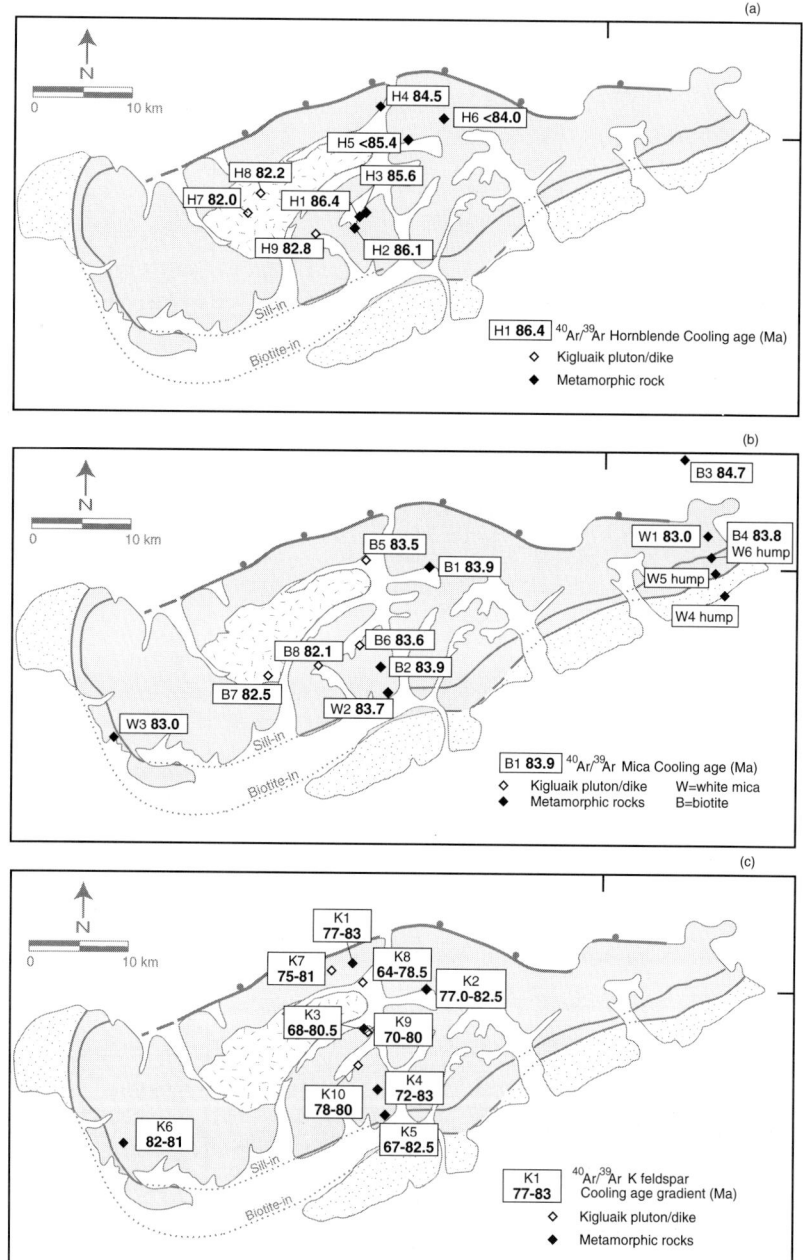

**Fig. 4.** (**a**) Sample locations and $^{40}$Ar/$^{39}$Ar ages of hornblende (H). (**b**) Sample locations and $^{40}$Ar/$^{39}$Ar ages of biotite (B) and white mica (W). (**c**) Sample locations and $^{40}$Ar/$^{39}$Ar maximum and minimum ages of K-feldspar samples. Sample locations from metamorphic rocks are black and intrusive samples are white.

old apparent ages were removed from the intermediate portions of biotite spectra because these steps generally have lower K/Ca ratios and indicate chlorite contamination (Lo & Onstott 1989).

Again, the isochron plots yield reasonable fits for small portions of the $^{39}$Ar released, but were generally poor for larger portions of the spectrum. Three white mica samples were deemed uninterpretable because of hump-shaped spectra. Detailed work by Hannula & McWilliams (1995) on other Nome Group samples shows that these ages have no significance.

K-feldspar samples were analysed over 47–104 step experiments with isolation times ranging from 15 min to 10 h. Initial steps commonly yielded very old apparent ages, but replicate steps at low temperatures allow an estimate of low-temperature ages. We were not able to use the Cl-correlated, excess argon correction technique of Harrison et al. (1994) because of low production of $^{38}$Ar from chlorine in the OSU reactor. Diffusion modelling of K-feldspar data is based on the theory of Lovera et al. (1989) that alkali feldspars have a distribution of diffusion domain sizes and that conventional step-heating experiments can be modelled to extract the diffusion parameters and reconstruct quantitative temperature–time histories. This powerful technique uses Arrhenius data to construct a log ($R/R_0$) plot (Richter et al. 1991), which yields information about diffusion domain sizes independent of heating schedule. Once diffusion domain sizes are understood, model age spectra can be synthesized by iterative calculation of different cooling histories (Lovera 1992) and matched to laboratory-obtained spectra. Temperature–time curves calculated from these model age spectra are presented below.

In presenting these data, we first discuss the cooling history of the metamorphic rocks from the sillimanite-grade core of the range by describing the data obtained from hornblende, white mica, biotite and K-feldspar. We then compare the cooling history of the metamorphic rocks with that of the Kigluaik intrusive complex. Finally, we describe thermochronological data from the transition between the gneiss dome and surrounding lower-grade rocks.

## Cooling history of upper amphibolite facies metamorphic rocks

The minimum age of peak metamorphism in the Kigluaik Mountains is interpreted from U/Pb monazite ages to be 91 ± 1 Ma (Amato & Wright 1998). The monazite data are complicated, but all ages are late Cretaceous and older than $^{40}$Ar/$^{39}$Ar hornblende data. This is a minimum age because monazite closes to lead diffusion at 725° ± 25°C (Parrish 1990) and peak temperatures for the complex are estimated to be >800°C (Lieberman 1988). Incompletely reset igneous monazites from the Cambrian Thompson Creek Orthogneiss also indicate that 91 Ma is a close approximation for peak temperatures in the range (Amato & Wright 1998). Thus all $^{40}$Ar/$^{39}$Ar ages obtained from these high-grade rocks must be cooling ages. As the Kigluaik pluton is 90 ± 1 Ma old, results from igneous minerals can also be used to define the cooling of the dome. The consistently small age differences between the monazite, hornblende and coexisting mica ages require relatively rapid cooling (c. 50°C Ma$^{-1}$) from >700°C to <300°C. For these relatively rapid cooling rates, we estimate closure temperatures based on published diffusion data as follows: hornblende 535° ± 50°C (Harrison 1981); muscovite 370° ± 50°C (Lister & Baldwin 1996); biotite 335° ± 50°C (Harrison et al. 1985; Grove & Harrison 1996).

Although closure temperatures calculated with diffusion data from hydrothermal experiments show considerable variability, empirical studies (Hanson & Gast 1967; Snee et al. 1988) yield results in reasonable agreement. K-feldspar data were modelled directly using the technique of Lovera (1992) to obtain quantitative medium- to low-temperature cooling histories.

### Hornblende ($T_c$ c. 535°C)

Hornblende is not a common phase in metamorphic rocks from the Kigluaik gneiss dome as bulk compositions tend to be too silicic, but samples were analysed from most mapped occurrences. At high structural levels, argon systematics were well behaved, yielding plateau ages for two of three samples; hornblendes from deeper structural levels showed signs of excess argon and yielded ages with lower precision (Fig. 5). By selecting steps from these spectra, we can obtain reasonable (generally younger) isochron ages; however, as these commonly represent only a small portion of the gas and different ages with small MSWDs can easily be fitted for most samples, we have little confidence in these ages. Our best interpretation of these complicated data is that the youngest ages in each spectrum are maximum ages for the samples. The structurally higher samples range from 2 to 2.5 km below the muscovite-out isograd. Samples H1 and H2 are from penetratively deformed amphibolite and H3 is from a melt-filled extension fracture along a syntectonic shear band in a granodioritic orthogneiss. H1, H2 and H3 yielded plateau or near-plateau ages of c. 86 Ma. Hornblende spectra from deeper structural levels are more disturbed, but are distinctively younger. Samples H4, H5 and H6 are projected onto the cross-section at 3.5, 4.1 and 5.1 km below the muscovite-out isograd, respectively. Sample H4 yielded a near-plateau spectrum with an interpreted age of 84.5 ± 0.3 Ma. H5 yielded a

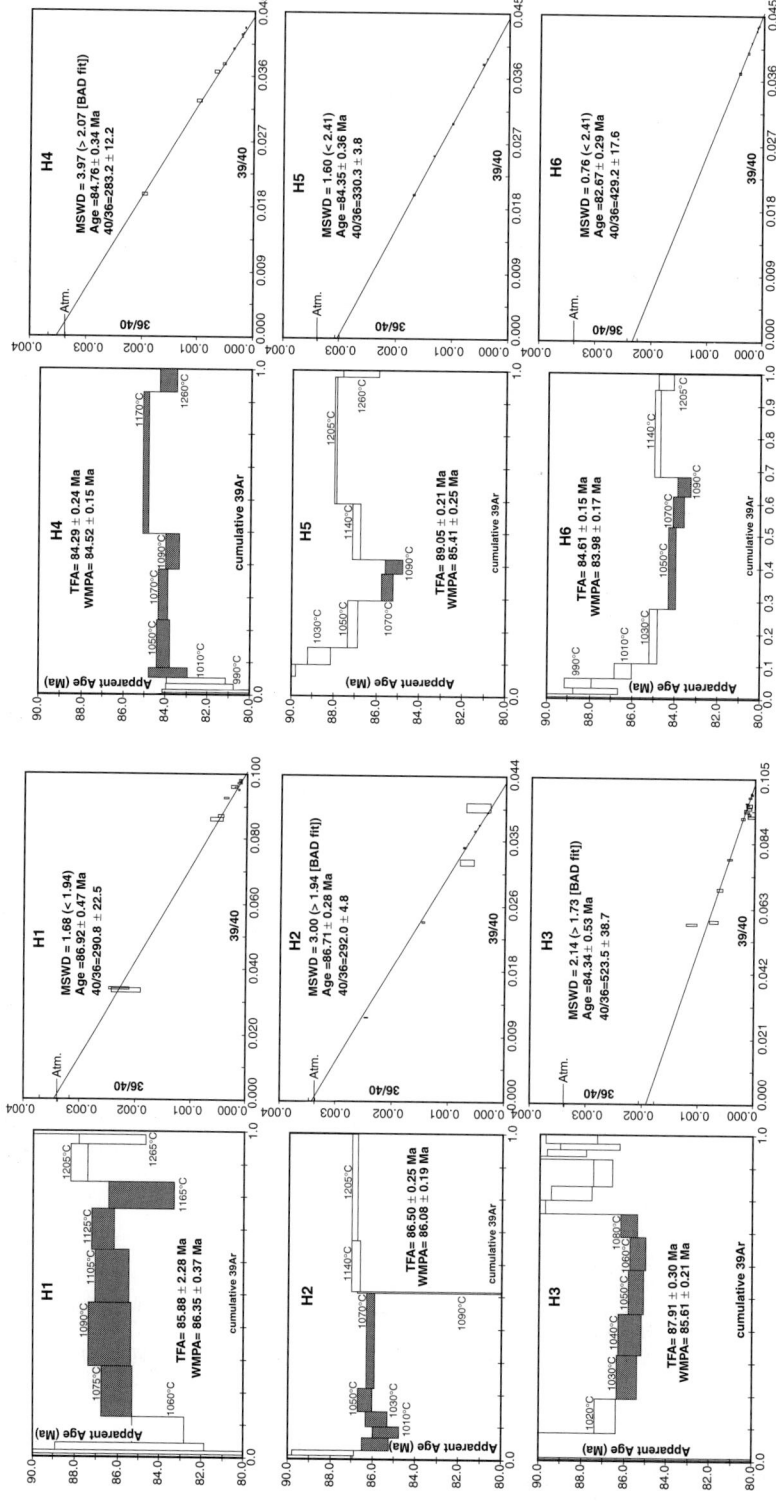

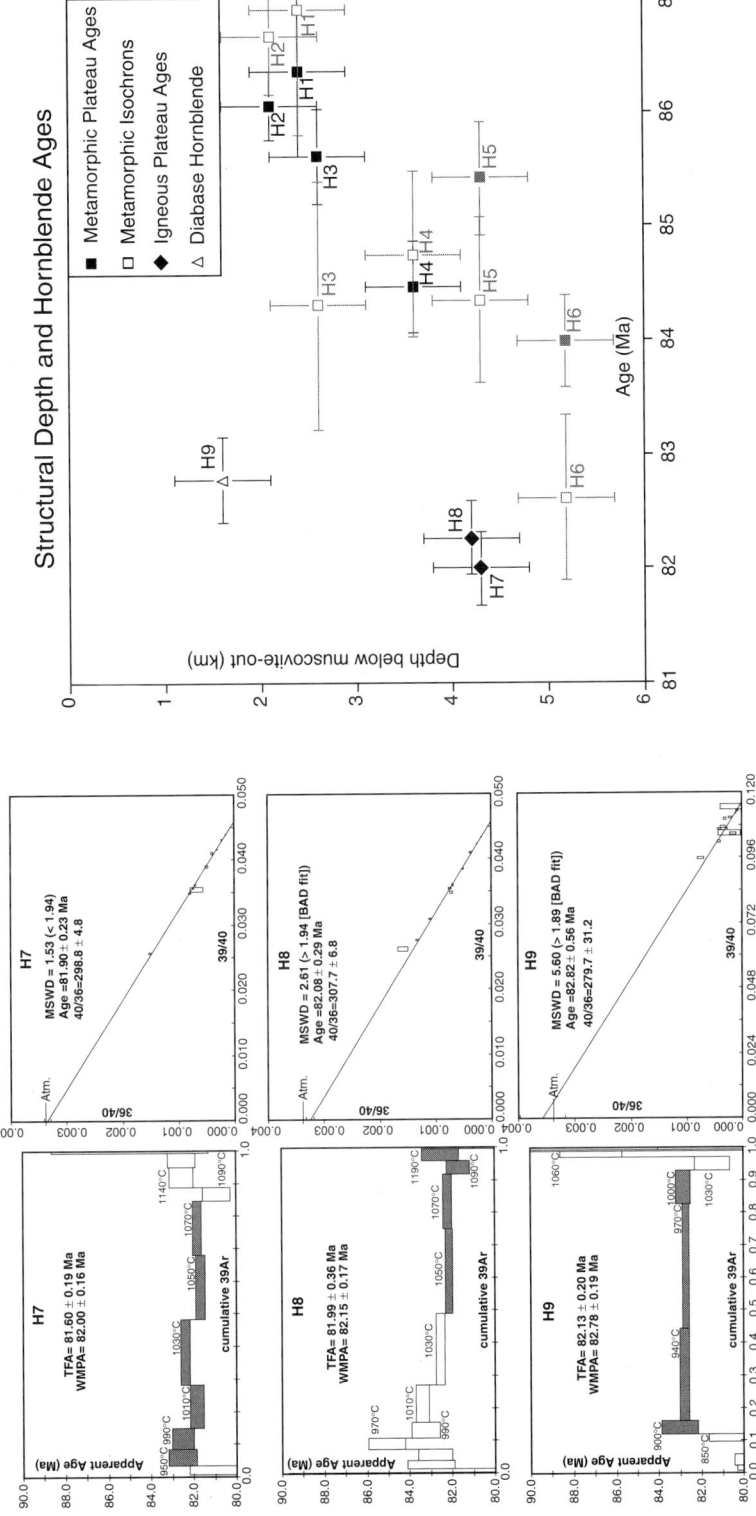

**Fig. 5.** (**a**) $^{40}$Ar/$^{39}$Ar metamorphic hornblende age spectra and their isochron plots. Shaded steps on spectra are those used in weighted mean plateau age (WMPA) calculations. Isochron plots are shown with steps used for the lowest MSWD. Goodness of fit index (MSWD) values are from Roddick (1978). Interpreted ages are tabulated in Table 1. (**b**) Hornblende plots from intrusive hornblendes. (**c**) Summary of the hornblende ages with 2σ errors versus their projected depth beneath the muscovite-out isograd. Greyed ages are disturbed samples.

U-shaped spectrum indicative of excess argon. Apparent ages of the youngest steps are 85.4 Ma, and the isochron age (MSWD = 1.60) for the first 42% of $^{39}$Ar released is 84.4 ± 0.7 Ma. Sample H6 is an amphibolite from nearly the deepest exposed portion of the gneiss dome and also yielded a moderately U-shaped spectrum suggestive of excess argon. The near-plateau age is 84.0 ± 0.3 and the interpreted isochron age (MSWD = 0.76) for 65% of the $^{39}$Ar is 82.7 ± 0.6. Both H5 and H6 apparently contain significant amounts of excess $^{40}$Ar and we interpret the youngest ages in the spectra (85.4 and 84.0, respectively) as maximum ages for the samples. As two hornblendes from intrusive rocks are from the same samples as give 90 Ma U/Pb zircon ages (Amato & Wright 1998), they should yield cooling ages reflecting the cooling of the gneiss dome. H7 and H8 are concordant with plateau ages of 82.0 ± 0.3 and 82.3 ± 0.3 Ma. They were collected at deep structural levels (c. 4.2 km below muscovite-out). Interpreted hornblende cooling ages (Fig. 5b, filled squares) generally decrease downward within the structural section, varying from 86–87 Ma at moderate structural levels to 82–83 Ma at deepest levels. We attribute this decrease in ages to downward movement of the isotherms during early high-temperature cooling before doming of the lithological layering. The rate at which the isotherms moved downward through the rock column is calculated to be at least 0.5 mm a$^{-1}$ (2.2 km per 4.2 Ma) and is closer to 1 mm a$^{-1}$ (2.5 km per 2.5 Ma) if only metamorphic samples (H1–H6) are compared. It is possible that this downward younging trend has been accentuated by thermal overprinting caused by c. 83 Ma dyke intrusions (see below), and this may explain the downward increase in excess argon (Fig. 5).

## White mica ($T_c$ c. 370°C)

Six white mica separates from metapelite and orthogneiss were analysed; three of these yielded well-behaved plateaux and three from isograd zone rocks yielded complex 'hump-shaped' spectra (Fig. 6) and will be discussed in the thermal overprint section below. The higher-grade samples were all collected from similar structural levels, near the sillimanite-in isograd (Fig. 4b) along the southern and western flanks of the range. The concordant ages of 83.0±0.8, 83.7 ± 0.5 and 83.1 ± 0.4 Ma (Figs 4b, 6; Table 1) come from widely separated localities, indicating essentially simultaneous cooling of this structural level through c. 370°C across the gneiss dome.

## Biotite ($T_c$ c. 335°C)

Four biotites from orthogneisses and paragneisses collected in the central and eastern portions of the range yielded nearly identical, but somewhat disturbed spectra. Interpreted ages for these biotites lie between 83.8 and 84.6 Ma with estimated errors of generally 0.4 Ma (Fig. 4b; Table 1). Ages obtained from biotite are within analytical error, but generally slightly older than ages obtained from white mica at the same locations. Several biotite spectra contain old apparent ages around the mid-point of the release spectrum and K/Ca ratios inversely track the apparent ages with lower ratios yielding older ages. This behaviour may be a consequence of excess argon or $^{39}$Ar recoil (Lo & Onstott 1989). To obtain interpreted ages from the complicated biotite spectra, we remove steps with low K/Ca ratios. Isochron plots of gas released during these experiments yield highly variable fits with ages generally within error of the discontinuous plateau ages.

There is no consistent variation in mica ages with either structural depth or geographical position, suggesting that rapid cooling through 335°C of the exposed, high-grade metamorphic complex occurred nearly simultaneously across the dome (Fig. 6). We interpret the nearly simultaneous cooling through mica closure irrespective of structural depth to indicate that doming of the lithological layering was largely completed by 83–84 Ma when the complex cooled through 300–350°C.

## Alkali feldspar ($T_c$ c. 300°C–150°C)

To obtain quantitative mid- to low-temperature cooling histories we ran detailed diffusion experiments on five K-feldspar samples from metamorphic rocks on the Mosquito Pass transect across the Kigluaik gneiss dome. All of the metamorphic K-feldspar yielded well-defined age gradients over most of the $^{39}$Ar released. Low-temperature apparent ages range from c. 65 to 75 Ma depending on the location of the sample, and high-temperature apparent ages consistently climb to 82 or 83 Ma (Fig. 7). The modelling results yielded generally straightforward Arrhenius plots and a good match between observed and modelled log $R/R_0$ plots (Table 2). Calculated activation energies ($E_a$) varied from 30.8 to 54.4 kcal/mol and frequency factors ($D_0$) varied from 2.25 to 9.51 s$^{-1}$, and reasonably good fits were obtained with 6–8 discrete diffusion domains (Table 2). Perhaps the biggest uncertainty in this modelling is in the absolute temperature assigned to the cooling

# DIAPIRIC ASCENT AND COOLING OF A GNEISS DOME

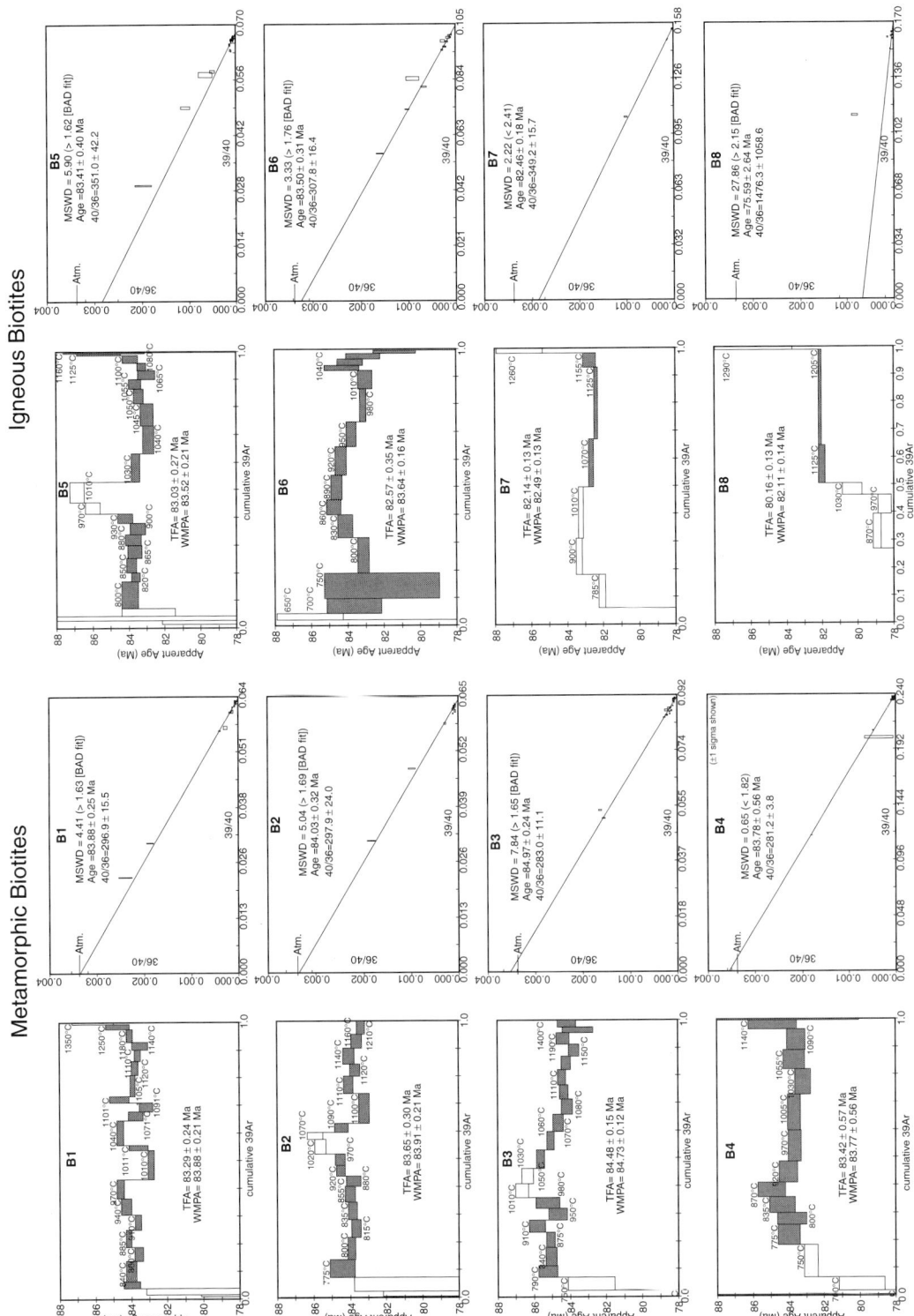

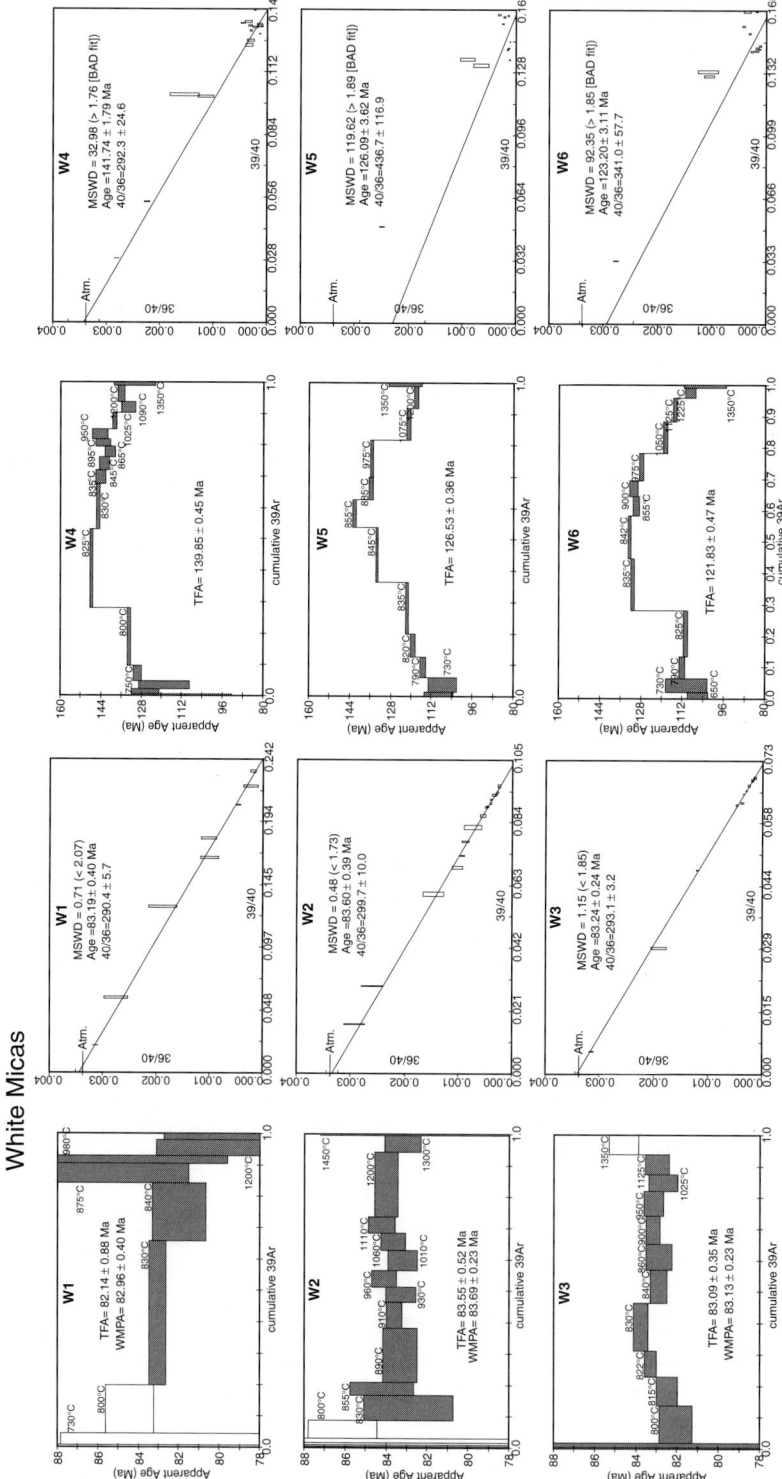

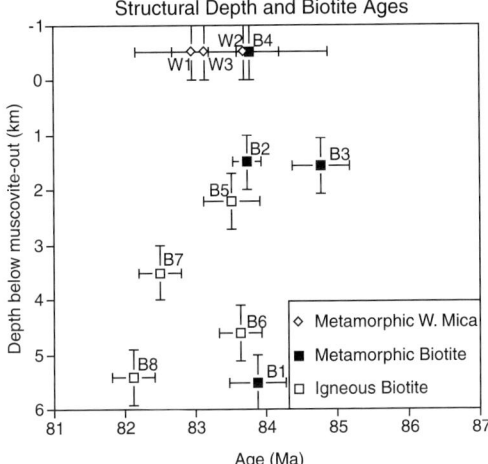

**Fig. 6.** (a) $^{40}Ar/^{39}Ar$ metamorphic and intrusive biotite age spectra and their isochron plots. Shaded steps are those used in weighted mean plateau age (WMPA) calculations. Isochron plots are shown with steps used for the lowest MSWD. Goodness of fit index (MSWD) values are from Roddick (1978). Interpreted ages are tabulated in Table 1. (b) Plots from white micas. (c) Summary of the mica ages with $2\sigma$ errors versus their projected depth beneath the muscovite-out isograd.

histories, as these are highly sensitive to the analytically measured temperatures of the resistance furnace. Several of the samples yielded closure temperatures higher or lower than those usually derived in these studies (>350°C or <150°C, Lovera et al. 1989) and we interpret those as contact problems between our thermocouple and crucible. As the behaviour is undoubtedly not linear across the temperature range, we did not attempt to correct any of the data. Despite these shortcomings, the modelling data generally fit well with hornblende and mica analyses and apatite fission-track results (Dumitru et al. 1995) from the same rocks. In addition, the trends we are attempting to quantify by generating cooling histories are apparent when comparing the unmodelled age spectra (Fig. 7).

K4, a sample from the Thompson Creek Orthogneiss (Fig. 2), is a representative example of our diffusion modelling experiments. This step-heating experiment comprised 104 steps between 15 min and 3 h long and yielded a detailed age spectrum with monotonically rising ages (Fig. 8). Some old apparent ages in the first 12% of the gas were quantitatively corrected by running replicate steps at those temperatures, to remove most of the excess argon. Using the crucible temperatures and the fraction of $^{39}Ar$ released for each step we fitted Arrhenius and log $(R/R_0)$ plots (Fig. 8) with seven discrete diffusion domains, $E_a = 46.0$ kcal mol$^{-1}$ and $D_0 = 6.83$ s$^{-1}$. To model the age spectrum, we ran 60 modelling cycles constraining the cooling to be monotonic, and the synthesized age spectra generally fitted the analytical data very well. The cooling history calculated from these synthesized spectra indicates moderate cooling rates from 330°C at 83 Ma to c. 170°C at 77 Ma followed by slower cooling to below 150°C at 73 Ma.

As is apparent by comparing the different age spectra, low-temperature portions of the cooling histories are a function of north–south position and proximity to the Kigluaik pluton, with older ages to the north and away from the pluton. All samples show fairly rapid (30–70°C Ma$^{-1}$) cooling between c. 83 and 78 Ma followed by much slower (<10°C Ma$^{-1}$) cooling younger than 78 or 76 Ma. Samples from the northern Kigluaik Mountains cooled 1–3 Ma sooner than those in the south, with the exception of K3, a sample subjected to reheating by a silicic intrusion at c. 83 Ma (Fig. 3; Table 1; see below).

*Summary of cooling history*

The combined thermochronological data from the metamorphic rocks in the Kigluaik gneiss dome reveal a fairly straightforward cooling history. Peak metamorphism is best dated by 91 Ma U/Pb monazite ages (Amato et al. 1994; Amato & Wright 1998). This was followed by a prolonged period of cooling to c. 550°C by 87–84 Ma, to c. 350°C by 83 Ma and to <200°C by c. 78 Ma (Fig. 4). Hornblende ages ($T_c$ c. 535°C) decrease with increasing structural depth (Fig. 5) and indicate that isotherms moved downward through the rock column at 0.5–1 mm a$^{-1}$. Samples at deepest structural levels are most disturbed, perhaps because of partial degassing during intrusion of dykes (see below) or because of higher ambient pressures of radiogenic argon. Mica ages are remarkably consistent throughout the gneiss dome and indicate that present levels of exposure all cooled through c. 350°C at 83–84 Ma. The three white mica ages are within analytical error and average 83.3 Ma whereas biotite results are weakly disturbed and ages vary from 83.8 to 84.6 Ma without regard to structural depth, present elevation or geographical coordinates (Fig. 6). Given the wide range of structural depth exposed across the range, we infer that doming was largely completed by this time. K-feldspar age gradients and modelled temperature–time histories demonstrate that rapid cooling continued until about 80 Ma and down to c. 200°C, and was followed by differential

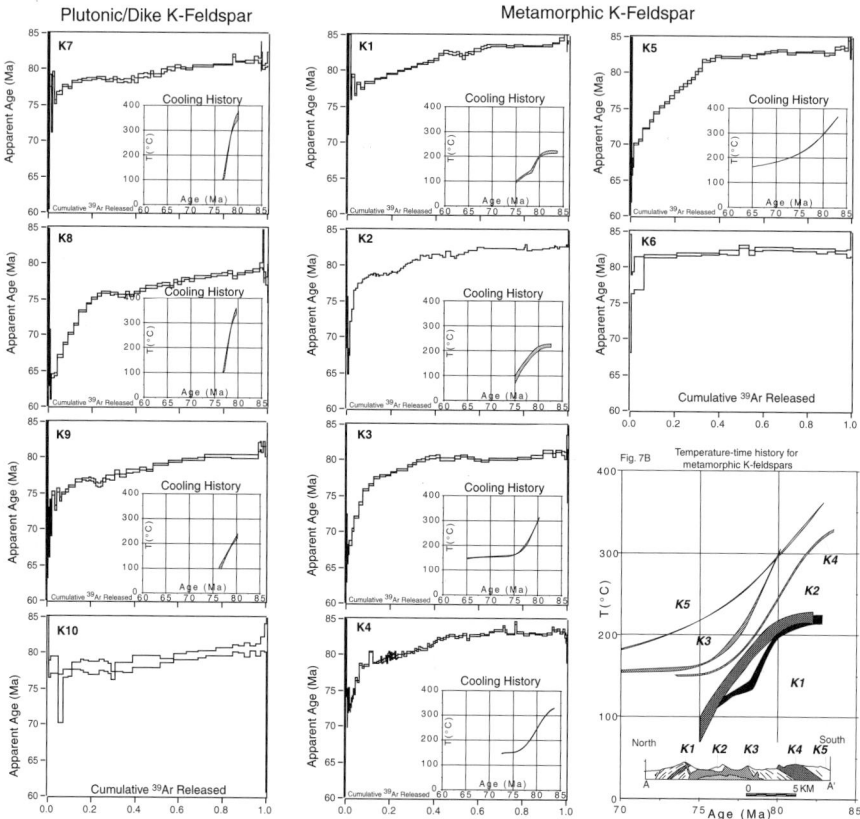

**Fig. 7.** (**a**) $^{40}Ar/^{39}Ar$ age spectra for K-feldspar with inset cooling histories determined by diffusion modelling of plutonic and metamorphic samples. Cooling histories are shown with 90% confidence intervals shaded. (**b**) K-feldspar cooling profiles shaded according to geographic position, with darkest to lightest from north to south. Sample locations are shown on inset cross-section. Samples cooled progressively from north to south except for K3, which we interpret was reheated by the silicic and diabase dykes.

cooling from north to south (Fig. 7). This low-temperature asymmetric cooling is attributed to differential exhumation and gentle southward tilting of the units, perhaps during early movement on the range-bounding Kigluaik fault (Fig. 2). This interpretation is supported by the fact that diabase dykes now dip steeply to the NW, compatible with a 10–20° southeastward tilt if they were intruded vertically. These lower-$T$ cooling histories from K-feldspar modelling are in accord with apatite fission-track results of Dumitru *et al.* (1995), which showed progressive cooling through 100°C from north to south in Eocene–Oligocene time.

## Age and cooling history of the Kigluaik pluton

One of the persistent questions concerning the origin of gneiss domes is why are they commonly associated with plutonic complexes? What role does magmatism play in their thermal and structural development? The Kigluaik gneiss dome is no exception, as it is cored by the Kigluaik plutonic complex, a large intrusion and associated dykes with a composition ranging from diorite to granite (Amato *et al.* 1994; Amato & Wright 1998). U/Pb ages of 90 Ma from the plutonic complex and 91 Ma from metamorphic rocks led Amato *et al.* (1994) to argue that it was emplaced in part during peak metamorphic conditions, and

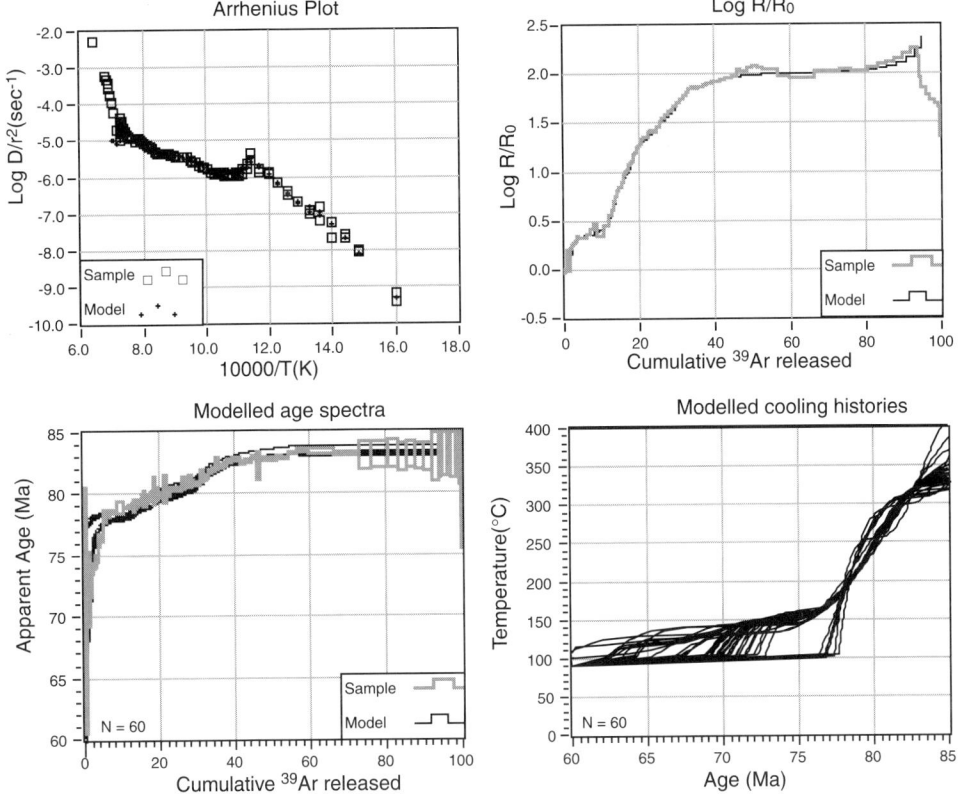

**Fig. 8.** Example of results of diffusion domain modelling (Lovera 1992) for K4 K-feldspar. (**a**) Arrhenius plot from the analysis (□) and the computer-fit model (+). (**b**) Log ($R/R_0$) plot for the sample (see Richter *et al.* 1991; Lovera 1992) (thick grey line) and the model (thin line). (**c**) Release spectrum for the sample with errors used to model (grey) and the 60 modelled spectra (black). (**d**) Range of cooling histories resulting from the modelled spectra.

they speculated that mantle-derived magmatism may have provided the heat for upper amphibolite facies conditions in the gneiss dome. As the pluton post-dates nearly all strain in the rocks, but was clearly intruded at high temperatures, we dated a suite of minerals from the plutonic complex and associated dykes to understand whether the plutonic and metamorphic country rocks have the same cooling history.

We analysed three hornblendes from the Kigluaik intrusive suite, two from the mafic root of the plutonic complex at its deepest levels and one from a fresh diabase dyke at high structural levels. The plutonic hornblendes were from the same samples that yielded U/Pb zircon ages of c. 90 Ma (Amato *et al.* 1994). These hornblendes yielded simple plateaus (Fig. 5) with indistinguishable ages of 82.2 ± 0.3 and 82.0 ± 0.3 Ma. The hornblende from the diabase dyke also yielded a plateau age of 82.8 ± 0.4 Ma.

Four biotites were dated from the intrusive rocks: B5 from a felsic intrusive mapped as the Kigluaik pluton, B6 from a felsic dyke associated with the pluton, and B7 and B8 from diabase dykes. B5 and B6 yielded similarly disturbed spectra with old apparent ages midway through the spectra and interpreted ages of 83.5 ± 0.4 and 83.6 ± 0.3 Ma. B7 and B8 from diabase dykes yielded 82.5 ± 0.3 and 82.1 ± 0.3 Ma plateau ages. B8 and H9 are from the same rock and show that the dyke cooled rapidly from >500°C to c. 335°C in less than 1 Ma.

Given the difference in ages of metamorphic and intrusive hornblendes and biotites, we dated K-feldspar samples from the Kigluaik pluton (K8) and from closely associated leucocratic dykes (K7, K9, K10) to determine if their low-temperature cooling profiles differed significantly from that of the metamorphic sequence. The three dyke samples and the plutonic sample yielded very similar argon release patterns and modelled cooling histories. All samples cooled rapidly from c. 300°C at 79–81 Ma to 150°C at 75 Ma.

These data do not define a simple cooling history for both metamorphic and plutonic rocks and we see three possible interpretations: (1) the U/Pb zircon ages may not date emplacement of the pluton, rather they could be inherited grains; (2) there may be a 7 Ma lag between zircon growth and pluton emplacement; (3) the pluton may be more complicated than we originally mapped. The U/Pb data are complicated (Amato & Wright 1998) and they used acid leaching experiments to remove high-uranium portions of the zircon to isolate a true age (Mattinson 1994). Scanning electron photomicrographs of leached zircons show that the portions dated to determine the 90 Ma age were euhedral igneous overgrowths on complicated, inclusion-rich zircons. As successive leach steps removed these cores and inclusions, we feel confident that the 90 Ma age is a zircon crystallization age. Although it seems farfetched that there is a 7 Ma lag between zircon growth and pluton emplacement, the two hornblendes from the intrusive rocks are slightly younger than the cooling trend of the metamorphic rocks (Fig. 5). However, the metamorphic hornblendes at deep structural levels yielded only maximum ages and may be closer in age to the 82 Ma plutonic hornblendes, or these plutonic hornblendes may have been slightly reheated by younger (c. 83 Ma) silicic and diabase intrusions. The hornblende from the dyke is clearly younger than the metamorphic hornblendes of its wall rocks, but is probably a younger dyke that post-dates the main plutonic complex.

Low-temperature thermochronology on biotite and K-feldspar from silicic dykes suggest that the cooling of some intrusive rocks is, in part, younger than cooling of the metamorphic complex. Intrusive rocks emplaced during peak metamorphic conditions should have low-temperature cooling histories comparable with those of metamorphic rocks. A comparison of intrusive and metamorphic K-feldspar from similar geographical positions indicates that the dated intrusive rocks consistently cooled at a younger time than the country rocks (Fig. 9). For example, K1 (metapelite) and K7 (leucocratic dyke) are separated by 2 km on the same ridge (Fig. 4c) and both show rapid cooling between >200° and 100°C, but the dyke cooled c. 5 Ma later. Similarly, samples K8 (silicic phase of Kigluaik pluton) and K2 (orthogneiss) are both from the core of the dome, but the plutonic rocks cooled through equivalent temperatures about 5–6 Ma after the metamorphic rocks. Samples K3 (metapelite) and K9 (leucocratic dyke) have more similar cooling histories, perhaps because K3 lies very close to and was reheated by the intrusion. However, even in this case, K3 is generally 50–100°C cooler than K9 until their cooling paths converge at 70 Ma (Fig. 7). We attribute the old apparent ages on biotites B5 and B6 to small amounts of excess argon, as detailed K-feldspar spectra from the same rocks (K8 and K10) reach only 78.5 and 80 Ma, and modelled temperatures from K8 show temperatures >300°C at 78 Ma. The combined geological and thermochronological data from the Kigluaik intrusive complex strongly suggest that it is a complicated unit; some portions were intruded at 90 Ma, shortly after peak metamorphic conditions, whereas some silicic and diabase dykes were intruded closer to 83 Ma.

## Thermal overprint of Kigluaik metamorphism on surrounding lower-grade rocks – the Big Creek transect

The nature of the contact between the sillimanite-grade rocks of the Kigluaik Mountains and surrounding blueschist–greenschist facies rocks of the Nome Group is crucial to understanding the origin of the gneiss dome. The contact is well exposed along the southern and western flank of the range (Fig. 2) and dips generally away from the Kigluaik Mountains, such that the Nome Group rocks structurally overlie the higher-grade rocks (Fig. 3). Importantly, this contact is not a discrete fault, but rather an abrupt transition or metamorphic field gradient defined by the sequential appearance of biotite, garnet, staurolite, sillimanite and finally sillimanite + K-feldspar. There is no fundamental break in rock type or evidence that any section has been omitted. Rather, the boundary between the Nome Group and Kigluaik Group is arbitrarily placed at the first appearance of biotite. These classic Barrovian isograds are remarkably closely spaced, as the entire transition from greenschist facies to upper amphibolite facies occurs over <2 km. It is likely that these isograds have been collapsed by strain that post-dates peak metamorphism (Miller et al. 1992), but there is no compelling sense of shear to the fabric. Instead, the dominant foliation and lineation of the polyphase fabrics in the Nome Group appear concordant with the relatively simple fabrics in the gneiss dome.

To better understand this boundary and directly compare the thermal history of the gneiss dome with that of the lower-grade Nome

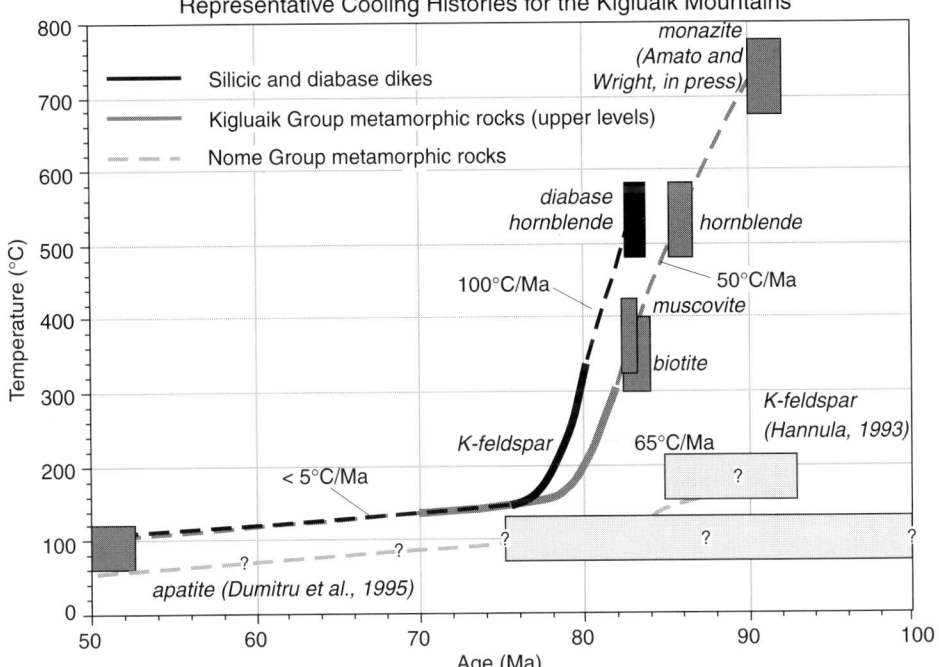

**Fig. 9.** Representative cooling histories for upper levels of the Kigluaik gneiss dome, silicic and diabase dykes, and Nome Group. Boxes show analytical data from c. 725°C (monazite from Amato & Wright (1998), 535°C (hornblende), 370°C (muscovite) and 335°C (biotite), and continuous lines show quantitative cooling histories from K-feldspar diffusion modelling. Apatite data are from Dumitru et al. (1995). Nome Group K-feldspar is an unpublished analysis from Hannula (1993).

Group rocks, we analysed a suite of white mica separates from samples that span the transition in the Big Creek area (Figs 2 and 3). We also hoped to evaluate at what metamorphic grade the older(?) micas of the Nome Group were completely reset by the Kigluaik metamorphism and to use these cooling ages from rocks that had only barely been reset as an indirect measure of the age of peak metamorphism. Although our results generally confirmed that the boundary is gradational with no discrete thermal break, they shed little light on either the age of metamorphism or the thermal history of the Nome Group. Instead, they provide important new insights into the retentivity of white mica (phengite) during thermal overprinting.

Earlier K–Ar dating of white micas in the Nome Group by Armstrong et al. (1986) yielded a broad range of ages from 122 to 194 Ma. A much more detailed $^{40}Ar/^{39}Ar$ investigation of these rocks found that the argon systematics were highly disturbed (Hannula et al. 1992; Hannula & McWilliams 1995). In particular, incremental heating analyses of Nome Group white micas consistently yielded convex upward, or 'hump-shaped' spectra, similar to mica spectra obtained from polymetamorphosed rocks in the Alps (Hammerschmidt & Frank 1991) and the Aegean (Wijbrans & McDougall 1986), with individual step ages as old as 334 Ma. Laser fusion analyses of single grains clearly demonstrated mixed populations of micas, with single grain ages ranging from 383 to 115 Ma. However, Hannula & McWilliams (1995) obtained relatively simple plateau ages of 116–125 Ma from the lowest-grade Nome Group rocks, so they concluded that all samples with older ages and complicated spectra or populations had incorporated excess argon.

We analysed four white micas and one biotite from a transect across the isograd interval between the Nome and Kigluaik Groups in the eastern Kigluaik Mountains (Figs 2 and 10; Table 1). Samples were collected from typical Nome Group metapelite (sample W4), and from the biotite zone (W5), the staurolite zone (W6, B4) and the first sillimanite zone (W1). The structurally highest sample (W4), yielded a disturbed

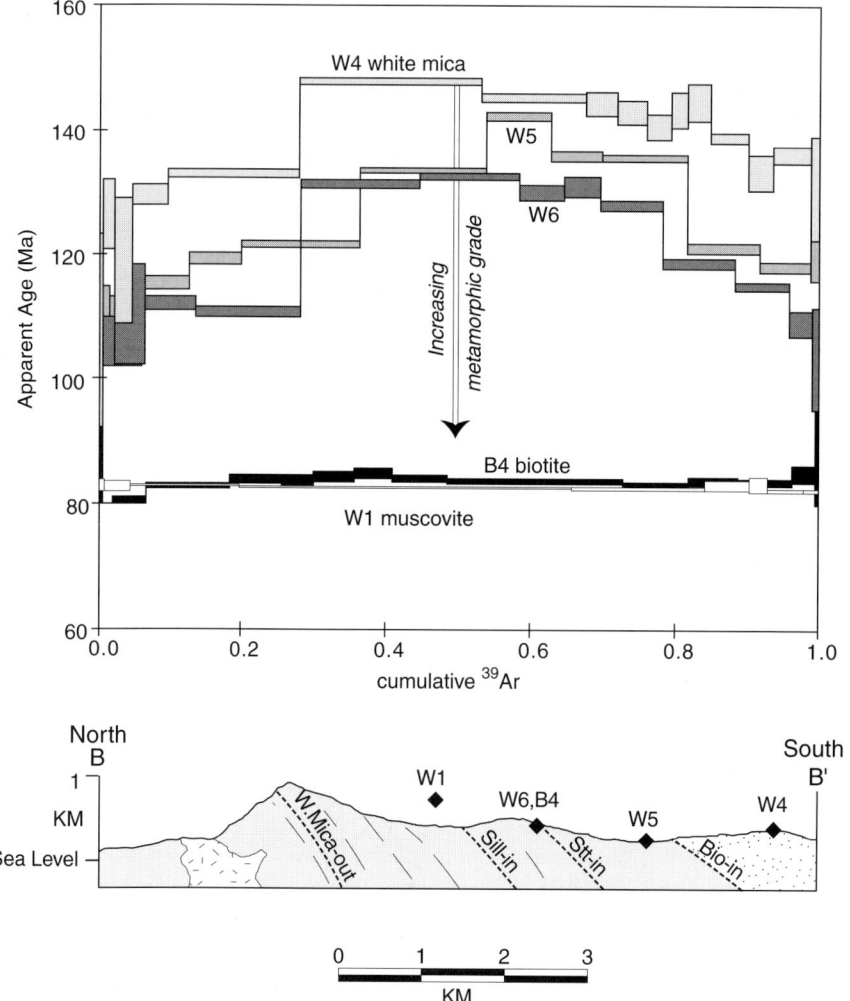

**Fig. 10.** $^{40}Ar/^{39}Ar$ white mica and biotite ages from a transect across Barrovian isograds in the eastern Kigluaik Mountains. White mica samples yield very disturbed spectra in all but the deepest rocks. Biotite results from staurolite grade and white mica results from sillimanite grade are interpreted as cooling ages.

hump-shaped spectrum with single step ages between 125 and 147 Ma and a total gas age of 140 Ma, similar to what had been obtained elsewhere in the Nome Group. The intermediate samples (W5 and W6) also yielded disturbed hump-shaped spectra, but the total gas ages decrease to 127 and 122 Ma, and both the maximum and minimum single-step ages decrease progressively into the higher-grade rocks (Fig. 10). These samples have two generations of mica clearly visible in thin section; one is a fine-grained white mica that is concordant with the younger dominant foliation and the other is a relict kinked and inclusion-ridden, coarser-grained mica. The same staurolite zone rock that yielded a white mica (W6) total gas age of 122 Ma yields a biotite (B4) plateau age of 83.8 ± 1.1 Ma. It is only at sillimanite grade, where there is incipient breakdown of white mica to K-feldspar + quartz, that white mica yields a well-behaved plateau age of 83.0 ± 0.8 Ma (W1).

The smooth gradient from old total gas ages and hump-shaped spectra in the Nome Group to young plateaux in the Kigluaik Group supports the idea that there is no structural break; rather there is a progressive metamorphic overprint of the Nome Group by the younger Kigluaik metamorphism. The most surprising result is that even at staurolite grade (i.e. metamorphism at temperatures >500°C), older white micas were apparently not completely reset. This implies a closure temperature of phengite that is well above the commonly quoted 350°C for muscovite (Snee *et al.* 1988) and the 370°C calculated from Lister & Baldwin (1996). As temperatures in excess of 500°C were needed to completely degas the older Nome Group white mica and, assuming these rocks subsequently cooled quickly, it is tempting to interpret the 83 Ma age as close to the age of peak metamorphism at this level, given an apparently high $T_c$ for white mica of >500°C. If so, the timing of peak metamorphism in the isograds must have lagged far behind the 91 Ma peak metamorphism at only slightly deeper structural levels in the range. This seems improbable because conductive heating rates for length scales of only a few kilometres are thought to be much faster. Our favoured interpretation is that the 83 Ma age reflects cooling of newly grown white mica through a lower $T_c$ (c. 370°C) well after peak metamorphism in the sillimanite-grade rocks, and that this cooling is recorded only in rocks where complete recrystallization of the more retentive phengite has occurred.

## Summary and discussion

The $^{40}Ar/^{39}Ar$ thermochronological data presented in this study in conjunction with the field relations, previous geochronology and metamorphic studies allow us to place important new constraints on the evolution of the Kigluaik Mountains. Any model for the origin of this gneiss dome and for the processes that unroofed the mid-crustal, sillimanite-grade rocks must account for the following observations.

(1) Upper amphibolite facies metamorphism and widespread partial melting of the Kigluaik Group occurred in late Cretaceous time and rocks first cooled through c. 725°C at c. 91 Ma (Amato & Wright 1998). The ambient pressures during peak metamorphism are poorly constrained, but limited barometric data from garnet and spinel lherzolites at the deepest structural levels (Lieberman 1988) and the local occurrence of kyanite and staurolite inclusions in garnet throughout the gneiss dome are compatible with peak metamorphism at moderately high pressures of 800–1200 MPa and temperatures in excess of 700°C (Lieberman 1988). Synmetamorphic deformational fabrics within the high-grade rocks consist of a high strain dominant foliation parallel to compositional layering, and well-developed stretching lineations that are oriented east–west in the core of the range. No consistent sense of shear is apparent in the amphibolite facies tectonites.

(2) After deformation had largely ended, but apparently before significant cooling of the complex took place, the mafic to silicic Kigluaik pluton was emplaced at 90 ± 1 Ma. Pressures and temperatures in the aureole of the pluton are estimated at 500–600 MPa and less than c. 780°C (Amato & Wright 1997).

(3) Following peak metamorphism and pluton intrusion, the complex cooled rapidly during late Cretaceous time, passing through c. 550°C at 85–86 Ma, 370°C at c. 83 Ma, and from 300 to 150°C between c. 82 Ma and 78 Ma. Thus, moderately rapid cooling rates of c. 35–75°C Ma$^{-1}$ were sustained for c. 12 Ma as the entire complex cooled by more than 550°C (Fig. 9). After 75–78 Ma, cooling rates abruptly decreased to <5°C Ma$^{-1}$ and the gneissic rocks resided at 150–100°C for a prolonged period. In Eocene time, renewed exhumation and moderately rapid cooling through 100°C is recorded by apatite fission-track dating (Dumitru *et al.* 1995), presumably related to the final exhumation of the complex.

(4) Spatial variations in cooling histories of the metamorphic rocks indicate that high-temperature cooling (≥500°C) probably occurred before doming of the complex. At 83–84 Ma all but the deepest levels of exposed metamorphic rocks cooled through c. 350°C simultaneously as subhorizontal isotherms passed downward through rocks that had already been strained and domed. Deepest structural levels cooled through hornblende closure at 82–84 Ma. Lower-temperature cooling (300–150°C) is somewhat asymmetrical and indicates that the rocks on the northern flank of the range cooled earlier, perhaps in conjunction with gentle southward tilting (Fig. 7).

(5) Diabase dykes cut across all units in the dome and yield an 82.8 Ma $^{40}Ar/^{39}Ar$ hornblende age. Lower-temperature cooling histories from at least some of the silicic intrusives are consistently younger than from their metamorphic country rocks and suggest that they were emplaced at c. 83 Ma, after the gneiss dome had already risen to shallow crustal levels and cooled to c. 300°C (Fig. 9). These intrusive rocks provided enough heat to locally perturb temperatures, but samples 3–4 km away from these intrusions do not show reheating.

(6) The contact between the sillimanite-grade rocks of the Kigluaik gneiss dome and older, much lower grade (blueschist–greenschist facies) rocks of the Nome Group is not a discrete fault; instead, it is a gradational metamorphic transition across very closely spaced isograds and there is no evidence for structural omission. It is

most reminiscent of a thermal aureole adjacent to a large plutonic complex. $^{40}Ar/^{39}Ar$ data from this transition appear to reflect progressive thermal overprinting of the Nome Group rocks as a consequence of the emplacement of the hot Kigluaik Group rocks (Fig. 10) The progressive rotation of lineations from E–W at deep structural levels to N–S in biotite-grade rocks is striking. We suggest that the lineations on the southern and western flanks record the emplacement of the dome at high structural levels and is the youngest deformational fabric in the area. Late-stage asymmetric folds and locally developed retrograde shear bands generally indicate radially directed transport, down-dip on the flanks of the present dome.

(7) Deformational fabrics that, at first inspection, appear to be continuous in the field are now interpreted to have formed at three distinct times and under very different conditions. Polyphase blueschist–greenschist facies penetrative fabrics in the Nome Group, including the dominant high strain foliation, may have formed during either regional shortening (Patrick 1988) or extension (Dumitru et al. 1995; Hannula et al. 1995), but in either case these rocks had finished deforming and lay at near-surface conditions (<200°C) by 90 Ma. The high-grade granoblastic gneissic foliation and associated E–W sillimanite lineation is apparently tightly constrained to be 91–90 Ma old, as it is associated with 91 Ma peak metamorphism but almost entirely pre-dates the crosscutting 90 Ma Kigluaik pluton. Finally, greenschist facies fabrics with a well-developed NW–SE to N–S stretching lineation in the isograd transition zone are interpreted to have continued to form until c. 83–82 Ma, when these rocks cooled quickly from >350° to <200°C.

Previous studies have called on regional shortening (Patrick & Lieberman 1988) and extension (Miller et al. 1992; Hannula et al. 1995) to exhume the gneissic complex, yet both of these models do not fit all of the field and timing relationships presented above. Implicit in both of these models is a unidirectional strain field (here c. N–S) that poorly explains E–W stretching lineations at deep structural levels and concentric foliations with down-dip shear indicators on the flanks of the dome. Patrick & Lieberman (1988) argued that all fabrics in the Nome and Kigluaik Groups are old (pre-late Cretaceous) and associated with shortening based on the apparent continuity of fabrics. They suggested that the high-temperature sillimanite metamorphism was a static event with mimetic growth of high-temperature minerals. We disagree with this conclusion because peak metamorphic minerals (i.e. sillimanite and hornblende) are clearly synkinematic. Furthermore, the timing of peak metamorphism and strain at deep structural levels is apparently 10–30 Ma younger than cooling in the Nome Group (Hannula et al. 1995). Regional extension is an attractive mechanism for juxtaposing rocks from different crustal levels, but it is difficult to reconcile E–W flow at one level with N–S flow at another. Brittle structures that are clearly extensional are not exposed in the cover rocks and there are no asymmetric cooling trends in the higher-grade rocks such as those normally identified in the footwalls of large normal faults (e.g. Hill et al. 1992; John & Foster 1993).

The combined structural and thermochronological data are most compatible with a model for the Kigluaik gneiss dome that invokes diapiric ascent of mid-crustal metamorphic rocks to shallow crustal levels followed by minor late-stage southward tilting. The overall development of the gneiss dome and the inferred position of the high-grade metamorphic rocks at various times during the unroofing are illustrated schematically in Fig. 11. This conceptual model attempts to explain several aspects of the Kigluaik Mountains that are not easily explained by other models.

The sustained, practically linear, rapid cooling history of the high-grade metamorphic rocks in the Kigluaik Mountains from temperatures in excess of 700°C to less than 200°C in less than 10 Ma is the single most important result of this study. As cooling rates are proportional to the thermal gradient (Spear 1993), reproducing our linear cooling history requires that the gradient remains constant during cooling. Thus, the Kigluaik Group must have been gradually exhumed and emplaced against progressively cooler rocks throughout the 90–80 Ma period. In contrast, a more rapid exhumation history followed by conductive relaxation of the isotherms would generate exponentially decreasing cooling rates (Spear 1993). Progressive unroofing in the footwall of a major, long-lived normal fault or in the hanging wall of a thrust fault could produce a similar linear cooling history, but no such structures are exposed. Furthermore, the symmetric high-temperature cooling history and the

**Fig. 11.** Time slices show the diapiric rise of the Kigluaik gneiss dome based on thermochronological data and thermobarometric work of Lieberman (1988). Shaded area shows the extent of rocks subjected to sillimanite-grade metamorphism. Stippled areas show deforming rocks at each time slice. Evolution of locations and temperatures of four levels (A–D) from the existing cross-section are shown on the time slices.

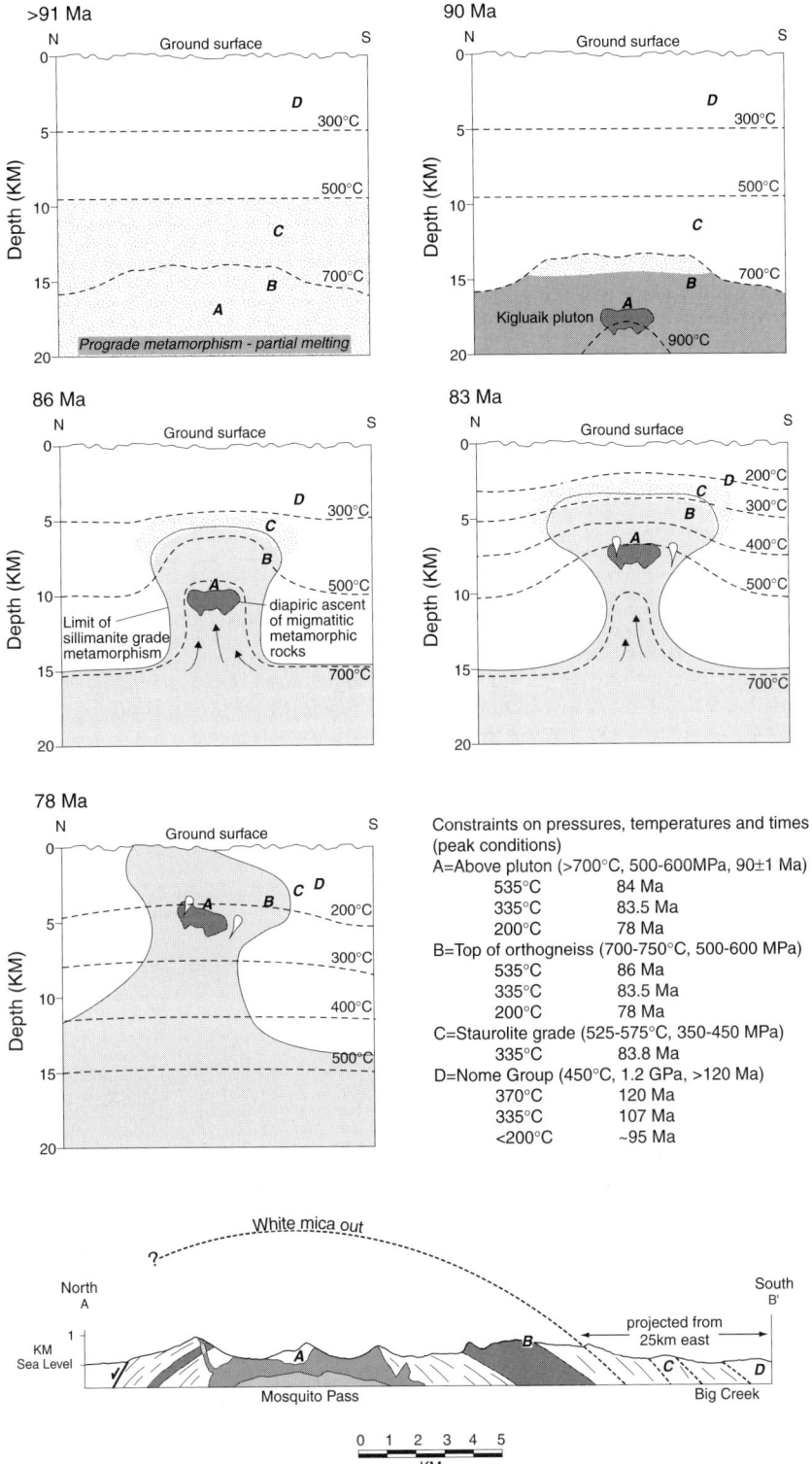

concentric distribution of tectonite fabrics (Fig. 2) around the high-grade rocks do not suggest the unidirectional shear that is implicit in such a tectonic model. We believe that the most straightforward explanation for the sustained rapid cooling of the Kigluaik Group rocks is provided by the exposed contact with the lower-grade Nome Group rocks. The resemblance of this contact to a thermal aureole with closely spaced isograds suggests that the Nome Group acted as the heat sink for cooling the Kigluaik gneiss dome as it rose to shallow crustal levels. The development of the domal geometry while still at reasonably elevated temperatures coupled with late-stage structures indicating down-dip transport and perhaps high strains along the steep margins of the dome add further credence to the diapiric model.

The ultimate cause of the high-grade metamorphism and the driving force for the ascent of the gneissic rocks are not clear. Metamorphism may have been the natural consequence of conductive relaxation of isotherms in crust that was previously thickened in the Brookian Orogeny (Patrick & Lieberman 1988), or it may have been triggered by the input of heat during widespread late Cretaceous mantle-derived magmatism (Amato et al. 1994). Once peak metamorphic conditions were achieved and widespread partial melting had occurred, ascent may have been driven mainly by the buoyancy inherent in metamorphic rocks that contain a large volume fraction of silicic melt segregations or it may have been carried up by an unexposed rising pluton. The cause of the asymmetry in the low-temperature cooling histories is unclear. It implies a southward tilting of the range in present geographic coordinates, and we speculate that this may reflect early movement on the Kigluaik Fault.

Perhaps the most controversial aspect of the evolutionary model that we propose for the Kigluaik Mountains is that it is entirely independent of tectonic setting. We propose that this large mass of high-grade metamorphic rocks was severely strained while at high temperature and then unroofed rapidly perhaps as much as 20 km to be juxtaposed against much lower grade rocks, yet we attribute this entirely to gravity, i.e. buoyancy forces acting on a low-density diapir of migmatitic rocks. The regional tectonic setting at the time may have been extensional, contractional or neutral.

This work constitutes a portion of the first author's PhD dissertation at UCSB. Field work and initial analytical work were funded by NSF grant EAR 90–18922 awarded to E. L. Miller. Remaining analytical work was funded by NSF EAR 93–17142 awarded to P. B. G. Field work for this study was conducted by the authors, K. Hannula and T. Little. E. Gans superbly drafted figures. O. Lovera and F. Spera generously provided diffusion modelling software and hardware, respectively. Discussions with J. Mattinson and W. McClelland helped us appreciate the complexities of U/Pb dating. U. Ring, P. Layer, V. Sisson and an anonymous reviewer improved the clarity and content of the manuscript.

## References

AMATO, J. M., 1995. *Tectonic evolution and petrogenesis of the Kigluaik Gneiss Dome, Seward Peninsula, Alaska: an integrated structural and geochemical study of extensional processes in mid-crustal rocks.* PhD thesis, Stanford University, CA.

—— & WRIGHT, J. E. 1997. Potassic mafic magmatism in the Kigluaik gneiss dome, Northern Alaska; a geochemical study of arc magmatism in an extensional tectonic setting. *Journal of Geophysical Research, B, Solid Earth and Planets*, **102**(4), 8065–8084.

—— & —— 1998. Geochronologic investigations of magmatism and metamorphism within the Kigluaik Mountains Gneiss dome, Seward Peninsula, Alaska. *In*: CLOUGH, J. G. & LARSON, F. (eds) *Short Notes on Alaska Geology, 1997*. Professional Report **118**. State of Alaska; Department of National Resources; Division of Geological and Geophysical Survey.

——, ——, GANS, P. B. & MILLER, E. L. 1994. Magmatically induced metamorphism and deformation in the Kigluaik Gneiss dome, Seward Peninsula, Alaska. *Tectonics*, **13**(3), 515–527.

ARMSTRONG, R. L., HARAKAL, J. E., FORBES, R. B., EVANS, B. W. & THURSTON, S. P. 1986. Rb–Sr and K–Ar study of metamorphic rocks of the Seward Peninsula and southern Brooks Range, Alaska. *In*: EVANS, R. B. & BROWN, E. H. (eds) *Blueschists and Eclogites*. Geological Society of America Memoir, **164**, 185–203.

BERNER, H., RAMBERG, H. & STEPHANSSON, O. 1972. Diapirism in theory and experiment. *Tectonophysics*, **15**, 197–218.

BURG, J. P., GUIRAUD, M., CHEN, G. M. & LI, G. C. 1984. Himalayan metamorphism and deformations in the north Himalayan Belt (southern Tibet, China). *Earth and Planetary Science Letters*, **69**, 391–400.

CRITTENDEN, M. D., CONEY, P. J. & DAVIS, G. H. (eds) 1980. *Cordilleran Metamorphic Core Complexes*. Geological Society of America Memoir, **153**.

DALRYMPLE, G. B. & DUFFIELD, W. A. 1988. High precision $^{40}Ar/^{39}Ar$ of Oligocene rhyolites from the Mogollon–Datil volcanic field using a continuous laser system. *Geophysical Research Letters*, **15**, 463–466.

DUMITRU, T. A., MILLER, E. L., O'SULLIVAN, P. B., AMATO, J. M., HANNULA, K. A., CALVERT, A. T. & GANS, P. B. 1995. Cretaceous to Recent extension in the Bering Strait region, Alaska. *Tectonics*, **14**(3), 549–563.

DUSEL-BACON, C., BROSGE, W. P., TILL, A. B., FITCH, M. R., MAYFIELD, C. F., REISER, H. N. & MILLER, T. P. 1989. *Distribution, Facies, Ages and Proposed Tectonic Associations of Regionally Metamorphosed Rocks in Northern Alaska*. US Geological Survey Professional Paper **1497A**.

EVANS, B. W. & PATRICK, B. E. 1987. Phengite-3T in high-pressure metamorphosed granitic orthogneisses, Seward Peninsula, Alaska. *Canadian Mineralogist*, **25**, 141–158.

GROVE, M. & HARRISON, T. M. 1996. $^{40}$Ar* diffusion in Fe-rich biotite. *American Mineralogist*, **81**, 940–951.

HAMMERSCHMIDT, K. & FRANK, E. 1991. Relics of high pressure metamorphism in the Lepontine Alps (Switzerland) – $^{40}$Ar/$^{39}$Ar and microprobe analyses on white K-micas. *Schweizerische Mineralogische und Petrographische Mitteilungen*, **71**, 261–274.

HANNULA, K. A. 1993. *Relations between deformation, metamorphism, and exhumation in the Nome Group blueschist–greenschist terrane, Seward Peninsula, Alaska*. PhD thesis, Stanford University, CA.

——, & McWILLIAMS, M. O. 1995. Reconsideration of the age of blueschist-facies metamorphism on the Seward Peninsula, Alaska, based on phengite $^{40}$Ar/$^{39}$Ar results. *Journal of Metamorphic Geology*, **13**, 125–139.

——, —— & GANS, P. B. 1992. $^{40}$Ar/$^{39}$Ar and compositional data on white micas, Seward Peninsula, Alaska: implications for the age of blueschist metamorphism. *Geological Society of America Abstracts with Programs*, **24**, 5.

——, MILLER, E. L., DUMITRU, T. A., LEE, J. & RUBIN, C. M. 1995. Structural and metamorphic relations in the southwest Seward Peninsula, Alaska: crustal extension and the unroofing of blueschists. *Geological Society of America Bulletin*, **107**(5), 536–553.

HANSON, G. N. & GAST, P. W. 1967. Kinetic studies in contact metamorphic zones. *Geochimica et Cosmochimica Acta*, **31**, 1119–1153.

HARRISON, T. M. 1981. Diffusion of $^{40}$Ar in hornblende. *Contributions to Mineralogy and Petrology*, **78**(3), 324–331.

——, DUNCAN, I. & McDOUGALL, I. 1985. Diffusion of $^{40}$Ar in biotite: temperature, pressure and compositional effects. *Geochimica et Cosmochimica Acta*, **49**, 2461–2468.

——, HEIZLER, M. T., LOVERA, O. M., CHEN, W. & GROVE, M. 1994. A chlorine disinfectant for excess argon released from K-feldspar during step heating. *Earth and Planetary Science Letters*, **123**(1–4), 95–104.

HILL, E. J., BALDWIN, S. L. & LISTER, G. S. 1992. Unroofing of active metamorphic core complexes in the D'Entrecasteaux Islands, Papua New Guinea. *Geology*, **20**, 907–910.

JOHN, B. E. & FOSTER, D. A. 1993. Structural and thermal constraints on the initiation angle of detachment faulting in the southern Basin and Range; the Chemehuevi Mountains case study. *Geological Society of American Bulletin*, **105**(8), 1091–1108.

LIEBERMAN, J. E. 1988. *Metamorphic and structural studies of the Kigluaik Mountains, Western Alaska*. PhD thesis, University of Washington.

LISTER, G. S. & BALDWIN, S. L. 1996. Modelling the effect of arbitrary P–T–t histories on argon diffusion in minerals using the MacArgon program for the Apple Macintosh. *Tectonophysics*, **253**(1–2), 83–109.

LO, C. H. & ONSTOTT, T. C. 1989. $^{39}$Ar recoil artifacts in chloritized biotite. *Geochimica et Cosmochimica Acta*, **53**, 2697–2711.

LOVERA, O. M. 1992. Computer programs to model $^{40}$Ar/$^{39}$Ar diffusion data from multidomain samples. *Computers in Geosciences*, **18**(7), 789–813.

——, RICHTER, F. M. & HARRISON, T. M. 1989. The $^{40}$Ar/$^{39}$Ar thermochronology for slowly cooled samples having a distribution of diffusion domain sizes. *Journal of Geophysical Research*, **94**(B12), 17917–17935.

MATTINSON, J. M. 1994. A study of complex discordance in zircons using step-wise dissolution techniques. *Contributions to Mineralogy and Petrology*, **116**, 117–129.

MAYFIELD, C. F., TAILLEUR, I. L. & ELLERSIECK, I. 1988. Stratigraphy, structure and palinspastic synthesis of the western Brooks Range, northwestern Alaska. *In:* GRYC, G. (ed.) *Geology and Exploration of the National Petroleum Reservoir in Alaska, 1974–1982*. US Geological Survey, Professional Paper, **1399**, 143–186.

MILLER, E. L., CALVERT, A. T. & LITTLE, T. A. 1992. Strain-collapsed metamorphic isograds in a sillimanite gneiss dome, Seward Peninsula, Alaska. *Geology*, **20**(6), 487–490.

MOORE, T. E., WALLACE, W. K., BIRD, K. J., KARL, S. M., MILL, C. G. & DILLON, J. T. 1992. *Stratigraphy, structure, and geologic synthesis of northern Alaska*. US Geological Survey Open File Report **92–330**.

PARRISH, R. R. 1990. U–Pb dating of monazite and its application to geologic problems. *Canadian Journal of Earth Science*, **27**, 1431–1450.

PATRICK, B. E. 1988. Synmetamorphic structural evolution of the Seward Peninsula blueschist terrane, Alaska. *Journal of Structural Geology*, **10**, 555–565.

—— & EVANS, B. W. 1989. Metamorphic evolution of the Seward Peninsula blueschist terrane. *Journal of Petrology*, **30**(3), 531–555.

—— & LIEBERMAN, J. E. 1988. Thermal overprint on the Seward Peninsula blueschist facies terrane: the Lepontine in Alaska. *Geology*, **16**(12), 1100–1103.

—— & McCLELLAND, W. C. 1995. Late Proterozoic granitic magmatism on Seward Peninsula and a Barentian origin for Arctic Alaska–Chukotka. *Geology*, **23**(1), 81–84.

RAMBERG, J. 1980. Diapirism and gravity collapse in the Scandinavian Caledonides. *Journal of the Geological Society, London*, **137**, 261–270.

RICHTER, F. M., LOVERA, O, HARRISON, T. M. & COPELAND, P. 1991. Tibetan tectonics from $^{40}$Ar/$^{39}$Ar analysis of a single K-feldspar sample. *Earth and Planetary Science Letters*, **105**(1–3), 266–278.

RODDICK, J. C. 1978. The application of isochron diagrams in $^{40}$Ar–$^{39}$Ar dating; a discussion. *Earth and Planetary Science Letters*, **41**(2), 233–244.

SAINSBURY, C. L., COLEMAN, R. G. & KACHADOORIAN, R. 1972. *Reconnaissance geological map of the Nome quadrangle, Seward Peninsula, Alaska*. U.S. Geological Survey Open-File Report **72-326**.

SNEE, L. W., SUTTER, J. F. & KELLY, W. C. 1988. Thermochronology of economic mineral deposits, dating the stages of mineralization at Panesqueira, Portugal, by high-precision $^{40}$Ar/$^{39}$Ar age spectrum techniques on muscovite. *Economic Geology*, **83**, 335–354.

SPEAR, F. S. 1993. *Metamorphic Phase Equilibria and Pressure Temperature Time Paths*. Mineralogical Society of America Monographs, **1**.

STURNICK, M. A. 1984. *Metamorphic petrology, geothermobarometry and geochronology of the eastern Kigluaik Mountains, Seward Peninsula, Alaska*. MS thesis, University of Alaska, Fairbanks.

TILL, A. B., DUMOULIN, J. A., GAMBLE, B., KAUFMAN, D. & CARROLL, P. I. 1986. *Preliminary geologic map and fossil data, Solomon, Bendeleben and southern Kotzebue quadrangles, Seward Peninsula, Alaska*. US Geological Survey Open File Report **86–276**.

TURNER, D. L. & SWANSON, S. E. 1981. Continental rifting; a new tectonic model for the central Seward Peninsula. *In*: WESCOTT, E. M. & TURNER, D. L. (eds) *Geothermal Reconnaissance Survey of the Central Seward Peninsula, Alaska*. University of Alaska Geophysical Institute, **UAG-R-284**, 7–36.

WIJBRANS, J. R. & MCDOUGALL, I. 1986. $^{40}$Ar/$^{39}$Ar dating of white micas from an Alpine high-pressure metamorphic belt on Naxos (Greece): the resetting of the argon isotopic system. *Contributions to Mineralogy and Petrology*, **93**, 187–194.

WORRALL, D. M. 1991. *Tectonic History of the Bering Sea and the Evolution of Tertiary strike-Slip Basins of the Bering Shelf*. Geological Society of America, Special Paper, **257**.

# Exposure of deep, dense rocks: interplay between erosion and sinking

ALLEN F. GLAZNER

*Department of Geology, CB# 3315, University of North Carolina, Chapel Hill, NC 27599, USA (e-mail: afg@unc.edu)*

**Abstract:** A numerical model of the interplay between viscous sinking and erosion predicts that the ultimate fate of dense rock bodies in the crust is a sensitive function of erosion rate and of the body's radius, density, and initial depth of emplacement. If all of these variables save one are held constant, the trajectories taken by the body on a depth–time plot diverge widely about a critical value of the remaining variable. For example, if erosion rate, depth, and density are held constant, there is a critical value of the body's radius below which it will be carried to the Earth's surface by erosion and above which it will plummet to neutral buoyancy in the deep crust. This critical sensitivity may explain, for example, why dense plutons larger than a few kilometres in diameter are rare in plutonic terranes. The calculations indicate that exceptionally high erosion or tectonic denudation rates, on the order of 10 km Ma$^{-1}$ or more, may be necessary to bring dense bodies to the surface before they sink through less dense crust.

Erosion is one mechanism by which deep-seated rocks may be exposed at the Earth's surface. Erosion rates estimated for tectonically active areas commonly exceed 1 km Ma$^{-1}$ (Benjamin *et al.* 1987; Burbank & Beck 1991; Brandon & Ring 1997), so deep rocks may be elevated at geologically rapid rates by this process. However, there is a competing effect that may keep rocks buried. Dense rock bodies (e.g. gabbro plutons or pods of eclogite) in less dense crust are gravitationally unstable and may sink if the enclosing rocks are warm enough or weak enough to be ductile at relevant time scales. For example, Glazner (1994) showed that the curious absence of large mafic bodies in mid-crustal plutonic areas may be explained by sinking of those plutons following the density increase that occurs during crystallization. Glazner & Miller (1997) gave field evidence that even granodiorite plutons in the middle crust sink when emplaced into relatively ductile, low-density wall rocks such as quartzite or calcite marble.

Therefore, a competition between elevation by erosion and sinking by negative buoyancy may control whether or not a dense rock body approaches the Earth's surface. If erosion is more rapid than sinking then the body will eventually reach the surface; if erosion is slower than sinking then the body will sink to a level of neutral buoyancy. This interplay is complicated by several important nonlinearities. In particular, the effective viscosity of the crust, which controls the rate of sinking, decreases exponentially with temperature. A rock body sinking through the crust will thus generally encounter progressively less viscous rocks with depth and will accelerate until achieving neutral buoyancy.

In this paper I consider the fate of dense plutons emplaced at mid-crustal levels. The results, however, are applicable to the general problem of a dense rock body (e.g. eclogite) emplaced in the crust tectonically, although the thermal problem will be somewhat different.

## General considerations and assumptions

### Viscosity needed for rapid pluton sinking

Figure 1 shows the crustal viscosity needed for a spherical body of given radius, 200 kg m$^{-3}$ denser than its surroundings, to sink at the geologically rapid rates of 1–100 km Ma$^{-1}$, assuming Stokes' Law. For plutons with radii of several kilometres, viscosities of 10$^{19}$–10$^{20}$ Pa s lead to rapid sinking. Thus, if crustal viscosity is so low, dense bodies will sink rapidly (assuming quasi-Newtonian viscous behaviour of the enclosing rock) and exposure at the surface will be difficult.

Geophysical studies indicate that crustal viscosities on the order of 10$^{20}$ Pa s are not unreasonable. For example, Wdowinski & Axen (1992) inferred lower-crustal viscosities of 10$^{19}$–10$^{21}$ Pa s from studies of isostatic rebound after tectonic denudation in the Basin and Range province. Masek *et al.* (1994) estimated maximum lower-crustal viscosities of 10$^{20}$–10$^{22}$ Pa s based on topographical analysis of rift flank

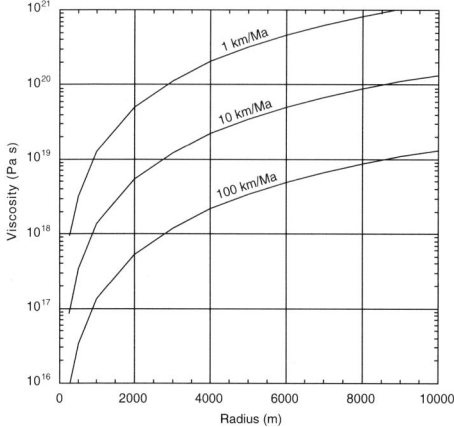

**Fig. 1.** Viscosity needed for a spherical body, 200 kg m$^{-3}$ denser than its surroundings, to fall at the indicated rate according to Stokes' Law. Geologically rapid rates (c. 10 km Ma$^{-1}$) require viscosities in the range $10^{19}$–$10^{20}$ Pa s, depending on radius.

uplift in Tibet. Hacker *et al.* (1992) used quartz grain-size piezometry to estimate viscosities in mid-crustal mylonite zones as low as $10^{18}$ Pa s in the Whipple Mountains of eastern California. Bills *et al.* (1994) estimated mid-crustal and lower-crustal viscosities in Utah to be <$10^{21}$ Pa s, on the basis of rebound of Lake Bonneville shorelines (this estimate is based upon loading on thousand-year time scales, whereas the other estimates, and this paper, deal with million-year time scales). Thus, it is possible that sinking of dense rock bodies is a significant process of mass redistribution in the middle and lower crust.

To investigate the interplay between pluton sinking and erosion I constructed a simple numerical model of the depth evolution of a dense rock body in an eroding, viscous crust.

*Assumptions*

In approaching this problem I make several important simplifying assumptions, as follows.

(1) The crust obeys a thermally activated power-law rheology. In such a crust the effective viscosity is a function of differential stress and temperature and decreases exponentially with temperature (Fig. 2). This is a justifiable model because (a) micromechanical theory predicts that steady-state dislocation creep, which microstructures of naturally deformed rocks indicate is the predominant solid-state flow mechanism in the crust, should be described by a thermally activated power-law, and (b) rock mechanics experiments producing microstructures that resemble those in naturally deformed rocks can be readily fitted using a thermally activated power-law (Twiss & Moores 1992, pp. 375–384). Power-law rheology is commonly used in modelling crustal behaviour (e.g. Kruse *et al.* 1991; Weinberg & Podladchikov 1995).

(2) The crust has an average composition of quartz diorite. Choosing weaker (e.g. granite or quartzite) or stronger (e.g. olivine) rheology will modify the results below but not change the general behaviour of the model.

(3) The temperature at a given depth is given by a steady-state, constant-gradient geotherm. For pluton emplacement, the particular case being studied here, this leads to temperatures that are too low in the upper part of the crust (because the thermal energy of the pluton is not accounted for) and too high in the lower part. Modelling the temperature field around a rising or sinking pluton is a difficult numerical problem (Ribe 1983; Mahon *et al.* 1988; Weinberg & Podladchikov 1995), which is beyond the scope of this paper. Assuming a linear geotherm will underestimate sinking rates of hot plutons because it does not account for local heating effects (Glazner & Miller 1997).

(4) Plutons sink at a rate given by Stokes' Law, assuming the relevant effective viscosity. Non-spherical geometry will modify sinking rates somewhat (e.g. McNown & Malaika 1950), but this effect is well within the uncertainty of the model.

Given these simplifying assumptions, the actual values for pluton capture depth, erosion rates necessary for pluton exposure, and so on are only rough approximations. The general behaviour exhibited by the model, however, is not strongly dependent on the assumptions or choices of variables.

## Model

### Choice of variables

The model comprises a specified crustal density profile, temperature profile, rheology, and erosion rate. The depths relative to the surface of spherical plutons of various radii and initial depths are then tracked through time.

*Coordinate system.* The depth origin is taken to be the eroding surface, such that wall rocks move to the surface at a constant velocity given by the erosion rate.

*Density.* Crustal density varies from 2600 kg m$^{-3}$ at the surface to 3000 kg m$^{-3}$ at 40 km depth. In

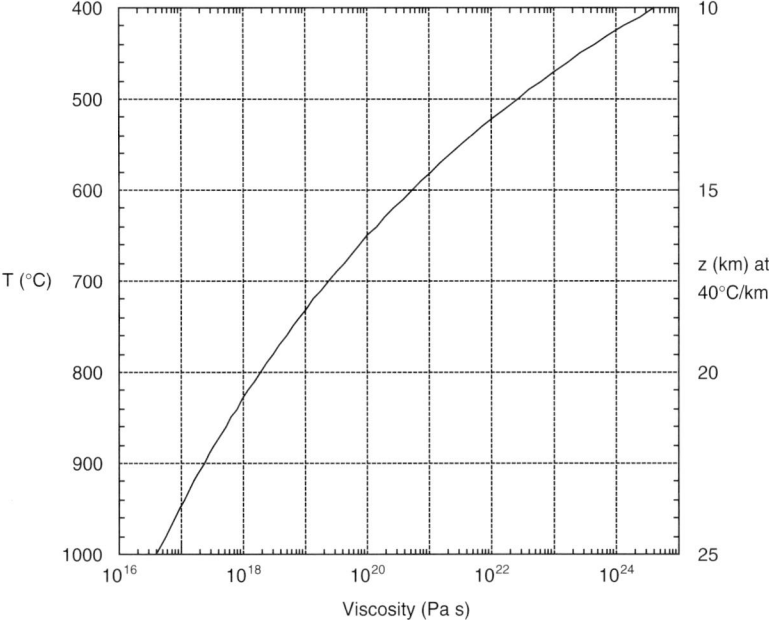

**Fig. 2.** Effective viscosity from equation (1) as a function of temperature, assuming a differential stress of 5 MPa. Viscosities around $10^{20}$ Pa s require temperatures in the 600–700°C range.

the zone of particular interest, 10–20 km depth, densities thus range from 2700 to 2800 kg m$^{-3}$. For simplicity of calculation, the density profile is fixed with respect to the eroding surface (dense wall rocks are not carried to the surface by erosion). This assumption has little effect on the results because most interesting effects happen within a few million years, before the density profile has time to change significantly.

*Temperature.* Temperature increases in the models at a constant gradient of 40°C km$^{-1}$. This is a gross oversimplification of reality, in which the geothermal gradient will vary with erosion rate, magmatic history, and several other important variables. However, even under high erosion rates (e.g. 1 km Ma$^{-1}$), a steady-state, hot geotherm, convex upward, will be established as hot material is advected to the surface (e.g. Glazner & Bartley 1985, equilibrium geotherm in their fig. 3). The relatively steep geothermal gradient assumed here is reasonable for many orogenic belts, especially those that are magmatically active (e.g. Griffin & O'Reilly 1987), but leads to unreasonably high temperatures in the lower crust. These excessive lower-crustal temperatures do not affect the results below, because dense rocks sink rapidly once their wall rocks attain temperatures around 600°C.

The assumption of a constant geotherm is perhaps the weakest assumption in the modelling below. However, a more accurate simulation of the geotherm will not affect the key result (that there is a sensitive interplay between model parameters); it will affect only the critical values of the variables.

*Rheology.* The power-law rheology for quartz diorite used by Kruse *et al.* (1991) was chosen for the model crust. In a power-law material the effective viscosity can be written in terms of the differential stress as

$$\mu_{eff} = \frac{(\sigma_1 - \sigma_3)^{1-n} e^{Q/RT}}{2A}$$

(Kruse *et al.* 1991, equation 9) where $n$, $Q$, and $A$ are experimental constants, $R$ is the universal gas constant, $T$ is the absolute temperature, and $\sigma_1-\sigma_3$ is the differential stress. For quartz diorite, the power-law parameters are $A = 19\,950$ GPa$^{-2.4}$/s$^{-1}$, $Q = 219\,000$ J mol$^{-1}$, and $n = 2.4$. For this model the differential stress is the stress imposed by the excess weight of the dense rock body. This value is set to a representative value

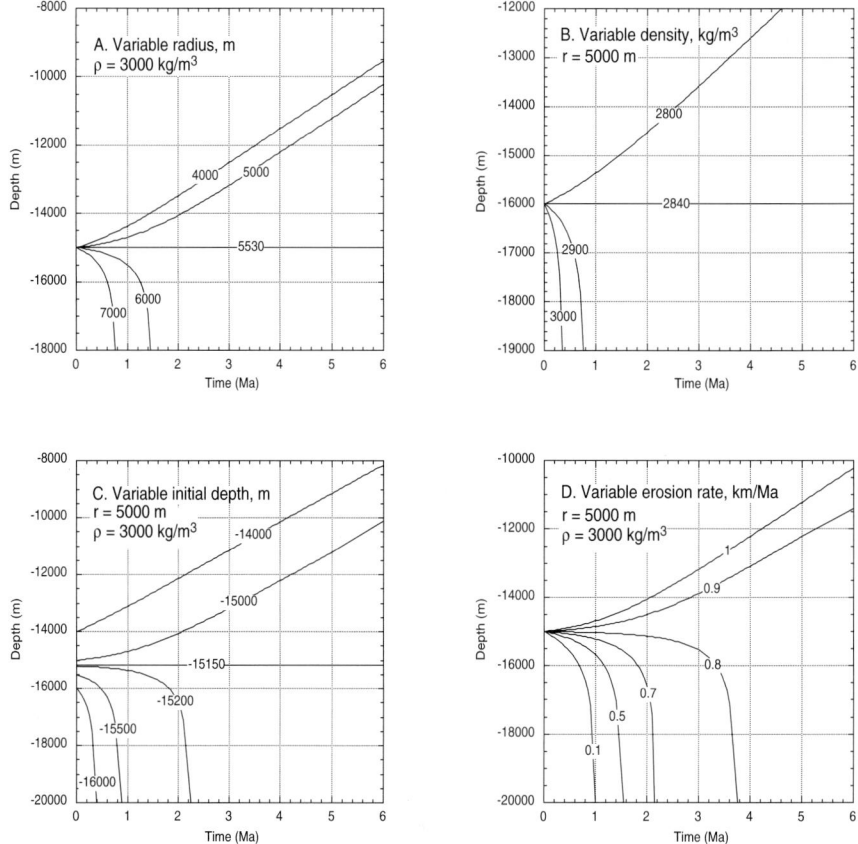

**Fig. 3.** Calculated depth of a dense rock body as a function of time for variable (**a**) radius, (**b**) density, (**c**) initial depth, and (**d**) erosion rate; typical values are radius 5000 m, density 3000 kg m$^{-3}$, initial depth 15 000 m, and erosion rate 1 km Ma$^{-1}$. It should be noted that for each variable there is a critical value at which the body 'hovers' (sinking balances erosion); for values only slightly different from the critical value the paths diverge widely. Initial depth in (**b**) is 16 000 m because for this model a pluton with radius 5000 m and density 3000 kg m$^{-3}$, emplaced at 15 000 m depth, will not sink (see (**c**), curve for –15 000 m).

of 5 MPa in all experiments, equivalent to the differential stress at the base of a cylindrical pluton 5 km tall that is 100 kg m$^{-3}$ denser than its wall rocks. Given these values, the effective viscosity is a function of $T$ and thus of depth in the crust. Varying the differential stress does not have a large effect on the results; for example, if the differential stress is changed by a factor of ten the critical depth in Fig. 3c (see below) changes by a few kilometres.

*Erosion rate.* Estimated erosion rates vary widely. In active orogenic areas rates may be as high as several kilometres per million years (Benjamin *et al.* 1987; Burbank & Beck 1991; Brandon & Ring 1997). Rates measured in less active drainage basins are much lower, typically <0.1 km Ma$^{-1}$ (Holland 1978). For most models a rate of 1 km Ma$^{-1}$ was used.

*Algorithm*

For each model the pluton density, pluton depth (taken as the centre of the pluton), pluton radius, and erosion rate were specified. At each time step the pluton moves up owing to erosion and down (if it is denser than its wall rocks) owing to negative buoyancy. The new position at each time step is calculated by summing the erosion rate and Stokes settling velocity, multiplying by the time step, and adding to the current position. A time step of 1000 years was sufficiently small to eliminate numerical errors (results using a time step of 100 years were not significantly different). A time step of $10^4$ years led to numerical instability for certain models. For example, if the time step is too large the pluton will overshoot its level of neutral buoyancy and then oscillate about it.

## Results

### Extreme sensitivity of results to variables

The four variables in the model are erosion rate, pluton density, initial depth, and pluton radius. The results show that if any three of these are fixed, the ultimate fate of the pluton depends critically on the value of the fourth. A small change in the fourth variable determines whether the pluton rises toward the surface or plummets to the deep crust.

Figure 3 demonstrates this sensitivity. Figure 3a shows how a pluton's trajectory depends on its radius. Erosion rate (1 km Ma$^{-1}$), pluton density (3000 kg m$^{-3}$), and initial depth (15 km) were held constant. In this model a pluton with a radius around 5000 m or less is carried to the surface by erosion, essentially at the full rate of 1 km Ma$^{-1}$, whereas a pluton with a radius of 6000 m plummets to the base of the crust in less than 2 Ma. The critical radius is 5530 m; a pluton of this radius will hover at a constant depth because the rate of sinking equals the erosion rate. The widely divergent curves for 5500 m and 5600 m radii illustrate how sensitive the model is to radius.

Figure 3b and c demonstrates the effects of pluton density and initial depth. Again, extreme sensitivity to model parameters is evident. For example, in Fig. 3c a pluton (radius 5000 m, density 3000 kg m$^{-3}$) emplaced at 15 km rises to the surface at nearly the full erosion rate of 1 km Ma$^{-1}$, one at 15.15 km hovers, and one at 15.2 km plummets to the deep crust about 2 Ma after emplacement.

Figure 3d illustrates the effect of variable erosion rate. Yet again, extreme sensitivity is the rule. For these model parameters (pluton density 3000 kg m$^{-3}$, depth 15 km, radius 5000 m), an erosion rate of 0.9 km Ma$^{-1}$ carries the pluton to the surface, but a rate of 0.8 km Ma$^{-1}$ allows it to plummet in 3.5 Ma.

### Capture depth

Figure 4 shows the maximum depth at which a 3000 kg m$^{-3}$ pluton of a given radius can be emplaced and not sink to the deep crust. The curves were calculated by setting the sum of the

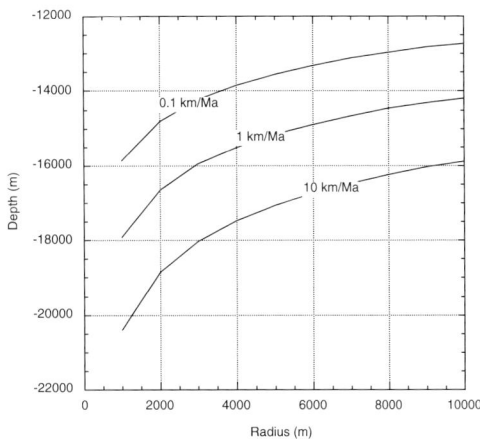

**Fig. 4.** Capture depth (greatest depth at which a dense body can start and still be brought to the surface by erosion) as a function of erosion rate. Curves calculated for a 3000 kg m$^{-3}$ pluton at 5 MPa differential stress.

erosion rate and Stokes settling velocity to zero. Plutons emplaced above these curves will be carried to the surface by erosion. The capture depth for plutons of several kilometres radius is in the range 12–17 km. This is the depth at which the model crust attains a temperature of >500°C and a viscosity in the range of $10^{20}$–$10^{22}$ Pa s. Dense plutons emplaced below the capture depth will sink at an accelerating rate, typically plummeting to the deep crust in a few million years.

## Discussion

The results above indicate that kilometre-scale rock bodies that are significantly denser than their surroundings will sink at geologically rapid rates if the crust behaves as a Newtonian fluid and if crustal viscosity is low enough, on the order of $10^{20}$ Pa s. The model results indicate that the ultimate fate of a dense rock body emplaced in the middle crust is an extremely sensitive function of the initial conditions to which the body is subjected. Key variables include erosion rate, initial depth, pluton radius, and pluton density. The calculations suggest that two plutons of similar characteristics can follow widely divergent paths if one is slightly denser, larger, or deeper than the other.

This process may explain some puzzling aspects of crustal evolution. For example, in the Sierra Nevada granite and granodiorite plutons are typically tens of kilometres in diameter in map view, whereas gabbro and diorite plutons only rarely exceed a few kilometres in maximum dimension (Moore & Sisson 1987; Ross 1989; Bateman 1992). This hundredfold discrepancy in area may be a consequence of sinking of larger mafic plutons. The calculations suggest that plutons more than a few kilometres in radius cannot be captured at mid-crustal levels, but small plutons can. Thus, large mafic plutons such as the Stillwater complex should only be found in the upper crust. Available data suggest that many large layered mafic complexes were indeed emplaced at shallow crustal levels (Glazner 1994).

The calculations also indicate that the common assumption that large, formerly deep-crustal gabbro bodies were emplaced by underplating of the crust near the crust–mantle boundary may not always be correct. Hamilton (1995) discussed several localities that expose the roots of subduction-related magma systems. In two of these, the Kohistan arc of Pakistan and the Western Fiordland region of New Zealand, thermobarometric and geochronological data indicate that the mafic bodies were emplaced at shallow levels but underwent high-pressure metamorphism shortly thereafter (Fig. 5). It is possible that these mafic bodies underwent the sinking mechanism proposed above, and such a mechanism should be examined for any proposed example of crustal underplating.

These calculations predict that it is difficult to elevate dense mid-crustal and lower-crustal rock bodies to shallow depths unless erosion or tectonic denudation rates are extremely high (see Brandon & Ring 1997). If the body lies at a depth such that the ambient temperature is high enough that the viscosity of the surrounding rocks is low, then it will sink faster than erosion brings it to the surface. The exceptionally high erosion or tectonic denudation rates inferred from some studies of dense ultra-high-pressure rocks may be necessary to bring such rocks to the surface before gravity can reclaim them.

Actual values of these variables calculated in this analysis are only rough approximations, given all the uncertainties in the input data. In particular, the critical radius of 5.5 km for a pluton emplaced at 15 km depth (Fig. 3a) is significantly greater than that of observed mafic plutons at any depth in the crust. The observed maximum radii of a few kilometres for mafic plutons in the Sierra Nevada suggest that the actual capture radius is on the order of a few kilometres.

This work is based on ideas generated at the 1996 Penrose Conference on exhumation of deep rocks; I thank the convenors and participants for their work in producing a stimulating conference. Constructive reviews by K. Furlong, M. Ducea, and an anonymous reviewer were helpful in clarifying the ideas and presentation. J. Bartley reviewed an early draft. I am especially grateful to the reviewers for allowing the simplified treatment above and agreeing that a far more sophisticated treatment of the problem would yield little additional insight. This study was supported by National Science Foundation grants EAR-9219521 and EAR-9526803.

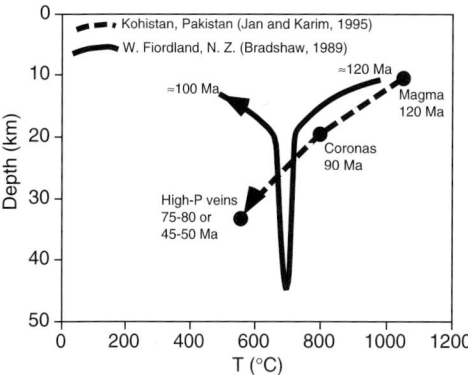

**Fig. 5.** Inferred depth–temperature–time trajectories for two deep-crustal mafic bodies from New Zealand and Pakistan (Bradshaw 1989; Jan & Karim 1995). Both mafic complexes were emplaced at mid-crustal depths and then underwent high-pressure, deep-crustal metamorphism. These $P$–$T$–$t$ paths could be a result of gravitational sinking followed by rapid exhumation.

## References

BATEMAN, P. C. 1992. *Plutonism in the Central Part of the Sierra Nevada Batholith, California*. US Geological Survey, Professional Paper, **1483**.

BENJAMIN, M. T., JOHNSON, N. M. & NAESER, C. W. 1987. Recent rapid uplift in the Bolivian Andes: evidence from fission-track dating. *Geology*, **15**, 680–683.

BILLS, B. G., CURREY, D. R. & MARSHALL, G. A. 1994. Viscosity estimates for the crust and upper mantle from patterns of lacustrine shoreline deformation in the eastern Great Basin. *Journal of Geophysical Research*, **99**, 22059–22086.

BRADSHAW, J. Y. 1989. Origin and metamorphic history of an Early Cretaceous polybaric granulite terrain, Fiordland, southwest New Zealand. *Contributions to Mineralogy and Petrology*, **103**, 346–360.

BRANDON, M. T. & RING, U. 1997. Penrose Conference report: exhumation processes: normal faulting, ductile flow, and erosion. *GSA Today*, **7**, 17–20.

BURBANK, D. W. & BECK, R. A. 1991. Rapid, long-term rates of denudation. *Geology*, **19**, 1169–1172.

GLAZNER, A. F. 1994. Foundering of mafic plutons and density stratification of continental crust. *Geology*, **22**, 435–438.

—— & BARTLEY, J. M. 1985. Evolution of lithospheric strength after thrusting. *Geology*, **13**, 42–45.

—— & MILLER, D. M., 1997. Late-stage sinking of plutons. *Geology*, **25**, 1099–1102.

GRIFFIN, W. L. & O'REILLY, S. Y. 1987. Is the continental Moho the crust–mantle boundary? *Geology*, **15**, 241–244.

HACKER, B. R., YIN, A., CHRISTIE, J. M. & DAVIS, G. A. 1992. Stress magnitude, strain rate, and rheology of extended middle continental crust inferred from quartz grain sizes in the Whipple Mountains, California. *Tectonics*, **11**, 36–46.

HAMILTON, W. B. 1995. Subduction systems and magmatism. *In*: SMELLIE, J. L. (ed.) *Volcanism Associated with Extension at Consuming Plate Margins.* Geological Society, London, Special Publication, **81**, 3–28.

HOLLAND, H. D. 1978. *The Chemistry of the Atmosphere and Oceans.* Wiley, New York.

JAN, M. Q. & KARIM, A. 1995. Coronas and high-$P$ veins in metagabbros of the Kohistan island arc, northern Pakistan: evidence for crustal thickening during cooling. *Journal of Metamorphic Geology*, **13**, 357–366.

KRUSE, S., MCNUTT, M., PHIPPS-MORGAN, J. & ROYDEN, L. 1991. Lithospheric extension near Lake Mead, Nevada – a model for ductile flow in the lower crust. *Journal of Geophysical Research*, **96**, 4435–4456.

MAHON, K. I., HARRISON, T. M. & DREW, D. A. 1988. Ascent of a granitoid diapir in a temperature varying medium. *Journal of Geophysical Research*, **93**, 1174–1188.

MASEK, J. G., ISACKS, B. L. & FIELDING, E. J. 1994. Rift flank uplift in Tibet: evidence for a viscous lower crust. *Tectonics*, **13**, 659–667.

MCNOWN, J. S. & MALAIKA, J. 1950. Effects of particle shape on settling velocity at low Reynolds numbers. *Transactions, American Geophysical Union*, **31**, 74–82.

MOORE, J. G. & SISSON, T. W. 1987. *Preliminary Geologic Map of Sequoia and Kings Canyon National Parks, California.* US Geological Survey Open-File Report **87–651**.

RIBE, N. M. 1983. Diapirism in the earth's mantle: experiments on the motion of a hot sphere in a fluid with temperature-dependent viscosity. *Journal of Volcanology and Geothermal Research*, **16**, 221–245.

ROSS, D. C. 1989. *The Metamorphic and Plutonic Rocks of the Southernmost Sierra Nevada, California, and their Tectonic Framework.* US Geological Survey Professional Paper, **1381**.

TWISS, R. J. & MOORES, E. M. 1992. *Structural Geology.* W. H. Freeman, New York.

WDOWINSKI, S. & AXEN, G. J. 1992. Isostatic rebound due to tectonic denudation – a viscous flow model of a layered lithosphere. *Tectonics*, **11**, 303–315.

WEINBERG, R. F. & PODLADCHIKOV, Y. Y. 1995. The rise of solid-state diapirs. *Journal of Structural Geology*, **17**, 1183–1195.

# Geological and geochronological constraints on the exhumation of a high-pressure metamorphic terrane, Oman

J. McL. MILLER[1,4], R. T. GREGORY[2], D. R. GRAY[1] & D. A. FOSTER[3]

[1]*Department of Earth Sciences, Monash University, Melbourne, Vic. 3168, Australia*
[2]*Department of Geological Sciences, Southern Methodist University, Dallas, TX 25275, USA*
[3]*Department of Earth Sciences, La Trobe University, Melbourne, Vic. 3083, Australia*
[4]*Present address: School of Earth Sciences, The University of Melbourne, Parkville, Vic. 3052, Australia*

**Abstract:** New $^{40}Ar$–$^{39}Ar$ age data from high-pressure former continental shelf rocks, now structurally beneath the Samail ophiolite, support a two-stage exhumation process for these rocks. High-$P$ rocks occur as stacked fold–nappes within a less deformed upper plate and a strong to intensely deformed lower plate separated by a major, ductile, crustal discontinuity. The highest-grade lower plate rocks yield $^{40}Ar$–$^{39}Ar$ ages that range from 82 to 131 Ma. The majority of these ages are inferred to represent cooling by the progressive emplacement of colder units during convergent margin tectonism, before emplacement to higher structural levels. In the lower plate, 82–79 Ma $^{40}Ar$–$^{39}Ar$ ages are associated with lower-grade assemblages defining both transposition fabrics and C'- type shear bands which overprint the high-$P$ assemblages during a NE-directed shearing event. Partial exhumation of these lower plate sequences resulted in the formation of regional closures by folding and transposition of earlier high-pressure fabrics. The lower plate closures are truncated by the structural break separating the two plates. Emplacement of the lower plate units to shallower crustal levels may have occurred before, or synchronously with, peak metamorphism of the upper plate units. The lower plate $^{40}Ar$–$^{39}Ar$ cooling(?) ages are older than mica crystallization ages of 76–70 Ma from the upper plate. These upper plate micas grew in axial surface fabrics associated with regional-scale fold–nappes which formed during the exhumation of the upper plate units and their transport over the higher-grade lower plate units. This nappe-forming event was possibly synchronous with structurally higher low-angle normal faulting and/or rapid erosion of the nappe pile before the deposition of Maastrichtian autochthonous units at *c.* 67–68 Ma.

A complex of variously deformed and metamorphosed pelagic and continental shelf sequences are superbly exposed in tectonic windows beneath the Samail ophiolite of the Oman mountains (Fig. 1). The northeastern sector of the Saih Hatat window (Figs 1 and 2) is unique in Oman, because these sequences have been subjected to blueschist–eclogite facies metamorphism (first described in detail by Lippard (1983) and Michard (1983)). The Oman high-pressure rocks are now exposed because of late Tertiary uplift accompanied by dome formation, but their formation and emplacement occurred during Cretaceous time. The deformation that resulted in their formation and exhumation affects rocks that are elsewhere in Oman considered part of the undeformed autochthon. The high-pressure rocks of Saih Hatat share many of the characteristics of A-type eclogites of Maruyama *et al.* (1996). The critical aspects of the tectonic history associated with exhumation are well preserved, because the Tertiary doming and uplift is the only post-emplacement event that affects these rocks.

In this paper, we first integrate the results of new $^{40}Ar$–$^{39}Ar$ geochronology with a new structural framework for the Saih Hatat window, and then consider the implications of these new geochronological data for the exhumation of the high-pressure rocks exposed in Oman and elsewhere. The new framework is based on 1:25 000 scale mapping utilizing Bureau de Recherches Géologiques et Minières (BRGM) stratigraphy (Le Metour *et al.* 1986) combined with detailed structural and microstructural analysis, and has been presented by Gregory *et al.* (1996, 1998) and Miller *et al.* (1998).

## Regional geological setting

The Oman mountains contain the Samail ophiolite (Fig. 1), which formed at a Tethyan spreading centre between 99 and 94 Ma (U–Pb zircon ages from plagiogranites (Tilton *et al.* 1981) and

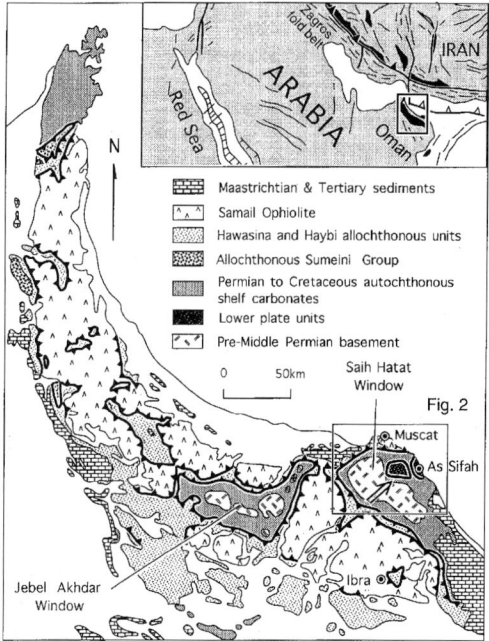

**Fig. 1.** Geological map of the Oman Mountains, showing the distribution of the Samail ophiolite, positions of windows containing pre-Permian basement, and the newly identified structural break within the Saih Hatat window outlining a structurally lower plate. Position of Fig. 2 is shown, arrow indicates deformation gradient across Saih Hatat (modified from Glennie et al. 1974). Inset is modified from Coleman (1981), dark lenses in the inset represent ophiolite bodies.

biostratigraphic data (Tippit et al. 1981)). The ophiolite was detached via intraoceanic thrusting along the metamorphic sole at c. 94–93 Ma ($^{40}$Ar–$^{39}$Ar on hornblende from amphibolite facies tectonite (Hacker et al. 1996; and references therein)), with later growth at 94–91 Ma of lower-grade phyllitic mica in cherts and calcsilicates ($^{40}$Ar–$^{39}$Ar, Hacker et al. 1996). Late Maastrichtian (c. 67–68 Ma) autochthonous sediments unconformably overlie all sequences; these units provide a minimum age for the ophiolite emplacement and exhumation of the sequences (Coleman 1981). The outcrop width of the Oman mountains and the stacking in the structural pile suggest that the ophiolite was transported several hundred kilometres (400–500 km, Glennie et al. 1974; Bechennec et al. 1990). A dominant set of north–south trending lineations within the metamorphic sole and mylonitic peridotite indicate that it was emplaced onto the margin from the north (Boudier & Coleman 1981; Boudier et al. 1985).

The ophiolite overlies allochthonous remnants of a telescoped marine basin (the Hawasina ocean basin) that was emplaced onto the passive Arabian margin. All of these rocks are exposed on the flanks of the Oman mountains, as well as within prominent tectonic windows (Fig. 1). Major units include: (1) mid-Permian to Cenomanian *Hawasina Complex* (remnants of an oceanic basin), *Haybi Complex* (volcanic rocks and 'exotic' limestones) and *Sumeini Group* (continental slope units), all of which were emplaced onto the passive continental margin during events which eventually culminated in the obduction of the Samail ophiolite (Glennie et al. 1974; Searle & Malpas 1980); (2) *Muti Formation* (part of the *Aruma Group*) comprising Cenomanian to Campanian conglomerates, shales, marls and limestones, believed to record the transition from a passive continental margin to a foreland basin (Glennie et al. 1974; Robertson 1987); (3) the *Hajar Super Group*, consisting of mid-Permian to Cenomanian continental shelf sequences (Glennie et al. 1974); (4) the *prePermian basement* (Glennie et al. 1974).

The timing of the high-pressure metamorphism, represented by the formation of garnet blueschist and eclogite assemblages at As Sifah (Fig. 2), is controversial. In particular, the question of whether all of the high-pressure metamorphism occurred during the emplacement of the ophiolite is still open. Lippard et al. (1986) obtained K–Ar ages of 80 and 100 Ma from garnet–glaucophane schists. $^{40}$Ar–$^{39}$Ar geochronology performed by three separate groups (Montigny et al. 1988; El-Shazly & Lanphere 1992; Searle et al. 1994) on white mica from high-pressure assemblages typically gives a range of ages between 131 and 86 Ma. Some of these ages are older than the plutonic crystallization age of c. 95 Ma for the Samail ophiolite (Tilton et al. 1981). Either the high-pressure metamorphism partly pre-dates the formation and obduction of the oceanic lithosphere, or all of the old ages have to be discounted because of the presence of excess argon in the white micas. In this study, we examine the Ar systematics of high-pressure rocks in the context of the fabric development and the cross-cutting relationships.

## Geology of Saih Hatat

The Saih Hatat window exposes strongly folded sequences of variously deformed and metamorphosed carbonates, quartzites, quartz–mica schists and mafic schists (Figs 1 and 2). During the production of regional 1:100 000 map sheets for Saih Hatat, BRGM geologists (Le Metour et al. 1986) recognized large tracts of overturned

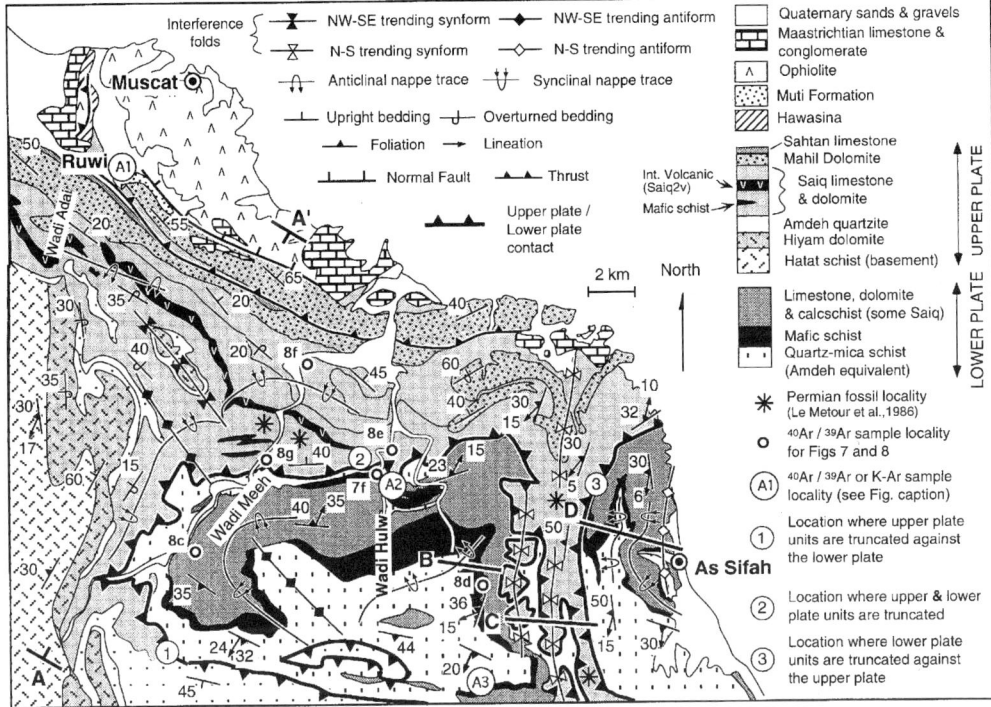

**Fig. 2.** Geological map of northeastern Saih Hatat (modified from Le Metour *et al.* 1986). Geochronological localities are: A1, 80 Ma, lawsonite schist ($^{40}$Ar–$^{39}$Ar step heating, El Shazly & Lanphere 1992); A2, 72 and 76 Ma ($^{40}$Ar–$^{39}$Ar total fusion and $^{40}$Ar–$^{39}$Ar step heating, respectively, El Shazly & Lanphere 1992); A3, 103 Ma (K–Ar, Montigny *et al.* 1988); some of the locations of the dated samples in Figs 7 and 8 are marked. (Note that nappe axial traces are complicated by the nappes becoming sheath-like in the vicinity of the upper plate–lower plate contact.)

rocks on the north limb of the dome; they identified these units as pre-Permian Hijam Formation and Amdeh Group now underlain by various members of the Permian Saiq Formation. This overturned section is part of the common limb shared between a regional anticlinal nappe recognized by Bailey (1981), and a large synformal nappe exposed in Wadi Meeh (Miller *et al.* 1998, fig. 2; Gregory *et al.* 1998).

In Wadi Meeh, the regional-scale synformal nappe is truncated by a ductile structural break that separates upper and lower 'plates' with differences in stratigraphy, metamorphic grade, structural style and strain (Figs 2–4 and Table 1). Various sections of this structural break have been identified by previous workers (e.g. Le Metour *et al.* 1986; Goffé *et al.* 1988; Montigny *et al.* 1988; Mann & Hanna 1990; El-Shazly 1994, 1995; Michard *et al.* 1994; Searle *et al.* 1994) but it has not been previously recognized as one major structure. The recognition of this structure simplifies the stratigraphy of Saih Hatat and is consistent with the existence of a major break required by discontinuities in pressure estimates derived from metamorphic assemblages (e.g. El-Shazly *et al.* 1990). The lower plate encompasses the metamorphic region III of El-Shazly *et al.* (1990).

## The upper plate

The core of the Saih Hatat dome exposes Hatat Schist, which contains a variety of lithologies from mafic schist to metasiliciclastic rocks (Figs 1 and 2). The rocks are strongly deformed with transposition layering and exhibit a well-developed lineation that parallels the regional Cretaceous lineation. Any pre-Permian deformation has been thoroughly overprinted by the Cretaceous event. The schist crops out in the core of the large regional antiformal nappe which closes to the north. In places, the Hatat Schist has overridden the Palaeozoic section.

Pre-Permian sedimentary rocks are represented by the Hijam Formation, consisting of limestone and dolomite with minor interlayered

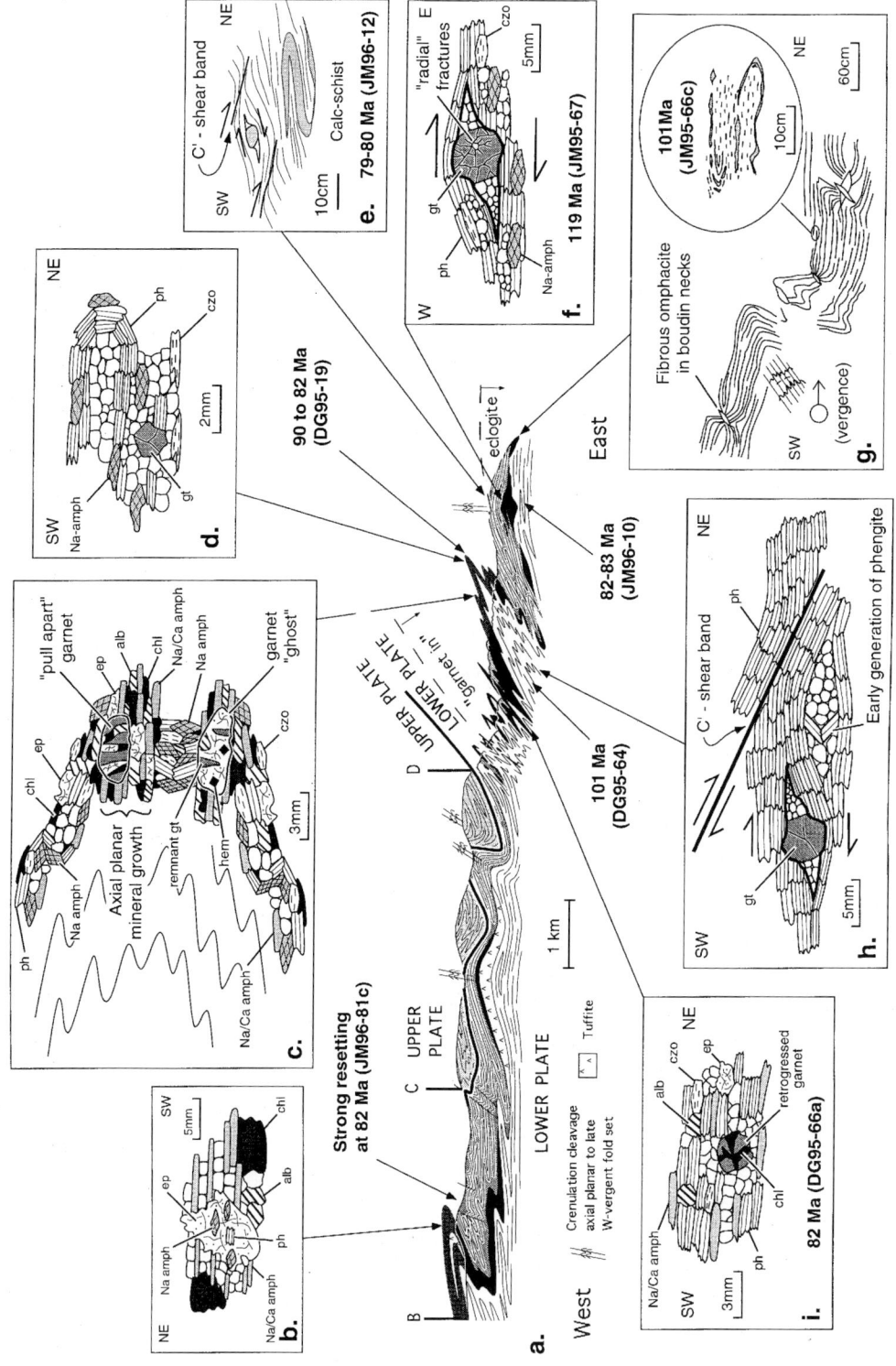

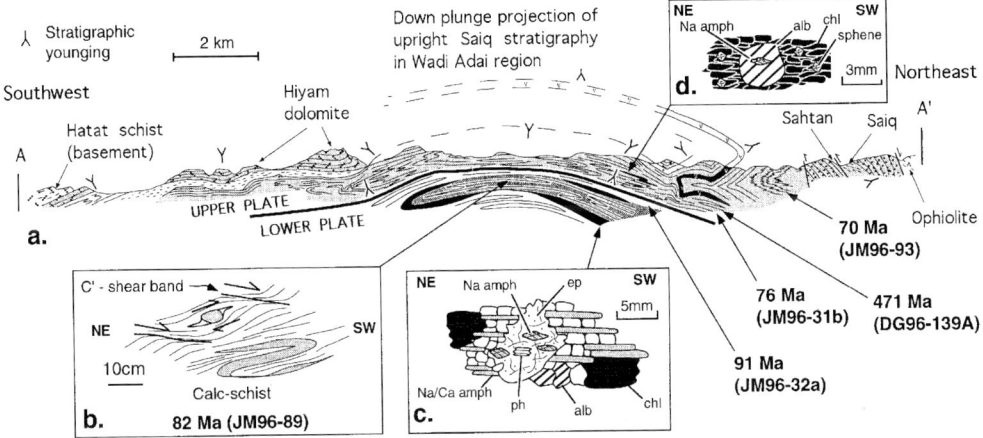

**Fig. 4.** (a) Structural profile along Wadi Meeh with $^{40}$Ar–$^{39}$Ar age data and key microstructural information (modified from fig. 2 of Miller *et al.* 1998). (b) Late calc-schist with strong to the northeast non-coaxial shear (C′-shear bands, asymmetrically sheared clasts). (c) Low-grade foliation within mafic schist axial planar to regional nappe; epidote porphyroblasts preserve higher-pressure sodic amphiboles (and phengite). It should be noted that the crosshatching across the amphiboles is for identification and does not represent amphiboles perpendicular to the c-axis. (d) Growth of albite, chlorite and sphene (mafic units). Chlorite fabric is axial planar to upper plate nappes. Abbreviations as in Fig. 3.

**Table 1.** *Structural comparison of upper and lower plates, northeastern Saih Hatat*

|  | Upper plate | Lower plate Wadi Huwl–Wadi Meeh | Lower plate As Sifah |
|---|---|---|---|
| Structural features | Regional closures Coaxial shear dominates strain gradient (cylindrical to sheath-like) | Regional sheath-like closures Non-coaxial shear (top to northeast) Large component of flattening | Regional shealth-like closures Non-coaxial shear (top to northeast) Large component of flattening |

**Fig. 3.** (a) Structural profile linking two exposed areas of lower plate at As Sifah and Wadis Meeh and Hulw, with $^{40}$Ar–$^{39}$Ar age data and key microstructural information. It should be noted that the section is drawn perpendicular to the transport direction because of the sheath-like nature of the regional closures (modified from fig. 2 of Miller *et al.* 1998). The sheath-like nature of the regional closures, extensive flattening and strong later fold interference by the N-trending fold set combine to make it difficult to ascertain whether the grade variation between the two exposed regions of the lower plate occurs up or down structural section. (b) Low-grade foliation within mafic schist axial planar to regional nappe; epidote porphyroblasts preserve higher-pressure sodic amphiboles (and phengite). It should be noted that the crosshatching across the amphiboles is for identification and does not represent amphiboles perpendicular to the c-axis. (c) Parasitic fold to a regional fold nappe which folds higher-grade fabrics. Epidote, chlorite, sodic–calcic amphiboles and albite have grown axial planar to the fold and in fractures formed during deformation of garnet porphyroblasts. Other garnets are almost completely replaced by these lower-grade fabrics. (d) Folded high-pressure fabric defined by garnet, phengite, glaucophane and clinozoisite. (e) Late calc-schist with strong to the northeast non-coaxial shear (C′-shear bands, asymmetrically sheared clasts). (f) Early east–west lineation (phengite, glaucophane and clinozoisite) from an eclogite-bearing mafic megaboudin. (g) Internal structure of mafic megaboudins. Internal boudins disrupt early isoclinal folds; the boudin necks sometimes contain fibrous omphacite. Later asymmetric folds overprint these earlier structures. (h) Shear sense indicators such as asymmetric pressure shadows around garnets and late shear bands give top to the northeast sense of shear. An earlier generation of phengite is preserved in microlithons. (i) Lineation defined by albite, epidote, phengite, chlorite, biotite and clinozoisite; garnet is retrogressed and not in equilibrium with the other minerals. Na-amph, glaucophane; Na–Ca-amph, sodic–calcic amphibole; ep, epidote; alb, albite; ph, phengite; rut, rutile; gt, garnet; hem, hematite; czo, clinozoisite; biot, biotite; act, actinolite.

clastic sediments, and the Amdeh Group, a succession of clastic rocks consisting of predominantly psammite at its base and becoming more pelitic towards its top. Overlying the pre-Permian rocks are the Saiq Formation, which is locally siliciclastic at its base, overlain by limestone and dolomite, fetid black limestone and thin-bedded limestone and dolomitic marl. Along much of the strike length of the northwestern upper plate units, the lower Saiq limestone and dolomite succession is separated from the fetid black limestone facies by mafic and felsic schist and minor clastic sediments. The Triassic Mahil and Sahtan formations overlie the Saiq succession. On the northern side of the Saih Hatat dome, Jurassic and Cretaceous stable platform sediments are absent, and the 'Muti' Formation or Hawasina mélange are in fault contact against the Sahtan or Mahil formations.

*The lower plate*

The lower plate consists of two subdomains (a window through the As Sifah dome and a window traversed by Wadis Meeh and Hulw), which are united by a common occurrence of metaquartzites overlain by mafic schist overlain by dolomite overlain by calc-schist. These rocks are more strongly deformed than the upper plate rocks. The window through the As Sifah dome exposes a major antiformal sheath-like closure with a core of metaquartzite and quartz–mica schist, overlain by mafic schist, dolomite and calc-schist. Eclogitic rocks are exposed in mafic boudins that define the lower limb of a synformal structure beneath the overlying antiformal closure.

The calc-schist and dolomite have negative $\delta^{13}C$ values inconsistent with a Permian age (R. T. Gregory, unpublished data). This suggests that the metaquartzite and quartz–mica schist may be correlated with Amdeh Group equivalents. If the dolomite and calc-schist are correlative with the Hijam Formation (which is older than the Ordovician Amdeh Quartzite), then the entire lower plate section may be overturned. Alternatively, if the section is upright and the quartzite is an Amdeh equivalent, then the calc-schist and dolomite succession represents a new facies of pre-Permian rocks, and perhaps a more distal portion of the Arabian platform.

In the window into the lower plate drained by the Wadis Meeh and Hulw ('Wadi' means dry river), quartz–mica schists and metaquartzites are overlain by mafic schists, felsic rocks, dolomite and calc-schist grading into carbonate rocks that have positive $\delta^{13}C$ values consistent with a Permian age assignment. With the exception of the calc-schists exposed near the upper plate–lower plate boundary, the rocks in this subdomain of the lower plate preserve more original features: relict pillows and deformed amygdules in the mafic schist, deformed fossils in the limestones, and some relic sedimentary features such as graded bedding in the clastic rocks.

## Structural constraints

*The lower plate*

The deepest levels in the eastern region of the lower plate at As Sifah expose eclogite megaboudins up to 1 km in length (Fig. 2). The lower plate tectonostratigraphy is folded by a series of sheath-like regional recumbent folds (Fig. 3a). In both exposed sections of the lower plate, these closures are associated with a strong northeast–southwest stretching lineation, which is marked by pull-apart limestone clasts, boudinage, vein sets orthogonal to the lineation, and pressure shadows on pyrite. The sheath-like nature of the regional closures has resulted in north–south outcrop traces for these closures in the As Sifah lower plate window, parallel to the stretching lineation (Miller *et al.* 1998). Outcrop traces of regional closures in the Wadi Meeh and Hulw subdomain of the lower plate reflect strong fold interference (Fig. 2).

The megaboudins at As Sifah have complicated internal structures and preserve earlier structural features and the highest-grade metamorphic assemblages (Table 1). The earliest structures are east–west oriented lineations with a west-over-east sense of shear, consistent in sense and direction within different, isolated, boudins. Internal boudins frequently disrupt early isoclinal folds within the megaboudins, and contain tension gashes with fibrous omphacite (Fig. 3g). All of these earlier structures within the megaboudins are frequently overprinted by a southwesterly inclined fold set, whose asymmetry consistently yields a southwest over northeast sense of shear and shear bands with a top to the northeast shear sense (Fig. 3g). They are bounded by intensely deformed calc-schist and quartz–mica schist which show intense transposition layering and a marked northeast–southwest stretching lineation (Fig. 3e).

Irrespective of the structural level, the lower plate in both subdomain windows shows the pervasive effects of intense, northeast-directed non-coaxial shear. Shear sense indicators include C- and C'-shear bands (see Passchier & Trouw 1996) (Figs 3e, h and 4b), asymmetrically sheared clasts and pressure shadows around porphyroblasts (Fig. 3h). Deformed conglomerates within the lower plate indicate a strong component of flattening accompanied by marked stretch in $X$, the maximum principal elongation direction, to produce flattened 'cigar-like' forms and the development of extensive shear bands. High strain ($XZ > 25:1$; Miller *et al.* 1998, fig. 2) has produced tight to isoclinal, sheath-like parasitic folds on the limbs of the closure, which makes using fold asymmetry to define the gross structure very unreliable.

*The upper plate*

Structures within the upper plate consist of a series of regional recumbent folds that initially substantially thicken the stratigraphic sequence, but at lower structural levels become more attenuated and sheath-like (Figs 2 and 4a). Folds within the upper plate limestones and dolomites have limbs extending for several kilometres, and a strong axial planar schistosity associated with a northeast–southwest trending stretching lineation. These folds have cylindrical form with fold axes at high angles to the stretching lineation at the highest structural levels, but the lower closures are sheath-like and have hinges sub-parallel to the stretching lineation; these changes in fold geometry and orientation are accompanied by a marked increase in $X/Z$ strain (see fig. 2 of Miller *et al.* 1988). Rare shear bands at the structurally deepest levels indicate top to the northeast transport, similar to that observed for the lower plate rocks. The deformed upper plate carbonates in the northeastern part of Wadi Adai are juxtaposed with small lenses of peridotite.

*The lower plate–upper plate discontinuity*

The map designated upper and lower plates of northern Saih Hatat (Fig. 2) are separated by a major ductile shear zone which truncates stratigraphic units of both plates, as well as the regional structures. Locally, there is imbrication and infolding of upper and lower plate sequences, resulting in fault-bounded slithers of upper plate units surrounded by lower plate units (compare the southern region of Fig. 2). This is associated with the development of a late, strong, asymmetric crenulation cleavage with top to the northeast shear (Fig. 3a). In some areas, there is clearly later brittle movement along this boundary (top to the south in the northern Hulw region of the lower plate). Hydrothermal activity along this break is reflected by the presence of gossan and/or small lead–zinc ore bodies (see Le Metour *et al.* 1986).

Exposures of this discontinuity and the two lower plate windows (As Sifah and Wadi Meeh–Wadi Hulw regions) are controlled by late dome and basin interference folds that are the result of fold interference between a northwest-trending, upright fold set and a north-trending, east-verging, westerly inclined fold set (Fig. 2).

## Metamorphic constraints

Problems exist with accurately defining peak metamorphic pressures and temperatures because of continuous recrystallization during exhumation (see Miller *et al.* 1998). The $P$–$T$ estimates will always be minima because the highest-grade minerals are preserved as relict inclusions in retrograded rocks or they were later pseudomorphed by other phases. The upper plate 'apparent' peak assemblages are distinguished by the presence of carpholite and inferred lower peak temperatures. Eclogite assemblages are only found as relicts in boudins that occur in the As Sifah subdomain of the lower plate. A pressure difference, possibly as large as 13 kbar, distinguishes the highest-pressure upper plate units and from the highest-pressure lower plate glaucophane eclogites (Fig. 5; see the references therein). Peak metamorphic conditions may have been different for the two subdomains within the lower plate.

*The upper plate*

The lower-grade assemblages, axial planar to recumbent folds in the upper plate, appear to post-date the peak of metamorphism. Chlorite–albite schists (mafic units) define the axial planar fabric to these regional nappes (Fig. 4d), with sodic amphiboles included as relicts within the albite porphyroblasts. Based on the absence of aragonite and the inferred stability fields of magnesio-carpholite, ferro-carpholite and chloritoid, El-Shazly (1994, 1995) argued that these upper plate units were metamorphosed at 6.8–9 kbar and 315–435°C. Alternatively, for most of the upper plate, Goffé *et al.* (1988) estimated $P$–$T$ conditions in the range 8–10 kbar and 180–250°C for carpholite–kaolinite assemblages and *c.* 6 to 8 kbar and 250–350°C for Fe–Mg carpholite–pyrophyllite assemblages (Fig. 5).

*The lower plate: As Sifah subdomain*

The eclogite facies metabasites are preserved in megaboudins exposed in the easternmost exposure of the lower plate at As Sifah (Fig. 3a). These record minimum pressures of *c.* 10–12 kbar

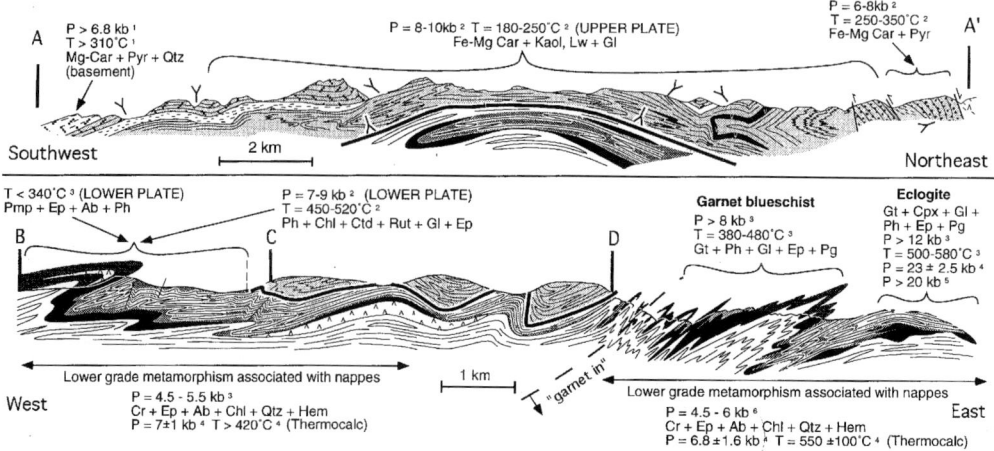

**Fig. 5.** Metamorphic summary for upper and lower plate localities in northeastern Saih Hatat (modified from fig. 2 of Miller *et al.* 1998). Car, Carpholite; Pyr, pyrophyllite; Qtz, quartz; Kaol, kaolinite; Lw, lawsonite; Gl, glaucophane; Pmp, pumpellyite; Ep, epidote; Ab, albite; Ph, phengite; Ctd, chloritoid; Rut, rutile; Pg, paragonite; Gt, garnet; Cpx, omphacitic pyroxene; Cr, crossite; Hem, hematite.

(jadeite = albite + quartz barometer of El-Shazly *et al.* (1990)). Peak temperatures were estimated at 500–580°C (garnet–pyroxene thermometer of El-Shazly *et al.* (1990)). Pure jadeite has not been documented at As Sifah and full petrological descriptions of these eclogites have been given by El-Shazly *et al.* (1990). Radial fractures around quartz inclusions in garnets (Fig. 3f) in the vicinity of the megaboudins suggest that very high pressures were attained (>20 kbar; Wendt *et al.* 1993). Similar 'radial' fractures are also found in garnets structurally higher in the lower plate at As Sifah. Application of a new garnet–pyroxene–phengite barometer by Wills *et al.* (1991) and Searle *et al.* (1994) resulted in calculated pressure values as high as 23 kbar for the lower plate eclogites. Alternatively, El-Shazly *et al.* (1997) argued that these estimates are too high and that metamorphic peak pressures were probably about 15 kbar.

Remnant jadeitic pyroxenes in structurally higher mafic schists in the lower plate have similar compositions to the jadeitic pyroxenes within the megaboudins (fig. 10 of El-Shazly *et al.* 1990), but have slightly lower minimum pressure estimates (>8 kbar, El-Shazly *et al.* 1990) because of the inferred lower temperature of metamorphism. Metamorphic temperatures of 380–480°C (zone B, fig. 10 of El-Shazly *et al.* 1990) were estimated for garnet blueschists, which are structurally higher than the eclogites (Fig. 5).

*The lower plate: Wadi Meeh and Wadi Hulw subdomain*

No garnet or pyroxene has been found within the lower plate window at Wadi Meeh and Wadi Hulw, in marked contrast to the As Sifah window (Fig. 5). Relict higher-pressure sodic amphiboles are, however, preserved in epidote porphyroblasts (Fig. 4c). El-Shazly *et al.* (1990) argued that the presence of pumpellyite indicates that temperatures may have been less than 340°C, although lawsonite has not been found. Alternatively, Goffé *et al.* (1988) argued that pseudomorphs of garnet exist in the Hulw region of the lower plate and that phengite, chlorite, chloritoid, rutile, glaucophane, quartz assemblages suggest pressures of 7–9 kbar and temperatures of 450–520°C. What is clear is that very high strain resulted in strong recrystallization of these units during exhumation, making determination of peak metamorphic pressures and temperatures for the Wadi Meeh and Hulw exposure of the lower plate difficult. However, there is a general lack of earlier coarse microstructures in this region, suggesting that early high-temperature metamorphism may not have occurred within the lower plate in the Hulw–Meeh region of the lower plate.

*Relationship of lower plate structures to the metamorphism*

Outside the megaboudins at As Sifah, peak metamorphic assemblages associated with the earlier structures are almost completely overprinted by intense deformation, which culminated in the formation of the regional recumbent closures (Fig. 5). A progressive decrease in metamorphic grade is recorded by assemblages that are consistently oriented around the northeast–southwest stretching lineation. Mafic lithologies show a change from garnet, phengite, sodic amphiboles and clinozoisite (Fig. 3d) to later fabrics defined by phengite, sodic–calcic amphiboles, epidote, clinozoisite and albite (+quartz) (Fig. 3h), indicating that pressures must have been at least below the jadeite = albite + quartz equilibrium line. Mafic units at the lowest grade show epidote, sodic–calcic amphiboles, albite, actinolite, hematite and chlorite assemblages, and are frequently associated with sodic–calcic amphibole and albite bearing veins that are orthogonal to the stretching lineation.

In the lower plate at As Sifah, the lowest-grade assemblages are synchronous with the formation of the regional sheath-like recumbent folds, which fold the earlier higher-grade assemblages (Fig. 3d). 'Pulled apart' remnant garnets in the hinges of parasitic folds associated with these recumbent folds have epidote, chlorite, hematite and albite in the resulting fractures (Fig. 3c). In the Wadi Meeh and Hulw region, similar low-grade assemblages (epidote, sodic–calcic amphiboles, albite, actinolite, hematite and chlorite) define axial planar fabrics to these parasitic folds in mafic units (Fig. 4c).

Application of the geobarometers of Brown (1977) and Maruyama *et al.* (1986) yields similar pressures (4.5–6 kbar) for amphiboles associated with the lower-grade stretching lineation in the lower plate at As Sifah, Wadi Meeh and Wadi Hulw. Using the thermodynamic data set of Holland & Powell (1990), Searle *et al.* (1994) calculated a slightly higher pressure of $7 \pm 1$ kbar for the low-grade assemblages at Wadi Hulw, but they also calculated a pressure of $6.8 \pm 1.6$ kbar for the lower-grade event in the As Sifah region of the lower plate. These results indicate that the lower plate appears to have eventually attained similar pressures as a result of the deformation associated with regional recumbent folding (Fig. 5).

## $^{40}$Ar–$^{39}$Ar geochronology

New $^{40}$Ar–$^{39}$Ar geochronology was performed on white mica populations selected from various fabric elements identified in the field. This has provided insight into the exhumation and cooling history of these sequences. The $^{40}$Ar–$^{39}$Ar results were interpreted in the context of the macro-, meso- or micro-scale structures (Figs 3, 4 and 6), and the silicon content of phengite was used systematically as a qualitative geobarometer.

Both laser and furnace $^{40}$Ar–$^{39}$Ar techniques were used here and the methods follow those described by Foster & Fanning (1997). Representative mica and amphibole analyses have been deposited with the Society Library and the British Library, Boston Spa, Wetherby, West Yorkshire LS23 3BQ, UK, as Supplementary Publication No. SUP18126 (9 pp.). All errors are $2\sigma$ unless otherwise stated. Mica size fractions and the mass of furnace samples are shown in Table 2. The micas were concentrated by magnetic separation and the final aliquots were hand picked under a binocular microscope to remove any impurities. Where there is a sufficient spread of $^{36}$Ar/$^{40}$Ar ratios, inverse isochron ages have been calculated (Table 2) using the method of York (1969). Calculated ages and associated mean square weighted deviations (MSWDs) for these isochrons are listed in Table 2. MSWD values over 2.5 generally suggest that the scatter of data is not necessarily a function of experimental error (e.g. Dalrymple & Lanphere 1974; Fleck *et al.* 1977; McDougall & Harrison 1988). It should be noted that the silicon content of phengite (higher Si indicates higher pressure) can be used as a barometer only if the limiting assemblage of K-feldspar + phlogopite + quartz is present (Massone & Schreyer 1987). None of the samples studied here have this limiting assemblage, and hence the Si content can be

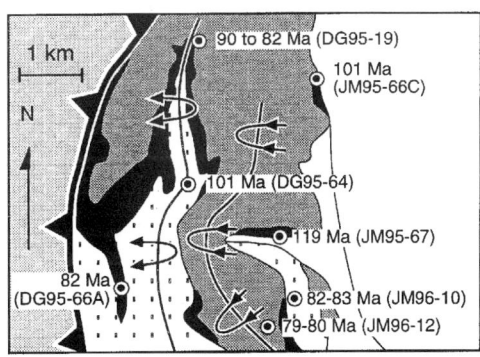

**Fig. 6.** Sample localities for $^{40}$Ar–$^{39}$Ar geochronology in the lower plate at As Sifah. Other samples are located in Fig. 2.

Table 2. Summary of $^{40}Ar-^{39}Ar$ age data; samples with an asterisk have slightly disturbed or discordant age spectra

| Sample no. | Mineral | Mass (mg), or no. of grains | Mica size fraction (μm) | $\%^{39}Ar$ | Steps used | Mean age (Ma) size of step weighting | Isochron age $^{36}Ar/^{40}Ar$ v. $^{39}Ar/^{40}Ar$ (Ma) | MSWD | $^{40}Ar/^{36}Ar$ ratio from isochron plot |
|---|---|---|---|---|---|---|---|---|---|
| JM95-67-L | phen. | 1 | 400–600 | 100 | 5/5 | 118.6 ± 0.8 | – | – | – |
| JM95-66c-F | phen. | 7.6 mg | 400–600 | 94 | 5–10/11 | 100.8 ± 0.9 | – | – | – |
| DG95-64-L* | mu. | 1 | 400–600 | 93 | 4–9/9 | 100.9 ± 0.9 | – | – | – |
| DG95-19-F* | phen. | 7.1 mg | 250–400 | 94 | 5–13/13 | 86.2 ± 0.6 | – | – | – |
| JM96-10-L.1 | phen. | 1 | 400–600 | 93 | 5–9/9 | 83.4 ± 0.3 | – | – | – |
| JM96-10-L.2 | phen. | 1 | 400–600 | 95 | 2–9/9 | 82.1 ± 0.5 | 82.2 ± 0.2 92% of gas (not step 8) | 1.8 | 293 ± 6 |
| JM96-32a-F | mu./phen | 7.5 mg | 250–400 | 52 | 6–11/12 | 91.0 ± 0.7 | 90.2 ± 0.2 | 1.4 | 321 ± 71 |
| DG95-66a-F | mu./phen | 8.0 mg | 400–600 | 75 | 1–10/12 | 82.7 ± 0.2 | 82.7 ± 0.2 | 1.9 | 303 ± 40 |
| JM96-12-L | mu./phen | 1 | 400–600 | 90 | 5–9/10 | 78.8 ± 0.5 | 79.6 ± 0.3 | 9.4 | 275 ± 21 |
| JM96-12-F* | mu./phen | 6.8 mg | 250–400 | 57 | 5–9/13 | 80.3 ± 0.8 | 80.7 ± 0.8 | 9.9 | 260 ± 41 |
| JM96-89-L* | mu./phen | 2–3 | 250–400 | 97 | 4–12/12 | 81.9 ± 1.1 | 79.8 ± 0.4 | 24.8 | 347 ± 24 |
| JM96-81c-F | mu. | 7.1 mg | 250–400 | – | – | – | – | – | – |
| JM96-31b-L | mu. | 4–5 | 250–400 | 74 | 4–8/9 | 75.7 ± 0.6 | – | – | – |
| JM96-93-L* | mu. | 4–5 | 250–400 | 100 | 1–8/8 | 70.0 ± 0.7 | 70.9 ± 0.4 | 6.0 | 278 ± 30 |
| JM96-93-L* | mu. | 4–5 | 250–400 | 60 | 5–8/8 | 70.0 ± 0.4 | 69.5 ± 0.6 | 2.6 | 340 ± 66 |
| JM96-139a-L | mu. | 1 | 250–400 | – | – | – | – | – | – |

used only qualitatively to determine relative pressures between samples having similar lithologies.

## $^{40}Ar/^{39}Ar$ results

Most of the age spectra exhibit concordant or nearly concordant steps over a significant fraction of the gas released (Figs 7 and 8). Many of the concordant samples fit the plateau criteria as defined by Fleck *et al.* (1977). However, several age spectra are partially disturbed or discordant and do not strictly satisfy these criteria. These discordant spectra may still give valuable information on the cooling–crystallization history of these samples.

### Lower plate: As Sifah (highest grade)

Sample JM95-67-L (Fig. 7a) is a phengite from a schist containing garnet, phengite, clinopyroxene, clinozoisite, glaucophane, and quartz. The phengite defines an early east–west lineation in one of the mafic boudins (Fig. 3f). Laser analysis of this sample gives a plateau age of $118.0 \pm 1.6$ Ma.

Phengite JM95-66c-F (Fig. 7b) was separated from the core of a mafic megaboudin (Figs 3g and 6) and furnace analysis of this sample gives a plateau age of $100.8 \pm 1.6$ Ma. In this sample, phengite coexists with omphacite, glaucophane, clinozoisite and quartz.

DG95-64-L (Fig. 7c) is a muscovite from a quartz–mica schist sampled high up in the structural pile (Figs 3a and 6). The age spectrum for this sample yields older initial low-temperature ages that step down to a mean age of $100.9 \pm 1.8$ Ma for the last six steps comprising 93% of the gas.

Phengite DG95-19-F (Fig. 7d) is from a mafic schist (phengite–glaucophane–epidote–quartz) structurally above the eclogite-bearing megaboudins (Figs 3a and 6). The phengite in this sample is in textural equilibrium with glaucophane. The last nine steps of the discordant age spectrum give ages ranging from 90 to 82 Ma.

JM96-10 (Fig. 7e) is a garnet–phengite–quartz schist with strongly retrogressed sodic amphiboles, which is structurally below the eclogite-bearing megaboudins (Figs 3a and 6). Two laser analyses on phengite from this sample gave mean ages of $83.4 \pm 0.6$ and $82.1 \pm 1.0$ Ma. The isochron age of $82.2 \pm 0.4$ Ma for JM96-10-run2 (Table 2) is within error of the plateau age ($82.1 \pm 1.0$ Ma), and provides an atmospheric value for the $^{40}Ar/^{36}Ar$ intercept.

### Lower plate: Wadi Hulw (early fabric elements)

JM96-32a (Fig. 7f) is a quartz–mica schist from the Hulw region of the lower plate, which is located just below the contact with the upper plate (Figs 2 and 4a). Apart from the first low-temperature step, the age spectrum for this sample displays an age gradient with values progressively increasing from 74 Ma to a plateau age of $91.0 \pm 1.4$ Ma (52% of gas).

### Lower plate: late fabric elements associated with regionally consistent stretching lineations

Late fabric elements (Fig. 8a–f) related to the exhumation of the high-pressure rocks have been identified using several criteria: (1) lower-pressure mineral assemblages, e.g. mica intergrown with sodic–calcic amphiboles and albite, as well as drops in the silicon content of phengite; (2) clear overprinting relationships in the field. The majority of plateau ages calculated for such samples correlate well with inverse isochron ages because they yield $^{40}Ar/^{36}Ar$ intercepts within $2\sigma$ error of the atmospheric value (see Table 2).

DG95–66A-F (Fig. 8a) is a furnace analysis from a strongly retrogressed schist (As Sifah region, Figs 3e and 6) which shows extensive growth of new micas characterized by lower Si content than peak metamorphic phengite. The mica is interlayered with biotite, sodic–calcic amphibole, albite, epidote and chlorite. This sample yields a plateau age of $82.3 \pm 1.6$ Ma for the first 75% of the gas released, whereas the high-temperature steps provide ages of 88–87 Ma.

JM96-12 (Fig. 8b) is a calc-schist with strong top-to-the-northeast shear indicators, which bounds the top of the eclogite-bearing megaboudins at As Sifah (Figs 3e and 6). The laser analysis JM96-12-L of a single mica grain yields a $78.8 \pm 1.0$ Ma plateau age. The furnace analysis JM96-12-F of a mica separate from the same sample gives apparent ages ranging from 83 to 77 Ma with a mean age of $80.3 \pm 0.6$ Ma for steps 5–9. Compositional data shows a clear drop in silicon content for mica associated with this late fabric element (see Fig. 8b).

JM96-89-L (Fig. 8c) is a laser analysis on mica from a calc-schist, which in terms of mesoscopic structure, is identical to sample JM96–12 (see Figs 3e and 4b), but was sampled in the Wadi Meeh region of the lower plate (Figs 2 and 4b). The last nine steps of the resulting partially

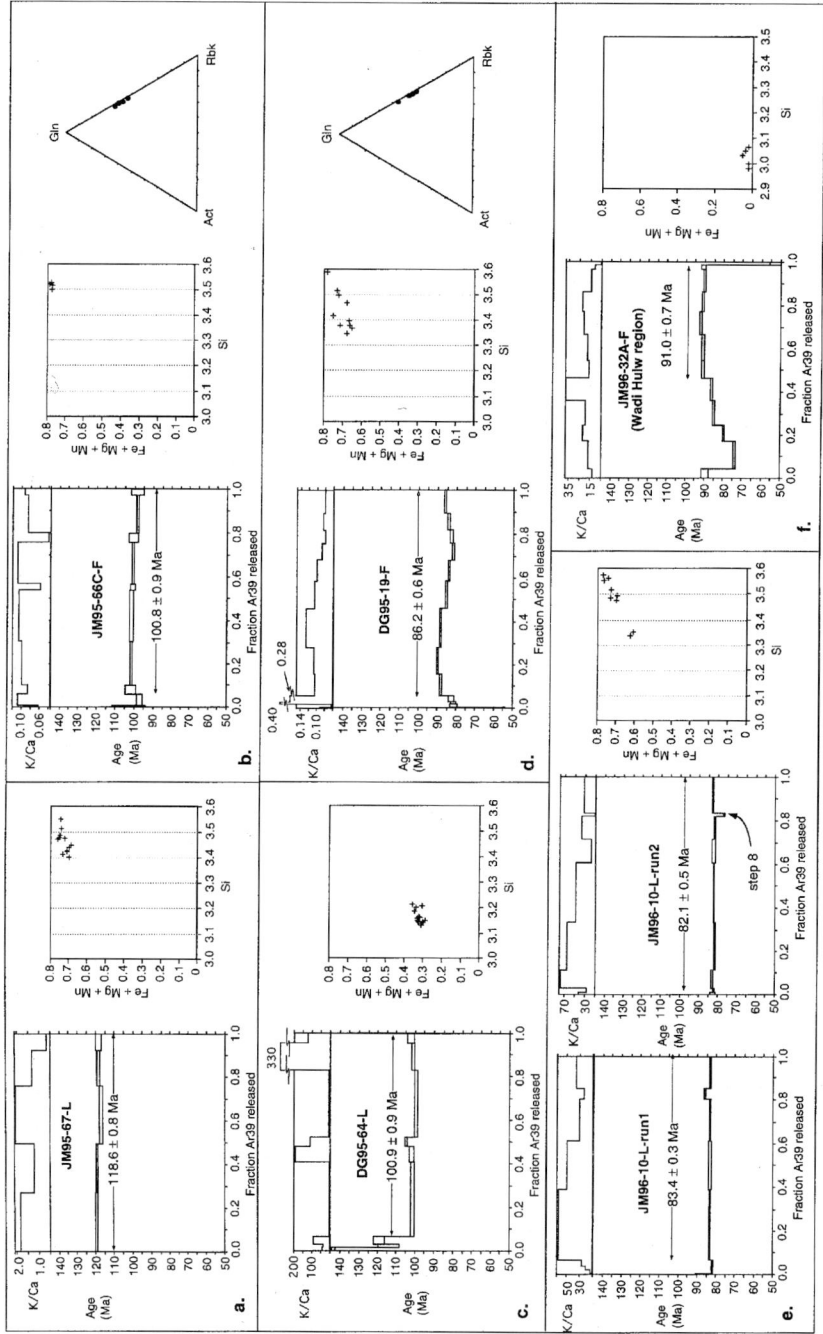

**Fig. 7.** $^{40}Ar-^{39}Ar$ age spectra and white mica compositional plots for samples from the lower plate (Group I ages). Analyses of amphiboles in textural equilibrium with white mica are plotted on a glaucophane (Gln)–actinolite (Act)–riebeckite (Rbk) ternary diagram. Mean ages (size of step weighting) have ± 2σ errors, individual age steps have ±1σ error bars.

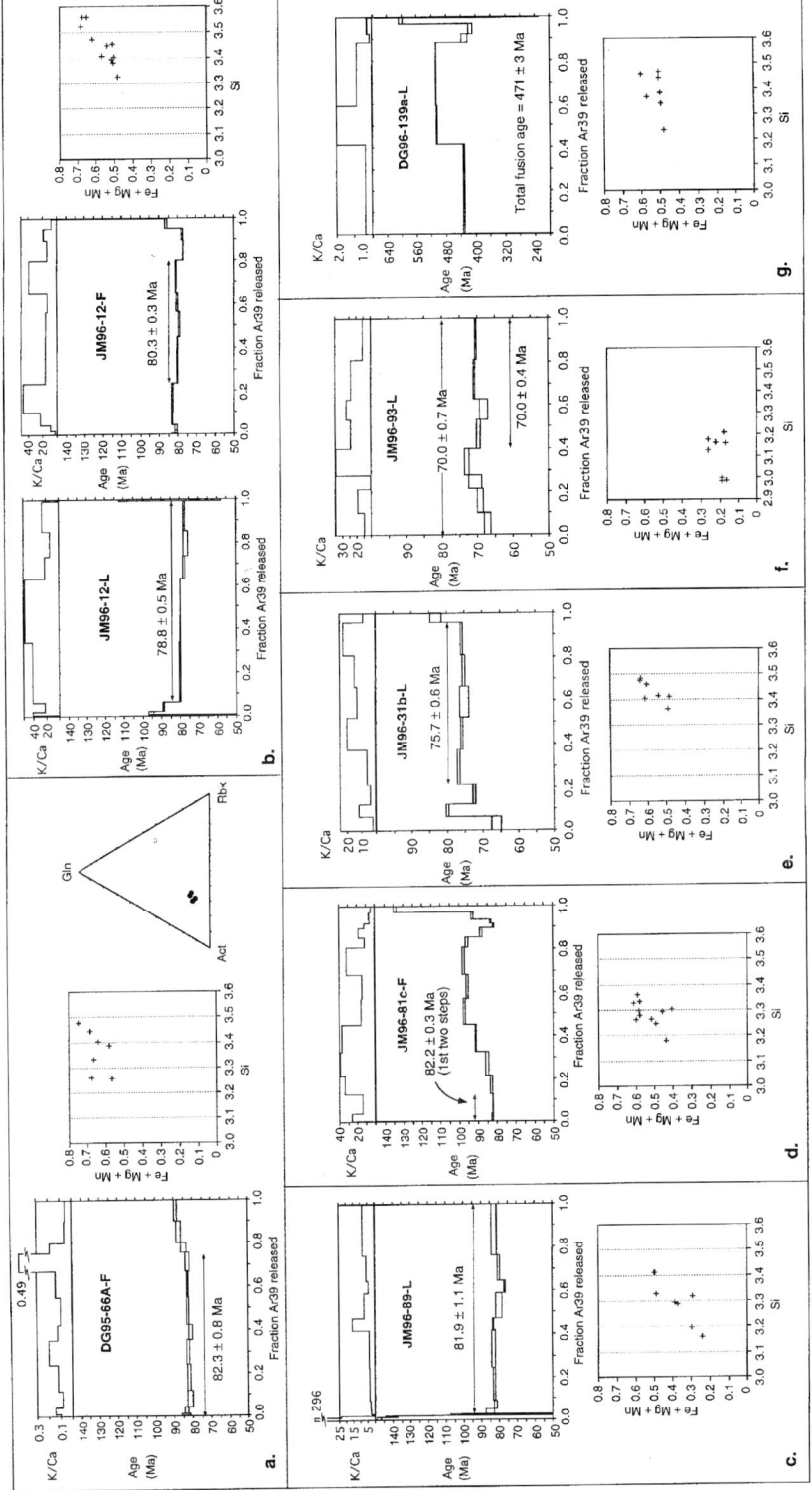

Fig. 8. $^{40}Ar-^{39}Ar$ age spectra and white mica compositional plots for late fabric elements in the lower plate (Group II ages, (a)–(d): sample localities shown in Figs 2 and 6) and upper plate (Group III ages, (e)–(g): sample localities shown in Fig. 2). Analyses of amphiboles in textural equilibrium with white mica are plotted on a glaucophane (Gln)–actinolite (Act)–riebeckite (Rbk) ternary diagram. Mean ages (size of step weighting) have ± 2σ errors, individual age steps have ± 1σ error bars.

discordant spectrum give a mean age of 81.9 ± 2.2 Ma. Like sample JM96-12, compositional data for mica JM96-89 also highlight a clear drop in silicon content for mica associated with this fabric element (see Fig. 8c), but with slightly lower initial Si values compared with JM96-12.

Mica JM96-81c-F (Fig. 8d) was separated from a pelitic schist, located in the hinge region of one of the regional nappes in the Hulw region of the lower plate (Figs 2 and 3a). The sample is strongly retrogressed, with chloritoid almost completely breaking down to biotite, chlorite and muscovite, and extensive new mica growth. This sample yields a strongly discordant saddle-shaped spectrum with ages ranging from 82 to 98 M. No meaningful plateau age can be calculated here. However, the mean of the first two steps is 82.2 ± 0.6 Ma.

## Upper plate

Mica JM96-31b-L (Fig. 8e) was extracted from a northeast–southwest lineation in a strongly foliated limestone sampled at the base of the upper plate nappes, just above the upper plate–lower plate discontinuity (Figs 2 and 4a). This sample is close to (<100 m) the lower plate sample JM96-32a (Fig. 7f). The age spectrum given by laser analysis JM96-31b-L exhibits three initial low-temperature steps with discordant ages, followed by five higher-temperature steps defining a plateau age of 75.6 ± 1.2 Ma (74% of the gas).

JM96-93-L (Fig. 8f) is a parasitic fold taken from the hinge region of an upper plate nappe which has muscovite growing axial planar to it (Figs 2 and 4a). The results from laser analysis of this sample show some age variation with K/Ca. The last four steps of the age spectrum are within $2\sigma$ error of each other and have a mean age of 70.0 ± 0.8 (61% of gas), indistinguishable from the total fusion age of 70.0 ± 1.4 Ma.

DG96-139a-L (Fig. 8g) is a laser analysis of muscovite from a Permian limestone unit (Saiq Formation) which was sampled in the upper plate just above the contact with the lower plate in the northern region of Wadi Meeh (Figs 2 and 4a). The age spectrum of this sample displays pre-Permian ages with a total fusion age of 471 ± 6 Ma, suggesting that the micas have retained part of their detrital age.

## Discussion of $^{40}Ar/^{39}Ar$ data

The age spectra obtained on white mica using both furnace and laser experiments can be divided into three main groups: (I) 119–82 Ma ages on phengites (high-Si phengite) and muscovites from older fabric elements in the lower plate (Figs 6 and 7); (II) 82–79 Ma ages on micas from late fabric elements; these micas are associated with lower-pressure mineral assemblages that overprint the older fabric elements during the exhumation of the lower plate sequences (Fig. 8); (III) 76–70 Ma ages on micas from fabric elements in the upper plate and pre-Permian detrital ages (Fig. 8).

## Lower plate: highest-grade micas (earliest fabrics)

The majority of Group I ages were obtained from high-pressure fabric elements formed at temperatures of 440–580°C (zone C, El-Shazly et al. 1990). Estimates of the closure temperature for Ar in muscovite and phengite range from 350 to 450°C (Purdy & Jager 1976; Sisson & Onstott 1986; Blankenburg et al. 1989; Kirschner et al. 1996; Lister & Baldwin 1996). Therefore, the majority of the Group I ages reflect the time of cooling below the closure temperature and cannot be interpreted as crystallization ages.

## Lower plate: retrograde assemblages (later fabrics)

At present, there are no well-constrained temperature estimates for the retrograde exhumation path of these sequences, and hence the Group II ages (82–79 Ma) may reflect either cooling or crystallization ages. If they represent cooling ages, then they give only a minimum value for the age of exhumation of these units to shallower crustal levels.

## Upper plate

The upper plate metamorphism occurred at a significantly lower temperature than that of the lower plate (< 350°), resulting in the growth of carpholite, lawsonite and sodic amphibole (Fig. 5). This difference in temperature, combined with the preservation of Hercynian ages in some age spectra, led El-Shazly & Lanphere (1992) to interpret $^{40}Ar–^{39}Ar$ ages from this region as crystallization and not cooling ages. The retrograde assemblage albite–chlorite–sphene in the upper plate mafic rocks (Fig. 4d) is also lower temperature than the retrograde assemblage epidote–sodic–calcic amphibole–chlorite–albite found in lower plate mafic units (Fig. 4c). Preservation of pre-Permian $^{40}Ar–^{39}Ar$ ages in detrital micas, the occurrence of high-pressure–low-temperature minerals such as carpholite and lawsonite, and the lack of textural evidence for temperatures exceeding the closure temperature of muscovite after the growth of these high-pressure minerals, all suggest that the upper plate $^{40}$Ar–$^{39}$Ar ages may represent mica crystallization ages rather than cooling ages.

## Exhumation history of the high-pressure rocks

The upper plate–lower plate discontinuity has transposed two terranes with different tectonometamorphic histories (Gregory et al. 1998; Miller et al. 1998). Basement rocks were emplaced over younger shelf sequences in an upper plate with a regional nappe-like geometry. Higher-grade rocks are exposed in the structurally lowest position and nappe structures are associated with the exhumation of this higher-grade terrane from subcrustal depths. The lower plate rocks record a longer history of fabric development and metamorphism, whereas the upper plate rocks record a simpler history. The current range of $^{40}$Ar–$^{39}$Ar age data is summarized in Fig. 9.

### Lower plate

The oldest apparent ages (i.e. >101 Ma) are restricted to eclogite localities exposed at the deepest structural level of the lower plate shear zone at As Sifah (Fig. 6). Eclogitic assemblages are not recorded by the rocks in the Wadi Hulw–Meeh subdomain. Otherwise, the age distributions for the lower plate fabric elements (i.e. Groups I and II) as described by this study are scattered throughout both lower plate domains (Wadi Hulw–Meeh and As Sifah windows), the key difference being that the lower plate sequences in the Meeh–Hulw region registered initially lower peak temperatures and pressures with respect to the As Sifah region. However, the entire lower plate has experienced intense deformation and retrogression associated with the exhumation of the eclogites and garnet blueschists at As Sifah.

The older ages in Group I (c. 130 Ma) have been used by El-Shazly & Lanphere (1992) to support a model of early crustal thickening or A-type subduction. The younger ages (≤80 Ma) were related to a separate second event, the emplacement of the Samail ophiolite. Alternatively, Searle et al. (1994) argued that any ages older than 95 Ma were the result of excess argon (i.e. ages below the thick dashed line in Fig. 9), and that all of the high-pressure assemblages could be related to the attempted northward subduction of the continental shelf, which eventually resulted in the emplacement of the Samail ophiolite (also the model of Michard et al. (1994)). Irrespective of which of these workers is correct, the $^{40}$Ar–$^{39}$Ar geochronology shows a series of cooling events in the 95–82 Ma period (and potentially from 131 to 82 Ma, Fig. 9). Only the youngest high-Si phengite cooling ages within Group I overlap with the Group II ages associated with lower-grade fabrics related to exhumation. This suggests that many of the older cooling ages (i.e. > 82 Ma) are not related to exhumation (Fig. 9), but may reflect cooling of the eclogite and garnet blueschists at depth, possibly by the progressive emplacement of colder footwalls, before emplacement to shallower crustal levels (Fig. 10a).

Structural and metamorphic evidence indicates that the exhumation of the lower plate units occurred during northeast-directed transport with the development of regional-scale, recumbent closures, that fold and transpose the earlier high-pressure fabrics. This strong non-coaxial shearing event pervasively affected the entire lower plate where a strong component of flattening has resulted in the development of extensive C′-shear bands (Fig. 10b). Calc-schists (with shear bands formed during this northeast-directed shearing event) have mica ages that range from 82 to 79 Ma (Fig. 8b and c), and other mica from the hinge zone of a regional closure in the Hulw region of the lower plate shows intense resetting by this event at c. 82 Ma (Fig. 8d). The youngest apparent ages from Group I (83–82 Ma, high-Si phengite from a retrogressed sample, Fig. 7e) may reflect cooling of high-pressure minerals during the initial stages of exhumation of the lower plate sequences. Mica with a lower silica content (with respect to the earlier fabric elements), which is also associated with retrograde sodic–calcic amphiboles and albite (indicating that pressures are below the albite = jadeite + quartz reaction line), has an 82 Ma plateau age (Fig. 8a). The new $^{40}$Ar–$^{39}$Ar ages combined with the metamorphic data suggest that partial exhumation of the lower plate sequences occurred between 82 and 79 Ma (Figs 9 and 10b).

### Upper plate

$^{40}$Ar–$^{39}$Ar mica crystallization ages at 80 Ma (El-Shazly & Lanphere 1992), 76 Ma, 70 Ma (Group III) are all younger than the cooling ages on the highest-pressure fabric elements in the lower plate (Group I).

Mica growth axial planar to upper plate nappes occurred at 70 Ma (Fig. 8f). Other muscovite from a northeast–southwest lineation at the base of the nappe pile in the upper plate gave a 76 Ma plateau age. Both ages are interpreted

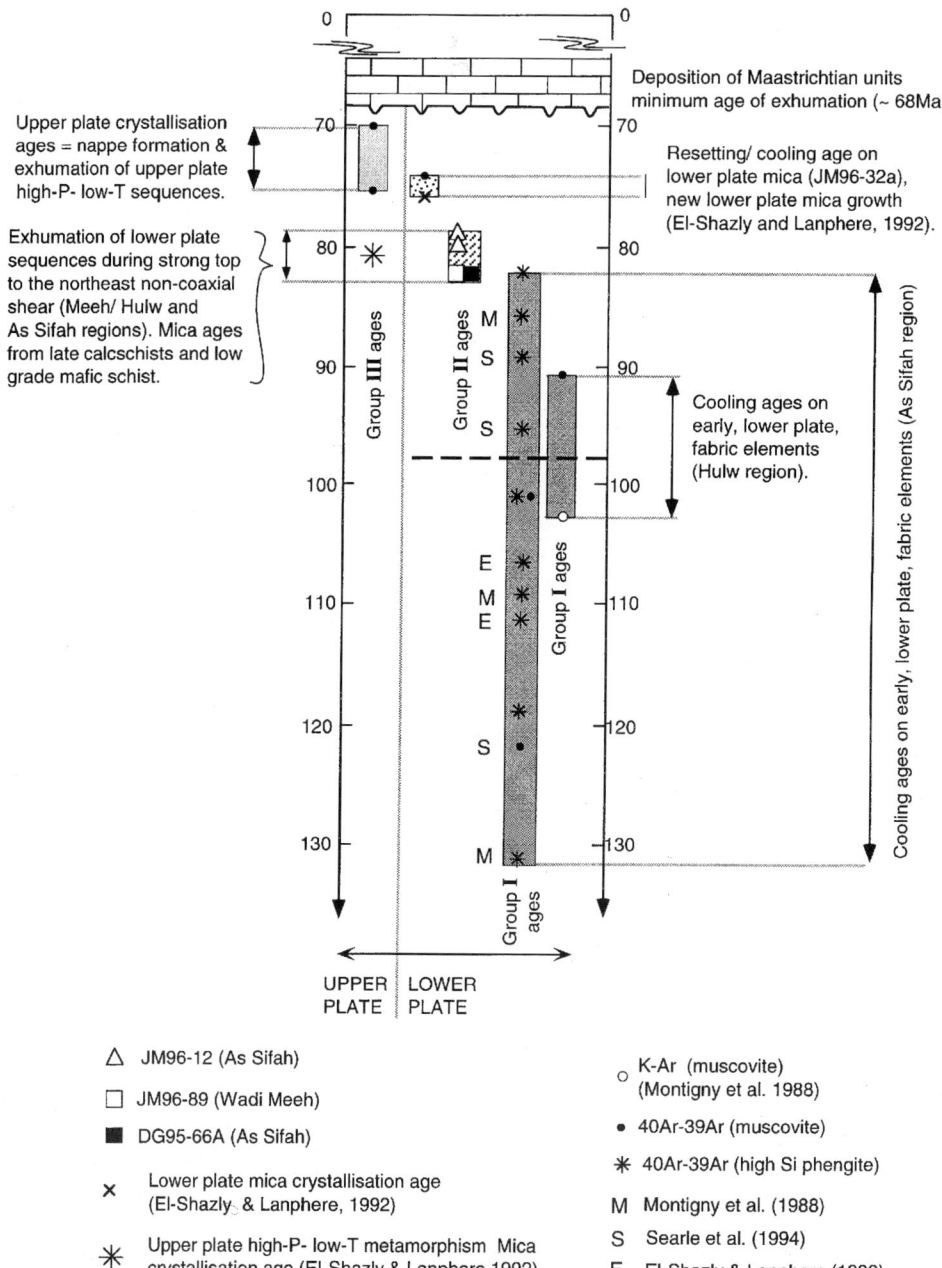

**Fig. 9.** Summary of geochronological data from northeastern Saih Hatat. It should be noted that the majority of Group I ages are 'discordant' age spectra (in particular the ages from El-Shazly & Lanphere (1992)) and the ages of Montigny et al. (1988) represent total fusion ages from discordant age spectra. The 95 Ma age from Searle et al. (1994) has an inverse isochron age of 96 Ma with an atmospheric $^{40}Ar-^{36}Ar$ intercept.

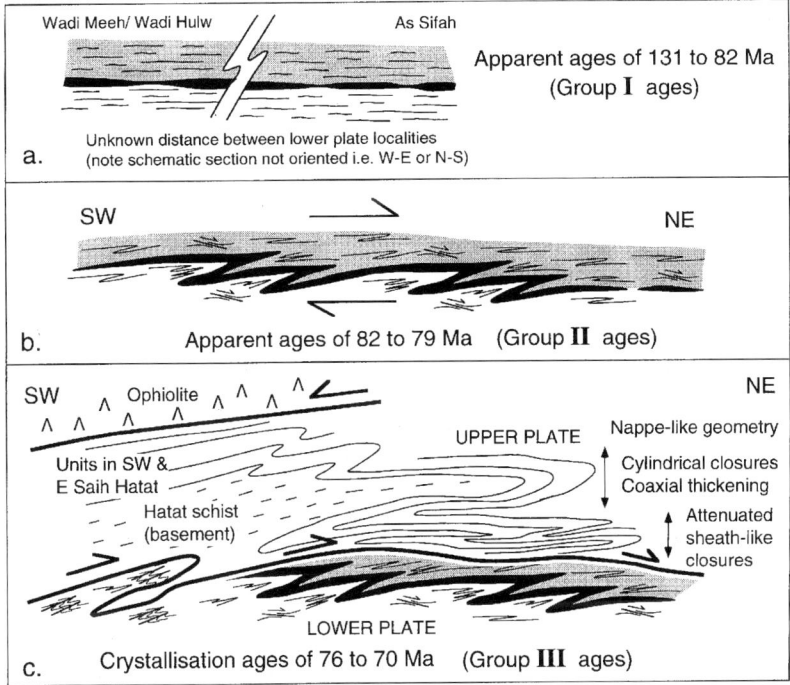

**Fig. 10.** Schematic evolutionary diagram for the high-pressure rocks in Saih Hatat based on structural and geochronological data. (**a**) Initial cooling events within deformed and metamorphosed lower plate tectonostratigraphic units of carbonate, mafic schist and quartz–mica schist. (**b**) Partial exhumation of lower plate sequences during strong top to the northeast shearing. This results in the formation of regional sheath-like closures, C- and C'-shear bands. (Note the orientation of the shear zone is unknown and is drawn horizontal for convenience.) (**c**) Juxtaposition of upper and lower plates via emplacement of upper plate nappes. (Note upper plate nappe-like geometry, truncation of lower plate closures and the infolding of upper and lower plate units during progressive deformation (at the southern end of the lower plate windows) with the development of a strong asymmetric crenulation cleavage.)

to represent mica growth synchronous with the upper plate nappe formation (i.e. from 76 to 70 Ma). Axial planar mineral assemblages to these upper plate nappes (chlorite–albite–sphene for mafic units) contain relict high-pressure minerals, which suggests that the nappes formed at lower pressure during the exhumation of the upper plate high-pressure units. This upper plate nappe-forming event has locally thickened the stratigraphy, and is at odds with them being related to extensional exhumation of these units. To exhume these upper plate units, there must have been either rapid erosion during the nappe-forming event, or structurally higher tectonic thinning (see Coleman 1981).

The youngest age constraint for the emplacement of the Samail ophiolite and exhumation of the high-pressure rocks is the deposition of autochthonous Maastrichtian limestones (c. 67–68 Ma, Fig. 9), which was pre-dated by extensive lateritic weathering and/or tectonic thinning of the Samail ophiolite (see Coleman 1981; Hopson et al. 1981). The inferred transport direction of the ophiolite is the exact opposite sense to transport direction inferred for the regional upper plate nappes and the earlier exhumation fabrics of the lower plate. A new structural section through the ophiolite (Gregory et al. 1998) suggests that these higher-level, low-angle faults represent deformation which extensionally thinned the ophiolitic structural pile. Other workers have also argued for extension along this structural break (e.g. Michard et al. 1994). The young $^{40}Ar-^{39}Ar$ ages in the upper plate (as young as 70 Ma, very close to the deposition age of the Maastrichtian units; Fig. 9) suggests this higher-level deformation that thins the structural pile may have been synchronous with northeast directed nappe formation in the upper plate (Fig. 10c).

The timing of the upper plate metamorphism and the latest retrogression of the lower plate rocks is associated with the arrival of the ophiolite. Igneous detritus from the ophiolite found within the autochthonous Mid- to Late Campanian (c. 74–70 Ma) Juweiza Formation (Glennie et al. 1974; Warbuton et al. 1990, fig. 2) suggests that the Samail ophiolite had been emplaced onto the margin, before extensive mica growth associated with regional nappe formation in the upper plate (76–70 Ma). Previously published $^{40}$Ar–$^{39}$Ar crystallization ages of $80.2 \pm 0.8$ Ma on mica within the Ruwi lawsonite schist (El-Shazly & Lanphere 1992) are potentially compatible with the 'overburden' of the Samail ophiolite causing the high-pressure metamorphism of the upper plate units. This $^{40}$Ar–$^{39}$Ar age from the Ruwi schist overlaps with the Group II ages related to the partial exhumation of the lower plate units (note large asterisk representing age of Ruwi schist in Fig. 9). This suggests the lower plate units may have been emplaced to shallower crustal levels before, or during, high-pressure metamorphism of the upper plate sequences.

The upper plate–lower plate shear zone developed in association with the deformation of the upper plate. Field data show a change in fold style (cylindrical to sheath-like) and an increase in strain in upper plate units with closer proximity to the upper plate–lower plate discontinuity (Miller et al. 1998), as well as lower plate closures that are clearly truncated by this break (see Fig. 2). The discontinuity has a strong control on the structural style within the upper plate, but it post-dates the regional closures in the lower plate associated with partial exhumation of the lower plate sequences from 82 to 79 Ma. The lower plate has not, however, behaved passively as a rigid unit during the emplacement and exhumation of the upper plate, as parts of the lower plate are infolded with upper plate units (Fig. 10c). Furthermore, previously published $^{40}$Ar–$^{39}$Ar crystallization ages (El-Shazly & Lanphere 1992) from strongly retrogressed lower plate localities in Wadi Hulw (location A2, Fig. 2) suggest that mica growth occurred in the lower plate in the 76–72 Ma period. There may also have been some resetting of lower plate age spectra, in relation to the juxtaposition of the two plates (c. 74 Ma, JM96-32a-F, Fig. 7f).

The style of deformation in the upper plate and the refolding of the ductile shear zone separating the upper and lower plates suggest that the lower plate–upper plate contact is not a low-angle normal fault, even though it places upper plate units on top of rocks that at one time resided at higher pressure, although the latter is often used as a criterion for recognizing extension exhumation regimes.

## Conclusions

This study has provided constraints on the timing and duration of exhumation of the high-pressure rocks in the Saih Hatat window. Importantly, the timing relationships evidenced by our $^{40}$Ar–$^{39}$Ar study match those obtained from the relative sequence of events deduced from field relationships. The higher-grade lower plate units have cooling ages on high-pressure fabric elements that are as young as 82 Ma. Fabrics associated with the exhumation and northward transport of the lower plate sequences have ages ranging from 82 to 79 Ma. Crystallization ages on mica associated with the northward emplacement of the upper plate nappes are younger (76–70 Ma) and suggest that nappe formation occurred synchronously with structurally higher tectonic thinning, erosion and southward transport of the overlying ophiolite, Hawasina and Haybi complexes.

We wish to thank the Ministry of Petroleum and Minerals for their kind support during field work, particularly M. Kassim (Director General of Minerals) and H. Al-Azry (Deputy Director General of Minerals). This study was supported by ARCSM 1995 and ARC A39601548 awarded to D. R. G. and NSF-EAR 91–06016 awarded to R. T. Gregory, A. El-Shazly and M. Holdaway. J. M. acknowledges support from a Monash Graduate Scholarship. Reviews by D. Waters, R. Oberhänsli and an anonymous reviewer improved the manuscript. S. Costa is thanked for improving the revised manuscript.

## References

BAILEY, E. H. 1981. Geologic map of the Muscat–Ibra area, Sultanate of Oman. *Journal of Geophysical Research*, **86**.

BECHENNEC, M., LE METOUR, J., RABU, VILLEY, M. & BEURRIER, M. 1990. The Hawasina Basin: a fragment of a starved continental margin, thrust over the Arabian Platform during obduction of the Sumail Nappe. *In*: ROBERTSON, A. H., SEARLE, M. P. & RIES, A. C. (eds) *The Geology and Tectonics of the Oman Region*. Geological Society, London, Special Publications, **49**, 323–343.

BLANKENBUREG, F. V., VILLA, A., BAUR, H., MORTEANI, G. & STEIGER, R. H. 1989. Time calibration of a *PT*-path from the Western Tauern Window, eastern Alps: the problem of closure temperatures. *Contributions to Mineralogy and Petrology*, **101**, 1–11.

BOUDIER, F. & COLEMAN, R. G. 1981. Cross section through the peridotite in the Samail ophiolite, southeastern Oman. *Journal of Geophysical Research*, **86**, 2573–2592.

——, BOUCHEZ, J. L., NICOLAS, A., CEULENEER, G., MISERI, M. & MONTIGNY, R. 1985. Kinematics of ocean thrusting in the Oman Ophiolites: model of plate convergence. *Earth and Planetary Science Letters*, **75**, 215–222.

BROWN, E. H. 1977. The crossite content of Ca-amphibole as a guide to the pressure of metamorphism. *Journal of Petrology*, **18**, 53–72.

COLEMAN, R. G. 1981. Tectonic setting for ophiolite obduction in Oman. *Journal of Geophysical Research*, **86**, 2497–2508.

DALRYMPLE, G. B. & LANPHERE, M. A. 1974. $^{40}Ar/^{39}Ar$ age spectra of some disturbed terrestrial samples. *Geochimica et Cosmochimica Acta*, **38**, 715–738.

EL-SHAZLY, A. K. 1994. Petrology of lawsonite-, pumpellyite- and sodic amphibole bearing metabasites from NE Oman. *Journal of Metamorphic Geology*, **12**, 23–48.

—— 1995. Petrology of Fe–Mg-carpholite-bearing metasediments from NE Oman. *Journal of Metamorphic Geology*, **13**, 379–396.

—— & LANPHERE, M. A. 1992. Two high pressure metamorphic events in NE Oman: evidence from $^{40}Ar–^{39}Ar$ dating and petrological data. *Journal of Geology*, **100**, 731–751.

——, COLEMAN, R. G. & LIOU, J. G. 1990. Eclogites and blueschists from NE Oman: petrology and P–T evolution. *Journal of Petrology*, **31**, 629–666.

——, WORTHING, M. A. & LIOU, J. G. 1997. Interlayered eclogites, blueschists, and epidote amphibolites from NE Oman: a record of protolith compositional control and limited fluid infiltration. *Journal of Petrology*, **38**, 1461–1487.

FLECK, R. J., SUTTER, J. F. & ELLIOT, D. H. 1977. Interpretation of discordant $^{40}Ar/^{39}Ar$ age-spectra of Mesozoic tholeiites from Antarctica. *Geochimica et Cosmochimica Acta*, **41**, 15–32.

FOSTER, D. A. & FANNING, C. M. 1997. Geochronology of the Idaho–Bitterroot batholith and Bitterroot Dome: an example of magmatism preceding and contemporaneous with metamorphic core complex extension. *Geological Society of America, Bulletin*, **109**, 379–394.

GLENNIE, K. W., HUGHES CLARKE, M. W., BOEUF, M. G., PILAAR, W. F. & REINHARDT, B. M. 1974. *Geology of the Oman Mountains*. Verhandelingen van het Koninklijk Nederlands geologisch mijnbouwkundig Genootsschap, Amsterdam, **31**.

GOFFÉ, B., MICHARD, A., KIENAST, J. R. & LE MER, O. 1988. A case of obduction-related high pressure, low temperature metamorphism in upper crustal nappes, Arabian continental margin, Oman. *Tectonophysics*, **151**, 363–386.

GREGORY, R. T., GRAY, D. R. & MILLER, J. McL. 1996. Tectonics of the Arabian margin associated with the emplacement of the Oman ophiolite, NE Saih Hatat. *Geological Society of America Abstracts with Programs*, **28**, A-369.

——, —— & —— 1998. Tectonics of the Arabian margin associated with the formation and exhumation of high-pressure rocks, Sultanate of Oman. *Tectonics*, **17**, 567–670.

HACKER, B. R., MOSENFELDER, J. L. & GNOS, E. 1996. Rapid emplacement of the Oman Ophiolite: thermal and geochronological constraints. *Tectonics*, **15**, 1230–1247.

HOLLAND, T. J. B. & POWELL, R. 1990. An enlarged and updated internally consistent thermodynamic dataset with uncertainties and correlations: the system $K_2O–Na_2O–CaO–MgO–MnO–FeO–Fe_2O_3–Al_2O_3–TiO_2–SiO_2–C–H_2–O_2$. *Journal of Metamorphic Geology*, **8**, 89–124.

HOPSON, C. A. HOPSON, C. A., COLEMAN, R. G., GREGORY, R. T., PALLISTER, J. S. & BAILEY, E. H. 1981. Geologic section through the Samail Ophiolite and associated rocks along the Muscat–Ibra transect, southeastern Oman Mountains. *Journal of Geophysical Research*, **86**, 2527–2544.

KIRSCHNER, D. L., COSCA, M. A., MASSON, H. & HUNZIKER, J. C. 1996. Staircase $^{40}Ar/^{39}Ar$ spectra of fine-grained white mica: timing and duration of deformation and empirical constraints on argon diffusion. *Geology*, **24**, 747–750.

LE METOUR, J., DE GRAMONT, X. & VILLEY, M. 1986. *Geological map of Masqat and Quryat*, Sheets **NF40-4A**, **NF40-4D**, Scale 1:100 000. Explanatory notes. Ministry of Petroleum and Minerals, Directorate General of Minerals, Sultanate of Oman.

LIPPARD, S. J. 1983. Cretaceous high pressure metamorphism in NE Oman and its relationship to subduction and ophiolite nappe emplacement. *Journal of the Geological Society, London*, **140**, 97–104.

——, SHELTON, A. W. & GASS, I. G. 1986. *The Ophiolite of Northern Oman*. Geological Society of London, Memoir **11**.

LISTER, G. S. & BALDWIN, S. L. 1996. Modelling the effect of arbitrary P–T–t histories on argon diffusion in minerals using the MacArgon program for the Apple Macintosh. *Tectonophysics*, **253**, 83–109.

MASSONE, H.-J. & SCHREYER, W. 1987. Phengite geobarometry based on the limiting assemblage with K-feldspar, phlogopite, and quartz. *Contributions to Mineralogy and Petrology*, **96**, 212–224.

MANN, A. & HANNA, S. S. 1990. The tectonic evolution of pre-Permian rocks, Central and Southeastern Oman Mountains. *In*: ROBERTSON, A. H., SEARLE, M. P. & RIES, A. C. (eds) *The Geology and Tectonics of the Oman Region*. Geological Society, London, Special Publications, **49**, 307–325.

MARUYAMA, S., CHO, M. & LIOU, J. G. 1986. Experimental investigations of blueschist greenschist equilibria: pressure dependence of $Al_2O_3$ contents in sodic amphiboles – a new geobarometer. *In*: EVANS, B. W. & BROWN, E. H. (eds) *Blueschists and Eclogites*. Geological Society of America Memoir, **164**, 1–16.

——, LIOU, J. G. & TERABAYASHI, M. 1996. Blueschists and eclogites of the world and their exhumation. *International Geology Review*, **38**, 485–594.

McDOUGALL, I. & HARRISON, T. M. 1988. *Geochronology and Thermochronology by the $^{40}Ar–^{39}Ar$ Method*. Oxford Monographs of Geology and Geophysics, 9. Oxford University Press, Oxford.

MICHARD, A. 1983. Les nappes des Mascate (Oman), rampe epicontinental d'obduction à facies schiste bleu, et la dualité apparente des ophiolites

Omanaises. *Sciences Géologique, Bulletin, Strasbourg,* **36**, 3–16.
——, GOFFÉ, B., SADDIQI, O., OBERHÄNSLI, R. & WENDT, A. S. 1994. Late Cretaceous exhumation of the Oman blueschists and eclogites: a two stage extensional mechanism. *Terra Nova,* **6**, 404–413.
MILLER, J. McL., GRAY, D. R. & GREGORY, R. T. 1998. Exhumation of high-pressure rocks, northeastern Oman. *Geology* **26**, 235–238.
MONTIGNY, R., LE MER, O. & WHITECHURCH, H. 1988. K–Ar and $^{40}$Ar/$^{39}$Ar study of metamorphic rocks associated with the Oman ophiolite: tectonic implications. *Tectonophysics,* **151**, 345–362.
PASSCHIER, C. W. & TROUW, R. A. J. 1996. *Microtectonics.* Springer, Berlin.
PURDY, J. W. & JAGER, E. 1976. *K–Ar Ages on Rock Forming Minerals from the Central Alps.* Memorie degli Istituti di Geologia e Mineralogia dell'Universita di Padova, **30**.
ROBERTSON, A. 1987. The transition from a passive margin to an Upper Cretaceous foreland basin related to ophiolite emplacement in the Oman Mountains. *Geological Society of America Bulletin,* **99**, 633–653.
SEARLE, M. P. & MALPAS, J. 1980. Structure and metamorphism of rocks beneath the Semail ophiolite of Oman and their tectonic significance in ophiolite obduction. *Transactions of the Royal Society of Edinburgh,* **71**, 247–262.
——, WATERS, D. J., MARTIN, H. N. & REX, D. C. 1994. Structure and metamorphism of blueschist–eclogite facies rocks from the NE Oman Mountains. *Journal of the Geological Society, London,* **151**, 555–576.

SISSON, V. B. & ONSTOTT, T. C. 1986. Dating blueschist metamorphism: a combined $^{40}$Ar/$^{39}$Ar and electron microprobe approach. *Geochimica et Cosmochimica Acta,* **50**, 2111–2117.
TILTON, G. R., HOPSON, C. A. & WRIGHT, J. E. 1981. Uranium–lead isotopic ages of the Semail Ophiolite, Oman, with applications to Tethyan ridge tectonics. *Journal of Geophysical Research,* **86**, 2763–2775.
TIPPIT, P. R., SMEWING, J. D. & PESSAGNO, E. A. 1981. The biostratigraphy of sediments in the volcanic unit of the Semail ophiolite. *Journal of Geophysical Research,* **86**, 2527–2544.
WARBUTON, J., BURNHILL, T. J., GRAHAM, R. H. & ISAAC, K. P. 1990. The evolution of the Oman Mountains Foreland basin. *In:* ROBERTSON, A. H., SEARLE, M. P. & RIES, A. C. (eds) *The Geology and Tectonics of the Oman Region.* Geological Society, London, Special Publications, **49**, 419–427.
WENDT, A. S., D'ARCO, P., GOFFÉ, B. & OBERHÄNSLI, R. 1993. Radial cracks around α-quartz inclusions in almandine: constraints on the metamorphic history of the Oman mountains. *Earth and Planetary Science Letters,* **114**, 449–461.
WILLS, H. N., WATERS, D. J. & SEARLE, M. P. 1991. A clockwise $P$–$T$ path at $P > 20$ kbar for eclogite and high-$P$ schist beneath the Semail ophiolite, Oman, *Terra Abstracts,* **3**, 98.
YORK, D. 1969. Least squares fitting of a straight line with correlated errors. *Earth and Planetary Science Letters,* **5**, 320–324.

# New insight into the dynamic development of the Southern Alps, New Zealand, from detailed thermochronological investigation of the Mataketake Range pegmatites

GEOFFREY E. BATT[1,3], BARRY P. KOHN[2], JEAN BRAUN[1], IAN McDOUGALL[1] & TREVOR R. IRELAND[1,4]

[1]*Research School of Earth Sciences, The Australian National University, Canberra, A.C.T. 0200, Australia*

[2]*Australian Geodynamics Cooperative Research Centre, School of Earth Sciences, La Trobe University, Melbourne, Vic. 3083, Australia*

[3]*Present address: Department of Geology and Geophysics, Yale University, P.O. Box 208109, New Haven, CT 06520–8109, USA*

[4]*Present address: Department of Geological and Environmental Sciences, Stanford University, Stanford, CA 94305-2215, USA*

**Abstract:** The Southern Alps are an actively developing mountain range resulting from continuing oblique continent–continent collision between the Australian and Pacific plates through the South Island of New Zealand. The thermal histories of five granite pegmatites from the Mataketake Range in the southwest of this orogen are revealed here by U–Pb dating of zircon, K–Ar, $^{40}Ar-^{39}Ar$ and Rb–Sr dating of muscovite, biotite and alkali feldspar, and fission-track dating of apatite and zircon. The constraint of these thermal histories, and in particular the high thermal resolution gained from analysis of K-feldspar by the $^{40}Ar-^{39}Ar$ dating method, offers a novel insight into the dynamic evolution of the Southern Alps. The pegmatites were probably formed by localized anatectic partial melting at temperatures of 620–680°C between 67 and 82 Ma. Other than initial cooling to the ambient temperature of the host schist following intrusion (estimated at c. 500°C), these bodies share an apparently uniform thermochronological record, and this thermal history is thus largely indicative of cooling during local exhumation. Between 10 and 5 Ma, the Mataketake Range area cooled from c. 350 to 260°C, at a rate of 18°C Ma$^{-1}$. This corresponds to unroofing at between 0.8 and 1 km Ma$^{-1}$ during this interval, and is probably a reflection of the inferred early history of the Australian–Pacific plate boundary through the South Island of New Zealand as a dominantly strike-slip feature. A reduction in cooling rate at 5 Ma, immediately followed by an episode of extremely rapid cooling at c. 350°C Ma$^{-1}$ marks the initiation of rapid exhumation and the development of the perturbed geothermal structure of the modern Southern Alps. This supports suggestions that the present tectonic regime of the orogen developed in a single episode of tectonic reorganization 5 Ma ago.

The current estimated position of the Euler pole for relative motion between the Australian and Pacific plates is at 62.9 ± 1.0°S, 179.9 ± 0. 4°W (Smith *et al.* 1996). Movement about this pole results in right lateral oblique convergence (Walcott 1984; Sutherland 1995) across the Alpine Fault, which marks the boundary between the Australian and Pacific plates in the South Island of New Zealand (Norris *et al.* 1990; Cooper & Norris 1994). The surface expression of this continuing convergence is the Southern Alps, a major mountain range produced by rapid uplift and distributed deformation within the Pacific Plate crust east of the Alpine Fault.

Despite their current geographical prominence, the Southern Alps are a relatively recent addition to the New Zealand landscape. Although the Alpine Fault has a history as a strike-slip feature extending back to c. 25 Ma (Kamp 1986; Cooper *et al.* 1987), the current tectonic regime, and thus the convergence responsible for the modern orogenic activity along this boundary, has developed within only the last c. 10 Ma (Field *et al.* 1989; Norris *et al.* 1990; Sutherland 1995, 1996).

Numerous attempts have been made to assess the exact timing of this major dynamical change using stratigraphy (Cutten 1979; Mildenhall & Pocknall 1984; Field *et al.* 1989; Sutherland 1996), regional tectonic reconstructions

(Walcott 1984; Sutherland 1995), and isotope geochronology (Adams 1981; Adams & Gabites 1985; Tippett & Kamp 1993). As illustrated by the detailed work of Tippett & Kamp (1993), much of the discordance in inferred timing of developments between these studies can be attributed to the significant regional variation in rates of uplift and exhumation observed across the Southern Alps. Different regions of the orogen have experienced widely differing histories of exhumation and cooling. Geographically distributed groups of samples thus do not preserve a unified geochronological response to any tectonic changes they may have experienced (Cliff 1985; Batt & Braun 1997).

Rather than trying to assess and correct for this spatial variability, this study approaches the thermochronological investigation of the Southern Alps through detailed analysis of one small area of the orogen, the Mataketake Range in South Westland (Fig. 1). Granite pegmatites from this area were dated by the U–Pb method by Chamberlain et al. (1995) as part of an assessment of the higher-temperature thermochronology of the Southern Alps. The study reported here advances on the work of Chamberlain et al. (1995) by applying multiple dating methods ($^{40}$Ar–$^{39}$Ar, K–Ar, fission-track, U–Pb, and Rb–Sr) to individual pegmatites, combining the high- and low-temperature portions of the cooling history of the region.

The thermal histories compiled using these data are internally consistent, obviating consideration of lateral variations in dynamics. An additional advantage is our use of $^{40}$Ar–$^{39}$Ar dating of K-feldspar. Unlike other isotopic dating methods, $^{40}$Ar–$^{39}$Ar dating of K-feldspar allows experimental measurement of the actual diffusion characteristics of a sample, giving much more detailed and precise thermochronological results (Lovera et al. 1991; Richter et al. 1991). Meaningful thermal histories can thus be extracted from such age spectra, on the assumption that the release of argon in the laboratory occurs by the same mechanism and is controlled by the same diffusion boundaries as in nature. Previous studies have shown this method to produce a good correlation between argon kinetics and age spectra for most plutonic K-feldspars that have not undergone low-temperature alteration (e.g. Lovera et al. 1991, 1993; Leloup et al. 1992), and thus the technique should apply well to the pegmatitic feldspars analysed in this study. Overall, the new thermochronological data presented here, and the level of detail they reveal, justifies a re-examination of the timing of the development of this orogen.

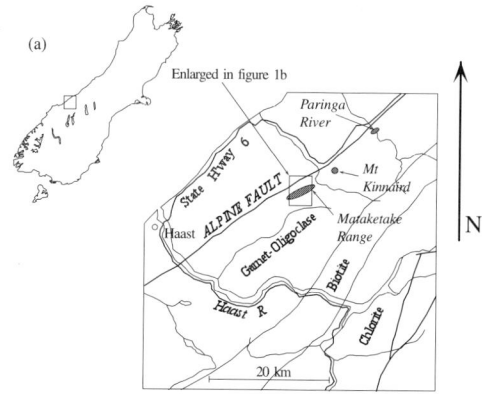

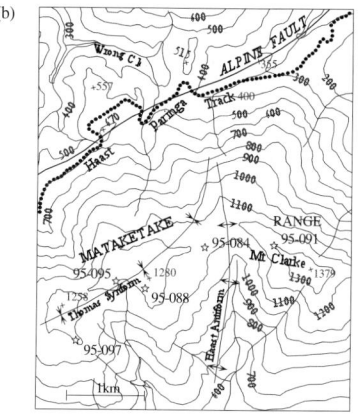

**Fig. 1.** (a) Regional geology of the Haast area in the southwest of the Southern Alps. Metamorphic isograds shown are after Mutch & McKellar (1964) and Gair (1967). Areas of granite pegmatites within the Haast Schist are marked by grey ellipses. Inset: reference map of the South Island showing area enlarged. (b) Local structural and topographical map showing the localities of the pegmatite samples analysed in this study. Contours drawn at 100 m intervals.

## Geological setting

Uplift in the Southern Alps is strongly focused against the Alpine Fault, with between 70 and 90% of the modern Pacific–Australian relative plate velocity accommodated on this structure (Norris et al. 1990; Sutherland 1994). Deformation associated with the continuing development of the Southern Alps is similarly concentrated adjacent to the Alpine Fault. In the region of this study (Fig. 1), the metamorphic grade of exposed rocks increases from biotite zone c. 25 km to the southeast of the fault to

K-feldspar bearing amphibolite facies (equivalent to the kyanite zone) adjacent to the Alpine Fault (Cooper 1972; Wallace 1974; Grapes 1995).

Also in this immediate area, a suite of granitic pegmatite bodies crop out within 2–3 km of the Alpine Fault in the Mataketake Range, Mount Kinnaird, and the Paringa River valley (Fig. 1) (see also Wallace (1974) and Chamberlain et al. (1995)). These granitic bodies, together with a series of lamprophyric dykes that also occur in this area and further to the southeast (Cooper et al. 1987; Adams & Cooper 1996), are the only intrusive rocks exposed in the Southern Alps, and as such represent an important geochronological resource for constraining the development of the orogen (Cliff 1985).

These pegmatites are best exposed and most abundant on the Mataketake Range (Fig. 1), where they are generally intruded parallel to foliation in the surrounding schist and have elongated podiform shapes up to 40 m thick by several hundred metres along strike. The pegmatites are dominated by quartz, muscovite, and K-feldspar ($Or_{92-95}$ composition (Wallace 1974)), with lesser biotite, plagioclase, and tourmaline. Accessory minerals include zircon, monazite, and apatite. Petrological and $^{87}Sr/^{86}Sr$ relationships suggest these intrusive rocks may have been formed as partial melts of the local schist country rock (Aronson 1965; Wallace 1974; Chamberlain et al. 1995).

The work presented here is based on the analysis of geochronological data from five of these granitic bodies from the Mataketake Range (Fig. 1b), mainly samples 95-084, 95-088 and 95-097, and to a lesser extent 95-091 and 95-095.

## Experimental results

### U–Pb ages

U–Pb zircon ages were obtained from four samples of granite pegmatite from the Mataketake Range (Fig. 1) on the Sensitive High Resolution Ion MicroProbe facility (SHRIMP I) at the Australian National University, using the techniques outlined by Muir et al. (1996). The 572 Ma SL13 standard was used for U–Pb ratio and U and Th concentration calibration. Samples 95-088, 95-095, and 95-097 all contain clear, euhedral zircon crystals. U–Pb data for these zircons are presented in Table 1. All of these zircons are characterized by extremely high U concentrations (all greater than 1000 ppm by weight, and extending as high as 2.0% by weight), and low Th/U ratios (<0.1).

The U–Pb and $^{207}Pb/^{206}Pb$ data presented in Table 1 are not corrected for common Pb. For these data, the common lead component (presented as the fraction of common Pb, $f^{206}$), was estimated from the $^{207}Pb/^{206}Pb$ ratio assuming each datum is a mixture between radiogenic and common lead. The age of the datum is then given by the intercept of the common lead mixing line with concordia. The high concentrations of uranium, and hence radiogenic lead in these samples result in common lead contributions that are generally low (<1%), despite the relatively young age of the material. The resulting $^{206}Pb/^{238}U$ age data are presented in Fig. 2 as probability density diagrams, where the curves represent the sum of individual unit-area Gaussian age distributions for individual analyses from each sample. The curves are normalized to the number of analyses in the mean. The main peaks in the data indicate that the crystallization ages are 68–80 Ma, corresponding to late Cretaceous crystallization. There is, however, a high degree of over-dispersion in all the samples analysed here, with mean square weighted deviation (MSWD) values of the weighted means being in excess of that expected for dispersion through analytical errors alone. In these circumstances, outliers are sought which do not belong to the main population. Younger analyses are generally ascribed to Pb loss, whereas older analyses may be the result of an inherited component.

The weighted mean of all analyses from sample 95-084 gives an MSWD of 3.6. Two analyses are more than $3\sigma$ removed from the mean, one high and one low, and rejecting these analyses gives a weighted mean of 69.8 ± 0.8 Ma ($1\sigma_m$). Similarly for 95-088, with the rejection of two outliers, both young in this case, the MSWD falls from 4.19 to 1.92 for the weighted mean of 81.5 ± 0.5 Ma ($1\sigma_m$). For sample 95-095, the rejection of three older analyses results in a lowering of the MSWD from 3.82 to 1.12 and a weighted mean of 67.3 ± 0.8 Ma ($1\sigma_m$). The MSWD of sample 95-097 is marginally high at 2.44 and rejecting the youngest analysis gives a mean age of 74.4 ± 0.8 Ma ($1\sigma_m$) with an MSWD of 1.84. The errors in these mean ages must be augmented by the systematic error component produced by the U–Pb calibration and including this component gives the following final age estimates: 95-084, 69.8 ± 2.2 Ma; 95-088, 81.5 ± 1.4 Ma; 95-095, 67.3 ± 2.1 Ma; and 95-097, 74.4 ± 2.2 Ma, where the cited errors are at the $2\sigma$ confidence limit for the means.

The outlier rejection incorporated here is purely statistical with the most deviant points being rejected sequentially. Rejection of young outliers is consistent with the high U concentrations and the

**Table 1.** *Analytical data fropm U–Pb SHRIMP analyses of samples*

| Grain | U (ppm) | Th (ppm) | Th/U | $f^{206}$ (%)* | $^{207}Pb/^{206}Pb$ | $^{238}U/^{206}Pb$ | Age† (Ma ± 1 SD) |
|---|---|---|---|---|---|---|---|
| *95–084 zircon, grid reference G37 142041* | | | | | | | |
| 1 | 2717 | 56 | 0.021 | 0.25 ± 0.31 | 0.04932 ± 0.00241 | 97.52 ± 3.27 | 65.6 ± 2.2 |
| 2 | 16258 | 835 | 0.051 | 0.04 ± 0.09 | 0.04777 ± 0.00074 | 87.75 ± 3.46 | 73.0 ± 2.9 |
| 3 | 18283 | 956 | 0.052 | −0.16 ± 0.11 | 0.04646 ± 0.00084 | 74.37 ± 4.15 | (86.2 ± 4.8) |
| 4 | 2808 | 59 | 0.021 | −0.38 ± 0.45 | 0.04450 ± 0.00354 | 83.87 ± 5.36 | 76.7 ± 4.9 |
| 5 | 2321 | 36 | 0.016 | 0.49 ± 0.28 | 0.05117 ± 0.00218 | 96.85 ± 8.26 | 65.9 ± 5.6 |
| 6 | 1978 | 9 | 0.005 | 0.62 ± 0.49 | 0.05222 ± 0.00386 | 100.58 ± 2.93 | (63.4 ± 1.9) |
| 7 | 10523 | 327 | 0.031 | −0.03 ± 0.13 | 0.04720 ± 0.00100 | 90.55 ± 3.18 | 70.8 ± 2.5 |
| 8 | 2064 | 34 | 0.017 | 0.84 ± 0.28 | 0.05400 ± 0.00218 | 92.91 ± 2.25 | 68.4 ± 1.7 |
| 9 | 6404 | 279 | 0.044 | 0.26 ± 0.24 | 0.04946 ± 0.00188 | 91.17 ± 4.42 | 70.2 ± 3.4 |
| 10 | 2373 | 69 | 0.029 | 0.19 ± 0.26 | 0.04897 ± 0.00204 | 88.48 ± 2.55 | 72.3 ± 2.1 |
| | | | | | | Weighted mean age (Ma ± s)‡ = | 69.8 ± 0.8 |
| *95–088 zircon, grid reference G37 133036* | | | | | | | |
| 1 | 3941 | 94 | 0.024 | 2.03 ± 0.25 | 0.06616 ± 0.00226 | 78.15 ± 2.07 | 80.3 ± 2.1 |
| 2 | 4015 | 97 | 0.024 | 0.21 ± 0.08 | 0.04959 ± 0.00072 | 78.20 ± 1.70 | 81.7 ± 1.8 |
| 3 | 3210 | 43 | 0.013 | 0.22 ± 0.08 | 0.04972 ± 0.00071 | 75.33 ± 2.16 | 84.8 ± 2.4 |
| 4 | 4128 | 116 | 0.028 | 0.41 ± 0.10 | 0.05149 ± 0.00093 | 75.65 ± 1.74 | 84.3 ± 1.9 |
| 5 | 3338 | 70 | 0.021 | 0.37 ± 0.12 | 0.05096 ± 0.00107 | 80.08 ± 1.06 | 79.7 ± 1.0 |
| 6 | 3541 | 73 | 0.021 | 0.19 ± 0.30 | 0.04934 ± 0.00273 | 80.13 ± 1.06 | 79.8 ± 1.1 |
| 7 | 4970 | 69 | 0.014 | 0.14 ± 0.05 | 0.04888 ± 0.00049 | 78.45 ± 1.76 | 81.5 ± 1.8 |
| 8 | 4557 | 147 | 0.032 | 0.10 ± 0.05 | 0.04860 ± 0.00050 | 75.97 ± 1.47 | 84.2 ± 1.6 |
| 9 | 3374 | 55 | 0.016 | 0.19 ± 0.13 | 0.04936 ± 0.00116 | 79.16 ± 1.82 | 80.8 ± 1.9 |
| 10 | 2702 | 48 | 0.018 | 0.16 ± 0.08 | 0.04901 ± 0.00071 | 82.72 ± 1.42 | (77.3 ± 1.3) |
| 11 | 5203 | 76 | 0.015 | 0.26 ± 0.09 | 0.05011 ± 0.00087 | 74.80 ± 1.59 | 85.4 ± 1.8 |
| 12 | 2784 | 12 | 0.004 | 0.72 ± 0.13 | 0.05405 ± 0.00118 | 88.17 ± 2.36 | (72.2 ± 1.9) |
| | | | | | | Weighted mean age (Ma ± s)‡ = | 81.5 ± 0.5 |
| *95–095 zircon, grid reference G37 132039* | | | | | | | |
| 1 | 12110 | 297 | 0.025 | −0.15 ± 0.07 | 0.04636 ± 0.00058 | 82.04 ± 3.15 | (78.2 ± 3.0) |
| 2 | 12102 | 479 | 0.040 | −0.39 ± 0.08 | 0.04439 ± 0.00064 | 85.42 ± 2.81 | (75.3 ± 2.5) |
| 3 | 3073 | 135 | 0.044 | 0.11 ± 0.22 | 0.04822 ± 0.00175 | 93.90 ± 6.01 | 68.2 ± 4.3 |
| 4 | 3790 | 146 | 0.038 | 0.27 ± 0.21 | 0.04953 ± 0.00164 | 93.88 ± 2.56 | 68.1 ± 1.9 |
| 5 | 4705 | 241 | 0.051 | 0.29 ± 0.17 | 0.04969 ± 0.00138 | 94.21 ± 1.98 | 67.9 ± 1.4 |
| 6 | 8012 | 536 | 0.067 | 0.14 ± 0.18 | 0.04849 ± 0.00141 | 92.58 ± 3.25 | 69.2 ± 2.4 |
| 7 | 1240 | 12 | 0.010 | 0.43 ± 0.25 | 0.05071 ± 0.00195 | 99.02 ± 2.87 | 64.5 ± 1.9 |
| 8 | 7946 | 525 | 0.066 | −0.16 ± 0.16 | 0.04602 ± 0.00123 | 99.55 ± 3.10 | 64.5 ± 2.0 |
| 9 | 3253 | 44 | 0.013 | 0.11 ± 0.17 | 0.04828 ± 0.00134 | 91.38 ± 2.92 | 70.1 ± 2.2 |
| 10 | 14176 | 474 | 0.033 | −0.10 ± 0.16 | 0.04669 ± 0.00131 | 86.67 ± 2.37 | (74.0 ± 2.0) |
| | | | | | | Weighted mean age (Ma ± s)‡ = | 67.3 ± 0.8 |
| *95–097 zircon, grid reference G37 125028* | | | | | | | |
| 1 | 789 | 9 | 0.012 | 1.20 ± 0.65 | 0.05697 ± 0.00514 | 85.04 ± 4.06 | 74.5 ± 3.6 |
| 2 | 18965 | 974 | 0.051 | −0.21 ± 0.10 | 0.04592 ± 0.00077 | 82.47 ± 2.39 | 77.9 ± 2.2 |
| 3 | 4934 | 158 | 0.032 | 0.39 ± 0.14 | 0.05034 ± 0.00112 | 102.43 ± 7.27 | (62.4 ± 4.4) |
| 4 | 3806 | 75 | 0.020 | 0.09 ± 0.23 | 0.04815 ± 0.00179 | 91.11 ± 3.23 | 70.3 ± 2.5 |
| 5 | 12446 | 474 | 0.038 | −0.05 ± 0.09 | 0.04707 ± 0.00069 | 87.41 ± 1.75 | 73.4 ± 1.5 |
| 6 | 5971 | 227 | 0.038 | −0.12 ± 0.14 | 0.04652 ± 0.00111 | 90.06 ± 3.19 | 71.3 ± 2.5 |
| 7 | 11952 | 486 | 0.041 | −0.11 ± 0.11 | 0.04669 ± 0.00088 | 82.46 ± 2.49 | 77.8 ± 2.3 |
| 8 | 5054 | 152 | 0.030 | 0.02 ± 0.12 | 0.04765 ± 0.00093 | 88.29 ± 2.21 | 72.6 ± 1.8 |
| 9 | 6512 | 209 | 0.032 | 0.11 ± 0.29 | 0.04851 ± 0.00230 | 80.21 ± 2.58 | 79.8 ± 2.6 |
| 10 | 2610 | 58 | 0.022 | 0.26 ± 0.24 | 0.04962 ± 0.00191 | 84.33 ± 5.82 | 75.8 ± 5.2 |
| | | | | | | Weighted mean age (Ma ± s)‡ = | 74.4 ± 0.8 |

* $f^{206}Pb$ is percentage of common $^{206}Pb$ relative to total $^{206}Pb$ measured based on $^{207}Pb/^{206}Pb$ expected for radiogenic composition.
† Radiogenic $^{206}Pb/^{238}U$ age.
‡ Weighted mean age excludes data in parentheses. σ does not include propagated error from SL13 standard.

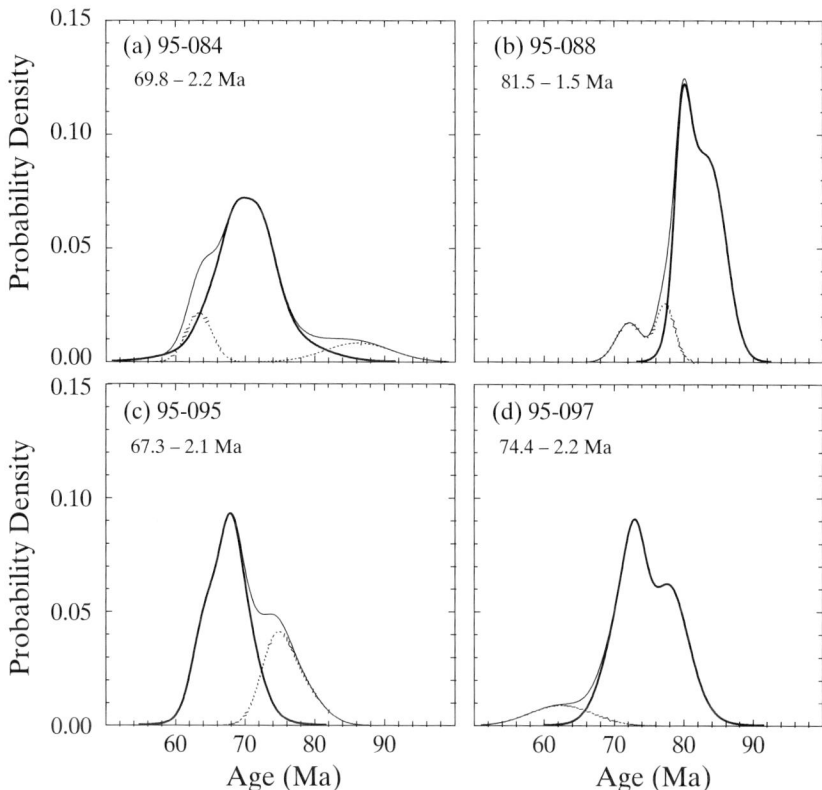

**Fig. 2.** Probability density diagrams for U–Pb ages for zircon from pegmatites analysed in this study (Table 1). The fine continuous line corresponds to all analyses (the sum of the outliers and included data), the heavy continuous line corresponds to data included in the final age calculation, and the broken line represents the probability density of the outliers alone. The final ages include the propagated error component from the SL13 standard calibration.

concomitant Pb loss through radiation damage and subsequent migration of lead. However, it is interesting to note that the analyses with the oldest ages also have the highest U concentrations, which is inconsistent with Pb loss solely as a response to high U concentration. Such trends have been occasionally observed during ion microprobe analysis and have been suspected as being caused by inappropriate U–Pb calibration for high U (>3000 ppm) zircons. However, it is difficult to differentiate between what might be a U–Pb calibration effect and the possibility that the older ages are indeed that of an inherited component coincidentally with higher U concentrations. It could be argued that the three older, and higher U, analyses in sample 95–095 are indeed an earlier crystallization event coincident with the crystallization of sample 95–097. Although there are some details in the U–Pb systematics of these pegmatitic zircons that warrant further study, in the context of this work the incorporation or rejection of the outliers has little bearing on the individual crystallization ages of the pegmatites, as opposed to the age range between them. The ages reported here for 95-084 and 95-095 (69.8 ± 2.2 and 67.3 ± 2.1 Ma, respectively) agree well with previously published U–Pb ages for zircon (67.9 ± 7.2 Ma) and monazite (67.8 ± 2.7 Ma and 69.2 ± 2.4 Ma) from a Mataketake Range granite pegmatite (Chamberlain et al. 1995). Although sample 95-097 at 74.4 ± 2.2 Ma is marginally older, the U–Pb age of zircon from 95-088, at 81.5 ± 1.5 Ma, clearly falls outside the uncertainty limits based on the other samples and establishes a protracted formation history for the Mataketake Range pegmatites from at least 82 to 67 Ma.

## Rb–Sr ages

Rb–Sr analysis of selected pegmatite samples was carried out at the Australian National University. All of the analysed mineral separates display significant enrichment in rubidium, with Rb contents ranging from 190 to 1020 ppm by weight (Table 2). Three-mineral analysis (K-feldspar, biotite, and muscovite) of sample 95-097 does not yield an isochron (Fig. 3). This pattern is consistent with protracted cooling of 95-097 through the respective strontium closure temperatures of these minerals. Muscovite and K-feldspar close to the diffusion of strontium at approximately the same temperature in gradually cooled systems (e.g. Wagner *et al.* 1977; Jäger 1979; Parrish *et al.* 1988; Giletti 1991), and thus should give a uniform apparent age irrespective of the $T$–$t$ path of a sample. A two-mineral isochron plotted through the K-feldspar and muscovite results for 95-097 gives a model initial $^{87}Sr/^{86}Sr$ ratio of 0.7060 ± 0.0002, and yields an age of 71.7 ± 0.7 Ma. This initial strontium isotope ratio agrees with the initial $^{87}Sr/^{86}Sr$ ratio of 0.7058 derived for a sample of a Mataketake Range pegmatite by Aronson (1965) through analysis of a plagioclase separate lacking a measurable $^{87}Rb$ content. Taking 0.7060 ± 0.0002 as the initial $^{87}Sr/^{86}Sr$ ratio for the biotite separate from 95-097 gives an apparent Rb–Sr age for this biotite of 11.3 ± 0.5 Ma (Fig. 3).

A two mineral (K-feldspar–muscovite) isochron plotted for sample 95-084 gives a model initial $^{87}Sr/^{86}Sr$ ratio of 0.7059 ± 0.0006, again in agreement with the determination of Aronson (1965), and with that derived for sample 95-097 above. This muscovite–K-feldspar isochron for 95-084 gives a model Rb–Sr age of 64.4 ± 2.3 Ma (Fig. 3 and Table 2).

Notably, the Rb–Sr ages of the two pegmatites are significantly different, and also differ from the age calculations of Aronson (1965), who arrived at model Rb–Sr ages of 60 Ma for muscovite and K-feldspar, and 25.8 Ma for biotite from his analysis of a granite pegmatite from the

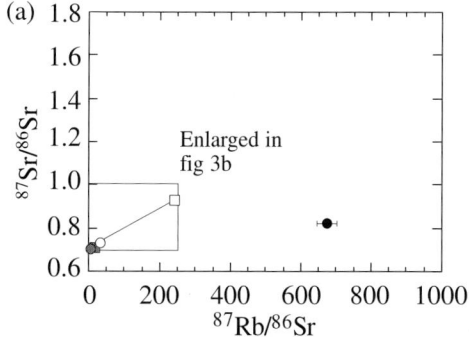
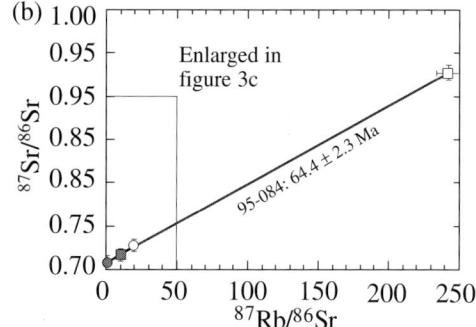
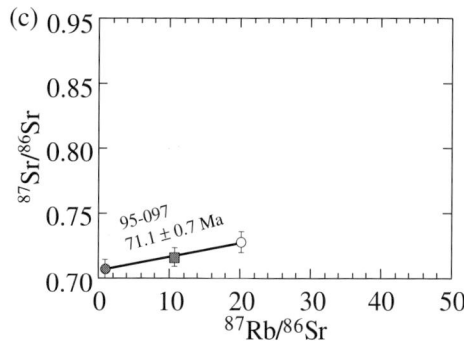

**Fig. 3.** Results of Rb–Sr analyses (Table 2) illustrating isochrons constructed between the data points. Square symbols mark the analyses of sample 95-084, circles mark those of 95-097. White symbols represent muscovite analyses, grey symbols represent K-feldspar, and black represent biotite.

**Table 2.** *Analytical data from Rb–Sr analyses of samples*

| Sample | Mineral | Sample weight (mg) | Rb (ppm) | Sr (ppm) | $^{87}Rb/^{86}Sr$ | ±2 SD (%) | $^{87}Sr/^{86}Sr$ | Age (Ma) |
|---|---|---|---|---|---|---|---|---|
| 95–097 | K-feldspar | 203.55 | 190.05 | 603.33 | 0.9119 | 2.1 | 0.70692 | 71.1 ± 0.7* |
| | Muscovite | 170.77 | 297.97 | 45.98 | 20.12 | 2.4 | 0.72633 | 71.1 ± 0.7* |
| | Biotite | 213.38 | 588.11 | 2.554 | 673.6 | 4.2 | 0.81471 | 11.3 ± 0.5* |
| 95–084 | K-feldspar | 202.01 | 497.97 | 85.08 | 10.69 | 2.2 | 0.71567 | 64.4 ± 2.3† |
| | Muscovite | 181.94 | 313.82 | 5.961 | 242.1 | 3.3 | 0.92736 | 64.4 ± 2.3† |

*Based on initial $^{87}Sr/^{86}Sr$ ratio of 0.7060 ± 0.0001.
† Based on initial $^{87}Sr/^{86}Sr$ ratio of 0.7059 ± 0.0006.

Mataketake Range.

## K–Ar ages

K–Ar and $^{40}$Ar–$^{39}$Ar dating of material were carried out using the techniques described by Batt (1997). A summary of the techniques used in this study is available from the British Library Document Supply Centre, Boston Spa, Wetherby, West Yorkshire LS23 7BQ, UK, and the Geological Society Library, as SUP 18127 (11 pp.). Figure 4a is an Arrhenius plot of the results of $^{40}$Ar–$^{39}$Ar analysis of K-feldspar sample K95-084, showing log ($D/r^2$) versus the reciprocal of absolute temperature, where $D$ is a frequency factor and $r$ the characteristic dimension of the diffusion domain. It is immediately apparent that these results do not fall on a single straight line, indicating that this sample has several discrete diffusion domains with varying argon retentivities (Richter et al. 1991; Lovera 1992). The measured $^{40}$Ar–$^{39}$Ar age spectrum (Fig. 4b) further supports the existence of multiple discrete diffusion domains in this sample, with a variety of different $^{40}$Ar–$^{39}$Ar ages revealed during progressive argon release with increasing temperature.

The progressive evolution of $^{39}$Ar is taken as a proxy for the progress of the experiment towards completion. The initially high ages decreasing rapidly through the first c. 5% of the gas release from this sample are a common feature of $^{40}$Ar–$^{39}$Ar analysis of K-feldspars. Such patterns are generally attributed to excess argon ($^{40}$Ar from a source other than in situ decay of $^{40}$K within a sample) in fluid inclusions (Harrison et al. 1994), or taken up at grain margins at low temperatures (Richter et al. 1991), and are omitted from the interpretation

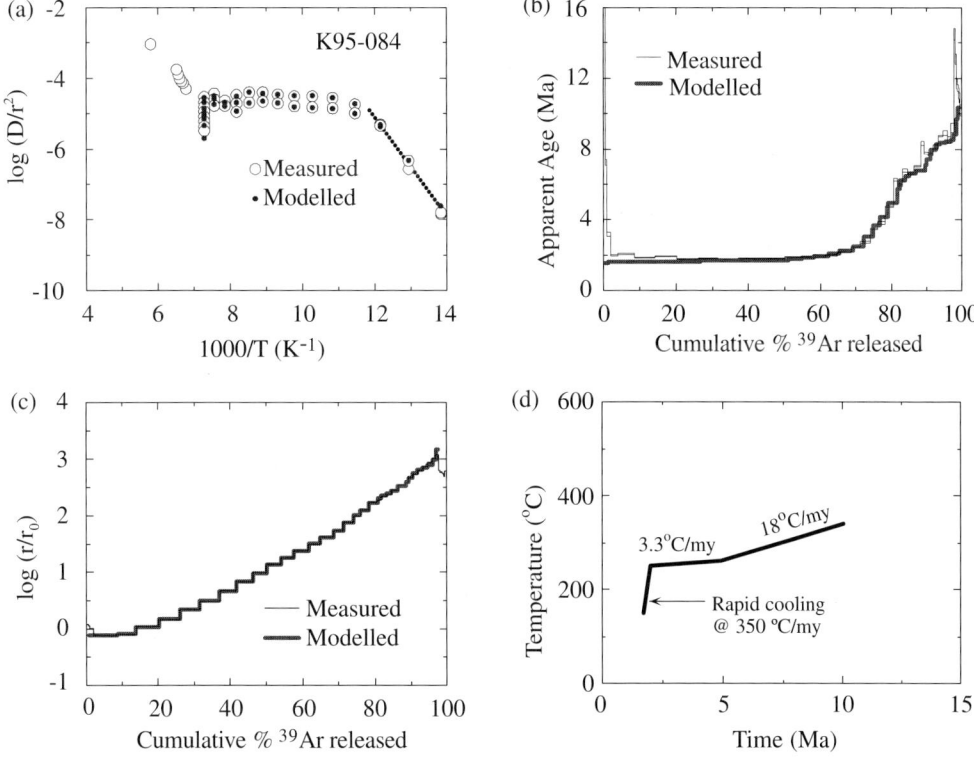

**Fig. 4.** Experimental $^{40}$Ar/$^{39}$Ar results and modelled characteristics of K-feldspar K95-084. (**a**) Arrhenius plot. Modelled $D/r^2$ is calculated using a planar geometry model, following the method outlined by Lovera et al. (1989). The dotted line through the low-temperature data points marks the fitted Arrhenius relationship used to calculate the activation energy and diffusion length scale for argon in the sample. This Arrhenius relationship has a value of $E_a$ of 54.6 ± 4.2 kcal mol$^{-1}$, and $D_0/a^2$ of 13.91. (**b**) Observed $^{40}$Ar–$^{39}$Ar age v. cumulative per cent $^{39}$Ar released, and best fit model age spectrum produced by the thermal history illustrated in (**d**) below. (**c**) Plot of log ($r/r_0$). Modelled curve is calculated as outlined for (**a**) above. (**d**) Thermal history producing the modelled age spectrum in (**b**). Uncertainty in the activation energy calculated from (**a**) gives uncertainty of c. ±30°C in the absolute temperature of this history, but does not affect the form of the $T$–$t$ curve.

of age spectra. Excess argon has previously been found to be abundant in rocks from South Westland (Adams & Gabites 1985; Adams & Cooper 1996), and K–Ar analysis of mica is interpreted to show its possible presence in the pegmatites analysed here (see below), making such an explanation for this early irregularity in the age spectrum likely in this case.

The last few per cent of gas released is also ignored in analysis of these results, because it was extracted above the incongruent melting temperature of K-feldspar (1050°C), and thus does not in any way reflect the diffusion properties which the sample had in the natural environment. The structural breakdown of the sample is reflected in the marked decrease seen in $\log(r/r_0)$ over the final 3% of gas release, and the anomalous decrease seen in the age determined from gas released over this interval (Fig. 4b).

K95-084 exhibits a plateau age of 1.8 Ma over virtually all of the first 60% of the $^{39}$Ar released (Fig. 4b). The remaining 40% of the gas release is characterized by a consistent sigmoidal increase to a maximum apparent age of c. 8.5 Ma (Fig. 4b). This aspect of the age spectrum is of particular interest, as it conforms closely with the behaviour predicted for diffusion from multiple discrete domains (Dodson 1973; Lovera et al. 1989), supporting the concept of gas release and retention in this sample being governed by volume diffusion processes.

The measured $^{40}$Ar–$^{39}$Ar age release data for samples K95-088 (Fig. 5) and K95-097 (Fig. 6) are similar to those determined for K95-084. Minor differences do arise, however, with K95-088 (Fig. 5b) displaying a second plateau towards the final stages of gas release. This secondary plateau has consistent ages of c. 9.5 Ma over the last 19% of gas released from this sample. Although displaying a broadly similar form, the age spectrum of K95-097 increases monotonically to c. 8 Ma before the onset of melting.

The variation observed in the older age component for these three samples is sufficient to give them significantly different K–Ar and $^{40}$Ar–$^{39}$Ar total fusion ages (Table 3). K95-084 has a K–Ar age of 3.64 ± 0.05 Ma and a total

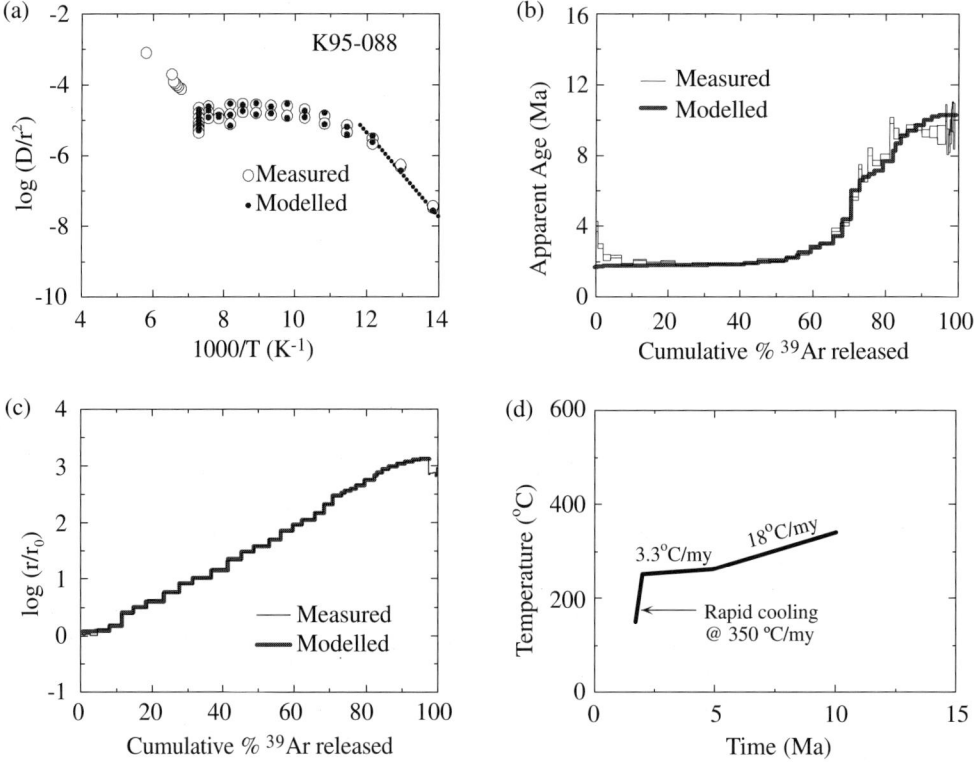

**Fig. 5.** As Fig. 4 for K-feldspar sample K95-088. The fitted Arrhenius relationship for this sample has a value of $E_a$ of 54.3 ± 3.8 kcal mol$^{-1}$, and $D_0/a^2$ of 11.95.

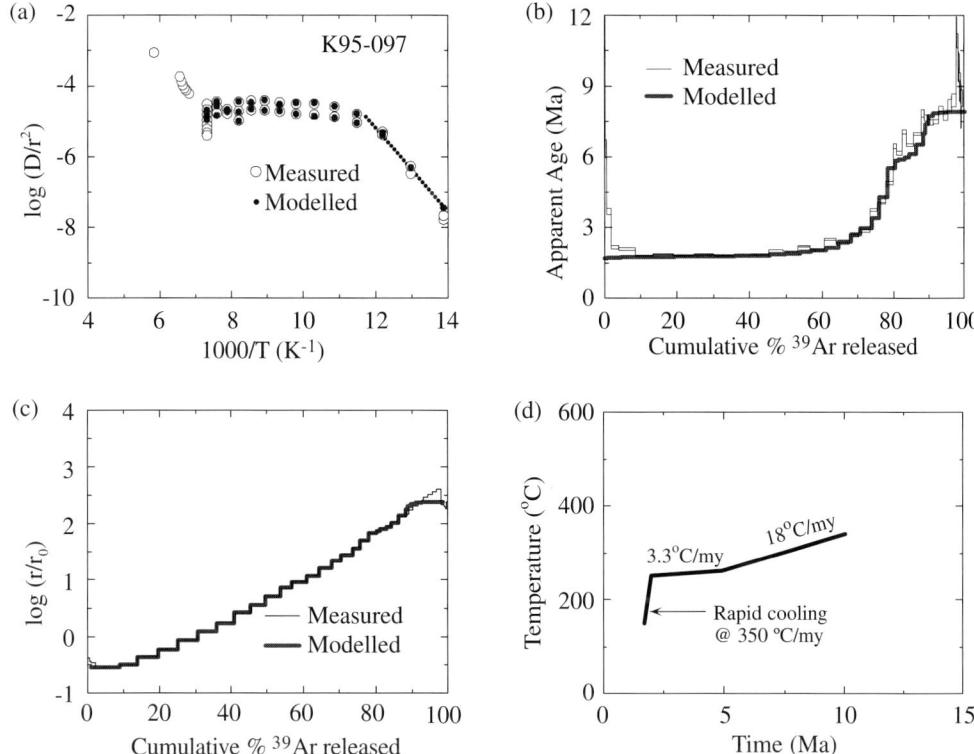

Fig. 6. As Fig. 4 for K-feldspar sample K95-097. The fitted Arrhenius relationship for this sample has a value of $E_a$ of $54.2 \pm 3.4$ kcal mol$^{-1}$, and $D_0/a^2$ of 11.78.

fusion $^{40}$Ar–$^{39}$Ar age (the mean age of gas released over the entire experiment) of $3.72 \pm 0.05$ Ma. In contrast, K95-088 and K95-097 yield K–Ar ages of $5.78 \pm 0.13$ Ma and $4.73 \pm 0.05$ Ma, and total fusion $^{40}$Ar–$^{39}$Ar ages of $4.65 \pm 0.12$ Ma and $4.10 \pm 0.07$ Ma, respectively.

None of these ages overlap between the three different pegmatites. This demonstrates that at the level of precision of these age measurements, the compositional and/or structural differences between the feldspars are sufficient to significantly affect their kinetic behaviour, and thus their ability to retain radiogenic argon. This variation in behaviour with composition is common to many isotopic dating systems (Cliff 1985; Lister & Baldwin 1996), and illustrates the weakness of assuming that several separate specimens of a given mineral will behave identically, even under the same tectonic, and later, experimental conditions.

This also displays the power of the $^{40}$Ar–$^{39}$Ar dating method applied to K-feldspars, in that this method has the capacity to overcome these behavioural differences and resolve the significance of the age variation between these samples. By independently assessing the diffusion behaviour of each sample, $^{40}$Ar–$^{39}$Ar dating allows much more accurate assessments to be made of argon retention over time.

The closure temperatures of argon and strontium in biotite are approximately equal (see below), and thus biotite should give the same age by both the Rb–Sr and K–Ar (or $^{40}$Ar–$^{39}$Ar) dating methods. Biotite was present in one only of the analysed pegmatites, B95-097, giving ages of $15.6 \pm 0.2$ Ma by K–Ar dating and $15.8 \pm 0.3$ Ma by $^{40}$Ar–$^{39}$Ar laser fusion. Multi-step $^{40}$Ar–$^{39}$Ar dating of biotite B95-097 (Fig. 7) yields a somewhat irregular, but largely flat age spectrum with a total fusion age of $16.5 \pm 0.3$ Ma. Despite the apparently simple behaviour exhibited by this age spectrum, which falls well within the established criteria for defining a 'meaningful' plateau age (Dalrymple & Lanphere 1974; Fleck et al. 1977; Berger & York 1981), this age of 16.5 Ma is significantly older than the Rb–Sr age of this biotite sample (as discussed above), indicating the likely presence of excess argon

**Table 3.** Summary of K–Ar and $^{40}$Ar–$^{39}$Ar age determinations of granite pegmatite samples from the Mataketake Range

| | | | | K–Ar age determinations[1] | | | $^{40}$Ar–$^{39}$Ar age determinations (Ma ± 1 SD) | | | |
|---|---|---|---|---|---|---|---|---|---|---|
| | | | | | | | | | Step heating experiments | |
| Sample | Grid ref. | Mineral | Wt % K | $^{40}$Ar* ($10^{-11}$ mol g$^{-1}$) | $^{40}$Ar* (% total) | K–Ar age (Ma ± 1 SD) | Laser total fusion | Isochron | Mean | Plateau |
| 95-084 | G37 142041 | Muscovite | 8.93, 8.89 | 40.4 | 42.8 | 26.0 ± 0.3 | – | – | – | – |
| | | K-feldspar | 12.1, 12.1 | 7.64 | 54.2 | 3.64 ± 0.05 | – | – | 3.72 ± 0.05 | – |
| 95-088 | G37 134036 | Muscovite | 8.92, 8.94 | 24.0 | 43.9 | 15.4 ± 0.2 | – | – | – | – |
| | | K-feldspar | 12.3, 12.2 | 12.3 | 15.8 | 5.78 ± 0.13 | – | – | 4.65 ± 0.12 | – |
| 95-097 | G37 125028 | K-feldspar | 10.9, 10.91 | 8.96 | 69.3 | 4.73 ± 0.05 | – | – | 4.10 ± 0.07 | – |
| | | Biotite | 7.96, 7.99 | 21.9 | 36.6 | 15.6 ± 0.2 | 15.78 ± 0.26 | 16.4 ± 0.1 | 16.5 ± 0.3 | 16.5 ± 0.3[2] |
| | | Muscovite | 8.87, 8.88 | 28.2 | 36.6 | 18.2 ± 0.2 | 19.69 ± 0.32 | 19.0 ± 0.3 | 19.6 ± 0.5 | 19.4 ± 0.4[2] |

[1] Decay constants are $\lambda_e = 0.581 \times 10^{-10}$ a$^{-1}$ and $\lambda_\beta = 4.962 \times 10^{-10}$ a$^{-1}$.
[2] See Fig. 7.

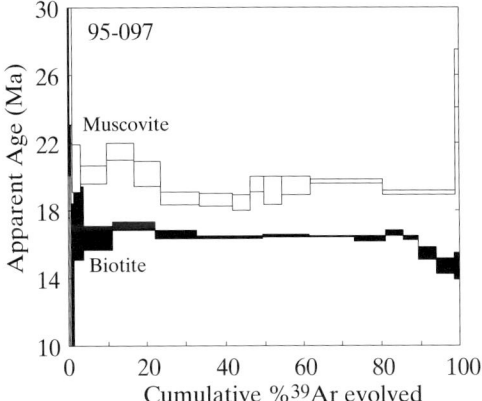

**Fig. 7.** $^{40}$Ar/$^{39}$Ar age spectra for muscovite sample M95-097 and biotite sample B95-097.

contamination. It has been well established from previous studies that biotite can yield flat age spectra such as that observed here while containing significant excess argon (Pankhurst et al. 1973; Roddick et al. 1980; Foland 1983). This ambiguity is believed to be due to structural changes caused in the biotite by dehydration during heating *in vacuo* (McDougall & Harrison 1988).

Marked variation is also observed in the K–Ar age of muscovite from these pegmatites. As documented in Table 3, muscovite sample M95-084 has a K–Ar age of 26.0 ± 0.3 Ma, muscovite sample M95-088 an age of 15.4 ± 0.2 Ma, and sample M95-097, 18.2 ± 0.2 Ma. The implications of this variation in age are discussed further below. Although slightly irregular in its form, the $^{40}$Ar–$^{39}$Ar age release spectrum for muscovite sample M95-097 (Fig. 7) is again approximately flat, yielding a total fusion age of 19.6 ± 0.5 Ma (Table 3).

*Fission-track ages*

Fission-track dating was carried out using the techniques described by Gleadow et al. (1976), Green (1981), and Hurford & Green (1983). A summary of the techniques used in this study is available as SUP 18127 (see p. 267). For all samples, the central age (Galbraith & Laslett 1993), which is essentially a weighted-mean age, is reported. Where appropriate, lengths of confined tracks were measured using the procedure outlined by Green (1986). Only fully etched and horizontal confined tracks in grains with polished faces parallel to prismatic crystal faces were measured (Laslett et al. 1982). The principles used to interpret the fission-track ages and fission-track length distributions have been detailed by Gleadow et al. (1986), Green et al. (1989), and Hasebe et al. (1994). Fission-track ages for five apatite and two zircon samples from the Mataketake Range pegmatites are presented in Table 4. The apatite ages range between 1.3 and 1.8 Ma and are concordant within analytical error, as are the zircon ages of 4.2 ± 0.3 and 4.5 ± 0.4 Ma. Overall weighted mean ages for the apatite and zircon are 1.4 ± 0.2 Ma, and 4.3 ± 0.3 Ma, respectively. Horizontal confined track lengths were measured in apatite samples 95-088, 95-091, and 95-095, but their low abundance prevents any rigorous statistical analysis of track length variation. Track lengths are generally long (>14 μm), with relatively narrow distributions. A relatively long mean confined track length (c. 10.2 μm) was also observed in zircon sample 95-084, with an additional 'tail' of shorter tracks between 8 and 10 μm in length.

The two zircon fission-track ages reported here (4.2 ± 0.3 Ma and 4.5 ± 0.4 Ma) agree with zircon ages of 3.6 ± 0.8 Ma (Tippett & Kamp 1993), 4.2 ± 1.6 Ma, and 4.8 ± 0.8 Ma (Kamp et al. 1989) reported at a comparable distance from the Alpine Fault in the Haast River valley 25 km to the southwest of the Mataketake Range (Fig. 1a). The apatite fission-track ages from these samples, however, were below measurable limits (Kamp et al. 1989; Tippett & Kamp 1993), in marked contrast to those measured here. Our ability to measure relatively youthful apatite ages, as well as significantly more precise zircon ages than previously reported for this region, results from the significantly higher uranium content of apatite and zircon in the granite pegmatites compared with that of apatite and zircon in the Alpine Schist (Table 4).

**Cooling history**

The granite pegmatites exposed on the Mataketake Range are interpreted to have most probably been formed by anatectic melting as a result of rapid exhumation or access of water to hot crustal rocks (Wallace 1974; Chamberlain et al. 1995). The petrology of the pegmatites indicates that they were formed by partial melt at 620–680°C (Wallace 1974). No indication was found in either this study or the work of Chamberlain et al. (1995) of older cores within zircons extracted from these pegmatites, and thus the U–Pb ages obtained from these zircons are inferred to date the intrusion and crystallization of the respective pegmatite bodies. This suggests

**Table 4.** Fission-track analytical data and age calculations

| Sample no. | Grid ref. | Mineral | No. of grains | Standard track density ($\times 10^6$ cm$^{-2}$)* | Fossil track density ($\times 10^4$ cm$^{-2}$)* | Induced track density ($\times 10^6$ cm$^{-2}$)* | Uranium content (ppm) | $\chi^2$ prob (%) | Age dispersion (%) | Central age (Ma) ($\pm 1\sigma$)† | Mean track length $\pm$SE ($\mu$m) | SD ($\mu$m) |
|---|---|---|---|---|---|---|---|---|---|---|---|---|
| 95-084 | 142041 | Apatite | 24 | 1.233 (4111) | 2.000 (32) | 3.067 (4907) | 31 | 78.1 | 2.09 | 1.5 ± 0.3 | – | – |
|  |  | Zircon | 13 | 0.451 (2479) | 240.1 (547) | 16.580 (3778) | 1471 | 4.5 | 12.88 | 4.2 ± 0.3 | 10.73 ± 0.08 (56) | 0.62 |
| 95-088 | 134036 | Apatite | 28 | 1.124 (4651) | 0.851 (15) | 1.027 (1810) | 11 | 76.5 | 9.86 | 1.8 ± 0.5 | 14.24 ± 0.23 (2) | 0.32 |
|  |  | Zircon | 5 | 0.550 (2185) | 291.5 (194) | 16.260 (1082) | 1537 | 30.9 | 6.13 | 4.5 ± 0.4 | – | – |
| 95-091 | 150042 | Apatite | 20 | 1.280 (4715) | 1.484 (19) | 2.437 (3119) | 24 | 95.5 | 0.11 | 1.5 ± 0.3 | 14.56 ± 0.57 (4) | 1.13 |
| 95-095 | 132039 | Apatite | 24 | 1.299 (4715) | 1.182 (17) | 2.263 (3256) | 22 | 97.9 | 0.01 | 1.3 ± 0.3 | 14.58 ± 0.67 (6) | 1.63 |
| 95-097 | 125028 | Apatite | 32 | 1.244 (4111) | 1.270 (26) | 2.422 (4960) | 24 | 98.2 | 0.23 | 1.3 ± 0.3 | – | – |

* Number of tracks counted or measured is given in parentheses. Standard and induced track densities are measured on external mica detectors ($g = 0.5$), fossil track densities are measured on internal mica surfaces.
† Ages calculated using zeta = 383.5 for dosimeter glass CN-5. All ages are central ages (Galbraith & Laslett 1993).

that the initial formation and intrusion of these pegmatites occurred over a protracted interval, ranging from c. 68 Ma to up to 81 Ma (Table 1).

The Rb–Sr muscovite–K-feldspar isochron ages of 95-084 and 95-097 differ markedly from one another, and are slightly younger than the U–Pb zircon ages of the same samples (Fig. 8). This suggests that these Rb–Sr ages reflect the post-emplacement cooling of individual bodies rather than any regional cooling event. The closure temperature for diffusion of strontium is c. 500 ± 50°C from muscovite (Jäger 1979;

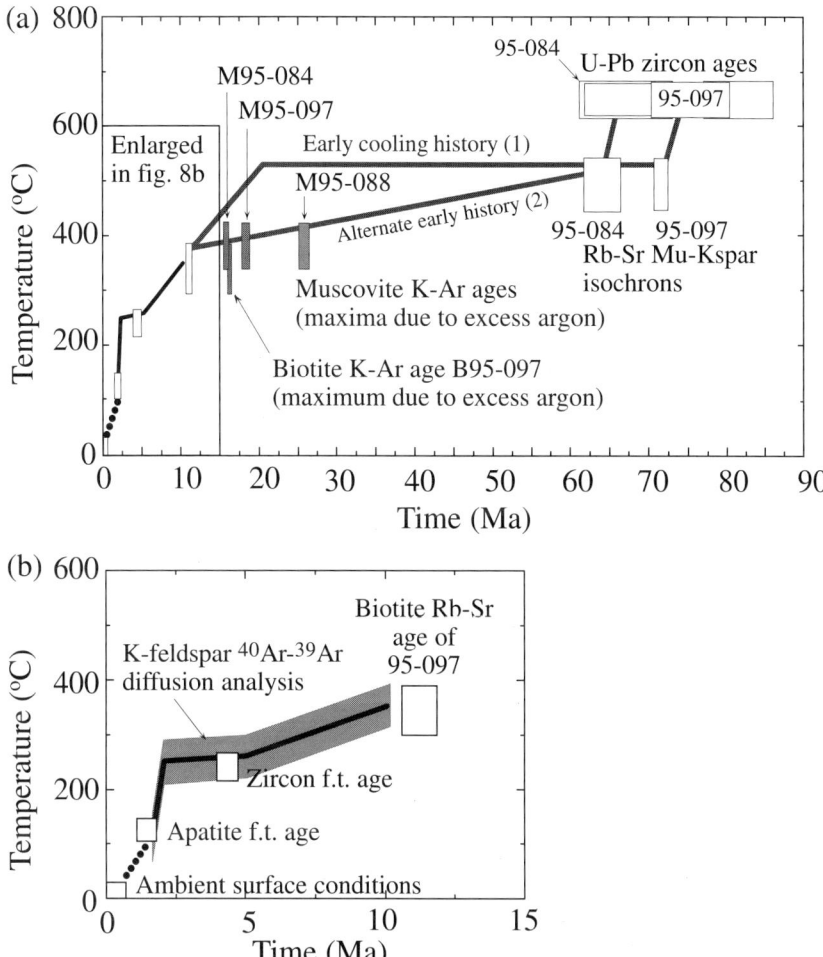

**Fig. 8.** Summary of cooling history for the analysed granite pegmatites. Boxes show the estimated closure temperature and error range for the respective thermochronometers used. (**a**) Illustration focusing on higher-temperature, less well-constrained end of the cooling history. The temperature of the U–Pb box assumes that emplacement of the pegmatites occurred at approximately their temperature of formation, as outlined in the text. Mica ages are included as grey boxes and treated only as absolute maximal age constraints because of contamination by excess argon. Early cooling history (1) assumes that the calculated cooling rate for the higher-temperature K-feldspar domains extends back to the inferred initiation of the Australia–Pacific plate boundary through New Zealand at c. 20 Ma. Alternative early history (2) assumes constant cooling rate between the emplacement constraints and the higher-temperature K-feldspar constraint. (**b**) Enlargement to show better constrained recent cooling history. Uncertainty on the absolute value of the calculated $T$–$t$ curve derived from $^{40}$Ar–$^{39}$Ar dating of K-feldspar is indicated by the grey shaded area. This uncertainty does not carry over into the form of the $T$–$t$ curve, which is analysed instead in Fig. 9. Ambient surface conditions are taken as 0–20°C. Because of the current rapid denudation of the Southern Alps, it is assumed that these samples can only have arrived at these surface conditions roughly at the present day.

Wagner et al. 1977; Parrish et al. 1988), and 500–550°C from K-feldspar (Giletti 1991). These Rb–Sr ages thus indicate that the individual pegmatite bodies cooled to below c. 500°C immediately following their initial emplacement. This implies that the host schist rocks must have been below 500°C at this time: significantly cooler than the estimated peak metamorphic temperature for the Haast area of 600–700°C (Cooper 1972).

K-feldspar, as discussed previously, contains multiple discrete argon diffusion domains of varying retentivity. Experimental investigation of these domains allows the direct calculation of the cooling history required to produce the observed age distribution of a sample (Lovera et al. 1989; Lovera 1992). Following the method of Lovera et al. (1989), this process first involves domain characterization. Activation energy ($E$) and $D_0/r^2$ (frequency factor over the square of the effective domain size) are calculated from the approximately linear low-temperature portion of the Arrhenius plot of the experimental data (Figs 4a, 5a and 6a). In the version of Lovera's modelling routines used in this study, activation energy remains fixed, and the program then determines the distribution of domain sizes that best reproduces the Arrhenius information from the entire step heating experiment. Scatter in the experimental data used to constrain the activation energy of the sample (Fig. 4a) translates into uncertainty in the absolute temperature calibration of the modelled thermal history. This uncertainty does not affect the form of the $T$–$t$ curve, however, which is dependent on the length scale distribution of diffusion domains in the sample, and is calculated by iterative fitting of the modelled and observed age spectra.

For K95-084, the $^{40}$Ar–$^{39}$Ar experimental results are fitted best by eight diffusion domains with uniform activation energies of 54.6 ± 4.2 kcal mol$^{-1}$ (Fig. 4) and the domain sizes shown in Table 5. Such a domain-size distribution corresponds to a range of effective closure temperatures from c. 150 to 350°C, and thus K95-084 offers constraint over this entire interval of the cooling history of this pegmatite.

Using these parameters, the experimental age release spectrum of 95-084 is best synthesized by a three-stage thermal history (Fig. 4d). The older ages seen in the final 10–15% of gas release are well modelled by cooling at 18°C Ma$^{-1}$ beginning at around 9–10 Ma. To produce the well-defined age plateau observed over the first 60% of gas release, in contrast, requires extremely rapid cooling beginning at around 2 Ma, approaching rates of 350°C Ma$^{-1}$ (Fig. 4d).

**Table 5.** *Domain size distributions determined for K-feldspar samples*

| Domain | $D_0/a^2$ (s$^{-1}$) | Volume fraction |
|---|---|---|
| K95-084, 8 domains, each with $E_a = 54.6 \pm 4.2$ kcal/mol$^{-1}$ | | |
| 1 | 13.91 | 0.1272 |
| 2 | 12.67 | 0.1544 |
| 3 | 11.43 | 0.1489 |
| 4 | 10.12 | 0.1275 |
| 5 | 9.01 | 0.1405 |
| 6 | 7.79 | 0.0932 |
| 7 | 6.56 | 0.1227 |
| 8 | 5.34 | 0.0854 |
| K95-088, 8 domains, each with $E_a = 54.3 \pm 3.8$ kcal/mol$^{-1}$ | | |
| 1 | 11.95 | 0.0884 |
| 2 | 10.44 | 0.1177 |
| 3 | 9.19 | 0.1634 |
| 4 | 8.04 | 0.1381 |
| 5 | 6.99 | 0.1493 |
| 6 | 5.71 | 0.1082 |
| 7 | 4.82 | 0.0737 |
| 8 | 4.02 | 0.1612 |
| K95-097, 8 domains, each with $E_a = 54.2 \pm 3.4$ kcal/mol$^{-1}$ | | |
| 1 | 11.78 | 0.0904 |
| 2 | 10.85 | 0.1185 |
| 3 | 9.92 | 0.1325 |
| 4 | 9.00 | 0.1239 |
| 5 | 7.93 | 0.1355 |
| 6 | 7.01 | 0.1247 |
| 7 | 5.77 | 0.1219 |
| 8 | 4.32 | 0.1527 |

Cooling at these extremely rapid rates continuously from 2 Ma onwards would have sample K95-084 arriving at ambient surface temperatures around 1–1.5 Ma ago (Fig. 4d), and require that it then remained exposed between that time and the present. This appears unlikely, given the present active erosion of South Westland (Griffiths & McSaveney 1986; Hicks et al. 1990; Hovius et al. 1997), suggesting instead that sample 95-084 may have experienced a reduction in cooling rate in Pleistocene time. This possibility is discussed further in the interpretation of fission-track results below.

The steep sigmoidal increase in age observed in the later released argon component of K95-084 requires an extensive period where the higher closure temperature domains accumulate argon while the lower-temperature domains are still open to complete diffusive loss. As there is no great discontinuity in the diffusion domain characteristics of the sample (Table 5), a hiatus in cooling is thus suggested between these two periods. This is best modelled by temperature dropping from 260 to 250°C between 5 Ma and 2 Ma (Fig. 4d), a cooling rate of some 3.3°C Ma$^{-1}$.

As noted above, the predicted age release distribution is sensitive to the chosen $T$–$t$ path. The robustness of this modelled thermal history can thus be assessed by examining the sensitivity of the fit between calculated and observed age spectra to variation in cooling history (Fig. 9). The age and form of the initial plateau in predicted age release for K95-084 permit only a limited range of $T$–$t$ histories during the 2 Ma event. Variation in the time of onset of this episode of rapid cooling directly affects the age of the modelled plateau (Fig. 9b). Given the consistency of this young age plateau in the observed age distribution of K95-084, significant variation in onset time is thus unlikely for this cooling event.

Figure 9c illustrates the influence that the rate of cooling at 2 Ma has on the form of the age plateau. Decreasing the modelled cooling rate to a still rapid 125°C Ma$^{-1}$ (sufficient to bring K95-084 to ambient surface conditions at a constant cooling rate from 2 Ma ago to the present) causes the lowest-temperature domain of this sample to remain open to diffusive loss of argon appreciably later than the other low closure temperature domains, giving the modelled age distribution a downwardly concave form. Although this disagrees with the observed age distribution for K95-084, this alternative cooling history is difficult to completely rule out, because of the potential of the low-temperature portion of the age release spectra to be affected by minor excess argon contamination, as discussed previously (Dunlap et al. 1991; Lovera et al. 1991).

The inferred cooling rate reduction at 5 Ma is investigated in Fig. 9d. A simple two-stage $T$–$t$ history lacking this cooling hiatus fails to produce the sharp rise between the initial plateau and the older ages in the later interval of gas release. As this steep rise is such a well-defined feature of the age release spectrum of K95-084, it appears that a reduction in cooling rate between c. 260 and 250°C is strongly supported by the data. Variations of 1 Ma to either side of the modelled 5 Ma timing of this cooling rate reduction produce significant deviations from the later released portion of the observed age spectrum for this sample (Fig. 9e). It is thus considered unlikely that this cooling rate reduction occurred more than 1 Ma to either side of the modelled time of 5 Ma.

The measured age spectrum of K95-084 can also be well reproduced by a significant pulse of heating at c. 1.8 Ma (Fig. 9f). This heating event partially outgasses the argon accumulated in the sample before this time, giving the distinctive young plateau and related features observed in this age spectrum (Fig. 9g). Such a heating event, however, is incompatible with the fission-track ages of this sample (as discussed below), so this solution is not pursued any further here.

K-feldspar samples K95-084, K95-088, and K95-097 differ from each other principally in the extent and age of the older gas component released late in their respective analyses. These samples also differ slightly in their diffusion characteristics (Table 5), such that K95-088 and K95-097 contain a greater proportion of gas in their larger (and thus higher effective closure temperature) domains. These differences in diffusion properties mean that the varying age spectra of all three samples can be modelled using the same cooling history (Figs 5d and 6d). This correlation suggests that the details of this cooling history are significant, and regional in their occurrence.

Assuming the 18°C Ma$^{-1}$ cooling rate calculated for the late gas release of the K-feldspars extends back to the time of closure of muscovite and biotite, the experimentally determined diffusion characteristics of muscovite (Robbins 1972; Lister & Baldwin 1996) give an inferred closure temperature of 380 ± 40°C (Dodson 1973; Lister & Baldwin 1996). For the same conditions, the diffusion characteristics of biotite (Harrison et al. 1985) yield a closure temperature of 340 ± 40°C (Dodson 1973; Lister & Baldwin 1996).

Because of the likely presence of excess argon, the K–Ar and $^{40}$Ar–$^{39}$Ar analyses of biotite from sample 95-097 give only a maximum possible time since this sample cooled through the closure temperatures of argon in biotite. As discussed previously, the 11.3 ± 0.5 Ma Rb–Sr age of biotite from sample 95-097 offers better constraint on this portion of the cooling history of the area, with biotite closing to the diffusion of both argon and strontium at the same temperature (Jäger et al. 1967; Dodson 1973; Del Moro et al. 1982; Dodson & McClelland-Brown 1985; Harrison et al. 1985; Yuhara & Kagami 1995).

No direct evidence of excess argon contamination is observed in the muscovite samples extracted from these pegmatites. The biotite Rb–Sr and K–Ar ages of sample 95-097, however, as noted above, indicate that excess argon is present, and $^{40}$Ar–$^{39}$Ar analysis of muscovite from a schist sample from the Mataketake Range (Batt 1997) shows that muscovite from this area also incorporates such contamination. It is thus our preferred interpretation that these

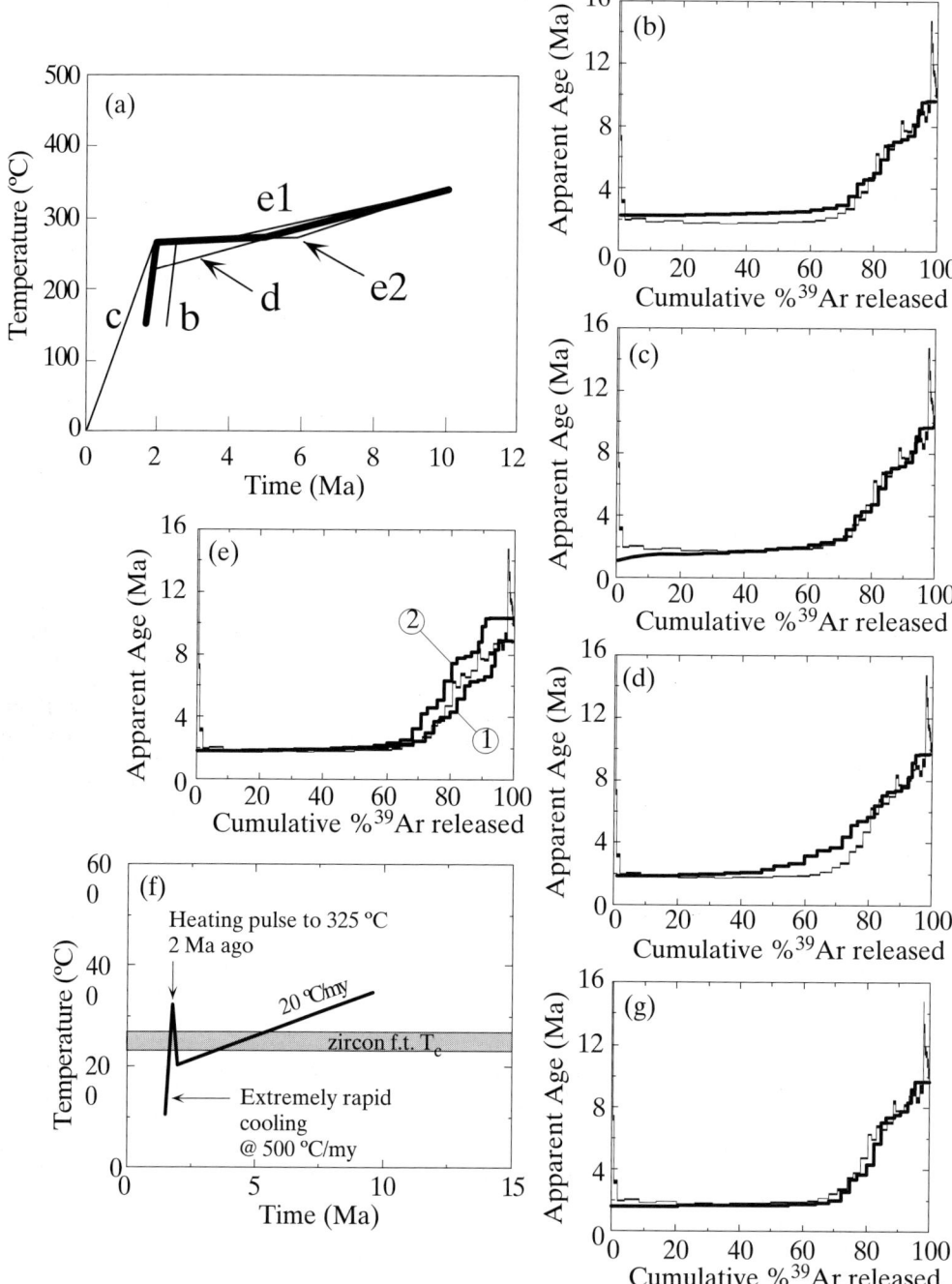

**Fig. 9.** Illustration of the sensitivity of the modelled age spectrum of K95-084 to alternative thermal histories. (**a**) Thermal histories employed to produce the model age spectra shown in (**b**)–(**e**). (**b**)–(**e**) each show the effects of the associated thermal history (black line) against the experimentally observed age spectrum for K95-084 (grey line). (**f**) Thermal history incorporating a heating pulse at 2 Ma. It should be noted that this pulse exceeds the annealing temperature of fission tracks in zircon (as discussed in the text) and thus would be expected to have reset that chronometer as well. (**g**) Model age spectrum produced by the thermal history illustrated in (**f**) (black line) compared with experimental age release spectrum for K95-084 (grey line).

muscovite ages do not directly constrain the time at which these samples cooled through the closure temperature of argon in this mineral, but rather reflect maximal constraints on this time, as discussed for biotite above. This interpretation allows a great deal of freedom in possible cooling histories between c. 350 and 500°C (Fig. 8).

Following Kamp et al. (1989) a fission-track closure temperature of 240 ± 25°C is adopted for zircon (see also Foster et al. (1996) and Brandon et al. (1998) for discussion), whereas for the extremely rapid cooling experienced within the last 2 Ma, the closure temperature for fission tracks in apatite would be of the order of 120 ± 20°C (Tippett & Kamp 1993). The zircon and apatite fission-track analyses of these pegmatites thus lie within the $T$–$t$ interval constrained by $^{40}Ar$–$^{39}Ar$ dating of K-feldspar (Fig. 8). The mean calculated age of zircon, at 4.3 ± 0.3 Ma, is consistent with the K-feldspar derived cooling curve, and the track length distribution of sample 95-084, which has a 'tail' of shorter tracks, indicates a moderate rate of cooling.

As discussed above, the zircon fission-track ages of these pegmatites are inconsistent with the prominent 1.8 Ma plateau age observed in the $^{40}Ar$–$^{39}Ar$ age spectra of these samples being produced by an episode of heating. This hypothetical heating event would take the area significantly above the inferred closure temperature of fission tracks in zircon (240 ± 25°C), resulting in fission-track annealing as well as partially degassing the accumulated argon within the K-feldspars (Fig. 9f). The well-constrained ages and dominant long track lengths of these zircons (as noted above) provide confidence that no such heating event occurred in this area.

The mean apatite fission-track age of these samples, at 1.4 ± 0.2 Ma, again falls on the K-feldspar derived cooling history. The apatite fission-track ages and length measurements support the extreme rapidity of cooling inferred for the pegmatites over more moderate alternatives (Fig. 8). In doing so, they also favour the possible occurrence of a reduction in the cooling rate experienced by these samples within the past 1–1.5 Ma, as discussed above.

In summary, the thermochronological data collected from the pegmatites of the Mataketake Range support a consistent, well-constrained cooling history for the area. The key features of this cooling history are a reduction in cooling rate from c. 18°C Ma$^{-1}$ to less than 5°C Ma$^{-1}$ at around 5 Ma, and a subsequent major increase in the rate of cooling to 350°C Ma$^{-1}$, beginning at 2 Ma. This then appears to have been followed in the recent past (1–1.5 Ma) by a further reduction in cooling rate. The precise timing of this postulated recent reduction in cooling rate is not well constrained by the available data, and cooling during this period may thus still have exceeded 100°C Ma$^{-1}$ (Fig. 8b).

## Interpretation

Between 10 and 5 Ma, the pegmatites of the Mataketake Range cooled from approximately 350 to 260°C, at a rate of 18°C Ma$^{-1}$ (Fig. 8). Interpretation of this history can be made on the basis of the geothermal gradient of the area before its exhumation. Kamp (1997) calculated this palaeo-geothermal gradient to be 20 ± 2°C/km, from fission-track analysis of the Hutt Range in central Canterbury. This geothermal gradient would suggest that the analysed pegmatites began this period between 16 and 19 km below the surface. From there, an elementary interpretation of the inferred cooling rate requires exhumation at between 0.8 and 1 km Ma$^{-1}$ over this 5 Ma interval.

For an orogen of the scale of the Southern Alps at these rates of exhumation, heat transfer by advection of material towards the surface occurs at a comparable rate to conduction of heat through typical crustal rock (Batt & Braun 1997). This relative efficiency of diffusive heat transfer acts to stabilize the geothermal structure of the region, preventing significant disturbance of crustal isotherms (Batt & Braun 1997). In such a situation, the inferred exhumation rate of 0.8–1 km Ma$^{-1}$ provides a dynamically reasonable explanation of the observed cooling rate. Unroofing at these moderate rates would be in keeping with the early history of the Alpine Fault as a dominantly strike-slip feature before the initiation of the present phase of rapid convergence and uplift in the Southern Alps (Kamp 1986; Sutherland 1995).

This behaviour is in marked contrast to that suggested by the rapid cooling experienced in the Mataketake Range beginning 2 Ma ago. The implied cooling rates for this period of c. 350°C Ma$^{-1}$ require exhumation rates far in excess of the 1–2 km Ma$^{-1}$ critical value for the instigation of dynamic thermal perturbation in a crustal-scale orogen (Batt & Braun 1997). Under these conditions, diffusion of heat into the surrounding crust cannot keep pace with the rate of advection of hot material. This leads to the presence of anomalously hot crustal material at shallow depth (Koons 1987; Shi et al. 1996; Batt & Braun 1997), and a correspondingly high near-surface geothermal gradient. Exhumation through this near-surface thermal boundary layer would result in a short episode of extremely rapid cooling to ambient surface conditions (Koons 1987; Shi et al. 1996; Batt & Braun 1997), as calculated for these pegmatites. Such behaviour would be in keeping with the rapid exhumation (Griffiths & McSaveney 1986; Hicks et al. 1990; Hovius et al. 1997) and high near-surface heat flow (Allis & Shi 1995; Ingham 1995; Shi et al. 1996) observed in the Southern Alps at the present day. The initiation of this rapid cooling 2 Ma ago, however, does not in itself provide a constraint on the beginning of this period of rapid exhumation (Batt & Braun 1997).

The recent tectonic reconstruction of Sutherland (1995) suggests that the Australian–Pacific instantaneous pole of rotation moved significantly at 5 Ma, and has not shifted appreciably since that time. The present dynamics of the South Island, with a significant component of convergence across the Alpine Fault resulting in the continuing rapid uplift of the Southern Alps must thus also have been established at c. 5 Ma. This suggestion is also supported by a change in the sedimentation rates and provenance of detritus in West Coast sedimentary basins at this time (Sutherland 1996).

In the context of this established history, the cooling hiatus experienced by these pegmatites at 5 Ma is taken to reflect the thermal reorganization and shear heating effects associated with the tectonic reorganization of the New Zealand region reported by Sutherland (1995). For material within the orogen at the time of such a dynamical change, the active migration of isotherms and consequent development of a perturbed thermal structure in response to the increased exhumation rate acts to prevent the immediate occurrence of a commensurate increase in cooling rate, and may in fact result in material deep within an orogen experiencing an initial decrease in cooling rate (Batt 1997). Although the contribution of shear heating to the thermal budget of the Southern Alps is a highly debated issue (e.g. Adams & Gabites 1985; Allis 1986; Shi et al. 1996), the possible short-term occurrence of such heating effects associated with this sudden change in tectonic style and rates in the South Island (Johnston & White 1983) would tend to maintain temperatures near the plate boundary at a higher level, possibly thus accounting for the observed decrease in cooling rate at this time.

Although similar variations in cooling rate could also be produced as a consequence of the three-dimensional transport and exhumation path of material with respect to the thermal structure of the orogen (Shi et al. 1996; Whittington 1996; Batt & Braun 1997), such behaviour is considered unlikely to have exerted a major effect on these pegmatite samples. If these pegmatites were to have been advected into the actively deforming zone of the Southern Alps at or shortly after 5 Ma, for example, and encountered a perturbed geothermal structure already established by the orogenesis, then their own initial exhumation within that zone would have been largely parallel to the regional isotherms, resulting in slow cooling despite possibly rapid rates of unroofing. Notably, however, these pegmatites experienced significant cooling (and inferred exhumation) between 5 and at least 10 Ma, as discussed above. The Mataketake Range thus appears to have been located in a tectonically active region before 5 Ma, suggesting that these pegmatites were already in close proximity to the plate boundary during this earlier interval, and thus ruling out the possibility of approach to a pre-existing Southern Alpine orogen as a feature of their dynamic history.

The K-feldspar $^{40}$Ar–$^{39}$Ar and apatite fission-track ages of these pegmatites also suggest a possible reduction in the cooling rate experienced by this area in the more recent past (1–1.5 Ma ago), implying the operation of additional dynamical phenomena. In this regard, Waschbusch et al. (1998) suggested that the focus of deformation, and hence the uplift rate, at the Alpine Fault in the southwest of the Southern Alps may have been reduced in the recent past because of the effects of 'subduction zone rollback' (Waschbusch & Beaumont 1996) beneath Otago and South Westland. Although this possibility would fit with the variation in surface heat flow observed close to the Alpine Fault by Shi et al. (1996) between the Franz Josef valley in the central region of the Southern Alps ($190 \pm 50$ mW m$^{-2}$) and the Haast valley in the south of the orogen ($90 \pm 25$ mW m$^{-2}$), it appears inconsistent with the rapid Holocene uplift rates of 7–8 mm a$^{-1}$ calculated for the Paringa region by Simpson et al. (1994). In light of such apparent disparity, the interpretation of this more recent reduction in the cooling rate experienced by these pegmatites remains open.

## Conclusions

The data reported in this study are consistent with the previously published U–Pb zircon and monazite ages of Chamberlain et al. (1995), and the zircon and apatite fission-track ages of Kamp et al. (1989) and Tippett & Kamp (1993) for the Mataketake Range and surrounding areas of South Westland. The combination of dating methods used herein yield cooling histories that are largely internally consistent, both within any given sample and between the different pegmatites analysed. These results combine to give the Mataketake Range an extremely well-constrained thermal history. Other than initial cooling to the ambient temperature of the host schist (estimated at c. 500°C) following intrusion in late Cretaceous time, these pegmatites share an apparently uniform thermochronological record, and this thermal history is thus likely to be largely indicative of regional cooling during uplift and exhumation, with most uplift in the late Miocene–Recent interval.

The key tectonic signal in the thermal history of the Mataketake Range pegmatites is seen at 5 Ma, when a marked decrease in cooling rate occurred. This hiatus in cooling is identifiable only because of the high resolution of the $^{40}$Ar–$^{39}$Ar method applied to the dating of K-feldspar. The consistent requirement of this event to explain the observed age spectra of the three K-feldspar samples analysed in this study, however, strongly confirms its significance.

Rather than indicating a decrease in tectonic activity, the short duration of this event and the ensuing occurrence of a pulse of extremely rapid cooling lead to the conclusion that this reduction in cooling rate marks the perturbation of the geothermal structure in response to a major increase in exhumation rate. As such, this event at 5 Ma is consistent with the established timing of the change from dominantly strike-slip motion on the Alpine Fault to the oblique convergence and rapid uplift of the present tectonic environment of the South Island of New Zealand, and thus the initiation of the development of the Southern Alps. Such timing for this development is consistent with recent tectonic reconstructions for the New Zealand region (Sutherland 1995), and also coincides with significant reorganization of Pacific plate motion at 5 Ma (Cox & Engebretson 1985; Pollitz 1986; Cande & Kent 1992).

This research was conducted while G. E. B. was funded by a John Conrad Jaeger PhD scholarship from the Australian National University, and funding for collaboration with La Trobe University was provided by a travel grant from the Research School of Earth Sciences at the ANU. G. E. B. also expresses his thanks to J. Dunlap for his initial introduction to the interpretation of K-feldspar $^{40}$Ar–$^{39}$Ar experiments. Rb–Sr analyses were carried out by M. McCulloch and V. Bennett at the Australian National University. Irradiation of samples for $^{40}$Ar–$^{39}$Ar and fission-track dating was undertaken in the HIFAR reactor of the Australian Nuclear Science and Technology Organisation through grants provided by the Australian Institute of Nuclear Science and Engineering. Some work reported here was conducted as part of the Australian Geodynamics Cooperative Research Centre program, and this paper is published with the permission of the Director, AGCRC. This manuscript has been greatly improved by constructive reviews from N. Mortimer, P. Koons, and M. Brandon.

## References

ADAMS, C. J. 1981. Uplift rates and thermal structure in the Alpine Fault zone and Alpine Schist, Southern Alps, New Zealand. *In*: MCCLAY, K. R. & NEVILLE, J. (eds) *Thrust and Nappe Tectonics*. Blackwell Scientific, Oxford, 211–222.

—— & COOPER, A. F. 1996. K–Ar age of a lamprophyre dike swarm near Lake Wanaka, west Otago, South Island, New Zealand. *New Zealand Journal of Geology and Geophysics*, **39**, 17–23.

—— & GABITES, J. E. 1985. Age of metamorphism and uplift in the Alpine Schist Group at Haast Pass, Lake Wanaka and Lake Hawea, South Island, New Zealand. *New Zealand Journal of Geology and Geophysics*, **28**, 85–96.

ALLIS, R. G. 1986. Mode of crustal shortening adjacent to the Alpine Fault, New Zealand. *Tectonics*, **5**, 15–32.

—— & SHI, Y. 1995. New insights to temperature and pressure beneath the central Southern Alps, New Zealand. *New Zealand Journal of Geology and Geophysics*, **38**, 585–592.

ARONSON, J. L. 1965. Reconnaissance rubidium–strontium geochronology of New Zealand plutonic and metamorphic rocks. *New Zealand Journal of Geology and Geophysics*, **8**, 401–423.

BATT, G. E. 1997. *The crustal dynamics and tectonic evolution of the Southern Alps, New Zealand: insights from new geochronological data and fully coupled thermodynamical finite element modelling*. PhD thesis, Australian National University, Canberra, A. C. T.

—— & BRAUN, J. 1997. On the thermo-mechanical evolution of compressional orogens. *Geophysical Journal International*, **128**, 364–382.

BERGER, G. W. & YORK, D. 1981. Geothermometry from $^{40}$Ar/$^{39}$Ar dating experiments. *Geochimica et Cosmochimica Acta*, **45**, 795–811.

BRANDON, M. T., RODEN-TICE, M. K. & GARVER, J. I. 1998. Late Cenozoic exhumation of the Cascadia accretionary wedge in the Olympic Mountains, NW Washington State. *Geological Society of America Bulletin*, **110**, 985–1009.

CANDE, S. C. & KENT, D. V. 1992. A new geomagnetic polarity time scale for the late Cretaceous and Cenozoic. *Journal of Geophysical Research*, **97**, 13917–13951.

CHAMBERLAIN, C. P., ZEITLER, P. K. & COOPER, A. F. 1995. Geochronologic constraints of the uplift and metamorphism along the Alpine Fault, South Island, New Zealand. *New Zealand Journal of Geology and Geophysics*, **38**, 515–523.

CLIFF, R. A. 1985. Isotopic dating in metamorphic belts. *Journal of the Geological Society, London*, **142**, 97–110.

COOPER, A. F. 1972. Progressive metamorphism of

metabasic rocks from the Alpine Schist Group of southern New Zealand. *Journal of Petrology*, **13**, 457–492.

—— & NORRIS, R. J. 1994. Anatomy, structural evolution, and slip rate of a plate-boundary thrust: the Alpine Fault at Gaunt Creek, Westland, New Zealand. *Geological Society of America Bulletin*, **106**, 627–633.

——, BARREIRO, B. A., KIMBROUGH, D. L. & MATTINSON, J. M. 1987. Lamprophyre dike intrusion and the age of the Alpine Fault, New Zealand. *Geology*, **15**, 941–944.

COX, A. & ENGEBRETSON, D. 1985. Change in motion of the Pacific plate at 5 Ma. *Nature*, **313**, 472–474.

CUTTEN, H. N. C. 1979. Rappahannock Group: Late Cenozoic sedimentation and tectonics contemporaneous with Alpine Fault movement. *New Zealand Journal of Geology and Geophysics*, **22**, 535–553.

DALRYMPLE, G. B. & LANPHERE, M. A. 1974. $^{40}$Ar/$^{39}$Ar age spectra of some undisturbed terrestrial samples. *Geochimica et Cosmochimica Acta*, **38**, 715–738.

DEL MORO, A., PUXEDDU, M., RADICATI DI BROZOLO, F. & VILLA, I. M. 1982. Rb–Sr and K–Ar ages on minerals at temperatures of 300°–400°C from deep wells in the Larderello geothermal field (Italy). *Contributions to Mineralogy and Petrology*, **81**, 340–349.

DODSON, M. H. 1973. Closure temperatures in cooling geochronological and petrological systems. *Contributions to Mineralogy and Petrology*, **40**, 259–274.

—— & MCCLELLAND-BROWN, E. 1985. Isotopic and palaeomagnetic evidence for rates of cooling, uplift and erosion. *In*: SNELLING, N. J. (ed.) *The Chronology of the Geological Record*. Geological Society, London, Memoirs, **10**, 315–325.

DUNLAP, W. J., TEYSSIER, C., MCDOUGALL, I. & BALDWIN, S. 1991. Thermal and structural evolution of the intracratonic Arltunga Nappe Complex, central Australia. *Tectonics*, **14**, 1182–1204.

FIELD, B. D., BROWNE, G. H., DAVY, B., et al. 1989. *Cretaceous and Cenozoic Sedimentary Basins and Geological Evolution of the Canterbury Region, South Island, New Zealand*. New Zealand Geological Survey Basin Studies 2. New Zealand Geological Survey, Lower Hutt.

FLECK, R. J., SUTTER, J. F. & ELLIOT, D. H. 1977. Interpretation of discordant $^{40}$Ar/$^{39}$Ar age-spectra of Mesozoic tholeiites from Antarctica. *Geochimica et Cosmochimica Acta*, **41**, 15–32.

FOLAND, K. A. 1983. $^{40}$Ar/$^{39}$Ar incremental heating plateaus for biotites with excess argon. *Isotopic Geoscience*, **1**, 3–21.

FOSTER, D. A., KOHN, B. P. & GLEADOW, A. J. W. 1996. Sphene and zircon closure temperatures revisited: empirical calibrations from $^{40}$Ar/$^{39}$Ar diffusion studies of K-feldspar and biotite. Conference abstract, International Workshop on Fission Track Dating, University of Gent, Gent, p. 37.

GAIR, H. S. 1967. *Geological Map of New Zealand 1:250 000, Sheet 20, Mt Cook*. Department of Scientific and Industrial Research, Wellington.

GALBRAITH, R. F. & LASLETT, G. M. 1993. Statistical models for mixed fission track ages. *Nuclear Tracks*, **21**, 459–470.

GILETTI, B. J. 1991. Rb and Sr diffusion in alkali feldspars, with implications for cooling histories of rocks. *Geochimica et Cosmochimica Acta*, **55**, 1331–1343.

GLEADOW, A. J. W., DUDDY, I. R., GREEN, P. F. & LOVERING, J. F. 1986. Confined fission track lengths in apatite; a diagnostic tool for thermal history analysis. *Contributions to Mineralogy and Petrology*, **94**, 405–415.

——, HURFORD, A. J. & QUAIFE, R. D. 1976. Fission track dating of zircon; improved etching techniques. *Earth and Planetary Science Letters*, **33**, 273–276.

GRAPES, R. H. 1995. Uplift and exhumation of Alpine Schist, Southern Alps, New Zealand: thermobarometric constraints. *New Zealand Journal of Geology and Geophysics*, **38**, 525–534.

GREEN, P. F. 1981. A new look at statistics in fission track dating. *Nuclear Tracks*, **5**, 77–86.

—— 1986. On the thermo-tectonic evolution of Northern England: evidence from fission track analysis. *Geological Magazine*, **123**, 493–506.

——, DUDDY, I. R., LASLETT, G. M., HEGARTY, K. A., GLEADOW, A. J. W. & LOVERING, J. F. 1989. Thermal annealing of fission tracks in apatite; 4, Quantitative modelling techniques and extension to geological timescales. *Chemical Geology*, **79**, 155–182.

GRIFFITHS, G. A. & MCSAVENEY, M. J. 1986. Sedimentation and river containment on Waitangitaona alluvial fan; South Westland, New Zealand. *Zeitschrift für Geomorphologie*, **30**, 215–230.

HARRISON, T. M., DUNCAN, I. & MCDOUGALL, I. 1985. Diffusion of $^{40}$Ar in biotite: temperature, pressure and compositional effects. *Geochimica et Cosmochimica Acta*, **49**, 2461–2468.

—— HEIZLER, M. T., LOVERA, O. M., CHEN, W. & GROVE, M. 1994. A chlorine disinfectant for excess argon released from K-feldspar during step heating. *Earth and Planetary Science Letters*, **123**, 95–104.

HASEBE, N., TAGAMI, T. & NISHIMURA, S. 1994. Towards fission-track thermochronology; reference framework for confined track length measurements. *Chemical Geology*, **112**, 169–178.

HICKS, D. M., MCSAVENEY, M. J. & CHINN, T. J. H. 1990. Sedimentation in proglacial Ivory Lake, Southern Alps, New Zealand. *Arctic and Alpine Research*, **22**, 26–42.

HOVIUS, N., STARK, C. P. & ALLEN, C. P. 1997. Sediment flux from a mountain belt derived by landslide mapping. *Geology*, **25**, 231–234.

HURFORD, A. J. & GREEN, P. F. 1983. The zeta age calibration of fission-track dating. *Isotope Geoscience*, **1**, 285–317.

INGHAM, M. R. 1995. Electrical structure along a transect of the central South Island, New Zealand. *New Zealand Journal of Geology and Geophysics*, **38**, 559–564.

JÄGER, E. 1979. Introduction to geochronology. *In*: JÄGER, E. & HUNZIGER, J. C. (eds) *Lectures in Isotope Geology*. Springer, Berlin, 1–12.

———, NIGGLI, E. & WENK, E. 1967. Rb–Sr Alterbestimmungen an Glimmern der Zentralalpern. Beiträge zur Geologie Karte der Schweiz, NF 134. Lieferungen, Kümmerly and Frey, Bern.

JOHNSTON, D. C. & WHITE, S. H. 1983. Shear heating associated with movement along the Alpine Fault, New Zealand. Tectonophysics, **92**, 241–251.

KAMP, P. J. J. 1986. The mid-Cenozoic Challenger Rift System of western New Zealand and its implications for the age of the Alpine Fault. Geological Society of America Bulletin, **97**, 255–281.

——— 1997. Paleogeothermal gradient and deformation style, Pacific front of the Southern Alps Orogen: constraints from fission track thermochronology. Tectonophysics, **274**, 37–58.

———, GREEN, P. F. & WHITE, S. H. 1989. Fission track analysis reveals character of collisional tectonics in New Zealand. Tectonics, **8**, 169–195.

KOONS, P. O. 1987. Some thermal and mechanical consequences of rapid uplift; an example from the Southern Alps, New Zealand. Earth and Planetary Science Letters, **86**, 307–319.

LASLETT, G. M., KENDALL, W. S., GLEADOW, A. S. W. & DUDDY, I. R. 1982. Bias in measurement of fission-track length distributions. Nuclear Tracks and Radiation Measurements, **6**, 79–85.

LELOUP, P. H., HARRISON, T. M. & RYERSON, F. 1992. Structural, petrological and thermal evolution of the Diancang Shan (PRC). Eos Transactions, American Geophysical Union, **73**, 310.

LISTER, G. S. & BALDWIN, S. L. 1996. Modelling the effect of arbitrary P–T–t histories on argon diffusion using the MacArgon program for the Apple Macintosh. Tectonophysics, **253**, 83–109.

LOVERA, O. M. 1992. Computer programs to model $^{40}Ar/^{39}Ar$ diffusion data from multi-domain samples. Computers and Geosciences, **18**, 789–813.

———, HEIZLER, M. & HARRISON, T. M. 1993. Argon diffusion domains in K-feldspar, II, kinetic properties of MH-10. Contributions to Mineralogy and Petrology, **113**, 381–393.

———, RICHTER, F. M. & HARRISON, T. M. 1989. The $^{40}Ar/^{39}Ar$ thermochronometry for slowly cooled samples having a distribution of diffusion domain sizes. Journal of Geophysical Research, **94**, 17917–17935.

———, ——— & ——— 1991. Diffusion domains determined by $^{39}Ar$ released during step heating. Journal of Geophysical Research, **96**, 2057–2069.

MCDOUGALL, I. & HARRISON, T. M. 1988. Geochronology and Thermochronology by the $^{40}Ar/^{39}Ar$ Method. Oxford University Press, New York.

MILDENHALL, D. C. & POCKNALL, D. T. 1984. Paleobotanical evidence for changes in Miocene and Pliocene climates in New Zealand. In: Late Cainozoic Palaeoclimates of the Southern Hemisphere. Rotterdam, A. A. Balkema, 159–171.

MUIR, R. J., IRELAND, T. R., WEAVER, S. D. & BRADSHAW, J. D. 1996. Ion microprobe dating of Paleozoic granitoids: Devonian magmatism in New Zealand and correlations with Australia and Antarctica. Chemical Geology (Isotope Geosciences Section), **127**, 191–210.

MUTCH, A. R. & MCKELLAR, I. C. 1964. Geological Map of New Zealand 1:250 000, Sheet 19, Haast. Department of Scientific and Industrial Research, Wellington.

NORRIS, R. J., KOONS, P. O. & COOPER, A. F. 1990. The obliquely convergent plate boundary in the South Island of New Zealand: implications for ancient collision zones. Journal of Structural Geology, **12**, 715–725.

PANKHURST, R. J., MOORBATH, S., REX, D. C. & TURNER, G. 1973. Mineral age patterns in a ca. 3700 Ma old rock from W. Greenland. Earth and Planetary Science Letters, **20**, 157–170.

PARRISH, R. R., CARR, S. D. & PARKINSON, D. L. 1988. Eocene extensional tectonics and geochronology of the Southern Ominica belt, British Columbia and Washington. Tectonics, **7**, 181–212.

POLLITZ, F. 1986. Pliocene changes in Pacific plate motion. Nature, **320**, 738–741.

RICHTER, F. M., LOVERA, O. M., HARRISON, T. M. & COPELAND, P. 1991. Tibetan tectonics from $^{40}Ar/^{39}Ar$ analysis of a single K-feldspar sample. Earth and Planetary Science Letters, **105**, 266–278.

ROBBINS, G. A. 1972. Radiogenic argon diffusion in muscovite under hydrothermal conditions. MS thesis, Brown University, Providence, RI.

RODDICK, J. C., CLIFF, R. A. & REX, D. C. 1980. The evolution of excess argon in alpine biotites – a $^{40}Ar$–$^{39}Ar$ analysis. Earth and Planetary Science Letters, **48**, 185–208.

SHI, Y., ALLIS, R. & DAVEY, F. 1996. Thermal modelling of the Southern Alps, New Zealand. Pure and Applied Geophysics, **146**, 469–501.

SIMPSON, G. D. H., COOPER, A. F. & NORRIS, R. J. 1994. Late Quaternary evolution of the Alpine Fault zone at Paringa, South Westland, New Zealand. New Zealand Journal of Geology and Geophysics, **37**, 49–58.

SMITH, D. E., KOLENKIEWICZ, R., ROBBINS, J. W., TORRENCE, M. H., HEFLIN, M. & SOUDARIN, L. 1996. A space geodetic plate motion model. Eos Transactions, American Geophysical Union, **77**, 1996 special supplement, S73.

SUTHERLAND, R. 1994. Displacement since the Pliocene along the southern section of the Alpine Fault, New Zealand. Geology, **22**, 327–330.

——— 1995. The Australia–Pacific boundary and Cenozoic plate motions in the SW Pacific: some constraints from Geosat data. Tectonics, **14**, 819–831.

——— 1996. Transpressional development of the Australia–Pacific boundary through southern South Island, New Zealand; constraints from Miocene–Pliocene sediments, Waiho-1 borehole, South Westland. New Zealand Journal of Geology and Geophysics, **39**, 251–264.

TIPPETT, J. M. & KAMP, P. J. J. 1993. Fission track analysis of the late Cenozoic vertical kinematics of continental Pacific crust, South Island, New Zealand. Journal of Geophysical Research, **98**, 16119–16148.

WAGNER, G. A., REIMER, G. M., & JÄGER, E. 1977. Cooling ages derived by apatite fission-track, mica Rb–Sr and K–Ar dating: the uplift and cooling history of the Central Alps. Memoire degli

*Istituti di Geologia e Mineralogia, dell'Universita di Padova*, **30**, 1–27.

WALCOTT, R. I. 1984. Reconstructions of the New Zealand region for the Neogene. *Palaeogeography, Palaeoclimatology, Palaeoecology*, **46**, 217–231.

WALLACE, R. C. 1974. Metamorphism of the Alpine Schist, Mataketake Range, South Westland, New Zealand. *Journal of the Royal Society of New Zealand*, **4**, 253–266.

WASCHBUSCH, P. & BEAUMONT, C. 1996. Effect of a retreating subduction zone on deformation in simple regions of plate convergence. *Journal of Geophysical Research*, **101**, 28133–28148.

——, BATT, G. E. & BEAUMONT, C. 1998. Subduction zone retreat and recent tectonics of the South Island of New Zealand. *Tectonics*, **17**, 267–284.

WHITTINGTON, A. G. 1996. Exhumation overrated at Nanga Parbat, northern Pakistan. *Tectonophysics*, **206**, 215–226.

YUHARA, M. & KAGAMI, H. 1995. Cooling history of the Katsuma quartz diorite in the Ina District of the Ryoke Belt, southwest Japan Arc. *Journal of the Geological Society of Japan*, **101**, 434–442.

# Exhumation history of orogenic highlands determined by detrital fission-track thermochronology

JOHN I. GARVER[1], MARK T. BRANDON[2], MARY RODEN-TICE[3] & PETER J. J. KAMP[4]

[1]*Geology Department, Union College, Schenectady, NY 12308–2311, USA (e-mail: garverj@union.edu)*

[2]*Department of Geology and Geophysics, Yale University, P.O. Box 208109, 210 Whitney Avenue, New Haven, CT 06520–8109, USA*

[3]*Earth and Environmental Sciences, State University of New York at Plattsburgh, Plattsburgh, NY, 12901, USA*

[4]*Geochronology Research Unit, Earth Sciences Department, University of Waikato, Private Bag 3105, Hamilton, New Zealand*

**Abstract:** A relatively new field in provenance analysis is detrital fission-track thermochronology which utilizes grain ages from sediment shed off an orogen to elucidate its exhumational history. Four examples highlight the approach and usefulness of the technique. (1) Fission-track grain age (FTGA) distribution of apatite from modern sediment of the Bergell region of the Italian Alps corresponds to ages obtained from bedrock studies. Two distinct peak-age populations at 14.8 Ma and 19.8 Ma give calculated erosion rates identical to *in situ* bedrock. (2) Zircon FTGA distribution from the modern Indus River in Pakistan is used to estimate the mean erosion rate for the Indus River drainage basin to be about 560 m Ma$^{-1}$, but locally it is in excess of 1000 m Ma$^{-1}$. (3) FTGA distribution of detrital apatite and zircon from the Tofino basin records exhumation of the Coast Mountains in the Canadian Cordillera. Comparison of detrital zircon and apatite FT ages gives exhumation rates of c. 200 m Ma$^{-1}$ during the interval between c. 34 and 54 Ma, but higher rates (c. 1500 m Ma$^{-1}$) at c. 56 Ma. (4) FTGA analysis of apatite grain ages from a young basin flanking Fiordland in New Zealand indicates that removal of cover strata was followed by profound exhumation at c. 30 Ma, which corresponds to plate reorganization at this time. Exhumation rates at the onset of exhumation were c. 2000–5000 m Ma$^{-1}$. These studies outline the technique of detrital FTGA applied to exhumation studies and highlight practical considerations: (1) well-dated, stratigraphically coordinated suites of samples that span the exhumation event provide the best long-term record; (2) strata from the basin perimeter are the most likely to retain unreset detrital ages; (3) the removal of 'cover rocks' precedes exhumation of deeply buried rocks, which retain a thermal signal of the exhumation event; (4) steady-state exhumation produces peak ages that progressively young with time and have a constant lag time; (5) same-sample comparison of zircon and apatite peak ages is best in sequences with high-uranium apatite grains (>50 ppm), and peak-ages statistics can be improved by counting numerous apatite grains (>100).

Orogenic sediments provide an integrated view of the tectonic and climatic factors that shape the evolution of mountainous regions. Stratigraphers and sedimentologists have spent considerable effort using the provenance of sediments to understand the evolution of adjacent source regions. A relatively new field in provenance analysis is detrital fission-track thermochronology, which utilizes the fission-track ages of single detrital grains to identify and characterize the source region and also to quantify its thermochronological evolution.

Unravelling the long-term evolution of mountain systems requires a detailed understanding of the relationship between uplift, erosion, and deposition. One approach has been to apply different mineral thermochronometers to bedrock samples for eroded landscapes, or to sedimentary detritus that collects in nearby basin sequences. Studies using bedrock exposures are limited to only those rocks exposed at the surface and therefore provide only limited information about the entire evolution of that orogenic system because overlying rock, which presumably had a record of earlier thermotectonic events, has been stripped away and lost to adjacent basins. On the other hand, sedimentary basins that surround orogenic belts contain an easily accessible, long-term orogenic record, but the evidence is fragmentary and needs to be

GARVER, J. I., BRANDON, M. T., RODEN-TICE, M. & KAMP, P. J. J. 1999. Exhumation history of orogenic highlands determined by detrital fission-track thermochronology. *In:* RING, U., BRANDON, M. T., LISTER, G. S. & WILLETT, S. D. (eds) *Exhumation Processes: Normal Faulting, Ductile Flow and Erosion.* Geological Society, London, Special Publications, **154**, 283–304.

reassembled. Sediments, however, commonly provide the only long-term record for an evolving orogenic system. In detrital FT thermochronology, the goal is to examine stratigraphic sequences, and relate upsection changes in FT grain ages to the thermal evolution of the source region. Another similar approach has been laser-probe $^{40}$Ar/$^{39}$Ar dating of detrital potassium-bearing minerals such as biotite, muscovite, amphibole, and feldspar (e.g. Horstmann 1987; Cohen et al. 1990; Copeland and Harrison 1990; Copeland et al. 1990; Grist et al. 1990; Renne et al. 1990; Turner et al. 1996; Dallmeyer et al. 1997).

FT dating of common detrital uranium-bearing minerals such as apatite, zircon and sphene has been used for some time now in provenance analysis, stratigraphic correlation, and dating sediments (McGoldrick & Gleadow 1977; Wagner et al. 1979; Zeitler et al. 1982, 1986; Hurford et al., 1984; Yim et al., 1985; Baldwin et al. 1986; Kowallis et al. 1986; Naeser et al. 1987, 1989; Cerveny et al. 1988; Garver 1988; Vance 1989; Corrigan & Crowley 1990; Brandon & Vance 1992; Hasebe et al. 1993; Maranville 1993; Garver & Brandon 1994a,b; Frisbee 1995; Carter 1996; Dunkl et al. 1996; Tagami & Dumitru 1996; Pereira et al. 1996; Rohrman et al. 1996). A more recent development has been to use FT ages of detrital zircon or apatite from stratigraphic sequences to reconstruct source region exhumation (Garver et al. 1993; Brown et al. 1994; Garver & Brandon 1994b; Johnson & Lonergan 1996; Rohrman & Andriessen 1996; Rohrman et al. 1996). It seems clear that this approach holds great promise for understanding the long-term relationship between tectonics, topography, climate, erosion, and sedimentation in orogenic settings.

Generally, source regions have distinct thermotectonic signatures, which are clearly retained in the erosional detritus. In our experience, most source regions in orogenic settings can be characterized as one of three important thermotectonic regimes: (a) source regions exhumed by progressive erosion; (b) source regions unroofed by tectonic processes; (c) source regions characterized by high geothermal gradients and active volcanism. The first case, which is highlighted in this paper, involves a source region that evolves through progressive erosion. In this case, rocks move upward in the crust through a closure isotherm, the FT clock starts, and eventually the rock reaches the surface, is eroded and the detritus is deposited in adjacent basins. All of these cases have distinct FT grain age (FTGA) patterns.

This paper reviews the basic concepts and techniques of detrital FT thermochronology, and highlights the strategy and implementation of this approach by looking at both the zircon and apatite systems. Then, to illustrate these concepts, we provide examples, from a small drainage basin in the centre of the Italian Alps, the Indus River, which drains the western Himalayan Mountains, a basin flanking the Coast Mountains of western Canada, and a basin flanking the Fiordland block in southern New Zealand.

## Erosion, transport, and deposition

In a simple setting, erosional exhumation progressively unroofs rocks in a source area and the sediment is shed into an adjacent basin (Fig. 1a and b). During exhumation, a rock passes through the closure temperature $T_c$ and closure depth $Z_c$ for the isotopic system of interest to the surface (which has an average temperature of $T_s$). Once at the surface, it is then eroded at time $t_e$, transported, and deposited at $t_d$ in adjacent sedimentary basins. A critical aspect of this system is the $\Delta T$ for the sample, which is given by $T_c - T_s$. Sediment storage for a certain length of time $(t_e - t_d)$ may occur sometime between erosion $(t_e)$ and deposition $(t_d)$. Ultimately, we are interested in determining the effective exhumation rate of the system from a well-dated stratigraphic sequence ($\dot{\varepsilon}$ in m Ma$^{-1}$). To do this, several important variables need to be estimated including, most importantly, the geothermal gradient ($G$) at the time of closure ($t_c$). Both of the variables are difficult to quantify accurately and as a result are the main limiting factors in using detrital FTGA analysis to estimate exhumation.

One can use FT ages from bedrock samples to estimate long-term exhumation rates. Use of this approach is routine in the FT analysis of bedrock samples, especially suites of samples from mountainous regions (see review by Wagner & Van den haute (1992)). A common approach is to assume a linear geotherm and a specific closure temperature. In this case, the average exhumation rate is given by the following equation:

$$\dot{\varepsilon}_m = [(T_c - T_s)/G] / \Delta t \qquad (1)$$

where $\dot{\varepsilon}_m$ is a model exhumation rate, $\Delta t = t_e - t_c$ which is the lag time, and $G$ is the geothermal gradient.

In detrital FTGA spectra, essentially the same approach is used for distinct grain-age populations. These populations are referred to as component 'populations' or 'peaks' and they are defined by either a Gaussian or binomial fit

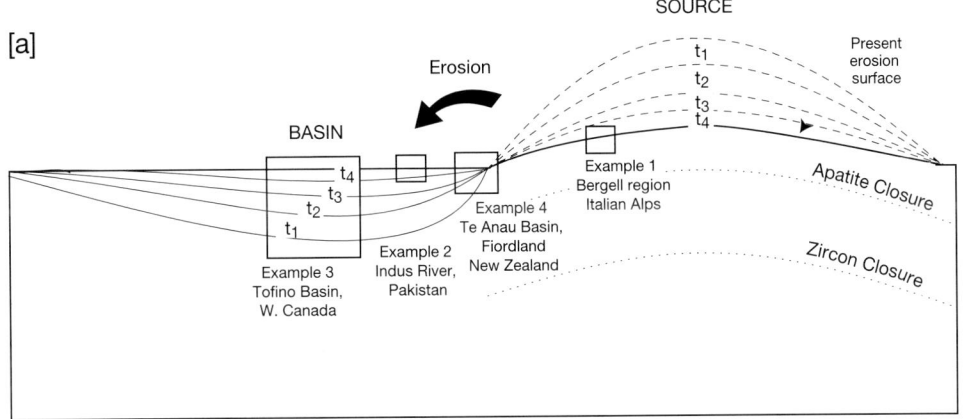

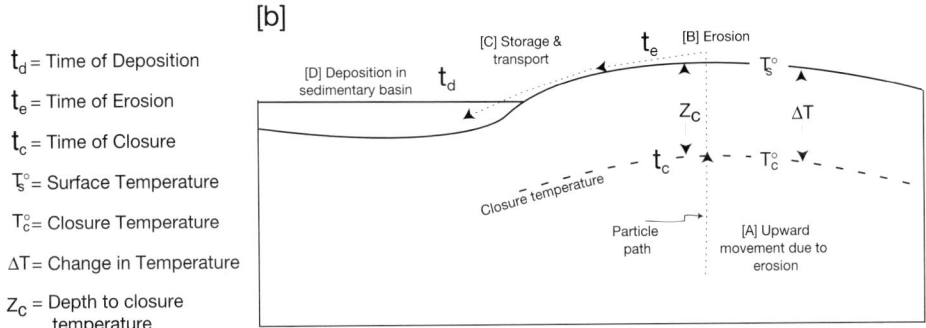

**Fig. 1.** Schematic representation of an idealized erosion of a mountain range and the variables used in FTGA calculations. (**a**) Schematic diagram showing development of an exhumational sequence through time. In this simple case, a source is progressively unroofed and the sedimentary detritus is deposited in an adjacent basin. It should be noted that the thickness of the strata varies in the basin, but locally, once the strata are thick enough, they enter the apatite annealing zone and in terms of FT ages, the provenance information in the apatite is lost. Time slices ($t_1$–$t_4$) correspond to four progressive and continuous intervals of erosion and deposition in which initial deepening in the basin was followed by basin infilling and erosional decay. The examples in this paper come from the different areas of the system denoted in the figure. (**b**) Variables used in model exhumation calculations. It should be noted that a particular rock follows the dotted lined marked as 'particle path'. As the rock moves upward toward the surface, it passed the critical closure isotherm and fission tracks begin to accumulate. Through exhumation of the region, that rock then passes upwards to the surface some distance $Z'_c$ which is estimated by making assumptions about the closure temperature, surface temperature and the geothermal gradient. The rock is then eroded and deposited in an adjacent basin, a part of the journey that is inferred to be geologically instantaneous.

to the composite dataset (Galbraith & Green 1990; Brandon 1992, 1996). For example, let us assume that a young population or 'peak' of apatite grain ages (see section on 'peak-fitting') $P_1$ from Recent river sediment occurs at 14 Ma. In this case the lag time or $\Delta t$ in equation (1) is given by $t_d - t_c$, which represents the time between passage through the partial annealing zone (PAZ) and deposition.

For simplicity, it should be noted that for detrital samples, lag time includes not only the time from closure to the surface, but also the time needed to erode and transport the sediment into nearby sedimentary basins, but we suspect that the transport time is fast in rapidly eroding mountain belts (Brandon & Vance 1992). We take the approach here that when viewed at the scale of the entire drainage system, sediment storage is part of the erosion process. In other words, the material has not been eroded from the drainage system until it has moved out of the entire system. This view becomes less credible when the distance between the erosional drainage and the sampled basinal setting become large. With large distances, there is a greater probability that transient basins may have existed along that transport path.

## FT dating

Naturally formed fission tracks in common minerals are generally the result of fission of $^{238}$U at a rate of about $10^{-16}$a$^{-1}$ (see Fleischer et al. 1975; Naeser 1976; Wagner & Van den haute 1992). These fission events result in the recoil of two highly charged sub-equal nuclei that create a damage zone in the crystal lattice, which can be easily enlarged by chemical etching and then viewed at high magnification with a microscope.

Annealing of fission tracks is a function of time and temperature. Although the annealing temperatures of apatite and zircon have been studied in the laboratory, difficulty arises when extrapolating these conditions to geologically reasonable conditions (see Wagner & Van den haute 1992). Resetting and closure of fission-track ages are commonly approximated using the concepts of a PAZ and an effective closure temperature, which are summarized in Fig. 2. The PAZ describes the temperature range in which existing fission tracks are partially annealed. In practice, the upper and lower limits of the PAZ can be defined by the temperatures needed to cause 10 and 90% annealing after a specified period of time (Fig. 2a). For natural α-damaged zircon, the PAZ is about 180–240°C for heating times of between 1 and 25 Ma, whereas for fluorapatite, it is about 40–120°C. Composition and radiation damage can influence annealing behaviour. For instance, fission tracks will persist to higher temperatures in zircons with little or no α-radiation damage (Kasuya & Naeser 1988) or in apatites with higher Cl contents (Green et al. 1985). Thus, it is important to keep these factors in mind when interpreting fission-track ages. We are particularly concerned with the lower-temperature side of the PAZ, which determines where grain ages start to reset. Assuming a typical continental geotherm of 25°C km$^{-1}$ and an average surface temperature of c. 10°C, we would expect to preserve unreset detrital grain ages at depths less than 1.2 km for fluorapatite and less than about 7 km for α-damaged zircons.

Closure of the fission-track system occurs as a sample cools through the PAZ. None the less, it is useful to define an effective closure temperature $T_c$, which is the temperature of the sample at the time indicated by the fission-track age. If

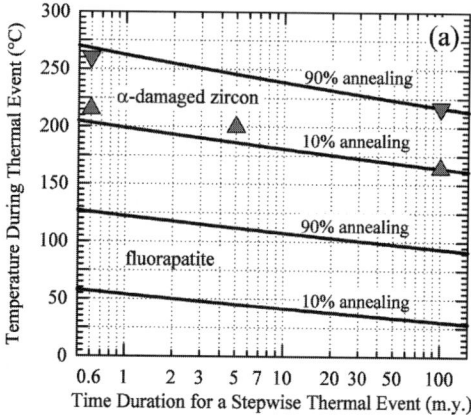

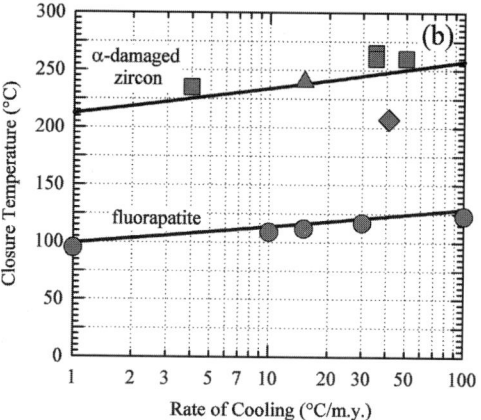

**Fig. 2.** The thermal stability and $T_c$ for fission tracks in fluorapatite (Durango) and natural α-damaged zircon (from Brandon et al. 1998). (**a**) The partial annealing zones (PAZs) are represented by the 10 and 90% annealing isopleths, which have an Arrhenius-like relationship. (**b**) $T_c$ is estimated for monotonic cooling through the PAZ at an approximately constant cooling rate. Symbols show other estimates for $T_c$: circles, predictions of a track-length annealing model; triangles, comparison of zircon FT ages and biotite K/Ar ages; squares, comparisons of zircon FT ages with $^{40}$Ar/$^{39}$Ar K-feldspar ages; diamonds, comparison of zircon FT ages with apatite FT and K/Ar biotite ages (see Brandon et al. (1998) for complete discussion).

one can assume that the sample cooled through the PAZ at an approximately steady rate, then $T_c$ can be calculated as a function of cooling rate (Fig. 2b). For typical geological cooling rates of 1–30°C Ma$^{-1}$, the $T_c$ for fluorapatite is about 100–120°C, and for α-damaged zircon about 215–240°C. Laboratory experiments by Yamada et al. (1995) suggest that $T_c$ for zircon might be

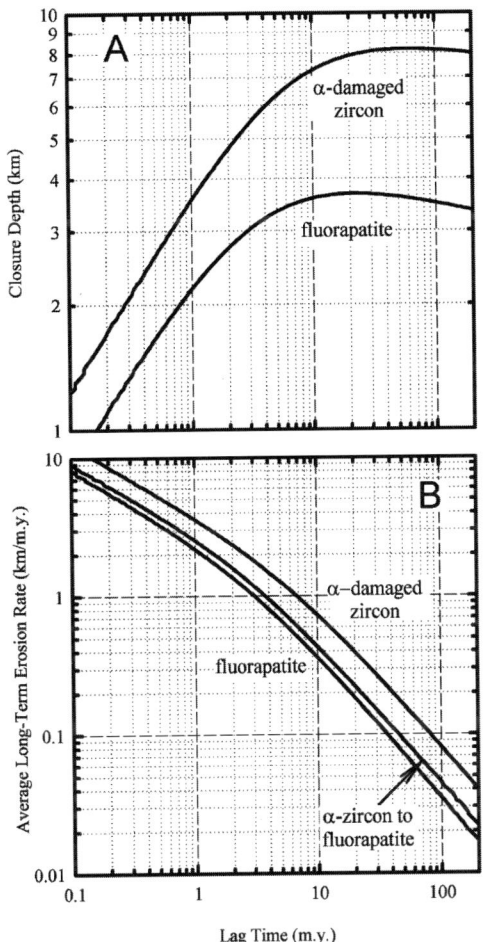

**Fig. 3.** The lag time shown as a function of closure depth and average long-term erosion rates. (**a**) Relationship between apatite and zircon closure depth with lag time. The lines makes approximate corrections for changes in closure depth caused by displacement of isotherms and by changes in cooling rates, both of which are a function of exhumation rates. (**b**) Calculations of averaged erosion rates and lag time. The influence of fast erosion rates on the thermal profile can be seen at lag time of about 5 Ma, below which correction must be made for the advection of heat by erosion. The closure depth for long lag times, takes on a constant value of c. 3 km for apatite with $t > 10$ Ma. and c. 6.5 km for zircon with $t > 20$ Ma (see Brandon *et al.* 1998). The relationship between cooling rate and closure temperature was modelled using the equation of Dodson (1979) with values reported by Brandon & Vance (1992) and Brandon *et al.* (1998). The thermal profile is specified using the solution for steady-state erosion given by Stüwe *et al.* (1994), which assumes a constant erosion rate and fixed temperatures at the base and top of an eroding infinite layer. For the Himalayan example shown in this paper, the layer was assumed to be 30 km thick and the base and top were held at 910°C and 10°C, which implies an initial linear thermal gradient of c.30°C km$^{-1}$. The model results are not very sensitive to the thickness of the layer for thicknesses greater than c.20 km. A lower thermal gradient or lower surface temperature will result in a deeper closure depth, all other factors being equal.

50–100°C greater than cited here, but this result is inconsistent with other estimates (see comparisons in Fig. 2). Brandon *et al.* (1998) have reviewed this problem, and suggested that the anomalously high thermal stability observed by Yamada *et al.* (1995) would be consistent for zircons with low α-radiation damage. The amount of α damage present in the zircons studied by Yamada *et al.* (1995) is not known.

In most cases, we are actually interested in estimating the closure depth $Z_c$ (Fig. 3). The closure depth is a function of the geotherm and the rate of cooling, both of which are directly affected by the exhumation process. Brandon *et al.* (1998) proposed an approximate method for estimating closure depth using a one-dimensional steady-state model where erosion at the surface is balanced by accretion at depth. The model assumes that temperatures at the base and surface of a one-dimensional layer are maintained fixed at their initial values. These fixed values can be viewed as describing the initial geotherm (i.e. the steady-state thermal profile before the onset of exhumation). The solution of this model for fission-track closure is not strongly dependent on the layer thickness so long as the thickness is greater than 20 km. The model assumes that the erosion rate has been operating for long enough for the thermal profile to reach a steady-state configuration, which usually requires several millions of years to occur (see Brandon *et al.* (1998) for details). A faster erosion rate will cause isotherms to migrate closer to the surface, but it will also result in a faster cooling rate.

An important aspect of the thermal annealing of apatite is that tracks are progressively shortened while in this PAZ, and the track length distribution can be used qualitatively to assess the rate of cooling through this zone (Gleadow *et al.* 1986). Where cooling is rapid, track lengths of c. 14–15 μm are typical.

Apatites that cooled more slowly (i.e. slower exhumation) will tend to have shorter track lengths because they spend more time in the PAZ. Unfortunately, individual apatite grains rarely have more than one or two tracks with appropriate orientations for measurement. In studies of reset samples, this problem is circumvented by collecting track-length measurements from the entire population of grains, but this approach is not appropriate for unreset detrital grain ages because each grain can have a different thermal history. The application of track-length analysis to zircon is still in its developmental phase, but the same problem of low numbers of horizontally confined tracks in single grains will preclude routine use for unreset detrital zircons.

## Detrital FT chronology

Unlike a conventional FT age where dated grains from a single source have a common thermal history, FTGA distributions contain many single grain ages, which originated from a variety of thermotectonic source terranes. Several techniques have been proposed to evaluate grain-age distributions, but the Gaussian and binomial peak-fitting methods are the most routinely used (see, e.g. Hurford *et al.* 1984; Seward & Rhodes 1986; Galbraith 1988; Galbraith & Green 1990; Brandon & Vance 1992; Brandon 1996). Here we employ both 'peak-fitting' methods. (Programs used to peak-fit the data in this paper are available by anonymous FTP from: love.geology.yale.edu//pub/brandon/ft/ft_peaks. For Gaussian peak-fitting, individual grain ages are first collectively approximated by a continuous probability density (PD) plot. Next, Gaussian peaks are fitted to the PD, and the sum of the fitted peaks is the model prediction. With an appropriate fit, individual peak ages (P1, P2, etc., in Fig. 4) can then be evaluated. For the binomial peak-fitting technique, populations are estimated directly from the data without the need to generate a PD plot. This latter method has some advantages in that it is based directly on the binomial distribution, which better represents counting statistics for FT dating. It also gives more precise estimates for the uncertainties of peak ages. In practice, however, the two methods commonly give similar results especially when the uncertainty for grain ages is small, which is typically the case for zircon. For both methods, individual fitted peaks have a mean age, size, and peak width ($W$). It should be noted that the peak width ($W$) is the relative standard deviation of a peak expressed as a fraction of the peak age.

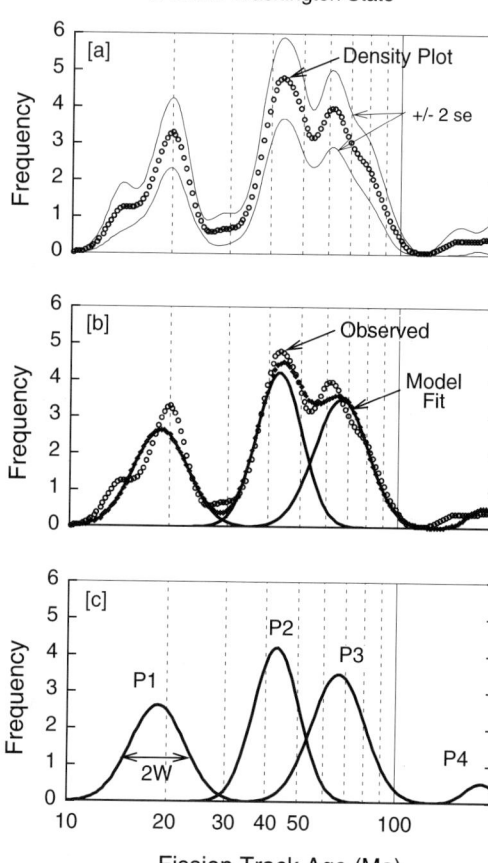

**Fig. 4.** Composite probability plots and fitted peaks for detrital zircon FT ages from rocks of the Olympic Subduction Complex (western Washington State, USA), shown here to illustrate the general procedure of peak-fitting detrital FT data (data from Brandon & Vance 1992). These rocks are part of the modern accretionary prism to the Cascadia subduction zone, and the zircons in these sandstones were derived from a variety of thermotectonic terranes behind and within the Cascade arc (see Brandon & Vance 1992; Maranville 1993; Frisbee 1995). (**a**) Density plot (±2 SE) showing the observed distribution of all grain ages ($n = 50$). (**b**) The density plot is then fitted by the Gaussian peak-fitting method (individual peaks) and the modelled fit is maximized to the observed density plot. (**c**) Fitted peak ages are then inferred to represent the age of component populations in the source region (see Brandon (1996) for full explanation).

The two main factors that influence counting statistics and therefore the precision of a single grain-age are uranium concentration and age; both directly affect the number of countable tracks. Zircon typically ranges between 5 and 4000 ppm uranium (see Speer 1980) and apatite mainly ranges between <1 and 300 ppm but typically falls well below 100 ppm (e.g. Wagner & Van den haute 1992; Dill 1994). For common detrital zircons, the mean uranium concentration is $c.$ 420 ppm with a mode of $c.$ 140 ppm and a range between 100 and 600 ppm (Garver & Brandon in preparation). Therefore, apatite grains range in composition from about 1 to 100 ppm whereas zircon grains typically range from 100 to 500 ppm. This difference in U content results in much lower precision for apatite FT grain ages than that for zircon FT ages.

To illustrate this important effect of uranium variation on FT age, we show a typical distribution of grain ages from the FT reference standard, the Oligocene Fish Canyon Tuff (Fig. 5). In both plots, about 45 typical single-grain ages are plotted. The peak width, $W$, for apatite is about 0.39, which is much greater than for zircon, where $W = 0.20$. For this example, both FT ages are identical, but the zircon has $c.$ 380 ppm U and apatite has $c.$ 13 ppm (Fig. 5). Given identical cooling ages, less uranium means fewer tracks, which results in less precise grain ages. Thus, high U concentrations are needed (>50 ppm) if apatite is to be used for detrital geochronological study. Our experience suggests that apatite suites from evolved continental crustal blocks tend to have higher uranium concentrations and are better suited for exhumation studies.

## Examples of detrital FT geochronology

In these examples, we show how detrital FTGA distributions can be used to study exhumation of mountainous source regions. In outlining these four cases, we pose several logical questions for the reader to consider: (1) Does the method faithfully capture the cooling ages of bedrock in a drainage basin? (2) Can the cooling ages be applied to understanding the exhumation of an orogenic system? (3) What problems, assumptions, and limitations are encountered? (4) Can detrital FTGA analysis be used to evaluate the temporal evolution of exhumation by using stratigraphically coordinated samples? (5) What approaches are most fruitful and what problems are encountered?

### Modern river sediment shed from the Bergell region, Central European Alps

In this first example, we show how FT grain ages from a small modern drainage reflect cooling in the bedrock as independently determined from

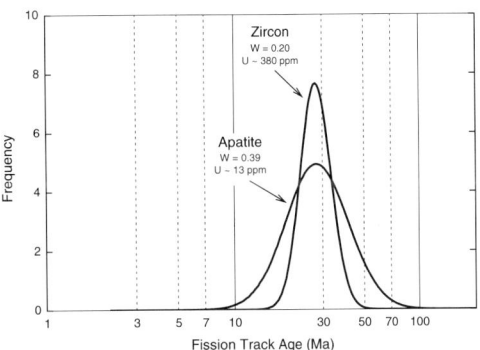

**Fig. 5.** Plot showing the probability distribution of zircon and apatite grain ages determined by fission-track dating for the 27.9 ± 0.3 Ma Fish Canyon Tuff age standard from Southfork, Colorado (Naeser et al. 1981). Minerals from this quickly cooled welded tuff are commonly used as an age standard in fission-track dating and in other geochronological techniques. For both zircon and apatite, $c.$ 45 grains are used to construct the plot. It should be noted that the peak is about two times wider for the apatite compared with the zircon, with the contrast ascribed to large differences in uranium concentrations between the two minerals. The lower precision of apatite is typical in most detrital suites, and as a result, apatite is less useful for provenance analysis, especially when the uranium concentration falls below $c.$ 50 ppm.

other studies (Fig. 6). Sediment was collected from a small river near the town of Novate in northernmost Italy near Lake Como to the south and the border between Italy and Switzerland to the north. The upstream drainage is entirely underlain by the Bergell pluton and adjacent country rocks, and it was chosen because the area has a very well-constrained thermal history representative of this part of the Alpine system. In this example we are interested in establishing the FTGA distribution of sediment with a well-defined provenance and thermal history. One hundred and fourteen grains with moderate to high uranium concentrations (mean of 58 ppm, but ranging between 50 and 200 ppm) define two distinct peak ages at 14.8 Ma and 19.8 Ma (Table 1; a full set of fission-track data is available from the Society Library and from the British Library Document Supply Centre, Boston Spa, Wetherby, West Yorkshire LS23 7BQ, UK, as Supplementary Publication No. SUP 18128 (10 pp.)).

The collision between Africa and the European plate has discontinuously affected rocks in this area from Cretaceous to late Tertiary time;

## Apatite fission track ages from the Bergell region, Italian Alps

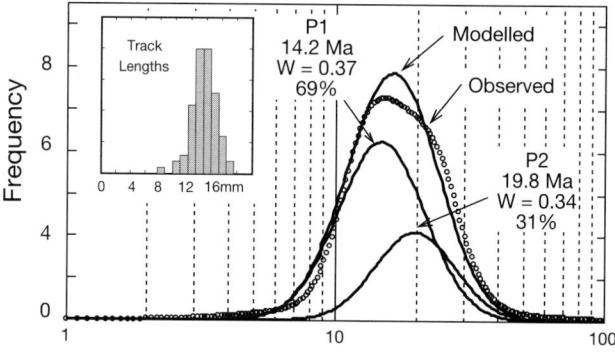

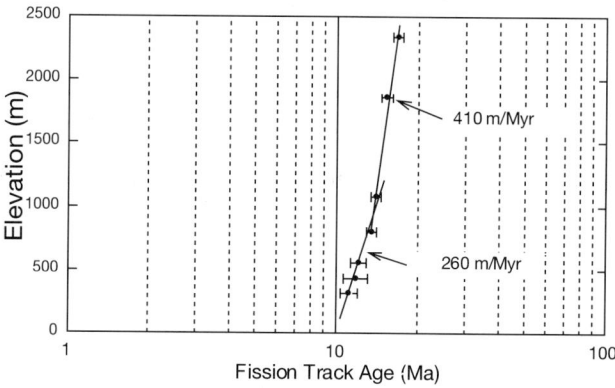

**Fig. 6.** Fission-track ages from the Bergell region in the Italian Alps: (**a**) detrital FT ages from apatite from river sediment (Table 1); and (**b**) apatite from bedrock samples as reported by Wagner *et al.* (1979). For the detrital sample, all 114 single grain ages were summed to form the probability density (PD) plot (shown as the line defined by small circles). Fitted peaks, P1 and P2, are a best fit using binomial statistics. These peaks sum to a model density plot, which is shown as the continuous line. Wagner *et al.* (1979) recognized two cooling episodes on the age–elevation plot. The first is defined by samples from c. 2500 to c. 1000 m and gives a exhumation rate of c. 410 m Ma$^{-1}$ at a mean age of c. 14.2 Ma. The second is defined by samples <1000 m, and gives a exhumation rates of 260 m Ma$^{-1}$ at a mean age of c. 11.7 Ma. Inset in (**a**) shows the track-length distribution for the river-sediment apatite. These track lengths (mean 13.98 μm, SD 1.56 μm, $n = 86$) were measured from detrital grains presumably with different thermal histories, but the relatively long mean length of c. 14 μm suggests rapid cooling of rocks in the source despite mixing.

here we are interested in the most recent episode of uplift and exhumation. Estimates of long-term exhumation rates for the Central Alps range between about 200 and 400 m Ma$^{-1}$ (see Clark & Jäger 1969; Schaer *et al.* 1975; Wagner *et al.* 1977, 1979; Wagner & Van den haute 1992). The Bergell region of the Central Alps is underlain by the Bergell plutonic suite that intrudes the Penninic Nappes (Schmid *et al.* 1996). Intrusion of the Bergell plutonic suite occurred at about 33–28 Ma, and the suite is notable because some 10 km of magmatic section are enclosed at present (Hansmann 1996; Schmid *et al.* 1996). Apatite fission-track dates from plutonic boulders in Late Oligocene strata of the Po Plain (to the south of the Alps in

**Table 1.** *Apatite FT peak ages for Bergell river sediments, Italian Alps*

| Unit | Age (Ma) | n | P1 | P2 | P3 | Track length | n |
|---|---|---|---|---|---|---|---|
| Modern river | 0 | 114 | 14.2 ± 1.2<br>W = 0.37<br>69% | 19.8 ± 3.6<br>W = 0.34<br>31% | – | 13.98 ± 0.1<br>1.56 | 86 |

Apatite grains were mounted in epoxy resin on glass slides and polished to expose internal grain surfaces. The mounts were then etched in 5 M $HNO_3$ for 20 s at room temperature. With low-uranium mica flakes affixed to the polished and etched mounts, the samples were irradiated at the nuclear reactor at Oregon State University with a nominal fluence of $8 \times 10^{15}$ neutrons $cm^{-2}$. Samples were irradiated with the glass dosimeter CN-1 as well, and the Fish Canyon Tuff and Durango age standards. Grains were counted using a Leitz Ortholux microscope at 1250× (100× dry objective, 1.25 tube factor, and 10× oculars). Ages were calculated using a mean weighted $zeta_{CN-1}$ of 107.0 ± 4.5 (± 1 SE; J.I.G.). Peak ages were calculated using the binomial peak-fitting method of Galbraith & Green (1990) (see Brandon 1996). The number of grains in each analysis is represented by '*n*'.

NE Italy) are between 23 and 28 Ma, indicating that the upper levels of the Bergell plutonic suite cooled rapidly after intrusion and was then brought to the surface, eroded, and deposited in flanking basins (Wagner *et al.* 1979; Giger & Hurford 1989). Fission-track dating of bedrock exposures in the Bergell pluton and adjacent rocks elucidate the Miocene exhumation history (Wagner *et al.* 1977, 1979). These data from elevation traverses indicate that the Miocene cooling history of the area was marked by continuous exhumation with rates of *c.* 410 m $Ma^{-1}$ between about 16 and 14 Ma followed by rates of about 260 m $Ma^{-1}$ between about 14 and 11 Ma (Fig. 6).

A comparison of the detrital FTGA distribution (Fig. 6a) with the results of FT dating of apatite from bedrock exposures (Fig. 6b; filled points, fitted with exhumation rates; from Wagner *et al.* 1977, 1979) provides an instructive comparison. On the plot of bedrock FT ages (Fig. 6b), it should be noted that the upper slope is 410 m $Ma^{-1}$ at an interval of about 14.2 Ma, and the lower slope shows exhumation of *c.* 260 m $Ma^{-1}$ at about 11.7 Ma. Several important features appear in this comparison. First, we note that the young peak age, P1 (14.8 ± 1.2 Ma), is not significantly different from the average of the high-elevation bedrock samples (mean of 14.2 ± 0.7 Ma), which suggests that the bulk of the sediment of this detrital sample was derived from high elevations (*c.* > 1000 m). Second, the older peak age is not well represented in the bedrock data, which could indicate that rocks that cooled at *c.* 20 Ma are not well sampled in this area (presumably these high-elevation rocks are more difficult to reach). Although the peak widths in this sample are wide, they are well fitted to the data because of the high average uranium content and the large number of grains counted (*n* = 115). In this case, we can see that the detrital grain ages provides an excellent sample of the bedrock in the drainage system.

*Erosion rates of the Indus River drainage basin, Pakistan*

In this example, we look at an emerging technique of using each single zircon to estimate long-term erosion rates. We apply this approach to modern sediment dated by Cerverny *et al.* (1998) from the upper part of the Indus River drainage basin in Pakistan, which here extends from the foothills of the Himalaya to the southern margin of the Tibetan Plateau. Although this sort of analysis is under development by our group, it promises to provide a simplified approach to interpreting FTGA. Upstream from the sample collection point, the Indus River drains the Nanga Parbat–Haramosh massif with zircon FT ages of <3.2 Ma, as well as a variety of terranes with much older cooling ages (Zeitler 1985; Cerveny *et al.* 1988). The highest exhumation rates for the Nanga Parbat–Haramosh massif are estimated to be *c.* 5000 m $Ma^{-1}$ but rates for surrounding areas are much lower (Zeitler 1985; Zeitler *et al.* 1993).

Our goal in this example is to estimate spatially averaged erosion rates over the entire drainage basin using a FTGA distribution. Modern erosion rates are commonly estimated from measurements of sediment yield, which is defined as the flux of sediment carried by a river, as normalized by the area of the upstream drainage system. The units used are volume of equivalent solid rock removed per time and area of the drainage system, with units of km $Ma^{-1}$ (i.e. mm $a^{-1}$).

Lag time measures the amount of time needed

to erode the source terrane to a depth equal to the closure depth $Z_c$, which depends on the temperature profile during erosion and the closure properties for the thermochronometer being used. Here we make the assumption that the basal heat flux is relatively constant and that steady erosion is the primary factor controlling the shape of the temperature profile and the rate of cooling of the rocks. There are a few relatively young intrusions in the Nanga Parbat area (Zeitler et al. 1993), erosion rates may be slightly overestimated for this part of the drainage (i.e. zircon FT ages reflect cooling of magma, and not cooling induced by exhumation).

The steps in our analysis are illustrated and explained in Fig. 7a–d. Ceverny et al. (1988) dated two zircon mounts with different etch times; we mixed the results from the mounts in a manner designed to correct for grain-age bias (Garver & Brandon in preparation). This figure emphasizes some of the general aspects involved in transforming the FTGA probability density plot (Fig. 7a) to a plot of the erosion-rate distribution as a function of fractional and cumulative (Fig. 7c and d) drainage area. The transformation of the lag time to erosion rate is based on the relationships shown in Fig. 3.

An additional transformation is needed to account for the fact that erosion rate influences the volume of sediment and ultimately the relative proportion of the dated detrital mineral present in the FTGA distribution. To illustrate this point, let us consider a hypothetical drainage where half the basin is eroding at 500 m Ma$^{-1}$ and the other half at 1000 m Ma$^{-1}$. The two source terranes yield zircon FT ages indicating lag times of 15 Ma and 5 Ma, respectively. The relative flux of sediment will be different by a factor of two because of the factor of two difference in erosion rate. Let us assume that the rocks in each source terrane have approximately the same density of datable zircons. If correct, then we would expect that a probability density plot constructed for a sample of zircon FT grain ages would have two peaks, with the young peak being twice the size of the old peak. This bias can be corrected by dividing the probability density by the associated erosion rate and then renormalizing the density to a probability mass of one. The peaks on this plot would have equal size, reflecting the fact that they each represent equal areas within the drainage system. Without this correction, the estimated average erosion rate would have a substantial upward bias.

Our findings suggest that the mean erosion rate for the Indus River (Fig. 7) is about 560 m Ma$^{-1}$, and a small part of the basin has erosion rates in excess of 1000 m Ma$^{-1}$. This result is surprising given that the drainage system is surrounded by the highest peaks in the world. We infer that high rates are only locally developed in the drainage.

## Exhumation of the Coast Plutonic Complex, Western Canada

Perhaps the most complicated yet most promising of detrital FT techniques involves calculating interval exhumation rates based on mineral pairs from detrital apatite and zircon. In particular, if we were able to look at small differences in $T_c$, we could reduce the time interval over which our long-term exhumation rates are averaged. In this example, we look at how apatite and zircon peak ages can be compared to estimate exhumation rates.

The Tofino basin in southwestern British Columbia (BC) and northwest Washington State contains a nearly continuous sequence of Eocene to Miocene marine strata derived from the Coast Mountains of British Columbia in western North America (Fig. 8). In a recent study of detrital zircon from strata of this basin, Garver & Brandon (1994b) showed that quartzofeldspathic sandstones from this basin record the erosional unroofing of the Coast Plutonic Complex (CPC) and related rocks in the Coast Range Mountains. The zircon fission-track record revealed that component populations became successively younger upsection, indicating progressive unroofing of a metaplutonic source terrane. This progressive unroofing and exhumation produces a well-developed forward moving peak. From these data, Garver & Brandon calculated constant long-term (30–40 Ma) exhumation rates of about 250 m Ma$^{-1}$ throughout much of Tertiary time, which suggests that the erosional system was in a steady state.

Here, we supplement this analysis by presenting apatite fission-track ages for some of the samples reported by Garver & Brandon (1994b) (a full set of data is available as SUP 18128, see p. 290). Then, we estimate exhumation rates using mineral pairs (apatite–surface and apatite–zircon). These model exhumation rates are used.

Strata of the Tofino basin are mostly offshore, but they are locally exposed as a tilted sequence of Eocene and younger marine sedimentary rocks on the Olympic Peninsula of Washington State (Tabor & Cady 1978; Muller et al. 1983; Brandon & Vance 1992; Garver & Brandon 1994b). The BC Coast Mountains are dominated by the CPC, which is composed of Jurassic to Tertiary quartz diorites to granodiorites that

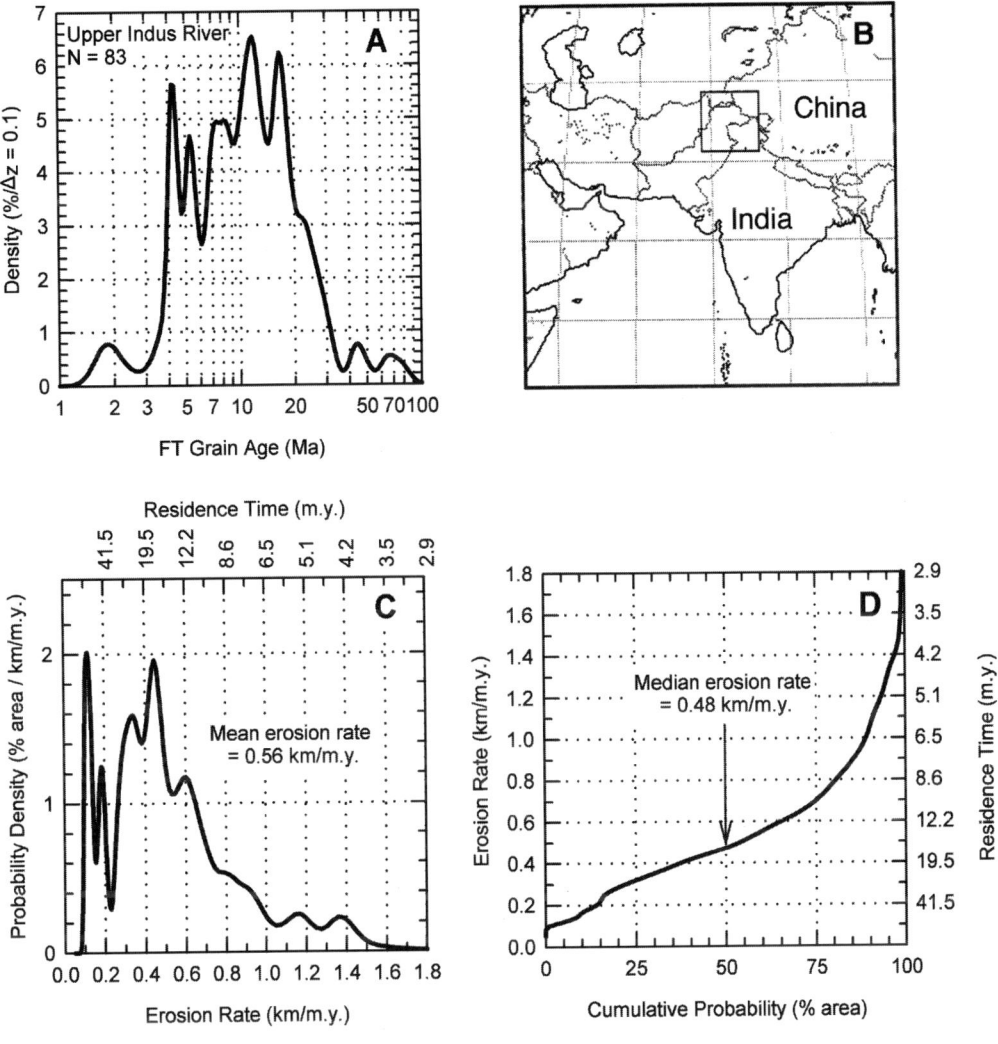

**Fig. 7.** Transformation of a detrital FTGA distribution to an erosion-rate distribution using data from the Indus River in Pakistan (data from Cerveny *et al.* 1988). (**a**) The FTGA distribution for detrital zircons from modern sands of the Indus River in Pakistan, as reported by Cerveny *et al.* (1988). The PD plot was constructed using the method of Brandon (1996). Probability density is given in number of grains per *c.* 10% increments on the grain-age scale. (**b**) Location map for sample. (**c**) The PD plot for erosion rate was converted from the FTGA PD plot using the erosion rate–lag time relationship for zircon FT ages present in Fig. 3. Probability density was normalized to relative area by scaling the transformed PD relative to the inverse of the erosion rate. The mean erosion rate is *c.* 560 m Ma$^{-1}$. (**d**) The cumulative probability of erosion rate by area was calculated by integrating the PD plot in (**c**). The median value (50%) is *c.* 480 m Ma$^{-1}$. This plot indicates that *c.* 10% of the basin has erosion rates faster than 1000 m Ma$^{-1}$.

extend nearly 1500 km from southernmost BC north to Alaska (Roddick 1983). Significant uplift and exhumation of the BC Coast Mountains, particularly pronounced and well documented in the north–central section, is known to have occurred after intense Paleocene–Eocene plutonic and orogenic activity (Fig. 8; Parrish 1983). Following this activity, rapid cooling and exhumation of the central and northern BC Coast Mountains occurred at *c.* 55 Ma (Cook & Crawford 1996) with exhumation rates that may have been as high as *c.* 2000 m Ma$^{-1}$ (Hollister

1982). Following this event, the north-central part of the BC Coast Mountains had apparent exhumation rates of $c.150$ m Ma$^{-1}$ from 25 to 15 Ma (Parrish 1983), but about 250–300 m Ma$^{-1}$ at 35–45 Ma and about 400 m Ma$^{-1}$ at 30–40 Ma. These studies indicate that sediment shed from this north–central source should have a distinct thermal history marked by rapid and widespread cooling at $c.55$ Ma and relatively slow cooling after this event.

We now consider the sedimentary record of this exhumation history (Figs 9 and 10; Tables 2–4). First, we correlate the main peak in each distribution based on the assumption that the main peak contains the majority of the grains, that it becomes younger with time, and that

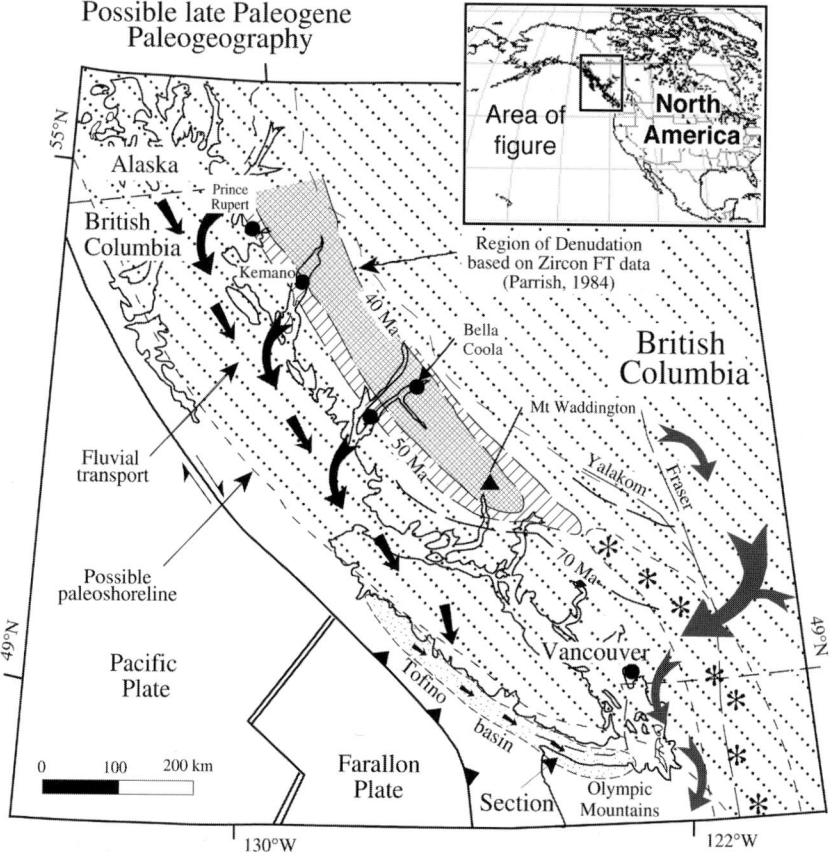

**Fig. 8.** Location map and inferred Paleogene palaeogeography of the Coast Plutonic Complex in western British Columbia, Canada. In this setting, two important sediment sources are identified. The first is sediment derived from the Cascade arc (shown schematically by '*') and a variety of thermotectonic source terranes in the back-arc region. The back-arc sediments mix with sediments from the arc, and ultimately are transported to the forearc. The second is sediment derived from the progressive uplifted and exhumation of the BC Coast Mountains, dominated by dioritic rocks of the Coast Plutonic Complex, which is the focus of our third example. Although most of this sediment from the BC Coast Ranges was shed westward, much of it flowed south along the continental margin and was ultimately deposited in the Tofino basin, which extends as far south as northern Washington State. In the BC Coast Ranges, FT data from bedrock give an important baseline cooling history with which we compare our results (see Parrish 1983). The occurrence of Coast Plutonic Complex-derived sediments in the Tofino basin suggests that the area was largely emergent and fluvial transport was responsible for much of the sediment movement (arrows) to the south. The stratigraphic section from which our samples were derived is labelled 'section', which occurs along the northern edge of the Olympic Mountains in the USA (Washington State).

apatite peaks are younger than the zircon peaks for a particular source. Using this approach, we can compare the main peak in the three uppermost samples (Table 3, Fig. 10). That the lag between the dominant apatite peak age and the dominant zircon peak age becomes smaller with time and is very close in age to the youngest unit examined (Clallam), indicates rapid exhumation at this time (here c. 56 Ma). Next, the peak lag and exhumation rate for these peaks can be calculated for all the peaks (Table 4). The peak lag is the difference between the peak ages (i.e. for apatite and zircon) or the difference between the peak age and deposition. Both estimates give us an indication how long it took for the source rock to move between one datum and the next (Table 4).

Calculated exhumation rates range between about 100 and 1500 m Ma$^{-1}$ for the main peak. This peak is attributed to unroofing of the Coast Plutonic Complex, which has estimated exhumation rates of c. 120 and 230 m Ma$^{-1}$ during the interval between c. 34 and 54 Ma (Parrish 1983). The zircon–apatite peaks in the Clallam Formation suggest exhumation rates of about 1500 m Ma$^{-1}$ at c. 56 Ma. This result is consistent with other studies that suggest rapid cooling occurred at about this time, or slightly younger, with exhumation rates as high as c. 2000 m Ma$^{-1}$ (Hollister 1982; Cook & Crawford 1996).

The preceding discussion demonstrates several important aspects of using apatite and zircon FT chronology together. Because peak widths vary according to age and uranium concentration, apatite peaks are almost always more poorly resolved than zircon peaks. Our experience suggests that this difference in precision makes comparisons between apatite and zircon FT ages possible only if (a) apatite statistics are improved by counting a large number of grains (100 or more) present in the distribution, (b) there are only a few peaks (one to two), and (c) peaks are well separated in time (by more than 20–30% of their peak ages).

## Exhumation of Fiordland, South Island, New Zealand

Rocks of Fiordland in New Zealand consist of a Palaeozoic to Mesozoic metaplutonic complex that has experienced episodic exhumation since Cretaceous time, which is well recorded in the flanking strata of the Te Anau basin (Figs 11 and 12). In this example, we show how fission-track ages of detrital apatite samples record the exhumation of Fiordland during the establishment of the modern plate regime which is dominated by the Alpine Fault, a major oblique-slip fault that separates the Pacific and Australian plates. This example illustrates how strata from the perimeter of a basin can be used to avoid problems with thermal resetting, which would be expected for detrital apatite in a central more deeply buried part of the basin (Fig. 1). In the case of the Te Anau basin, most of the basin fill has been buried deeply enough to have thermally affected apatite, and most of the provenance information has been lost because of thermal resetting.

The Te Anau basin has been a long-lived depocentre from Eocene to Recent time, and as a result it has received some 8000 m of sediments from adjacent terranes (Turnbull et al. 1993). The Fiordland block, in which plutonic and metaplutonic rocks form the basement to the western edge of the basin, has been a major contributor of detritus to the basin. One of the main source regions for sediment in the Te Anau basin was the Fiordland block. In Late Eocene time, subsidence rates accelerated, but in Early Oligocene time, the basin received a thick sequence of marine sediments derived from Fiordland (Turnbull et al. 1993). It is widely thought that the Eocene to Oligocene phase of sedimentation was driven by a new transtensional regime set up during plate reorganization at this time (e.g. Stock & Molnar 1982). Here and elsewhere in New Zealand, Middle to Upper Oligocene strata are dominated by carbonate rocks, which suggests a period of tectonic quiescence. Lower to Middle Miocene sediments of the Te Anau basin record deep-marine conditions in the centre of the basin but shallow marine facies along its western margin. Although much of the detritus in this interval is mudstone, some units on the western margin, including the Borland Formation (Fig. 12), are coarse grained and derived from Fiordland (Turnbull et al. 1993). In late Mid-Miocene to Pliocene time, the Prospect Formation records an enormous increase in sediment derived from adjacent terranes. Palaeocurrents and pebble types suggest that for the Prospect Formation, palaeoflow was generally to the south. The source region included terranes to the northeast (Caples), but also Fiordland rocks (Turnbull et al. 1993). This final phase of sedimentation is related to contraction and uplift along the Alpine strike-slip fault system.

An important aspect of our analysis here, and the reason that we selected samples along the basin periphery, is that although the total fill of the Te Anau basin is thick in the centre of the basin, successive phases of uplift have resulted in onlap of the strata onto the edge of the

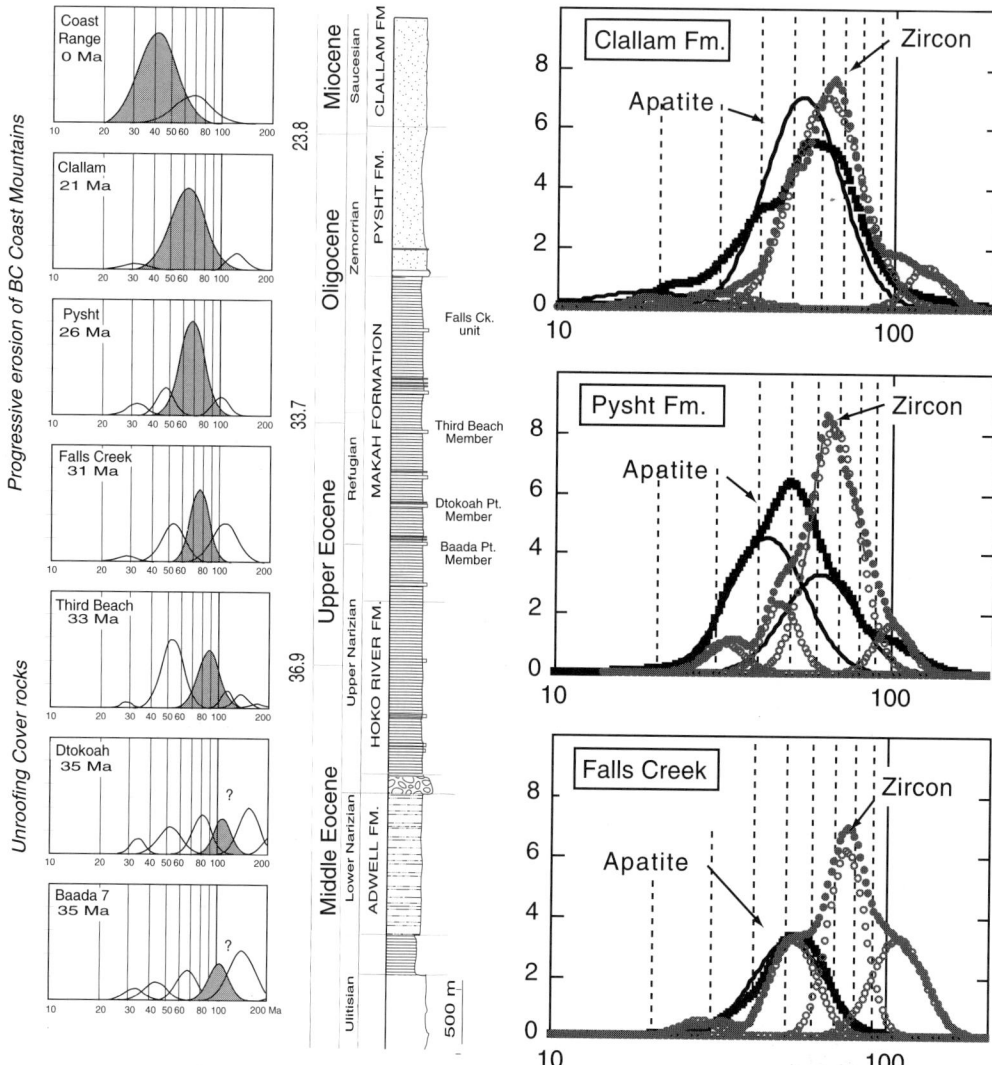

**Fig. 9.** Stratigraphy of part of the Tofino basin (in Washington State, USA) showing the forward moving peaks of FT grain ages of detrital zircon (modified from Garver & Brandon 1994b). It should be noted that the uppermost plot shows all data from the northern Coast Plutonic Complex (bedrock, from Parrish 1983), treated as if all dated grains were detrital (see Garver & Brandon 1994b). A comparison of apatite and zircon peak ages for the upper three units is shown in Fig. 10.

**Fig. 10.** Comparison of zircon and apatite peak ages from the three upper units of the Tofino basin exposed on the northern flanks of the Olympic Peninsula in Washington State (USA). Samples shown from oldest at the bottom to youngest at the top. Sediments in these units are inferred to have been derived from the Coast Plutonic Complex. The FT ages record the progressive erosion of this area through time (Garver & Brandon 1994b). Shown are the PD plot for zircon (○) and apatite (■) along with the fitted peak ages for zircon (●) and apatite (continuous lines). It should be noted that the difference in the main peak age between apatite and zircon becomes shorter with time, which is inferred to represent the exhumation of a rapidly cooled section of upper crust at c. 56 Ma.

**Table 2.** Comparison of zircon and apatite FT peak ages from detritus shed from the BC Coast Ranges, Canada

| Mineral | Age range (Ma) | P1 | P2 | P3 | P4 | P5 |
|---|---|---|---|---|---|---|
| *Clallam Formation (depositional age is 17.7–24.1 Ma)* | | | | | | |
| Zircon ($N_t = 51$) | 19–147 | 22.5 ± 1.9 (4%) $W = 0.25$ | 58.9 ± 4.7 (68%) $W = 0.25$ | 77.3 ± 6.8 (15%) $W = 0.25$ | 117.8 ± 9.3 (26%) $W = 0.25$ | – |
| Apatite ($N_t = 50$) | 08–127 | 18.1 ± 3.4 (8%) $W = 0.33$ | 53.3 ± 2.5 (92%) $W = 0.26$ | – | – | – |
| *Pysht Formation (depositional age is 24.1–28.5 Ma)* | | | | | | |
| Zircon ($N_t = 50$) | 29–110 | 37.3 ± 2.8 (9%) $W = 0.22$ | 63.8 ± 4.6 (74%) $W = 0.22$ | 92.9 ± 6.8 (18%) $W = 0.22$ | – | – |
| Apatite ($N_t = 50$) | 25–109 | – | 42.9 ± 4.9 (59%) $W = 0.26$ | 62.1 ± 10.9 (41%) $W = 0.25$ | – | – |
| *Falls Creek Unit (depositional age is 28.5–32.7 Ma)* | | | | | | |
| Zircon ($N_t = 50$) | 26–136 | 28.7 ± 2.1 (4%) $W = 0.21$ | 50.5 ± 3.6 (17%) $W = 0.21$ | 72.1 ± 5.2 (46%) $W = 0.21$ | 103.4 ± 7.4 (27%) $W = 0.21$ | 135.4 ± 9.9 (6%) |
| Apatite ($N_t = 20$) | 18–110 | – | – | 51.5 ± 2.9 (100%) $W = 0.25$ | – | – |

For zircon (see Garver & Brandon 1994b), fractions were hand picked, and then two equal portions were mounted in FEP Teflon, polished, and etched in a KOH–NaOH eutectic at 225°C for 12–20 h; one mount was given a 'long' etch, and the other was given a 'short' etch (see Naeser *et al.*, 1987). The etched mounts were irradiated at the Oregon State nuclear reactor using a nominal fluence of $2 \times 10^{15}$ neutrons cm$^{-2}$. The internal gradient within the irradiated package was estimated by interpolating track densities for fluence monitors (CN-5) placed within of the irradiation tube. Fission tracks were counted at 1250× (100× oil immersion objective, 1.0 tube factor, 12.5× oculars) on an Olympus BH-2 microscope. Ages were calculated using an mean weighted zeta$_{CN-5}$ of 323.5 ± 9 (±1 SE; J.I.G. at Union). Apatite grains were mounted in epoxy resin on glass slides and polished to expose internal grain surfaces. The mounts were then etched in 5 M HNO$_3$ for 20 s at 21°C. Samples were irradiated at the Oregon State nuclear reactor with a nominal fluence of $5 \times 10^{15}$ neutrons cm$^{-2}$. Samples were irradiated with the glass dosimeter CN-1 as well and the Fish Canyon Tuff, Durango and Mt Dromedary apatite age standards. Apatite grains were counted using a Leitz Ortholux microscope at 1600× (160× dry objective and 10× oculars). Ages were calculated using a zeta$_{CN-1}$ of 105.8 ± 2.5 (±1 SE; M.R.-T). FT data were decomposed using a peak-fitting routine of Brandon (1992). In this routine, grain ages are calculated, the grains are ordered in increasing age and the resultant distribution of grain ages is fitted with a probability density distribution. The probability distribution invariably represents several component populations, which must be extracted using a peak-fitting routine, which fits individual peaks to the probability density plot. For zircon we use the Gaussian peak-fitting routine of Brandon (1992), and for apatite we use the binomial fitting routine of Galbraith & Green (1990).

**Table 3.** *Summary of principle detrital peaks used in model calculations, detritus from the BC Coast Ranges, Canada*

|  |  | Peak 1 |  |
|---|---|---|---|
| Clallam | Z | 58.9 | (68) |
|  | A | 53.3 | (92) |
|  | S | 21.0 |  |
| Pysht | Z | 63.8 | (74) |
|  | A | 42.9 | (49) |
|  | S | 26.0 |  |
| Falls Ck | Z | 72.1 | (46) |
|  | A | 51.5 | (100) |
|  | S | 31.0 |  |

Z, Zircon; A, apatite; S, surface or depositional age. Numbers in parentheses are the percentage of grains that make up the peak.

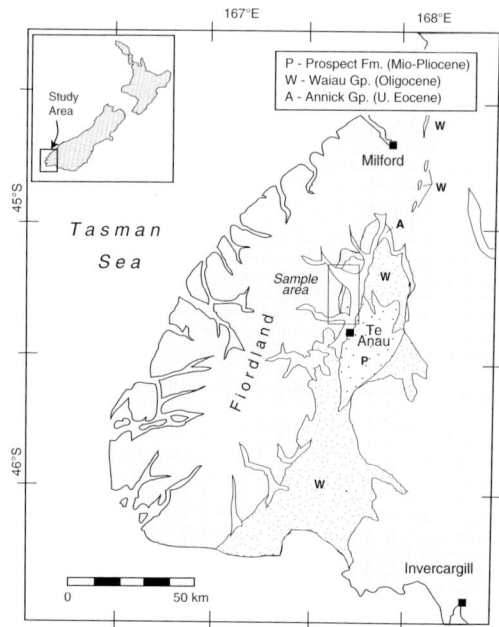

**Fig. 11.** Simplified geological map of the Fiordland area in the southern part of the South Island, New Zealand, showing basement rocks and adjacent basin strata of the Te Anau basin. P, Prospect Formation; W, Waiau Group throughout most of the basin but also includes the Nightcaps Group along the SE edge of the exposed basin; A, Annick Group; B, basement terranes including only locally minor exposures of Eocene to Cretaceous strata (southern tip of Fiordland; see Fig. 12 for ages and stratigraphy). (Modified from Turnbull *et al.* 1993.)

Fiordland block. The result is that basin-margin strata are relatively condensed in thickness. Thus the section there preserves a long record without the complication of thermal annealing of the apatites.

Density plots with fitted peaks are shown for samples with depositional ages spanning some 15 Ma from *c.* 34 to 20 Ma (Fig. 12; Table 5). This suite of samples records a 15 Ma interval that includes initial rifting, subsequent waning of tectonic activity, and finally the establishment of the Alpine Fault system at *c.* 23 Ma. The data have several important features and the results can be evaluated into two groups; a group deposited between *c.* 35 and 32 Ma and one deposited between *c.* 29 and 20 Ma. The older samples contain two peaks (P1 and P2) that become older with time (Fig. 13). This is an unusual situation because generally FT peak ages move forward as a source is exhumed with time. Progressively older component ages in

**Table 4.** *Model exhumation rates based on comparing main peak age in apatite and zircon from detritus from the BC Coast Ranges, Canada*

| Unit | Age (or $t_d$) (Ma) | Pair | Lag time (Ma) | Midpoint (Ma) | Z (km) | $\dot{\varepsilon}_{model}$ (m Ma$^{-1}$) |
|---|---|---|---|---|---|---|
| Pysht | 26 | AS | 16.8 | 34.5 | 4.5 | 266 |
| Clallam | 21 | AS | 32.3 | 37.5 | 4.5 | 140 |
| Falls Creek | 31 | AS | 20.5 | 41.3 | 4.5 | 220 |
| Pysht | 26 | ZA | 20.9 | 53.4 | 4.9 | 234 |
| Clallam | 21 | ZA | 2.8 | 56.1 | 4.9 | 1600 |
| Falls Creek | 31 | ZA | 20.6 | 61.8 | 4.9 | 238 |

Pair refers to the thermochronometers used (ZA, zircon–apatite pair; AS, apatite–surface pair). Lag time is the peak lag (PL$_1$ Ma) (peak age – depositional age) for samples going to the surface; peak age minus peak age for mineral pairs) for individual samples. This number represents the time it took rock to pass from one known isotherm to either the surface or another known isotherm. Midpoint ages represent the time between closure of each system, and the table is sorted on midpoint ages because these show model erosion rates through time. Depth is the crustal thickness based on model assumptions as outlined in text. $\dot{\varepsilon}_{model}$ represents the model rate calculated for the interval; based on a generic geotherm of 25°C km$^{-1}$ as outlined in text.

the these samples were deposited about 9 Ma apart. The occurrence of a peak age in strata that have essentially the same age (lag time c. 1 Ma or less; see Table 5) suggests that exhumation rates were extremely rapid at this time (c. 29 Ma), probably in excess of 2000 m Ma$^{-1}$ (Fig. 3). In the youngest strata, the peak age remains about the same (c. 29 Ma) and the peak lag is about 7 Ma. In this case, exhumation of the source terrane must have slowed abruptly to below 500 m Ma$^{-1}$ as this would be the average rate over this interval (between 29 and 20 Ma); otherwise, if exhumation was progressive through this time, the peak age would have progressively younged with time (a typical 'moving peak'). In previous work, we have referred to this as a static peak and have interpreted the peak age to represent the time of rapid crustal quenching (Brandon & Vance 1992; Garver & Brandon 1994a,b). This pattern is the hallmark of extensional exhumation where a substantial thickness of crustal section is rapidly cooled because of normal faulting.

The reduction in exhumation rates of P1 is supported by the fact that during this interval (c. 27–23 Ma; Late Oligocene) most of New Zealand was covered by shallow-marine limestones. Only after initiation of the Alpine Fault (c. 23 Ma) did rocks retaining this 'static peak' become uplifted, deeply eroded, and shed into adjacent basins. The onset of motion on the Alpine Fault is marked by widespread clastic sedimentation in New Zealand. Therefore, the erosion of rocks with this cooling age appears to have been episodic. For example, at c. 20 Ma, the lag time for P1 is about 7 Ma, giving an average exhumation rate over this entire interval of about 500 m Ma$^{-1}$. None the less it seems likely that the instantaneous exhumation rate during this interval could have been much greater (> 1000 m Ma$^{-1}$) sometime during this 7 Ma interval. Based on what we know about the tectonic history of New Zealand, rapid exhumation was likely at the c. 29 Ma because of rifting and at c. 23–20 Ma because of inception of the Alpine Fault.

In this example, we have shown several important aspects of using detrital apatite to understanding exhumation. First, it is important to sample rocks that have not been thermally annealed. In this regard, condensed sections around the perimeter of a basin are very attractive candidates. Second, peak ages with short lag times indicate rapid exhumation. Third, erosion of 'cover rocks' or the 'dead zone' must precede exhumation of deeply buried rocks (sub-PAZ). The dead zone does not contain a thermal signal of the exhumation event and several possibilities

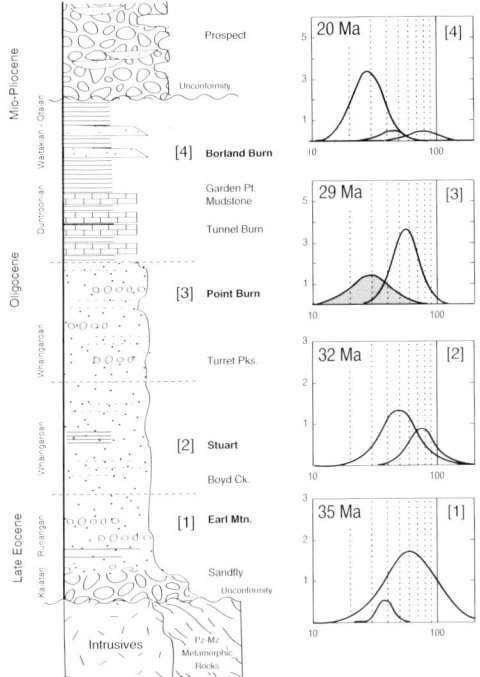

**Fig 12.** Stratigraphy and FT peak ages for detrital apatite from strata on the flanks of the Fiordland block in the southern part of the South Island, New Zealand (see Table 5). In this upsection progression, we recognize two suites of samples. The first suite (35 and 32 Ma) is inferred to represent the initial unroofing of cover strata, and it is possible that this group has backward-moving peaks as explained in the text. The second suite (29–20 Ma) shows the emergence and then dominance of a young peak that we infer to be associated with exposure of reset rocks that resided below the partial annealing zone before exhumation. This young peak (P1, shaded, in both the 29 and 20 Ma samples) does not become significantly younger with time, which suggests that rapid exhumation occurred at c. 30 Ma (>5000 m Ma$^{-1}$) followed by slow exhumation between 29 and 20 Ma (c. <500 m Ma$^{-1}$). Stratigraphy is a schematic representation derived from data and descriptions of Turnbull (1985), and Turnbull et al. (1993).

younger and younger strata, referred to as a backward moving peak, may represent the downward removal of cover strata that have an unroofing sequence preserved in their layers.

The younger two samples record very rapid exhumation of Fiordland rocks at about 30 Ma, followed by slow and continuous erosion. In these two younger samples, the young (P1) age is about the same (c. 29 Ma) despite the fact that

**Table 5.** *Apatite FT peak ages for the Te Anau basin, Fiordland, New Zealand*

| Unit | NZ stage | Age (Ma) | Range (Ma) | n | P1 | P2 | P3 | Track length | SD | (n) |
|---|---|---|---|---|---|---|---|---|---|---|
| Waiau Gp | Otaian | c. 20.5 | (18.7–22.0) | 32 | 28.5 ± 1.7<br>$W = 0.30$<br>79% | 45.7 ± 8.4<br>$W = 0.28$<br>11% | 77.8 ± 12.4<br>$W = 0.28$<br>10% | 13.36 ± 0.58<br>13.20 ± 0.28 | 1.15<br>1.41 | (05)<br>(61) |
| Waiau Gp | Upper Whaingaroan | c. 29.0 | (28.0–30.0) | 36 | 29.4 ± 3.2<br>$W = 0.37$<br>37% | 56.4 ± 3.2<br>$W = 0.25$<br>63% | – | 11.54 ± 0.40<br>11.43 ± 0.28 | 2.18<br>2.18 | (29)<br>(60) |
| Waiau Gp | Lower Whaingaroan | c. 32.0 | (30.0–34.0) | 17 | 49.1 ± 5.2<br>$W = 0.35$<br>68% | 75.9 ± 5.2<br>$W = 0.25$<br>32% | – | 11.40 ± 0.54<br>11.49 ± 0.47 | 1.85<br>1.15 | (29)<br>(07) |
| Annick Gp | Runangan | 35.0 | (34.0–36.0) | 22 | 38.1 ± 2.8<br>$W = 0.14$<br>8% | 61.9 ± 2.7<br>$W = 0.47$<br>91% | – | 11.95 ± 0.27 | 1.76 | (42) |

Apatite grains were mounted in epoxy resin on glass slides and polished to expose internal grain surfaces. The mounts were then etched in 5 M $HNO_3$ for 20 s at room temperature. With low-uranium mica flakes affixed to the polished and etched mounts, the samples were irradiated at the X-7 facility of the HIFAR reactor, Lucas Heights, Australia, (following procedures outlined by Kamp *et al.* (1992)) with a nominal fluence of $10 \times 10^{15}$ neutrons $cm^{-2}$. Samples were irradiated with the glass dosimeter SRM 612 as well, and the Fish Canyon Tuff, Durango and Mt Dromedary age standards. Grains were counted using a Leitz Ortholux microscope at $1250\times$ ($100\times$ dry objective, 1.25 tube factor, and $10\times$ oculars). Ages were calculated using a mean weighted zeta$_{SRM612}$ of 348.7 ± 4.3 (±1 SE; I.W.S.W.) or 343.5 ± 4.5 (±1 SE; I.J.L.). Data from two separate samples were combined for each analysis, and track lengths are shown for each sample, with the exception of one sample of the Annick Group in which no lengths were measured. Peak ages calculated using a binomial statistical analysis, with the exception of the oldest sample, which was fitted using a Gaussian approximation because it is poorly approximated by binomial statistics (for procedure and program see Brandon (1996)). Depositional age approximated from regional stratigraphy. $W$ is the fraction of the half-width of the peak. The percentage of grains that compose each peak are shown below $W$. The number of grains in each analysis is represented by $n$.

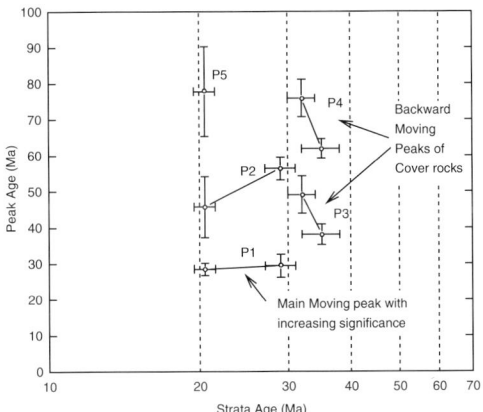

**Fig. 13.** Relationship between peak age and strata age for apatite FT ages of the Te Anau basin, Fiordland, New Zealand. It should be noted that the older strata contain older peak ages inferred to have been derived from reworking of sedimentary rocks. The younger peaks, especially P1, are inferred to represent the exhumation of plutonic rocks of Fiordland.

exist as the detrital FT pattern that may result from erosion of this crustal layer. In some cases, older PAZs fossilized in previous exhumation events may give younger and younger peak ages as erosion uncovers deeper levels. This situation results in forward moving peaks, which track successive incision into the dead zone. In other cases, the dead zone may consist of basin strata that have not been thermally annealed. In this case, the resulting pattern of FT grain ages can be complicated because of recycling of older unreset detritus. It is possible that in the case of exhumation of basin strata, backward moving peaks may result from erosion of cover strata that have a pre-existing unroofing signature (Garver & Brandon 1994b).

## Conclusions and future directions

These studies highlight the advantages and approach of detrital FTGA analysis. A few general observations can be made about application of this approach to understanding exhumation. First, the evolution of exhumation is best evaluated when samples are collected from a well-dated stratigraphic section that spans the exhumation event. Second, a condensed section from the perimeter of a basin is most likely to preserve unreset detrital grains and therefore give provenance information. This problem is especially acute for apatite. Third, during erosion, the removal of cover rocks or a 'dead zone' precedes the exposure of rocks with young cooling ages. Fourth, fast exhumation results in short lag times and slow exhumation results in a long lag time. Steady-state conditions are reflected in a constant lag time. Fifth, for most source regions with several distinct cooling ages, about 50 grain FT grain ages will capture the principle grain-age populations, but more grain ages will give better resolution. In cases where zircon and apatite ages from the same sample are compared, the comparison is easiest when samples have high-uranium apatite grains (>50 ppm), and numerous apatite grains have been counted (>100).

A variety of studies are required to advance detrital FT thermochronology. To quantitatively evaluate FT ages of different mineral systems, future work needs to address the relative proportion of apatite and zircon grains derived from various source rock lithologies, as well as the fate of these minerals in transport, burial, and diagenesis. In particular, there is a need to obtain a full quantitative sampling of FT grain ages free from procedural biases. Although probably small, there are several important aspects of etching and counting that need to be evaluated in this context. Finally, a better understanding of the thermal history of individual grains and effective closure temperature of each individual grain will also advance this approach. This avenue of research involves understanding the annealing characteristics of individual grains through crystal-specific constraints based on chemistry, track lengths, or etch pit size in apatite. In zircon, there is a need to resolve the effective closure temperature of zircon as well as the relationship between alpha damage and closure temperature.

This work was funded directly and indirectly by National Science Foundation grant EAR 9614730 and EAR 9418989 (to J.I.G.), EAR 9005777 and EAR 9418989 to M.T.B., and EAR 9117941 to M.R.-T. Irradiation was funded, in part, by the Reactor Use Sharing (R.U.S.) programme awarded to B. Dodd and the Oregon State Reactor Facility. M. E. Bullen collected and separated the sample from the Italian Alps. J.I.G. gratefully acknowledges logistical support of the Department of Earth Sciences at the University of Waikato, New Zealand, during a 1997 sabbatical leave. This paper benefited from helpful review by G. Lister and three anonymous reviewers.

## References

BALDWIN, S. L., HARRISON, T. M. & BURKE, K. 1986. Fission-track evidence for the source of accreted sandstones, Barbados. *Tectonics*, 5, 457–468.

BRANDON, M. T. 1992. Decomposition of fission-track grain age distributions. *American Journal of Science*, **292**, 535–564.

—— 1996. Probability density plot for fission-track grain-age samples. *Radiation Measurements*, **26**, 663–676.

—— & VANCE, J. A. 1992. New statistical methods for analysis of fission-track grain-age distributions with applications to detrital zircon ages from the Olympic subduction complex, western Washington State. *American Journal of Science*, **292**, 565–636.

——, RODEN-TICE, M. K., & GARVER, J. I. 1998. Late Cenozoic exhumation of the Cascadia accretionary wedge in the Olympic Mountains, northwest Washington State. *Geological Society of America Bulletin*, **110**, 985–1009.

BROWN, R., SUMMERFIELD, M. A. & GLEADOW, A. J. W. 1994. Apatite fission-track analysis: its potential for the estimation of exhumation rates and implications for models of long-term landscape development. *In:* KIRKBY, M. J. (ed.) *Process Models and Theoretical Geomorphology.* Wiley, New York, 23–53.

CARTER, A. 1996. The application of FT provenance studies to paleogeographic reconstructions: an example from Thailand (abstract). *In: Abstracts for the International Workshop on Fission-track dating, Gent 1996.* Geological Institute and Institute for Nuclear Sciences of the University of Gent, 19.

CERVENY, P. F. 1988. *Uplift and erosion of the Himalaya over the past 18 million years: evidence from fission track dating of detrital zircons and heavy mineral analysis.* MS thesis, Dartmouth College, Hanover, NH.

——, NAESER, N. D., ZEITLER, P. K., NAESER, C. W. & JOHNSON, C. W. 1988. History of uplift and relief of the Himalaya during the past 18 million years: evidence from fission-track ages of detrital zircons from sandstones of the Siwalik Group. In: KLEINSPEHN, K. & PAOLA, C. (eds) *New Perspectives in Basin Analysis,* Springer, New York, 43–61.

CLARK, S. P., JR & JÄGER, E. 1969. Exhumation rates in the Alps from geochronologic and heat flow data. *American Journal of Science*, **267**, 1143–1160.

COHEN, H. A., ONSTOTT, T. C., LUNDBERG, N. & HALL, C. M. 1990. 40/39 Ar laser probe dating of detrital phenocrysts to constrain the age of volcanism, Gravina Belt, SE Alaska. *EOS Transactions, American Geophysical Union*, **71**, 1617.

COOK, R. D. & CRAWFORD, M. L. 1996. Exhumation and tilting of the western metamorphic belt of the Coast Orogen in southern southeastern Alaska. *Tectonics.*

COPELAND, P. & HARRISON, T. M. 1990. Episodic rapid uplift in the Himalaya revealed by $^{40}Ar/^{39}Ar$ analysis of detrital K-feldspar and muscovite, Bengal Fan. *Geology*, **18**, 354–357.

——, —— & HEIZLER, M. T. 1990. $^{40}Ar/^{39}Ar$ single-crystal dating of detrital muscovite and K-feldspar from Leg 116, southern Bengal Fan: implications for the uplift and erosion of the Himalayas. *In:* COCHRAN, J. R. & STOW, D. A. V. (eds) *Proceedings of the Ocean Drilling Program, Scientific Results*, **116**, Ocean Drilling Program, College Station, TX, 93–114.

CORRIGAN, J. D. & CROWLEY, K. D. 1990. Fission-track analysis of detrital apatites from sites 717 and 718, Leg 116, Central Indian Ocean. *In:* COCHRAN, J. R. & STOW, D. A. V. (eds.) *Proceedings of the Ocean Drilling Program, Scientific Results*, **116**, 75–87.

DALLMEYER, R. D., KEPPIE, J. D. & NANCE, R. D. 1997. $^{40}Ar/^{39}Ar$ ages of detrital muscovite within Lower Cambrian and Carboniferous clastic sequences in northern Nova Scotia and southern New Brunswick: applications for provenance regions. *Canadian Journal of Earth Sciences*, **34**, 156–168.

DILL, H.-G. 1994. Can REE patterns and U-Th variations be used as a tool determine the origin of apatite in clastic rocks? *Sedimentary Geology*, **92**, 175–196.

DODSON, M. E. 1979, Theory of cooling ages. *In:* JAEGER, E. & HUNZIKER, J. C. (eds) *Lectures in Isotope Geology.* Springer, Berlin, 194–202.

DUNKL, I., FRISCH, W., KUHLEMANN, J. & BRÜGEL, A. 1996. 'Combined pebble dating': a new tool for provenance analysis and for estimating alpine exhumation (abstract). *In: Abstracts for the International Workshop on Fission-track Dating, Gent 1996.* Geological Institute and Institute for Nuclear Sciences of the University of Gent, 32.

FLEISCHER, R. L., PRICE, P. B. & WALKER, R. M. 1975. *Nuclear Tracks in Solids.* University of California Press, Berkeley, CA.

FRISBEE, A. J. 1995. *A fission-track study of detrital zircons from Pacific Northwest river sediments.* BSc thesis, Union College, Schenectady, NY.

GALBRAITH, R. F. 1988. Graphical display of estimates having differing standard errors. *Technometrics*, **30**, 271–281.

—— & GREEN, P. F. 1990. Estimating the component ages in a finite mixture. *Nuclear Tracks and Radiation Measurements*, **17**, 197–206.

GARVER, J. I. 1988. Stratigraphy and tectonic significance of the clastic cover to the Fidalgo Ophiolite, San Juan Islands, Washington. *Canadian Journal of Earth Sciences*, **25**, 417–423.

—— & BRANDON, M. T. 1994a. Fission-track ages of detrital zircon from mid-Cretaceous sediments of the Methow–Tyaughton basin, southern Canadian Cordillera. *Tectonics*, **13**, 401–420.

—— & —— 1994b. Erosional exhumation of the British Columbia Coast Ranges as determined from fission-track ages of detrital zircon from the Tofino basin, Olympic Peninsula, Washington. *Geological Society of America Bulletin*, **106**, 1398–1412.

——, ——, RODEN, M. K. & ARCHIBALD, D. A. 1993. Time-integrated history of cooling and exhumation of the Coast Plutonic Complex, B.C., based on isotopic ages of detrital minerals. *Geological Society of America Abstracts with Programs*, **25**, 172–173.

GIGER, M. & HURFORD, A. J. 1989. Tertiary intrusives of the Central Alps: their Tertiary uplift, erosion, redeposition and burial in the south Alpine foreland. *Eclogae Geologicae Helvetiae*, **82**, 857–866.

GLEADOW, A. J. W., DUDDY, I. R., GREEN, P. F. &

LOVERING, J. F. 1986. Confined fission track lengths in apatite: a diagnostic tool for thermal history analysis. *Contributions to Mineralogy and Petrology*, **94**, 405–415.

GREEN, P. F., DUDDY, I. R., GLEADOW, A. J. W. & TINGATE, P. R. 1985. Fission track annealing in apatite: track length measurements and the form of the Arrhenius plot. *Nuclear Tracks and Radiation Measurements*, **10**, 323–328.

——, ——, LASLETT, G. M., HEGARTY, K. A., GLEADOW, A. J. W. & LOVERING, J. F. 1989. Thermal annealing of fission tracks in apatite: 4. Quantitative modelling techniques and extension to geological timescales. *Chemical Geology (Isotope Geosciences Section)*, **79**, 155–182.

GRIST, A. M., REYNOLDS, P. H. & ZENTILLI, M. 1990. Provenance and thermal history of detrital sandstones of the Scotian Basin, offshore Nova Scotia, using apatite fission track and $^{40}Ar/^{39}Ar$ methods. *Atlantic Geology*, **26**, 171.

HANSMANN, W. 1996. Age determinations of the Tertiary Masino–Bregaglia (Bergell) intrusives (Italy, Switzerland) – a review. *Schweizerische Mineralogische und Petrographische Mitteilungen*, **76**, 421–451.

HASEBE, N., TAGAMI, T. & NISHIMURA, S. 1993. Evolution of the Shimanto accretionary complex: a fission-track thermochronological study. *In*: UNDERWOOD, M. B. (ed.) *Thermal Evolution of the Tertiary Shimanto Belt, Southwest Japan: an example of Ridge–Trench Interaction*. Geological Society of America, Special Paper, **273**, 121–136.

HOLLISTER, L. S. 1982. Metamorphic evidence for rapid (2mm/yr) uplift of a portion of the Central Gneiss complex, Coast Mountains, B.C. *Canadian Mineralogist*, **20**, 319–332.

HORSTMANN, U. E. 1987. *The metamorphic evolution of the Damara Orogeny, Namibia, as deduced from K/Ar dating of detrital white mica from molasse sediments of the Nama Group* (in German). Goettinger Arbeiten zur Geologie und Palaeontologie, **32**.

HURFORD, A. J. 1986. Cooling and uplift patterns in the Lepontine Alps, south central Switzerland, and an age of vertical movement on the Insubric fault line. *Contributions to Mineralogy and Petrology*, **92**, 413–427.

——, FITCH, F. J. & CLARKE, A. 1984. Resolution of the age structure of the detrital zircon populations of two Lower Cretaceous sandstones from the Weald of England by fission-track dating. *Geological Magazine*, **121**, 269–277.

JOHNSON, C. & LONERGAN, L. 1996. Quantifying the exhumation history of Tertiary mountain belts from stratigraphic, provenance, and fission-track studies (abstract). *In*: *Abstracts for the International Workshop on Fission-track Dating, Gent 1996*. Geological Institute and Institute for Nuclear Sciences of the University of Gent, 61.

KAMP, P. J. J., GREEN, P. F. & TIPPETT, J. M. 1989. Tectonic architecture of the Mountain front–foreland basin transition, assessed by fission track analysis; fission track analysis reveals character of collisional tectonics in New Zealand. *Tectonics*, **8**, 169–195.

KASUYA, M. & NAESER, C. W. 1988, The effect of a-damage on fission-track annealing in zircon. *Nuclear Tracks and Radiation Measurements*, **14**, 477–480.

KOWALLIS, B. J., HEATON, J. S. & BRINGHURST, K. 1986. Fission-track dating of volcanically derived sedimentary rocks. *Geology*, **14**, 19–22.

MCGOLDRICK, P. J. & GLEADOW, A. J. W. 1977. Fission track dating of lower Palaeozoic Sandstones at Tatong, North Central Victoria. *Journal of the Geological Society of Australia*, **24**, 461–464.

MARANVILLE, R. E. 1993. Fission-track dating of detrital zircons from modern rivers in the Pacific Northwest: implications for the provenance of the Olympic subduction complex. *Green Mountain Geologist*, **20**, 6.

MULLER, J. E., SNAVELY, P. D., JR & TABOR, R. W. 1983. *The Tertiary Olympic Terrane, Southwest Vancouver Island and Northwest Washington*. Geological Association of Canada, Mineralogical Association of Canada, Canadian Geophysical Union, Field Trip Guidebook, Trip 12.

NAESER, C. W. 1976. *Fission-track Dating*. US Geological Survey Open File Report **76–190**.

——, ZIMMERMAN, R. A. & CEBULA, G. T. 1981. Fission track dating of apatite and zircon: an interlaboratory comparison. *Nuclear Tracks*, **5**, 65–72.

NAESER, N. D., NAESER, C. W. & MCCULLOH, T. H. 1989. The application of fission-track dating to the depositional and thermal history of rocks in sedimentary basins. *In*: NAESER, N. D. & MCCULLOH, T. H. (eds) *Thermal History of Sedimentary Basins, Methods and Case Histories*. Springer, New York, 157–180.

——, ZEITLER, P. K., NAESER, C. W. & CERVENY, P. F. 1987. Provenance studies by fission-track dating – etching and counting procedures. *Nuclear Tracks and Radiation Measurements*, **13**, 121–126.

PARRISH, R. R. 1983. Cenozoic thermal evolution and tectonics of the Coast Mountains of British Columbia 1: Fission-track dating, apparent uplift rates and patterns of uplift. *Tectonics*, **2**, 601–631.

PEREIRA, A. J. S. C., CARTER, A., HURFORD, A. J., NEVES, L. J. P. F. & GODINHO, H. M. 1996. Evidence for the unroofing history of Hercynian granitoids in central Portugal derived from Mesozoic sedimentary zircons (abstract). *In*: *Abstracts for the International Workshop on Fission-track dating, Gent 1996*. Geological Institute and Institute for Nuclear Sciences of the University of Gent, 86.

RENNE, P. R., BECKER, T. A. & SWAPP, S. M. 1990. $^{40}Ar/^{39}Ar$ laser-probe dating of detrital micas from the Montgomery Creek Formation, Northern California: clues to provenance, tectonics, and weathering processes. *Geology*, **18**, 563–566.

RODDICK, J. 1983. Geophysical review and composition of the Coast Plutonic Complex, south of latitude 55°N. *In*: RODDICK, J. A. (ed.) *Circum-Pacific Plutonic Terranes*, Geological Society of America, Memoir, **159**, 195–211.

ROHRMAN, M. & ANDRIESSEN, P. 1996. The relationship between sedimentary source and depositional area: a new challenge for AFT thermochronology (abstract). *In*: *Abstracts for the*

*International Workshop on Fission-track Dating, Gent 1996*. Geological Institute and Institute for Nuclear Sciences of the University of Gent, 94.

——, —— & VAN DER BEEK, P. 1996. The relationship between basin and margin thermal evolution assessed by fission track thermochronology: and application to offshore southern Norway. *Basin Research*, **8**, 45–63.

SCHAER, J. P., REIMER, G. M. & WAGNER, G. A. 1975. Actual and ancient uplift rate in the Gotthard region, Swiss Alps: A comparison between precise levelling and fission-track apatite age. *Tectonophysics*, **29**, 293–300.

SCHMID, S.M., BERGER, A., DAVIDSON, C., ET AL. 1996. The Bergell pluton (southern Switzerland, northern Italy) – overview accompanying a geological–tectonic map of the intrusion and surrounding country rocks (review). *Schweizerische Mineralogische und Petrographische Mitteilungen*, **76**, 329–355.

SEWARD, D. & RHOADES, D. A. 1986. A clustering technique for fission-track dating of fully to partially annealed minerals and other non-unique populations. *Nuclear Tracks and Radiation Measurements*, **11**, 259–268.

SPEER, J. A. 1980. Zircon. *In*: RIBBE, P. H. (ed.) *Orthosilicates*. Mineralogical Society of America, Reviews in Mineralogy, **5**, 67–112.

STOCK, J. & MOLNAR, P. 1982. Uncertainties in the relative plate positions of the Australia, Antarctica, Lord Howe, and Pacific Plates since the Late Cretaceous. *Journal of Geophysical Research*, **87**, 4697–4717.

STÜWE, K., WHITE, L. & BROWN, R. 1994. The influence of eroding topography on steady-state isotherms: application to fission track analysis. *Earth and Planetary Science Letters*, **124**, 63–74.

TABOR, R. W. & CADY, W. H. 1978. *The Structure of the Olympic Mountains, Washington – Analysis of a Subduction zone*. US Geological Survey, Professional Paper, **1033**.

TAGAMI, T. & DUMITRU, T. A. 1996. Provenance and thermal history of the Franciscan accretionary complex: constraints from zircon fission track thermochronology. *Journal of Geophysical Research*, **101**, 11353–11364.

TURNBULL, I. M. 1985. *Sheet D42AC and part sheet D 43 – Te Anau Downs Geologic map of New Zealand*, 1:50 000 (1 sheet) and notes, 1st edn. Department of Scientific and Industrial Research, Wellington.

——, URUSKI, C. I. ET AL. 1993. *Cretaceous and Cenozoic Sedimentary Basins of Western Southland, South Island, New Zealand*. Institute of Geological and Nuclear Science, Monograph, **1**.

TURNER, S. P., KELLEY, S. P., VANDENBERG, A. H. M., FODEN, J. D., SANDIFORD, M. & FLOETTMANN, T. 1996. Source of the Lachlan fold belt flysch linked to convective removal of the lithospheric mantle and rapid exhumation of the Delamerian–Ross fold belt. *Geology*, **24**, 941–944.

VANCE, J. A. 1989. Detrital kyanite and zircon: provenance and sediment dispersal in the Middle and Late Eocene Puget and Cowlitz groups, SW Washington and NW Oregon. *Geological Society of America, Abstracts with Programs*, **21**, 153.

WAGNER, G. A. & VAN DEN HAUTE, P. 1992. *Fission-Track Dating*. Kluwer Academic Publishers, Dordrecht.

——, MILLER, D. S. & JÄGER, E. 1979. Fission track ages on apatite of Bergell rocks from Central Alps and Bergell boulders in Oligocene sediments. *Earth and Planetary Science Letters*, **45**, 355–360.

——, REIMER, G. M. & JÄGER, E. 1977. *Cooling ages derived by apatite fission-track, mica Rb–Sr and K–Ar dating: the uplift and cooling history of the Central Alps*. Memorie degli Istituti di Geologia and Mineralogia dell'Universita di Padova, **30**.

YAMADA, R., TAGAMI, T., NISHIMURA, S. & ITO, H. 1995, Annealing kinetics of fission tracks in zircon: An experimental study. *Chemical Geology (Isotope Geosciences Section)*, **122**, 249–258

YIM, W. W. S., GLEADOW, A. J. W. & VAN MOORT, J. C. 1985. Fission-track dating of alluvial zircons and heavy minerals. *Journal of the Geological Society, London*, **142**, 351–356.

ZEITLER, P. K. 1985. Cooling history of the NW Himalaya, Pakistan. *Tectonics*, **4**, 127–151.

——, CHAMBERLAIN, C. P., & SMITH, H. A. 1993. Synchronous anatexis, metamorphism, and rapid denudation at Nanga Parbat (Pakistan Himalaya). *Geology*, **21**, 347–350.

——, JOHNSON, N. M., BRIGGS, N. D. & NAESER, C. W. 1982. History of uplifts in northwestern Himalayas using study of ages by fission tracks of detrital Siwalik zircons. *In*: *Symposium on Mesozoic and Cenozoic Geology in Celebration of the 60th Anniversary of the Geological Society of China, Beidaihe, China*. Geological Society of China, Beidaihe, 108–109.

——, ——, —— & —— 1986. Uplift history of the NW Himalaya as recorded by fission-track ages on detrital Siwalik zircons. *In*: JIQING, H. (ed.) *Proceedings of the Symposium on Mesozoic and Cenozoic Geology in Connection of the 60th Anniversary of the Geological Society of China*. Geological Publishing House, Beijing, 481–494.

# Detachment faults in the Aegean core complex of Ios, Cyclades, Greece

M. A. FORSTER & G. S. LISTER

*Australian Crustal Research Centre, Monash University, Melbourne, Vic. 3168, Australia*
*(e-mail: mforster@earth.monash.edu.au)*

**Abstract:** Several generations of detachment fault systems have been recognized in the Aegean metamorphic core complex of Ios, Cyclades, Greece. Multiple strands occur in each fault system, with fault-bounded slices often characterized by distinctive lithologies. These faults accomplished significant exhumation, and telescoped the crustal section. Different slices have been deformed at different crustal levels, and then juxtaposed, so that adjacent fault slices display different deformation styles. The Aegean detachment systems differ from equivalent faults observed in the US core complexes in that they separate relatively cohesive tectonic slices. However, the Aegean detachments are dissected by the multiple systems of later high-angle normal faults that define the current geomorphology, and this raises significant difficulties in correlating detachment systems between the different islands of the Cycladic archipelago. A number of distinct extensional events can be defined on Ios. The Cycladic blueschist belt was exhumed relatively early during the history of Alpine orogeny, after it had been thrust over a Hercynian 'basement' terrane, with the oldest recognized detachment faults associated with Miocene N–S extension. This led to the exhumation of the 'basement' terrane from beneath the blueschist 'series' rocks. The Ios Detachment Fault system now separates the metasediments of the Cycladic blueschist belt from the mylonitized 'basement' complex. It was associated with the operation of the south-directed, crustal-scale, middle to upper greenschist facies, South Cyclades Shear Zone. The youngest detachment faults formed during late-stage W-directed extension, and formed a prominent sequence of low-angle normal faults termed the Coastal Fault system. These were associated with spectacular breccias, and relatively narrow ductile shear zones which formed during and subsequent to the formation of the Ios dome, under significantly lower grade conditions (lower-greenschist facies assemblages). Final exhumation of the Cycladic blueschist belt appears to have been associated with the detachment faults that led to its exposure beneath non-metamorphic sediments, and remnants of the Cyclades 'ophiolite nappe' as observed on adjacent islands.

Much of the Aegean microplate (Fig. 1) was affected by a major period of extensional tectonism during Oligo-Miocene time, with an associated period of regional greenschist to amphibolite facies metamorphism, and plutonism (Altherr *et al.* 1982; Angelier *et al.* 1982; Lister *et al.* 1984; Avigad & Garfunkel 1989, 1991; Avigad *et al.* 1992). This period of extension was accomplished by the operation of high-angle normal faults, low-angle normal faults, and crustal-scale ductile shear zones (e.g. Lister *et al.* 1984; Ridley 1984; Faure & Bonneau 1988; Avigad & Garfunkel 1989; Urai *et al.* 1990; Avigad *et al.* 1992; Gautier *et al.* 1993; Fig. 2). Specific examples of listric normal faults and/or low-angle normal (detachment) faults have been documented on Paros and Naxos (Lister *et al.* 1984; Urai *et al.* 1990; Gautier *et al.* 1993; Gautier & Brun 1994a; Howard & John 1995), Tinos (Avigad & Garfunkel 1989), Sifnos (Avigad *et al.* 1992; Avigad 1993), Mykonos (Lee & Lister 1992), Kos (Altherr *et al.* 1982), Ikaria (Lister, unpublished data), Syros (Ridley 1984) and Ios (Lister *et al.* 1984; Forster 1996; Forster & Lister 1996a; Vandenberg & Lister 1996).

Although there are numerous examples of low-angle normal faults throughout the Aegean region (see above), evidence for regional-scale detachments has only recently begun to emerge (Fig. 2). Island-scale faults have been noted on Tinos, Andros (Jolivet, pers. comm.), and Sifnos (Avigad 1993). However, data for the existence of regional-scale low-angle faults have mostly been based on observations at single outcrops (e.g. Ios, Naxos, Paros). Gautier *et al.* (1993) suggested the existence of a single detachment fault extending from the eastern coast of Naxos to the eastern coast of Paros (their fig. 2, p. 1182). The existence of this single detachment fault (with a strike length c. 100 km) is extrapolated from relatively small outcrops in the east and west of Naxos and in the east of Paros. However, this might not be a valid extrapolation if these are segments of different detachment faults that have formed at different times throughout the history of continental extension.

FORSTER, M. A. & LISTER, G. S. 1999. Detachment faults in the Aegean core complex of Ios, Cyclades, Greece. *In:* RING, U., BRANDON, M. T., LISTER, G. S. & WILLETT, S. D. (eds) *Exhumation Processes: Normal Faulting, Ductile Flow and Erosion.* Geological Society, London, Special Publications, **154,** 305–323.

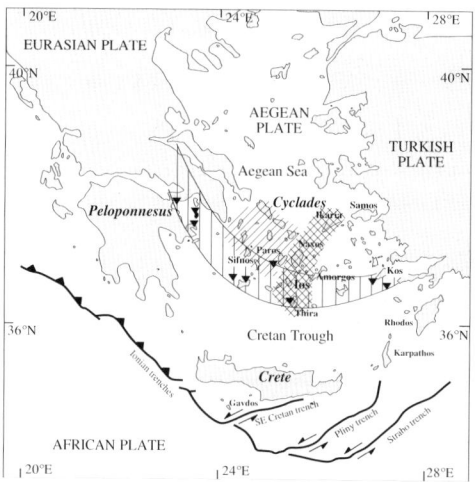

**Fig. 1.** Cycladic archipelago in its regional tectonic context. Locations of islands mentioned in the text are shown. There are two Miocene structural domains (shown hatched). The central domain was affected by roughly N–S-directed extension. The extension direction in the northwest domain was more heterogeneous, with the lineations in Miocene shear zones varying from SE–NW to NE–SW trends.

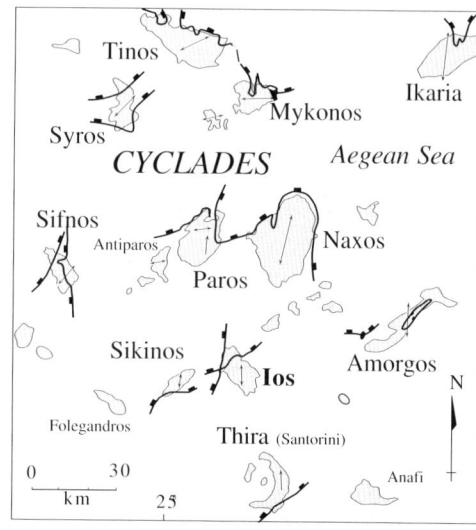

**Fig. 2.** Low-angle normal faults (LANFs) in the central Aegean. Ios: Van der Maar (1980a), Forster (1996); Sifnos: Avigad et al. (1992), Avigad (1993); Tinos: Avigad & Garfunkel (1989), Gautier & Brun (1994a); Paros: Gautier et al. (1993); Paros: Gautier et al. (1993), Forster & Lister (unpublished data); Syros: Altherr et al. (1982); Ikaria: Lister (unpublished data); Mykonos: Altherr et al. (1982); Naxos: Gautier et al. (1993); Amorgos: Forster (unpublished data).

In this paper we report multiple generations of island-scale low-angle normal faults on the Aegean metamorphic core complex of Ios (Fig. 3). One set of faults can be traced across the island, from east to west, for a strike length of c. 12 km (Figs 3–5) with mostly continuous outcrop. The roughly N–S-striking faults can be followed through a sequence of peninsulas along the NW coast of Ios, and can be confidently extrapolated over a strike length of 5–6 km (Figs 6–8).

The fault systems on Ios can be compared with those on Paros, Naxos, Amorgos, Sifnos, Sikinos, Tinos and Thira. There are many similarities as well as differences, suggesting that the low-angle faults recognized throughout the Cyclades may be part of a number of linked detachment system of differing generations, formed at a range of structural levels. The key issues are to determine how to correlate between different islands, and to time the formation of the different detachments, and thus to constrain the exhumation history of the Cyclades blueschist belt. Ios was chosen for a detailed study because it lies close to the western boundary of the central structural domain of the Cyclades (Blake et al. 1981; Lister & Forster 1996a) and thus might reveal critical relations between the two domains.

## The Ios Detachment Fault system

The upper levels of the Hercynian basement complex on Ios (Henjes-Kunst & Kreuzer 1982; Van der Maar & Jansen 1983; Andriessen et al. 1987; Keay & Lister 1996a; Keay 1998) have been intensely deformed by a crustal-scale, top-to-the-south shear zone, termed the South Cyclades Shear Zone (SCSZ) (Banga 1983; Lister et al. 1984; Vandenberg & Lister 1996). This shear zone is now arched from north to south, in a manner analogous to the shear zones in the US core complexes (e.g. Wernicke 1981, 1985; Wernicke & Burchfiel 1982; Spencer 1984; Davis 1988; Reynolds & Lister 1990). The lower structural levels of the SCSZ are cut by localized top-to-the-north shear zones (Lister & Keay 1996), which appear to form a different generation of shear zones from those that were previously recognized. These overprinting top-to-the-north shear zones were not recognized by Vandenberg & Lister (1996), but can be clearly observed in the headland south of Mylopotas Beach (Lister & Keay 1996). The actual contact of the mantle garnet–mica schist with the underlying augen-gneiss core is marked by such a north-directed shear zone, with

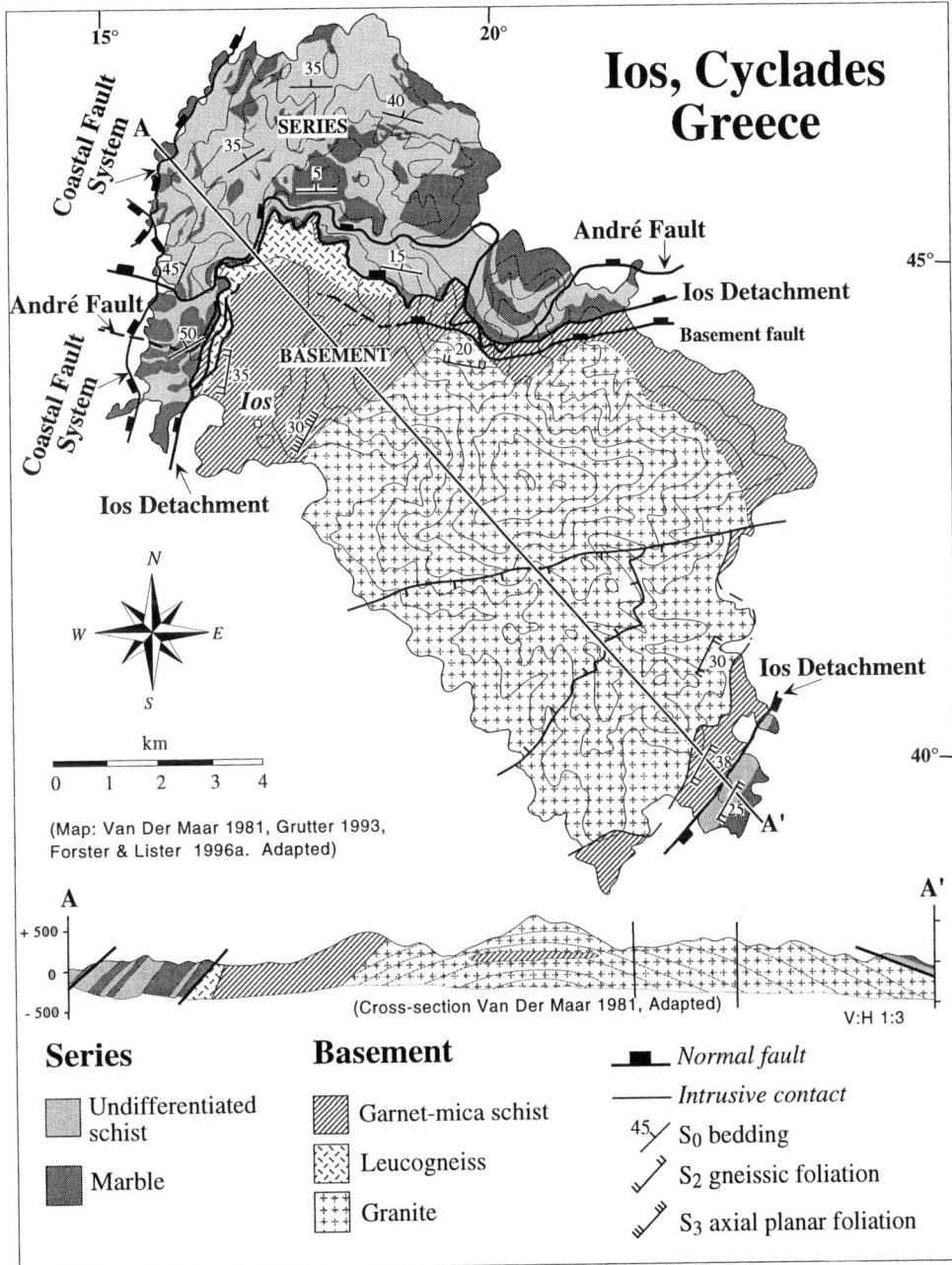

**Fig. 3.** Geology of Ios showing multiple faults of the Ios Detachment Fault system, which can be traced in an arch across the island, cutting both the basement and the overlying series (Forster 1996). Part of the Coastal Fault system along the northwest coast of Ios is also shown.

intensely developed fabrics and lineations in a zone 1–5 m wide. These north-directed shear zones may be analogous to the antithetic shear zones of the mylonite front recognized in the US core complexes (Reynolds & Lister 1990) and therefore may be related to doming.

The upper levels of the SCSZ are truncated by a system of low-angle faults (the Ios Detachment

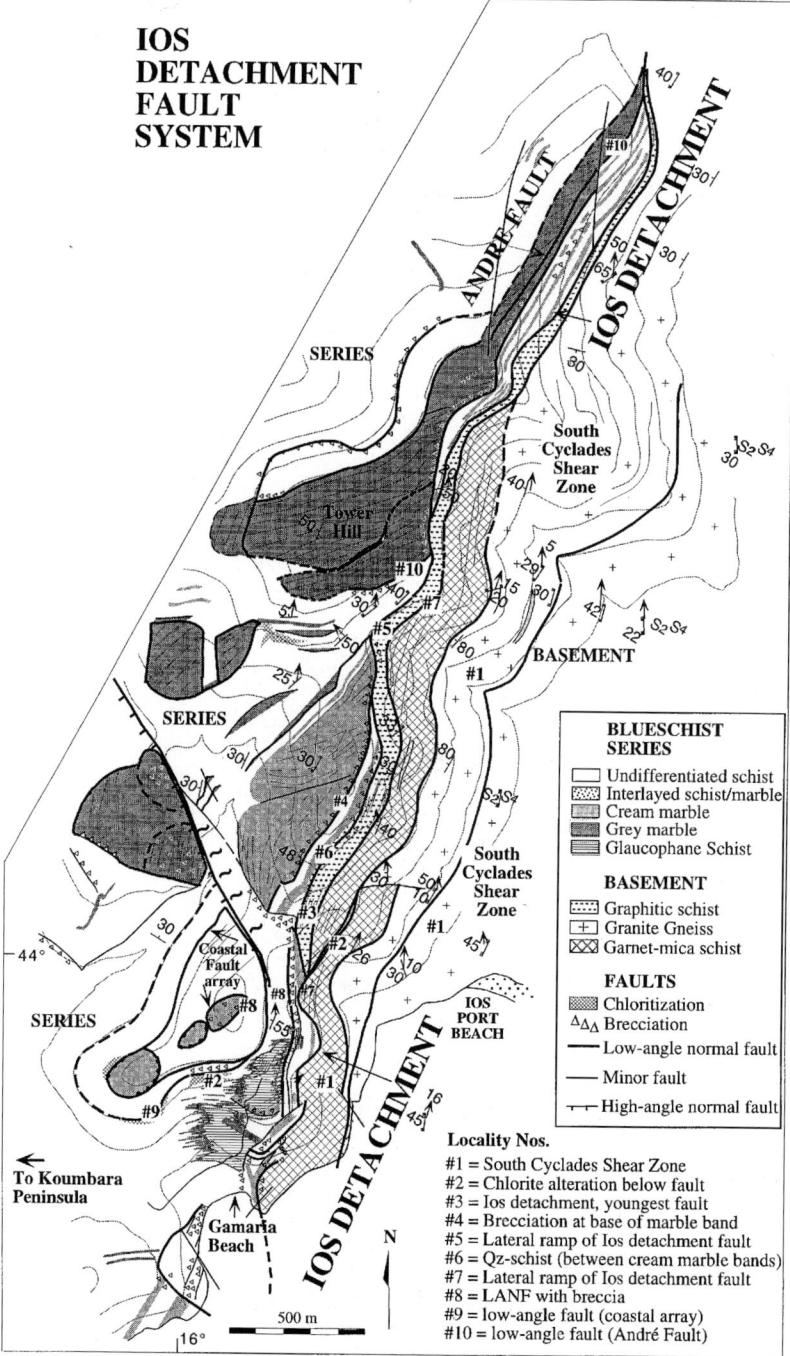

**Fig. 4.** Map of the Ios Detachment Fault system adjacent to Ios Port. (Note multiple faults in basement lithologies, dissecting the South Cyclades Shear Zone (SCSZ).)

**Fig. 5.** View of the Ios Detachment Fault system, taken from the hill at Ios Chora looking northwestward. Multiple faults occur in the basement and series, above and below the Ios Detachment.

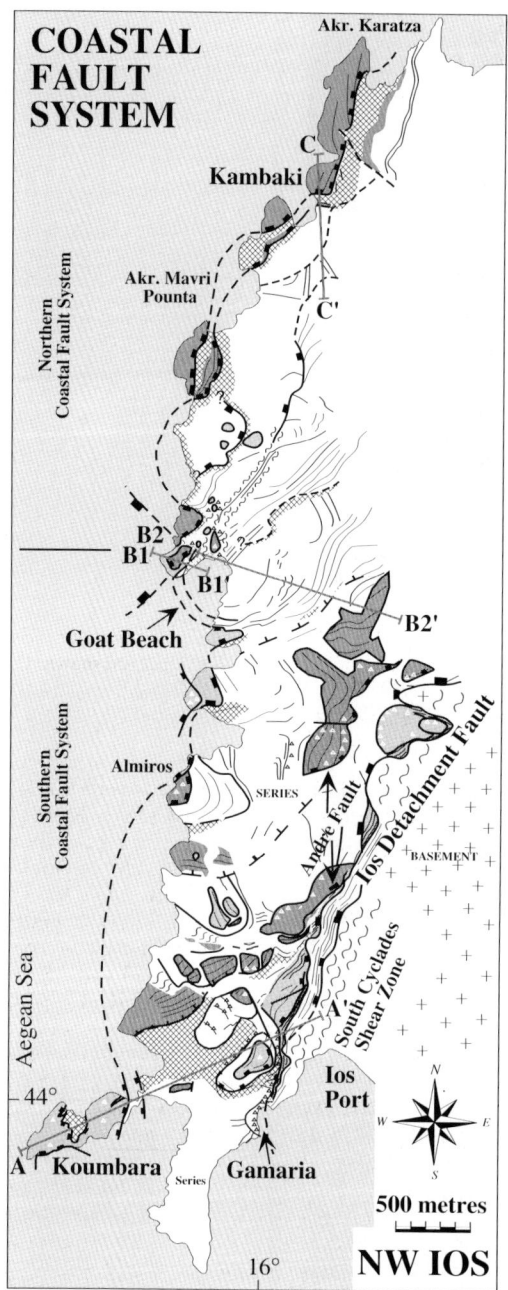
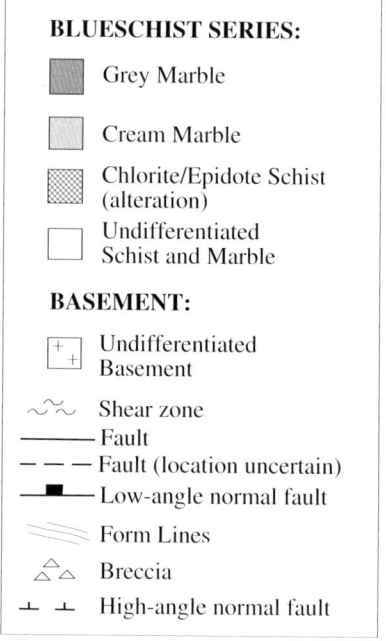

**Fig. 6.** Map of the Coastal Fault system along the northwest coast of Ios (showing locations of cross-sections).

Fault system). These faults accomplish the final juxtaposition of the basement and overlying blueschist sequence (named the 'series' by Van der Maar (1980a, b, 1981)). The Ios Detachment is most obvious where it truncates cream marbles of the series adjacent to basement schists (Fig. 4). However, the Ios Detachment Fault is only one of a sequence of low-angle faults which anastomose subparallel to the basement–series contact.

Multiple low-angle normal faults (LANFs) occur both in the series and the basement above and below the Ios Detachment. The most obvious and laterally continuous of these fault

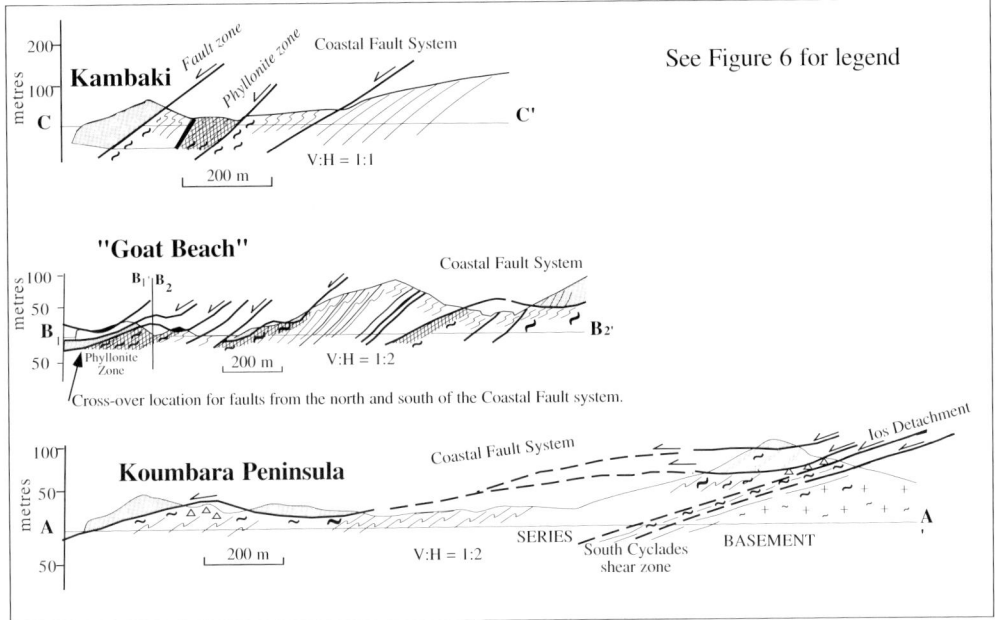

**Fig. 7.** Cross-sections through the Coastal Fault system along the north-west coast of Ios (locations shown in Fig. 6).

lies structurally above the Ios Detachment, and is here named the André Fault (Fig. 3). This younger island-scale LANF juxtaposes grey marble bands against the domain of cream marbles. However, at several locations the André Fault incises into the basement, across the Ios Detachment.

The Ios Detachment Fault system is exposed north of Ios town, and in the southern part of the Ios dome (near Manganari Beach). It represents the earliest formed detachment fault system that has been recognized on Ios and we suggest (see below) that it formed during N–S directed extension. Equivalent low-angle faults in the south of the island dip (both bedding and foliation) less steeply than in the northern part of the dome.

## Location of the Ios Detachment

The anastomosing LANFs of the Ios Detachment Fault system cause considerable lateral variation in lithology. This explains why the location of the series–basement contact has been a contentious issue between previous researchers working in this area (Van der Maar 1980a, b; Van der Maar & Jansen 1983; Grütter 1993; Vandenberg & Lister 1996).

Van der Maar (1980b) and Van der Maar & Jansen (1983) suggested that the contact is a pre- to syn-blueschist facies thrust plane, localized along an older erosional surface or unconformity. The supposed thrust surface is defined by a layer of actinolite schist in the upper level of the basement complex. These workers suggested that the actinolite schist is the remnant of a retrograded ophiolite, and hence identified this surface as a thrust plane. Van der Maar (1980a, b) and Van der Maar & Jansen (1983) suggested that the series–basement contact is located up to c. 100 m below the lowest cream marble band. The schist under the lowest marble bed was thought to be part of the overlying series.

In contrast, Grütter (1993) suggested that the basal cream marble units, as well as the intervening schist and metabasic rock, are part of an older stratigraphic sequence that unconformably overlies the basement. He suggested that there is a thrust at a higher structural level that juxtaposed series against series. Vandenberg & Lister (1996) suggested that a late single north-dipping normal fault (not a detachment fault) had cut through the SCSZ and was located at the base of the cream marbles, at Ios Port and below grey marble at Plakato.

The Ios Detachment can be difficult to locate because of: (a) the similarities of lithology between the series and basement; (b) outcrop

**Fig. 8.** Low-angle normal faults in the Cyclades. (**a**) Fault in the Coastal Fault system on NW Ios, at Kambaki. Grey marble breccia (without ductile fabrics) overlies phyllonites formed by intense deformation and chlorite–epidote alteration in a 100 m thick shear zone. (**b**) Fault plane exposed across a topographic high on the east coast of Ios. (**c**) Detachment faults on the northwest coast of Paros, with similar characteristics to the Ios Detachment. A fault separates the cream marble band and underlying basement schist. An additional fault plane truncates the top of the cream marble band. HANF, high-angle normal fault.

that is not always laterally continuous and/or is poorly outcropping; (c) incision of younger, structurally higher low-angle faults through the Ios Detachment Fault; (d) the overprinting of the fault zone by later high-angle normal faults. The greatest difficulty is caused by the similarity between the lithology of basement and series schists, because the structurally lowest schists of the series have no diagnostic features that allow them to be readily distinguished from basement schist. Where series quartz–mica schist includes small garnets (c. 1–1.5 mm), the unit is indistinguishable from basement garnet–mica schist units. Where lithologies are not diagnostic, and the foliation of the basement and series is subparallel, the only manifestation of the fault plane is a distinct change in topography (a break in slope).

In most cases, however, the position of the fault plane can be readily determined; for example, where the foliation in the basement and series is truncated by the fault, or where diagnostic mineral assemblages are present in either the basement or the series. Diagnostic mineralogy in the basement garnet–mica schist includes such minerals as biotite and/or hornblende, which are not observed in the series schists. Similarly, a diagnostic mineralogy for a series schist is the presence of sodic amphiboles.

The location of the boundary between the basement and series is important as it reveals the processes and the sequence of events that have been involved in the exhumation of the basement terrane.

## Movement sense on the Ios Detachment

Movement direction, or sense of movement, on the Ios Detachment system has been determined by the ductile structures in the footwall. The mylonites and phyllonites immediately adjacent to the fault are more intensely foliated, with well-developed shear bands and mylonitic lineations. This suggests that they were affected by drag on the fault, with top-to-the-south movement taking place.

There are several examples of extensional 'ramp and flat' geometry (Scott & Lister 1992). Along strike, the Ios Detachment cuts to different levels in the series, as can be seen by variation in the number of cream marble beds that crop out along the contact zone. Mylonitized cream marbles are intersected at a low angle (up to c. 10–20°) on what are essentially ramp geometries (Figs 5 and 9). The ramps produce a characteristic morphology in the trace of the fault strands.

The geometry of these ramps determines the movement sense for the Ios Detachment. If the fault was originally a north-dipping normal fault or a north-dipping thrust fault, the fault should ramp structurally upwards towards the south. This geometry is not observed. Therefore the fault is either a reoriented (originally south-dipping) normal fault or a reoriented (originally south-dipping) thrust fault. If it was an originally south-dipping thrust, the predicted sense-of-shear in the underlying shear zone would be north directed, not south directed as observed.

The remaining alternative is that the fault is a reoriented (originally south-dipping) normal fault that was arched upwards so that it now appears to have a thrust geometry. Movement on the regional fault was therefore extensional. This fits with other observations, for example that the cooling history of the basement terrane is dominated by the youngest thermal events (Baldwin & Lister 1998).

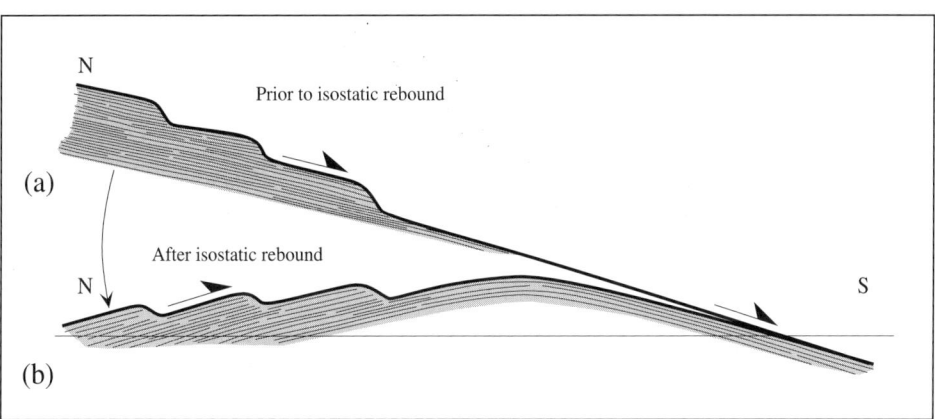

**Fig. 9.** Evolution of the Ios Detachment Fault, from (**a**) a south-directed normal fault (ramping to the south); to (**b**) a fault with an apparently thrust geometry, but with a normal fault character revealed by the characteristic geometry of the ramp structures. An originally northward-dipping fault (either a reverse fault or a normal fault) would not have developed the observed ramp geometries.

## The coastal low-angle fault system

Structurally higher in the series another array of faults has been discovered (named in this study the Coastal Fault System). Typically, these faults juxtapose massive marble units against structurally lower blueschist units. These faults dissect the series, and cut down through almost the entire preserved section (Figs 6 and 7), from the highest structural levels (at Kambaki), to the lowest structural levels (at Koumbara). These are exposed on the flanks of the Ios dome, along the NW and NE coasts, although the faults on the NE coast display a less complex deformational history and are not discussed in detail in this paper.

On the northwest coast narrow shear zones beneath the faults display localized roughly E–W-trending phyllonitic stretching lineations. This E–W extensional system overprints the fabrics and lineations formed during earlier and dominant N–S extension observed over the rest of the island. Considerable variation in deformational style occurs between juxtaposed units. This is in respect to both the overall characteristics of the fault slices and the immediate footwall and hanging wall adjacent to a particular fault.

The faults can be best observed in a number of coastal outcrops in a sequence of peninsulas along the northwest coast of Ios (e.g. Fig. 8a). The traces of these faults anastomose from N–S to roughly NE–SW. They dip shallowly (0–35°) to the west (see cross-section, Fig. 7), but the overall dip of the faults increases towards the north. The fault at Koumbara near the basement boundary dips from 0° to $c.$ 10° west, whereas at Kambaki the fault dips $c.$ 35° NNW. These low-angle faults appear to be normal faults, and small-scale normal faults (<10 m strike length) can often be observed in the hanging walls.

Small-scale asymmetrical folding and boudinage are abundant in the altered zone of the footwall. These folds occur below all faults in the Coastal Fault system. Bedding and foliation are generally more steeply dipping than the faults and are thus truncated by these faults. These folds thus appear to have formed during extensional deformation. Fault movement has distorted the more steeply dipping fabrics in the blueschist, demonstrating that a normal sense of movement has occurred.

A marked contrast in the deformational style of adjacent fault slices can be observed in the Coastal Fault system. The hanging wall is often pervasively brecciated whereas the footwall immediately below the fault displays intensely lineated phyllonitic fabrics in shear zones. These have been imprinted on the older ductile fabrics. The hanging walls on the coastal peninsulas display evidence of an origin at more surficial crustal levels than footwall, as they are relatively undeformed or brecciated, and generally do not display ductile features such as fabrics or foliations. In contrast, footwall rocks display ductile deformation and blueschist facies metamorphic mineral parageneses. Movement along these low-angle faults has thus juxtaposed deeper level metamorphic tectonites (in the footwall of each fault) against metamorphic rocks from significantly higher crustal levels (in the hanging wall).

Over a distance of $c.$ 100 m several low-angle faults can be found, and one can walk down the structural section from brecciated carbonate, through transitional brittlely to ductilely deformed carbonate and schist, into schist and carbonate that deformed in a completely ductile manner. A general decrease in the degree of ductility is evident across many of these faults, from footwall to hanging wall, suggesting that, in total, they have produced a telescoping of rocks from different structural depths.

### Two generations of LANFs in the Coastal Fault array

The fault array which occurs on the northwest coast of Ios can be divided into two distinct arrays of LANFs. Features which distinguish the two arrays are (a) different lithologies in the hanging walls and (b) different deformation characteristics in both the footwalls and hanging walls. The youngest fault array occurs in the south, from Koumbara to Goat Beach. The older fault array crops out in the northwest of the island, from Goat Beach to Kambaki.

It is possible that these two fault arrays may represent a variation in structural depth during a single faulting event or continuing faulting with a change in deformation character during the event. The older faults are steeper and may have reoriented as the Ios dome formed. The timing of the Coastal Fault system can thus be constrained to occur progressively during formation of the Ios dome.

The two different types of faults of the Coastal Fault system appear to cross at Goat Beach (Fig. 6). At this location, the more steeply dipping faults seen in the north are cut by more shallowly dipping faults with brecciated cream marbles in their hanging walls. These breccias are fine- to coarse-grained angular cataclasites (<3 cm) and appear to vary from matrix supported to not matrix supported. The cataclasite

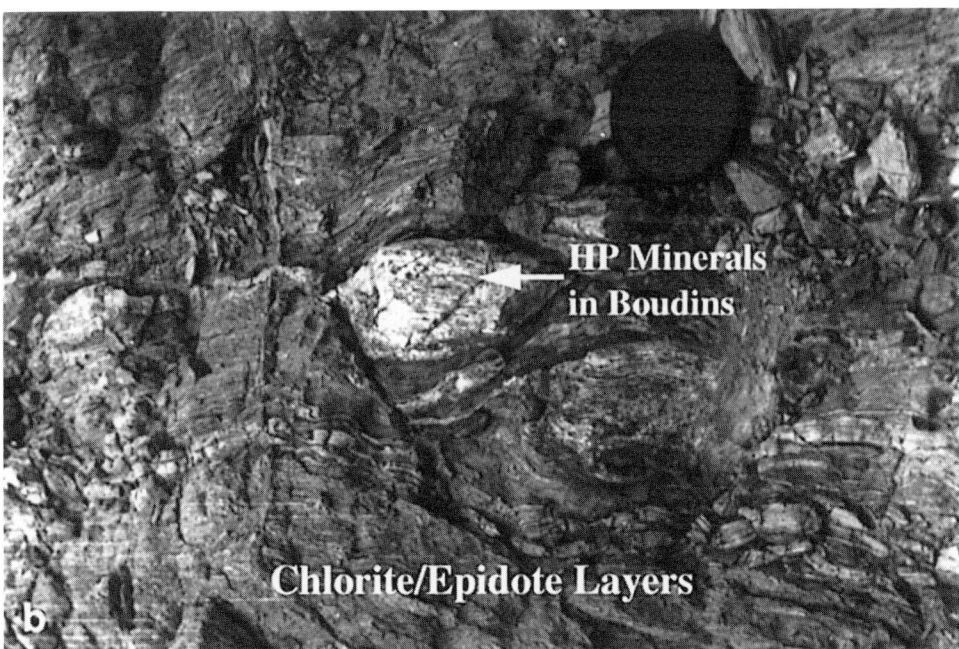

**Fig. 10.** Contrast in the rheology of rocks above the low-angle normal faults of the Coastal Fault system. (**a**) Brecciated cream marble above an LANF in the Coastal Fault array. The breccia fragments are indurated, with a wide range in grain size being present. There are fragments of the grey marble units included in this breccia. (**b**) Pervasive deformation and alteration of high-pressure metamorphic rocks immediately below an LANF. Layers of quartz and epidote are folded and boudinaged with relict glaucophane remaining in small boudins. Layering has been reoriented by the shear zone beneath the normal fault. This late deformation is also responsible for the formation of the boudin.

has been indurated as the result of later fluid flow and cementation. Fragments of the veined grey marble breccia (up to 30 cm) from the underlying fault slice occur within the cream marble groundmass (Fig. 10a).

The faults located between Koumbara and Goat Beach characteristically juxtapose recrystallized massive cream marble beds against ductilely deformed epidote–quartz–chlorite rocks derived from a blueschist protolith (Fig. 10b), but faults can also occur with brecciation in the footwall. For example, at Koumbara an incohesive brecciated zone occurs at the fault contact in chloritized volcanic material. There is no fabric in this breccia. However, immediately below the breccia in an epidote–quartz–chlorite assemblage, an approximately E–W extension direction is inferred from stretching lineations.

The faults located between Goat Beach and Kambaki display distinctly different lithologies in their hanging walls, located in the peninsulas, from those found in the hanging walls of the coastal faults near Koumbara. In this northern region the peninsula hanging wall units consist of grey marbles which display no ductile deformation or obvious recrystallization. Instead, these units are pervasively cataclased. The other distinguishing features of this northern section are the zones of intense mylonitization in the retrograded blueschist units of the footwall.

The coastal faults found on the east coast of Ios display similar characteristics to those found near Koumbara, where similar lithologies are juxtaposed, i.e. cream marble units in the hanging wall defining the peninsulas, and glaucophane–albite blueschist assemblages in the footwall. However, in some cases the juxtaposition of these lithologies appears to have been accomplished by narrow ductile shear zones rather than faults.

The most significant difference between the faults on the east and west coasts is that no E–W lineations are observed along the east coast (or in the centre of the island, where LANFs occur at the very highest topographic points). The blueschist assemblages adjacent to the faults on the east coast (and central region) have not been retrogressed or affected by alteration. Fluid-related retrogression occurred only during E–W extension, under lower-greenschist facies conditions, and overprinted the earlier formed fabrics. This occurred at shallow crustal levels.

Geomorphology reflects the existence of fault truncations at the top of many of the large marble units associated with the hanging wall of the Coastal Fault system (Fig. 8b). This has been noted to also occur on the NW coast of Paros, where similar faults between large carbonate units and underlying schists are observed (Fig. 8c).

## Shear zones associated with low-angle faulting

Shear zones from micro- to macro-scale occur in association with the Coastal Fault system. These shear zones display either N–S- or roughly E–W-trending mineral lineations. The shear zones contain mostly phyllonite, often intensely lineated, and can occur through individual fault slices and but are generally associated with alteration zones below the faults.

Phyllonites are formed where the most intense retrogression has occurred, in shear zones dominated by epidote–quartz–chlorite assemblages. Phyllonite shear zones may occur up to 10–50 m below a particular LANF. In some cases an intense roughly E–W-trending lineation in a phyllonite-bearing shear zone occurs directly below the fault. Some of the narrow shear zones (<10 m) splay into the actual low-angle fault itself, but then diverge into the underlying footwall. This geometry is similar to that observed in the 'mylonite front' (Davis 1988; Reynolds & Lister 1990) in the Colorado extensional corridor, in the western US core complexes.

The intensity of the fabric in the phyllonite zones is reflected in the degree of reorientation of the dominantly N–S-trending lineations. The most intensely foliated phyllonites have lineation trends that are roughly E–W, whereas the less intensely foliated zones have NE–SW lineation trends.

The E–W lineated fabrics always occur in an epidote–quartz–chlorite or graphitic zone in the footwall. These E–W lineated fabrics overprint the structurally higher and lower N–S extensional features. Parts of the Coastal Fault system may have formed during the N–S extensional event, and been reactivated and overprinted during the later E–W extensional episode.

## Breccia zones

Breccia zones are generally relatively narrow (1–5 m wide), affecting the altered zone, as well as localized areas of the phyllonites. This suggests a time sequence as well as specifying the deformational style of continuing extension. This evolutionary sequence is observed beneath the detachment faults of the Colorado River extensional corridor (Davis *et al.* 1986; Reynolds & Lister 1990).

The style of breccias varies considerably from the north to the south of the series, as well as locally. In the north (e.g. at Kambaki) the hanging walls of grey marble are fine (2–5 mm) angular cataclasites which are pervasive throughout the rock mass. The breccias are apparently unrelated to the location of the underlying fault, and have been formed by penetrative fracturing and some brecciation, but the fragments are jigsaw breccias with small relative displacement.

The hanging walls of the structurally lower faults in the southern areas display distinctly lineated ductilely deformed grey marbles, which have also been brecciated. These are observed only as remnants of the hanging wall on the side of the dome (e.g. between Koumbara and Gamaria, Figs 6 and 7). These grey marbles are unlike the brecciated massive cream carbonates on the peninsulas, which display evidence of a ductile history overprinted by brecciation. The brecciation generally extends c. 1 m up into the hanging wall. However, as only the base of some of these remnants is preserved breccia occurs over the entire outcrop.

## Discussion

The effects of extension in the Aegean microplate are recorded by high-angle normal faults (e.g. Drooger & Meulenkamp 1973; Angelier 1978; Jackson & McKenzie 1983), gently dipping listric normal faults (Ridley 1984), LANFs (Avigad & Garfunkel 1989, 1991; Lee & Lister 1992; Gautier et al. 1993; Gautier & Brun 1994a & b) and by major crustal normal-sense ductile shear zones (e.g. Lister et al. 1984; Buick 1988, 1991; Urai et al. 1990; Gautier et al. 1993). Extensional features recognized on Ios now include the top-to-the-south SCSZ (Lister et al. 1984; Vandenberg & Lister 1996), multiple LANFs that juxtapose the basement and blueschist series, as well as multiple LANFs that dissect the series and are largely responsible for the map pattern (Forster & Lister 1996a–d). In addition, geomorphology suggests the existence of several high-angle normal fault scarps.

The Cyclades is not affected by one single a really extensive detachment fault system. LANFs of a similar character to the Ios Detachment Fault are exposed on Paros, where basement schists are juxtaposed against cream marble beds by an LANF (Fig. 8c). This fault system may therefore be of regional extent. LANFs similar to the Coastal Fault system may also exist regionally. The LANFs that have drawn most attention to date are those that separate the non-metamorphic rocks and remnants of the Cyclades ophiolite nappe (Dürr et al. 1978; Lee & Lister 1992; Gautier & Brun 1994a,b) from the underlying metamorphic rocks. However, these higher-level faults, which separate the metamorphic rocks from the overlying non-metamorphic rocks, are either poorly exposed or non-existent on Ios (Lister & Forster 1996a).

### Bowing-up of the Ios Detachment

This study suggests that the faults that separate the series from the basement in the south of Ios are from the same fault system as the faults that separate the series from the basement in the north. These bounding faults occur at the same structural levels, and they juxtapose the same lithologies (e.g. garnet–mica schist against retrograded blueschist). We have presented data that suggest the northern bounding fault has been bowed upwards until its dip reversed.

In the north, the Ios Detachment system cuts structurally downwards into the lower plate, producing slices of mylonitized basement, and truncating the fabrics of the SCSZ. The fault array also cuts structurally upwards into the upper plate, so that the mylonitized cream marbles are intersected at a low angle along ramps. These geometries are consistent with south-directed motion on a subsequently bowed-up detachment fault. Normal faults at Manganari also cut into the upper plate, with the same south-directed sense. This implies one bowed-up detachment fault over the whole island. The Ios Detachment thus appears to have behaved in the same manner as equivalent detachment faults recognized in the US Cordillera (e.g. Spencer 1984).

The origin of the N–S arch may be related simply to isostatic adjustment after tectonic denudation (Spencer 1984). This would explain why the dips in the south are less steep than those in the north, and is consistent with the notion that the north-directed overprinting shear zones are antithetic shear zones related to a mylonite front (see Davis 1988; Reynolds & Lister 1990). It is also possible that part of the dome formation is related to the effect of magma emplacement at depth, but no direct evidence for Miocene plutonism has been found on Ios (see Baldwin & Lister 1998). The LANFs on Ios are not reoriented high-angle normal faults, because they formed at low angles to overall shallowly dipping pre-existing fabrics (see Wernicke & Axen 1987).

An additional variable to consider is the effect of later-formed high-angle normal faults that cut the detachment faults. These are responsible for many complexities in the map pattern, particularly in respect to the way they offset the detachment faults on fault scarps. There may also be a systematic decrease in dip from north to south, related to the rotation of tilt blocks that have cut through the dome.

## Relation between the South Cyclades Shear Zone and the Ios Detachment

Detailed mapping shows that the single fault which has previously been associated with the juxtaposition of the lower and the upper plate (Van der Maar 1980a,b; Vandenberg & Lister 1966) is in fact only one of a complex array of faults. This complex array of anastomosing ramped and fluted normal faults has led to confusion surrounding the previous attempts to define the actual location of the contact. All of the faults are younger than the SCSZ, as they all cut the mylonites of the SCSZ.

This study suggests that, in some areas, the cream marble bands immediately above the Ios Detachment were also intensely deformed in the SCSZ. The anastomosing fault array that defines the boundary between the basement and the series has in fact dissected the SCSZ, and fault slices comprise rocks from the basement as well as from the series that were caught up in this shear zone.

The SCSZ may have extended beneath the Sea of Crete (as shown in fig. 2.4 of Lister & Forster 1996a). The Ios Detachment may have accommodated the final movements in this shear zone. At that time, however, the mylonites now exposed on Ios would have been kinematically inactive, whereas mylonites deep beneath the Sea of Crete in the southern extension of the SCSZ would have been actively deforming. This model for the evolution of the Ios core complex is similar to that proposed by Wernicke (1981, 1985), Wernicke & Burchfiel (1982), and Lister & Davis (1989).

The significance of the late overprinting top-to-the-north shear zones on Ios are, as yet, not understood. Top-to-the-north overprinting shear zones also occur through the series, defined by button schist and S–C fabrics.

## Variation in time and space of the kinematics of extension

At least two distinct episodes of extension can be inferred on Ios based on the change in kinematics from N–S to roughly E–W extension, and the observed overprinting of the different shear zones (see above). The key question is why the direction of extension should change in the second period of extensional faulting, and why these faults are confined to a localized region on the NW coast of the island.

Ios lies on the western edge of the central structural domain of the Cyclades (Lister & Forster 1996b). There is an abrupt change in the orientation of Miocene stretching lineations on adjacent islands (e.g. Gautier & Brun 1994a). The roughly E–W-directed extensional overprint may indicate that Ios in fact lies on the eastern flank of the central to northwestern structural domain (see Lister & Forster 1996a, their fig. 2.1). E–W-directed extension appears to be associated with extension at shallower crustal levels than during the period of N–S extension. Variation in the direction of extension in the different structural domains may be a natural consequence of the evolution of the geometry of the Aegean, with N–S extension reflecting the earlier effects of slab retreat, before significant curvature had been induced into the Hellenic Arc (see Kissel & Laj 1988).

## Timing of detachment faulting on Ios

Detachment faulting took place at different times in the Miocene history of the Aegean region. Avigad & Garfunkel (1989) reported detachments on Tinos which must have operated before the emplacement of granitoid plutons at 18 Ma. Some of the faults on Naxos formed subsequent to emplacement of the granodiorite (e.g. Gautier & Brun 1994a) before 12 Ma (Keay & Lister 1996b). New detachment faults may form in each episode, but individual detachment faults may exhibit a complex history of reactivation.

The first period of extensional faulting appears to be associated with the operation of the SCSZ, as outlined above, although the brittle faults of the Ios Detachment clearly post-date the ductile fabrics of the SCSZ. Operation of the SCSZ, and the formation of the Ios Detachment appears to (at least in part) pre-date dome formation, although the development of the dome may have been broadly synchronous with operation of the fault.

The Ios Detachment system is distinctly arched, with approximately the same degree of curvature as the SCSZ, whereas the structurally higher André Fault displays a flatter geometry across the island. This suggests that the André Fault was formed later in the doming process than the structurally lower Ios Detachment.

The Coastal Fault system developed during doming, and faults were forming while N–S extension was still taking place. These faults are best preserved on the NW flanks of the dome, where they have dominantly roughly N–S trends. However, here the roughly E–W-trending lineations and refolded fabrics in localized shear zones beneath the faults suggest that faulting either continued or was reactivated late to post-doming while roughly E–W-directed extension was taking place.

These timing relationships are confirmed by the mineral parageneses preserved in shear zones associated with fault movement. Shear zones with roughly E–W-trending lineations are associated with lower-greenschist facies assemblages whereas those with N–S-trending lineations preserve blueschist to greenschist facies assemblages. This suggests that these shear zones operated over a substantial period of the denudational history, accomplishing significant exhumation of the Cyclades blueschist belt.

## Telescoping of the crustal section

The footwalls of faults of the Coastal Fault system were considerably more ductile than the hanging walls. In some cases, the hanging wall is brecciated whereas the footwall has well-developed ductile fabrics. This seems to imply that the footwall and the hanging wall come from different crustal levels, and that significant relative motion must have occurred on the extensional low-angle faults that juxtaposed these lithologies. It is difficult to argue that rocks at the same structural level would develop such different deformational styles.

Brittle behaviour is typical of more shallow crustal levels, and/or is the result of high fluid pressures or high strain rates, and some lithologies are more brittle in their response to deformation than others. However, in this case, lithology does not seem to be the explanation for the different behaviour. Lowermost carbonate units have undergone ductile deformation and are overprinted by later brecciation, whereas structurally higher units appear to have undergone mostly brittle deformation throughout their history. There are no veins in the brecciated marbles, and thus there is no evidence for enhanced fluid pressure. Brittle behaviour of the carbonates thus can be taken to imply that they were derived from shallower crustal levels.

The schist in the footwalls apparently comes from deeper structural levels than the hanging walls, because ductile behaviour is pervasive and is only locally overprinted by brecciation at the fault zone. Boudinaged albite–quartz bands in the footwall contain remnants of high-pressure mineralogies (e.g. glaucophane), but outside the boudins there is pervasive chlorite–epidote alteration. The phyllonitic shear zones immediately beneath the low-angle faults have been cut by the low-angle faults. Together these data imply that the faults that juxtapose the schists below the brecciated marbles are normal faults that have accomplished considerable telescoping of the structural section.

Such behaviour is typical of detachment faults (Wernicke 1981, 1985; Wernicke & Burchfiel 1982; Davis et al. 1986; Davis & Lister 1988), and can be explained by progressive evolution from ductile behaviour to brittle behaviour in a single movement zone during uplift. For this reason the mylonitic or phyllonitic zone does not exactly coincide with the fault zone. The later-formed fault has cut the ductilely deformed rocks during the progressive evolution of a single movement zone.

On a broad scale, there is an overall decrease in ductility from the lowermost structural levels of the series to its uppermost structural levels. The low-angle faults seem to mark locations where there is an abrupt increase (downwards) in apparent ductility, and/or metamorphic grade. At the very top of the pile, the rocks are pervasively brecciated with no evidence of ductility. At the base of the pile there are well-preserved blueschists, with a strongly developed schistose fabrics.

The variation in metamorphic grade within the schist units appears to be related to the degree of retrogression of older mineral parageneses. At the top of the pile, the only relics of the blueschists are found in centimetre-scale boudins (Fig. 10b), whereas at the base of the pile, there are pods of well-preserved eclogites (Forster & Lister 1996d, 1999) that reveal some of the oldest and highest-pressure parts of the $P$–$T$–$t$ history.

The blueschists must have already been exhumed to relatively shallow crustal levels by the time fluid movement along these faults caused a chlorite–epidote overprint of the blueschist facies mineral assemblages. However, these faults were responsible for considerable exhumation, as different structural and metamorphic conditions in the hanging walls and footwalls imply significant relative movement to have occurred. Similar arguments for substantial fault movement have been made by Enrile (1991).

## How was exhumation accomplished?

Several workers have commented that the Cycladic eclogite–blueschist belt was (at least partially) exhumed before the onset of Miocene extensional tectonism (e.g. Wijbrans & McDougall 1986, 1988). This conclusion is based on the occurrence of Oligocene and Eocene $^{40}Ar/^{39}Ar$ cooling ages, obtained from phengitic white mica and coexisting K-feldspar (Wijbrans

&McDougall 1986, 1988; Maluski et al. 1987; Wijbrans et al. 1990; Bröcker et al. 1993; Baldwin 1996; Baldwin & Lister 1998). This (partial) exhumation must have taken place after the Cycladic eclogite–blueschist belt had been thrust over the Hercynian basement terrane, and after the period of high-pressure metamorphism (but see Forster & Lister 1999). No information is available as to the nature of the structures that accomplished such exhumation, however.

Exhumation may have initially taken place as the result of the operation of major extensional shear zones. Ductile fabrics formed in these shear zones indicate that deformation commenced under high-pressure conditions in which glaucophane remained stable, and continued until lower-greenschist facies conditions. By the time that Oligo-Miocene Barrovian facies amphibolite–greenschist regional metamorphism took place (Kreuzer et al. 1978; Altherr et al. 1982; Andriessen et al. 1987) the rocks of the Cycladic blueschist belt were not much deeper than 15–20 km. A significant part of the exhumation history had already been accomplished.

The LANFs that separate the non-metamorphic rocks from the underlying metamorphic tectonites on Naxos, Paros and Mykonos in fact define a distinct late generation of detachment faults (additional to those described for Ios). The upper plates of these extensional systems comprise non-metamorphic rocks inferred to be fragments of the Cyclades ophiolite nappe, emplaced early during the Alpine collision (Dürr et al. 1978). The upper plates include fragments of serpentinized ultrabasic bodies, and immature boulder conglomerates derived from sources unrelated to the underlying igneous and metamorphic rocks in the lower plates. These were responsible for the last stages of exhumation of the Aegean metamorphic core complexes.

## Conclusions

This study changes our view in respect to the way extensional tectonism has taken place in the Aegean region. LANFs have been documented throughout the Cyclades that represent fragments of a number of once linked detachment systems, formed at a range of structural levels and at different times, with different kinematic frameworks operating.

The early LANFs define the Ios Detachment Fault system, and these juxtapose the series and the basement. This anastomosing fault array dissects the earlier formed (south-directed) SCSZ, and fault slices comprise rocks from both the basement and the series that had been earlier caught up in this shear zone.

The Ios Detachment Fault system is younger than the SCSZ but it follows the same domal form, and it apparently has the same south-directed sense of motion. The Ios Detachment may be kinematically coordinated to the SCSZ, although the currently exposed mylonites were kinematically inactive when they were cut by the Ios Detachment. The Ios Detachment formed as a south-directed LANF that was subsequently bowed over the length of the Ios dome. The SCSZ may continue beneath the Sea of Crete, and may contain younger mylonites that were coeval with the Ios Detachment.

The Coastal Fault system is defined by sub-horizontal to gently dipping normal faults that are younger faults than those of the Ios Detachment system. This fault system is associated with N–S extension which is overprinted by an approximately E–W- to NE–SW-trending lineations in W-directed narrow phyllonitic shear zones. This change in kinematics from N–S to roughly E–W may be related to extension in an adjacent (western) domain of the Cyclades, where E–W- to NE–SW-trending extension occurred.

The Coastal Fault system formed subsequent to doming and/or during the late stages of the formation of the Ios dome. This system cuts progressively into deeper structural levels of the series. Near Ios Port the Coastal Fault system has been observed to intersect older faults associated with the Ios Detachment. The Coastal Fault system has attenuated the crustal section, juxtaposing slices from distinctly different crustal levels.

This project was supported by an Australian Research Council Grant, in collaboration with the Institute of Geological and Mineral Exploration (Athens, Greece). In this respect the help and support of V. Avdis is gratefully acknowledged. On Ios our work was made pleasant and enjoyable through the generous assistance and friendship offered by V. Daskalakis and G. Gana, as well as by K. and O. Batsalis. These individuals greatly assisted the progress of this study.

## References

ALTHERR, R., KREUZER, H., WENDT, I., LENZ, H., WAGNER, G.A., KELLER, J., HARRE, W. & HÖHNDORF, A. 1982. A Late Oligocene/Early Miocene high temperature belt in the Attic–Cycladic crystalline complex (SE Pelagonian, Greece). *Geologisches Jahrbuch*, **E23**, 97–164.

ANDRIESSEN, P. A. M., BANGA, G. & HEBEDA, E. H. 1987. Isotopic age study of pre-Alpine rocks in the basal units on Naxos, Sikinos and Ios, Greek Cyclades. *Geologie en Mijnbouw*, **66**, 3–14.

ANGELIER, J. 1978. Tectonic evolution of the Hellenic arc since the late Miocene. *Tectonophysics*, **4**, 23–36.

——, LYBERIS, N., LE PICHON, X., BARRIER, E. & HUCHON, P. 1982. The tectonic development of the Hellenic Arc and the Sea of Crete: a synthesis. *Tectonophysics*, **86**, 159–196.

AVIGAD, D. 1993. Tectonic juxtaposition of blueschists and greenschists in Sifnos Island (Aegean Sea); implications for the structure of the Cycladic blueschist belt. *Journal of Structural Geology*, **15**, 1459–1469.

—— & GARFUNKEL, Z. 1989. Low-angle faults above and below a blueschist belt, Tinos Island, Cyclades, Greece. *Terra Nova*, **1**, 182–187.

—— & GARFUNKEL, Z. 1991. Uplift and exhumation of high pressure metamorphic terrains: the example of the Cycladic blueschist belt (Aegean). *Tectonophysics*, **188**, 357–372.

——, MATTHEWS, A., EVANS, B. W. & GARFUNKEL, Z. 1992. Cooling during exhumation of a blueschist terrane: Sifnos (Cyclades) Greece. *European Journal of Mineralogy*, **14**, 619–634.

BALDWIN, S. L. B., 1996. Contrasting P–T–t histories for blueschists from the western Baja terrane and the Aegean: effects of synsubduction exhumation and backarc extension. *In*: BEBOUT, G. E., SCHOLL, D. W., KIRBY, S. H. & PLATT, J. P. (eds) *Subduction Top to Bottom*. Geophysical Monograph, American Geophysical Union, **96**, 135–141.

—— & LISTER, G. S. 1998. Thermochronology of the South Cyclades Shear Zone, Ios, Greece: the effects of ductile shear in the argon partial retention zone (PRZ). *Journal of Geophysical Research – Solid Earth*, **103 (B4)**, 7315–7336.

BANGA, G. 1983. *A major crustal shear zone in the Greek Islands Ios and Sikinos*. MSc thesis, University of Utrecht.

BLAKE, M. C., BONNEAU, M., GEYSSANT, J., KIENAST, J. R., LEPRVRIER, C., MALUSKI, H. & PAPANIKOLAOU, D. 1981. A geologic reconnaissance of the Cycladic blueschist belt, Greece. *Geological Society of America Bulletin*, **92**, 247–254.

BRÖCKER, M., KREUZER, H., MATTHEWS, A. & OKRUSCH, M. 1993. $^{40}Ar/^{39}Ar$ and oxygen isotope studies of polymetamorphism from Tinos Island, Cycladic blueschist belt, Greece. *Journal of Metamorphic Geology*, **11**, 223–240.

BUICK, I. S. 1988. *The metamorphic and structural evolution of the Barrovian overprint, Naxos, Cyclades, Greece*. PhD thesis, University of Cambridge.

—— 1991. Mylonite fabric development on Naxos, Greece. *Journal of Structural Geology*, **6**, 643–655

DAVIS, G. A. 1988. Rapid upward transport of mylonitic gneisses in the footwall of a Miocene detachment fault, Whipple Mountains, southeastern California. *Geologische Rundschau*, **77**, 191–209.

—— & LISTER, G. S. 1988. Detachment Faulting in Continental Extension: Perspectives from the Southwestern U.S. Cordillera, *In*: CLARK, S. P. JR, BURCHFIEL, B. D. & SUPPE, J. (eds) *Processes in continental lithospheric deformation*. Geological Society of America, Special Papers, **218**, 133–159.

——, REYNOLDS, S. J. & LISTER, G. S. 1986. Structural evolution of the Whipple and South mountains shear zones, southwestern United States. *Geology*, **14**, 7–10.

DROOGER, C. W. & MEULENKAMP, J. E. 1973. Stratigraphic contribution to geodynamics in the Mediterranean area: Crete as a case history. *Société Géologique de Grèce Bulletin*, **10**, 193–200.

DÜRR, S., ALTHERR, R., KELLER, J., OKRUSCH, M. & SEIDEL, E. 1978. The median Aegean crystalline belt: stratigraphy, structure, metamorphism, magmatism. *In*: CLOOS, H., ROEDER, D. & SCHMIDT, K. (eds) *Alps, Apennines, Hellenides*. International Union of Geological Sciences, Report, **38**, 455–477.

ENRILE, J. L. H. 1991. Extensional tectonics of the Toledo ductile–brittle shear zone, central Iberian Massif. *Tectonophysics*, **191**, 311–324.

FAURE, M. & BONNEAU, M. 1988. Données nouvelles sur l'extension néogène de l'Égée: la déformation ductile du granite miocène de Mykonos (Cyclades, Grèce). *Comptes Rendus de l'Académie des Sciences, Série, II*, **307**, 1553–1559.

FORSTER, M. A. 1996. *Deformation and metamorphism of the upper plate, Ios, Cyclades, Greece*. MSc report, Monash University, Melbourne, Vic.

—— & LISTER, G. S. 1996a. Evolution of the Ios upper plate. *In*: LISTER, G. S. & FORSTER, M. A. (eds) *Inside the Aegean Metamorphic Core Complexes: a Field Trip Guide Illustrating the Geology of the Aegean Metamorphic Core Complexes*. Australian Crustal Research Centre, Technical Publication, **45**, 51–60.

—— & —— 1996b. Detachment fault systems on Ios: Traverse 3 – Tower Hill. *In*: LISTER, G. S. & FORSTER, M. A. (eds) *Inside the Aegean Metamorphic Core Complexes: a Field Trip Guide Illustrating the Geology of the Aegean Metamorphic Core Complexes*. Australian Crustal Research Centre, Technical Publication, **45**, 41–46.

—— & —— 1996c. Detachment fault systems on Ios: Traverse 4 – Port Beach shear zone to Koumbara Peninsula. *In*: LISTER, G. S. & FORSTER, M. A. (eds) *Inside the Aegean Metamorphic Core Complexes: a Field Trip Guide Illustrating the Geology of the Aegean Metamorphic Core Complexes*. Australian Crustal Research Centre, Technical Publication, **45**, 47–52.

—— & —— 1996d. Detachment fault systems on Ios: Traverse 5 – Varvara Boudin to the 'Goat Beach'. *In*: LISTER, G. S. & FORSTER, M. A. (eds) *Inside the Aegean Metamorphic Core Complexes: a Field Trip Guide Illustrating the Geology of the Aegean Metamorphic Core Complexes*. Australian Crustal Research Centre, Technical Publication, **45**, 53–60.

—— & —— 1999. Separate episodes of eclogite and blueschist facies metamorphism in the Aegean metamorphic core complex of Ios, Cyclades, Greece. *In*: MACNIOCAILL, C. & RYAN, P. (eds) *Continental Tectonics* Geological Society, London, Special Publications, in press.

GAUTIER, P. & BRUN, J. P. 1994a. Ductile crust exhumation and extensional detachments in the central Aegean (Cyclades and Evvia Islands). *Geodinamica Acta*, **7**, 57–85.

—— & —— 1994b. Crustal-scale geometry and kinematics of late-orogenic extension in the central Aegean (Cyclades and Evvia Island). *Tectonophysics*, **238**, 399–424.

——, —— & JOLIVET, L. 1993. Structure and kinematics of Upper Cenozoic extensional detachment on Naxos and Paros (Cyclades Islands, Greece). *Tectonics*, **12**, 1180–1194.

GRÜTTER, H. S. 1993. *Structural and metamorphic studies on Ios, Cyclades, Greece*. PhD thesis, University of Cambridge.

HENJES-KUNST, F. & KREUZER, H. 1982. Isotopic data of pre-Alpidic rocks from the island of Ios (Cyclades, Greece). *Contributions to Mineralogy and Petrology*, **80**, 245–253.

HOWARD, K. A. & JOHN, B. E. 1995. Rapid extension recorded by cooling age patterns and brittle deformation, Naxos, Greece. *Journal of Geophysical Research*, **100**, 9969–9979.

JACKSON, J. A. & MCKENZIE, D. 1983. The geometrical evolution of normal fault systems. *Journal of Structural Geology*, **5**, 471–482.

KEAY, S. M. 1998. *The geological evolution of the Cyclades, Greece: constraints from SHRIMP U–Pb geochronology*. PhD thesis, Australian National University, Canberra, A. C. T.

—— & LISTER, G. S. 1996a. Evolution of the Naxos core complex. *In*: LISTER, G. S. & FORSTER, M. A. (eds) *Inside the Aegean Metamorphic Core Complexes: a Field Trip Guide Illustrating the Geology of the Aegean Metamorphic Core Complexes*. Australian Crustal Research Centre, Technical Publication, **45**, 61–74.

—— & —— 1996b. The Naxos Detachment Fault. *In*: LISTER, G. S. & FORSTER, M. A. (eds) *Inside the Aegean Metamorphic Core Complexes: a Field Trip Guide Illustrating the Geology of the Aegean Metamorphic Core Complexes*. Australian Crustal Research Centre, Technical Publication, **45**, 89–94.

KISSELL, K. & LAJ, C. 1988. The Tertiary geodynamical evolution of the Aegean arc: a paleomagnetic reconstruction. *Tectonophysics*, **146**, 183–201.

KREUZER, H., HARRE, W., LENZ, H., WENDT, I., & HENJES-KUNST, F. 1978. K/Ar and Rb/Sr Daten von Mineralen aus dem polymetamorphen Kristallin der Kykladen, Insel Ios (Griechenland). *Fortschritte der Mineralogie*, **56**, 69–70.

LEE, J. & LISTER, G. S. 1992. Late Miocene ductile extension and detachment faulting, Mykonos, Greece. *Geology*, **20**, 121–124.

LISTER, G. S. & DAVIS, G. A. 1989. The origin of metamorphic core complexes and detachment faults formed during Tertiary continental extension in the northern Colorado River region, U.S.A. *Journal of Structural Geology*, **11**(1–2), 65–94.

—— & FORSTER, M. A. (eds) 1996a. *Inside the Aegean Metamorphic Core Complexes: a Field Trip Guide Illustrating the Geology of the Aegean Metamorphic Core Complexes*. Australian Crustal Research Centre, Technical Publication, **45**.

—— & —— 1996b. The nature and origin of the Aegean core complexes. *In*: LISTER, G. S. & FORSTER, M. A. (eds) *Inside the Aegean Metamorphic Core Complexes. Field Guide to the islands of Thera, Ios, Naxos and Paros*. Australian Crustal Research Centre, Technical Publication, **45**, 11–20.

—— & KEAY, S. M. 1996. The lower plate of Ios Core Complex. *In*: LISTER, G. S. & FORSTER, M. A. (eds) *Inside the Aegean Metamorphic Core Complexes. Field Guide to the islands of Thera, Ios, Naxos and Paros*. Australian Crustal Research Centre, Technical Publication, **45**.

——, BANGA, G. & FEENSTRA, A. 1984. Metamorphic core complexes of Cordilleran type in the Cyclades, Aegean Sea, Greece. *Geology*, **12**, 221–225.

MALUSKI, H., BONNEAU, M. & KIENAST, J. R. 1987. Dating the metamorphic events in the Cycladic area: $^{40}Ar/^{39}Ar$ data from metamorphic rocks of the island of Syros (Greece). *Bulletin de la Société Géologique de France*, **8**, 833–842.

REYNOLDS, G. S. & LISTER, G. S. 1990. Folding of mylonitic zones in Cordilleran metamorphic core complexes: evidence from near the mylonitic front. *Geology*, **18**, 216–219.

RIDLEY, J. 1984. Listric normal faulting and the reconstruction of the synmetamorphic structural pile of the Cyclades. *In*: DIXON, J. E. & ROBERTSON, A. H. F. (eds) *The Geological Evolution of the Eastern Mediterranean*. Geological Society, London, Special Publications, **17**, 755–761.

SCOTT, R. J. & LISTER, G. S. 1992. Detachment faults: evidence for a low-angle origin. *Geology*, **20**, 833–836.

SPENCER, J. E. 1984. Role of tectonic denudation in warping and uplift of low-angle normal faults. *Geology*, **12**, 95–98.

URAI, J. L., SCHUILING, R. D. & JANSEN, J. B. H. 1990. Alpine deformation on Naxos (Greece). *In*: KNIPE, R. J. & RUTTER, E. H. (eds) *Deformation Mechanisms, Rheology and Tectonics*. Geological Societym, London, Special Publication, **54**, 509–522.

VAN DER MAAR, P. A. 1980a. The geology and petrology of Ios, Cyclades, Greece. *Annales Géologiques des Pays Helléniques*, **30**, 206–224.

—— 1980b. *Metamorphism on Ios and the geological history of the southern Cyclades*. PhD thesis, University of Utrecht.

—— 1981. *Geological map of Greece, Island of Ios*, 1:50 000 scale. Institute for Geology and Mineral Exploration (I.G.M.E), Athens.

—— & JANSEN, J. B. H. 1983. The geology of the polymetamorphic complex on Ios, Cyclades, Greece, and its significance for the Cycladic massif. *Geologische Rundschau*, **72**, 283–299.

VANDENBERG, L. C. & LISTER, G. S. 1996. Structural analysis of basement tectonites from the Aegean metamorphic core complex of Ios, Cyclades, Greece. *Journal of Structural Geology*, **18**(12), 1437–1454.

WERNICKE, B. 1981. Low-angle normal faults in the Basin and Range Province: nappe tectonics in an extending orogen. *Nature*, **291**, 645–648.

—— 1985. Uniform-sense normal simple shear of the continental lithosphere. *Canadian Journal of Earth Sciences*, **22**, 108–126.

—— & AXEN, G. J. 1987. On the role of isostasy in the evolution of normal fault systems. Geology, **16**, 848–851.

—— & BURCHFEIL B. C. 1982. Modes of extensional tectonics. *Journal of Structural Geology*, **4**, 105–115.

WIJBRANS, J. R. & MCDOUGALL, I. 1986. $^{40}Ar/^{39}Ar$ dating of white micas from an Alpine high pressure metamorphic belt on Naxos (Greece); the resetting of the argon isotopic system. *Contributions to Mineralogy and Petrology*, **93**, 187–194.

—— & —— 1988. Metamorphic evolution of the Attic Cycladic metamorphic belt on Naxos (Cyclades, Greece) utilising the $^{40}Ar/^{39}Ar$ age spectrum measurements. *Journal of Metamorphic Geology*, **6**, 571–594.

——, SCHLIESTEDT, M. & YORK, D. 1990. Single grain argon laser probe dating of phengites from the blueschist to greenschist transition on Sifnos (Cyclades, Greece). *Contributions to Mineralogy and Petrology*, **104**, 582–594.

# Controls on pseudotachylyte formation during tectonic exhumation in the South Mountains metamorphic core complex, Arizona

LAUREL B. GOODWIN

*Department of Earth and Environmental Science, New Mexico Tech, Socorro, NM 87801–4796, USA (e-mail: lgoodwin@nmt.edu)*

**Abstract:** Pseudotachylyte formed during the later stages of extension in the South Mountains metamorphic core complex. Microstructural and microchemical evidence of pseudotachylyte formation processes is well preserved in granitoid mylonite host rocks because of rapid cooling during tectonic exhumation. Petrographic studies indicate a cataclasite precursor to pseudotachylyte. Transmission electron microscope analysis shows glass in pseudotachylyte, providing evidence of melting. Electron microprobe analyses demonstrate variations in glass, microlite, and crystallite composition with proximity to quartz clasts, indicating reaction between clasts and melt. Amphibole crystallites exhibit sieve textures, interpreted as recording rapid crystallization and subsequent decompression. Decompression is postulated to result from drops in fluid pressure along jogs in the slip surface or following formation of injection veins. Both pseudotachylyte and cataclasite are typically subparallel to C-surfaces in the host mylonite. Cataclasite not associated with pseudotachylyte is composed of quartz + feldspars ± biotite. Pseudotachylyte everywhere contains biotite. The model proposed for formation of pseudotachylyte and cataclasite in foliated granitoid protoliths includes seismic failure along C-surfaces. Where biotite is absent, slip produces cataclasite. Where biotite is present, initial cataclasis may be followed by frictional melting, facilitated by biotite's low melting temperature and lowering of the melting temperatures of other phases by release of water during biotite breakdown.

Formation of pseudotachylyte is not in itself an exhumation process. However, pseudotachylyte commonly forms during exhumation and cooling of deep crustal shear zones. In particular, it appears to be an integral part of the deformation that accommodated extension and exhumation of metamorphic core complexes in the eastern Cordilleran metamorphic belt (S. J. Reynolds, pers. comm. 1992). This paper is focused on formation of pseudotachylyte in the South Mountains metamorphic core complex of Arizona. A key point developed here is the important mechanical weakening effect created by concentrating phyllosilicate phases along foliation planes. In short, the development of foliations at depth can influence subsequent deformation at shallower crustal levels as rocks are exhumed.

Pseudotachylyte is considered one of the few unambiguous indicators of palaeoseismic activity (see Magloughlin 1993; Means 1993; Swanson 1993; Tullis 1993). Yet there is ample evidence of episodic pseudotachylyte formation within ductile shear zones, and mylonitic overprinting of pseudotachylyte veins has been well documented (e.g. Sibson 1980; Passchier 1982, 1984; Wenk & Weiss 1982; Hobbs et al. 1986; Swanson 1988; Reynolds et al. 1998). The deformation path by which pseudotachylyte develops in ductile shear zones is a point of discussion.

Three hypotheses have been proposed: (1) pseudotachylyte within mylonite zones records deformation at the brittle–ductile transition (e.g. Sibson 1977; Passchier 1982, 1984; White & White 1983; Koch & Masch 1992); (2) brittle failure occurs below the seismogenic zone because faults nucleate within the brittle regime and propagate downward into mylonite zones or fractures initiate at rheological boundaries that act as stress risers (Sibson 1980); (3) pseudotachylyte in mylonite zones records transient plastic instabilities that locally raise temperature and result in melting (Sibson 1980; Hobbs et al. 1986; White 1996).

As White (1996) noted, each of these paths constrains the conditions of pseudotachylyte formation. The first path requires ambient temperatures of less than 400°C (Sibson 1984). The second path is limited by the distance over which faults can propagate or the magnitude of stress concentration possible during ductile flow. Finally, conditions under which the third path can be followed are defined by a limiting critical upper temperature for a given rock above which thermo-mechanical instabilities cannot be achieved (Hobbs et al. 1986; Ord & Hobbs 1989).

Geological evidence for pseudotachylyte formation by each of these paths will, of course, be different. Until recently, workers advocating production of pseudotachylyte by ductile instabilities relied on theoretical analyses. White (1996), however, provided microstructural and mineralogical evidence supporting this process. Specifically, he demonstrated ambient pressure and temperature conditions that, combined with apparently low fluid pressures, are sufficient for the generation of plastic instabilities. In addition, he showed that strain was accommodated in mylonites largely by dislocation glide; fractures are notably absent, even adjacent to and within porphyroclasts contained in pseudotachylyte.

Distinction between the first two deformation paths by which pseudotachylyte may form must be made by considering whether or not deformation occurred under greenschist facies conditions. In both cases, however, the emerging model for pseudotachylyte formation incorporates both extreme comminution through cataclasis and frictional melting (Magloughlin 1989, 1992) and explains clast distribution and composition in terms of the mechanical characteristics of the phases in the protolith to the pseudotachylyte (Spray 1992). Magloughlin (1989, 1992) documented evidence of a cataclasite precursor to frictional melt in pseudotachylyte. Shimamoto & Nagahama (1992) studied clast size distributions in pseudotachylyte and found that the larger grains showed a range in size predicted by a cataclasis model, but the smaller grains exhibited less variation in size, and smaller grain sizes, than anticipated. Their observations are consistent with initial cataclasis followed by frictional melting of the finer size fraction of clasts. Spray (1992) presented convincing evidence that phases with lower shear yield strengths, fracture toughnesses, and thermal conductivities would be susceptible to relatively greater grain size reduction (through comminution and thermal shock) than other phases. Subsequent frictional melting would occur preferentially in the finer-grained phases. Maddock (1992) emphasized the importance of localization of hydrous mafic phases along foliation planes. Finally, Spray (1995) conducted high-speed slip experiments on Westerly granite and documented the following sequence of events: fracture, progressive comminution, surface melting of mineral fragments, fragment–fragment adhesion, and production of a fragment-laden, melt-supported suspension. He subsequently suggested that whether frictional melting follows cataclasis, and the amount of cataclasite melted, will depend on strain rate, with the percentage of melt increasing with increasing strain rate.

Pseudotachylyte is found in all of the metamorphic core complexes of Arizona (S. J. Reynolds, pers. comm. 1992), and has been reported in a metamorphic core complex in southeastern California (John 1987). Despite this widespread distribution, detailed study of pseudotachylyte formed in this extensional tectonic regime has been limited to preliminary work by Goodwin et al. (1998) and Reynolds et al. (1998). In this paper, I report on the characteristics of pseudotachylyte formed in granitoid mylonites of the South Mountains metamorphic core complex, demonstrate that they formed at the brittle–ductile transition, and document controls on pseudotachylyte formation in this tectonic environment. It is hypothesized that a combination of mineralogy (i.e. the presence of a hydrous mafic phase, such as biotite), structural and mineralogical anisotropy (strong foliation(s) and segregation of phases into mafic and felsic domains), and an appropriately oriented stress field (so that slip was localized along pre-existing foliation planes) were required to form pseudotachylyte in the South Mountains. Where this combination was not present at least locally (i.e. biotite in folia within felsic areas) cataclasite could form, but frictional melting did not follow cataclasis.

## Geological setting

Considerable attention has been focused on low-angle extensional fault systems since the recognition of a belt of metamorphic core complexes spanning the length of the eastern Cordillera of North America (Coney 1980a,b). These complexes are characterized by the association of extensive mylonite zones with relatively younger low-angle normal faults (Davis 1980; Rehrig & Reynolds 1980). These structures record the transition from ductile to brittle deformation during crustal thinning.

The South Mountains metamorphic core complex is one of the most comprehensively studied (Reynolds 1985; Reynolds et al. 1986; Reynolds & Lister 1987, 1990; Smith et al. 1991) and arguably the simplest metamorphic core complex in the eastern Cordillera. It can be divided into lithologically distinct segments: a western Precambrian half comprising gneiss and minor granite and an eastern half which consists largely of a composite middle Tertiary pluton (Fig. 1). Both areas record core complex deformation and both are cut by late- to post-mylonitic dykes; many of these dykes are chemically and lithologically similar to phases of the pluton. Cross-cutting relationships between the mylonitic fabric, the pluton in the eastern half of the South Mountains, and the dykes allow relative age relationships to be determined.

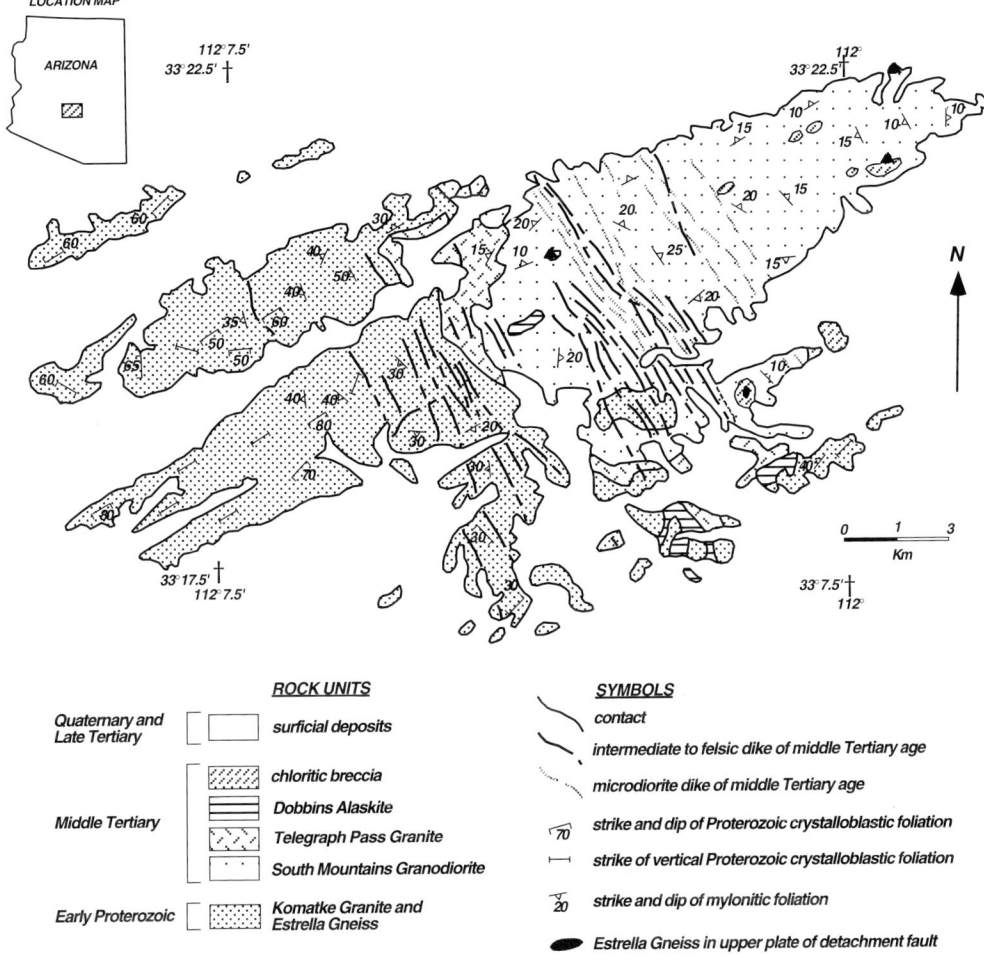

**Fig. 1.** Simplified geological map of the South Mountains metamorphic core complex, adapted from Reynolds (1985). ▲, Sample localities.

These field observations clearly indicate that mylonitization followed each pulse of Tertiary magmatism, with the exception of the youngest dykes (Reynolds 1985).

The composite pluton is dated at 22 ± 4 Ma by U–Pb (zircon) methods (Reynolds 1985; Reynolds *et al.* 1986). Stable isotopic, fluid inclusion, and petrographic data indicate that ductile deformation was taking place at *c.* 600°C, whereas brittle deformation began to dominate after cooling to *c.* 400°C (Smith *et al.* 1991). Fission-track dates on apatite record cooling to *c.* 110°C by *c.* 17.5 Ma (Fitzgerald *et al.* 1994). Enough thermochronological data have been obtained from the South Mountains granodiorite (Fitzgerald *et al.* 1994) to construct a cooling curve, on which deformational events may be plotted (Fig. 2). Although deformation took place at elevated temperatures, it should be noted that the dykes and plutons of the area show evidence of shallow to moderate depths of intrusion (Reynolds 1985). As deformation was broadly syn-intrusive, this suggests that deformation (including pseudotachylyte formation) occurred under low confining pressure, estimated at less than 300 MPa. High-quality thermochronological data are not available for the late- to post-mylonitic dykes. The one K–Ar date on hornblende from a younger dioritic dyke, 28.4 ± 0.5 Ma, is clearly inaccurate, as it suggests that

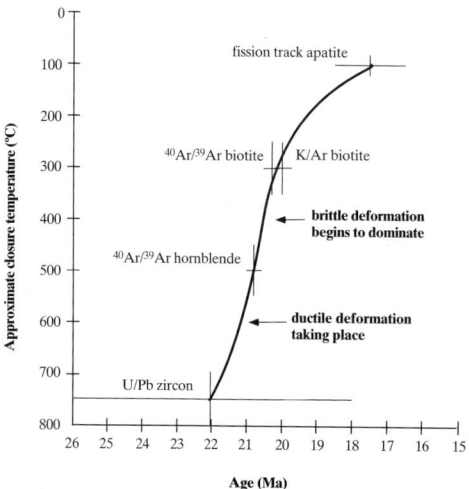

**Fig. 2.** Cooling curve for South Mountains granodiorite. Data from Reynolds (1985), Reynolds et al. (1986) and Fitzgerald et al. (1994). Temperatures indicated by arrows determined by Smith et al. (1991). Approximate closure temperatures from Harrison et al. (1979, 1985), Harrison (1981) and Ghent et al. (1988).

the dyke is older than the pluton it cuts; this anomalously old date is believed to reflect excess argon (Reynolds et al. 1986).

The South Mountains metamorphic core complex is typical of such complexes in Arizona: in addition to mylonitic rocks and a detachment fault, it contains pseudotachylyte veins. The veins are associated with the mylonite; they are generally subparallel to and locally cross-cut the mylonitic fabric and range from undeformed to strongly mylonitized. The eastern half of the South Mountains offers the perfect opportunity to investigate pseudotachylyte formation because: (1) it is possible to consider these studies in the broader context of a well-established structural history; (2) unlike Precambrian rocks to the west, the middle Tertiary pluton has experienced only one deformational event, so structural complexities caused by overprinting are absent; (3) the protolith to the pseudotachylyte fits the profile of a 'typical' pseudotachylyte-bearing mylonite, both structurally and mineralogically; (4) fluids present during deformation were related to intrusion, and remained close to chemical and oxygen or hydrogen isotope equilibrium with the composite pluton throughout the ductile-to-brittle deformation of the rock (Smith et al. 1991); the system to be studied is therefore as close to chemical equilibrium as one could hope to find in nature; (5) the rapid cooling history (Fig. 2), and lack of introduction of fluids from an external source, represent ideal conditions for preservation of microstructures that record the deformation history.

## Pseudotachylyte from the South Mountains metamorphic core complex

Five samples of pseudotachylyte, collected from three areas of optimum exposure (Fig. 1), were analysed in detail. Four of these samples were collected from the volumetrically dominant South Mountains granodiorite; one was obtained from the Dobbins Alaskite. The host rock in all cases is an S–C mylonite (Reynolds 1985). One sample of the granodiorite mylonite was situated directly beneath a chlorite breccia developed in footwall rocks beneath the main detachment surface. The remaining samples were collected from the footwall further from the detachment fault. Observations about lateral variations in pseudotachylyte geometry v. foliation geometry were limited to the size of the samples, 10–15 cm, but are confirmed by earlier field and petrographic observations (Reynolds et al. 1998). Details of hand-sample and petrographic study, back-scattered electron (BSE) imaging and electron microprobe analyses, and investigations by transmission electron microscopy (TEM) are elaborated in the following sections.

### Host rock: structural and mineralogical heterogeneity

The host rock to pseudotachylyte in the South Mountains is a variably mylonitized, composite granitoid pluton. Point counts of three of the host mylonites indicate that the samples vary in the range 55–63% plagioclase and potassium feldspar, 19–37% quartz, 6–16% biotite, and 1–2% trace minerals (largely titanite, amphibole, and opaque phases) consistent with the variation noted by Reynolds (1985). The rocks are S–C mylonites. S-surfaces are generally more pervasive than C-surfaces, and are defined in part by mineralogical domains dominated, in order of abundance, by: (1) feldspar, (2) quartz, and (3) biotite (Fig. 3a and b). All minerals are elongate parallel to S-surfaces. Biotite is localized along both S- and C-surfaces, with cleavage planes typically subparallel to the foliations, and is generally finer grained along C-surfaces. C-surfaces are typically discontinuous. The mylonites include a mineral lineation that lies

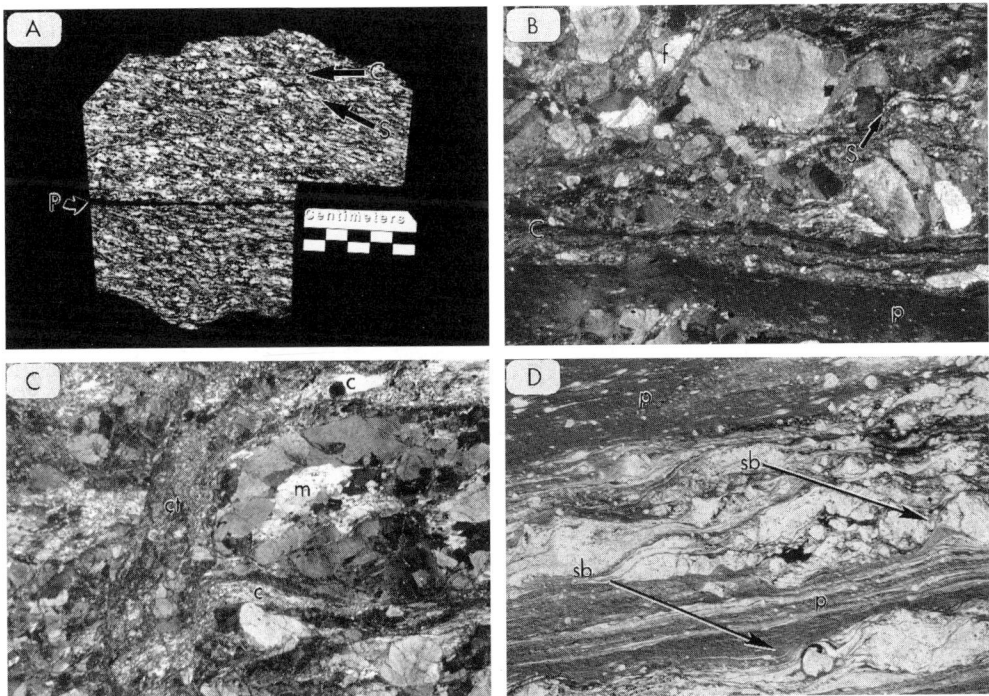

**Fig. 3.** (**a**) Photograph of hand sample of mylonite and pseudotachylyte (P). Labelled arrows indicate S- and C-surfaces. (Note that pseudotachylyte is subparallel to C-surfaces.) Photomicrographs were taken with crossed polars; the field of view of each is 6.5 mm. (**b**) Section cut perpendicular to foliations and parallel to lineations in host rock. Pseudotachylyte (p) is parallel to C-surfaces. (Note fracturing of feldspars (e.g. 'f').) Labelled arrow indicates monocrystalline quartz ribbons defining S-surfaces. 'C' marks C-surface, defined by concentration of biotite and exhibiting sharp boundaries with adjacent quartz and feldspar. (**c**) Section cut perpendicular to C-surfaces and lineation in host rock. Quartz–feldspar cataclasite (ct), locally including small amounts of biotite. Foliations in mylonite are horizontal in this view. 'm' marks mylonite preserved between branches of cataclasite labelled 'c'. Branches parallel to foliations (c) record ductile deformation of cataclasite (note monocrystalline quartz ribbons). (**d**) Section cut parallel to lineations and perpendicular to foliations in host rock. Sample containing ultramylonite formed from pseudotachylyte veins (p) bracketing mylonite. (Note pervasive shear band foliation developed in ultramylonite formed from pseudotachylyte (sb, arrow parallel to shear bands).) Shear bands in mylonite (sb) are more widely spaced than those in ultramylonite.

within S-surfaces, and ridge-in-groove slickenside striae on C-surfaces (see Lin & Williams 1992). The intersection between S- and C-surfaces is roughly at right angles to the lineations, indicating that the lineations were perpendicular to the vorticity axis during flow. Samples were therefore examined in sections cut perpendicular to C-surfaces and both parallel and perpendicular to slickenside striae.

Microstructures in the host rock record non-uniform flow and suggest that deformation was accommodated by different mechanisms in different minerals. Quartz exhibits both monocrystalline and type 1 polycrystalline ribbons (Fig. 3b; see Boullier & Bouchez 1978). Fracturing of feldspar is widespread and individual feldspar grains locally include quartz-filled extension fractures, though some small feldspar porphyroclasts show evidence of subgrain formation. Biotite fish locally exhibit folds and kinks; evidence for intracrystalline cataclasis (see Goodwin & Wenk 1990) is limited. Most biotite is so fine grained that deformation mechanisms cannot be elucidated. Collectively, these observations suggest deformation under greenschist facies conditions (see Simpson 1985). This interpretation is consistent with the thermal history of the South Mountains metamorphic core complex (Fig. 2), which indicates that although mylonitization began under amphibolite facies conditions, it continued through the greenschist facies as the core of the complex was exhumed and cooled. Thus, as the rock will preferentially record its most recent experiences, we would expect greenschist facies microstructures to be preserved in the mylonites. It should be noted, however, that similar microstructures might be produced at higher-grade conditions with faster strain rates.

In thin section, C-surfaces are defined by discrete, very fine-grained, typically biotite-rich but locally quartz-rich zones (Fig. 3b). In many areas, these zones exhibit sharp boundaries with quartz and feldspar grains. C-surfaces are more closely spaced adjacent to some, though not all, pseudotachylyte veins. Fracture densities in the mylonite increase with increasing proximity to pseudotachylyte veins. Transgranular fractures subparallel to pseudotachylyte margins are also common (Fig. 4a).

## Cataclasite

Cataclasite is distinguished from pseudotachylyte by the following characteristics:

(1) Pseudotachylyte includes relatively large polycrystalline quartz clasts and smaller, less common, feldspar clasts. Rock fragments including biotite are locally evident, particularly at branches between generation and injection veins, but biotite clasts are absent. The largest clasts in cataclasite are also typically polycrystalline quartz and feldspar clasts are common but, unlike pseudotachylyte, biotite clasts and clasts containing pseudotachylyte are found locally.

(2) Clasts in pseudotachylyte are generally rounded; clasts in cataclasite are typically angular.

(3) Pseudotachylyte matrix material appears brown in regular thin sections. In ultra-thin sections, it is colourless and includes biotite and hornblende crystals ≤10 μm in size, discussed in detail in later sections. With crossed polars, this matrix appears largely isotropic; only the largest of the crystals are birefringent. The matrix to cataclasite consists of very small clasts, which largely exhibit birefringence with crossed polars.

**Fig. 4.** Samples cut perpendicular to C-surfaces and ridge-in-groove slickenside striae in host mylonite. The field of view of the largest photo is 6.5 mm; all photos are the same scale. (**a**) Crossed polars. Cataclasite (ct) cut by pseudotachylyte (p). (Note sharp boundaries with quartz-rich domain (above) and feldspar-rich domain (below).) Arrow marks transgranular fracture subparallel to cataclasite–pseudotachylyte boundary. (**b**) Mutual cross-cutting relationships between pseudotachylyte (p) and cataclasite (ct). Ductile deformation of pseudotachylyte is evidenced by the development of a foliation and extension of clasts. The upper right-hand corner of the cataclasite includes pseudotachylyte clasts. Pseudotachylyte is injected into anastomosing foliations (a) as well as across foliations (i).

Both light microscope and electron microprobe examination indicate that these grains are dominantly quartz and feldspar, with biotite only locally evident. Where biotite is present, it typically has a 'shredded' appearance.

Cataclasite typically is localized in zones parallel or subparallel to C-surfaces. Locally, cataclasite cuts the mylonite at a high angle to C-surfaces; these high-angle cataclasite zones generally are connected to low-angle zones (Fig. 3c). Boundaries between cataclasite and mylonite are sharp. Polycrystalline quartz clasts record evidence of ductile flow before fragmentation. Some zones of cataclasite contain biotite; others, or parts of others, do not. Where biotite is present, it is typically very fine grained, generally lacks a preferred crystallographic orientation, and may be localized along cataclasite margins or the boundaries of large quartz clasts.

Cataclasite without biotite is never observed associated with pseudotachylyte. Cataclasite with biotite locally grades into pseudotachylyte, as illustrated in Fig. 4, where mutually cross-cutting relationships are evident. In this example, the cataclasite includes porphyroclasts of green biotite surrounded by fine-grained fragments of green biotite in one area. Pseudotachylyte with fine-grained brown biotite crystals cross-cuts and fingers into the cataclasite. The pseudotachylyte records evidence of subsequent ductile flow (i.e. a solid-state foliation; see following sections), and ductilely deformed pseudotachylyte is present in clasts in overprinting cataclasite.

Where cataclasite branches, with segments both parallel and perpendicular to foliations, there is evidence of ductile overprinting of segments parallel to the foliations (Fig. 3c). Elongation of quartz grains and development of monocrystalline quartz ribbons resulted in formation of a strong foliation in these segments. The foliation is absent or less well developed where the cataclasite cross-cuts foliations. Locally, quartz-rich domains in the mylonite are strewn with small feldspar clasts. Both the mean and the maximum grain size of feldspar clasts in these domains are significantly smaller than elsewhere in the mylonite. These domains may also represent areas in which ductile deformation superseded cataclasite formation.

## Pseudotachylyte

Pseudotachylyte from the South Mountains metamorphic core complex is grey to black and largely aphanitic in hand sample, though clasts are locally macroscopically visible. The pseudotachylyte cross-cuts the mylonite, but is itself variably mylonitized (Figs 3d and 5; Reynolds *et al.* 1998). Pseudotachylyte fault or generation veins (criteria for classification have been summarized by Sibson (1975) and Magloughlin & Spray (1992)) range from 1 mm to 1.5 cm in width. Viewed in sections cut parallel to lineations and perpendicular to foliations in the host rock, these veins are roughly planar and are typically parallel or subparallel to C-surfaces (Fig. 3a and b). Examined in detail (sub-millimetre scale), the margins of the veins are locally irregular; feldspar porphyroclasts locally protrude into the veins, and veins in places bulge into biotite-rich S-surfaces (see Sibson 1975). These generation veins occur as isolated features, paired in generation zones (see Grocott 1981; Swanson 1992), or in more complex extensional ladder networks (see Sibson 1975; Swanson 1992). Locally, pseudotachylyte forms the matrix to fine-grained breccias in pull-apart structures. In these more complex geometries, generation veins are linked with injection veins, which exhibit a variety of orientations. Injection veins are most commonly subparallel to S-surfaces, but occur in other orientations, including at a high angle to the C-surfaces. The least deformed pseudotachylyte vein studied is locally subparallel to C-surfaces, but part of the vein cuts across foliations and terminates within the hand sample. These observations suggest that at least part of the sample is an injection vein. Viewed in sections cut at right angles to C-surfaces and ridge-in-groove slickenside striae (Fig. 4b), both the foliations and the pseudotachylyte veins are less planar than they appear in sections cut parallel to lineations. Injection veins both cut across foliations and are preferentially localized along anastomosing foliation planes. Thus, in three dimensions, both foliations and pseudotachylyte veins locally anastomose around lens-shaped, feldspar-rich mineralogical domains.

Pseudotachylyte locally grades into cataclasite; cataclasite is also found in pockets within and along the margins of pseudotachylyte (Fig. 4). Most of the cataclasite preserved along the margins of the pseudotachylyte is composed of quartz and feldspars. Large clasts in pseudotachylyte are typically rounded, may have irregular margins, and are dominantly polycrystalline quartz. Polycrystalline quartz clasts exhibit ductile microstructures, consistent with observations made at all scales that pseudotachylyte cross-cuts mylonite.

Pseudotachylyte veins in the South Mountains exhibit varying degrees of ductile overprint, from relatively undeformed veins to ultramylonite. This overprint may vary within a single vein, or between veins in a single hand sample. Photomicrographs of ultra-thin sections of the least deformed vein studied are shown in Fig. 5a and b. The sample illustrated experienced

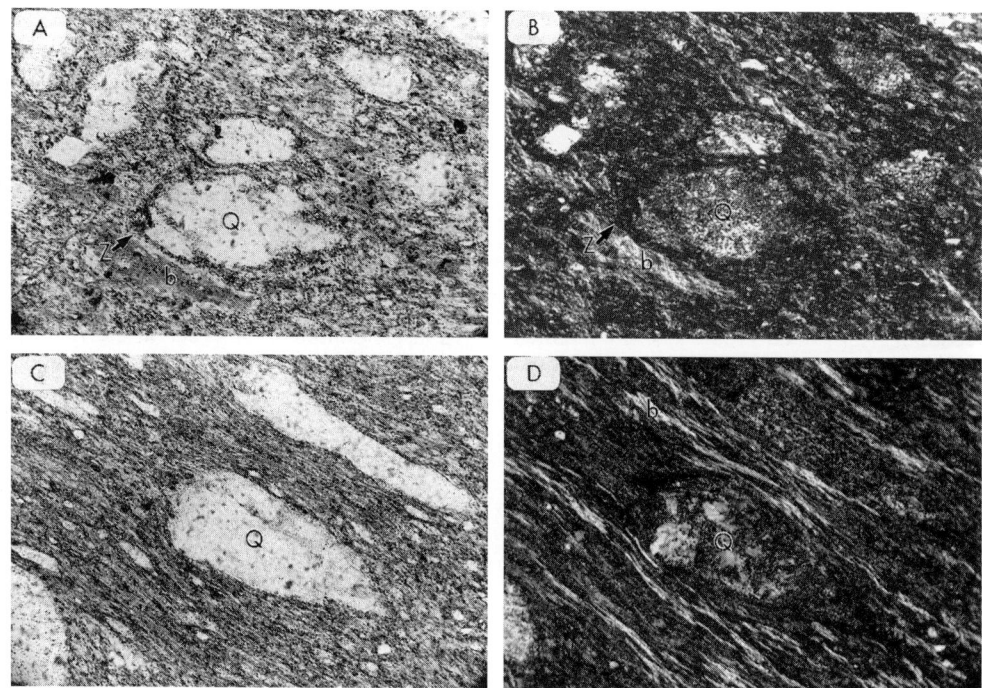

**Fig. 5.** Photomicrographs of ultra-thin sections of variably deformed pseudotachylyte, taken from sections cut parallel to foliations and lineations in the host rock. C-surface traces trend roughly from the upper left to the lower right of each photo. The field of view of each photomicrograph is 1.3 mm. (**a**) Plane-polarized light. Distribution of phases in pseudotachylyte. Zones (Z) including tiny acicular dark crystals rim quartz clasts (e.g. Q). 'b' indicates biotite-rich area. (**b**) Crossed polars. Zones (Z) rimming quartz clasts (e.g. Q) are isotropic. Irregular areas between rimmed clasts are rich in biotite (e.g. 'b') and include patches of isotropic material. (**c**) Plane-polarized light. Quartz clasts (Q), zones rimming clasts, and areas between these zones are all extended to form a foliation parallel to C-surfaces. (**d**) Crossed polars. Birefringent biotite (b) is concentrated along foliation planes in ductilely deformed pseudotachylyte. (Note fine-grained portions of quartz clasts in (**b**) and (**d**).)

relatively minor deformation subsequent to formation of pseudotachylyte. Minor ductile deformation is indicated by a moderately developed crystallographic preferred orientation of euhedral biotite grains, and by elongation of quartz clasts parallel to the C-surfaces and associated ridge-in-groove slickenside striae in the host mylonite. The quartz clasts include fine-grained, irregularly shaped, apparently recrystallized areas; these are found at the margins of large clasts, and pervade most small clasts (Fig. 5). In addition, this weakly deformed sample exhibits a regular distribution of phases within the pseudotachylyte. Quartz clasts have irregular, rounded margins, and are rimmed by tiny (c. 10 μm or less in size), acicular, dark crystals in a colourless matrix. Both the crystals and the matrix are optically isotropic (compare Fig. 5a and b). Between these rimmed clasts are irregular-shaped areas rich in <10 μm long biotite crystals in a fine-grained matrix of colourless material. The biotite crystals in these areas are birefringent, as much of the fine-grained matrix, though patches of the matrix are isotropic.

The distribution of phases shown in Fig. 5a and b is modified through subsequent ductile deformation. Zones around and between quartz clasts are extended into lenses and laminae with deformation, the clasts themselves are stretched, and biotite and the dark acicular phase become very well aligned (Fig. 5c and d). Biotite grains are segregated into foliation planes. Margins with surrounding mylonite remain sharp after ductile overprinting, but the pseudotachylyte develops into ultramylonite (Fig. 3b and d). Ultramylonite developed after pseudotachylyte typically contains C-surfaces, a pervasive shear band (or C′) foliation, and quartz clasts strongly elongate parallel to the slickenside striae in the host mylonite. All of these features record solid-state deformation. Compositional banding in strongly deformed pseudotachylyte (e.g. Fig. 3b) may result from extension of initial heterogeneities, as described above, or be a relict feature of flow banding in the pseudotachylyte. Flow bands have been noted in a little deformed injection vein (Reynolds *et al.* 1998). At this stage in the deformation, biotite grains are aligned in both C-surfaces and shear bands, and are relatively evenly distributed within the ultramylonite. Areas of mylonite sandwiched between pseudotachylyte veins (Fig. 3d) are generally finer grained and more highly deformed than elsewhere in the rock. Shear bands present in such areas are more widely spaced and may be inclined at a different angle from those in ultramylonite.

The sample collected adjacent to the chlorite breccia experienced a greater degree of post-mylonitic microfracturing than the other samples, but a mylonitic foliation can still be distinguished. Pseudotachylyte in this sample exhibits relationships and features similar to those of samples collected elsewhere: pseudotachylyte veins are subparallel to C-surfaces and exhibit varying degrees of ductile overprint, including local folding. An important difference is that significant cataclasis superseded pseudotachylyte formation, though one pseudotachylyte vein cuts all other structures in the rock.

## TEM analysis

TEM studies were focused on determining whether or not glass is present in pseudotachylyte. Glass is characterized in TEM by diffuse ring patterns in selected area diffraction and the absence of Bragg contours in bright field images (see Wenk 1978). These features, which result from the lack of a crystal lattice, are both evident in areas in all four samples of pseudotachylyte examined by TEM. For example, Fig. 6 shows tiny, roughly rectangular biotite crystals surrounded by glass. The composite diffraction pattern in Fig. 6 is produced by diffraction through both glass (diffuse ring pattern) and biotite included within the glass (spot pattern).

## Electron microprobe analysis

BSE images of the least-deformed pseudotachylyte (Fig. 5a and b) were taken with a JEOL JSM-6300 scanning electron microscope with 8 kV accelerating voltage to optimize resolution. Three zones can be distinguished based on the

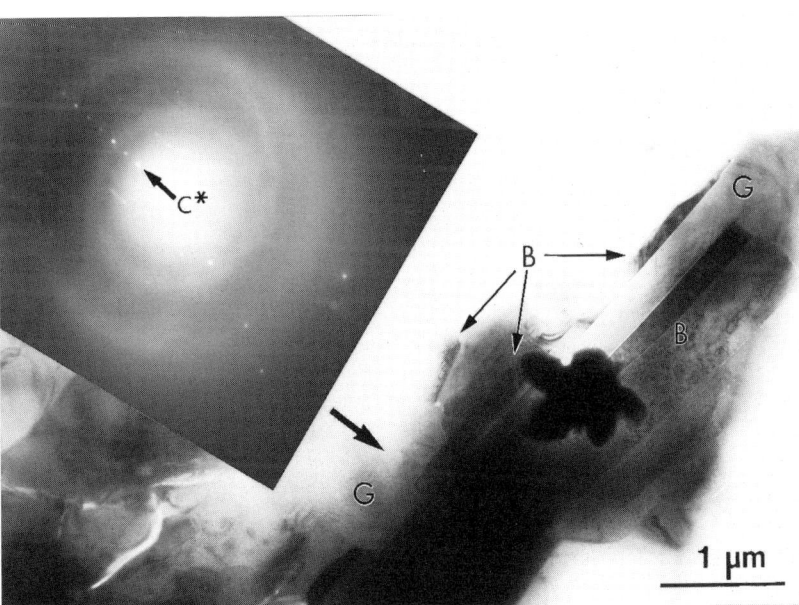

**Fig. 6.** TEM image of biotite crystallites in glass. Boundaries between biotite grains (B) and adjacent glass (G) are sharp and straight. (Note that glass extends between large grain on right and grains marked by arrows to the left.) Glass surrounds the large grain on three sides. Selected area diffraction pattern is a composite of a diffuse ring pattern from the glass and a spot pattern resulting from biotite diffraction. Starfish-shaped 'blob' is contaminant deposited on TEM foil

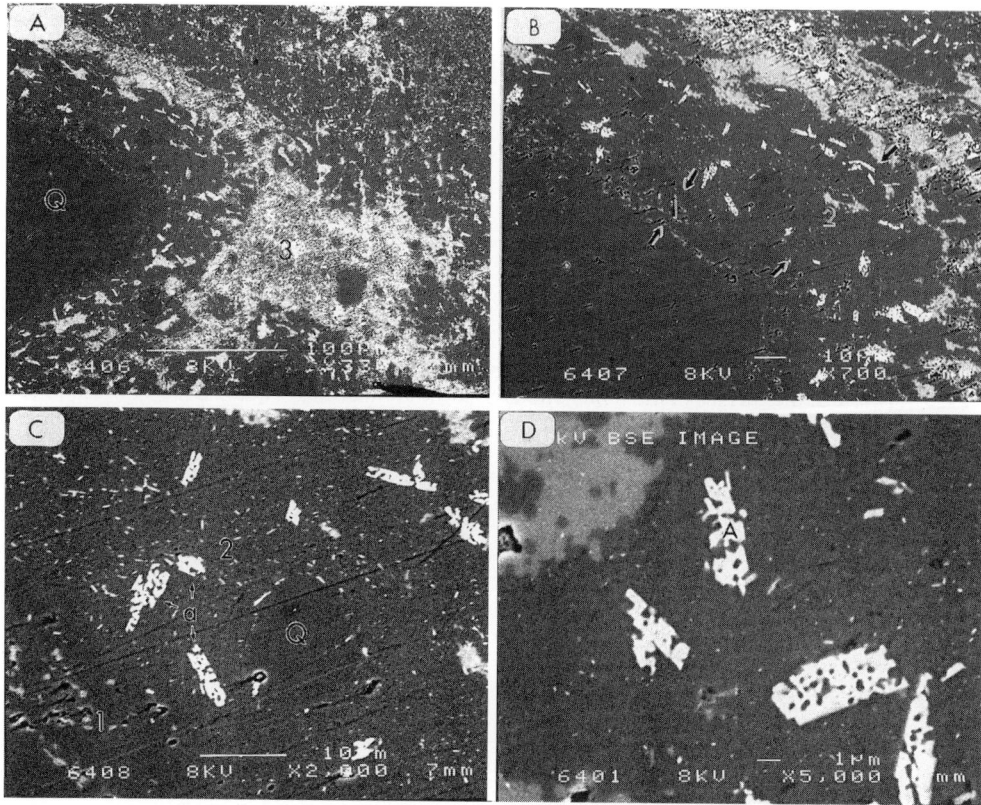

**Fig. 7.** Back-scattered electron images of crystalline and non-crystalline phases in pseudotachylyte. Zones 1, 2, and 3, described in text, are numbered. Relative brightness of phases imaged is as follows: amphibole > biotite > K-feldspar > glass in Zone 2 > glass in Zone 1 > quartz. (**a**) Distribution of phases adjacent to quartz clast (Q) mirrors clast margin. Irregular-shaped regions between quartz clasts (Zone 3) are rich in K-feldspar and biotite. Smaller, black quartz clasts are evident in Zone 2. (**b**) Enlargement showing positions of zones 1 and 2. (**c**) Enlargement of dark area adjacent to quartz clast, showing Zones 1 and 2. Zone 1, bordering the large quartz clast (lower left-hand corner), is Si-rich glass with biotite and K-feldspar crystallites. Zone 2 is glass with amphibole and biotite crystallites. Larger amphibole crystallites (a) exhibit distinctive sieve texture. (**d**) Close-up of sieve texture in amphibole crystallites (A). Zone 2–Zone 3 contact is imaged in upper left-hand corner.

distribution of phases and spatial proximity to quartz clasts (Fig. 7). Boundaries between zones are irregular and locally interfingering. Zones 1 and 2 represent areas in which a clear, colourless, isotropic matrix is visible in thin section. Zone 3 represents the biotite-rich region seen with a light microscope.

The margins of quartz clasts, viewed in BSE images and photomicrographs, are rounded and embayed. Zone 1 rims on quartz clasts show little contrast with the quartz itself in BSE, but they include small, boxy and platy crystals and clusters of crystals (Fig. 7c) that are shown with X-ray mapping to be biotite and K-feldspar. X-ray mapping of Zone 2 indicates the presence of small, platy biotite crystals and acicular amphibole crystals that vary in longest dimension from <1 μm to c. 10 μm. The larger of these crystals exhibit interesting sieve textures, which may be present but not visible in the tiny amphibole crystals (Fig. 7d). Zone 3 is shown by X-ray mapping and spot analyses to be dominantly K-feldspar and biotite.

Analyses of each zone were obtained by rastering the beam over areas between $10 \times 10$ μm and $40 \times 40$ μm in size, depending on the width of the zone at any given point, with a JEOL 733 electron microprobe. Na loss was minimized by the use of a 10 nA beam current; the accelerating voltage was 15 kV. Results are shown in Table 1, in which analyses in each zone have been averaged, and Fig. 8, in which $Al_2O_3$, $K_2O$, and $Na_2O$ have been plotted with respect to $SiO_2$. In addition, spot analyses were made of intercrystalline regions of Zone 2 (Table 1, Zone 2 glass). The beam diameter was varied from 2 to 5 μm to avoid crystals as much as possible, while at the same time maintaining a somewhat defocused beam to minimize sodium loss. For both spot and raster analyses of Zone 1, care was taken to avoid quartz clasts. Areas to be analysed were located with BSE images; examination of BSE and secondary electron images (SEI) subsequent to analysis indicated that quartz clasts were largely sidestepped. Some overlap in the third dimension (i.e. of a clast not visible at the surface of the thin section) is nevertheless possible, but is not deemed sufficient to produce the variations shown here. One suspect point, with 90 wt % $SiO_2$, is evident in Fig. 8a. It should be noted that the intercrystalline regions of Zone 2 have compositions similar to albite with variable amounts of $SiO_2$ added (stoichiometric albite would have roughly 68% $SiO_2$ and comparably greater amounts of $Al_2O_3$ and $Na_2O$). All of these data record trends of increasing $SiO_2$ with increasing proximity to quartz clasts, with a concomitant decrease in $Al_2O_3$, FeO, MgO, and $TiO_2$. $Na_2O$, CaO, and $K_2O$ generally decrease with increasing $SiO_2$. With $SiO_2$ contents typically of c. 69% or less, however, $Na_2O$ and CaO locally drop sharply, and $K_2O$ increases equally abruptly. These represent areas dominated by K-feldspar, largely in Zone 3 but locally present in Zone 2.

Although the analyses as plotted in Fig. 8 suggest smooth variations in composition, a great deal of variation is evident within each zone. This variation is related in part to the heterogeneous distribution of quartz clasts of different sizes, both within and outside the plane of the thin section (Fig. 5). In addition, the nature of the processes that produced the microstructures illustrated appears to be such that the boundaries between zones are not smooth and gradational. Finally, the zones themselves vary in width and include heterogeneous distributions of crystals.

It is evident that the totals of analyses recorded in Table 1 vary from zone to zone. This variation appears to reflect, at least in part, selective plucking during polishing of the slide. Both BSE and SEI photographs document extensive damage of the surface of Zone 3, and less dramatic damage of the other two zones. Areas analysed were chosen to avoid holes as much as possible, though it was impossible to analyse representative portions of Zone 3 without including damaged areas. Variations in water content may also influence totals, as may the presence of trace elements that were not analysed. Small amounts of BaO, for example, were detected locally, but were not deemed sufficiently significant to analyse. Na loss should have been greatest for spot analyses, as the beam rested in one place rather than scanning. In addition, the intercrystalline regions analysed are shown below to be glass, which is more susceptible to Na loss than is crystalline material. Totals of these analyses, however, are very high (mean of 99.61; Table 1, Zone 2 glass), suggesting that Na loss was successfully minimized.

## Discussion and conclusions

### Significance of zones in pseudotachylyte

The least deformed sample of pseudotachylyte from the South Mountains metamorphic core complex includes optically isotropic zones that rim quartz clasts. The fact that the zones are isotropic suggests that they are either not crystalline or contain only crystallites too small to exhibit birefringence. Analyses of these zones (Table 1, Fig. 8) indicate that they have compositions that vary with proximity to the clasts. This

**Table 1.** *Electron microprobe analyses of zones in pseudotachylyte, given as weight per cent oxide ± 1σ*

|  | Zone 1 (12 analyses) | Zone 2 (19 analyses) | Zone 2 Glass (26 analyses) | Zone 3 (9 analyses) |
|---|---|---|---|---|
| $SiO_2$ | 79.44 ± 4.50 | 73.10 ± 4.35 | 72.03 ± 2.23 | 69.88 ± 2.77 |
| $TiO_2$ | 0.06 ± 0.04 | 0.10 ± 0.02 | NA | 0.12 ± 0.02 |
| $Al_2O_3$ | 11.69 ± 2.88 | 14.89 ± 2.45 | 17.03 ± 1.53 | 16.08 ± 1.48 |
| FeO | 0.35 ± 0.22 | 0.56 ± 0.05 | 0.18 ± 0.25 | 0.69 ± 0.36 |
| MgO | 0.11 ± 0.05 | 0.17 ± 0.08 | 0.07 ± 0.09 | 0.19 ± 0.06 |
| CaO | 0.97 ± 0.47 | 1.18 ± 0.22 | 0.89 ± 0.27 | 0.93 ± 0.65 |
| $Na_2O$ | 6.02 ± 1.63 | 6.94 ± 1.86 | 9.24 ± 1.97 | 4.49 ± 2.50 |
| $K_2O$ | 0.66 ± 0.78 | 1.41 ± 2.08 | 0.26 ± 0.38 | 5.70 ± 3.97 |
| Total | 99.28 | 98.35 | 99.61 | 98.06 |

NA, not analysed.

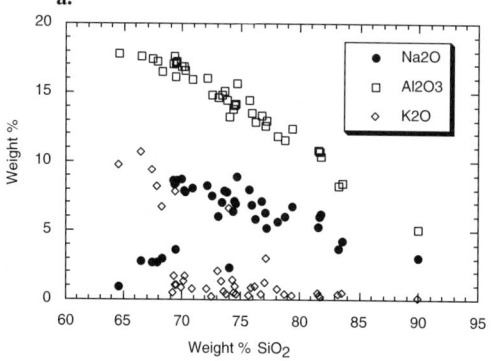

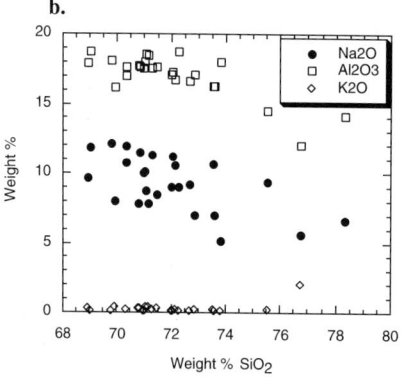

**Fig. 8.** Plots of $Na_2O$, $Al_2O_3$, and $K_2O$ v. $SiO_2$ from analyses averaged for Table 1. (**a**) Scanned areas of Zones 1, 2, and 3. (**b**) Spot analyses of intercrystalline glass from Zone 2. (See text for discussion.)

variation in composition, the fact that Zones 1 and 2 mimic the morphology of clast margins, and the rounded and irregular character of clast margins, all suggest that Zones 1 and 2 are composed largely of glass, and that the melt from which the glass was quenched reacted with quartz clasts. TEM analyses, which indicate that glass is present in all four samples of pseudotachylyte analysed to date, including this little deformed sample, support this interpretation. In particular, Fig. 6 shows biotite crystals surrounded by glass. The tiny crystals of biotite and amphibole found in all three zones are therefore interpreted to be microlites and crystallites that crystallized subsequent to melting. Fine-grained K-feldspar that composes much of Zone 3 is similarly interpreted to have crystallized from a melt. Other workers have also found reaction rims around clasts in pseudotachylyte, which may include glass, devitrified glass and crystals (e.g. Ermanovics et al. 1972; Lin 1994).

These relationships suggest melting of all major phases in the protolith except quartz, which shows evidence of reaction with the melt but not wholesale melting. K-feldspar has the maximum melting temperatures of the phases melted (see table 1 of Spray 1992). Orthoclase melts at $1150 \pm 20°C$ dry at 0.1 MPa (Shairer & Bowen 1955), providing a constraint on the temperature at which pseudotachylyte in the South Mountains formed. Water released by the breakdown of biotite (c. 650°C) would lower this melting point a certain amount, depending on how much water was given off.

Following melting of other phases, quartz crystals partially dissolved. The compositional heterogeneity of the veins suggests that no convection took place following melting. Partial dissolution therefore involved two processes: reaction at the crystal–melt interface and diffusion. Zhang et al. (1989) conducted experimental and theoretical studies of diffusive dissolution of a number of phases, including quartz, in an andesitic melt. Quartz was partially dissolved in the melt at c. 1300°C and 550 MPa. Zhang et al. found that, without convection, the interface melt rimming the dissolving crystal reached a constant 'saturation' composition in less than a second. Concentration profiles of all components subsequently propagated into the melt at a rate proportional to the square root of the run duration; diffusion is thus the rate-determining step for crystal dissolution. Diffusion rates are influenced by a number of factors. At temperatures close to the liquidus of the crystal, diffusion rates slow because the melt and crystal are near equilibrium. In addition, the higher the $SiO_2$ content of the melt, the more polymerized it becomes, slowing diffusion through the melt (Hofmann 1980; Watson 1982; Lesher & Walker 1986). Quartz generally exhibits normal diffusion profiles, with $SiO_2$ increasing with proximity to the crystal, and most other components decreasing (Zhang et al. 1989). $K_2O$ is an exception, showing uphill diffusion into the more polymerized melt. The dissolution distance, measured as the difference in thickness of the partially dissolved quartz crystal before and after the experiment, was $18 \pm 14$ μm in 1 h; the calculated dissolution distance is $18 \pm 4$ μm (Zhang et al. 1989).

How can this work be applied to pseudotachylyte from the South Mountains? The composition of the pseudotachylyte melt, inferred from analyses presented in Table 1, is significantly different from the andesitic melt studied by Zhang et al. (1989); however, diffusivities measured during the latter experiments agree well with those determined for quartz dissolution in a basaltic melt at 1300°C (Watson 1982) and rhyolite–dacite diffusion couples, where the interface melt composition is rhyolite (Baker 1988). If we consider the zones around quartz to represent 'frozen' compositional gradients in the original melt, then all components, including $K_2O$, exhibit normal diffusion profiles (Fig. 8). It is not clear why $K_2O$ does not show evidence of uphill diffusion toward the more $SiO_2$-rich melt. Dissolution distances reported by Zhang et al. (1989) seem reasonable for the South Mountains given the scale of the microstructural and microchemical heterogeneities observed (Table 1; Figs 7 and 8) and the short time scales required for cooling of thin pseudotachylyte veins.

## Crystallites and microlites

Crystal morphologies provide information about growth rates and the degree of supercooling of the liquid in which the crystal grew (Lofgren 1980). Rapid crystallization from a melt typically results in skeletal crystals in all minerals studied (Lofgren 1980). Plagioclase shows changes from tabular crystals, to elongate skeletal crystals, to dendritic crystals, and finally to spherulitic crystals with increasing degree of supercooling (Lofgren 1974). In the case of pseudotachylyte from the South Mountains, BSE and TEM analyses show that biotite crystallites and microlites are tabular in form and amphibole crystallites have unusual sieve textures. These sieve textures are distinct from skeletal textures (see Lofgren 1980). Sieve textures in amphibole of the South Mountains could, however, form through partial dissolution of initially tabular or skeletal crystals. If we consider amphibole crystallites to have had an initially tabular or skeletal form, the sieve textures suggest that amphibole and biotite crystallized rapidly before substantial supercooling of the melt.

Sieve textures in plagioclase from volcanic rocks are believed to record disequilibrium between crystals and melt, resulting in partial resorption of the plagioclase crystals (Nelson & Montana 1992). Two mechanisms by which disequilibrium can be produced have been proposed to explain these textures: (1) magma mixing (e.g. Dungan & Rhodes 1978; Tsuchiyama 1985); (2) magmatic decompression (e.g. Nelson & Montana 1992). The former has been demonstrated in theory only, and seems an unlikely mechanism to explain formation of sieve textures in amphibole in pseudotachylyte of the South Mountains. This mechanism would require introduction of a melt of significantly different composition subsequent to amphibole crystallization. No evidence for mixing of melts of different composition has been found in the South Mountains. Magma mixing also would have to occur in a way that did not physically disturb zones around quartz clasts, or zones would have to develop subsequent to mixing. Neither of these possibilities seems likely, particularly in the short time frames required for crystallization (see below).

Both theory and experiment demonstrate that sieve textures similar to those seen in amphibole (Fig. 7) can form in plagioclase through rapid decompression following crystallization (Nelson & Montana 1992). Essentially, the pressure drop changes the composition of the stable plagioclase phase, resulting in partial dissolution of the crystallites. If amphibole sieve textures were generated in a similar fashion, then pseudotachylyte records a pressure drop following melting and subsequent crystallization. Fluid pressure drops in pseudotachylyte might be created by (1) irregularities, or jogs, on the slip surface (S. Cox, pers. comm. 1995) and/or (2) bleeding of pseudotachylyte into injection veins through hydraulic fracturing following melt generation. Either of these processes would happen within short time spans. It is therefore worthwhile to consider how rapidly crystallization took place.

Most of the data we have on crystallization processes and times come from work on plagioclase (see Lofgren (1980) and Cashman (1993) and references therein). However, crystallization processes and times for different silicate minerals are believed to be similar, so it is worth considering the results of these studies here. Data on plagioclase crystallization rates have, in general, been obtained from mafic dykes and volcanic rocks (e.g. Ikeda 1977; Cashman 1993). Ikeda (1977) focused mainly on basaltic andesite dykes ranging from 5 to 1500 cm wide, and compared the widths of tabular plagioclase grains in individual dykes with cooling rates calculated using Jaeger's (1968) dyke model. He also, however, considered plagioclase growth in felsic plutonic rocks by looking at grain size distributions in the Iritono granodiorite at 300 m and 2 km from its contact with country rock, and in the Uwajima granite at 2 km from its margin. Cooling rates were calculated using the sphere model of Jaeger (1968). Grain size modes were then plotted against crystallization rates in both cases.

Ikeda's (1977) data can be used to provide constraints on rates of crystallization of amphibole in pseudotachylyte from the South Mountains. By considering the maximum width of the most regularly shaped amphibole crystallites, rather than modal width, a maximum estimate of the time required for crystallization can be obtained. The maximum width of the most tabular grains shown in Fig. 7 is roughly 2 μm. An empirical fit through Ikeda's (1977) data from dykes, which have similar morphology to pseudotachylyte veins but more mafic compositions, suggests a crystallization time for these grains of 17 s. An empirical fit through the data from granitoid plutons, which are similar in composition but very different in morphology from pseudotachylyte veins, indicates a crystallization time of less than 10 s. Although these comparisons are by no means ideal, I believe they offer a reasonable order of magnitude estimation of crystallization time. Decompression must therefore have taken place rapidly following frictional melting.

To freeze in these microstructures and preserve glass, pseudotachylyte must have been quenched soon after decompression. The sample collected near the chlorite breccia (formed during the last stage of core complex deformation) exhibits cataclastic deformation of pseudotachylyte, as well as late-stage pseudotachylyte cross-cutting fractures and cataclasite generated during brecciation. Pseudotachylyte veins cut mylonite, but are locally mylonitized. These observations suggest that pseudotachylyte formed in the late stages of mylonitization, when brittle deformation began to dominate, beginning at about 400°C (Fig. 2). Cooling of thin pseudotachylyte veins to ambient temperatures is expected to have occurred rapidly. Although 400°C may at first seem too hot for glass preservation, consideration of the glass transformation temperature ($T_g$, at which the transition from supercooled liquid to glass occurs) for albite, which is a reasonable analogue for much of the glass in pseudotachylyte from the South Mountains (e.g. Table 1), indicates otherwise. The $T_g$ of albite is 763°C (Arndt & Haeberle 1973).

## *Relationship between pseudotachylyte and cataclasite*

It has been established that pseudotachylyte in the South Mountains metamorphic core complex formed by melting. Several lines of evidence suggest that this was frictional melting, rather than melting related to transient ductile instabilities. Microstructures in the host mylonite are consistent with deformation at greenschist facies conditions at the time of pseudotachylyte generation (Fig. 3b). Microfractures are qualitatively observed to increase with increasing proximity to pseudotachylyte.

Transgranular fractures subparallel to pseudotachylyte vein boundaries are common (Fig. 4a). The sharp margins that many C-surfaces make with adjacent quartz and feldspar grains (Fig. 3b) suggest the possibility of frictional sliding on C-surfaces. Finally, and perhaps most important, both cataclasite and pseudotachylyte are found in the South Mountains metamorphic core complex. Both cataclasite and pseudotachylyte cut mylonite, and evidence has been presented of both frictional melting of cataclasite and cataclasis of frictional melt. The two are spatially associated (e.g. Fig. 4) and both show evidence of local ductile overprinting (Fig. 3). These observations suggest broadly contemporaneous brittle and ductile deformation, consistent with greenschist facies deformation.

Zones of cataclasite are generally subparallel to C-surfaces, but locally cut C-surfaces at angles up to 90°. Cataclasite is in many places intimately associated with pseudotachylyte, occurring as pockets in and along the margins of, or interfingered with, pseudotachylyte. Cataclasite that is not associated with pseudotachylyte includes little to no biotite; cataclasite associated with pseudotachylyte always contains biotite. Biotite is the main mafic phase present in the host rock; amphibole is locally present only as an accessory mineral. All pseudotachylyte contains biotite microlites and crystallites, suggesting that the protolith to the melt also contained biotite. As mentioned above, some of the host mylonites contain as little as 6% biotite, but this biotite is typically localized along S- and C-surfaces, particularly C-surfaces. Pseudotachylyte generation veins are generally subparallel to C-surfaces, whereas injection veins are most commonly located along S-surfaces.

These observations collectively indicate that cataclasite with biotite was a precursor to pseudotachylyte in the South Mountains metamorphic core complex. In rocks in which there is little biotite, pseudotachylyte was generated only where that biotite was concentrated, generally along C-surfaces.

## *Model for pseudotachylyte formation*

Based on all of the observations presented, a model is suggested for the generation of pseudotachylyte in the South Mountains metamorphic core complex. In this model, footwall rocks develop a strong mylonitic fabric during deformation from amphibolite through greenschist facies conditions. Strain hardening as the rock cooled is postulated to result in transient brittle failure, causing cataclasis. Brittle deformation is largely localized along C-surfaces in the mylonite. Where the resulting cataclasite contains biotite, frictional melting may, but does not always, follow cataclasis.

Spray (1995) has proposed that frictional melting of cataclasite, and the degree of melting of cataclasite, is dependent on strain rate. At strain rates less than $10^{-2}$ s$^{-1}$, he suggested that only cataclasite will form. For strain rates between $10^{-2}$ and $10^6$ s$^{-1}$, he suggested that the percentage of frictional melt will increase with increasing strain rates. The data presented here provide some support for Spray's model, in that frictional melting clearly did not always follow cataclasis; however, these data also suggest that the model needs to be modified to consider rock composition. Whether or not temperatures reached at a given strain rate are sufficient to melt the rock will depend on the minerals present. This point is emphasized by a literature search, which indicates that pseudotachylyte appears to be generated in quartzite only in impact structures such as Meteor Crater or Vredefort Dome. As shown earlier by Allen (1979), Spray (1992), and Maddock (1992), frictional melting is facilitated by the presence of a hydrous mafic phase.

The following factors contribute to localization of pseudotachylyte along C-surfaces:

(a) C-surfaces are optimally oriented for slip. Pseudotachylyte formed late to post-mylonitization. Therefore, mylonitization and post-mylonitic intrusion of dykes bracket the formation of pseudotachylyte. Extension directions within the South Mountains metamorphic core complex remained constant throughout mylonitization, as indicated by the consistent orientations of extension lineations and slickenside striae on C-surfaces (see Lin & Williams 1992). Post-mylonite dykes are oriented at right angles to lineations in the mylonites (Reynolds 1985). Thus, it is likely that surfaces optimally oriented for ductile shear would be similarly well oriented for brittle shear.

(b) C-surfaces are mechanically weak because of the concentration of crystallographically aligned biotite (see Sibson 1975; Allen 1979; Shea & Kronenberg 1993; Rawling 1997).

(c) Grain-size reduction and subsequent melting of biotite is facilitated by its relatively low strength, thermal conductivity, and melting temperature (Spray 1992).

(d) Water released during melting of biotite decreases melting temperatures of other phases, aiding further melting (Allen 1979).

## Ductile overprinting of pseudotachylyte

Ductile strain was localized in pseudotachylyte after its formation, as indicated by (1) ductile overprinting of pseudotachylyte, (2) evidence of higher strain in mylonite lenses and layers between pseudotachylyte veins (Fig. 3d), and (3) evidence indicating that the veins formed during the waning stages of mylonitization. Passchier (1982), Hobbs et al. (1986), Koch & Masch (1992), and White (1996) have all suggested that grain-size sensitive flow mechanisms operate during ductile deformation of pseudotachylyte. Passchier (1982) and Hobbs et al. (1986) used the activity of grain-size sensitive processes to explain macroscopic superplasticity of deformed pseudotachylyte; White (1996) presented TEM evidence of deformation mechanisms consistent with grain-size sensitive flow. Koch & Masch (1992) pointed out that glasses are known by metallurgists to deform easily by superplastic flow (Haasen 1974). Both grain-size reduction associated with cataclasis and subsequent melting and partial recrystallization are deemed favourable to the operation of grain-size sensitive mechanisms in pseudotachylyte of the South Mountains. Perhaps more important than grain-size reduction is the preservation of glass, which is rheologically weak relative to holocrystalline rock.

Two microscopically pervasive foliations are evident in deformed pseudotachylyte: one is parallel to C-surfaces in the host rock and the other is a shear band (or C') foliation (Fig. 3b and d). A similar fabric was described in phyllonite formed from cataclasite in the Santa Rosa mylonite zone (Goodwin & Wenk 1995). The mean grain size of the phyllonite is c. 3 µm, similar to the mean grain size of pseudotachylyte from the South Mountains. Goodwin & Wenk (1995) suggested a link between the dominant deformation mechanisms operating in a given zone and the character of the fabric that develops. The deformation mechanisms they described in phyllonite, dominantly grain-boundary sliding in quartz, dynamic recrystallization and/or intracrystalline slip in plagioclase, and dynamic recrystallization in concert with intracrystalline slip and mechanical rotation of biotite, are consistent with grain-size sensitive deformation. Thus, pervasive C-surfaces and shear band folia may record grain-size sensitive flow.

From a broader perspective, generation of rheologically weak zones of cataclasite and pseudotachylyte may have allowed ductile deformation to continue under $P$–$T$ conditions in which coarser-grained rocks would have failed brittlely (Reynolds et al. 1998).

*Applicability to other areas*

The protoliths to pseudotachylyte formed in shear zones tend to be strongly foliated, and (where noted) pseudotachylyte generation veins commonly form parallel, or at a small angle (typically <20–30°, occasionally as high as 45°), to the main foliation (e.g. Sibson 1975; Grocott 1981; Passchier 1982, 1984; Swanson 1988; Bossière 1991; Maddock 1992; McNulty 1995). However, there are important instances in which pseudotachylyte is not subparallel to foliations (e.g. Techmer et al. 1992; McNulty 1995). In the cases listed in which the protoliths are well described, the foliation is typically defined in part by hydrous phases, generally biotite and hornblende, which are segregated along foliation planes. In most of these sites, where the metamorphic histories of the rocks are well constrained, pseudotachylyte formed in rocks initially deformed under amphibolite or granulite facies metamorphic conditions. Does this mean that the proposed model for formation of pseudotachylyte in the South Mountains metamorphic core complex can be applied to other pseudotachylytes?

The Outer Hebrides Thrust pseudotachylyte, first described by Sibson (1975) and most recently studied by White (1996), exhibits many of the geometric relationships described here. Evidence for brittle failure associated with pseudotachylyte formation is, however, notably absent, and this led White (1996) to propose that it formed instead through transient plastic instabilities. On the other hand, controls on pseudotachylyte formation in terranes deformed under greenschist facies conditions may well be similar to those described here. In particular, the deformation and cooling history typical of metamorphic core complexes is ideal for the formation of pseudotachylyte in this manner. Initiation of deformation under relatively high-grade metamorphic conditions ensures the formation of a well-developed fabric, in which foliations are defined in part by mineralogical domains. Syntectonic intrusion of granitoids, typical of metamorphic core complexes, increases the likelihood that mylonites with foliations defined by hydrous mafic phases form. Continuation of deformation through the brittle–ductile transition during exhumation and cooling provides the appropriate metamorphic conditions for formation of cataclasite and pseudotachylyte. Instances where pseudotachylyte cross-cuts foliations (e.g. Techmer et al. 1992; McNulty 1995) may be confined to circumstances in which greenschist facies or lower-grade deformation overprints foliations formed during an earlier tectonic regime.

This work would not have been possible without the generous assistance of S. J. Reynolds, who took me to appropriate sample sites in the South Mountains and provided me with samples from his own collection. I thank C. J. Ferranti for point counts of mylonites, and G. Bond for assistance in the search for glass by TEM. M. Spilde bailed me out when the University of New Mexico's JEOL 733 electron microprobe acted up. V. Robertson of JEOL produced BSE images of selected areas of pseudotachylyte during a demonstration of the JEOL JSM-6300. C. Passchier, S. Ralser, and J. Spray provided thoughtful reviews, which resulted in a better paper. Work was supported by the Research Council of New Mexico Tech and by NSF Grant 9304973.

## References

ALLEN, A. R. 1979. Mechanism of frictional fusion in fault zones. *Journal of Structural Geology*, **1**, 231–243.

ARNDT, J. & HAEBERLE, F. 1973. Thermal expansion and glass transition temperatures of synthetic glasses of plagioclase-like compositions. *Contributions to Mineralogy and Petrology*, **39**, 175–183.

BAKER, D.R. 1988. Chemical diffusion in intermediate to silicic melts. *Eos Transactions, American Geophysical Union*, 69, 511.

BOSSIÈRE, G. 1991. Petrology of pseudotachylytes from the Alpine Fault of New Zealand. *Tectonophysics*, **196**, 173–193.

BOULLIER, A.-M. & BOUCHEZ, J.-L. 1978. Le quartz en rubans dans les mylonites. *Bulletin du Société Géologique de France*, **7**, 253–262.

CASHMAN, K. V. 1993. Relationship between plagioclase crystallization and cooling rate in basaltic melts. *Contributions to Mineralogy and Petrology*, **113**, 126–142.

CONEY, P. J. 1980a. Cordilleran metamorphic core complexes: an overview. *In*: CRITTENDEN, M. D. JR., CONEY, P. J. & DAVIS, G. H. (eds) *Cordilleran Metamorphic Core Complexes*. Geological Society of America, Memoir, **153**, 7–31.

—— 1980b. Introduction. *In*: CRITTENDEN, M. D. JR., CONEY, P. J. & DAVIS, G. H. (eds) *Cordilleran Metamorphic Core Complexes*. Geological Society of America, Memoir, **153**, 1–6.

DAVIS, G. H. 1980. Structural characteristics of metamorphic core complexes, southern Arizona. Geological Society of America, Memoir, **153**, 35–78.

DUNGAN, M. A. & RHODES, M. J. 1978. Residual glasses and melt inclusions in basalts from DSDP legs 45 and 46: evidence for magma mixing. *Contributions to Mineralogy and Petrology*, **67**, 417–431.

ERMANOVICS, J. F., HELMSTAEDT, H. & PLANT, A. G. 1972. An occurrence of Archean pseudotachylyte from southeastern Manitoba. *Canadian Journal of Earth Science*, **9**, 257–265.

FITZGERALD, P. G., REYNOLDS, S. J., STUMP, E., FOSTER,

D. A. & GLEADOW, A. J. W. 1994. Thermochronologic evidence for timing of denudation and rate of crustal extension of the South Mountains metamorphic core complex and Sierra Estrella, Arizona. *Nuclear Tracks and Radiation Measurements*, **21**, 555–563.

GHENT, E. D., STOUT, M. Z. & PARRISH, R. R. 1988. Determination of metamorphic pressure–temperature–time ($P$–$T$–$t$) paths. *In*: NISBET, E. G. & FOWLER, C. M. R. (eds) *Short Course on Heat, Metamorphism, and Tectonics*. Mineralogical Association of Canada, 155–188.

GOODWIN, L. B. & WENK., H. R. 1990. Intracrystalline folding and cataclasis in biotite of the Santa Rosa mylonite zone: HUEM and TEM observations. *Tectonophysics*, **172**, 201–214.

—— & —— 1995. Development of phyllonite from granodiorite: mechanisms of grain-size reduction in the Santa Rosa mylonite zone, California. *Journal of Structural Geology*, **17**, 689–707.

——, REYNOLDS, S. J., FERRANTI, C. J., ELLZEY, P. D. & LISTER, G. S. 1998. Pseudotachylyte from the South Mountains metamorphic core complex, Arizona. *In*: SNOKE, A. W., TULLIS, J. A. & TODD, V. R. (eds) *Fault-Related Rocks: A Photographic Atlas*. Princeton University Press, Princeton, NJ, 122–123.

GROCOTT, J. 1981. Fracture geometry of pseudotachylyte generation zones: a study of shear fractures formed during seismic events. *Journal of Structural Geology*, **3**, 169–179.

HAASEN, P. 1974. *Physikalische Metallkunde*. Springer, New York.

HARRISON, T. M. 1981. Diffusion of $^{40}$Ar in hornblende. *Contributions to Mineralogy and Petrology*, **78**, 324–331.

——, ARMSTRONG, R. L., NAESER, C. W. & HARAKAL, J. E. 1979. Geochronology and thermal history of the Coast Plutonic complex, near Prince Rupert, BC. *Canadian Journal of Earth Science*, **16**, 400–410.

——, DUNCAN, I. & MCDOUGALL, I. 1985. Diffusion of $^{40}$Ar in biotite: temperature, pressure, and compositional effects. *Geochimica et Cosmochimica Acta*, **49**, 2461–2468.

HOBBS, B. E., ORD, A. & TEYSSIER, C. 1986. Earthquakes in the ductile regime? *Pure and Applied Geophysics*, **124**, 309–336.

HOFMAN, A. W. 1980. Diffusion in natural silicate melts: a critical review. *In*: HARGRAVES, R. B. (ed.) *Physics of Magmatic Processes*. Princeton University Press, Princeton, NJ, 385–418.

IKEDA, Y. 1977. Grain size of plagioclase of the basaltic andesite dikes, Iritono, central Abakuma plateau. *Canadian Journal of Earth Science*, **14**, 1860–1866.

JAEGER, J. C. 1968. Cooling and solidification of igneous rocks. *In*: HESS, H. H. & POLDERVAART, A. (eds) *Basalts*, 2 Interscience (Wiley), New York, 503–536.

JOHN, B. E. 1987. Geometry and evolution of a mid-crustal extensional fault system: Chemehuevi Mountains, southeastern California. *In*: COWARD, M. P., DEWEY, J. F. & HANCOCK, P. L. (eds) *Continental Extensional Tectonics*. Geological Society, London, Special Publications, **28**, 313–335.

KOCH, N. & MASCH, L. 1992. Formation of alpine mylonites and pseudotachylytes at the base of the Silvretta nappe, Eastern Alps. *In*: MAGLOUGHLIN, J. F. & SPRAY, J. G. (eds) *Frictional Melting Processes and Products in Geological Materials*. Tectonophysics, **204**, 289–306.

LESHER, C. E. & WALKER, D. 1986. Solution properties of silicate liquids from thermal diffusion experiments. *Geochimica et Cosmochimica Acta*, **50**, 1397–1411.

LIN, A. 1994. Glassy pseudotachylyte veins from the Fuyun fault zone, northwest China. *Journal of Structural Geology*, **16**, 71–83.

LIN, S. & WILLIAMS, P. F. 1992. The origin of ridge-in-groove slickenside striae and associated steps in an S-C mylonite. *Journal of Structural Geology*, **14**, 315–321.

LOFGREN, G. 1980. Experimental studies on the dynamic crystallization of silicate melts. *In*: HARGRAVES, R. B. (ed.) *Physics of Magmatic Processes*. Princeton University Press, Princeton, NJ, 487–551.

LOFGREN, G. A. 1974. An experimental study of plagioclase crystal morphology; isothermal crystallization. *American Journal of Science*, **274**, 243–273.

MADDOCK, R. H. 1992. Effects of lithology, cataclasis and melting on the composition of fault-generated pseudotachylytes in Lewisian gneiss, Scotland. *In*: MAGLOUGHLIN, J. F. & SPRAY, J. G. (eds) *Frictional Melting Processes and Products in Geological Materials*. Tectonophysics, **204**, 261–278.

MAGLOUGHLIN, J. F. 1989. The nature and significance of pseudotachylite from the Nason terrane, North Cascade Mountains, Washington. *Journal of Structural Geology*, **11**, 907–917.

—— 1992. Microstructural and chemical changes associated with cataclasis and frictional melting at shallow crustal levels: the cataclasite–pseudotachylite connection. *In*: MAGLOUGHLIN, J. F. & SPRAY, J. G. (eds) *Frictional Melting Processes and Products in Geological Materials*. Tectonophysics, **204**, 243–260.

—— 1993. Identification criteria and evidence for melting in pseudotachylyte, and the relationship to faulting, seismic slip and ultracataclasite. *Geological Society of America Abstracts with Programs*, **25**, A115.

—— & SPRAY, J. G. 1992. Frictional melting processes and products in geological materials: introduction and discussion. *In*: MAGLOUGHLIN, J. F. & SPRAY, J. G. (eds) *Frictional Melting Processes and Products in Geological Materials*. Tectonophysics, **204**, 197–206.

MCNULTY, B. A. 1995. Pseudotachylyte generated in the semi-brittle and brittle regimes, Bench Canyon shear zone, central Sierra Nevada. *Journal of Structural Geology*, **17**, 1507–1521.

MEANS, W. D. 1993. Slickensides as paleoseismographs. *Geological Society of American Abstracts with Programs*, **25**, A115.

NELSON, S. T. & MONTANA, A. 1992. Sieve-textured plagioclase in volcanic rocks produced by rapid decompression. *American Mineralogist*, **77**, 1242–1249.

ORD, A. & HOBBS, B. E. 1989. The strength of the continental crust, detachment zones and the development of plastic instabilities. *Tectonophysics*, **158**, 269–289.

PASSCHIER, C. W. 1982. Pseudotachylite and the development of ultramylonite bands in the Saint-Barthélemy Massif, French Pyrenees. *Journal of Structural Geology*, **4**, 69–79.

—— 1984. The generation of ductile and brittle shear bands in a low-angle mylonite zone. *Journal of Structural Geology*, **6**, 273–281.

RAWLING, G. C. 1997. *Dilatancy and anisotropy during brittle deformation of gneiss*. MS thesis, State University of New York, Stony Brook.

REHRIG, W. A. & REYNOLDS, S. J. 1980. Geologic and geochronologic reconnaissance of a northwest-trending zone of metamorphic core complexes in southern and western Arizona. *Geological Society of America, Memoir*, **153**, 131–157.

REYNOLDS, S. J. 1985. *Geology of the South Mountains, central Arizona*. Arizona Bureau of Geology and Mineral Technology Bulletin, **195**.

—— & LISTER, G. 1987. Structural aspects of fluid–rock interactions in detachment zones. *Geology*, **15**, 362–366.

—— & —— 1990. Folding of mylonite zones in Cordilleran metamorphic core complexes; evidence from near the mylonitic front. *Geology*, **18**, 216–219.

——, GOODWIN, L. B., LISTER, G. S., ELLZEY, P. D. & FERRANTI, C. J. 1998. Development of ultramylonite from pseudotachylyte, South Mountains metamorphic core complex, Arizona. *In*: SNOKE, A. W., TULLIS, J. A. & TODD, V. R. (eds) *Fault-Related Rocks: A Photographic Atlas*. Princeton University Press, Princeton, NJ, 124–125.

——, SHAFIQULLAH, M., DAMON, P. E. & DEWITT, D. 1986. Early Miocene mylonitization and detachment faulting, South Mountains, central Arizona. *Geology*, **14**, 283–286.

SCHAIRER, J. F. & BOWEN, N. L. 1955. The system $K_2O-Al_2O_3-SiO_2$. *American Journal of Science*, **253**, 681.

SHEA, W. T. & KRONENBERG, A. K. 1993. Strength and anisotropy of foliated rocks with varied mica contents. *Journal of Structural Geology*, **15**, 1097–1121.

SHIMAMOTO, T. & NAGAHAMA, H. 1992. An argument against the crush origin of pseudotachylites based on the analysis of clast-size distribution. *Journal of Structural Geology*, **14**, 999–1006.

SIBSON, R. H. 1975. Generation of pseudotachylyte by ancient seismic faulting. *Geophysical Journal of the Royal Astronomical Society*, **43**, 775–794.

—— 1977. Fault rocks and fault mechanisms. *Journal of the Geological Society, London*, **133**, 191–213.

—— 1980. Transient discontinuities in ductile shear zones. *Journal of Structural Geology*, **2**, 165–171.

—— 1984. Roughness at the base of the seismogenic zone: contributing factors. *Journal of Geophysical Research*, **89**, 5791–5799.

SIMPSON, C. 1985. Deformation of granitic rocks across the brittle–ductile transition. *Journal of Structural Geology*, **7**, 503–511.

SMITH, B. M., REYNOLDS, S. J., DAY, H. W. & BODNAR, R. J. 1991. Deep-seated fluid involvement in ductile–brittle deformation and mineralization, South Mountains metamorphic core complex, Arizona. *Geological Society of America Bulletin*, **103**, 559–569.

SPRAY, J.G. 1992. A physical basis for the frictional melting of some rock-forming minerals. *Tectonophysics*, **204**, 205–221.

—— 1995. Pseudotachylyte controversy: fact or friction? *Geology*, **23**, 1119–1122.

SWANSON, M. T. 1988. Pseudotachylyte-bearing strike-slip duplex structures in the Fort Foster Brittle Zone of southern Maine. *Journal of Structural Geology*, **10**, 813–828.

—— 1992. Fault structure, wear mechanisms and rupture processes in pseudotachylite generation. *Tectonophysics*, **204**, 223–242.

—— 1993. Seismo-structural deformation features in dextral strike-slip faults of coastal Maine. *Geological Society of America, Abstracts with Programs*, **25**, A115.

TECHMER, K. S., AHRENDT, H. & WEBER, K. 1992. The development of pseudotachylyte in the Ivrea–Verbano Zone of the Italian Alps. *In*: MAGLOUGHLIN, J. F. & SPRAY, J. G. (eds) *Frictional Melting Processes and Products in Geological Materials*. *Tectonophysics*, **204**, 307–322.

TSUCHIYAMA, A. 1985. Dissolution kinetics of plagioclase in the melt system diopside–albite–anorthite, and origin of dusty plagioclase in andesites. *Contributions to Mineralogy and Petrology*, **89**, 1–16.

TULLIS, T. E. 1993. The difficulty of determining whether slip on an ancient fault was seismic or not. *Geological Society of America, Abstracts with Programs*, **25**, A114.

WATSON, E. B. 1982. Basalt contamination by continental crust: some experiments and models. *Contributions to Mineralogy and Petrology*, **80**, 73–87.

WENK, H.-R. 1978. Are pseudotachylites products of fracture or fusion? *Geology*, **6**, 507–511.

—— & WEISS, L. E. 1982. Al-rich calcic pyroxene in pseudotachylite: an indicator of high pressure and high temperature? *Tectonophysics*, **84**, 329–341.

WHITE, J. C. 1996. Transient discontinuities revisited; pseudotachylyte, plastic instability and the influence of low pore fluid pressure on deformation processes in the mid-crust. *Journal of Structural Geology*, **18**, 1471–1486.

—— & WHITE, S. H. 1983. Semi-brittle deformation within the Alpine fault zone, New Zealand. *Journal of Structural Geology*, **5**, 579–589.

ZHANG, Y., WALKER, D. & LESHER, C. E. 1989. Diffusive crystal dissolution. *Contributions to Mineralogy and Petrology*, **102**, 492–513.

# Quantifying tectonic exhumation in an extensional orogen with thermochronology: examples from the southern Basin and Range Province

DAVID A. FOSTER[1,2] & BARBARA E. JOHN[3]

[1] *Department of Earth Sciences, La Trobe University, Melbourne, Vic. 3083, Australia*
[2] *Present address: Department of Geology, University of Florida, Gainesville, FL 32611, USA (e-mail: dfoster@geology.ufl.edu)*
[3] *Department of Geology and Geophysics, University of Wyoming, Laramie, WY 82071, USA*

**Abstract:** The Colorado River extensional corridor is a region of large-scale crustal attenuation, where the integration of thermochronological techniques illuminates processes of extension and exhumation. Biotite and K-feldspar $^{40}Ar-^{39}Ar$, and zircon and apatite fission-track ages from metamorphic core complexes in the corridor consistently young in the displacement directions of detachment faults. The onset of rapid extension is denoted by an abrupt change, at c. 22–21 Ma, in cooling rate derived from age–closure temperature plots, and in slope of age–distance gradients in footwalls. Linear age–distance trends with slopes of c. 3–8 km Ma$^{-1}$, for low-temperature thermochronometers, give time-averaged displacement rates on detachment faults. Gradients of mineral age across thick tilted hanging wall fault blocks reveal palaeotemperature gradients when the angle of tilting is known. Pre-extensional thermal gradients for tilted blocks in the southern Basin and Range province were c. 20–25°C km$^{-1}$, 17° ± 5°C km$^{-1}$, and c. 20–24°C km$^{-1}$. Thermochronological data from the footwall of the Chemehuevi detachment fault reveal a trend of increasing temperature, in the known slip direction, from ≤200° to 350–400°C over an exposed distance of 23 km, at the onset of extension. The gradual increase in temperature constrains the Chemehuevi detachment fault to have had an original dip of c. 15–30°, within the brittle upper crust.

The Colorado River extensional corridor (Howard & John 1987) of the southern Basin and Range province (Fig. 1) is one of the classical extensional orogens where models for asymmetric extension, detachment faulting, and metamorphic core complex formation were developed (e.g. Davis et al. 1980; Wernicke 1981; Reynolds & Spencer 1985; Howard & John 1987; Lister & Davis 1989). After nearly 20 years of intensive study important aspects of the evolution of this rapidly extended terrane, and the mechanisms for widespread tectonic unroofing of mid-crustal rocks remain unresolved. For example, the initial dip of the low-angle normal (detachment) faults, dip of these faults during movement in the seismogenic upper crust, how detachment faults were modified with displacement, and the relationship between detachment faults and the metamorphic tectonites and plutons exposed in the footwalls are still controversial (e.g. Scott & Lister 1992; John & Foster 1993; Wernicke 1995; Axen & Bartley 1997). In metamorphic core complexes, the majority of exhumation of the footwall rocks can be shown to have been tectonic, but the erosional component of final unroofing also needs to be defined to understand the evolution of these highly extended regions.

This paper aims to define the nature of the tectonic exhumation of mid-crustal rocks exposed in the metamorphic core complexes in the central part of the Colorado River extensional corridor, southeastern California and western Arizona (Figs 1 and 2). We review detailed thermochronological results from a series of our recent publications on specific areas in the extensional corridor, as well as new data from the region. We then show how the integration of those results with structural data helps define rapid extension and tectonic exhumation of continental crust. A number of recent publications present thermochronological datasets from detachment terranes, most of which are restricted to analyses of footwall rocks (e.g. Foster et al. 1990b; Richard et al. 1990; Holm et al. 1992; Dokka 1993; Holm & Dokka 1993; John & Foster 1993; Lee 1994). We demonstrate here the utility of accurately determining the temperature–time histories of rocks from a spectrum of palaeodepths, and from both hanging wall and

FOSTER, D. A. & JOHN, B. E. 1999. Quantifying tectonic exhumation in an extensional orogen with thermochronology: examples from the southern Basin and Range Province. *In:* RING, U., BRANDON, M. T., LISTER, G. S. & WILLETT, S. D. (eds) *Exhumation Processes: Normal Faulting, Ductile Flow and Erosion.* Geological Society, London, Special Publications, **154**, 343–364.

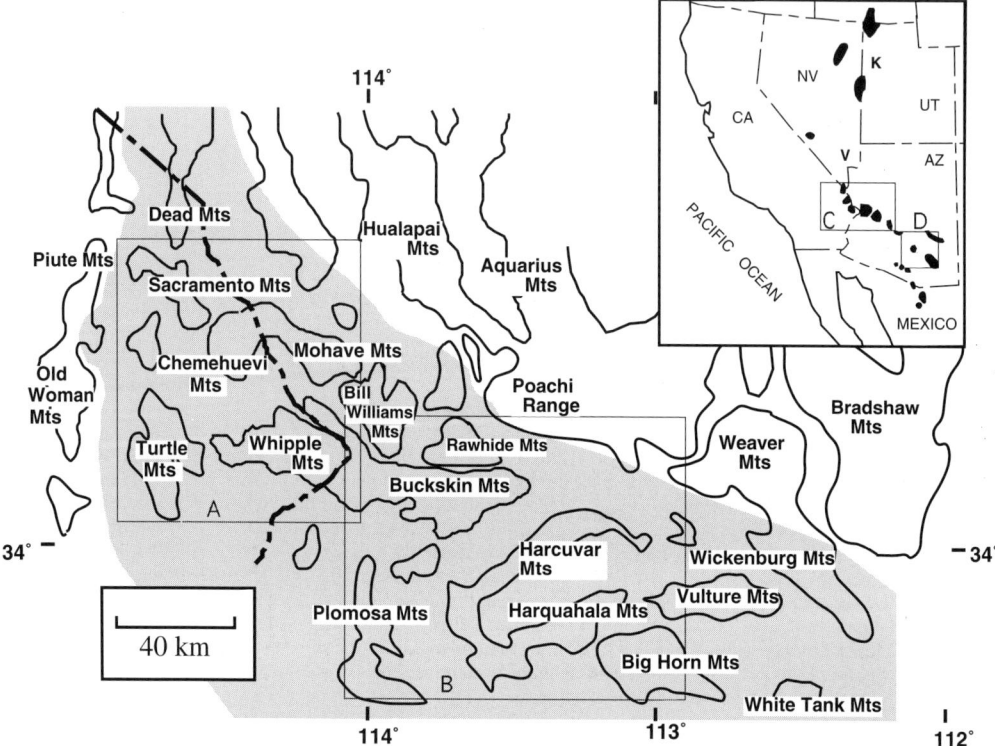

**Fig. 1.** Map showing mountain ranges in western Arizona and southeastern California. The shaded area shows the Colorado River extensional corridor, a region of large magnitude extension west and south of the Colorado Plateau. Boxes A and B show the areas of Figs 2 and 7, box C on inset map shows the location of the map, and box D on the inset map shows the location of Fig. 12. The black areas on the inset map indicate exposed footwall rocks of metamorphic core complexes. Abbreviations on the inset: V, Virgin Mountains; K, Kern–Deep Creek Mountains.

footwall rocks. Specifically, the Colorado River extensional corridor thermochronological results allow the following variables to be assessed: (1) timing and duration of rapid crustal extension; (2) rates of cooling and exhumation of mid-crustal rocks; (3) time-averaged displacement rates on detachment faults; (4) palaeo-geothermal gradients before and during rapid extension; (5) initial dips of detachment faults bounding core complexes and tilted hanging wall crustal sections.

Thermochronology, the application of temperature-sensitive rock dating techniques to the reconstruction of rock thermal histories, is dominated by the $^{40}Ar-^{39}Ar$ and fission-track dating techniques. Fission-track dating is particularly useful for thermal history studies at relatively low temperatures, below about 110–130°C using apatite, and up to about 240–280°C using sphene and zircon (e.g. Gleadow & Duddy 1981; Green et al. 1989; Brandon & Vance 1992; Foster et al. 1996b; Brandon et al. 1998). Higher temperatures (about 150° to ≥500°C) are accessible with $^{40}Ar-^{39}Ar$ analysis of various potassium-bearing minerals (for a review, see McDougall & Harrison (1988)), and developments in the thermochronological applications of this field in the last 15 years have been extremely significant. A recent advance in the ability to define segments of time–temperature histories using K-feldspars is particularly relevant to this study (Lovera et al. 1989, 1991, 1993). Integrating these methods allows investigation of both the thermal and tectonic evolution of rock bodies over the temperature regimes prevailing to significant depths in the Earth's crust.

Thermochronological data reveal detailed information about tectonic exhumation of mid-crustal rocks when the datasets are comprehensive enough to give the pre- and syn-extension thermal conditions of hanging wall and footwall

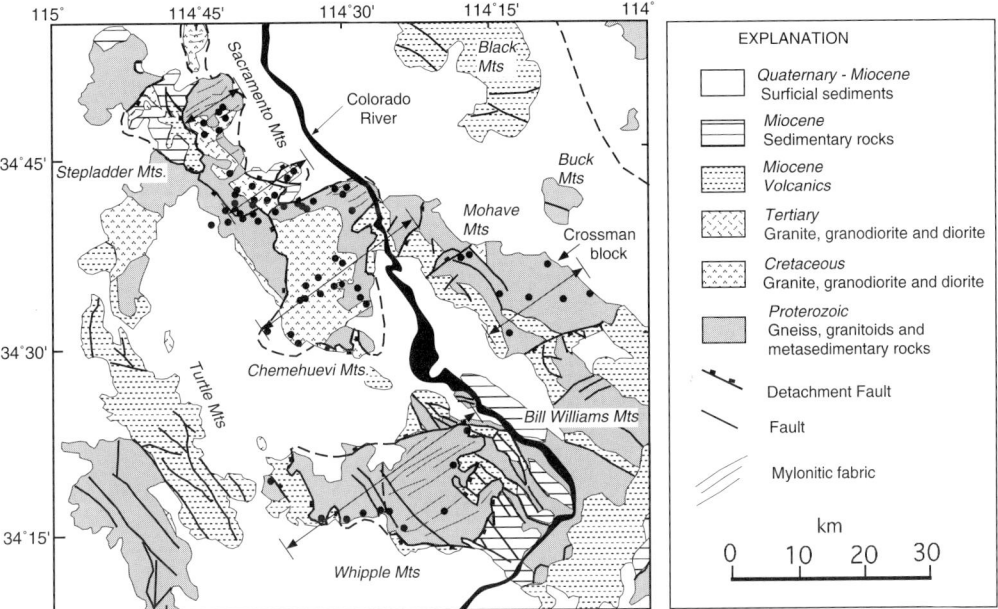

**Fig. 2.** Geological map of the central Colorado River extensional corridor (modified from Campbell & John 1996b). ●, Sample locations for thermochronological data. Arrows show approximate locations of sample transects discussed in the text.

rocks. In these cases, the data and their implications are consistent with thermal models of detachment faulting (e.g. Ketchum 1996) and other geological and geophysical data. With more limited thermochronological data under- or overestimations can be made about the timing of tectonic exhumation, rates of slip, and palaeothermal field gradient (Ruppel et al. 1988; House & Hodges 1994; Foster 1995; Ketchum 1996). In this review we demonstrate applications of thermochronology successful in defining one or more aspects of the thermal–extensional history.

## Geological framework

The Cenozoic deformation history of the eastern Mojave and western Sonoran Deserts is dominated by extreme crustal extension along the 50–100 km wide Colorado River extensional corridor (Fig. 1; Howard & John 1987). Regionally, major middle Tertiary extension involving the upper and middle crust was accomplished along brittle northeast-dipping detachment faults. These faults cut gently down-section in the direction of tectonic transport from a headwall breakaway between the Turtle and Old Woman Mountains (Howard & John 1987). Low-angle normal or detachment faults are exposed around the domal core complexes in the central part of the corridor, including the Whipple Mountains (Davis et al. 1980; Davis & Lister 1988), Buckskin Mountains (Reynolds & Spencer 1985), Chemehuevi Mountains (John 1982, 1987a), Sacramento Mountains (McClelland 1982; Spencer 1985; Campbell & John 1996b), and the Dead Mountains (Spencer 1985). Transport of the hanging wall of each fault, where known, was to the northeast (Davis et al. 1980; Howard et al. 1982; John 1982; Spencer 1985; Campbell & John 1996a). This transport direction holds even for the structurally deepest exposed faults in the system. Cumulative slip on the fault system increases to the northeast across the corridor and totals an estimated 50 km in the central part of the belt (Howard & John 1987), with even greater total slip in the southern part (Spencer & Reynolds 1991). Regional field relations indicate that the basal fault(s) cut initially to depths of 10–15 km, the palaeothickness of the Mohave Mountains block (Fig. 3), above the regionally developed Chemehuevi–Whipple detachment fault system. Farther east, geological and seismic-refraction data (McCarthy et al. 1991; Wilson et al. 1991) suggest that the fault system at present lies at a depth of less than 3 km under the Mohave Mountains and is rooted farther east under the relatively unfaulted and untilted Hualapai Mountains and Colorado Plateau. Middle Tertiary crustal extension in this region therefore occurred along an asymmetric shear system (Howard & John 1987).

As a consequence of extension, Proterozoic, Mesozoic and Cenozoic crystalline rocks that initially resided at palaeodepths of at least 10–15 km in the crust were juxtaposed against volcanic and sedimentary rocks deposited at the surface at less than $c.18$ Ma. Slip on the faults at all exposed levels of the crust was unidirectional. Brittle thinning above the basal detachment fault(s) affected the entire upper crust, and wholly removed it in the region of maximum extension, on the California side of the Colorado River, in the central part of the corridor (Fig. 3).

## Timing of extension

The onset and cessation of rapid extension and exhumation in the highly extended orogen may be defined by detailed documentation of the oldest and youngest tilted strata, crystallization age of pre-, syn- and post-extensional plutons, and the timing of facies changes and the appearance of footwall-sourced rocks in hanging wall basins. The oldest tilted sedimentary and volcanic rocks in the Colorado River extensional corridor, determined by dating of the volcanic horizons, decreases from 26–21 Ma in the Buckskin and Harcuvar metamorphic core complexes (Spencer & Reynolds 1989, 1991, 1993; Lucchitta & Suneson 1993), to about 23–21 Ma around the Whipple Mountains (e.g. Beratan 1993; Nielson 1993; Dorsey & Becker 1995; Nielson & Beratan 1995), to about 22–20 Ma around the Chemehuevi Mountains (Howard et al. 1993), and to about 18 Ma near the Dead and Newberry Mountains (Howard et al. 1994). This suggests that the onset of crustal stretching in the extensional corridor decreases in age from south to north (see Armstrong & Ward 1991; Howard et al. 1994).

The oldest Tertiary plutons in the core complexes also decrease in age from south to north along strike of the Colorado River extensional corridor. In the Buckskin Mountains the pre- to syn-extensional Swansea suite was intruded at 21.7 ± 0.7 Ma (Bryant & Wooden 1989), based on U–Pb zircon dates, or is perhaps as old as c. 26 Ma based on $^{40}Ar-^{39}Ar$ hornblende data, which may be slightly biased to older ages by excess argon (Richard et al. 1990). In the Whipple Mountains syn-kinematic tonalite gives ages of 26 ± 5 Ma (Wright et al. 1986) and 24.1 ± 0.5 Ma (new U–Pb zircon data from the same sample; Foster, unpublished data). Plutons that were emplaced just after the onset of extension in the Chemehuevi Mountains give $^{40}Ar-^{39}Ar$ hornblende and U–Pb zircon ages of 21–19 Ma (Foster et al. 1990b; John & Foster 1993; Foster et al. 1996a). Syn-extensional plutons in the Sacramento Mountains were intruded at 19 ± 1 Ma (Foster et al. 1996b; Pease et al. 1995, 1999). Finally, further north in the Newberry and Dead Mountains, the oldest Miocene plutons give ages of c. 18 Ma (Howard et al. 1996).

These data indicate that the onset of extension was progressive in the extensional corridor from c.26–22 Ma in the south to c.20–18 Ma in the north. However, there is some uncertainty about the exact onset of extension at specific locations and in some of the individual core

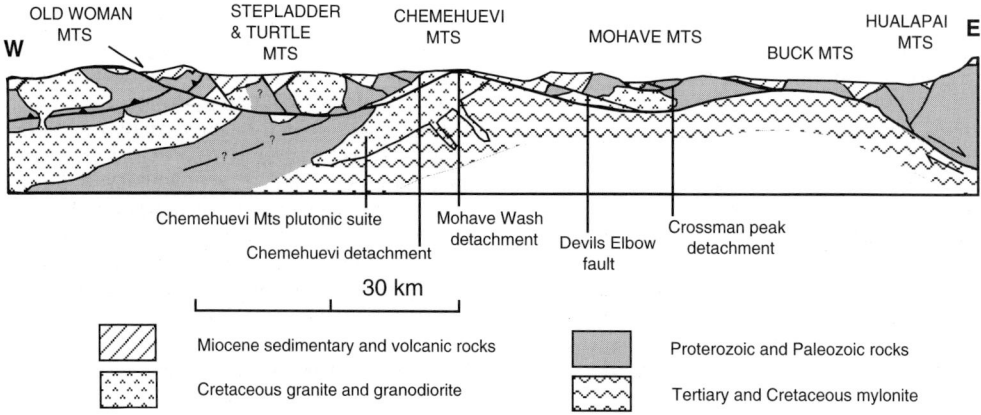

**Fig. 3.** Cross-section of the central Colorado River extensional corridor showing the footwall and hanging wall rocks to the detachment fault system exposed in the Chemehuevi Mountains (modified from Howard et al. 1987).

complexes, because of (1) limited reliable dates for the oldest syn-extensional sedimentary deposits and a non-uniform distribution of these deposits, and (2) a lack of syn-kinematic plutons in all core complexes and the uncertainty of when they were intruded with respect to the onset of extension.

In addition to dating syn-extensional plutons and volcanic rocks, thermochronological data from footwall and hanging wall rocks reveal the timing and duration of extension (1) by showing dramatic increases in cooling rate at specific times, and (2) when breaks in slope in trends of mineral age v. slip distance for sample transects can be identified.

## Accelerated cooling rates

To accurately determine the onset of extension with thermochronological data requires information from a spectrum of palaeodepths and from isotopic systems with different temperature sensitivities. With sufficient data, locations in metamorphic core complexes can be identified where specific isotopic systems change from recording mixed ages, older than the onset of extension, to cooling ages related to tectonic exhumation. Samples that yield the timing of the onset of extension are those that were at or slightly deeper than the depth of the high-temperature boundary of the partial closure interval (Dodson 1973, 1979) for the particular thermochronometric system. Lower-temperature (i.e. <350°C) thermochronological data from relatively shallow palaeodepths along detachment faults (less than c. 5–15 km depth) are those that retain information about the onset of extension and cooling associated with tectonic exhumation. Those samples from higher temperatures and deeper structural levels undergo rapid cooling after those at shallow levels and give younger ages because of (1) having resided at temperatures significantly above the closure temperature of the various isotopic systems, (2) a delay in cooling of the deeper parts of the footwall, because of advection of isotherms so that cooling does not keep pace with extension (Ruppel *et al.* 1988; House & Hodges 1994; Scott *et al.* 1998), and (3) minor heating of the deepest parts of the footwall because isotherms become compressed at the detachment fault (Ketchum 1996).

Temperature–time paths for samples from footwalls of the Chemehuevi and Sacramento core complexes, which were exhumed by the same detachment fault system, are shown in Figs 4 and 5. The individual sample cooling curves show breaks in slope that reflect the onset of rapid cooling. The onset of accelerated cooling rates recorded by these samples becomes younger with position along the detachment fault, with the deeper-level samples recording very rapid cooling up to 5 Ma after those from shallower structural depths. Cooling rates also change with time along the detachment as the rocks are transported closer to the surface. The rates between temperatures of 400° and 300°C (15–40°C Ma$^{-1}$) are significantly less than rates between c. 300°C and the surface (>100°C Ma$^{-1}$) (see also a detailed example of this effect for the Buckskin Mountains given by Scott *et al.* (1998)). The cooling rate acceleration is related to (1) the decreasing amount of cover with time, (2) the linked increase in cooling effects from the surface, a cold hanging wall and associated flow of meteoric water, and (3) the fact that cooling does not keep pace with rapid exhumation. This change in cooling rate is predicted by conductive models of detachment faulting (e.g. Ruppel *et al.* 1988).

In the Chemehuevi Mountains example, the onset of extension is indicated by the break in slope of the cooling curve for only the structurally most shallow sample (CHM10), which occurs at c. 22–21 Ma. Samples from progressively deeper structural levels show a lag in the onset of rapid cooling. The 22–21 Ma increase in cooling rate for CHM10 is similar to time estimated for the onset of extension from the adjacent sedimentary basin sequences (Miller & John 1988).

Rapid increase in cooling rates is characteristic of tectonic exhumation in metamorphic core complexes (see also discussion by Scott *et al.* (1998)). The consistently rapid cooling rates (>40–100°C Ma$^{-1}$) for all of the core complexes studied in the Colorado River extensional corridor suggests that progressive tectonic exhumation was the most important process in unroofing. However, final exhumation included a significant contribution from erosion (Miller & John 1988). We estimate that well over 75% of the exhumation of the footwall rocks in these metamorphic core complexes was due to detachment faulting. Regionally, rapid erosion was relatively more important for exhuming hanging wall rocks, because of the greater amount of topographical relief around the higher-angle fault blocks. One way we are attempting to quantify the relative importance of erosional processes is by documenting the thermal histories and timing of deposition of distinctive landslide megabreccias and blocks in the sedimentary basin sections that can be traced to footwall sources (e.g. Miller & John 1988, 1993, 1995; Yin & Dunn 1992; Nielson & Beratan 1995; Campbell & John 1996*a*). These data allow

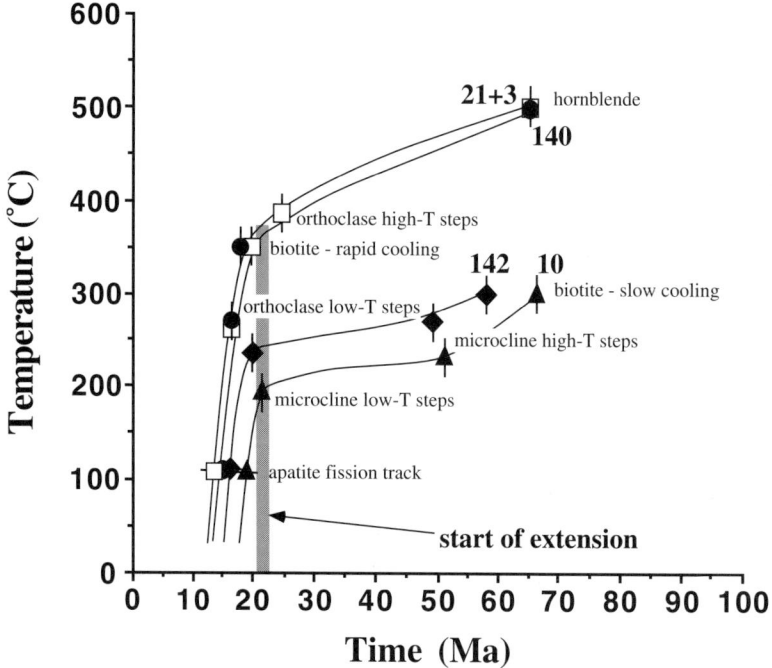

**Fig. 4.** Cooling curves for selected samples from footwall rocks in the Chemehuevi Mountains. The thermal history curves are defined by $^{40}$Ar–$^{39}$Ar (hornblende, biotite and K-feldspar) and apatite fission-track thermochronological data (Foster et al. 1990b; John & Foster 1993). High- and low-temperature ($T$) microcline and orthoclase closure temperatures are calculated for each sample based on Arrhenius data (e.g. Lovera et al. 1989). The break in slope of the curves, indicating a change in cooling rate, for the samples that started cooling from lowest temperatures (10 and 142) gives the time at which rapid extension started (shaded bar). Samples from deeper structural depths (21 and 140) start to cool rapidly sometime after extension started. Error bars are smaller than the symbols when not shown.

estimates of time-averaged erosional exhumation for the rocks from the megabreccias, which can then be compared with those estimated for footwall rocks (John & Foster, in preparation).

## Exhumed palaeoisotherms

A second approach for estimating the onset of extension uses sample traverses to identify positions of palaeoisotherms that were 'quenched' when fault slip started. The critical footwall palaeoisotherms are those that correspond to the pre-extension boundaries between the partial closure intervals and effectively open-system behaviour, for respective thermochronological systems. These threshold temperatures represent the approximate closure isotherms for the K–Ar and fission-track isotopic systems at rapid cooling rates $\gg 10°C$ Ma$^{-1}$ (e.g. c. 300–350°C for biotite K–Ar). This method requires information from samples over a spectrum of palaeodepths and from several thermochronometric systems, and can be used when locations, within the footwall, are identified where specific isotopic systems change from recording mixed ages, older than the onset of extension, to cooling ages the same as or younger than the time of tectonic exhumation. Such exhumed zero-age isotherms are commonly marked by a well-defined break in slope in plots of age against distance in the slip direction (Fitzgerald et al. 1991; John & Foster 1993; Foster et al. 1994).

A plot showing the variation in age for four different thermochronometers along a projection in the slip direction of the Chemehuevi detachment fault is shown in Fig. 6. All samples included here were collected from beneath the zone of cataclasis associated with the detachment in Cretaceous and Proterozoic granitic or gneissic rocks lacking Tertiary fabric. Therefore, Tertiary partial recrystallization of the minerals is unlikely to have affected the isotopic systems (e.g. Reddy et al. 1996). On this plot, all of the

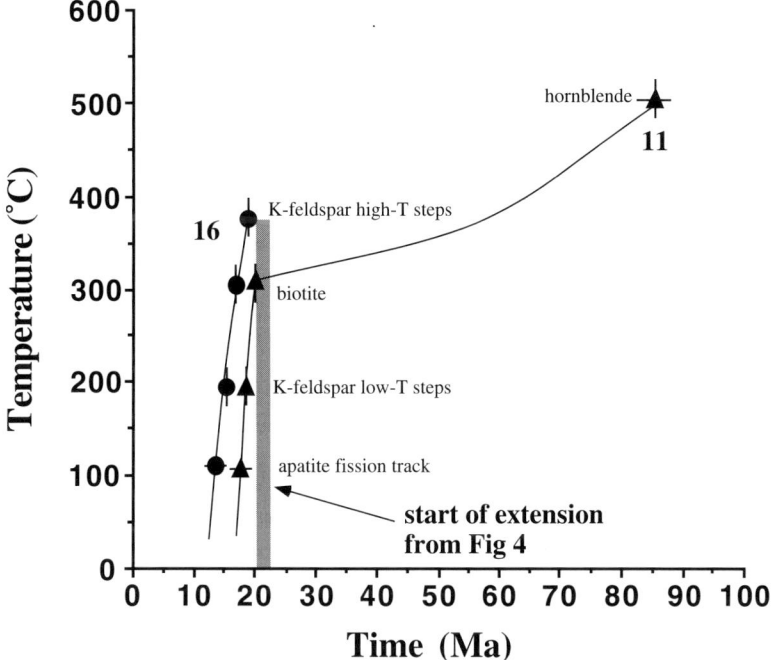

**Fig. 5.** Cooling curves for two samples from different palaeodepths within the Sacramento Mountains footwall (data from Foster *et al.* 1990*b*, 1991). High- and low-temperature (*T*) K-feldspar closure temperatures are calculated for each sample based on Arrhenius data (e.g. Lovera *et al.* 1989). As shown in Fig. 4, the sample from shallow structural levels, which resided at relatively lower temperature before extension, shows a break in slope at the time at which extension started. Error bars are smaller than the symbols when not shown.

mineral ages decrease to the northeast in the direction of slip, and increasing palaeodepth and pre-extension temperature. The curves defined for the minerals with higher closure temperatures (biotite and the high-temperature plateaux of the K-feldspar age spectra) have three distinct parts. At the shallowest structural levels, these ages are Late Cretaceous to early Tertiary, and reflect cooling after emplacement of the Chemehuevi plutonic suite. The biotite and high-*T* K-feldspar ages then decrease gradually with distance toward the northeast, with their curves becoming much steeper at distances between *c.* 9 and 13 km, along the projected section. At this point, there is a break in slope below which the-values decrease more gradually to the eastern boundary of the footwall. The curve for sphene fission-track ages is defined by only a limited number of data, but is probably similar to biotite and high-*T* K-feldspar curves. Low-temperature thermochronometers (minimum ages or low-*T* plateaux of K-feldspar age spectra and apatite fission track) give approximately linear trends over most of the footwall. In the Chemehuevi Mountains the break in slope for the biotite and K-feldspar system and the inferred location for the sphene fission-track system is 20–22 Ma.

Biotite ages are older than coexisting K-feldspar high-*T* plateau ages in the southwest part of the exposed footwall, where the K-feldspars are microcline. The biotite ages record cooling through about 300–350°C (Harrison *et al.* 1985), depending on cooling rate, and the microclines cooling through 280–300°C (Lovera *et al.* 1989). At deeper palaeodepths along the footwall, starting near the break in slope, the K-feldspars are orthoclase, which record maximum closure temperatures of about 340–380°C (Lovera *et al.* 1989; Foster *et al.* 1990*a*). This is why at the deeper structural levels the K-feldspar maximum ages become equal to or older than the biotite ages. The form of the individual K-feldspar age spectra also changes from southwest–northeast in the slip direction. In the southwest, the spectra exhibit steep age gradients from Oligocene–Miocene dates for low-temperature steps to Paleocene–Eocene ages for high-temperature steps (Foster *et al.* 1990*b*;

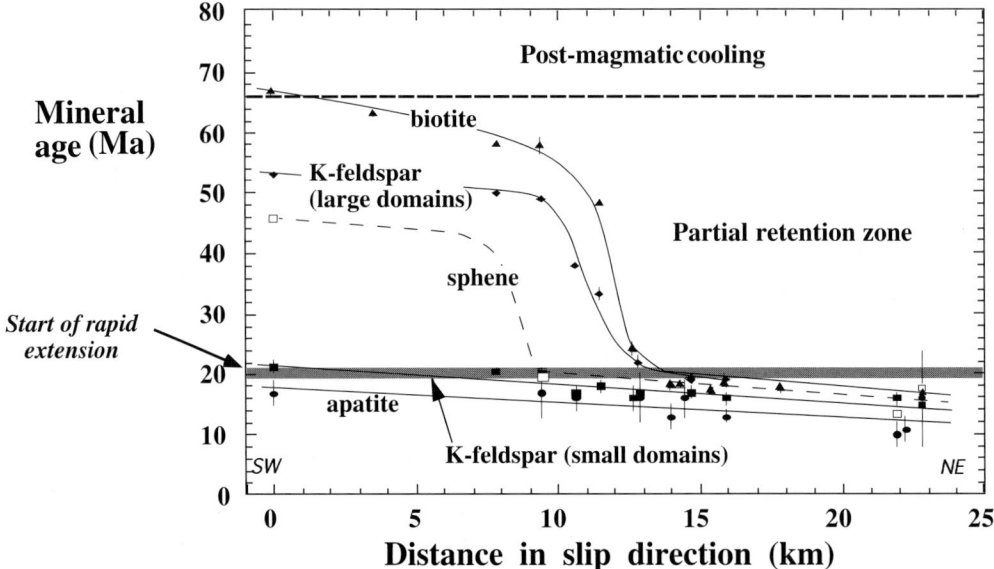

**Fig. 6.** Plot showing the relationship between age for different thermochronometers and distance in the known slip direction (southwest to northeast) of the Chemehuevi detachment fault. The curves defined by the higher-temperature thermochronometers start with mixed ages at shallow structural levels, and give way to rapidly cooled samples, at deeper structural levels, that reflect the time at which they were tectonically exhumed from higher temperatures. The break in slope in these curves marks the time at which rapid extension started and the first 'zero aged' samples were rapidly cooled. Error bars are smaller than the symbols when not shown.

Foster & John, unpublished data). The age spectra become progressively flatter to the northeast, to below the break in slope, where only a few million years separate Miocene maximum and minimum ages. The flatter age spectra indicate more rapid Miocene cooling through a greater range of palaeo-temperatures for samples from deeper palaeodepths.

The shape of the mineral age–palaeodepth curves is analogous to the variation in mineral age with depth observed for thermochronological data from deep boreholes or uplifted mountain blocks (e.g. Gleadow & Duddy 1981; Foster & Gleadow 1996). The interval between 9 and 13 km distance along the traverse, where the curves steepen abruptly, represents the former depths of the partial closure intervals where ambient temperatures were sufficiently high to cause significant but not complete resetting, before Miocene unroofing. Below the break in slope, at c. 13 km distance, all of the thermochronometers were open systems with 'zero age' before extension. The break in slope for the curves gives the time at which the tectonic exhumation began. At this position in the footwall, the effect of extension and slip on the detachment fault was to rapidly cool the samples (Foster 1995; Ketchum 1996). At any location down-dip of these isotherms the isotopic systems for respective minerals record progressively younger ages, which, taken in isolation, would underestimate the timing of extension. This approach for determining initial exhumation appears valid for the thermochronometers shown here, where closure temperatures are up to c.350 °C, but should be applicable for higher-temperature systems in situations where the detachment does not flatten with depth before the isotopic closure temperature (e.g. Ketchum 1996), or the thermal history of the deeper crust is not complicated by plutonism.

## Rates of slip on detachment faults

Rate of slip on detachment fault systems may be estimated from the inverse slope of mineral age with distance, in the slip direction, for thermochronological systems that had zero age before extension (Foster et al. 1993; John & Foster 1993). This is most applicable to low-temperature thermochronological (K-feldspar $^{40}Ar-^{39}Ar$ and fission-track) data because (1) it is difficult to determine if higher-temperature thermochronometers yield simple cooling ages, and (2) the effect of thermal pulses on mineral ages is a potential problem at deeper crustal levels. For accurate slip rate estimates the closure isotherm for the thermochronometer must have remained approximately horizontal during the interval of slip revealed by the data. This assumption was investigated using 2D conductive cooling models by Ketchum (1996), and found to be reasonable with a few million years after the onset of extension, because the isotherms reach a steady-state position by that time. Before this time the isotherms advance along the detachment surface, causing uncertainties in the slip rate (Ketchum 1996). These data give time-averaged slip rates on time scales of about a million years and do not rule out significantly faster or slower rates of detachment slip over shorter time scales.

## Slip rate examples

Average slip rates on detachments in the Colorado River extensional corridor indicated by the thermochronological data (Foster *et al.* 1993; John & Foster 1993, 1996*a*), are similar to slip rates estimated by restoring structural markers of known age from the footwalls of the Whipple, Buckskin, and Harcuvar Mountains core complexes with hanging wall equivalents (e.g. Davis 1988; Spencer & Reynolds 1991; Scott *et al.* 1998). Plots of the variation of apatite fission-track age with distance in the Buckskin–Rawhide and Harcuvar metamorphic core complexes (Fig. 7) are shown in Fig. 8 (Foster *et al.* 1993). The inverse slope of the results from the Buckskin core complex suggests a rate of 7.7 ± 3.6 km Ma$^{-1}$ (±2 $\sigma$). Regressions of the slip rate and 2$\sigma$ errors, for these and other slip rates, were calculated using methods of York (1969) for noncorrelated errors, considering 2$\sigma$ errors in age and sample location–projection. The plots are shown as inverse slopes for the ease of comparing the data with the map cross-sections. For the regressions of the apatite fission-track data, in those cases where the number of data points is relatively small (Figs 8 and 9), the 2$\sigma$ errors are considered to approximate the uncertainties. All of the samples from the Buckskin footwall give ages at least 5 or 6 Ma younger than the onset of extension, and have long mean track lengths, indicating rapid cooling. A value of 6.5 ± 3.0 km Ma$^{-1}$ is given by results from the adjacent Harcuvar Mountains core complex. The trend in the Harcuvar Mountains is influenced by one relatively precise apatite age of *c.* 21 Ma from the structurally shallowest sample (from the Granite Wash Mountains). Removing this one sample, because of the possibility it may have cooled before the depth of the 110°C isotherm was stationary (e.g. Ketchum 1996), gives a slip rate of 7.7 ± 3.1 Ma. As expected, this rate is similar to that from the Buckskin

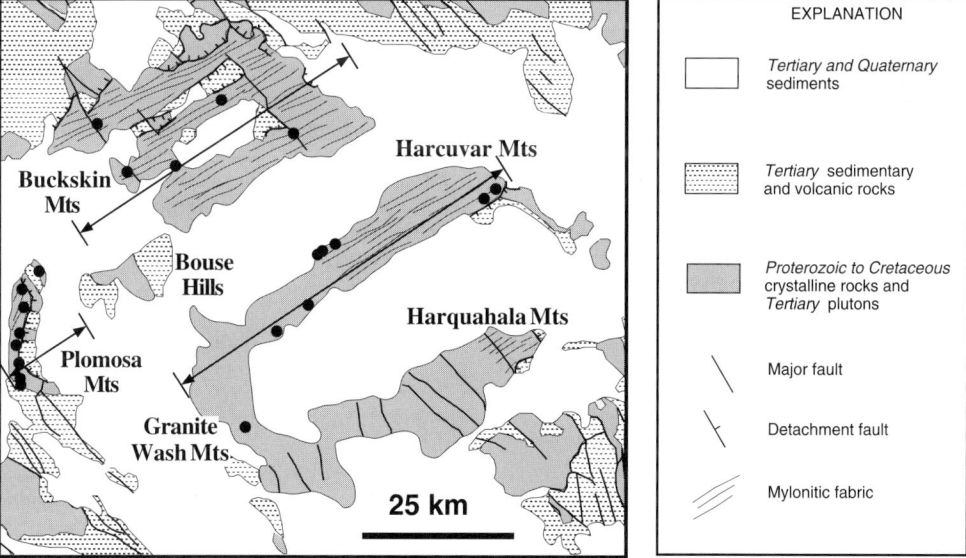

**Fig. 7.** Simplified sample and geological map of the Buckskin, Harcuvar and north Plomosa Mountains modified from Foster *et al.* (1993). ●, Sample locations.

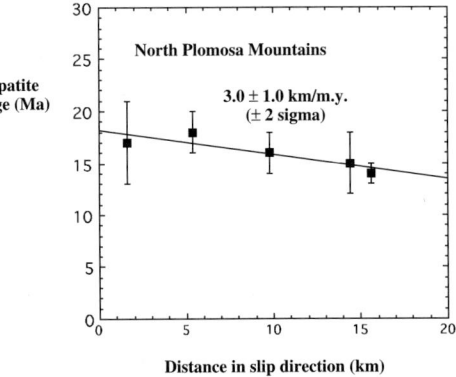

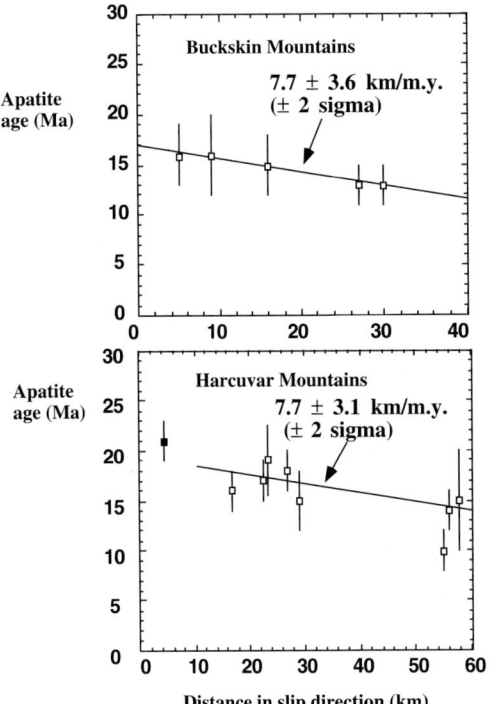

Fig. 8. Plot of apatite fission-track age against distance in the slip direction for the Buckskin and Harcuvar detachment faults (Foster et al. 1993).

Fig. 9. Plot of apatite fission-track age against distance in the slip direction for the Plomosa detachment fault (data from Foster & Spencer 1992).

Mountains because both footwalls were unroofed from beneath the same detachment system, at approximately the same time (Spencer & Reynolds 1991).

Spencer & Reynolds (1991) estimated an extension rate of 8–9 km Ma$^{-1}$ along the Buckskin–Rawhide detachment, based on the amount of slip (c. 90 km) that had occurred between 23–25 Ma (when syn-extensional basins started to form) and c. 15 Ma (when the lower plate rocks cooled >100°C). The timing of initial exposure of footwall rocks at the southwestern and northeastern ends of the core complex, based on distinctive clasts in conglomerates, also indicates a slip rate of c. 7 km Ma$^{-1}$ for this fault (Scott et al. 1998).

South of the Buckskin Mountains, less extension was accommodated by slip on the Plomosa detachment fault (Fig. 7) in the North Plomosa Mountains (Spencer & Reynolds 1991). The Plomosa Mountains footwall was probably tilted about 45° to the south during extension, by movement on an underlying fault, based on the dip of the mid-Tertiary strata on top of the block. Five apatite fission-track ages, all <20 Ma, occur over a slip distance of about 15 km in this block (Foster & Spencer 1992). A regression of these data gives a slip rate estimate of 3.0 ± 1.0 km Ma$^{-1}$ (Fig. 9), which is about half of that in the Buckskin Mountains, and consistent with the more limited extension on this fault.

Minimum ages of K-feldspar age spectra from the Whipple Mountains (Fig. 2) decrease from about 21 Ma to 15 Ma over an exposed slip distance of about 30 km (Foster 1994). These data give a slip rate of 7.8 ± 4.4 km Ma$^{-1}$ (±2σ) (Fig. 10). Four apatite fission-track ages, for samples with long track lengths, suggest a similar slip rate (Fig. 10). Davis & Lister (1988) estimated that the slip rate on the Whipple detachment was c. 8.2 km Ma$^{-1}$, based on the 40 km of horizontal offset the Chambers Well dyke swarm, in the footwall, and equivalent dykes in the Mohave Mountains hanging wall (Howard et al. 1982), between 20 and 15 Ma (Fig. 2).

Results of similar studies from the central part of the Colorado River extensional corridor, where the total amount of slip on the detachment faults is less, and over a shorter time interval, than in the south, reveal consistently lower average slip rate estimates. Figure 11 shows plots of K-feldspar minimum age and apatite fission-track age against slip distance for the Chemehuevi detachment. A regression of the 15 K-feldspar ages gives a time-averaged rate of 3.8 ± 1.0 km Ma$^{-1}$. Extension probably started here at about 21–22 Ma (see above), so if we use only those sample with dates younger than 19 Ma the estimated slip rate is 6.25 ± 1.4 km Ma$^{-1}$. The apatite fission-track results for this detachment give a similar overall rate of 3.3 ± 0.9 km Ma$^{-1}$. A slip rate of c. 3 km Ma$^{-1}$ is also indicated by biotite and hornblende $^{40}$Ar–$^{39}$Ar data from the

QUANTIFYING TECTONIC EXHUMATION 353

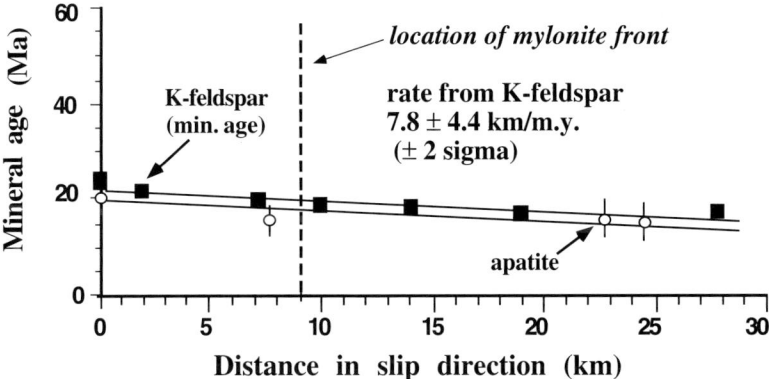

**Fig. 10.** Plot of the minimum ages of K-feldspar $^{40}$Ar–$^{39}$Ar age spectra (plateau or isochron ages of low-temperature steps), against distance in the slip direction, for footwall rocks to the Whipple detachment fault. Error bars are smaller than the symbols when not shown.

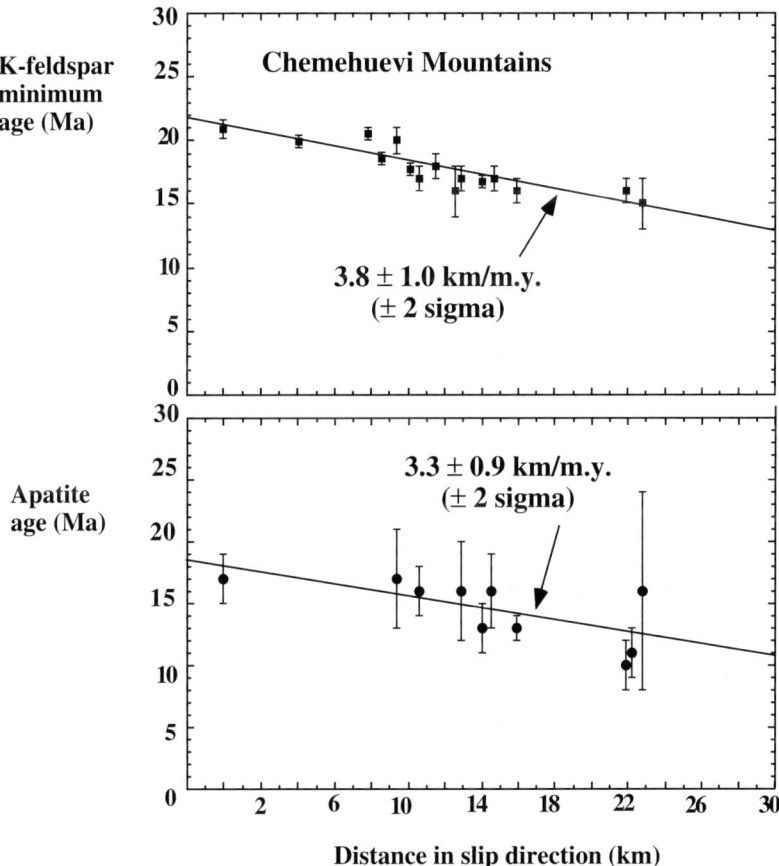

**Fig. 11.** Plots of the minimum ages of K-feldspar $^{40}$Ar–$^{39}$Ar age spectra (plateau or isochron ages of low-temperature steps) and apatite fission-track age against distance in the known slip direction for footwall rocks to the Chemehuevi detachment fault (data from Foster *et al.* 1990*b*; John & Foster 1993; Foster & John unpublished data).

northern Sacramento Mountains footwall, to the north of the Chemehuevi Mountains (Pease *et al.* 1999).

As all slip rate estimates discussed here use only low-temperature thermochronological data, and samples giving dates a few million years after the onset of extension, there is probably very little error because of the movement of isotherms during evolution of the detachment faults (see Ketchum 1996). It is also significant that the results from the low-temperature thermochronological transects are similar to rates suggested by independent geological data, and that results from adjacent metamorphic core complexes are similar.

## Syn-extensional geothermal gradient

Tilted upper plate fault blocks, up to 15 km thick, in highly extended parts of the Basin and Range province, preserve sections of what was the upper crust before extension. These tilted crustal sections arise when blocks are progressively tilted along with their bounding normal faults (Davis *et al.* 1980; Howard *et al.* 1982; Holm & Wernicke 1990; Fryxell *et al.* 1992; Mueller & Snoke 1993; Howard & Foster 1996). Where such blocks are structurally intact they provide unique insight into the structural and thermal histories of rocks with depth (Howard *et al.* 1982; Foster *et al.* 1990b, 1991; Fitzgerald *et al.* 1991; Gans *et al.* 1991; Fryxell *et al.* 1992; Howard & Foster 1996). In sufficiently thick hanging wall blocks, where palaeohorizontal is known from the dips of horizontal sheet intrusions (see evidence given by Howard (1991)) or from Tertiary or older unconformities, the variation in mineral age with palaeodepth can be used constrain the geothermal gradient at the onset of extension. Knowledge of the geothermal gradient before extension is critical for evaluating models of crustal extension involving magmatic input into the crust, variations in crustal strength, fluid flow, and heat advection.

Samples collected from transects in the movement direction of the bounding normal faults, of the tilted blocks, reveal thermal histories from increasingly deeper palaeodepths (Foster *et al.* 1991, 1994; Gans *et al.* 1991; Howard & Foster 1996). Palaeoisotherms can be identified when isotopic systems record pre-extension ages at shallow depths, and when at deeper levels the mineral age v. palaeodepth curve (at the break in slope) intersects the time at which extension started (Fitzgerald *et al.* 1991; Howard & Foster 1996). When two palaeoisotherms are identified, or one palaeoisotherm combined with the location of the Tertiary unconformity (i.e. the surface temperature), a geothermal gradient can be calculated. A geothermal gradient was calculated for the Grayback fault block in the Tortilla Mountains, Arizona, by Howard & Foster (1996). Other areas where this approach was applied include the Mohave Mountains (Foster *et al.* 1994), the Piute Mountains (Fig. 1) (Foster *et al.* 1991), the Gold Butte block of the South Virgin Mountains (Fig. 1) (Fitzgerald *et al.* 1991), and the Kern–Deep Creek Mountains, Nevada (Fig. 1) (Gans *et al.* 1991).

## Tortilla Mountains, palaeogeothermal gradient

The Grayback fault block in the Tortilla Mountains, Arizona (Figs 12 and 13), exposes a Proterozoic to Paleocene granitic crustal section about 12 km thick (Howard & Foster 1996). The crustal section was tilted eastward during Oligocene to Miocene extension, which led to the exhumation of core complexes in the Santa Catalina, Rincon, Tortolita and Picacho

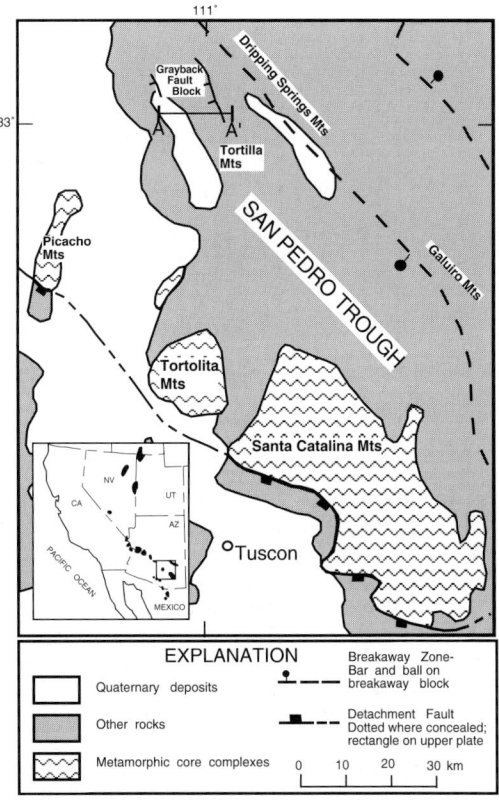

**Fig. 12.** Map of part of south–central Arizona showing location of the Tortilla Mountains and Grayback fault block and their proximity to the Santa Catalina Mountains metamorphic core complex (after Howard & Foster 1996).

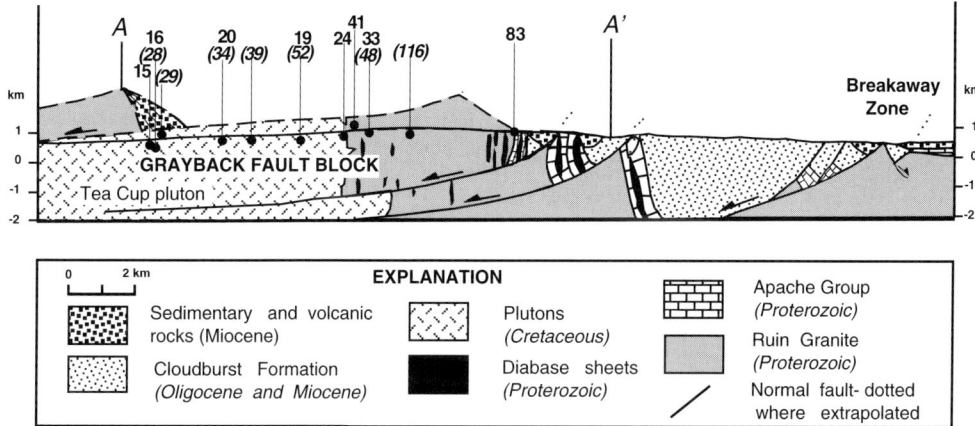

**Fig. 13.** Cross-section of the Grayback fault block (A–A' in Fig. 12) showing the tilted crustal section and bounding normal faults (modified from Howard & Foster 1996). Sample locations and apatite and zircon (in parentheses) fission-track ages are also shown.

Mountains (Fig. 12) (Dickinson 1991). Stratigraphy of the Tertiary rocks indicates that tilting of the Grayback block took place between 25 and 16 Ma. The block was homoclinally tilted about 90° based on the orientation of Proterozoic Apache Group that overlies the granitoids, which dip c. 80° overturned, and vertical dips for Proterozoic diabase sheets (Howard 1991). These diabase sheets are approximately parallel to Proterozoic and younger strata throughout Arizona and California and thus were horizontal before tilting (Howard 1991; Howard & Foster 1996).

Apatite and zircon fission-track ages from the Grayback fault block are shown in the cross-section (Fig. 13) and in a plot of age against palaeodepth (Fig. 14). The apatite ages decrease westward (deeper palaeodepths) from c. 83 Ma at the unconformity to a break in slope at c.24 Ma and c.5–6 km depth. Mean track lengths for samples between 0 and 6 km depth are <13 μm, indicating relatively slow cooling through the apatite partial annealing zone. Below c. 5–6 km depth, the apatite ages decrease from c. 24 to 15 Ma and have long mean track lengths (>14 μm), indicating more rapid cooling. The zircon fission-track ages also decrease to the west and become concordant with the initiation of extension at palaeodepths of 12.1–12.3 km.

The break in slope in the apatite age transect represents the position of the base of the apatite partial annealing zone (c. 110°C) (Gleadow & Fitzgerald 1987; Foster & Gleadow 1996) before Tertiary tilting. The break in slope is concordant with independent stratigraphic constraints on the initiation of tilting. All of the apatite samples below the break in slope were cooled rapidly

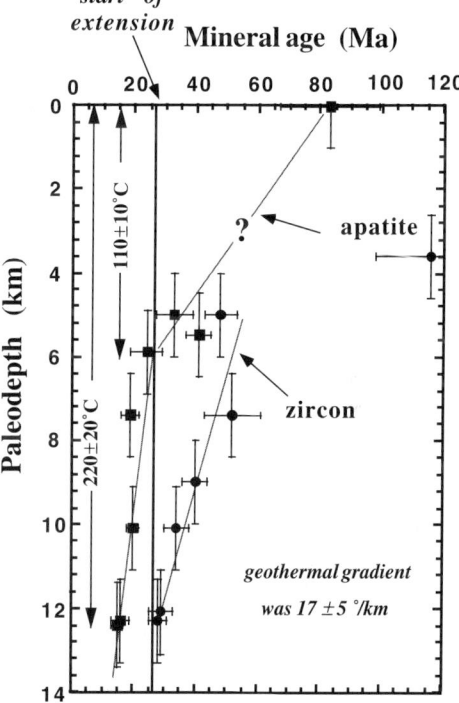

**Fig. 14.** Plot of apatite and zircon fission-track ages against palaeodepth for the Grayback fault block (Howard & Foster 1996). This block was tilted west c. 90° during extension, exposing a thick crustal section that records the positions of two isotherms (c. 110° and c. 220°C), before extension starting at 25 Ma. The depths of these isotherms allow a geothermal gradient of 17° ± 5°C km$^{-1}$ to be calculated for the time just before tilting and tectonic exhumation.

from temperatures where tracks were totally annealed, based on the long mean track lengths. The form of the zircon age profile with no Oligocene–Miocene break in slope suggests that all of the samples were at or colder than and structurally above the temperature of total annealing (c. 230–250 °C for $10^7$ year time scales; Brandon & Vance 1992; Foster et al. 1996b; Brandon et al. 1998). However, the fact that the dates of the two deepest samples are concordant with, but not younger than, start of tilting suggests that the samples at 12.3 km were at c. 220 ± 30°C at c. 25 Ma.

Howard & Foster (1996) calculated the palaeogeothermal gradient for the Grayback block from the difference in depth of the palaeoisotherms at 5.7 ± 0.4 km (110 ± 10°C) and 12.15 ± 0.7 km (220 ± 30°C). This gives a gradient of 17.1 ± 5.3 °C km$^{-1}$. The errors include values of known and estimated errors in annealing temperatures and the projections (see Howard & Foster 1996). A gradient can also be calculated between the surface and the 110 ± 10°C isotherm. Assuming a surface temperature of 15 ± 10°C for the late Oligocene time gives a gradient of 16.7 ± 4.9°C km$^{-1}$. The mean of these two estimates gives a palaeogeothermal gradient of 17 ± 5 °C km$^{-1}$.

## Other palaeogeothermal gradient examples

The palaeogeothermal gradient for the Crossman block in the Mohave Mountains (Fig. 2) is more crudely defined than that for the Tortilla Mountains (Foster et al. 1994; Howard & Foster 1996). This greater error is partly due to more scattered sampling and therefore a larger projection error, as well as the presence of a Miocene dyke swarm. A plot of apatite fission-track age, microcline minimum age, and biotite K–Ar age against palaeodepth for the Crossman block (data from Foster et al. 1990a; Nakata et al. 1990) reveals about 4–5 km between the intersections of the microcline and biotite curves with the time at which extension started (Fig. 15). The difference in closure temperature between the low-temperature ages of the microcline (c. 180–200°C, Lovera et al. 1989) and biotite (c. 280–320°C, Harrison et al. 1985) suggest a geothermal gradient of c. 20–25°C km$^{-1}$. However, values as high as c. 35°C km$^{-1}$ and as low as c. 16°C km$^{-1}$ are allowed by the present data.

A palaeogeothermal gradient for the South Virgin Mountains, Nevada (Fig. 1), of c. 20–24°C km$^{-1}$ can be derived from the c. 4.6 km depth to the 110°C base of the palaeo-apatite partial annealing zone in the Gold Butte tilt

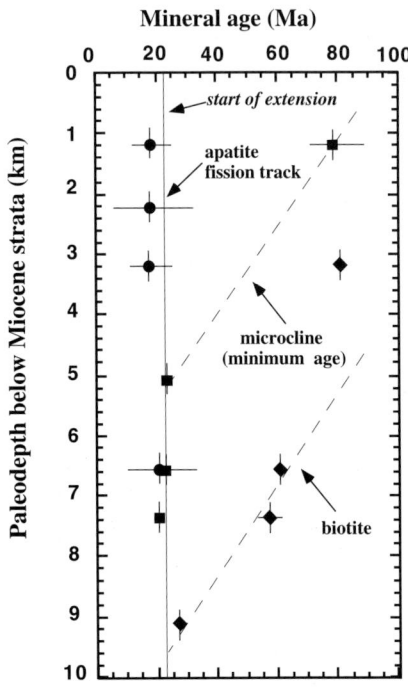

**Fig. 15.** Plot of mineral age against palaeodepth for the Crossman block of the Mohave Mountains. This block exposed a thick crustal section that was tilted c. 90° at c. 20–22 Ma (Howard et al. 1982). Two palaeoisotherms are crudely identified in this block corresponding to the depths where biotite and K-feldspar (low-temperature steps) record zero-ages before tilting and exhumation (data from Foster et al. 1990b; Nakata et al. 1990). These results suggest a pre-extension geothermal gradient of c. 20–25°C km$^{-1}$, with values between c. 16° and c. 35°C km$^{-1}$ allowed by the data (Foster et al. 1994; Howard & Foster 1996).

block (Howard & Foster 1996). The c. 4.6 km depth of the c. 110°C isotherm is based on the apatite age traverse of Fitzgerald et al. (1991) and the sum of the 1.6 km of overlying basement and c. 3 km of sedimentary cover before extension.

## Implications of the thermal gradients

The relatively low geothermal gradients for these three fault blocks have implications for the mechanisms leading to the initiation of extension. The Tortilla and South Virgin Mountains underwent high degrees of extension despite the relatively cold thermal structure in these areas before extension. Models of continental extension commonly call upon magmatism and mantle heat input for the initiation of extension (e.g. Gans et al. 1989; Armstrong & Ward 1991). The low initial geotherms are not consistent with these models. Preservation of tilt blocks with large thicknesses may be enhanced by low geothermal gradients, where the thickness may reflect the depth to which the brittle upper crust can fracture and tilt above a ductile lower crust (Howard & Foster 1996).

Geothermal gradients in the highly extended area undoubtedly increased after the onset of extension, to values of 30–50°C km$^{-1}$ or even higher (Foster et al. 1991; Lund et al. 1993), as a result of rapid exhumation of mid-crustal rocks and the presence of voluminous plutonism and volcanism. Plutonism and volcanism are approximately contemporaneous in and adjacent to many metamorphic core complexes, leading to refined models linking magmatism with exhumation of the footwall rocks and development of low-angle detachment faults (Gans et al. 1989; Thompson & McCarthy 1990; Armstrong & Ward 1991; Foster et al. 1992; Lister & Baldwin 1993; Parsons & Thompson 1993). Magmatism and high heat flow may well be important for the exhumation of metamorphic core complexes, but if the low initial geothermal gradients were widespread during Oligocene–Miocene extension, the crust around the core complexes may have been relatively hotter than surrounding areas.

## Initial dips of detachment faults

### Background

Much research in extensional tectonics has focused on the existence and nature of low-angle normal or detachment faults. Controversy has centred on the question of whether gently dipping normal faults were initiated and moved in their present low-angle configuration, or were active with relatively high dips, and have been rotated passively to a more gentle orientation through time (see review by Wernicke (1995)). The variation in palaeotemperature in the slip direction of detachment faults can be helpful in constraining their initial dips through the seismogenic crust provided that (1) the thermal gradient is known or can be calculated from hanging wall tilt blocks, (2) major lateral variations in the thermal gradient can be ruled out in the area of study, and (3) the samples are from below a single fault system. Those on both sides of this debate recognize that the thermochronological data have the potential to discriminate between the two end-member models. Nevertheless, thermochronology has been used to support the entire range of models (e.g. Foster et al. 1990b; Holm et al. 1992; Dokka 1993; John & Foster 1993; Lee 1995). This is partly due to relatively small datasets used in some studies and uncertainties in some mineral closure temperatures used. In this section we summarize a relatively large dataset, utilizing a range of thermochronometers from the Chemehuevi Mountains (Figs 2 and 16), to demonstrate how temperature-time data can be used to define the initial geometry of these faults.

### Chemehuevi Mountains example

Extension involving the upper and middle crust in the Chemehuevi Mountains area was accomplished along a stacked, anastomosing sequence of brittle, northeast-dipping low-angle normal faults discordantly cutting heterogeneous quartzofeldspathic basement (John 1987a,b). The large displacement (>18 km) Chemehuevi detachment fault separates footwall and hanging wall rocks from different structural levels in the crust, commonly with gross lithological 'mismatch'. Above the Chemehuevi detachment fault, the hanging wall was distended by innumerable high-angle faults, accommodating up to 100% average extension. Footwall accommodation to movement on the fault system was small. An older, small displacement (≤2 km) Mohave Wash fault lies structurally below the Chemehuevi detachment. Moderately to steeply dipping (40°–80°) NW-trending (110–170°) normal faults, and NE-trending (050°–060°) strike-slip faults truncate and are truncated by the Mohave Wash fault. These faults have tens to hundreds of metres separation, but are nowhere known to cut the structurally higher Chemehuevi detachment fault. Throughout the area, the youngest fault is demonstrably the regionally developed Chemehuevi detachment. This relationship precludes the possibility of rotations of the low-angle normal faults to their gentle orientation by younger higher-angle faults.

Fault rock type and the mineral deformation mechanisms associated with movement on the detachment faults suggest that they were initiated and moved within the brittle and transitional brittle–ductile regimes. Fault rocks produced by slip include incoherent gouge, breccia, rocks of the cataclasite series, and rare protomylonite and pseudotachylite. Thickness of these fault rocks varies from <1 m to >200 m. Fault rocks associated with the small displacement Mohave Wash fault are characterized by microfracturing over broad regions (up to 200 m), with evidence for repeated fracturing and fluid flow events. Those associated with the Chemehuevi detachment show overprinting relations that imply an evolution from wide

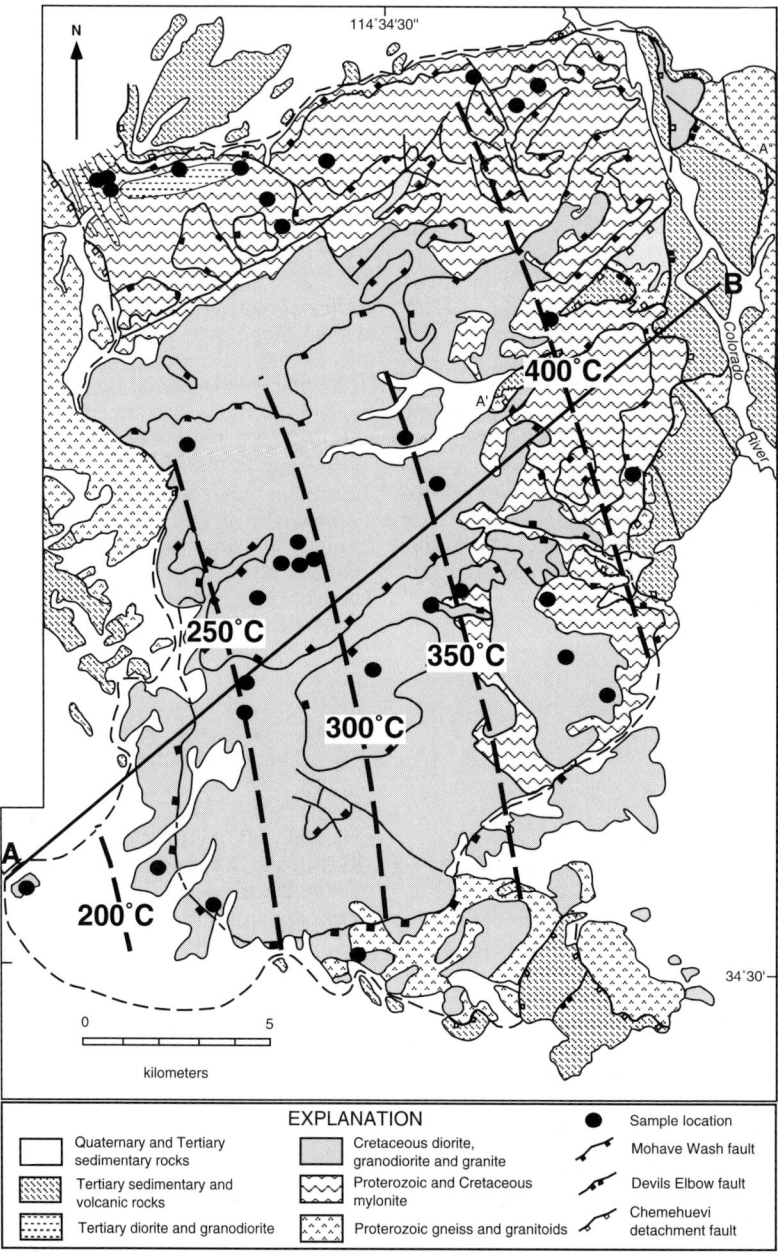

**Fig. 16.** Geological and sample location map of the Chemehuevi Mountains (modified from John & Foster (1993), with new data). The section A–B is the transect through the southern part of the footwall discussed in Fig. 17. The thick dashed lines are palaeoisotherms for the southern and central parts of the footwall to the Chemehuevi detachment fault when extension started at c.22 Ma. Paleotemperatures for sample points constraining the isotherms, were calculated from thermal histories of the footwall rocks obtained from a large database of $^{40}$Ar–$^{39}$Ar and fission-track measurements, where three to five minerals with different closure temperatures were analysed from each sample (after Foster et al. 1990; John & Foster 1993; Foster & John unpublished data). The isotherms show a gradual increase in temperature to the northeast in the known direction of tectonic transport. Isotherms are not shown for the northern part of the footwall because of the presence of Tertiary syn-extensional plutons in that area (Foster et al. 1990b, 1996b; John & Foster 1993).

zones of hydrothermally altered cataclasite (as the Mohave Wash fault) at moderate levels in the crust to a narrower zone of breccia within the upper crust and to a sharp, planar discontinuity marked by breccia and locally gouge at very shallow crustal levels. Examination of these fault rocks indicate that cataclastic flow and frictional sliding were the dominant deformation mechanisms. The presence of elongate quartz and potassium feldspar suggests that in the structurally deepest exposures of the fault system, incipient crystal plastic behaviour in quartz and pressure solution became important deformation mechanisms.

Minor cross-cutting veins of pseudotachylite with subhorizontal generating surfaces occur in, and adjacent to the detachment fault zone(s). Elsewhere the faults are characterized by a high concentration of cross-cutting mineralized veins and fractures that imply episodic fracturing and fluid flow associated with detachment faulting. The presence of pseudotachylite suggests that the fault system was seismically active during at least part of its movement history.

Application of $^{40}Ar-^{39}Ar$ and fission-track thermochronology to rocks in the footwall of the Chemehuevi fault system provides further constraints on the timing, initiation angle and evolution of regional detachment faulting (Foster et al. 1990b; John & Foster 1993; Foster & John, unpublished data). Geographical patterns of cooling ages in the footwall provide a basis for interpreting the unroofing history of the domed footwall as it moved from under the hanging wall. Contoured values of mineral age, for biotite, K-feldspar, apatite, zircon, and sphene, decrease northeastward in the slip direction and provide information about palaeotemperatures of footwall locations before, during, and after crustal extension (Fig. 16). These thermochronological data indicate a moderate palaeotemperature field gradient across the footwall before faulting.

At c.22 Ma, granitic rocks exposed in the southwestern and northeastern portions of the footwall were at ≤200°C and ≥400°C, respectively, and were separated by a distance of some 23 km down the known slip direction (Fig. 16).

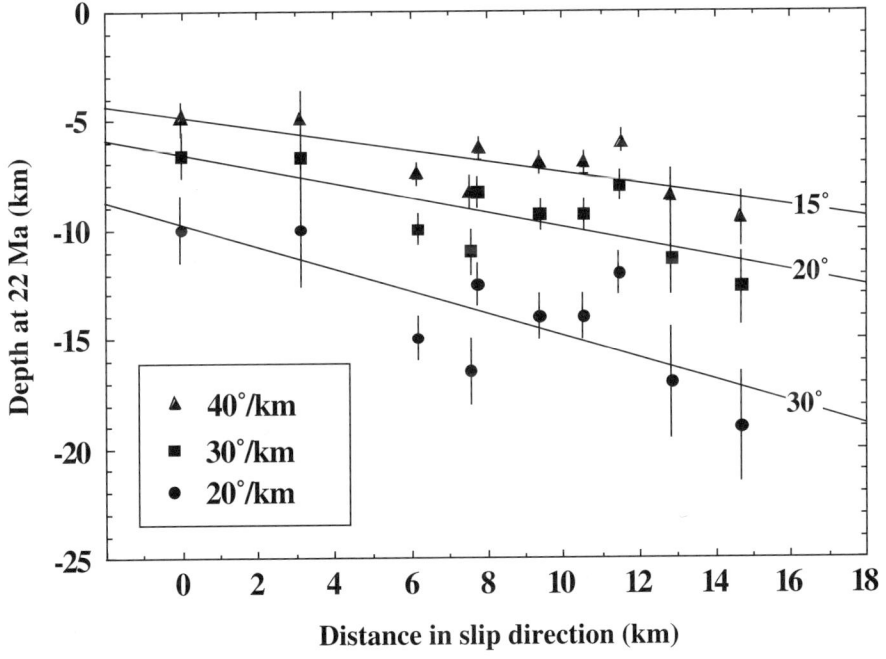

Fig. 17. Plot of calculated palaeodepth for samples from the southern Chemehuevi Mountains projected to a southwest–northeast cross-section in the direction of slip on the Chemehuevi detachment fault. Palaeodepth for each point is calculated from the temperature of each sample based on the thermochronological data, at 22 Ma, and the three geothermal gradients shown (after Foster et al. (1990) and John & Foster (1993), with new data). Lines indicate regressions of the data for each gradient and give regional average initial dips of the Chemehuevi detachment. The low angles of dip are consistent with structural data (reviewed by John & Foster (1993)). The geothermal gradients used are within the range of those estimated from upper plate blocks in the Colorado River extensional corridor.

This gradual increase in temperature with depth is attributed to the gentle warping of originally subhorizontal isothermal surfaces, and constrains the exposed part of the Chemehuevi detachment fault to have had a regional dip initially of 15–30° using a range of geothermal gradients (Fig. 17), including those obtained from adjacent tilted crustal sections (Mohave and Piute Mountains). A similar initial dip <30° was calculated for the Harquahala Mountains detachment (Fig. 7), where it developed within the brittle upper crust, from a more limited thermochronological dataset (Richard et al. 1990; Foster et al. 1993).

The significance of Cenozoic normal faulting exposed in the Chemehuevi Mountains cannot be over-emphasized. In contrast to many core complexes, the footwall to the Chemehuevi Mountains detachment fault system is underlain by few demonstrably Tertiary mylonitic rocks, apart from minor, narrow greenschist facies shear zones (John 1987a,b; John & Mukasa 1990). This relationship implies that deformation was dominantly within the brittle regime. Detailed geological mapping (John 1987b, 1999) demonstrates contemporaneous high-angle and low-angle normal faulting, with the regionally developed low-angle normal faults showing the most recent movement (John 1986, 1987a,b). This relationship precludes the possibility of rotations of the low-angle normal faults to their gentle orientation by younger higher-angle faults. Documented slip on the fault system exposed in the Chemehuevi Mountains accommodated large strains (>18 km), at least locally accompanied by the presence of pseudotachylite. The fault system apparently cut gently down through the upper crust, to a minimum depth of roughly 12–15 km, the deepest exposed parts of the system today, and was domed from mid-crustal depths and locally denuded during continued slip. Together these relations indicate that the detachment fault system accommodated large magnitude crustal extension along gently dipping (≤30°), in part seismically active normal faults.

## Conclusions

The examples presented here constrain many aspects of the evolution of the southern Basin and Range province and Colorado River extensional corridor. It should be emphasized that relatively large datasets were needed to define aspects of the extensional evolution, and more detailed thermochronological and geochronological studies are needed for a full understanding of the thermal evolution of the area. In our present level of understanding significant advances have been made in elucidating some factors including the following:

(1) Rapid cooling of rocks in metamorphic core complexes is the direct result of tectonic exhumation with a component of this cooling possibly related to contact with colder hanging wall rocks and fluid flow, with only a minor contribution from erosion.

(2) The onset of extension can be determined by thermochronological measurements of minerals that were at or very near the transition between totally open systems ('zero age') and the interval of partial closure, just before tectonic exhumation. Analyses from samples that were at higher temperatures before extension will reveal information about only the cooling–unroofing history, and will underestimate the time at which extension and exhumation started.

(3) Estimates of average rates of detachment fault slip derived from plots of age against distance for low-temperature thermochronometers provide independent constraints that can be directly compared with geological evidence. Potential complications caused by increases in geothermal gradients during rapid extension and plutonism need to be assessed, but these can be minimized by using minerals with very low closure temperature and only those samples that record ages >2–3 Ma after the onset of slip.

(4) Regional geothermal gradients may have been rather low to moderate immediately before onset of Oligocene–Miocene extension in the southern Basin and Range province. These relatively low geothermal gradients run counter to some ideas that extension in this area was initiated by magmatism alone. However, magmatism may still have been an important factor in the development of metamorphic core complexes, and/or the rapid exhumation of mid-crustal rocks, with the colder crust of the tilted crustal sections representing passive failure toward a growing lower area of actively thinning hotter crust.

(5) The syn-extensional geothermal gradients provide useful constraints on palaeodepth estimates for footwall rock samples at the onset of extension, which in turn can define the geometry of detachment faults within the brittle upper crust. A large, multi-mineral thermochronological dataset from the Chemehuevi footwall reveals a gradual palaeotemperature field gradient at the onset of extension at 22 Ma. This indicates that the exposed part of the Chemehuevi detachment fault initiated at low angles of 15–30°.

This paper arose from a number of collaborative studies and helpful discussions with E. Campbell, G. Davis, M. Fanning, P. Fitzgerald, A. Gleadow, M. Harrison, K. Howard, B. Kohn, G. Lister, C. Miller, J. Miller, V. Pease, S. Reynolds, R. Scott, and J. Spencer. Reviews by S. Inger and S. Reddy, and editorial comments by M. Brandon were helpful in improving the manuscript. This work was supported by grants from the Australian Research Council (D.A.F.), La Trobe University (D.A.F.), the Australian Institute of Nuclear Science and Engineering (D.A.F.), the US Geological Survey (B.E.J.), and National Science Foundation (B.E.J.).

## References

ARMSTRONG, R. L. & WARD, P. 1991. Evolving geographic patterns of Cenozoic magmatism in the North American Cordillera: the temporal and spatial association of magmatism and metamorphic core complexes. *Journal of Geophysical Research*, **96**, 13201–13224.

AXEN, G. L. & BARTLEY, J. M. 1997. Field-tests of rolling hinge existence, mechanical types, and implications for extensional tectonics. *Journal of Geophysical Research*, **102**, 20515–20537.

BERATAN, K. K. 1993. Tertiary stratigraphy of the southern Whipple Mountains, southeastern California. *US Geological Survey Bulletin*, **2053**, 151–157.

BRANDON, M. T. & VANCE, J. A. 1992. Tectonic evolution of the Cenozoic Olympic subduction complex, Washington State, as deduced from fission-track ages for detrital zircons. *American Journal of Science*, **292**, 565–636.

——, RODEN-TICE, M. K. & GARVER, J. I. 1998. Late Cenozoic exhumation of the Cascadia accretionary wedge in the Olympic Mountains, NW Washington State. *Geological Society of America Bulletin*, **110**, 985–1009.

BRYANT, B. & WOODEN, J. L. 1989. Lower-plate rocks of the Buckskin Mountains, Arizona: a progress report. *In*: SPENCER, J. W. & REYNOLDS, S. J. (eds) Geology and mineral resources of the Buckskin and Rawhide Mountains, west-central Arizona. *Arizona Geological Survey Bulletin*, **198**, 47–50.

CAMPBELL, E. A. & JOHN, B. E. 1996a. Magmatic vs. amagmatic accommodation of Miocene extension in the southernmost Sacramento Mountains, southern Basin and Range. *Geological Society of America, Abstracts with Programs*, **28**(7), A450.

—— & —— 1996b. Constraints on extension-related plutonism from modeling of the Colorado River gravity high. *Geological Society of America Bulletin*, **108**, 1242–1255.

DAVIS, G. A. 1988. Rapid upward transport of mid-crustal mylonitic gneisses in the footwall of a Miocene detachment fault, Whipple Mountains, southeastern California. *Geologische Rundschau*, **77**, 191–209.

—— & LISTER, G. S. 1988. Detachment faulting in continental extension: perspectives from the southwestern U.S. Cordillera. *In*: CLARK, S. P. JR, BURCHFIEL, B. C. & SUPPE, J. (eds) Processes *in continental lithospheric deformation*. Geological Society of America, Special Paper, **218**, 133–159.

——, ANDERSON, J. L., FROST, E. G. & SHACKLEFORD, T. J. 1980. Mylonitization and detachment faulting in the Whipple–Buckskin–Rawhide Mountains terrane, southeastern California and western Arizona. *In*: CRITTENDEN, M. D. JR, CONEY, P. J. & DAVIS, G. H. (eds) *Cordilleran Metamorphic Core Complexes*. Geological Society of America, Memoir, **153**, 79–129.

DICKENSON, W. R. 1991. Tectonic setting of faulted Tertiary strata associated with the Catalina core complex in southern Arizona. *Geological Society of America Special Paper*, **264**.

DODSON, M. E. 1973. Closure temperature in cooling geochronological and petrological systems. *Contributions to Mineralogy and Petrology*, **40**, 259–274.

—— 1979. Theory of cooling ages. *In*: JAEGER, E. & HUNZIKER, J. C. (eds) *Lectures in Isotope Geology*. Springer, Berlin, 194–202.

DOKKA, R. K. 1993. Original dip and subsequent modification of a Cordilleran detachment fault, Mojave extensional belt, California. *Geology*, **21**, 711–714.

DORSEY, R. J. & BECKER, U. 1995. Evolution of a large Miocene growth structure in the upper plate of the Whipple detachment fault, northeastern Whipple Mountains, California. *Basin Research*, **7**, 151–163.

FITZGERALD, P. G., FRYXELL, J. E. & WERNICKE, B. P. 1991. Miocene crustal extension and uplift in southeastern Nevada: constraints from fission track analysis. *Geology*, **19**, 1013–1016.

FOSTER, D. A. 1994. The denudation of metamorphic rocks in the Basin and Range (USA) core complexes. *Geological Society of Australia Abstracts*, **36**, 47–48.

—— 1995. Limits on the tectonic significance of rapid cooling events in extensional settings. Insights from the Bitterroot metamorphic core complex, Idaho–Montana: Comment. *Geology*, **23**, 1051–1052.

—— & GLEADOW, A. J. W. 1996. Structural framework and denudation history of the flanks of the Kenya and Anza Rifts, East Africa. *Tectonics*, **15**, 258–271.

—— & SPENCER, J. E. 1992. *Apatite and zircon fission-track dates from the Northern Plomosa Mountains, western Arizona*. Open-File Report, Arizona Geological Survey, 92–9, 11.

——, GLEADOW, A. J. W., REYNOLDS, S. J. & FITZGERALD, P. G. 1993. The denudation of metamorphic core complexes and the reconstruction of the Transition Zone, west-central Arizona: constraints from apatite fission-track thermochronology. *Journal of Geophysical Research*, **98**, 2167–2185.

——, HARRISON, T. M., COPELAND, P. & HEIZLER, M. T. 1990a. Effects of excess argon on K-feldspar age spectra in the presence of large diffusion domains and plagioclase inclusions. *Geochimica et Cosmochimica Acta*, **54**, 1699–1708.

——, ——, MILLER, C. F. & HOWARD, K. A. 1990b. The

$^{40}$Ar/$^{39}$Ar thermochronology of the eastern Mojave Desert, California, and adjacent western Arizona with implications for the evolution of metamorphic core complexes. *Journal of Geophysical Research*, **95**, 20005–20024.

——, HOWARD, K. A. & JOHN, B. E. 1994. Thermochronologic constraints on the development of metamorphic core complexes in the lower Colorado River area. *US Geological Survey Circular*, **1107**, 103.

——, JOHN, B., CAMPBELL, E. A. & FANNING, C. M. 1996a. The role of syntectonic plutonism in development of the Colorado River Extensional Corridor: Sacramento Mountains example. *Geological Society of America Abstracts with Programs*, **28**, A450.

——, KOHN, B. P. & GLEADOW, A. J. W. 1996b. Sphene and zircon fission track closure temperatures revisited: empirical calibrations from $^{40}$Ar/$^{39}$Ar diffusion studies of K-feldspar and biotite. *International Workshop on Fission Track Dating Abstracts*, August 26–30, University of Gent, Gent, 37.

——, MILLER, C. F., HARRISON, T. M. & HOISCH, T. D. 1992. The $^{40}$Ar/$^{39}$Ar thermochronology and thermobarometry of metamorphism, plutonism, and tectonic denudation in the Old Woman Mountains area, California. *Geological Society of America Bulletin*, **104**, 176–191.

——, MILLER, D. S. & MILLER, C. F. 1991. Tertiary extension in the Old Woman Mountains area, California: evidence from apatite fission track analysis. *Tectonics*, **10**, 875–886.

FRYXELL, J. E., SALTON, G. G., SELVERSTON, J. & WERNICKE, B. 1992. Gold Butte crustal section, South Virgin Mountains, Nevada. *Tectonics*, **11**, 1099–1120.

GANS, P. B., MAHOOD, G. A. & SC HERMER, E. 1989. Synextensional Magmatism in the Basin and Range Province: a Case Study from the Eastern Great Basin. Geological Society of America, Special Paper **233**.

——, MILLER, E. L., BROWN, R., HOUSEMAN, G. & LISTER, G. S. 1991. Assessing the amount, rate, and timing of tilting in normal fault blocks: a case study for tilted granites in the Kern-Deep Creek Mountains, Utah. *Geological Society of America Abstracts with Programs*, **23**, 28.

GLEADOW, A. J. W. & DUDDY, I. R. 1981. A natural long-term track annealing experiment for apatite. *Nuclear Tracks*, **5**(1/2), 169–174.

—— & FITZGERALD, P. G. 1987. Uplift history and structure of the Transantarctic Mountains: new evidence from fission track dating of basement apatites in the Dry Valleys area, southern Victoria Land. *Earth and Planetary Science Letters*, **82**, 1–14.

GREEN, P. F., DUDDY, I. R., LASLETT, G. M., HAGERTY, K. A., GLEADOW, A. J. W. & LOVERING, J. F. 1989. Thermal annealing of fission tracks in apatite 4. Quantitative modelling techniques and extension to geological timescales. *Chemical Geology*, **79**, 155–182.

HARRISON, T. M., DUNCAN, I & MCDOUGALL, I. 1985. Diffusion of $^{40}$Ar in biotite: Temperature, pressure and compositional effects. *Geochimica et Cosmochimica Acta*, **49**, 2461–2468.

HOLM, D. K. & DOKKA, R. K. 1993. Interpretation and tectonic implications of cooling histories: an example from the Black Mountains, Death Valley extended terrane, California. *Earth and Planetary Science Letters*, **116**, 63–80.

—— & WERNICKE, B. 1990. Black Mountains crustal section, Death Valley extended terrain, California. *Geology*, **18**, 520–523.

——, SNOW, J. K. & LUX, D. R. 1992. Thermal and barometric constraints on the intrusive and unroofing history of the Black Mountains: implications for timing, initial dip, and kinematics of detachment faulting in the Death Valley region. *Tectonics*, **11**, 507–522.

HOUSE, M. A. & HODGES, K. V. 1994. Limits on the tectonic significance of rapid cooling events in extensional settings. Insights from the Bitterroot metamorphic core complex, Idaho-Montana. *Geology*, **22**, 1007–1010.

HOWARD, K. A. 1991. Intrusion of horizontal dikes: tectonic significance of Middle Proterozoic diabase sheets widespread in the upper crust throughout the southwestern US. *Journal of Geophysical Research*, **96**, 12461–12478.

—— & FOSTER, D. A. 1996. Thermal and unroofing history of a thick, tilted Basin and Range crustal section Tortilla Mountains, Arizona. *Journal of Geophysical Research*, **101**, 511–522.

—— & JOHN, B. E. 1987. Crustal extension along a rooted system of imbricate low angle faults, Colorado River extensional corridor, California and Arizona. *In*: COWARD, M. P., DEWEY, J. F. & HANCOCK, P. L. (eds) *Continental Extensional Tectonics*. Geological Society, London, Special Publications, **28**, 299–311.

——, CHRISTIANSEN, P. P. & JOHN, B. E. 1993. Cenozoic stratigraphy of northern Chemehuevi Valley and flanking Stepladder Mountains and Sawtooth Range, southeastern California. *US Geological Survey Bulletin*, **2053**, 123–125.

——, GOODGE, J. & JOHN, B. E. 1982. Detached crystalline rocks of the Mohave, Buck and Bill Williams Mountains, western Arizona. *In*: FROST, E. G. & MARTIN, D. L. (eds) *Mesozoic–Cenozoic Tectonic Evolution of the Colorado River Region, California, Arizona, and Nevada*. Cordilleran Publishers, San Diego, CA, 377–392.

——, JOHN, B. E., DAVIS, G. A., ANDERSON, J. L. & GANS, P. B. 1994. A Guide to Miocene extension in the lower Colorado River region, Nevada, Arizona, and California. *US Geological Survey Open-File Report* **94–246**.

——, —— & MILLER, C. F. 1987. Metamorphic core complexes, Mesozoic ductile thrusts, and Cenozoic detachments: Old Woman Mountains–Chemehuevi Mountains transect, California and Arizona. *In*: DAVIS, G. H. & VAN DEN DOLDER, E. H. (eds) *Geological Diversity of Arizona and its margins; excursions to choice areas*. Arizona Bureau of Geology and Mineral Technology Special Paper, **5**, 365–382.

——, WOODEN, J. L., SIMPSON, R. W. & PEASE, V. L. 1996. Extension-related plutonism along the Colorado River extensional corridor. *Geological Society of America Abstracts with Programs*, **28**, A450.

JOHN, B. E. 1982. Geologic framework of the Chemehuevi Mountains, southeastern California. *In*: FROST, E. G. & MARTIN, D. L. (eds) *Mesozoic-Cenozoic Tectonic Evolution of the Colorado River Region, California, Arizona, and Nevada*. Cordilleran Publishers, San Diego, CA, 317–325.

—— 1986. *Structural and intrusive history of the Chemehuevi Mountains area, southeastern California and western Arizona*. PhD thesis, University of California, Santa Barbara.

—— 1987a. Geometry and evolution of a mid-crustal extensional fault system: Chemehuevi Mountains, southeastern California. *In*: COWARD, M. P., DEWEY, J. F. & HANDCOCK, P. L. (eds) *Continental Extensional Tectonics*. Geological Society, London, Special Publications, **28**, 313–335.

—— 1987b. *Geologic map of the Chemehuevi Mountains area, San Bernardino County, California and Mohave County, Arizona*. U.S. Geological Survey Open-File Report, **87–666**.

—— 1999. *Geologic map of the Chemehuevi Mountains area, San Bernardino County, California, and Mohave County, Arizona*. US Geological Survey Miscellaneous Field Investigation Map, in press.

—— & FOSTER, D. A. 1993. Structural and thermal constraints on the initiation angle of detachment faulting in the southern Basin and Range: the Chemehuevi Mountains case study. *Geological Society of America Bulletin*, **105**, 1091–1108.

—— & MUKASA, S. B. 1990. Footwall rocks to the mid-Tertiary Chemehuevi detachment fault: a window into the Late Cretaceous middle crust. *Journal of Geophysical Research*, **95**, 463–485.

KETCHUM, R. A. 1996. Thermal models of core-complex evolution in Arizona and New Guinea: implications for ancient cooling paths and present-day heat flow. *Tectonics*, **15**, 933–951.

LEE, J. 1995. Rapid uplift and rotation of mylonitic rocks from beneath a detachment fault: insights from potassium feldspar $^{40}Ar/^{39}Ar$ thermochronology, northern Snake Range, Nevada. *Tectonics*, **14**, 54–77.

LISTER, G. S. & BALDWIN, S. L. 1993. Plutonism and the origin of metamorphic core complexes. *Geology*, **21**, 607–610.

—— & DAVIS, G. A. 1989. The origin of metamorphic core complexes and detachment faults formed during Tertiary continental extension in the Colorado River region, U.S.A. *Journal of Structural Geology*, **11**, 65–93.

LOVERA, O. M., HEIZLER, M. T. & HARRISON, T.M. 1993. Argon diffusion domains in K-feldspar II: Kinetic properties of MH-10. *Contributions to Mineralogy and Petrology*, **113**, 381–393.

——, RICHTER, F. M. & HARRISON, T. M. 1989. $^{40}Ar/^{39}Ar$ geothermometry for slowly cooled samples having a distribution of diffusion domain size. *Journal of Geophysical Research*, **94**, 17,917–17,936.

——, RICHTER, F. M. & HARRISON, T. M. 1991. Diffusion domains determined by $^{39}Ar$ released during step heating. *Journal of Geophysical Research*, **96**, 2057–2069.

LUCCHITTA, I. & SUNESON, N. H. 1993. Stratigraphic section of the Castaneda Hills–Signal area, Arizona. *US Geological Survey Bulletin*, **2053**, 139–144.

LUND, K., BEARD, L. S. & PERRY, W. J., JR 1993. Relation between extensional geometry of the northern Grant Range and oil occurrences in Railroad Valley, east-central Nevada. *Bulletin, American Association of Petroleum Geologists*, **77**, 945–962.

MCCARTHY, J., LARKIN, S. P., FUIS, G .S., SIMPSON, R. W. & HOWARD, K. A. 1991. Anatomy of a metamorphic core complex: seismic refraction/wide angle reflection profiling in southeastern California and western Arizona. *Journal of Geophysical Research*, **96**, 12259–12291.

MCCLELLAND, W. C. 1982. Structural geology of the central Sacramento Mountains, San Bernadino County, California. *In*: FROST, E. G. & MARTIN, D. L. (eds) *Mesozoic-Cenozoic Tectonic Evolution of the Colorado River Region, California, Arizona & Nevada*. Cordilleran Publishers, San Diego, 401–406.

MCDOUGALL, I. & HARRISON, T. M. 1988. *Geochronology and Thermochronology by the $^{40}Ar/^{39}Ar$ Method*. Oxford University Press, New York.

MILLER, J. M. G. & JOHN, B. E. 1988. Detached strata in a Tertiary low-angle normal fault terrane, southeastern California: A sedimentary record of unroofing, breaching, and continued slip. *Geology*, **16**, 645–648.

—— & —— 1993. Tertiary stratigraphy of the Chemehuevi Mountains, southeastern California and western Arizona. *US Geological Survey Bulletin*, **2053**, 119–121.

—— & —— 1995. Sedimentation patterns support active low-angle normal faulting, S.E. California. *Geological Society of America Abstracts with Programs*, **27**, A69–70.

MUELLER, K. J. & SNOKE, A. W. 1993. Progressive overprinting of normal fault systems and their role in Tertiary exhumation of the East Humbolt–Wood Hills metamorphic complex, northeast Nevada. *Tectonics*, **12**, 361–371.

NAKATA, J. K., PERNOKAS, M. A., HOWARD, K. A, NIELSON, J. E. & SHANNON, J. 1990. K–Ar and fission track ages (dates) of volcanic, intrusive, altered and metamorphic rocks in the Mohave Mountains area, west–central Arizona. *Isochron West*, **56**, 8–20.

NIELSON, J. E. 1993. Stratigraphic and structural correlation of Tertiary strata of the Mohave Mountains and Aubrey Hills, Arizona. *US Geological Survey Bulletin*, **2053**, 133–138.

—— & BERATAN, K. K. 1995. Stratigraphic and structural synthesis of a Miocene extensional terrane, southeast California and west-central Arizona. *Geological Society of America Bulletin*, **107**, 241–252.

PARSONS, T. & THOMPSON, G. A. 1993. Does magmatism

influence low-angle normal faulting? *Geology*, **21**, 247–250.

PEASE, V., FOSTER, D., O'SULLIVAN, P., WOODEN, J., ARGENT, J. & FANNING, C. 1999. The Northern Sacramento Mountains, part II: exhumation history and detachment faulting. *In*: MAC NIOCAILL, C. & RYAN, P. (eds) *Continental Tectonics*, Geological Society, London, Special Publications, in press.

——, ——, WOODEN, J., O'SULLIVAN, P. & ARGENT, J. 1995. Tertiary plutonism and extension in the Sacramento Mountains, SE California, U.S.A. *Eos Transactions, American Geophysical Union*, **46**, F369.

REDDY, S. M., KELLY, S. P. & WHEELER, J. 1996. A $^{40}Ar–^{39}Ar$ laser probe study of micas from the Seria Zone, Italian Alps; implications for metamorphism and deformation histories. *Journal of Metamorphic Geology*, **14**, 493–508.

REYNOLD, S. J. & SPENCER, J. E. 1985. Evidence for large-scale transport on the Bullard detachment fault, west-central Arizona. *Geology*, **13**, 353–356.

RICHARD, S. M., FRYXELL, J. E. & SUTTER, J. F. 1990. Tertiary structure and thermal history of the Harquahala and Buckskin Mountains, west–central Arizona: implications for denudation by a major detachment fault. *Journal of Geophysical Research*, **95**, 19973–19988.

RUPPEL, C. L., ROYDEN, L. H. & HODGES, K. V. 1988. Thermal modeling of extensional tectonics: application to pressure–temperature–time histories of metamorphic rocks. *Tectonics*, **7**, 947–958.

SCOTT, R. S. & LISTER, G. S. 1992. Detachment faults: evidence for a low-angle origin. *Geology*, **20**, 833–836.

——, FOSTER, D. A. & LISTER, G. S. 1998. Rapid cooling of denuded lower plate rocks from the Buckskin–Rawhide metamorphic core complex, west-central Arizona. *Geological Society of America Bulletin*, **110**, 588–614.

SPENCER, J. E. 1985. Miocene low-angle normal faulting and dike emplacement, Homer Mountain and surrounding areas, southeastern California and southernmost Nevada. *Geological Society of America Bulletin*, **96**, 1140–1155.

—— & REYNOLDS, S. J. 1989. Middle Tertiary tectonics of Arizona and adjacent areas. *In*: JENNY, J. P. & REYNOLDS, S. J. (eds) *Geologic Evolution of Arizona*. Arizona Geological Society Digest, **17**, 539–574.

—— & —— 1991. Tectonics of mid-Tertiary extension along a transect through west-central Arizona. *Tectonics*, **10**, 1204–1221.

—— & —— 1993. Startigraphy of Middle Tertiary rocks in the central and eastern Buckskin Mountains, west–central Arizona. *US Geological Survey Bulletin,* **2053**, 1149–1150.

THOMPSON G. A. & MCCARTHY, J. 1990. A gravity constraint on the origin of highly extended terranes. *Tectonophysics*, **174**, 197–206.

WERNICKE, B. 1981. Low-angle normal faults in the Basin and Range Province: nappe tectonics in an extending orogen. *Nature*, **291**, 645–647.

—— 1995. Low-angle normal faults and seismicity: a review. *Journal of Geophysical Research*, **100**, 20159–20174.

WILSON, J. M., MCCARTHY, J. & JOHNSON, R. A. 1991. An axial view of a metamorphic core complex: crustal structure of the Whipple and Chemehuevi Mountains, southeastern California. *Journal of Geophysical Research*, **96**, 12293–12311.

WRIGHT, J. E., ANDERSON, J. L. & DAVIS, G. A. 1986. Timing of plutonism, mylonitization, and decompression in a metamorphic core complex, Whipple Mountains, CA. *Geological Society of America Abstracts with Programs*, **18**, 201.

YIN, A. N. & DUNN, J. F. 1992. Structural and stratigraphic development of the Whipple–Chemehuevi detachment fault system, southeastern California: implications for the geometrical evolution of domal and basinal low-angle normal faults. *Geological Society of America Bulletin*, **104**, 659–674.

YORK, D. 1969. Least-squares fitting of a straight line with correlated errors. *Earth and Planetary Science Letters*, **5**, 320–324.

# Index

Note: Page numbers in *italics* refer to illustrations; those in **bold type** refer to tables.

Aar massif 158, 159, 163, 164
   cooling anomaly 175
   exhumation 167
accommodation 160
accretionary fluxes 7
accretionary wedges
   B-type subduction 29
   décollement 80
   deformation 78
   Franciscan Complex 55
   pressure gradients 149
   steady-state 80–81
   tapering 79
   thickening 148
   viscous flow 150
actinolite 9, 133, 137, 139–141
   precipitation 140
   schists 37, 49, 311
actinolite–tremolite phase 37
actinolite–winchite–riebeckite solid solutions *136*
Adria microcontinent 87
   collision 88, 102
   thickness 104
Adriatic plate 7, 171
Adriatic promontory 158
advection 162, 166, 171, 277
Aegean core complex 305
Aegean microplate 305, 317
Aegean Sea 151
African–Eurasian plate collision 289
Aiguilles Rouge 164, 165
Akaishi Mountains 143
Al-rich phases, metamorphic fluids 78
Alaska
   erosion rates 10, 11
   map *206*
Alberta basin 184, 194
albite
   porphyroblasts 208, 247
   veins 249
Aleutian, subduction zones 147
Alisitos–Santiago Peak arc 32
alkali feldspar, cooling ages 218, *222*
Alpe Arami 9
Alpine Fault
   establishment 298
   Fiordland 295
   plate movement accommodation 262
   strike-slip 261, 277, 279
   uplift rate 278
Alpine Schist 271
Alpine–Himalayan collisional system 120, 320
Alps, *see also* Southern Alps
Alps, European 2, 6, 151, 290
Alps, Swiss 157

deformation and erosion *172, 173*
evolution *160*
exhumation models 166–174
geology 158–161
isothermal maps *165*
map *159*
Amdeh Group 243, 246
Amorgos 306
amphiboles
   analyses **138**
   compositions *136, 137*
   crystallites 337
   micrographs *133, 134*
   Sanbagawa 131–132, *135*
   sodic 247, 249, 313
amphibolite facies 37, 39, 45, 50
   Alaska 207
   Alps 158, 159
   Canadian Cordillera 186
   core complexes 183
   Franciscan complex 57
   Massif Central 190, 191
   New Zealand 263
   Oman 242
   South Mountains 338
anatexis 190, 191, 198, 271
Anatolian Push 105
Andean orogen 6
Andes 151
andesites 337
André Fault 311, 318
Andros 305
annealing model, fission-track dating 92
antigorite 37
antitaxial fibre method 56
Apache Group 355
apatite, chlorine content 286
aplite 187
Appenines 3, 6
   thrust belt 7
Ar/Ar ages
   Alps 162
   Baja metamorphism 37, 38, 39, 43, 47
   Canadian Cordillera 190
   Colorado River corridor 344, 346, *353*, 359
   Crete 90, 96
   Cyclades 319
   Franciscan complex 57
   Kigluaik Mountains 205, 206
   Massif Central 190
   New Caledonia 120
   New Zealand 262, 267–271, *267, 268, 269,* **270**, *271,* 277
Nome Group 208
Oman 242

Ar/Ar ages *continued*
  resolution 279
  Saih Hatat 249–255, **250**, *252*, *253*
  Sanbagawa 131, 132, 141
Ar/Ar thermochronology
  alkali feldspar age spectra 218, *222*
  biotite age spectra 218, *219*, *221*
  data **212**, **213**
  hornblende age spectra *216*, *217*
  Kigluaik Mountains 211–215
  methods 211–215
  mica age spectra 254
  phengite age spectra 254
  sample locations *214*
  white mica age spectra 218, *220*, *226*
Arabian platform 246
aragonite 57, 63, 95
arc–continent collision
  Southwest Pacific 123
  Taiwan 6, 131
Arctic Alaska terrane 206
argon
  excess 96, 218, 242, 255, 275, 328
  systematics 225, 242
arid landscapes 11
Arizona, map *354*
Arrhenius plots 215, 218, 221, 274
Aruma Group 242
As Sifah 242, 246, 247
  Ar/Ar data 251
  eclogites 255
  sample localities *249*
asthenosphere, upwelling 198
attenuation, sections 12, *13*
augen-gneiss 306
Australian Plate, compression 122
Australian–Pacific plate motions 261
Austroalpine nappes 158, 164

back-arc basins, formation of 7, 122, 145
back-arc imbrication 123
back-scattered electron images *334*, 335
backthrusting 158, 173, 175
Baja California 29
  cross-section *47*
  eclogites 37
  exhumation age 46–47
  extension age 41–44
  faults *40*, *41*
  greenschists 37
  major faults 44–47
  maps *33*, *34*, *35*, *36*
  mélange 37–39
  oceanic rocks 32, 33–39
  regional geology 32–33
Ballantrae Complex 122, 123
Banda Arc 122
barometric data 4
barroisite 139
Barrovian metamorphism
  Canadian Cordillera 184, 194
  Cyclades 320
  French Variscides 190
  Lepontine dome 158, 162

basal accretion 7
Basin-and-Range Province 2
  evolution 360
  exhumation 343
  metamorphism 8
  normal faulting 12, 13
  shear zones 123
basins, piggy-back 12
Bay of Islands 122, 123
Benioff zones 9
Bergell pluton 161, 164, 174
  sediments 289–291
Bering Sea, magmatism 207
Bering Strait, map *206*
Besshi unit 131, 132, 141
Betic Cordillera 3
  erosion rates 11
  normal faulting 12, 32
Betic-Rif orogen 9
Big Creek 225
biotite, cooling ages 218, *219*
biotite isograds 208
blackwall effect 44
blueschist facies 8
blueschists
  Baja California 29, 33–37, 44–47
  in collisional belts 9
  Cyclades 8, 306, 319
  densities 49
  footwall 45
  Franciscan Complex 56, 57
  hanging-wall 44, 45
  Kigluaik Mountains 208, 210
  lateral flow 130
  New Caledonia 111, 116, 119
  Oman 242
  overprinting 37, 47, 50
  protoliths 33
  Sanbagawa belt 129, 146
  types 38, 45, 50
Borland Formation 295
boudinage
  Baja California 36
  Canadian Cordillera 186
  Crete 98
  crustal scale 198, 200
  Ios 314, 319
  New Caledonia 117
  Saih Hatat 246, 251
Bouehndep 116
boundary type change, Baja California 33
bow-tie geometry 64
Bowser–Nechako basin 194
breccias
  Chemehuevi Mountains 359
  fault scarp 88
  Ios fault system 314, 316
  South Mountains 328
brittle extension 30, 198
  Baja California 36, 39, 46, 48
  Crete 97, 98
  Cyclades 319
brittle failure 325, 338
brittle shear 339

brittle–ductile transition 30, 36, 200, 325, 326
Brookian orogeny 207, 230
Brooks Range 122, 206
Brossasco–Isasca unit 20, 21
Buckskin Mountains 345, 346, 347, *351*, 352
bulk strength, crust 198
buoyancy
    Crete 87, 104
    as exhumation process 32, 49, 120
    Kigluaik dome 230
    negative 233, 236
    New Caledonia 111
button schist 318

$^{13}C$, Saih Hatat 246
Cache Creek terrane 184
calc-alkaline rocks 181
Canadian Cordillera 181
    exhumation 193–196
    map *184*
    section *185*
capture depth 237, 238
carapace shear zone 112, 118, 119
carbonates
    deep-water 88
    platform 88
    Wadi Adai 247
Caroline basin 110
carpholites 248
Cascadia 2, 56
cataclasites
    Alps 158, 161
    Canadian Cordillera 186
    Chemehuevi Mountains 359
    Ios 314, 317
    South Mountains 324, *330*, 338
Catalina Schist 47
Cedros Island
    Ar/Ar ages 39
    blueschists 37, 38
    fault zones 44, 49
    oceanic rocks 32, 41
Cévennes basin 191, 196
Chambers Well, dykes 352
charnockite 193
Chemehuevi Mountains
    cooling 347, *348*, 349
    extension 346, 357
    faults 345, 352
    map *358*
chert
    Franciscan Complex 56
    radiolarian 33, 131, 141
chlorite 118, 208, 319, 328
chloritoid 247, 248
chromitite 37
chrysotile 37
Chukotka Peninsula 206, 207
Cima Lunga 9, 162
cleavage
    SMT 62, 77, 79
    *see also* crenulation cleavage
climate change 11, 175
clinopyroxenites 37

clinozoisite 249
closure depths
    erosion 292
    thermochronometry 4, 5
closure temperature 15, 90, 91, 162
    isotherms 348
    Kigluaik Mountains 206, 215, 221, 227
    New Zealand 273, 277
    Saih Hatat 254
Coast Plutonic Complex
    exhumation 292–295
    map *294*
Coast Range 50
    fault zone 60
    ophiolite 56
Coastal Fault System 314, 318
    rheology *315*
Coble creep mechanism 78
Cochimí terrane 32
coesite, stability field 9, 17, 19
Col d'Amoss 116, 118, 119, 120
collision, syn-subduction 183
collisional belts 157
Colorado Plateau 346
Colorado River extensional corridor 343
    maps *344*, *345*
    section *346*
Columbia River detachment 186, 190, 196
compositional zoning 135, 141
compressional orogens 3
computer programs, strain measurement 65
condensed sections 301
continental collision 6, 183
    high-temperature metamorphism 198
    Mediterranean-style 17
    metamorphism 9
    metapelite accretion 198
continental crust, thickening 181
continental margins, active 181
continental rifts, metamorphism 8
contractional pure shear 30
convergence
    continental 157
    oblique 31, 131, 145, *146*, 147–148, 261
convergence rates 173
convergent boundaries 3, 6
    Baja California 33
    Hellenic margin 56, 87
    Pacific 181
convergent wedges *5*
    flow-fields *3*
cooling ages
    isotopic 11–12, 16
    Kigluaik 215
    and normal faulting 12
cooling anomalies, Alps 165
cooling curves
    Colorado River corridor *348*, *349*
    South Mountains *328*
cooling history
    Alps 162
    Kigluaik Mountains 215, 221–222, 224, *225*, 227
    New Zealand *273*, 278–279
cooling patterns, Alps 164–166, 174

cooling rates
    Alps 161–166, *166*
    Bergell pluton 291
    Colorado River corridor 347–348
    Crete 90, 91, 95
    and erosion rates 287
    Kigluaik Mountains 205
    New Zealand 275, 277
    Sanbagawa 142
Coral Sea 109
Cordilleran thrust belt, Washington 12
core complexes 183
    Arizona 325, 326
    Colorado River corridor 343, 346, 354–355
    Ios 306
    models 198–200, *199*
    New Caledonia 119, 122, 124
corner flow 104, 120, 150
Cortes terrane 32
Coulomb wedge model 79–80
crenulation cleavage 97, 131, 247
Crete
    geology 88
    map *89*, *92*
    normal faulting 12, 32, 56
    rapid exhumation 87
    sections *103*
critical-wedge model 61
Crossman block 356
crust
    anomalously hot 277
    brittle 196, 197–199
crustal extension, Alps 157, 158
crustal models, two-layer 181, 198, 200
crustal thickening
    Alps 168, 176
    Canadian Cordillera 194
    collisional belts 9, 183
    and erosion 30
    Saih Hatat 255
    thrust slices 123
crustal thinning
    faults 39
    Kigluaik Mountains 208
cryptic terranes 50
crystalline basement, Crete 102
crystallites 336, 337–338
culminations, domed 186, 190
Cyclades 305
    map *306*
    metamorphism 8
    normal faulting 32

Dabie Shan
    erosion 10, 30
    metamorphics 21
    oxygen isotope studies 20
Dead Mountains 345, 346
décollement
    accretionary wedges 80
    Canadian Cordillera 196
    Shimanto belt 143
decompression
    and crystallization 337, 338

metamorphic 194, 196
deep rocks, exposure 233
deformation
    graphical summary *76*
    methods of measurement 63–72
    migration 183
    mixed-mode 3
    partitioning 150–151
    timing 62
deformation fabrics, 14, 228
deformation rates 56
Del Puerto Canyon 60
    ductile fabrics 79
Delamerian orogen 9
dense rocks
    depth *236*
    sinking 233
densities
    blueschists 49
    crustal 234
    plutons 237
density gradients, inverted 198
denudation, term 4
depth, dense rocks *236*
detachment faults
    Alps 158–161, 174
    Canadian Cordillera 186, 196
    Colorado River corridor 345, 347, 352, 357
    core complexes 183
    extensional 87, 88, 99, 103
    footwalls 343
    Ios 305
    Massif Central 191
    slip rates 350–354
detrital grains, fission-track ages 283
diabase 210, 222, 223, 227
Diablo Range 57, 60, 62
    map *73*
    sampling 72
diagenesis, Alps 158
Diahot terrane 111, 113
diamond, stability field 9, 17, 19
diapirism
    Kigluaik dome *229*
    serpentinite 8
diaspore 95
diatexites 190, 191
diffusion domain modelling *223*
diorite, plutons 237
dip-slip faults 46, 113
discontinuity, metamorphic 164
dislocation creep 234
dislocation glide 326
dissolution, diffusion 336
dissolution event 140
divergent boundaries 3, *5*, 6
doming 218, 221, 227, 241, 307, 317
Dora-Maira massif 9, 20, 21, 31, 162
ductile crust, upwelling 194–196, 197
ductile fabrics, Del Puerto Canyon 79
ductile flow 6, 14
    brittle-plastic transition 30
    Crete 96
    as exhumation process 2, 30, 48

lower crust 198
within-wedge 61
ductile shear zones
   Ios 316
   New Caledonia 111, 123
   pseudotachylites in 325
   Saih Hatat 247, 258
ductile strain
   Franciscan complex 81
   margin-parallel 78
ductile thinning
   exhumation process 80
   pervasive 14, 97
dunite 37
Dunnage Zone 123
duplexing 34
dykes 187, 191
   andesite 337
   Chambers Well 352
   diabase 210, 222, 223, 227
   lamprophyre 43, 46, 263
   leucocratic 223, 224
   South Mountains 326
dynamic wedge tectonics 120

earthquakes, Pacific 147
East African rift, exhumation 8
eclogites
   As Sifah 255
   Baja California 37, 50
   Cima Lunga 162
   in collisional belts 9
   Cyclades 8, 319
   Erzgebirge 20
   exotic blocks 131
   Franciscan complex 8–9, 57
   glaucophane 247
   Massif Central 190
   New Caledonia 111, 116, 119
   Oman 242, 246
   omphacite structures 21
   overprinting 37
   Samail 241
ECORS profiles 193
electron microscopy 328
Enderby basin 186, 194
epidote 113, 249, 319
epidote-amphibolite facies 131, 132, 133
erosion
   Alps 158, 167, 169, 170, 175
   Canadian Cordillera 194
   Crete 105
   as exhumation process 2, 21, 30, 48, 284
   Franciscan complex 81
   Himalayas 10
   idealized *285*
   Massif Central 196
   New Zealand 274
   Saih Hatat 258
   surficial 10
   term 4
   erosion rates 233
   Alps 171, 175
   average *10*

and climate 11
and cooling rates 287
drainage-scale 4
Indus River 291–292, *293*
and lag times *287*, 291, 295, 299, 301
model 236
and vegetation cover 11
and vertical extension 7
erosion-mechanical model 168
erosional fluxes 7
Erzgebirge, eclogites 20
Eurasian Plate 87, 88, 102, 145
evaporites, Crete 88
exhumation
   Basin-and-Range Province 343
   buoyancy-driven 10, 87, 104
   calculations **81**
   Canadian Cordillera 193–196
   Coast Plutonic Complex 292–295
   Cyclades 319–320
   extensional 299
   Fiordland 295–301
   Massif Central 196–197
   mechanisms *193*
   model *80*
   rapid *16*, 87, 278
   Saih Hatat 255–258
   Shuswap Metamorphic Core Complex *195*
   tectonic 7, 99, 168
   term 4
   transient processes 21
   Velay Dome *197*
exhumation age, Baja California 46–47
exhumation history
   orogenic highlands 283
   preservation of 20–21
exhumation models, Alps 168–174
exhumation patterns *12*
exhumation processes 1, *2*
   accretionary prisms 29
   Baja California 47–49
   buoyancy 32
   observations **31**
   strike-slip faults 32
exhumation rates 4
   Alps 157, 166
   Coast Mountains 292, 293–295, **298**
   Crete 96
   and exhumation processes 21
   Fiordland 299
   fission-track analyses 284
   Southern Alps 277, 279
exotic blocks
   Baja mélange 37, 45, 49
   eclogite facies 131
   Sanbagawa 131
extension
   Colorado River corridor 346, 356
   late-orogenic 181
   synorogenic 174
extension age, Baja California 41–44
extensional orogens 3
extensional shear zones
   Crete 97, 98

extensional shear zones *continued*
  New Caledonia 116–118, 120, 123
extrusion processes 120, 146, 147, 149, 150–151

fabrics
  asymmetric 60
  fault-related 113
  Nome Group 208
  Saih Hatat 251
fanglomerates 186
Farallon lithosphere 47
fault arrays, Ios 314
fault gouge 39, 196, 357
fault imbrication 17
fault orientations, Baja California *42–43*
fault scarps 115
fault splays 45
fault zones, blueschist complex 44
faults
  Baja California *40, 41*
  folded 49
K-feldspars, domain sizes **274**
fibre overgrowths
  precipitation 78
  strain measure 63, 65
finite strain, Franciscan complex 77
Fiordland
  exhumation 295–301
  map *298*
Fish Canyon tuff 289
fission-track ages
  Alps 162
  Bergell *290*, **291**
  Coast Ranges **297**, *298*
  Colorado River corridor 344, 351, *352*, *355*
  Crete 90, 92, **93–94**, *95*, *97*
  detrital grains 283, 284, 286–288
  Fiordland *299*, **300**, *301*
  Fish Canyon tuff *289*
  Franciscan complex 59, 75
  Nome Group 208
  Olympic Mountains *288*
  Sanbagawa 131, 141, 143
  South Mountains 327
  Southern Alps 262, 271, **272**, 277
  Tofino basin *296*
flexural basins 183, 191, 194
flexural slip 76
fluid flux, solute mass transfer 78
fluorapatite 286
fluxes, distribution of 7
flysch
  Crete 88, 91, 102
  Songpan-Ganzi basin 10, 30
fold geometry, Franciscan complex 76
folds
  asymmetric 228
  Baja California 36
  Crete 96
  New Caledonia 116
  parasitic 254
  recumbent 117, 131, 247, 249
foliation
  granoblastic 208, 228

  layer-parallel 36
  phyllosilicate phases 325
  pseudotachylites 339
  recrystallized 117
  Sanbagawa 142, 147
  solid state 331
  sub-horizontal 14
  Velay Dome 191
Forcola fault 161, 164
forearc
  emergence 81
  Great Valley 56
  North American 33
  refrigerated 46
  submarine 48
  turbidites 41
  underplating 31
foreland basins
  Canadian Cordillera 194
  sediment capacity 11
Franciscan complex
  blueschists 29, 50
  ductile flow 30, 55
  eclogites 8–9
  high-pressure rocks 60–61
  map *58*
  normal faulting 31
  orogenic parameters **59**
  residence times 59–60
  setting 56–59
Franciscan Subduction Complex 55
Franz Josef valley 278
French Variscides 181
Frenchman's Cap 190
frictional melting 326, 338, 339
frontal accretion 7
frontal extrusion 151

gabbro
  ophiolitic 40
  Ortigalita Peak 57, 60
  plutons 237
Galicia–Newfoundland, sea-floor spreading 8
Mn-garnet 113
garnet isograds 208
garnet peridotites 9
garnet-glaucophane schist 113, 116, 242
garnet-oligoclase phase 15
generation veins 331
geobaric gradient 132, 167
geobarometry 249
geothermal gradient
  accretionary prisms 30, 36
  Canadian Cordillera 186
  closure temperature 284
  Colorado River corridor 354, 356
  Massif Central 191
  New Zealand 277, 278
  Sanbagawa 140
geotherms
  Alps *167*
  initial 287
  model 234
glass 339

glaucophane 57, 111, 117, 118, 139, 319
*Globigerina ampliapetura* zone 90
gneiss domes 207, 210, 222, 224, 227
gneissic layering 190
Gold Butte 354, 356
Gondwana, breakup 109, 121
gossan 247
Gotthard massif 161, 164, 165, 174
grain cores, zonation 138, *139*
grain dimensions, strain measures 63
grain populations 284, 285, 288
grain selvages 61
grain-size sensitive flow 339
Granite Wash Mountain 351
granites, core complexes 183
granitoids
    Canadian Cordillera 190
    I-type 181, 210
    Massif Central 191
    P–t data 16
granodiorite 190, 328
granulite facies
    Alps 158
    Massif Central 190
granulites, in collisional belts 9
gravitational collapse 30, 31, 102, 104, 105, 198, 233
gravity anomalies, Baja California 32
Grayback fault block 354, *355*
Great Valley basin 56, 81
Great Valley Group 50
greenschist facies
    accretionary wedges 56, 56
    Alps 159
    Canadian Cordillera 186, 187
    Crete 97
    Kigluaik Mountains 208
    New Caledonia 118
    pseudotachlyites 326, 329, 338, 340
    Sanbagawa 141, 151
    in UHP rocks 20
greenschists, Baja California 37, 45
Gulf of Mexico 194

Haast River valley 271, 278
Hajar Super Group 242
hanging wall
    as eclogite source 9
    thinning 149
Harcuvar Mountains 346, 351
Harquahala Mountains 360
Hatat Schist 243
Hawasina Complex 242, 246
Haybi Complex 242
heat production, radiogenic decay 198
heat transfer 166, 277, 278, 357
Hellenic Arc 318
Hellenic margin 7, 56, 102
Hellenic Subduction Zone 87
Helvetic nappes 158, 159, 163, 164
high-angle normal faults 317, 360
high-pressure rocks
    Cyclades 320
    Franciscan complex 60–61
    Saih Hatat 255

high-pressure-low-temperature rocks 56, 87, 88
    deformation history 96–98
    metamorphic history 95–96
    New Caledonia 113
    Nome Group 210
high-temperature-low-pressure rocks 88, 181
Hijam Formation 243, 246
Himalayas 3, 151
    erosion 10, 11
    lineations 145
horizontal extension
    brittle field 98, 100
    slab retreat 7
    and tectonic exhumation 4, 6
hornblende
    closure 227
    inclusions 137–138
    Kigluaik 215–218, 223
    plutonic 223–224
    Sanbagawa 132, 139
Hornelen Basin 174
Hualapai Mountains 346
Hutt Range 277
hydrothermal activity 247

Ikaria 305
illite crystallinity 143
impact structures 339
inclusions, zoned 139
INDEPTH traverse 198
Indo-Australian Plate 109, 121
Indus River, erosion rates 291–292, *293*
injection veins 331
*Inoceramus* 57
Insubric Line 158, 164, 169, 171, 173, 175
Intermontane superterrane 184, 194
internal rotation 78
Ionian Zone 88
Ios 305
    map *307*
    section *311*
Ios detachment fault system *308*, *309*, *310*, 317
    bowing 317
    evolution *313*
Iritono 337
island-arc terranes, Baja California 32, 39
island-arcs, magmatic 181
isostatic rebound 233
isotopic ages, footwall rocks 2
Italian Alps 9
Izanagi plate 145
Izumi Group 146

jadeite 57, 62, 63, 248
Japan, plate motions *145*
Japan Sea, opening 145
Jaujac basin 191, 196
Juweiza Formation 258

K–Ar ages
    Alaska 207
    Alps 161–162
    Baja California 43
    Canadian Cordillera 184, 190

K–Ar ages *continued*
  Crete 90, 91, 96
  Franciscan complex 57
  New Zealand 262, 267–271, **270**
  Oman 242
  South Mountains 327
K/Ca ratios 218
Kamuikotan 32
Kermadec-Tonga Trench 110
Kern-Deep Creek Mountains 354
Kigluaik dome, diapirism *229*
Kigluaik Group 207, 210
Kigluaik Mountains 205
  field relations 208–211
  map *207*
  section *209*
Kigluaik Pluton 208, 222–224, 227
Kii Peninsula 143
kinematic analysis, brittle extension 40
kinematic numbers 65, 66, 71, 73
kink bands 98, 113
klippen 45, 164
knockers 9, 56, 118, 119, 120
Kohistan arc 238
Koumac terrane 111, 116
Kula lithosphere 47
Kuma Group 132, 141
kyanite zone 263
Kyushu 143

laccoliths 187, 191
Ladybird leucogranite 187
lag times 11
  and erosion rates *287*, 291, 295, 299, 301
Lake Bonneville 234
Lake Como 289
lamprophyre dykes 43, 46, 263
landslides, bedrock 11
Laramide orogeny 79
lateral flow 132
lawsonite
  Crete 95
  New Caledonia 111, 113, 116
  Yolla Bolly 62
lawsonite–albite facies 56, 57
lead–zinc ores 247
Leech Lake Mountain 60
  map *72*
  sampling *72*
Lendas 95
Lepontine dome 158, 160, 162, 164, 167
  exhumation 167, 168, 174, 176
  thermal history *169*, *170*
leptynite-amphibolite gneiss 190
leucogranites 186, 191, 198
leucosomes 191
lherzolite 210, 227
limestone
  black 246
  interpillow deposits 33
  recrystallized 115
lineation
  Canadian Cordillera 190
  Ios 316

Kigluaik Mountains 208, 228
  mylonitic 79
  rotation 228
  Saih Hatat 249
  Sanbagawa 143
  stretching 129, 131, 132, 148, 249, 318
  strike-parallel 150
listric faults 305, 317
Lithoprobe seismic profiles 196
lithosphere, weakening 105
lithospheric delamination 104
lithospheric extension 305
lizardite 37
Lord Howe Rise 109
low-angle normal faults
  Aegean *306*
  Colorado River corridor 357, 360
  Cyclades *312*
  Ios 310–311, 314
  North American Cordillera 326
  *see also* high-angle normal faults
lower crust
  ductile flow 198
  Massif Central 191–193
Loyalty Basin 110, 111

mafic bodies
  in plutonic areas 233
  trajectories *238*
mafic complexes 238
Magdalena Island 33, 39
magmatic arcs 33, 50, 132
magmatism
  Bering Sea 207
  descending slabs 181
  and gneiss domes 222–223
  and ultrahigh-pressure rocks 20, 162
magnesio-carpholite 95
Mahil Formation 246
major faults, Baja California 44–47, **46**
Malton 190
mantle inclusion, in orogenic roots 17
mantle rocks, unroofing 8
mantle–crust detachment 104
Manus–Solomon–Vitiaz region 110
marbles
  aragonite 95
  Ios 310, 316, 317
margins, underthrust 183
Mariana margin 7
  metamorphism 8
mass flux 170, 171, 173
mass loss, volume strain 77, 78
mass transfer
  diffusional 48
  *see also* solution mass transfer
Massif Central 183, 190
  exhumation 196–197
  map *187*
  section *188*
Mataketake Range 261, 263, 275, 279
Median Tectonic Line 142, 146
megaboudins 246, 248, 251
megabreccias 347

megafolds, New Caledonia 116, 122
mélange
   Cima Lunga 162
   Crete 88
   New Caledonia 111, 118
   Sanbagawa 143
   serpentinite 32, 37–39, 41, 44, 45
   shear bands 143
   see also serpentinite
Menderes Massif 13, 105
metabasalts, Zermatt 9
metabasites 247
metabauxite 95
metacherts
   Sanbagawa 141
   Zermatt 9
metamorphic indicators, vein minerals 34
metamorphic transition 227
metamorphism
   peak 37
   settings 7
metapelites 95, 186, 191, 198
metaquartzites 246
metasomatism 44, 49
metatexites 190
Meteor Crater 339
microboudinage 140–141
microfossils, Sanbagawa 131
microlites 336, 337–338
microprobe analysis 333, *334*, **335**, *336*
migmatites 181
   Canadian Cordillera 190
   core complexes 183
   exhumation 193–197
   Massif Central 191, 196
mineral ages
   and palaeodepth *356*
   and slip distance *350*
miscibility gap 137
mode method 63
Mohave Mountains 345, 352, 354, 356, 360
Mohave Wash fault 357
Moho
   Canadian Cordillera 196
   as eclogite facies transition 19
   ECORS profile 193
   in forearcs 10
   UHP metamorphism 17
Mojave Desert 345
Molasse Basin 174, 175
Mont Blanc 164
Monte Cedros 45
Monte Santa Margarita 45
monzogranites 190
Mosquito Pass 218
Mount Isa orogen 9
Mount Kinnaird 263
mudstone, deformation 78
Mulhacen nappe 15
multiple deformations 130
Muti Formation 242, 246
Mykonos 305, 320
mylonites
   Alps 158, 159

Canadian Cordillera 186, 187
Cyclades 307, 313, 316, 317, 318
Massif Central 191
New Caledonia 112, 118, 119, 120
overprinting 325
reworking 161
Sanbagawa 141
South Mountains 327, 328, *329*, 338
viscosity 234

Nanga Parbat–Haramosh massif 291, 292
nappes
   deformation in 14
   Eastern Belt 57
   Massif Central 190
   ophiolitic 122
   Saih Hatat 255, 257
   Sanbagawa 132
Naxos 305, 306, 318, 320
Nazca plate 6
Neotethys Ocean 88, 104
New Caledonia 109
   Basin 109
   map *112*
   regional geology 111
   schist belt *115*, *117*
   structural relations *114*
   tectonic evolution 118–119, *119*, 120–123, *121*
New Guinea 121
New Zealand 110
Newberry Mountains 346
Nome Group 207, 208, 210, 224, 225
Norfolk Ridge 109, 110
normal faulting
   amount of slip 45–46, 49
   brittle 112, 123
   diagnosis 12
   episodic 8
   and exhumation 12
   as exhumation process 2, 30–32, 48
   orogen-parallel 113
   quantification problems 13
   sense of slip 32, 45–46
normal faults
   Alps 161, 174, 175
   Baja California *38*, 44
   Betic Cordillera 32
   Canadian Cordillera 186, 194
   Colorado River corridor 354
   Crete 32, 98
   Cyclades 32, 305
   Franciscan complex 31, 60
   Kigluaik Mountains 208, 222
   New Caledonia 111, 122
   rotation 199, 357
   Sanbagawa 149
   slip rates 13
   vergence 6
   see also low-angle normal faults
North American Cordillera 56, *182*, 317, 325
   geology 326–328
Norwegian Caledonides 8

obduction
    New Caledonia 111, 120, 122, 123
    Oman 242
oblique buoyant escape 104
oblique-slip faults 46
Oboke unit 131, 132, 141, 145
ocean-ocean collision 111, 123
oceanic assemblages, in collisional belts 9
oceanic crust, protoliths 33
oceanic rifts, metamorphism 8
oceanic rocks, Baja California 32, 33–39
offscraping 7
Okanagan–Eagle River–Adams detachment 186, 194
Old Woman Mountain 345
olistostromes
    Baja California 43
    Crete 88
Olympic Mountains 7
    erosion rates 11, 81
    fission track ages *288*
Olympic Peninsula 292
Oman 122, 241
Oman Mountains, map *242*
omphacite 113
    in eclogites 21
onlap 295
Ontong Java Plateau 110
open folds 98
ophicalcites 8
ophiolites
    Baja California 39, 44
    Coast Range 56, 60
    Crete 91
    Cyclades 317, 320
    formation of 122–123
    Oman 241, 242, 255, 257, 258
    Southwest Pacific 110, 120, 122
    in UHP metamorphics 17
organic matter, maturation 191
orogenic collapse 198
orogenic deformation 5
orogenic highlands, exhumation history 283
orogenic parameters, Franciscan complex **59**
orogenic roots 17–19
orogenic wedges 169, 170
orogenies, oscillating 109, 123
orogens
    asymmetric 171
    collapsed 181, 183–193
    terminology 2–3
orthogneiss 208, 215, 221, 224
Otago wedge 14, 278
Ouégoa 116
Outer Hebrides Thrust 340
overprinting
    Alps 163
    Canadian Cordillera 186
    ductile 339
    eclogites 37, 119, 120
    Ios fault zone 313, 318
    Kigluaik 224, 226, 228
Owen Stanley belt 122
oxygen isotope studies, Dabie Shan 20

$P$–$t$ data 16–17
$P$–$T$ paths *15*
    Saih Hatat 247
    Sanbagawa 131, 139, *140*
$P$–$T$–$t$ history, Crete 95, 96, *98*
$P$–$T$–$t$ paths 174
Pacific plate 33, 109, 279
paired metamorphic belts 129
palaeoisotherms 348–350, 354
palaeoseismicity 325
palaeostresses, brittle extension 40–41
palaeotemperatures, Crete 95
Pam Peninsula 116
Papua New Guinea 110
Papuan–New Caledonia–Norfolk Trench 121
parageneses 161, 162, 319
Paringa River valley 263, 278
Paros 305, 306, 316, 320
partial annealing zones 286, 287, 299, 301, 355
partial melting 186, 191, 198
passive margin, North American 186
PDS, *see* projected dimension strain
peak metamorphism
    Alps 158, 160, 161
    Franciscan complex 59
    Kigluaik Mountains 210, 215, 222, 227, 230
    Saih Hatat 247
    Sanbagawa 131, 132, 143
peak temperatures, Kigluaik 215
pegmatites 187, 190
    intrusion 271
    New Zealand 261, 262
Pelagonian Plate 102
penetrative fabric
    and ductile flow 30
    Nome Group 208
    Sanbagawa 132
penetrative strain 210
Peninsular Ranges batholith 32
Penninic Klippen belt 163, 164
Pennnic nappes 158, 161, 164, 167, 290
    exhumation 167
per-aluminous melts 181
peridotites
    mylonitic 242
    serpentinized 8, 49
    Wadi Adai 247
phengite
    closure temperature 227
    crystallization age 16
    excess argon 96
    New Caledonia 118
    Saih Hatat 248, 249
    Si content 62, 249
    stabilization 20
Philippine Sea plate 147
Phyllite–Quartz unit 88
    mesoscopic structures *100*
    relationships *101*
    structural orientation *99*
phyllonites 313, 314, 316, 319, 339
Pickett Peak terrane 57, 72
Pilat detachment 191

Pindos Ocean 88, 102
Pindos unit 88
Piute Mountains 354, 360
plagiogranites 241
Plakias, map *92*
plane strain 4, 141
plastic behaviour 30
plate boundaries, terminology 3
plate divergence 305
plate reconstructions, Japan 145
Plattenkalk 88, 90, 104
Plomosa Mountains 352
plutons
    Colorado River corridor 346
    diapiric emplacement 2, 210
    dimensions 237
    modelling 237–238
    sinking 233, 234
Po Plain 290
Poindimié 116
porosity, Franciscan wedge 61
porphyroblasts
    albite 208, 247
    pressure shadows 247
Pouébo terrane 111, 115
precipitation 170
prehnite–pumpellyite facies 56
pressure differences, Baja faults 46
pressure estimation 16
pressure shadows
    chlorite 208
    Saih Hatat 246
pro-deformation 169, 171
pro-shear zones 169
prograde metamorphism 131, 139, 140
projected dimension strain 56, 63–64, **64**
Prospect Formation 295
protoliths
    Alaska 207
    blueschists 33
    pseudotachylites 340
    Sanbagawa 131, 141
protomylonite 357
provenance problems 295
psammite 246
pseudotachylite 328–335
    biotite in *333*
    Canadian Cordillera 186
    Chemehuevi Mountains 359, 360
    clasts 330, 331, 334
    ductile deformation 332
    formation 338–339
    palaeoseismicity 325
    photographs *329*, *332*
    zones 335–337
Puerto Nuevo 41
pumpellyite, Saih Hatat 247
pumpellyite–actinolite facies 56, 131
Pyrenees 6
pyrophyllite 95, 247

quartz diorite 234, 235
quartzofeldpathic rock, buoyant rise of 2
Quesnel arc 184

radiogenic decay, heat production 198
radiolaria 33, 131, 141
Raleigh-Taylor instability 19
Rawhide detachment 352
Rb–Sr ages
    Alps 162, 163
    Canadian Cordillera 184, 190
    Franciscan complex 57
    Massif Central 191
    New Zealand 262, **266**, *266*, 273, 274
recrystallization
    Colorado River corridor 348
    Saih Hatat 247, 254
relief, increased 170
Rennell Arc 110
residence times, Franciscan complex 59–60
retreating plate boundary 6
retro-deformation 169, 171
retro-shear zones 169
retrograde metamorphism
    Crete 97
    New Caledonia 118
    Saih Hatat 254
    Sanbagawa 131, 132, 138, 140, 143
return flow, as exhumation process 149
rheology
    Coastal Fault System *315*
    power-law 234, 235
Rhône–Simplon Line 158, 160, 161, 164
rhyolite 186, 337
ribbon chert 143
ribbon quartz 208, 329
ridge crests, under-fed 8
riebeckite 133, 135, 137, 139–141
rifting 305
rock mechanics 234
rock uplift 4
Rocky Mountains 9, 184, 196
Rocles laccolith 191
runoff 170
rutile 248
Ruwi schist 258
Ryoke belt 132

Sacramento Mountains 345, 346, 347, 354
Sahtan Formation 246
Saih Hatat 241
    evolution *257*
    geochronology *256*
    geology 242–246
    map *243*
    metamorphic summary *248*
    structural profile *244*, *245*
    structure **245**
Saint Etienne basin 191, 196
Saiq Formation 243, 246, 254
Salt Range 11
Samail ophiolite 241, 255, 257, 258
sampling, deformation 72
San Andreas system 50, 57
San Benito Islands 33, 37, 39, 44, 49
San Carlos fault 44
San Juan–Cascades, exhumation 30
Sanbagawa belt 129

Sanbagawa belt *continued*
  block diagram *148*
  deformation history 140–141
  exhumation model 150
  structural geology 132, *144*
  thinning 149
sandstones, strain measures 64
Santa Ana terrane 32
Santa Margarita Island
  age data 39
  blueschists 37
  dykes 43
  fault zones 40, 45
  oceanic rocks 33
schistosity
  axial planar 247
  re-oriented 113
Sea of Crete 318
sea-floor spreading, Galicia–Newfoundland 8
secondary electron images 335
sediment accretion 6
sediment dispersion 11
sediment transport 194
sediment yield 291
sedimentary basins, orogenic records 283
sediments
  eroded 30
  synorogenic 11
seismic profiles
  ECORS 193
  Lithoprobe 196
  Tibetan plateau 198
seismic tomography 90, 104
Selkirk allochthon 184
semi-deformable antitaxial method 56, 64–72
Serí terrane 32
serpentinite
  diapirism 8, 32
  emplacement 44, 49
serpentinite mélange
  significance 49
  *see also* mélange
Sesia zone 9
Seward Peninsula 205, 207
shear
  heating 278
  pure 148
  sense of 208, 227, 246, 313
shear fabric, New Caledonia 116
shear indicators 132
shear zones
  Alps 158–161
  Cyclades 306, 314, 319
  in downgoing plates 6
  fluid in 20
  Massif Central 191
  New Caledonia 111, 116, 118, 122, 123, 124
  pressure break 13
  Sanbagawa 132
  at translational boundaries 4
sheath folds 131, 132
Shikoku 143
  map *130*
Shimanto Complex 131, 132, 142–143
SHRIMP facility 263

Shuswap Metamorphic Core Complex 183
  evolution *189*
  exhumation *195*
  formation 194
  map *184*
  overlying sediments 200
  section *186*
Sierra Nevada 56, 237, 238
sieve textures 337
Sifnos 305, 306
Sikinos 306
sillimanite isograds 208, 227
sillimanite-gneiss 205
sillon 111, 116
sills 187, 191
simple shear 181, 198
Simplon Fault Zone 158, 161, 164, 175
Simplon Shear Zone 159, *160*, 165
sinking, dense rocks 233, 237–238
slab break-off 104
slab pull 101
slab rollback 6, 104, 121
slickensides 329, 331, 332
Slide Mountain terrane 184
slip distance, and palaeodepth *359*
slip magnitudes 45–46, 49
slip rates
  detachment faults 350–354
  normal faults 13–14
slope, surface 170
Sm–Nd ages, Alps 162
SMT, *see* solution mass transfer
sole, metamorphic 50
Solomon–New Hebrides system 110
solution mass transfer 56, 61, 82–83
  analysis 73–77
  fabrics *66*
  measurements **67–70**, **74**
Songpan–Ganzi basin, flysch 10, 30
Sonoran Desert 345
South Cyclades Shear Zone 306, 318
South Fiji Basin 110
South Fork Mountain Schist 57, 72
South Loyalty Basin 116
South Mountains 325, 326–328, *327*
South Virgin Mountains 354, 356
South Westland 274, 278
Southern Alps, New Zealand 4, 261
  erosion rates 10, 11
  geology 262–263
  map *262*
Southwest Pacific 109
  map *110*
staurolite isograds 162, 208, 227
Stillwater complex 238
strain
  *see* plane, projected dimension, vertical, volume strain
  high temperature 208
strain directions, Franciscan complex *75*
strain gradients 44
strain hardening 338
strain partitioning 147
strain paths, Yolla Bolly *82*
strain rates

and cataclasites 326, 339
  Sanbagawa 140
  Yolla Bolly **82**
strain symmetry 73
strain types 73
stretching
  asymmetric 6
  symmetric 6
striation orientation, on foliation planes 44
strike-slip faults
  Alps 161
  Canadian Cordillera 186
  as exhumation process 32, 48
  New Caledonia 111
  New Zealand 261, 295
  Sanbagawa 145, 151
subduction
  A-type 157, 255
  B-type 29
  oblique 145
  oceanic 6
  polarity 171
  syn-collision 183
subduction channels 148, 149, 151
subduction zones
  advancing 6
  Aleutian 147
  Baja California 29, 33–37, 46, 48
  Crete 99
  metamorphism 8–9
  New Caledonia 121
  refrigeration 46
  retreating 6, 7, 56, 102, 121, 278
  sediment recycling 11
  Southwest Pacific 109–110
  Sumatra 147
subduction-collision transition 169
submarine fan sediments 131
Sumatra, subduction zones 147
Sumeini Group 242
surface processes, Alps 175
surface uplift 4
Syros 305

T–t paths 15–16
  Colorado River corridor 347
  Crete *91, 96*
  New Zealand 274, 275
  Sanbagawa 141–142, *142*
Taiwan 6
  arc-continent collision 131
  erosion rates 10
  map *147*
talc, stabilization 20
Tasman Sea 109
Tauern window 13
Te Anau basin 295
tectonic denudation 2, 4, 194, 196, 200
tectonic exhumation
  Alps 158
  Arizona 325
  Colorado River corridor 347
  Hellenic margin 87
  importance of 21
tectonic settings, exhumation 7

tectonic thinning 46, 132, 258
tectonic wedging 60, 79
tectonites 118, 242
TEM images, pseudotachylite *333, 334*
temperature, crustal 235
temperature maxima, Alps *163*
terminological problems 2
Tethys, spreading centre 241
thermal aureole 228, 230
thermal equilibration 181
thermal history
  Alps *168, 169, 170*
  Colorado River corridor 360
  Crete 90–95
  New Zealand 262, *276*
  South Mountains 329
thermal inversions 129
thermal relaxation 200
thermal resetting 295, 350
thermal stability, fission tracks *286*
thermochronology
  Alps 157, 161–164
  Canadian Cordillera 194
  Colorado River 344
  Crete 90–95
  Kigluaik Mountains 205, 211
  Sanbagawa 132, 143
thermochronometers, Colorado River corridor *350*
thermochronometric data 4, 167
thermomechanical instabilities 325
thermomechanical model *171, 172, 173*
thermotectonic regimes 284
thickening
  continental crust 181
  lithospheric 6
  use of term 4
thinning
  tectonic **46**
  use of term 4
Thira 306
Thompson Creek 215, 221
Thor-Odin dome 190, 194
Three Kings Arc 110
thrust faults
  and attenuation 13
  Cyclades 313
  structural levels 149
thrust planes, Ios 311
thrust sheets, Crete 87
thrust slices, New Caledonia 111, 120, 122, 123
thrusting
  Coast range 60
  role in crustal thickening 30
Tibetan plateau
  seismic profiles 198
  uplift 234
Ticino culmination 174, 175
Tinos 305, 306, 318
Toce culmination 174, 175
Tofino basin 292, *296*
Tormes Gneiss 174
Tortilla Mountains 354
Tosco–Abrejos fault zone 48
transform boundaries
  Baja California 33, 48

transform boundaries *continued*
  metamorphism 8
translational boundaries 3
  shear zones at 4
transtensional regime 295
transverse faults, Southwest Pacific 110
tremolite 137
trench sediments 56, 131, 143
Trinity Hills basin 186, 194
Tripolitza unit 88, 103
Turba fault 161, 174
turbidites
  Baja California 41
  blueschist protoliths 33
  forearc 41
  Shimanto 142
Turtle Mountain 345

U–Pb ages
  Alps 162
  Canadian Cordillera 187, 190, 198
  Colorado River corridor 346
  Franciscan complex 57
  Kigluaik 210, 215, 222, 223
  Massif Central 191, 198
  New Zealand 262, 263–265, **264**, *265*
  Nome Group 208
  Oman 241
  South Mountains 327
ultrahigh-pressure metamorphism 9
  fluid phase in 20
ultrahigh-pressure rocks 17–21
  Alps 162
  buoyancy 32
  eclogitized crust model *19*
  formation 22
  and magmatism 20
  thickened mantle model *18*
  thickness 20
  undeformed 21
ultramafics
  exotic blocks 49
  New Caledonia 118
ultramafite, Zermatt 9
ultramylonite 331, 332
underplating 7, 30, 31, 120, 238
  Baja California 34, 48
  Franciscan complex 60, 79
  Sanbagawa 145, 148, 150
underthrusting, Sanbagawa 146, 148
unroofing
  Coast Plutonic Complex 295
  Kigluaik 228, 230
  New Zealand 277
  term 4
upper plate rocks, Baja California 39
Uppermost unit, Crete 88, 90–95, 101, 105
upwelling
  asthenosphere 198
  ductile crust 194–196
uranium, in zircons 289
Uwajima 337

Valentine Springs Formation 57, 60, 72
Valhalla 190

Variscan belt *183*, 196
Variscides, French 190–193
vegetation cover, and erosion rates 11
vein minerals
  metamorphic indicators 34
  Yolla Bolly 62
vein systems
  Baja California 39
  Sanbagawa 132
  South Mountains 328, 331
Velay Dome 183, 191, 196
  evolution *192*
  exhumation *197*, 200
velocity gradients, at plate boundaries 6
vergence, folds 116
Vernon basin 186, 194
vertical extension 7
vertical shortening 83
vertical strain 4, 14
viscosity, crustal 233, *235*
viscous wedge model 79–80, 169
vitrinite reflectance 143, 186
Vizcaíno Peninsula 33, 37, 38, 39, 41
  fault zone 40, 45
volcanics
  Baja California 33
  Canadian Cordillera 194
volcanism, Melanesia 122
volume strain 4, 56, 66
  Franciscan complex 77–78
  Sanbagawa 132
Vredefort Dome 339

Wadi Adai, carbonates 247
Wadi Hulw 246, 248, 251
Wadi Meeh 243, 246, 248, 254
water, in melts 339
weathering, Samail ophiolite 257
wehrlite 37
Western Baja terrane 33
Western Fiordland region 238
Western Gneiss, Norway 9, 14
Whipple Mountains 234, 345, 346, 351
white mica
  cooling ages 218, *220*, 225
  excess argon 242
winchite 133, 137, 139, 140

xenoliths 193

Yolla Bolly Mountains 57
  map *71*
  metamorphic age 60
  sampling 72
  sandstone photomicrographs *62*
  section *59*
Yuma terrane 32

zeolite facies 56
Zermatt–Saas region 9
zircon crystallization age 224
zircons
  U content 289
  U–Pb calibration 265
zoisite 118